OCEAN

BARENTS
SEA

CHUKCHI
SEA

BERING STRAIT
(WEST)

Kara Strait

BALTIC
SEA

OKHOTSK
SEA

Kuril

Agattu Strait

Adak Strait

Samalga Pass

BOSPORUS
STRAIT OF
OTRANTO

BLACK
SEA

La Perouse
Strait

Kunashiri
Strait

DARDANELLES

Pohai
Strait

Tsugaru
Strait

Shikotan Strait

Strait
of
Messina

SEA OF JAPAN

Strait of
Carpathos

YELLOW
SEA

Kithera
Strait

STRAIT OF
HORMUZ

TSUSHIMA
STRAIT

WESTERN
CHOSEN

Tiran
Strait

PERSIAN
GULF

Suwanose
Strait

RED
SEA

Pescadores
Channel

PACIFIC

Hainan
Strait

ARABIAN
SEA

BAY
OF
BENGAL

FORMOSA
STRAIT

OCEAN

BAB EL MANDEB

Palk
Strait

Mindoro
Strait

PHILIPPINE
SEA

Balabac
Strait

Surigao
Strait

STRAIT OF
MALACCA

Koti Passage

Pemba
Channel

MAKASAR
STRAIT

Dampier
Strait

Zanzibar
Strait

Gaspar
Strait

Manipa
Strait

Bougainville Strait

SUNDA
STRAIT

Torres Strait

LOMBOK
STRAIT

OMBAI
STRAIT

MOZAMBIQUE CHANNEL

CORAL
SEA

INDIAN

OCEAN

Bass Strait

Cook
Strait

TASMAN
SEA

SELECTED STRAITS USED FOR INTERNATIONAL NAVIGATION

– – – 200 Mile Exclusive Economic Zones

——— Approximate Outer Edge of the Continental Shelf

◀— Warm Ocean Currents

⇢ Cold Ocean Currents

Major Straits in Capital Letters

SYSTEMATIC POLITICAL GEOGRAPHY

FOURTH EDITION

Martin Ira Glassner
Southern Connecticut State University

Harm J. de Blij
University of Miami

WILEY

JOHN WILEY & SONS
New York • Chichester • Brisbane • Toronto • Singapore

334850

LIBRARY OF CONGRESS
Library of Congress Cataloging-in-Publication Data

Glassner, Martin Ira, 1932–
   Systematic political geography / Martin Ira Glassner, Harm J. de Blij.—4th ed.
      p.   cm.
   Includes bibliographies and indexes.
   ISBN 0-471-63583-9

   1. Geography, Political.   2. Geopolitics.   I. De Blij, Harm J. II. Title.
JC319.G59 1989
320.9—dc 19   88-2655
                                                                                CIP

Printed in the United States of America
10 9 8 7 6 5 4 3 2 1

To Renée

Preface

In 1978 Harm de Blij and Wiley asked me to prepare a third edition of Harm's *Systematic Political Geography*. They gave me total freedom to rewrite the second edition and I made extensive changes to it. It was thoroughly reorganized and revised; had a greatly reduced number of reprinted articles; nearly all the maps, diagrams, and photographs were new; and it had a number of other new features. It was, in fact, essentially a new book, one that gave me an opportunity to set down on paper some of the stories, ideas, and other material I had been using in the classroom for nearly 15 years. In the preface to the Third Edition I expressed my gratitude to Harm and to Wiley for this freedom to make radical changes in an already fine introductory text. I do so again.

After they asked me late in 1986 to revise this text again, one of the first things I did was to send a form letter to all members of the Political Geography Specialty Group of the Association of American Geographers, some five hundred all told. In this letter, I solicited informal, candid comments on the text and suggestions for improvement. The responses were most gratifying. Not only did I receive a larger number of responses than I had expected, but some of them were quite thoughtful and detailed. One respondent, Emanuel Maier, even took the trouble to come to New Haven and give me over an hour of his time and several useful ideas. Later, Wiley sent out a formal questionnaire for

the same purpose, and the responses to it were also helpful. All responses were carefully considered in designing this edition, and many respondents will be pleased to find their suggestions incorporated in it. In addition, I have included many ideas that I have been pondering and material I have been collecting since the appearance of the Third Edition in 1980. The result is a greatly improved book, and I thank the many people who helped to make it so, including Christine Drake, who made detailed comments on the galleys.

The overall format and organization of the Third Edition remain basically unchanged, but a number of important adjustments have been made. Everything has, of course, been updated. The chapters on colonialism and international organizations have been condensed from two each down to one each, whereas the single chapter on Antarctica and outer space has now been split into two separate chapters. There are three completely new chapters: (1) on the geography of elections, (2) on the geography of war and peace, and (3) on outlaws and merchants of death, which covers piracy, drug trafficking, the arms trade, and terrorism. The many new sections scattered through other chapters include those on international economic sanctions, transnational corporations, refugees, and pollution across international boundaries. The last chapter contains a number of new topics suggested for further study.

By far the most difficult decision to make for this edition was whether to retain the complete texts of selected articles that was such an important feature of the first three editions. I finally decided, after much agonizing over the matter, to drop them and to replace them with vastly expanded sections of references after each group of chapters.

The references have been carefully chosen to include only items both useful and reasonably accessible to students, though they can be helpful to their professors and others as well. Only books, monographs, and articles in English are included. Regretfully excluded for purely pragmatic reasons are immensely valuable materials in other languages, documents and publications of governments and intergovernmental organizations, atlases, and unpublished items, such as dissertations and papers read at professional meetings. Articles appearing in anthologies or other edited volumes are not listed separately if the books themselves are listed. Nor are articles appearing in specialized journals listed; only the journals themselves are, as each is a treasure trove of material on the topic it covers. General texts in political geography, whether edited or single-author volumes, are listed with the references for Part One; studies of individual countries and regions are listed with Part Nine. There are some exceptions to most of these principles.

Placement of many items was difficult because some cover many topics and some topics could logically fall under many of the book's headings, so placement was sometimes arbitrary and not always consistent. No item is intentionally listed more than once. Readers are advised to scan *all* of the references and footnotes that might have any relation to a topic of interest *and* to use these references and footnote citations only as a springboard for research in the materials omitted. With all of its limitations, however, the collection of references and citations herein constitutes the largest bibliography of political geography in print in the world.

Another improvement over the Third Edition is the graphics. There are 37 new photographs and 16 new maps and graphs, and all of the original maps and graphs from the Third Edition have been redrafted. Most of the maps and graphs are original; the others have been drawn from a variety of sources. Of these, the most useful has been *World Eagle*, a marvelous monthly publication full of authoritative facts, statistics, and illustrations on a great many topics of current interest. It is produced at 64 Washburn Avenue, Wellesley, Massachusetts 02181. The index is also greatly expanded and improved.

Finally, as in the Third Edition, the word "we" is used throughout this book in the editorial sense since I had full responsibility for it. Harm de Blij cannot be blamed for any of its faults.

Martin Ira Glassner
New Haven, Connecticut
August 1987

Contents

PART ONE

INTRODUCTION TO POLITICAL
GEOGRAPHY   **1**

Chapter 1   The Field of Political
        Geography   3
     2   Personal Space and
        Territoriality   9
     3   Perceptions of the Political
        World   16
References for Part One   27

PART TWO

THE STATE   **35**

Chapter 4   State, Nation, and Nation-
        State   37
     5   The Emergence of States   46
     6   Modern Theories About
        States   60
     7   The Territory of the State   66
     8   Frontiers and Boundaries   78
     9   Core Areas and Capitals   94
    10   Unitary, Federal, and
        Regional States   106
    11   Anomalous Political Units   117
    12   Power Analysis   126
References for Part Two   134

PART THREE

POLITICAL GEOGRAPHY
WITHIN THE STATE   **155**

Chapter 13   First Order Civil Divisions   157
    14   Civil Divisions of the
        United States   170
    15   Special Purpose Districts   187
    16   The Geography of
        Elections   200
References for Part Three   208

PART FOUR

GEOPOLITICS   **221**

Chapter 17   Development Through
        World War II   223
    18   Geopolitics Since World
        War II   231
    19   The Geography of War
        and Peace   243
References for Part Four   259

PART FIVE

IMPERIALISM, COLONIALISM
AND DECOLONIZATION   **281**

Chapter 20   Colonial Empires   283
    21   The Dismantling of
        Empires   296
    22   The New States   309
References for Part Five   320

PART SIX

CONTEMPORARY
INTERNATIONAL RELATIONS   **343**

Chapter 23   International Law   345
    24   International Trade   357
    25   Economic Integration   374
    26   Land-locked States   386
    27   Intergovernmental
        Organizations   398
    28   Outlaws and Merchants of
        Death   423
References for Part Six   435

PART SEVEN

OUR LAST FRONTIERS   **459**

Chapter 29   The Traditional Law of the
        Sea   461

30 The Developing Law of the
Sea   473
31 Antarctica   485
32 Outer Space   499
References for Part Seven   508

PART EIGHT
THE POLITICAL GEOGRAPHY
OF EVERYDAY LIFE   523

Chapter 33 The Politics of Religion,
Language, and Ethnic
Diversity   525
34 The Politics of
Transportation and
Communications   537
35 The Politics of Population,
Migration, and Food   552
36 The Politics of Ecology,
Energy, and Land Use   568
References for Part Eight   582

PART NINE
LOOKING AHEAD   595

Chapter 37 Political Geography for the
Future   597
The United States   597
Very Small Places   598
States with Centrally
Planned Economies   601
Nongovernmental
Organizations   601
Public Policy   601
Individual Countries and
Regions   602
Indigenous Peoples   602
A New International
Economic Order   603
The Role of Political
Geographers in the Future   607
References for Part Nine   608
Name Index   I-1
Subject Index   I-3

Maps

Front End Papers—World Oceans
The Cuban Missile Crisis (1962) 20
North Africa from the Middle East 22
Deliberate Distortion of a Soviet Map 23
The Treason Cycle—A Modern
  Propaganda Map 24
The Map War in Indochina 25
Bantustans in South Africa 38
Somalia and the Horn of Africa 43
Pakhtunistan 44
Culture Hearths 48
Early African States 52
Early American States 55
Jordan—Saudi Arabia Boundary
  Settlement (1965) 68
Chamizal Dispute Settlement (1963) 68
Shapes of States 73
Exclaves 75
Evolution of Bulgarian Boundaries 81
Genetic Classification of Boundaries 84
East and South China Seas Disputes 88
Venezuela/Guianas Territorial Disputes 90
European Core Areas 96
African Core Areas 97
Asian Core Areas 98
France—Administrative Districts 108
Italy—Administrative Districts 115
The Republic of the South Moluccas 120
The United Arab Emirates 122
Namibia 123
Strategic Position of Ports of Northern
  Chile 127
Liechtenstein: Civil Divisions 158
United Kingdom: Reorganization of
  1974 160
China: First Order Civil Divisions 162
Nepal: Districts and Zones 163
Language and Administrative Divisions
  of India—1948 and 1971 167

Nigeria: Changes in Civil Divisions 169
Economic Development Regions (USA) 174
The Port Authority of New York and
  New Jersey 175
The Delaware River Basin Commission 177
The Tennessee Valley Authority 178
Albuquerque, NM.—Annexation 182
Los Angeles: A Patchwork City 183
Metropolitan Toronto 185
Federal Lands of the United States 193
Indian Reservations of the United
  States 195
The Navajo-Hopi Territorial Dispute 198
Canadian Elections 202
The Original Gerrymander 204
Contemporary Gerrymanders 205
German Propaganda Map of
  Czechoslovakia 229
Airman's View: De Seversky (1949) 232
The Decay of the Buffer Zone in
  Southern Africa 237
New Zealand—World Perspective 240
Who Is Surrounding Whom? 244
Strategic Mineral Production 248
The Middle East 250
The Indian Ocean 255
Japanese Colonial Empire 295
League of Nations Mandates 299
Partition Plans for Palestine 303
Morocco 318
Division of the Indus Rivers 351
The Río Lauca 353
The Lower Mekong Basin Scheme 354
Principal Ocean Trade Routes 362
European Community Trade Links 379
Latin American Integration
  Organizations 382
African Economic Groups 384
Land-locked States 387

Territorial Corridors   391
Kashmir—A Divided Land   401
The Jurisdiction of the South Pacific
  Commission   416
Military Alliances of the Post-World
  War II World   421
International Drug Trafficking   428
The North Sea Continental Shelf   466
Ocean Fisheries   469
Baselines of the Territorial Sea   475
The United States Exclusive Economic
  Zone   478
Mineral Resources of the Sea   480
Antarctica   486
Potential Ecuadorian Antarctic Claim   490
Proposed French Airstrip in Antarctica   494
Sikh Separatist Claims in Punjab   526

Tamil Separatist Claim in Sri Lanka   531
Ethnic Groups of Africa   532
Cyprus   533
Switzerland—Religions and Languages   535
Major International Highways   539
Major International Railroads   542
Maps Are Liars   546
The Politics of International Air
  Transport   547
Current Refugee Movements   561
World Nuclear Energy Generation   576
United States Foreign Trade Zones   599
Per Capita Gross National Product by
  Country   605
The Physical Quality of Life Index   606
Back End Papers—Major World
  Political Units

Part One

INTRODUCTION TO POLITICAL GEOGRAPHY

Chapter 1

THE FIELD OF POLITICAL GEOGRAPHY

Political geography is a varied and wide-ranging field of learning and research, exciting and endlessly fascinating to the student and useful to the practitioner in many fields. Its roots go back to Aristotle's model of an ideal or perfect State. It developed in spurts over the next two thousand years and appears now to be in the early stages of a new efflorescence. Throughout the history of political geography, interest in it has waxed and waned, and each period of development featured a particular emphasis without discarding the sound ideas that had emerged earlier. Despite this constant change, there is also continuity. In the words of the old gospel hymn, *When the Saints Go Marching In,* "We are traveling in the footsteps of those who've gone before."

As a preface to our brief review of the development of and current trends in political geography, we present a random selection of definitions of the field offered by some of our more prominent contemporary political geographers.

. . . the study of the variation of political phenomena from place to place in interconnection with variations in other features of the earth as the home of man. *Hartshorne*

. . . Political Geography, a subdivision of human geography, is concerned with a particular aspect of earth-man relationships and with a special kind of emphasis . . . the relationship between geographical factors and political entities. *Weigert*

. . . the study of political regions or features of the earth's surface. *Alexander*

. . . the geographical nature, the policy, and the power of the State. *Pounds*

. . . the study of political phenomena in their areal context. *Jackson*

. . . the study of the interaction of geographical area and political process. *Ad Hoc Committee on Geography, Association of American Geographers*

. . . the study of the spatial and areal structures and interactions between political process and systems, or simply, the spatial analysis of political phenomena. *Kasperson and Minghi*

. . . political geography is concerned with the spatial attributes of political process. *Cohen and Rosenthal*

These few quotations demonstrate that there is no generally accepted definition of the field. Although many critics, both within and outside geography, assail this vagueness as a fault, in reality it might be considered a virtue. The lack of a rigidly defined focus, of precise boundaries, has enabled political geographers to investigate various phenomena that exhibit both political and spatial, or geographic, characteristics without being concerned about straying from the central theme of the field or about intruding into someone else's territory. (We will have more to say about territory in the next chapter.) It has also enabled political scientists, sociologists, and other social scientists to "do" political geography—indeed, to make valuable contributions to the field without incurring the displeasure of the "insiders." The backgrounds of political geographers, in fact, are as varied as the subjects they study, and their ranks are frequently enriched by converts from other fields. All this contributes to the variety, fascination, and utility of political geography.

HIGHLIGHTS OF THE HISTORY OF POLITICAL GEOGRAPHY

The definitions just cited, different as they may seem, really fall into only two categories. The first five emphasize the study of political areas, two focusing specifically on the State, while the last three, the most recent, introduce the study of political processes, a reflection of the behavioral approach so popular in the social sciences in the 1960s and 1970s. Even in a period of only some 40 years, we have seen a shift in the orientation of many teachers and students of political geography; the changes over two thousand years have been much greater.

The single long-term interest underlying the field of political geography from its origins to the present is in the interrelationships between politics in all its manifestations and the physical environment. When Aristotle fashioned his model State, he examined the ratio between population and territory—and the qualities of both—as fundamental. He examined the requirements for the capital city, the composition of the army and navy, the boundaries, and other factors, in all of which the nature of the physical environment, especially climate, was seen as the determining factor. Although his deterministic approach has little support today (we discuss environmental determinism in more detail later), many of his ideas have been incorporated into basic concepts in political geography and are currently very much alive.

Strabo, a Greco-Roman geographer, wrote a seventeen-volume description of the entire known world more than three centuries after Aristotle, following this tradition but basing his work in considerable part on his own field observations. Although his work was designed for practical use by soldiers, administrators, and travelers, he did consider the factors necessary for the functioning of a political unit the size of the Roman Empire. Like Aristotle, he tended to be quite ethnocentric, believing that his homeland had the "ideal" climate, the best governing techniques, and so on.

After the decline of Rome and the rise of Christianity, science and scholarship were nearly eclipsed in Europe. But geography flowered anew in the Muslim world, particularly among Arab merchants, historians, travelers, and philosophers. One of the most widely traveled, observant, and prolific of these writers was Abd-al Rahman Ibn Khaldun. He wrote a detailed autobiography late in the fourteenth century in which he proposed theses about the most powerful political units of his time, the tribe and the city, and related them to the occupations determined by the physical environments of their regions. His investigations into the nature of political units led to theories about political integration and disintegration and ultimately to the concept of a cyclical nature of a State's existence.

French writers of the sixteenth and seventeenth centuries followed in this tradition. Montesquieu, for example, emphasized the role of the landscape and climate in determining systems of government. He observed many complex relationships between agriculture and population, political systems and occupations and land area, soil fertility and government. In the mid-seventeenth century William Petty—physician, teacher of anatomy at Oxford, cartographer, political scientist, and economist—published works discussing the territorial and demographic elements in the power of States, the concept of the sphere of influence, the role of capital cities, and a host of other politicogeographical issues. Petty's major works were *Political Anatomy of Ireland* and *Political Arithmetic*. Carl Ritter in Berlin, writing in the late nineteenth century, developed a cycle theory of State growth similar to Ibn Khaldun's but based more on analogy with organisms. Ritter was more scientific than his predecessors but still somewhat ethnocentric and deterministic. His *General Comparative Geography*, in a sense, was the first real attempt at the creation of models in political geography.

A new phase in the evolution of political geography began with Friedrich Ratzel, commonly referred to as the "father" of the discipline because of his innovations in both concept and methodology. Much of his work derived from that of Ritter and other intellectual forebears, and their tradition continues right through to the present. The political geography of the late nineteenth century and the first half of the twentieth was dominated by geopolitics, rooted in some of Ratzel's voluminous work. We devote two chapters to Ratzel and geopolitics not only because of their influence on both thought and political action during the period, but also because some elements of the geopolitical approach can still be useful today.

Concurrent with the growth of geopolitics and with the later stages of the previous period, and still quite vigorous, is the period of concentration on the study of political areas, especially the State. Most of the prominent political geographers of the twentieth century have been vital contributors to this trend. Isaiah Bowman concentrated on practical problems of political geography: State, regional, and global. Derwent Whittlesey adopted a historical approach, viewing the State genetically. Richard Hartshorne and Stephen B. Jones developed two of the basic theories of modern political geography: the "functional approach" and the "unified field theory," both discussed later.

A content analysis of the political geographic literature from the beginning to the present would undoubtedly show that the overwhelming bulk fits into this category. Over the past two thousand years, the great majority of people who may be

Friedrich Ratzel, 1844–1904. (Courtesy of Geoffrey J. Martin)

Isaiah Bowman, 1878–1950. (Courtesy of Geoffrey J. Martin)

considered political geographers have found areal studies—of States, boundaries, capitals, civil divisions, territorial disputes, federalism, colonialism, international rivers, and others—not only engaging and provocative, but serviceable as well. They have contributed immeasurably to our understanding of the development of political units on the landscape and to the understanding and solution of real political and environmental problems. A great deal of work of this type is still to be done, and there is no reason to believe that the appearance of new problems will in any way diminish in the future. Areal studies will always be needed, even as new approaches and new techniques become available for application to them and as many political geographers channel their efforts into other directions.

CONTEMPORARY POLITICAL GEOGRAPHY

Beginning in the early 1960s, several new trends emerged in political geography. These contain diverse elements and overlap considerably, making categorization difficult. Some classification is helpful, however, and so we attempt it here by grouping the contemporary emphases in the field into four broad areas.

First, there is a notable change in the *scale* of subjects studied. Traditionally, political geographers and their kin in other disciplines have tended to work at the meso- and the macroscale, essentially concentrating on the State and on interstate (international) or global problems. Today there is a higher proportion of work being done at the microscale; that is, at levels below that of the State. The study of administrative districts and systems of all kinds within States has spawned a new subspecialization—electoral geography—which uses the electoral district as its field. Other studies cover urban areas, cities and sections of cities, and the supply of municipal services. This seems to be a reflection of a general turning inward, a concern with ever-smaller social as well as

political units, a withdrawal from the universalist perspective of the period of World War II and the 15 or 20 years following it.

Second, there has been a shift in the *subject matter* itself, with increasing emphasis on spatial interactional factors (flow of goods, services, and ideas over space). Associated with this have been studies of the effects of spatial structural elements on this spatial interaction. Geographers have also been analyzing political power, conflict resolution, value systems, territoriality, and public policy—all subjects previously the domain of the political scientist, the sociologist, and the psychologist—but applying geographic attitudes, approaches, and techniques.

The new emphasis on such subjects has both encouraged and been stimulated by the adoption of new *methodologies*. Prominent among them have been quantification and behavioralism. In both cases, political geography was among the last of the major branches of geography to adopt them, just as geography lagged behind economics, psychology, and political science in adopting them. This is merely an observation, not a criticism, for there is no particular virtue in jumping on a bandwagon merely because it happens to be passing by and the music is enticing.

The advent of computers has encouraged people to find ways to use them, and geography, including political geography, offers some opportunities. New discoveries in psychology have led to analyses of decision making, ideologies, and perception—all with applications in political geography. Similarly, the systems approach, which combines mathematical and behavioral techniques, is being used experimentally but has not yet proven useful for general application. It remains to be seen whether the new techniques, particularly the "mathematization" of political geography, is really a momentous advance in our field or merely a passing fancy; whether numbers and formulas and models really help us to understand our world better than words and maps and photographs or drawings; whether computers are really an adequate substitute for fieldwork. We may very well find that the new methodologies are useful tools to add to the geographer's box of tested and proven tools but cannot replace any of them.

One of the chronic criticisms of political geography, even in comparison with other branches of geography, has been that it is long on facts and short on ideas. We seem to be on the way to rectifying that weakness, if indeed it is a weakness, by developing *new theories*. The tendency in our field has been to deal with the empirical and neglect the theoretical. A number of theories have been proposed over the centuries, particularly in this century, and some of them, concerning primarily the State and geopolitics, have achieved some currency, but the theoretical underpinnings of our field do need shoring up. Attempts are being made in this direction, but so far they have resulted in more themes than theories, and certainly no one theory is generally accepted by political geographers.

Some of the major themes that may lead to new theories are: areal or spatial association, which tries to explain the patterns of distributions of various phenomena occupying an area; attempts to link generating processes with the forms of spatial distribution so that the type of pattern that will be generated in an area at a particular time can be inferred if not predicted; spatial interaction, which provides the basis for many hypotheses about flows of phenomena; distance decay, which proposes that the effect of an event or activity tends to decrease with increasing distance from the site of its occurrence; territoriality and perception; a number of concepts from economics, such as friction of space, localized costs and benefits, externalities, and *"homo economicus"*; and, of course, the well-established concept of the political region.

A newer version of the geopolitical concept of "heartland-rimland" (discussed in Chapter Seventeen) is that of *core (center)-periphery*, derived from dependency theory, which visualizes the postcolonial world as one still characterized by a core of rich and dominant former colonial States surrounded by a periphery of poor and dependent former colonies that share but little in the wealth of the center. Many varieties of Marxist (or radical or socialist) theory have been applied in political geography in the past two decades and some of them are slowly gaining respectability, if not wide acceptance. *Antipode,* the journal of radical geography, is now well established and almost mainstream; it has come far from its origins as a distinctly marginal project of a very few British and American geographers.

There are many signs of a revival of interest around the world in political geography besides these new approaches being advanced in various quarters. The Association of American Geographers (AAG) established a Political Geography Specialty Group in 1979, and it has become one of the larger and more active of the some 30 speciality groups in the AAG. The journal *Political Geography Quarterly* was founded in 1982 and is similarly flourishing. The Institute of British Geographers now has a Political Geography Working Party and the International Geographical Union has a Study Group on the World Political Map. The Office of the Geographer in the U.S. Department of State has greatly expanded both its size and the scope of its operations and serves many government agencies. New centers of political geography have developed outside of the traditional ones in Western Europe and North America. The most notable of these is in Israel, where there was not even one course being taught in the subject anywhere in the country in 1970. Now, it is taught in several of the universities and is especially strong at the University of Haifa. Moshe Brawer, Yehuda Gradus, Nurit Kliot, Arnon Soffer, and Stanley Waterman are among the Israeli political geographers who have already earned international reputations.

In Part Nine we will present briefly some of the topics of current interest in political geography that we have been unable to discuss in detail earlier in the text. We also suggest other topics that need studying but that are currently being neglected. The references at the end of each Part of the book give some indication of the vast literature available to any seeker of additional material, including much that is new.

But we wish to reiterate that the current phase of the evolution of political geography is not the first one, but the fourth and it will very likely not be the last. There is, moreover, little in this phase that is truly new or truly unique. Most of it is derived from the work done in previous phases, though with fresh approaches and techniques, and most of it is common both to geography in general and to the other social sciences.

As the current scholars and practitioners in the field continue to develop their own interests, ideas, and methodology and as today's students join the ranks and make their own contributions, the field will continue to evolve, perhaps into a new phase we cannot yet perceive. Ours is a dynamic field, and a useful one, with unlimited opportunities for newcomers to express themselves and to attain recognition from their peers. We hope that many students will accept the challenge and join us.

OUR APPROACH

In this book we make no attempt to treat every aspect of political geography equally nor to incorporate all of the very latest ideas and materials. Essentially what we try to do is to develop the theme of our opening sentence: "Political geography is a varied and wide-ranging field of learning and research, exciting and endlessly fascinating to the student and useful to the practitioner in many fields."

We retain much of what has long been the core interest of our field, the study of the State. We add to this, and interweave with it, examinations of politicogeographical phenomena at both the micro- and the macroscale. We are also concerned with the kinds of problems, experiences, and relationships experienced by ordinary people in their everyday lives, and we examine some of them from the standpoint of political geography. Finally, we go beyond the very latest in the field to suggest some of the topics that may attract the attention of political geographers in the near future.

The organization of the book is orderly and rational, but arbitrary. One of the most distressing trends in modern scholarship and education is the increasing compartmentalization of human knowledge and experience. But our perceptual universe is expanding far more rapidly than our mental abilities, and so some division and classification is necessary. Although we must concede to reality and discuss one topic after another, we constantly remind the reader in various ways that everything in the book is related to everything else, not only in the book but also in human knowledge and experience. We cannot emphasize too strongly the unity of our field and its interconnections.

Another theme running through the book is one emphasized in this chapter: the continuity and incompleteness of our field. Every change, every development, every advance raises more questions than it answers. There is far more to be done in political geography than has already been done, and we ask questions throughout and suggest approaches and topics for future investigation as examples of the work that still needs doing.

We hope this book will serve not only as an introduction to the field of political geography, but also as a guide through it and an inducement to enter it.

Chapter 2

PERSONAL SPACE AND TERRITORIALITY

The subject of territoriality could well be discussed under the heading of perception, but both topics have been developed so much in recent decades that each deserves a separate chapter. Of the two, territoriality is the more controversial. We have already emphasized the interdisciplinary nature of political geography, and nowhere is this more evident than in consideration of territoriality, based as it is primarily on research in ethology (the study of animal behavior) and psychology. From this growing body of empirical and theoretical work, we can extract some ideas that may be helpful in understanding human political behavior. We must be extremely careful in doing so, however, for political parties and States do not necessarily behave in the same manner as animals or individual humans.

PERSONAL SPACE

One promising area of research ought to be familiar to all of us. It relates to the way we spread ourselves out in classrooms and theaters, restaurants and libraries. If we do not *have* to be crowded, we like to put a little distance between ourselves and the next person. This is *personal space*—an envelope of territory we carry about with us as an extension of ourselves. Humans as well as animals carry such envelopes around, but the shape and size of the envelope vary in humans from culture to culture. In terms of shape, we can tol-

erate greater proximity in front of us than beside us, and less behind us. Thus our "portable territory" is not symmetrical around our body. And, as Robert Sommer remarks, Englishmen keep farther apart than Frenchmen or South Americans. This subject was first touched on by Edward T. Hall in *The Silent Language* and later more fully developed in *The Hidden Dimension*. In the second book Hall introduced the term *proxemics* for the study of the way people perceive space and use it in various cultures.

How can we interpret the territorial behavior of societies and cultures from what we know about personal space and small-group proxemics? Psychologists and geographers have had occasion to refer to ranking and hierarchies in human society. Walter Christaller made this a central issue in his work on urban centers; in Allan K. Philbrick's book *This Human World*, it is a dominant theme. In Sommer's 1969 book the idea that human territorial organization relates to rank and hierarchy is explored in some detail.

Perhaps an example from Africa will be useful. In a study of the city of Mombasa, where more than a dozen of Kenya's tribal peoples were represented by substantial numbers in the population of about two hundred thousand, it was noted that the people of the "stronger" tribal groups generally possessed better residential locations than those of "weaker" tribal groups. The strengths of several of the major tribal peoples in Kenya may be a matter for debate, but their influence in the government is a good indicator. In any case few would argue that the Kikuyu and the Luo hold positions of power compared to, say, the Nandi and the Gyriama. Now, this indefinite power hierarchy is somehow reflected by the urban residential pattern in Mombasa—the spatial adjustment those peoples have come to within the confines of that city. This is an example of *dominance behavior,* and it may apply not just to tribal peoples living in urban centers, but to States and even power blocs as well. We shall return to this point shortly.

While the minimum acceptable personal space around an individual is chiefly determined by culture, reactions to invasions of this "forbidden zone" vary with the circumstances and the individuals involved. They range from "cocooning," withdrawing within oneself, through displays of discomfort to flight, leaving the place to the intruder, the most extreme nonviolent reaction. Regardless of the degree of tolerance for closeness of physical presence of others in various cultures and individuals, there is a limit for everyone. In some animals behavior is grossly distorted whenever there is extreme crowding of members of the same species. Similar reactions have been noted among people, but social adaptations sometimes work to prevent violent reactions among them. A great deal of research remains to be done on the effects of crowding on people, but we already know enough about their seriousness to make us add crowding to our list of adverse effects of excessive population growth.

ANIMAL TERRITORIALITY

Territoriality has been investigated since the seventeenth century and was first described in detail in 1920. Many vertebrates exhibit it in varying degrees, including cats, some birds, some fish, and monkeys. Robert Ardrey first brought the subject forcefully to public attention in his book

The Territorial Imperative. Ardrey makes some statements political geographers can hardly ignore. "A territory," he writes, "is an area of space, whether of water or earth or air, which an animal or group of animals defends as an exclusive preserve. The word is also used to describe the inward compulsion in animate beings to possess and defend such a space." He concludes that man "is as much a territorial animal as a mockingbird singing in the clear California night" and that "the territorial nature of man is genetic and ineradicable."

Our knowledge of animal territoriality is still rather rudimentary, but we do have detailed information to supplement and support many of Ardrey's observations. We know, for example, that territorial animals mark out their territories in various ways, commonly by a glandular secretion or urination. Territories are rarely rigidly bounded and exclusive; they generally overlap and sometimes overlap *almost* completely. They have gradations of sensitivity so that the animal will react differently according to species, the type of intruder, and the distance the intruder penetrates into the animal's territory. Territory is not immutable; it tends to get larger when food is scarce and shrink when food is plentiful. Finally, territory serves the animals that mark and defend it in several ways. It functions, for example, to regulate population by limiting breeding, reduce frictions among members of the species, protect feeding and nesting sites, reduce the rate of spread of diseases and parasites, and afford some security for weak or subordinate animals.

If we define the concept of territoriality as a pattern of behavior whereby living space is fragmented into more or less well-defined territories whose limits are viewed as inviolable by their occupants, it becomes clear that political geographers have for some time recognized this phenomenon. Any of these territories, after all, will acquire certain particular characteristics. Gottmann called it *iconography,* and we discuss this idea later. The Ratzel-Kjellén-Haushofer school of geopoliticians was in the process of defining a form of territoriality when they conceptualized the State as an organic being. Hartshorne and Jones urged that the State be viewed as an entity whose characteristics could be linked, ultimately, to the behavior of the individuals constituting it.

And therein lies one of the problems with the concept of territoriality. It seems inappropriate to make inferences about aggregate or group behavior from research dealing mainly with the behavior of individuals; it is also questionable to what extent human behavior may safely be equated with animal behavior. Ardrey may not have been entirely wrong, but caution is warranted in accepting sweeping generalizations.

HUMAN TERRITORIALITY

By combining the concepts of personal space, including dominance behavior, and animal territory we can interpret many aspects of human behavior as indicating some form of territoriality. We begin with very small units of territory and move to progressively larger ones, at each stage making some observations and reserving our analysis for the end.*

Obviously, individuals exhibit territorial behavior in those small, confined places where they spend most of their time—home, office, factory floor, automobile—whether or not they own these places legally. The home, in its interior furnishings and decor especially, expresses an individual's (or family's) personality. It is more than just a residence; it is a refuge, a fortress shielding the individual from the

*Much of the human behavior that we consider under the headings of personal space and human territoriality—those gestures and signals that, though differing from one culture to another, communicate messages to other people about ownership of space—is included in the studies known variously as psycholinguistics, paralinguistics, and nonverbal communication, which have been popularized recently as "body language."

problems of the world outside. There is certainly a strong desire to defend this home from intruders and even to discourage intruders from a favorite armchair or "my" room or "my" place at the dining table. People who have grown up in the United Kingdom or North America tend to idealize the single-family detached house on its own lot, perhaps with a fence or shrubbery around it. Some observers speculate that this dream and its realization respond to territorial needs. Many people, however, prefer apartments instead of houses and many prefer to rent rather than buy; perhaps other needs are more strongly felt than territoriality.

Bureaucrats (in government, business, or university) have similar feelings about "their" offices, parking places, and similarly assigned or adopted spaces. Many of the perquisites of office, the coveted status symbols, have little intrinsic or functional value but do enhance the space occupied by the individual or simply symbolize his or her right to use it. The elegant desk, the carpet or rug on the floor, the key to the executive washroom, the view from the window, all go with the position and are defended by the possessor. In our culture and in many others, there is a distinct correlation between status and size of territory. Dominant people tend to have more and larger territories. "Important" people generally are expected to—and desire to—have big homes, big lots, big offices, big cars, and often several homes, offices, cars. These symbols of status also reinforce status itself, so someone with a big home, office, or car must be "important."

Even in temporary situations, in the diplomatic world, for example, status is symbolized by the space assigned to or occupied by a person. It really does make a difference whether one sits at a formal dinner to the left or the right of the host, or above or below the salt. In formal business sessions, the presiding officer sits in a prominent place with the rapporteur and other officials nearby in a particular order.

These and many other sensitive matters relating to rank and precedence have been painfully worked out over centuries and are summarized in the term *protocol*, a prerequisite for conducting diplomatic business effectively.* In most cultures there is a less formal but equally effective protocol that regulates the temporary possession of territory, such as in libraries, cafeterias, and classrooms. A seat is known to "belong" to a particular person by prescription, even if not assigned, or a seat may be territorially claimed in the "owner's" absence by markers, such as books, coats, or purses.

On a larger scale, territoriality is manifested in neighborhoods. Street gangs often claim particular "turf" as their own, sometimes marking it out with graffiti or other symbols. Ethnic groups tend to congregate in particular areas within cities, frequently creating a replica (or a pale imitation) of a neighborhood in "the old country." If groups are forced by law or custom into segregated areas, the territorial feeling is unlikely to be so strong but such enforced segregation, as in the Jewish ghettos of Europe and most Muslim countries, may help to preserve a culture that would tend to change along with the general culture if place of residence were unrestricted. A new kind of segregated neighborhood is developing in our society: the specialized housing area designed for particular groups of people. Examples would be the retirement communities for "senior citizens," summer beach colonies, and garden apartments for "swinging singles," all similar in some ways to the old "company towns," but more likely to inspire territorial feelings.

Depending on where people live and their length of residence as well as personal factors, they may extend the territoriality they exhibit toward their homes and neighborhoods to their towns, counties, and states. They may take pride in these places and advertise their virtues everywhere they go. This kind of "boosterism" seems to be fading with increasing mobility and standardized entertainment. Even regional loyalties have lost much of their importance in the United States; one seldom hears heated arguments any more over the respective merits and vices of New England, the South, the Midwest, the Far West, or other regions. The Civil War is no longer a topic to inspire passions as it was when the authors were in high school. Yet people still do identify themselves as being from a particular town, county, state, or region with a touch of pride.

People generally feel more comfortable and perform better at home, among familiar surroundings. Sports teams are well aware of the "home court (or field or ice or arena) advantage," even if the advantage is only psychological. Performers put out extra effort for the hometown folks. A person will be more confident in a confrontation with another in his own home, office, or other territory than in the adversary's territory. This principle does not seem to be applicable, however, to individuals or groups that either have no territory or habitually operate outside their territory. In international affairs this could include transnational corporations, international terrorist groups, international political movements, and perhaps religious movements as well. It does apply to farmland one owns and land tenure is an important factor in economic development, even in a communist country such as the Soviet Union, where agricultural production is highest on the small private plots

---

*Who sits where at international conferences can be an important substantive problem to be resolved before negotiations actually begin because the seating symbolizes the relationships among the participants' countries or bear on the issue itself. Recent examples are the Allied–Soviet negotiations over the status of Berlin in the early 1950s and the US–North Vietnamese peace talks in Paris 20 years later. In both cases even the shape of the negotiating table was an issue. At UN meetings, States are normally seated in English alphabetical order to avoid status disputes and generally seating is rotated regularly so no country is permanently in the front or rear of the hall. This system deliberately diminishes the value of each territory (seat) so it is not worth fighting for.

of land rather than State-owned farms (*sovkhozi*) or collectively owned farms (*kolkhozi*). But it does not seem to apply to continents; people rarely have territorial feelings toward South America or Asia or North America.

This background helps us to examine that aspect of territoriality that most interests political geographers: political territoriality, particularly as expressed in the State. As we will see in more detail in subsequent chapters, the modern nation-state based on nationalism is quite a recent phenomenon; in its most important form, it dates back not more than about four centuries. It emerged out of feudalism in Europe as the concept of *regnum*, or personal sovereignty, was gradually replaced by that of *dominium*, or national sovereignty. That is, people gradually transferred their allegiance from an individual sovereign (king, duke, prince) to an intangible but territorial political entity, the State. It is no longer "Mary, Queen of Scots" but Elizabeth II, Queen "of the United Kingdom of Great Britain and Northern Ireland. . . ."*

Nationalism is probably the strongest political force in the world today, and nationalism is territorially based. People identify not only with others of their group but with a particular portion of the earth's surface. A person is French not only because he or she speaks French and identifies with other Frenchmen, but also because France is home. A person of French descent living abroad may still speak French and feel a kind of kinship toward Frenchmen, yet owe loyalty and feel warmly toward the country in which he or she was born and raised. In the mod-

ern world the ideal political ties are those of place, not descent (although, as we will see later, there are many places where this ideal has not yet been attained). Most countries confer nationality on the basis of *jus soli*, or place of birth, instead of *jus sanguinis*, or parentage (although the United States and some other countries recognize both).

Sometimes a people or a nation will be sustained in its travail by a mystical concept of territory, a dream of a land of their own once possessed but now lost or of a land never possessed but longed for. This mystical territoriality is best exemplified by the Jewish people, who endured 40 years in the wilderness until they reached the Promised Land, who longed for Eretz Yisrael during the Babylonian exile, who through 2000 years of dispersion in nearly every country in the world repeated every single year in their Passover prayers, "Next year in Jerusalem." Zionism, the national liberation movement of the Jewish people, developed in the nineteenth century to convert the spiritual dream into political reality. The Jewish nation was to return home, to Zion. They rejected offers of homelands in Uganda, British Guiana (now Guyana), Sinai, Cyprus, Angola, and other countries. Finally, in 1948, they reestablished the Jewish State on its original territory and called it Israel.

We will examine other examples of homeless nationalism, but none can compare with this one which inspired countless generations of people to dream of dying in the ancient homeland or at least of being buried there, inspired them to walk from Russian Poland to Turkish Palestine and rebuild the wilderness there with their hands, to live and work and die on the land.

Was this an expression of some animal instinct? Do the other examples of human territoriality mean that it is instinctive, or has it been acquired and modified by learning—through cultural evolution? In other words, Is it really an "imperative"? And the concomitant, human aggres-

---

*There are still some technical survivors of *regnum*. Baudouin, for example, is still "King of the Belgians" and a citizen of England or Scotland is a British *subject*. But these anachronisms are purely symbolic, part of the national culture. Even the very few remaining absolute monarchs in the world demand loyalty not as individuals but as Chiefs of State. The Chief of State is still recognized in international law as the sovereign, regardless of domestic title, but the sovereign is now the personification of the State.

sion: Is it similarly universal? Is it genetic in its origins? Or, is it neither?

Edward Soja writes

Only when human society began to increase significantly in scale and complexity did territoriality reassert itself as a powerful behavioral and organizational phenomenon. But this was a cultural and symbolic territoriality, not the primitive territoriality of the primates and other animals. . . . Thus, although "cultural" territoriality fundamentally begins with the origins of the cultured primate, man, it achieves a central prominence in society only with the emergence of the state. And it probably attains its fullest flowering as an organizational basis for society in the formally structured, rigidly compartmentalized, and fiercely defended nation-state of the present day.*

What Soja is suggesting, of course, is that territoriality as it is expressed by our complex societies is not the same phenomenon as that which commanded our distant ancestors—or our animal contemporaries. Nor, according to some anthropologists, is man the "naked ape," driven by savagery and killer instincts. Alland argues forcefully that aggressiveness and territoriality are not universal human "imperatives" at all; in fact, he goes so far as to say that territoriality is born with and nurtured by culture. Viewing some "primitive" peoples like those still living a hunting-and-gathering existence, Alland notes that these have the weakest of territorial imperatives; supposedly they are the ones who should have the strongest instinct of this sort if territoriality is indeed a biological urge. And Alland is not so convinced either about the vigor with which nation-states are inevitably defended; he views draft laws, anthem singing, pledge reciting, and flag-waving as evidence that States do not find defenders so easy to come by.

Torsten Malmberg, however, a Swedish biologist and geographer, accepts territoriality as innate in humans, and says, in

fact, "Territoriality probably reaches its highest development in the human species. . . ." He reviews much of what we have just presented but from a different perspective and in much greater detail. He warns strongly about the very serious consequences for the human species of unchecked population growth. Then he summarizes:

Behavioural territoriality has an important place in the man-environment systems of evolutionary adaptation. What evolved from environmental demands on our ancestors is now required by the individual from the milieu. Thus the only relevant criterion by which we can consider the natural adaptedness of any particular part of present-day-man's behavioural equipment is the degree to which and the way in which it can contribute to population survival in his primeval environment.*

Even if we accept Malmberg's argument, however, must we also accept the argument that countries go to *war* over resources, strategic positions, or *Lebensraum*? Not even Malmberg claims that! More recently, *Robert Sack* has developed a rather elaborate theory about territoriality based on two basic forms of spatial relation: action by contact and territoriality. Essentially, he conceives of human territoriality as only a strategy for access and control. While this concept can be applied to political actions of States, it leaves unanswered—and Sack deliberately avoids—the question of whether territoriality in human beings is a biological drive or instinct.

It would seem that people desire territory in general or a particular piece of territory only if and when they perceive that territory to have some real or potential value for them—economic, strategic, or psychological. Remote and barren areas of the earth, even when well known, were seldom claimed as anyone's owned territory and remained *res nullius* until nationalism (a cultural phenomenon) arose so recently in the western peninsula of Asia

*E.W. Soja, *The Political Organization of Space*, Association of American Geographers Resource Paper No. 8, Washington, 1971, p. 30.

*T. Malmberg *Human Territoriality*, The Hague, Mouton, 1980, pp. 319–320.

and led some people there to imperialism.

The debate over human territoriality will continue to rage in the professional literature and provide diversion at cocktail parties, but as the evidence continues to mount to support the arguments of each side, it is unlikely to be settled very soon. Meanwhile, we would be wise to be skeptical about it. The reality, however, may be less important than the perception, and it is to perception that we now turn our attention.

Chapter 3

PERCEPTIONS OF THE POLITICAL WORLD

Political geographers have tried to gain deeper insight into the processes whereby decision making takes place. What were the elements that went into Malawi's decision to relocate its capital at Lilongwe? What leads to the voters' decision in a particular area to select one candidate over another? How does a government choose an aggressive course of action toward a neighboring State?

For answers to these questions, it is necessary to turn to the work of psychologists and other social scientists who have studied the behavior of decision making. Decisions emanate from a complex set of processes. They are made because the person or group of persons making them desires a state of affairs different from the one that prevails. They are based on information, but in order to assess the process of decision making in a given context, we must know the amount of information the decision makers had and the degree of accuracy of that information. And that is not all. There are different *kinds* of decisions; they are made as a matter of habit or they may be made subconsciously or they are controlled—that is, they are made in response to a perceived choice of alternatives.

Fortunately, geographers other than political geographers have also been interested in research on decision making, and political geography has benefited from this work. Economic geographers have focused on the element of rationality as part of a decision-making model. This is the concept of the "intendedly rational man," who may have an incomplete and imperfect information set but who makes rational decisions when confronted with alternative courses of action. Julian Wolpert incorporated this factor into his search for the behavioral elements involved in people's decision to move elsewhere—the decision to migrate. In several articles of great relevance to political geography, Wolpert stresses a number of dimensions of decision making. Place utility has to do with a decision maker's relationship with the locale where the decision is being made. The *decision environment* is constituted by the decision-maker's perceptions of alternatives, information, cues, and stresses. This brings forward another related item: *threat*. Not infrequently, decisions must be made in haste, in a crisis atmosphere. This tends to reduce not only the time involved, but also the range of choices that might have been considered. This has obvious relevance to strategic decision making. In the face of armed hostilities or under actual conflict, decisions often must be made rapidly. Again, this leads into another dimension of decision making: *risk and uncertainty*. Voters, too, face uncertainty when they select among presidential candidates; they cannot be sure platform promises will be kept.

The literature on decision making in political geography is still limited. Noteworthy are two articles by Roger Kasperson, one that deals with the interplay of geography and politics in the city of Chicago (but with important implications of a wider nature) and another, involving the allocation of water resources, focusing on political decision making under crisis conditions.*

MENTAL MAPS

One of the most important factors influencing us as decision makers is our images of places, our *mental maps* of our immediate surroundings, our country, the world. When we are asked to draw our "mental map" of an area, we will distort it in a way that may significantly reflect our misconceptions. In other words, we hold in our mind a "model" of the spatial environment involving notions of distance, direction, shape, accessibility, and so forth—and on the basis of this model, which is at variance with reality, we operate.

So, too, do politicians. Complicated issues are simplified and thus distorted, and then decisions are made (and opinions swayed) accordingly. Out of distorted perceptions of the situation in Southeast Asia came policies that, had their consequences been known beforehand, would not have been followed. Ask the next person who raises the "Domino Theory" to draw you a map, by heart, to show you how it will work.

Spatial perception, of course, is only one dimension of a complex of images we hold of the world around us (including ourselves), a totality that has been called the *perceptual field*. This perceptual field is affected by numerous conditions: our cultural conditioning and the values attached to it, our attitudes, motivations,

goals. In the spatial context, at least within cultures, it may be possible to discover broad patterns of common behavior. Sonnenfeld addresses himself to these matters. He recognizes geographical, operational, perceptual, and behavior environments.

These generalizations deserve some elaboration.

In Chapter Two we pointed out that home is the most important place to people. It is, in fact, central in their world. This *centrality* of home—the most familiar, the most important place in the world—manifests itself clearly in children's maps of their neighborhoods or hometowns, in which their homes, schools, and other special places are shown as both larger and more central than they really are according to the scale of the drawing. It may be observed in the familiar humorous, but quite serious, maps of the United States as seen by Texans, New Yorkers, Bostonians, or others, in which home is enormous and fairly detailed with the rest of the country fading vaguely into the distance. And again by the all-too-common maps showing Alabama or Jordan or Podunk surrounded by concentric circles proclaiming its centrality in the universe and, hence, its ideal location for a vacation or for new industry. A view with profound political implications is the traditional Chinese perception of the world with China as "The Middle Kingdom" surrounded by barbarians.

In fact, every point on the globe is central, and concentric rings around it could demonstrate this clearly. A place is central only because individuals or groups perceive it that way, based on subjective criteria that are subject to change. Read "strategic" for "central" and you are thinking geopolitically.

Mental, or *cognitive*, *maps* are based on individual perceptions of the world, nearly always influenced strongly by location and culture. Mountain-dwellers and valley-dwellers have different surroundings, see the world differently, and hence have different attitudes in environ-

*R.E. Kasperson, "Toward a Geography of Urban Politics: Chicago, A Case Study," *Economic Geography, 41,* 2 (April 1965), 95–107; R.E. Kasperson, "Political Behavior and the Decision-making Process in the Allocation of Water Resources Between Recreational and Municipal Use," *Natural Resources Journal, 8,* 2 (April 1969), 7–18.

mental, social, and political affairs, among others. Similarly, seafarers and landlubbers have different perspectives; so do desert nomads and city residents. Perceptions of space are conditioned by occupation, by technology, and possibly by numerous other factors not fully understood.

Distance is also a matter of perception. It is a cliché by now to observe that the world is shrinking. It isn't actually, it just seems that way because circulation (transportation and communications) is so much faster, safer, and more efficient now and places are closer together in time than formerly. Americans, in fact, have become accustomed to measuring distance in time—minutes', hours', or days' travel time—or in money—bus zones.

But our perceptions of distance still vary from one individual to another. Is a place 200 miles away far away? That depends. People in Rhode Island and Texas would very likely have quite different answers. Imagine the difference in attitude between a citizen of a country in which it is physically or politically impossible to travel overland more than a few hours in any direction within the State's boundaries and one who lives in a country as huge as the United States or the Soviet Union— or one who lives in Micronesia and customarily sails in an open canoe 150 kilometers just to get a particular brand of tobacco.

Cognitive distance affects the orientation of peoples. Until the 1970s, Australians felt closer to England than to the Philippines, and Jamaicans still feel closer to England or Canada than to Antigua only 1600 km (1000 miles) away. These perceptions help explain why Australia has only

---

COMPARISON OF SPATIAL PERCEPTIONS

A good way to understand how spatial perspectives are created, reinforced, and expressed is to compare atlases produced in different parts of the world. Here is a summary of two student atlases produced in the United States and Australia, with the maps listed in the order in which they appear.

Goode's World Atlas, 17th ed. *Chicago: Rand McNally, 1986*		*The Jacaranda Atlas of the World* *Milton, Queensland: Jacaranda Press, 1986*	
Area Covered	**Pages of Maps**	**Area Covered**	**Pages of Maps**
World	53	World	32
Metropolitan Areas		Australia	23
(5 U.S., 13 foreign)	18	Papua New Guinea	2
United States and Canada	53	New Zealand	2
Northern Lands and Seas	1	South Pacific	2
Middle America	8	Asia	3
South America	9	USSR	1
Europe	29	South and Southeast Asia	4
USSR	9	Middle East	2
Asia	9	Europe	14
Middle East and Southwest Asia	4	Africa	8
South and East Asia	12	United States and Canada	10
Pacific Ocean	2	Middle America	2
Australia and New Zealand	8	South America	6
Africa	14	Arctic	2
Southern Lands and Seas	1	Antarctic	2
Ocean Floor	6		
Total	226	Total	114

weak links with the Philippines and why the West Indies Federation was doomed to failure.

Direction is a matter of cognition also, conditioned largely by culture and in particular by frequent exposure to maps oriented in a particular direction and by the kind of ethnocentrism already discussed under centrality. Our geographic vocabulary is full of expressions such as Near, Middle, and Far East; Down Under; Top of the World; and Subsaharan Africa. In the United States, the four cardinal points of the compass are not north, south, east, and west, but "Up North," "Down South," "Out West," and "Back East." We commonly speak of going up to Canada or down to Baja, of above the equator or below the Rio Grande.

None of these terms has any meaning whatever on a globe or on the surface of the earth itself. They are all based on the perspective of the viewers who coined the phrases and helped diffuse them around the world to distort other peoples' perceptions. Indonesia may be the Far East to a Britisher, but to an Australian it's the Near North. This was brought forcefully to the attention of Australians when Japanese troops occupied Indonesia during World War II, invaded eastern New Guinea, and threatened their homeland. Australia had to reorient her attention away from Britain to the west and toward Japan to the north and the United States to the northeast.

Cognitive direction is also linked with centrality; that is, a single situation viewed from two perspectives can look very different. During the Cold War in the 1950s, Americans were told that the country was surrounded by a chain of their own military bases protecting them from attack by the Soviet Union. But a map appearing in a popular weekly news magazine was centered on the Soviet Union rather than the United States and showed unmistakably that it was the Soviet Union that was surrounded—and by hostile, not friendly forces. Then, in October 1962, when the United States revealed that the USSR had placed intermediate-range ballistic missiles and long-range bombers in Cuba that were capable of delivering nuclear warheads to nearly every major city in North America, many Latin Americans chuckled about a small Latin American country (Cuba) pulling the tail feathers of the Yanqui eagle. The smiles faded, however, when a wire service map appeared in local newspapers showing concentric rings around Havana representing the ranges of the weapons. To their surprise and dismay, they saw that those very same nuclear warheads could be delivered anywhere in Latin America as far south as Lima. Fidel Castro lost many friends and admirers in Latin America that day. A final example illustrates the changing role of Mexico in the world as she develops economically and begins investing in Central America. Mexicans used to refer to the United States as "The Colossus of the North." They seldom use that sobriquet any more since they no longer feel so inferior to the United States, but now Central Americans are using it to refer to Mexico!*

Another type of cognitive map is largely *imaginary*, an idealized or stereotyped vision of a place that bears little resemblance to the *real* place but nevertheless influences decisions and actions. Soldiers go off to war to fight and die for the green fields of England or the Volga steppe or Main Street, USA. In truth, these images have come to them only through folklore and the soldiers may really live in the industrial slums of Birmingham or Kiev or Pittsburgh. But, be it ever so humble, a slum tenement is hardly worth dying for 10,000 miles away, and governments foster images of picket fences and pastures of plenty as morale boosters.

Boosterism has a long and honored history in the United States, and it has helped

*Then there is the classic story of the headline not long ago in The Times (London), that symbolizes why Britain has been so reluctant to join Europe: "Terrible Gale in the Channel—Continent Isolated."

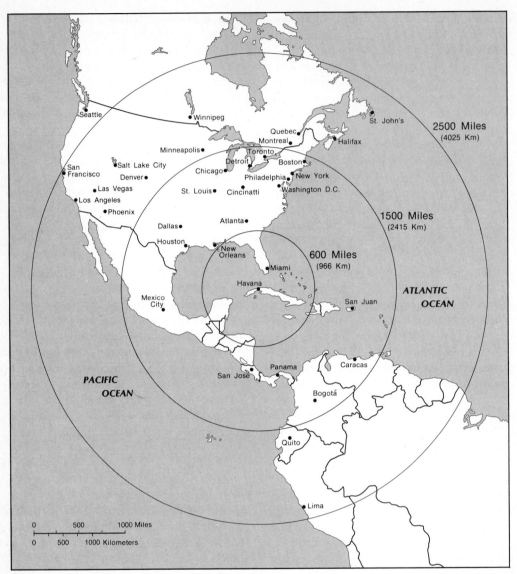

The Cuban Missile Crisis of 1962.   This map, centered on Havana, shows clearly the potential threat to Latin America as well as North America posed by Soviet offensive weapons based in Cuba. The inner circle represents the approximate range of the smaller missiles, the second that of the nuclear weapon-carrying bombers and the medium-range ballistic missiles, while the outer circle shows the approximate range of the intermediate-range missiles.

to shape the iconography that inspires loyalty. We discuss iconography in more detail in a later chapter, but here's an example designed to create a mental image of the United States. It's a song from the late nineteenth century, typical of the poems, advertisements, and songs created to entice immigrants to settle the new lands and build industries.

Bounding the U.S.

Of all the mighty nations in the East and in the West,
Why, the glorious Yankee nation is the wisest and the best.
There is room for all creation, and the banner is unfurled,
It's a general invitation to the people of the world.

Chorus

Come along, come along, make no delay;
Come from every nation, come from every
way,
Our land it is broad enough, and don't you
be alarmed,
Uncle Sam is rich enough, he'll give you all a
farm.

The St. Lawrence bounds our Northern line,
the crystal waters flow,
And the Rio Grande, the southern bound,
way down in Mexico.
From the great Atlantic Ocean where the sun
begins to dawn,
It peaks the Rocky Mountains, clear away in
Oregon.

Chorus

In the south, they raise the cotton, in the
West, the corn and pork,
While New England's manufactury, they do
the finer work,
And the little creeks and waterfalls, that force
along our hills,
Are just the thing for washing sheep and
driving cotton mills.

Chorus

This magnificent vision of America as a Garden of Eden was replaced not long after by one less attractive as settlers began to move out of the prairies of the Old Northwest and the Mississippi Valley, out onto the Great Plains where rainfall was scantier and the sod would not yield to the simple plows of the period. Then developed the legend of the Great American Desert blocking the way to California and Oregon, a region fit only for grazing, not for large-scale settlement. This legend, like the song, influenced migration patterns profoundly.

MAP DISTORTIONS

All maps are distortions of reality. They must be, for no flat map can accurately portray even a small portion of the earth's surface, to say nothing of the entire earth. The problem for the cartographer is how to manipulate the distortion so as to minimize it in those characteristics that are most important for the purpose of the map. The three essential qualities of every map are *projection*, *scale*, and *symbolization*.

In order to control distortion, the cartographer chooses the projection most appropriate for the purpose: an equal-area projection if the map is to show comparisons of areas in different parts of it or an equidistant projection if distance is the critical factor. The wrong projection can convey the wrong impression. The scale must be large enough for the amount of detail in the map to be seen clearly, but not so large that the map wastes space or becomes unwieldy. Symbols must fit the map scale and be easy for the reader to interpret, with the aid of a legend or key if necessary. A responsible cartographer will select and combine these qualities in various ways in order to transmit information about a part or all of the earth's surface. But even a map produced with the best of intentions is not always perfect; it is subject to errors in the original data on which it is based, the inexperience or subconscious biases of the cartographer, and the interpretations of the reader.

Even accurate maps can be misleading. The projection with which we are most familiar is Mercator's. It is a splendid projection for navigation, which is what Mercator designed it for in the sixteenth century, and is widely used today by aerial and maritime navigators all over the world. Unfortunately, however, it is all too commonly used for general-purpose or reference maps and, especially on small-scale maps, presents a grossly distorted picture of the world. Yet this is the map of the world most of us have indelibly printed on our minds. Wrong as it is, it still shapes our thinking. The original maps in this book have been designed and executed with the greatest care, each one individually, so as to be as useful as possible within the technical limitations im-

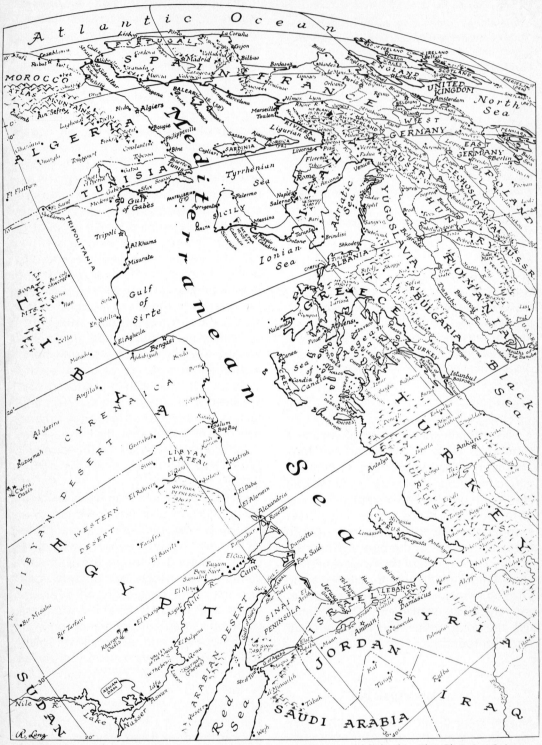

North Africa as seen from Mideast.   One of a series of 10 maps published in *The Christian Science Monitor* in late 1968, designed to show how people in various parts of the world perceive nearby areas that look very different when viewed from other perspectives. An understanding of the other person's perspective of the world can help us to appreciate his or her attitudes and decision-making. (© 1968 Russel H. Lenz. All rights reserved)

posed by the book itself. Nevertheless, to derive the fullest value from them, the reader must study each map carefully and try to understand its message.

Our Mercator view of the world prior to World War II led us to exaggerate the size of areas in the high latitudes and ignore the polar areas entirely, except when reading or hearing accounts of the exploits of polar explorers. But the global nature of the war plus the extensive use of aircraft made polar projection maps and other azimuthal (or zenithal) maps popular. American insularity had been reinforced by such historical and cultural factors as the central role of the Monroe Doctrine in our foreign policy, the location of the Pan American Union headquarters in Washington, Roosevelt's "Good Neighbor" policy toward Latin America, and the desire of many European immigrants to turn away from the misery and oppression they had left behind and look into the interior of North America for hope and opportunity. All this was changed by the war when Asia, the Pacific islands, North Africa, and the Soviet Union suddenly became our neighbors.

While we have tended to lapse back into our more traditional view of the world, we are now more accustomed to seeing maps not drawn to cylindrical projections and not oriented with north at the top.* During the war, Richard Edes Harrison produced a number of such unconventional maps and J. Paul Goode's homolosine projection became popular. Both helped us to develop mental maps somewhat more realistic than a Mercator-view. In the 1960s, *The Christian Science Monitor* ran a series of full-page maps that showed

portions of the earth as viewed from different points above it. The series was called "Global Perspectives" and was an attempt to help us understand the mental maps of the world carried around by people living in Canada, the Soviet Union, China, Africa, and other places. More of these maps more widely used would help us develop a more accurate mental image of the world.

Cartographers use graphic images resembling maps but deliberately drawn out of scale and not even drawn to a projection, but emphasizing size or shape as the primary symbol. Pictograms, cartograms, and block diagrams are the most common. We use a cartogram in Part Nine to emphasize the gross disparity in per capita incomes around the world. Maps and other graphic devices can be used to inform, as in most of the cases cited so far, or to persuade. Maps designed deliberately to channel the reader's thinking along certain lines are called *propaganda maps* and are far more common than we realize.

Deliberate distortion of a Soviet map. In the late 1960s the Soviet Union, which has had a reputation for cartographic excellence, began producing maps such as this one (redrawn for clarity), which distort many areas of the country. Apparently two techniques were used: creation of a pseudoprojection on which to draw the map and deliberate shifting of features from their actual positions. Such distortions are relatively easy to detect by comparison with older, accurate maps and through remote sensing from aircraft or space vehicles.

*There is no particular reason why north should be placed at the top of the map, except perhaps that most of the world's land area, people, States, and economic activity are located in the Northern Hemisphere. Medieval European maps were oriented (from *orient*—east) with east at the top, the general direction of Jerusalem, and some even had Jerusalem in the center with the rest of the world around it, and all were perfectly well understood by their readers.

PROPAGANDA MAPS

Some propaganda maps are relatively benign, such as the maps mentioned earlier that show the United States or the world revolving around Alabama or Jordan or Podunk. Others are found widely in commercial advertising: they depict how *our* bank blankets the state with its branch offices, or that *our* restaurant chain has 7479 locations in 59 states and, therefore, it must be better (or at least nearer) than the competition, or that *our* hotel is located precisely halfway between here and there and is therefore the best possible place to spend the night. We are concerned here, however, with maps that carry political messages.

Political propaganda maps, often in cartoon form, are not new. The late nineteenth and early twentieth centuries saw a spate of them, particularly in Europe, depicting countries in shapes or caricatures purporting to picture the national character. Prussian, Russian, Turk and Englishman were the favorite cartoon-map characters portraying stereotypes of the respective States. With changing tastes and circumstances and growing sophistication among propagandists and cartographers, propaganda maps became more serious and perhaps more persuasive.

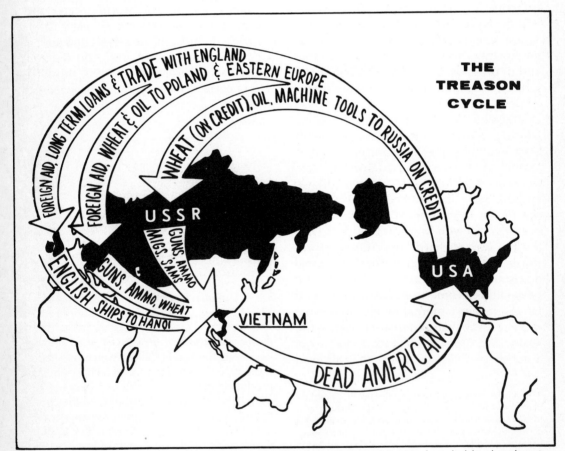

A modern propaganda map.   This map was distributed in Connecticut (and probably elsewhere) by the John Birch Society at the height of the Vietnam War in 1968. It uses a number of standard propaganda techniques to convey the message that dealing with those who engage in any way with our enemy of the moment is treason. An older example of a propaganda map may be found in Chapter 17.

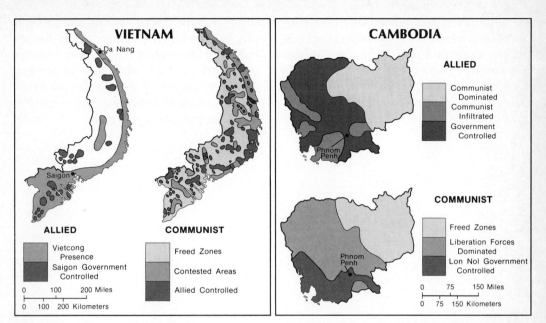

The map war in Indochina. These two pairs of maps represent versions of the military situation in South Vietnam and Cambodia (now Kampuchea) toward the end of 1970, produced by the United States and her allies on one hand and the Communist-led rebels on the other. Each side was trying to convince the world that it controlled the most territory, in order to win support and establish credibility. Similar map wars to influence public opinion were waged in Portuguese Guinea and elsewhere.

During the 1930s, the German "geopoliticans" made propaganda maps one of their principal weapons in generating support among intellectuals and the general public at home and abroad for the Nazi policy of German expansion.

The Germans clearly used projection, scale, and symbol to distort reality, always retaining enough truth in their maps to lend them credibility, and appealing to the emotions rather than the intellect of their readers. It is difficult to gauge their effectiveness, even with hindsight, but they did leave us a legacy of greater understanding of the utility of maps in persuasion as well as education. Even during World War II, the Allies used maps to dramatize the threat posed by the enemy or to justify strategies such as attacking the "soft underbelly of Europe" and invading one island in the Pacific after another in stepping-stone fashion toward Japan.

Both sides, but particularly the Allies, deliberately falsified maps that the enemy was allowed to "capture." The maps misled the enemy about planned military operations, part of the intelligence/counterintelligence battles that are so much a part of modern warfare. Even well after the war, the Soviet Union produced falsified maps of its own territory in apparent attempts to mislead NATO strategic planners about the locations of cities and natural features. These are not really propaganda maps but specialized types of military maps.

Maps are also used extensively in psychological warfare. During the "Cold War" of the 1950s and 1960s, both the United States and the USSR used maps to impress their peoples with the danger of imminent attack by the other, across Europe or over the north polar region. We have mentioned the maps of US bases surrounding the USSR; there were many more.

Maps continue to play an important role in attempts to mold public opinion on political issues. The Vietnam War, the Panama Canal treaties, the Egyptian blockade of the Strait of Tiran, "wars of national liberation," the Argentine–Chilean dispute over the Beagle Channel, and many other battles have been fought in maps as well as on the ground. There is no reason to believe that maps will be used in this way any less in the future. Examples are scattered through this book, and so are other instances of perception in political decision making. The concept could well be kept in mind as we explore some elements of political geography.

Books and Monographs

A

Alexander, Lewis M., *World Political Patterns*. Chicago: Rand McNally, 1963.

Alland, A., *The Human Imperative*. New York: Columbia Univ. Press, 1972.

Amadeo, Douglas, and Reginald Golledge, *Introduction to Scientific Reasoning in Geography*. New York: Wiley, 1975.

Ardrey, Robert, *The Territorial Imperative*. New York: Atheneum, 1966.

B

Bennett, D. Gordon (ed.), *Tension Areas of the World*. Delray Beach, FL: Park Press, 1982.

Bergman, Edward F., *Modern Political Geography*. Boston: Little, Brown, 1975.

Berry, Brian J.L., *The Nature of Change in Geographical Ideas*. De Kalb: Northern Illinois Univ. Press, 1979.

Bohannan, Paul, "Space and Territoriality," in *Africa and Africans*. Garden City, NY: Doubleday, Natural History Press, 1964.

Bowman, Isaiah, *The New World: Problems in Political Geography*. Yonkers, NY: World Book Co., 1921.

Burnett, Alan D. and Peter J. Taylor (eds.), *Political Studies from Spatial Perspectives*. Chichester, Sussex, Eng.; New York; and elsewhere: Wiley, 1981.

Busteed, Mervyn (ed.), *Developments in Political Geography*. London and elsewhere: Academic Press, 1983.

C

Carlson, Lucille, *Geography and World Politics*. Englewood Cliffs, NJ: Prentice-Hall, 1958.

Carpenter, C.R., "Territoriality: A Review of Concepts and Problems," in Anne Roe and George Simpson (eds.), *Behavior and Evolution*. New Haven, CT: Yale Univ. Press, 1958.

Coates, B.E. and others, *Geography and Inequality*. Oxford: Oxford Univ. Press, 1977.

Corbridge, Stuart. *Capitalist World Development; A Critique of Radical Development Geography*. Totowa, NJ: Rowman & Littlefield, 1986.

Cornish, G.A. and D. Whyte, *Geographical Changes Due to the Great War*. Toronto: A.T. Wilgress, 1920.

Cox, Kevin, *Conflict, Power and Politics in the City: A Geographic View*. New York: McGraw-Hill, 1973.

Cox, Kevin and others, *Locational Approaches to Power and Conflict*. New York: Halsted Press, 1974.

Crone, Gerald Roe, *Background to Political Geography*. Chester Springs, PA: Dufour, 1967.

D

Diesing, Paul, *Science and Ideology in the Policy Sciences*. New York: Aldine, 1982.

Dikshit, Ramesh K., *Political Geography: A Contemporary Perspective*. New Delhi: Tata McGraw-Hill, 1982.

Downs, Roger M. and David Stea, *Image and Environment*. Chicago: Aldine, 1973.

———, *Maps in Minds; Reflections on Cognitive Mapping*. New York: Harper & Row, 1977.

E

East, W. Gordon and A.E. Moodie (eds.), *The Changing World: Studies in Political Geography*. London: Methuen, 1968.

East, W. Gordon and J.R.V. Prescott, *Our Fragmented World*. New York: Macmillan, 1975.

F

Fisher, Charles A. (ed.), *Essays in Political Geography*. London: Methuen, 1968.

Fitzgerald, Charles P., *The Chinese View of Their Place in the World*. Oxford, England: Oxford Univ. Press, 1969.

Freeman, T.W., "Political Geography," in *A Hundred Years of Geography*. London: Duckworth, 1961; Chicago: Aldine, 1963.

G

Glacken, Clarence, *Traces on the Rhodian Shore*. Berkeley and Los Angeles: Univ. of California Press, 1967.

Gladwin, Thomas, *East Is a Big Bird*. Cambridge: Harvard Univ. Press, 1970.

Goblet, Y.M., *Political Geography and the World Map*. New York: Praeger, 1955.

Gottmann, Jean, *The Significance of Territory*. Charlottesville: Univ. Press of Virginia, 1973.

Gould, Peter, *People in Information Space: The Mental Maps and Information Surfaces of Sweden*, Lund Studies in Geography, Series

B, Human Geography, No. 42. Lund, Swed.: Royal Univ. of Lund, 1975.

Gould, Peter, and Rodney White, *Mental Maps*, 2nd. ed. Boston: Allen & Unwin, 1986.

H

Hall, E.T., *The Silent Language*, Garden City, NY: Doubleday, 1959.

———, *The Hidden Dimension*. Garden City, NY: Doubleday, 1966.

Harrison, Richard Edes, *Look at the World: The Fortune Atlas for World Strategy*. New York: Knopf, 1944.

Hartshorne, Richard, "Political Geography," in Clarence F. Jones and Preston James (eds.), *American Geography: Inventory and Prospect*. Syracuse, NY: Syracuse Univ. Press, 1954, 167–225.

Harvey, David, *Explanation in Geography*. New York: St. Martin's Press, 1969.

Huff, Darrell, *How to Lie with Statistics*. New York: Norton, 1954.

I

Isaacs, Harold R., *Images of Asia: American Views of China and India*. New York: Putnam, Capricorn Books, 1962.

J

Jackson, W.A. Douglas (ed.), *Politics and Geographic Relationships*, 2nd ed. Englewood Cliffs, NJ: Prentice-Hall, 1971.

Jackson, W.A. Douglas and Edward F. Bergman, *A Geography of Politics*. Dubuque, IA: W.C. Brown, 1973.

Johnston, Ronald J., *Geography and the State: An Essay in Political Geography*. London: Macmillan; New York, St. Martin's Press, 1982.

Jones, Stephen B. and Marion F. Murphy, *Geography and World Affairs*. Chicago: Rand McNally, 1971.

K

Kasperson, Roger E., *Frontiers of Political Geography*. Englewood Cliffs, NJ: Prentice-Hall, 1973.

Kasperson, Roger E. and Julian Minghi (eds.), *The Structure of Political Geography*. Chicago: Aldine, 1969.

Kates, Robert W., *Hazard and Choice Perception in Flood Plain Analysis*. Research Paper No. 78. Chicago: Univ. of Chicago, Dept. of Geography, 1962.

Kelman, H.C. (ed.), *International Behavior: A Social-Psychological Analysis*. New York: Holt, Rinehart & Winston, 1965.

Kidron, Michael and Ronald Segal, *The New State of the World Atlas*. London: Pluto Press; New York: Simon & Schuster, 1987.

Kirby, Andrew, *The Politics of Location; An Introduction*. London and New York: Methuen, 1982.

Klineberg, O., *The Human Dimension in International Relations*. New York: Holt, Rinehart & Winston, 1964.

L

Lewis, D., *We, the Navigators*. Honolulu: University Press of Hawaii, 1972.

Lorenz, Konrad, *On Agression*. New York: Harcourt, Brace, 1966.

Lowenthal, David (ed.), *Environmental Perception and Behavior*. Research Paper No. 109. Chicago: Univ. of Chicago, Dept. of Geography, 1967.

Lowenthal, David and Martyn J. Bowden (eds.), *Geographies of the Mind*. New York: Oxford Univ. Press, 1976.

Lynch, Kevin, *The Image of the City*. Cambridge: MIT Press, 1960.

M

Martin, Geoffrey J., *The Life and Thought of Isaiah Bowman*. Hamden, CT: Shoe String Press, Archon Books, 1980.

Monmonier, Mark S., *Maps, Distortion and Meaning*. Washington: AAG Resource Paper No. 75–4. 1977.

Montagu, M.F.A., *Man and Agression*. New York: Oxford Univ. Press, 1968.

Moodie, A.E., *Geography Behind Politics*. London: Hutchinson's Univ. Library, 1949.

Muir, Richard, *Modern Political Geography*. New York: Halsted Press, 1975.

Muir, Richard and Ronan Paddison, *Politics, Geography and Behaviour*. London and New York: Methuen, 1981.

N

National Academy of Sciences, *Studies in Political Geography*. Washington: Committee on Geography, 1965.

P

Pacione, Michael (ed.), *Progress in Political Geography*. London and elsewhere: Croom Helm, 1985.

Pearcy, George Etzel and others, *World Political Geography*, 2nd ed. New York: T.Y. Crowell, 1957.

Peattie, R., *Look to the Frontier: Geography for the Peace Table*. New York: Harper & Bros., 1944.

Peet, Richard, *Radical Geography*. London: Methuen, 1978.

Philbrick, Allen K., *This Human World*. New York: Wiley, 1963.

Pounds, Norman J.G., *Political Geography*, 2nd ed. New York: McGraw-Hill, 1972.

Preiswerk, Roy and Dominique Perrot, *Ethnocentrism and History: Africa, Asia and Indian America in Western Textbooks*. New York: NOK Publishers International, 1978.

Prestwick, Roger, "Maps and the Perception of Space," in David A. Lanegran and Risa Palm, *An Invitation to Geography*, 2nd ed. New York: McGraw-Hill, 1978, 12–37.

R

Renner, George T., *Political Geography and Its Point of View*. New York: Crowell, 1948.

Roucek, Joseph S. (ed.) *Contemporary Political Ideologies*. New York: Philosophical Library, 1961.

S

Saarinen, Thomas F., *Environmental Planning; Perception and Behavior*. Prospect Heights, IL: Waveland Press, 1976.

Saarinen, Thomas F., *Perception of Environment*, Washington, AAG, Resource Paper No. 5, 1969.

Sack, Robert David, *Human Territoriality*. New York: Cambridge Univ. Press, 1986.

Shafi, Mohammad and Mehdi Raza, *Spectrum of Modern Geography; Essays in Memory of Professor Mohammad Anas*. New Delhi: Concept, 1986.

Short, John R., *An Introduction to Political Geography*. London and elsewhere: Routledge & Kegan Paul, 1982.

Smith, D.M., *Human Geography: A Welfare Approach*. London: Arnold, 1977.

Soja, Edward W., *The Geography of Modernization in Kenya*. Syracuse, NY: Syracuse Univ. Press, 1968.

Sommer, R., *Personal Space: The Behavioral Basis of Design*. Englewood Cliffs, NJ: Prentice-Hall, 1969.

Sonnenfeld, J. "Geography, Perception, and the Behavioral Environment," in Paul W. English and R.C. Mayfield (eds.), *Man, Space, and Environment*. New York: Oxford Univ. Press, 1972, 244–251.

T

Taylor, Peter, J. and John House, *Political Geography; Recent Advances and Future Directions*. London and Sydney: Croom Helm; Totowa, NJ: Barnes & Noble, 1984.

Taylor, Peter J., *Political Geography; World-Economy, Nation-State and Locality*. London and New York: Longman, 1985.

Thompson, W.R. (ed.), *Contending Approaches to World System Analysis*. Beverly Hills, CA: Sage, 1983.

Tuan, Yi-Fu, *Topophilia: A Study of Environmental Perception, Attitudes and Values*. Englewood Cliffs, NJ: Prentice-Hall, 1974.

V

Van Valkenburg, Samuel, *Elements of Political Geography*. New York: Prentice-Hall, 1939.

Van Valkenburg, Samuel and Carl L. Stotz, *Elements of Political Geography*. New York: Prentice-Hall, 1954.

W

Weigert H. and Vilhjalmur Stefansson (eds.), *Compass of the World: A Symposium on Political Geography*. New York: Macmillan, 1945.

Weigert H. and Richard E. Harrison (eds.), *New Compass of the World: A Symposium on Political Geography*. New York: Macmillan, 1949.

Weigert H. and others, *Principles of Political Geography*. New York: Appleton-Century-Crofts, 1957.

Whittlesey, Derwent, *The Earth and the State*. Madison, WI: Published for the U.S. Armed Forces Institute by Holt, 1944.

Periodicals

A

Adams, D.K., "A Note: Geopolitics and Political Geography in the United States Between the Wars," *Australian Journal of Politics and History*, 6, 1 (May 1960), 77–82.

Ager, J., "Maps and Propaganda," *Society of University Cartographers Bulletin*, 11 (1978), 1–15.

Agnew, John A., "Socializing the Geographical Imagination: Spatial Concepts in the World-System Perspective," *Political Geography Quarterly*, 1, 2 (April 1982), 159–166.

——, "An Excess of 'National Exceptionalism': Towards a New Political Geography of American Foreign Policy," *Political Geography Quarterly*, 2, 2 (April 1983), 151–166.

Alexander, Lewis M., "Major Trends in the World Political Patterns," *Institute of Indian Geographers* (Bihar), Publ. No. 1 (1954), 6–12.

Anderson, J., "Geography, Political Economy and the State," *Antipode*, 10, (1978), 87–93.

B

Baker, J.N.L., "Geography and Politics: The Geographical Doctrine of Balance," *Transactions and Papers, Institute of British Geographers*, (1947), 1–15.

Barnett, J.R., "Scale, Process, and the Diffusion of a Political Movement," *Proceedings, AAG*, 4 (1972), 9–14.

Boggs, S. Whittemore, "Mapping the Changing World: Suggested Developments in Maps," *Annals, AAG*, 31, 2 (June 1941), 119–128.

——, "Carthohypnosis," *Scientific Monthly*, 64 (1947), 469–476.

——, "Geographic and Other Scientific Techniques for Political Science," *American Political Science Review*, 42 (1948), 223–238.

Boulding, Kenneth E., "National Images and International Systems," *Journal of Conflict Resolution*, 3, 2 (June 1959), 120–131.

Bunge, William W., "Geography of Human Survival," *Annals, AAG*, 63, 3 (September 1973), 275–295.

Burghardt, Andrew, "The Dimensions of Political Geography: Some Recent Texts," *Canadian Geographer*, 9 (1965), 229–233.

——, "The Core Concept in Political Geography: A Definition of Terms," *Canadian Geographer*, 13, 4 (1969), 349–353.

C

Carr, S. and D. Schlissler, "The City as a Trip: Perceptual Selection and Memory in the View from the Road," *Environment and Behavior*, 1, 1 (1969), 7–36.

Chappell, John E., Jr., "Marxism and Geography," *Problems of Communism*, 14 (1965), 12–22.

Chaudhuri, Manoranjan, "Evolution of Ideas in Political Geography," *Calcutta Geographical Review*, 9, 1–4 (1947; pub. 1949), 23–31.

D

DeBres, Karen, "Political Geographers of the Past IV: George Renner and the Great Map Scandal of 1942," *Political Geography Quarterly*, 5, 4 (October 1986), 385–394.

DeJonge, Derek, "Images of Urban Areas: Their Struggle of Psychological Foundations," *Journal of the American Institute of Planners*, 28 (November 1962), 266–276.

Dikshit, R.D., "Toward a Generic Approach in Political Geography," *Tijdschrift voor Economische en Sociale Geografie*, 61, 4 (July/August 1970), 242–245.

E

East, W. Gordon, "The Nature of Political Geography," *Politica*, 2, (1937), 259–263.

F

Fairbank, John K., "The Chinese World Order," *Encounter*, 27 (December 1966), 14–20.

Fischer, Eric, "German Geographical Literature, 1940–1945," *Geographical Review*, 36, 1 (January 1946), 92–100.

Fleming, D., "Cartographic Strategies for Airline Advertising," *Geographical Review*, 74, 1 (January 1984), 76–93.

Fleure, H.J., "The Geographic Study of Society and World Problems," *Scottish Geographical Magazine*, 48 (1932), 257–274.

G

Getis, Arthur and B.N. Boots, "Spatial Behavior: Rats and Man," *Professional Geographer*, 23, 1 (January 1971), 11–14.

Golledge, R.G., "Behavioral Approaches in Geography," *Australian Geographer*, 12 (1972), 59–79.

Gould, Peter and N. Lafond, "Mental Maps and Information Surfaces in Quebec and Ontario," *Cahiers de géographie de Québec*, 23, 60 (1979), 371–398.

Gulick, John, "Images of an Arab City," *Journal of the American Institute of Planners*, 29 (1963), 179–198.

H

Hall, E.T., "Proxemics," *Current Anthropology*, 9 (1968), 83–108.

Hall, Peter, "The New Political Geography," *Transactions, Institute of British Geographers*, Publ. No. 63 (1974), 48–52.

Harrison, Richard Edes, "The War of the Maps," *Saturday Review of Literature*, 26 (August 7, 1943), 24–27.

——, "The Face of the World: Five Perspectives for an Understanding of the Air Age," *Saturday Review of Literature*, 27 (1944), 5–6.

Harrison, Richard Edes and Robert Strauz-Hupe, "Maps, Strategy, and World Politics," *Smithsonian Institution Report*, (1943–1944), 253–258.

Hartshorne, Richard, "Recent Developments in Political Geography," *American Political Science Review,* 29, (1935), 785–804, 943–966.

———, "The Politico-Geographic Pattern of the World," *Annals, American Academy of Political and Social Science,* 218 (1941), 45–57.

———, "Political Geography in the Modern World," *Journal of Conflict Resolution,* 4, 1 (March 1960), 52–66.

Henrickson, Alan K., "The Map as an 'Idea': The Role of Cartographic Images During the Second World War," *American Cartographer,* 2, 1 (1975), 19–53.

Herman, T., "Group Values Toward the National Space: The Case of China," *Geographical Review,* 49, 1 (January 1959), 164–182.

Heske, Henning, "Political Geographers of the Past III: German Geographic Research in the Nazi Period—A Content Analysis of the Major Geography Journals 1925–1945," *Political Geography Quarterly,* 5, 3 (July 1986), 267–281.

Holdich, Thomas H., "Some Aspects of Political Geography," *Geographical Journal,* 34, 6 (December 1909), 593–607.

Horton, F.E. and David R. Reynolds, "An Investigation of Individual Action Spaces: A Progress Report." *Proceedings, AAG,* 1 (1969), 43–47.

I

Iwata, Kozo, "The Recent Trend of Political Geography," *Jimbun-chiri,* (Human Geography) 8, 3, (1956), 165–175.

J

Jackson, W.A. Douglas, "Whither Political Geography?" *Annals, AAG,* 48, 2 (June 1958), 178–183.

Johnson, Douglas Wilson, "A Geographer at the Front and at the Peace Conference," *Natural History,* 19 (1919), 511–521.

Johnston, Ronald J., "Activity Spaces and Residential Preferences: Some Tests of the Hypothesis of Sectoral Mental Maps," *Economic Geography,* 48, 2 (April 1972), 199–211.

———, "Texts, Actors and Higher Managers: Judges, Bureaucrats and the Political Organization of Space," *Political Geography Quarterly,* 2, 1 (January 1983), 3–19.

Johnston, Ronald J. and Mickey Lauria, Review Essay: "Political Geography: Improving the Theoretical Base," *Political Geography Quarterly,* 4, 3 (July 1985), 251–257.

Jones, Stephen B., "Field Geography and Post-war Political Problems," *Geographical Review,* 33, 4 (October 1943), 446–456.

———, "Views of the Political World," *Geographical Review,* 45, 3 (July 1955), 309–326.

K

Kirby, Andrew, "Pseudo-random Thoughts on Space, Scale, and Ideology in Political Geography," *Political Geography Quarterly,* 4, 1 (January 1985), 5–18.

Kliot, Nurit, "Contemporary Trends in Political Geography: A Critical Review," *Monadnock,* 50 (1976), 54–70.

———, "Contemporary Trends in Political Geography," *Tijdschrift voor Economische en Sociale Geografie* 73, (1982), 270–279.

Kriesel, Karl Marcus, "Montesquieu: Possibilistic Political Geographer," *Annals, AAG,* 58, 3 (September 1968), 557–574.

L

Ley, David and Roman Cybriewsky, "Urban Graffiti as Territorial Markers," *Annals, AAG,* 64, 4 (December 1974), 491–505.

Little, Kenneth B., "Personal Space," *Journal of Experimental Social Psychology,* 1, 237–247.

Lundén, Thomas, "Political Geography Around the World VI: Swedish Contributions to Political Geography," *Political Geography Quarterly,* 5, 2 (October 1986), 181–186.

Lyman, Stanford and Marvin B. Scott, "Territoriality: A Neglected Sociological Dimension," *Social Problems,* 15 (1967), 236–249.

M

Martin, Geoffrey J., "Ellsworth Huntington and the Pace of History," *Connecticut Review,* 5, 1 (October 1971), 83–123.

McCleary, George F., "Beyond Simple Psychophysics: Approaches to the Understanding of Map Perception," *Proceedings, American Congress of Surveying and Mapping* (1970), 189–209.

McColl, Robert W., "Political Geography as Political Ecology," *Professional Geographer,* 18, 3 (May 1966), 143–145.

Mercer, John, Review Essay: "Diversity as Strength," *Political Geography Quarterly,* 2, 1 (January 1983), 81–87.

Meyer, H.C., "The Chorographic Commentator—A Geographic Guide to Relationship Thinking," *Journal of Geography,* 40, 4 (April 1941), 145–152.

Mojumbar, K.K., "A Case for Specific Studies in Political Geography," *Geographic Observer,* 5 (1969), 50–61.

Muehrcke, Phillip C., and Juliana O. Muehrcke, "Maps in Literature," *Geographical Review*, 64, 3 (July 1974), 317–338.

Mookerjee, Sitanshu, "A Review of Political Geography," *Geographical Review of India*, 18, 1 (March 1956), 34–46.

O

Oatley, K., "Mental Maps for Navigation," *New Scientist*, 19 (1974), 863–866.

Obenbrugge, Jurgen, "Political Geography Around the World: West Germany," *Political Geography Quarterly*, 2, 1 (January 1983), 71–80.

Offe, C., "Advanced Capitalism and the Welfare State," *Politics and Society*, 2 (1972), 479–488.

Ogata, Sadako, "Japanese Attitudes Toward China," *Asian Survey*, (1965), 389–398.

Ormeling, F.J., Jr., "Cartographic Consequences of a Planned Economy—50 Years of Soviet Cartography," *American Cartographer*, 1, 1 (1974), 39–50.

Ormsby, H., "The Definition of Mitteleuropa and Its Relation to the Conception of Deutschland in the Writings of Modern German Geographers," *Scottish Geographical Magazine*, 51 (1935), 337–453.

Otok, Stanislaw, "Political Geography Around the World: Poland," *Political Geography Quarterly*, 4, 4 (October 1985), 321–327.

O'Tuathail, Gearóid, "Beyond Empiricist Political Geography: A Comment on van der Wusten and O'Loughlin," *Professional Geographer*, 39, 2 (May 1987), 196–197.

P

Politics and Geography. Special Issue of *International Political Science Review* 1, 4 (1980).

Porteous, J. Douglas, "Home: The Territorial Base," *Geographical Review*, 66, 4 (October 1976), 383–390.

Pounds, Norman J.G., "History and Geography: A Perspective in Partition," *Journal of International Affairs*, 18, 2 (1964), 161–172.

Prescott, John Robert Victor and J.G. Hajdu, "A Review of Some German Post-War Contributions to Political Geography," *Australian Geographical Studies*, 2, 1 (1964), 35–46.

Q

Quam, Louis O., "The Use of Maps in Propaganda," *Journal of Geography*, 42, 1 (January 1943), 21–32.

R

Rumley, D. and others, "The Content of Ratzel's *Politische Geographie*," *Professional Geographer*, 25, 3 (1973), 271–276.

Rushton, G., "Analysis of Spatial Behavior by Revealed Space Preference," *Annals, AAG*, 59, 2 (June 1969), 391–400.

S

Salter, P.S. and R.C. Mings, "A Geographic Aspect of the 1968 Miami Racial Disturbance: A Preliminary Investigation," *Professional Geographer* 21, 2, (March 1969), 79–86.

Sanguin, André-Louis, "Political Geographers of the Past II: André Siegfried, An Unconventional French Political Geographer," *Political Geography Quarterly*, 4, 1 (January 1985), 79–83.

Sauer, Carl O., "The Formative Years of Ratzel in the United States," *Annals, AAG*, 61, 2 (June 1971), 245–254.

Schat, P., "Political Geography: A Review," *Tijdschrift voor Economische en Sociale Geografie*, 60 (1969), 255–260.

Schofer, Jerry P., "Territoriality at the Micro-, Meso- & Macro-Scales," *Journal of Geography*, 74, 3 (March 1975), 151–158.

Scholler, Peter, "Advances and Aberrations of Political Geography and Geopolitics," *Erdkunde*, 11, 1 (1957), 1–20.

Schwarz, Henry G., "America Faces Asia: The Problem of Image Projection," *Journal of Politics*, 26 (1964), 532–549.

Sevrin, Robert, "Political Geography Around the World IV: Research Themes in Political Geography—A French Perspective," *Political Geography Quarterly*, 4, 1 (June 1985), 67–78.

Sillitoe, A., "A Sense of Place: One Man's Maps," *Geographical Magazine*, 47 (1975), 685–689.

Sinnhuber, K.A., "The Representation of Disputed Political Boundaries in General Atlases," *Cartographic Journal*, 2 (1964), 20–28.

Smith, H.R. and J.F. Hart, "The American Tariff Map," *Geographical Review*, 65, 3 (July 1955), 327–346.

Smith, Neil, "Political Geographers of the Past. Isaiah Bowman: Political Geography and Geopolitics," *Political Geography Quarterly*, 3, 1 (January 1984), 69–76.

Smith, Thomas R. and Lloyd D. Black, "German Geography: War Work and Present," *Geographical Review*, 36, 3 (July 1946), 398–408.

Sonnenfeld, J., "The Geopolitical Environment as a Perceptual Environment," *Journal of Geography*, 69, 7 (October 1970), 415–419.

Spate, O.H.K., "Toynbee and Huntington: A Study in Determinism," *Geographical Journal*, 118, 4 (December 1952), 406–428.

Sprout, Harold, "Political Geography as a Political Science Field," *American Political Science Review*, 25, 2 (May 1931), 439–442.

Sprout, Harold H. and Margaret T. Sprout, "Geography and International Politics in an Era of Revolutionary Change," *Journal of Conflict Resolution*, 4, 1 (March 1960), 145–161.

Stea, David, "Space, Territoriality and Human Movements," *Landscape*, 15 (1965), 13–16.

T

Tao, Jay, "Mao's World Outlook: Vietnam and the Revolution in China," *Asian Survey*, 8, 5 (May 1968), 416–432.

Thrift, Nigel and Dean Forbes, Review Essay: "A Landscape with Figures: Political Geography with Human Conflict," *Political Geography Quarterly*, 2, 3 (July 1983), 247–263.

Troll, Carl, "Geographic Science in Germany During the Period 1933–1945," *Annals, AAG*, 39, 2 (June 1949), 99–137.

Tuan, Yi-fu, "Environmental Psychology: A Review," *Geographical Review*, 62, 2 (April 1972), 245–256.

Tyner, Judith, "Persuasive Cartography," *Journal of Geography*, 81, 4 (July–August 1982), 140–144.

V

van der Wusten, Herman and John O'Loughlin, "Back to the Future of Political Geography: A Rejoinder to O'Tuathail," *Professional Geographer*, 39, 2 (May 1987), 198–199.

Van Valkenburg, Samuel, "A Political Geographer Looks at the World," *Professional Geographer*, 12, 4 (July 1960), 6–8.

W

Wallerstein, Immanuel, "The Rise and Future Demise of the World Capitalist System: Concepts for Comparative Analysis," *Comparative Studies in Society and History*, 16 (1974), 387–415.

Weigert, Hans W., "Maps Are Weapons," *Survey Geographic*, 30, 10 (October 1940), 528–530.

Whittlesey, Derwent, "The Horizon of Geography," *Annals, AAG*, 35, 1 (March 1945), 1–36.

Wolpert, Julian, "The Decision Process in Spatial Context," *Annals, AAG*, 54, 4 (October 1964), 537–558.

———, "Behavioral Aspects of the Decision to Migrate," *Papers, Regional Science Association*, 15 (1965), 17–29.

———, "Departures from the Usual Environment in Locational Analysis," *Annals, AAG*, 60, 2 (April 1970), 220–229.

Wright, John Kirtland, "Voting Habits in the United States," *Geographical Review*, 22, 3 (October 1932), 666–672.

———, "Geography and the Study of Foreign Affairs," *Foreign Affairs*, 17, 1 (October 1938), 153–163.

———, "Training for Research in Political Geography," *Annals, AAG*, 34, 4 (December 1944), 190–201.

Part Two

THE STATE

Chapter 4

STATE, NATION, AND NATION-STATE

Traditionally, political geographers have devoted more of their attention to study of the State than to any other topic. The State has been described, classified, analyzed, discussed, and argued about from the time of Aristotle and Plato—if not earlier. Today it is still the focus of our attention, and despite many other important and alluring subjects to investigate, some of them relatively new, this should not be surprising. The State has become in the past two centuries the dominant form of political organization in the world and is likely to remain so for some time to come.

In much of the discussion, not only among nonspecialists but even among scholars in many disciplines, the term "State" is confused with "state" or "nation." Often "State" or "state" or "nation" is used interchangeably with "nation-state," and sometimes the hyphenated word is used rather pompously in place of a simpler one. Before we proceed further, we should clarify the proper meanings of these terms.

STATE

A State is a place. It is also a concept represented by certain symbols and demanding (though not always obtaining) the loyalty of people. Its intangible qualities are discussed later, but here we can examine some of its more measurable qualities.

In order for a place to be considered a State in the strictest sense, it must possess to a reasonable degree certain characteristics:

1. **Land territory.** A State must occupy a definite portion of the Earth's land surface and should have more or less generally recognized limits, even if some of its boundaries are undefined or disputed.

2. **Permanent resident population.** An area devoid of people altogether, no matter how large, cannot be a State. An area only traversed by nomads or occupied seasonally by hunters or scientists similarly cannot be a State. A State is a human institution created by people to serve some of their particular needs.

3. **Government.** The people living within a territory must have some sort of administrative system to perform functions needed or desired by the people. Without political organization, there can be no State.

4. **Organized economy.** While every society has some form of economic system, a State invariably has responsibility for many economic activities, even if they include little more than the issuance and supervision of money and the regulation of foreign trade, and even if economic activities are managed badly.

5. **Circulation system.** In order for a State to function, there must be some or-

37

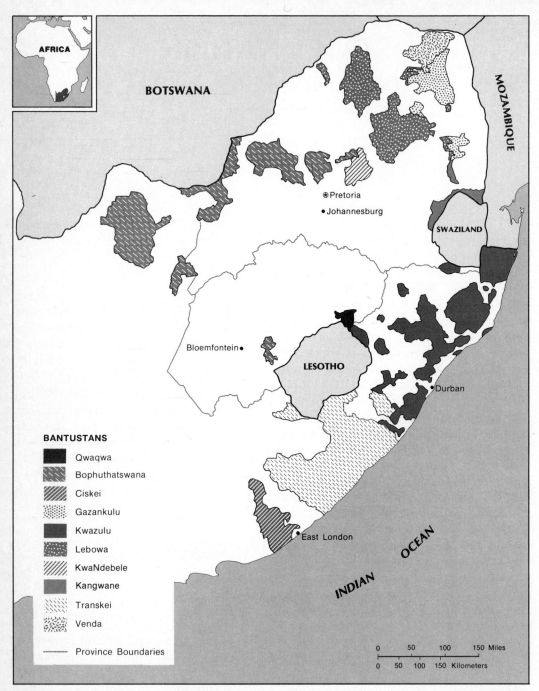

Bantustans or "homelands" in South Africa. The South African government has declared that the ultimate goal of its policy of *apartheid*, or "separate development" of the "races," is the creation of a separate State for each major group. Despite some consolidation in recent years, the "homelands" that are to evolve into these States are still badly fragmented. They total about 14 percent of the land area of the republic, generally the least desirable land, for about 80 percent of the population. By 1982 Transkei, Ciskei, Venda, and Bophuthatswana had been declared "independent" by South Africa, but this claim is taken seriously by few and officially recognized by none outside the republic.

ganized means of transmitting goods, people, and ideas from one part of the territory to another. All forms of transportation and communication are included within the term "circulation," but a modern State must have more sophisticated forms available to it than runners or the "bush telegraph."

These requirements for a State are all geographic, and political geographers have long been studying them. Two other requirements, however, have traditionally been in the realms of political science and international law. If we are to be realistic in any attempt to classify the political units of the world, we must give careful consideration to them. In many cases the geographic criteria may be quite clear, but only the political criteria can be decisive in determining whether or not a political unit is a State. These criteria are:

1. **"Sovereignty."** There is general agreement that a State is "sovereign" or "possesses sovereignty," but no agreement on precisely what constitutes "sovereignty." In a general sense it means power over the people of an area unrestrained by laws originating outside the area, or independence completely free of direct external control. If we try to get more specific than this, however, we encounter serious problems. There are so many qualifications and exceptions to the general concept that it would seem wise to use quotation marks around "sovereign" or "sovereignty" in most cases. However, that is often awkward and we may omit them as long as we understand that the term is imprecise and often used too loosely.

2. **Recognition.** For a political unit to be accepted as a State with "an international personality" of its own, it must be recognized as such by a significant portion of the international community, the existing States. In effect, it must be voted into the club by the existing members. Specialists in inter-

national law disagree about whether such recognition is "declarative" or "constitutive"; that is, whether the formal act of recognition simply puts the label of "State" on a political unit that has already demonstrated that it possesses most or all of the six criteria already mentioned, or whether it is the act of recognition itself that brings the State into being for the first time. There is also disagreement over what proportion of the club members must extend such a welcome and even over whether numbers alone have any value or whether affirmative acts by most of the "great powers" are necessary. These details need not concern us here, but we cannot forget that recognition is important, perhaps decisive, in determining the proper application of the word "State."

Among all the political units in the world, many are clearly States. The United States, for example, and Austria and Thailand, Venezuela and Tunisia. Many others are not: Hong Kong, Puerto Rico, Greenland, and French Polynesia, for example, are all dependent territories, even though they possess many of the attributes of statehood, including varying degrees of self-government. Still others are questionable. Is Andorra a State? Or the South African Bantustans? Or Taiwan? Or Vatican City? Or were a number of other entities whose status has become clarified only recently, such as Biafra, East Germany, Rhodesia, or Sikkim? We need not answer any of these questions. For our purposes, it is enough to understand that the term "State" has a fairly precise meaning and should be used with great care.

Nearly all States have civil divisions; that is, administrative districts within them which perform governmental functions and even, in the case of federal States, possess some measure of sovereignty. These civil divisions are frequently ranked in a hierarchy of size and/or authority, from first order civil divisions on down. States give many names to their first order civil divisions. Department, province, and county are among the most common.* In some States, all of which have at least nominally federal systems, the first order civil divisions are called states. These include Mexico, Brazil, India, and, of course, the United States. In this book, we follow the United Nations practice of capitalizing State when we mean an independent country and using a small "s" when we mean a first order civil division, such as Nuevo León, Mato Grosso, West Bengal, or South Carolina.†

NATION

Unlike a State, a nation is a group of people. Although definitions of the term vary considerably, political scientists and political geographers generally use it to mean a reasonably large group of people with a common culture, sharing one or more important culture traits, such as religion, language, political institutions, values, and historical experience. They tend to identify with one another, feel closer to one another than to outsiders, believe that they belong together. They are clearly distinguishable from others who do not share their culture.

This may seem rather vague, but it may help a bit if we point out that the term "people" when used as a singular noun is roughly equivalent to "nation," just as "country" is roughly equivalent to "State." Thus, we may properly speak of the Polish nation or the Navajo nation or the Ewe nation because each is a people distinguished from all others by its distinctive culture.

How small a group may be considered

*Others include krai, prefecture, division, parish, governorate, commune, circonscription, republic, region, wilaya, and district.

†It should be noted that the word is customarily capitalized when giving the official name of the division, as in State of New Hampshire or State of New South Wales. On the other hand, some countries use the word in their official long-form names: some are the State of Bahrain, State of Israel, State of Kuwait, Spanish State, and Independent State of Western Samoa.

a nation? There is no answer to this question. Certainly the group has to be larger than a family, a clan, or even a tribe, but how much larger no one can say. This is one of the dilemmas in the application of the principle of "national self-determination" or "self-determination of peoples," which we discuss later. We have in the world today some very large nations and some very small ones. Compiling a list would be an impossible task. Nevertheless, the concept of the nation as a cultural entity is very old, and it has been very important in the development of the modern political world. Perhaps the critical factor is whether the group in question considers itself to be a nation.

The feeling among a people that they constitute a nation may be carried further, into the political realm. They may begin to feel that they ought to have a State of their own. Or an individual may identify with the State in which he or she resides— or with another State. This linkage of emotion with a political entity called the State is termed, for lack of a better word, "nationalism." Nationalism is an old sentiment, apparently dating from our earliest civilizations, but we are primarily concerned here with its modern manifestations and its effects on the political organization of space.

Nationalism can be a very healthy emotion. It can encourage people to identify with a group larger than their family, clan, or tribe. It can get them interested, even involved, in the affairs of modern society. It can stimulate them to create new ideas and institutions that will advance society as a whole, in partnership with others and other nations. but if the emotion becomes more powerful, it can lead to a demand for self-determination among ever-smaller groups and to the fragmentation of society instead of its unification. Carried further, the emotion can lead to *irredentism*, the desire to incorporate within the State all areas that had once been part of it or in which live people who belong to the nation but not to the State. A still more extreme form of nationalism is *chauvinism*,* which may be defined as "superpatriotism," excessive and bellicose feelings of superiority over all other peoples and countries. Clearly this is unhealthy; it has frequently led to wars of aggression and to imperialism.

NATION-STATE

Just as the State is the dominant form of political organization today, the nation-state is the ideal form to which most nations and States aspire. That this ideal is very difficult to attain becomes evident when we try to count the number of true nation-states in the world today. Put in the simplest terms, a nation-state is a nation with a State wrapped around it. That is, it is a nation with its own State, a State in which there is no significant group that is not part of the nation. This does not mean simply a minority ethnic group, but a nationalistic group that either wants its own State or wants to be part of another State or wants at least a large measure of autonomy within the State in which it lives.

Despite being composed of a large number of minorities, the United States is arguably a nation-state. Japan certainly is. Sweden, Uruguay, Egypt, and New Zealand are all nation-states, though they have very different histories and demographic characteristics. The Soviet Union, on the other hand, is an empire, composed of many "nationalities," most of which have retained much of their cultural heritage in their ancestral homelands. Canada is frequently referred to as "two nations in one State," even as Germany is still sometimes called "one nation in two (or three or four) States." Belgium, South Africa, Afghanistan, and China are other examples of older countries that are not nation-states, and very few countries that have become independent since World War II can be considered nation-states. As

usual, other States fall in the question-mark column; it would take precise definitions, adequate data, and careful analysis to determine whether Israel, Jordan, Bolivia, or Liberia, for example, can truly be called nation-states.

In recent times we have seen a number of nations that are split among two or more States attempt to break away and form States of their own. Usually, those in one country are more enthusiastic about independence than their compatriots in adjacent countries. In no case yet has the group succeeded in assembling its politically separated parts and creating from them a new nation-state. Some examples are demands for independent States of Macedonians living in Bulgaria, Greece, Yugoslavia, and perhaps Albania; Basques living in Spain and France; and Kurds of Turkey, Iraq, Iran, Syria, and possibly the Soviet Union.

In the first case the issue seems to be raised from time to time primarily by Bulgaria, and there appears to be little enthusiasm for it among those who may be considered Macedonian. The great majority of Spanish Basques are probably content with a large measure of autonomy within Spain, which most of them received in 1977. Only a small minority is actively fighting for independence, while across the border, the French Basques seem to be content with the cultural autonomy they have long enjoyed. The Kurdish rebellion against Iraq has been prolonged and frequently intense, while the Kurds in Turkey and Iran have also rebelled at times, demanding autonomy within these countries—at the least. These examples indicate the very great difficulty today of detaching a portion of a State to form a new one on ethnic grounds. The creation of Bangladesh is a special case in which ethnic considerations were important but not decisive.

The other method of trying to create nation-states from existing States is the secession of areas within a State that are inhabited largely or exclusively by minority ethnic groups. We have seen numer-

---

*From Nicolas Chauvin, a French soldier wounded and decorated many times during the Revolutionary and Napoleonic wars but who retained a simple-minded devotion to Napoleon. He came to typify the cult of military glory and ultranationalism.

ous examples of this in the past few decades alone, and we are likely to see more in the future. Indonesia, for example, has had serious rebellions in the Riau Archipelago, in Southern Sulawesi (Celebes), and in the Southern Maluku (Molucca) Islands. Minority ethnic, or national, groups have engaged in full-scale rebellion or at least terrorism in order to obtain independence or autonomy in such countries as Chad (Muslim Toubou tribesmen), India (Mizos, Nagas), France (Bretons, Corsicans), Burma (Karens, Kachins, Shans), Yugoslavia (Croats), Ethiopia (Eritreans, Somalis), Nigeria (Ibos), China (Uigurs, Tibetans, others), Canada (Quebecois), Vietnam ("Montagnards"), Philippines (Moros), Sudan (southern blacks), Sri Lanka (Tamils), and many others. None of these rebellions has been any more successful in creating new States than the attempts to create new States from parts of several others.

IRREDENTISM

Another manifestation of nationalism, mentioned earlier, is irredentism. The term derives from an area in northern Italy that was still part of Austria in 1861 after the rest of Italy had been unified into a nation-state. Young Italian nationalists referred to it as *Italia irredenta,* or unredeemed Italy. Today there are a number of irredentist movements around the world, although it is often difficult to distinguish them from relatively routine border disputes, ideological conflicts, or simple aggression. There are also many cases in which people who speak the language or share other cultural attributes with those of a particular State live across the border of a neighboring State but are not involved in irredentist claims. They live in peace as citizens of their State and the neighboring States are on good terms.

On occasion, however, such circumstances lead to political problems. Sometimes it is felt that members of a nation living in an adjoining State are being mis-

treated and the State embodying their nation tries to defend them. This may lead to more vigorous action on their behalf, claims to the territory in which they live, and ultimately to military action designed to incorporate the territory into its own. Strong factions in Greece and Cyprus, for example, have long campaigned for *enosis,* or union of the two countries. Turkey has tended in the same direction to protect the Turkish Cypriot minority and in 1974 actually invaded Cyprus, ultimately occupying some 40 percent of the island. While they have not annexed the occupied area, they have sponsored therein a "Turkish Federated State of Cyprus."

Indonesia in the 1950s waged a vigorous campaign for the incorporation into her national territory of Dutch-held western New Guinea. She was finally successful in 1962 in forcing out the Dutch and in May 1963 obtained control after a nine-month period of United Nations administration. Indonesia then turned her attention increasingly to her "confrontation" with the new State of Malaysia, demanding all the former British territories on the island of Borneo. In this endeavor she received no support from the United Nations and temporarily withdrew from that body in January 1965. A change in government in late 1965 and the forced retirement of President Sukarno led to the abandonment of the irredentist (or imperialist) campaign in August 1966.

The longest-running, most violent, and most clear-cut irredentist campaign in our time has been the Somali drive for a "Greater Somalia." The Somali Democratic Republic, itself the product of a merger of the former British Somaliland and Italian Somaliland, has since independence in 1960 been waging a propaganda and guerrilla war against Ethiopia and less violent, but occasionally intense, battles with Kenya and Djibouti for control of the Somali-inhabited portions of those countries. The campaign reached a crescendo with a full-scale Somali invasion of the Ogaden region of Ethiopia in the summer of 1977. By October, the Somalis had

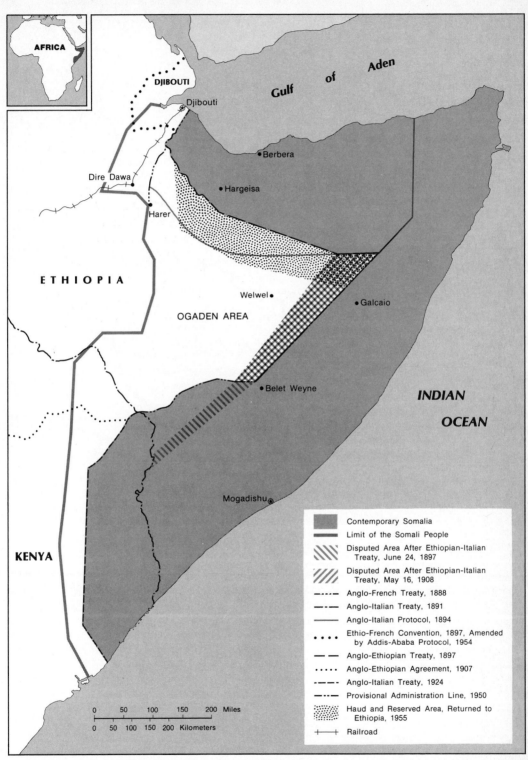

Somalia and the Horn of Africa.   Somali irredentism is based on the distribution of ethnic Somalis throughout a large contiguous area governed by four countries, and on the succession of conflicting political boundaries in the Horn arranged by foreign powers without Somali participation.

One version of the proposed State of Pushtunistan. This crude map claims as the territory for a new Pushtu (Pakhtun) State nearly all of Pakistan west of the Indus River, including all of Baluchistan, but none of the adjacent area of Afghanistan occupied by Pushtu-speaking peoples. Note the absence of such major cities as Peshawar and Sukkur, whereas many quite unimportant places are shown. (Source of original map: Rahman Pazhwak, *Pakhtunistan; The Khyber Pass as The Focus of the New State of Pakhtunistan*, London: Afghan Information Bureau, [1951]).

occupied most of the Ogaden, including all but one of its important cities, and cut the vital Ethiopian railway to the port of Djibouti. This aggression received little support from abroad and within a few months the Ethiopians, with considerable Cuban help, had driven the Somali regulars and most of the guerrillas back across the border. Since then the Ogaden has been relatively peaceful, but Somali irredentism has not been extinguished.

A different and rather strange case is that of Pushtunistan (also called Pakhtoonistan and other varieties of the name). Afghanistan was not happy about the creation of Pakistan in 1947 and late that year made her first demand of the new State for the formation of an independent entity in its Northwest Frontier Province. The people (called Pathans in Pakistan) on both sides of the border are closely related, but the Afghans have never suggested either adding the Pathan-inhabited area of Pakistan to their own State or ceding the Pushtu-inhabited area of Afghanistan to a new State carved out of Pakistan. Thus, there is no attempt to reunite a people divided by a colonial boundary. It seems instead to be an artificial political issue created by Afghanistan for her own purposes, waxing and waning according to political conditions in the two countries and relations with the Soviet Union, Iran, Arab States, and others. It may also be related to the Afghan desire for a seacoast, as evidenced by a map of the proposed new country produced by an Afghan diplomat. It has resulted in border closings, broken relations between Afghanistan and Pakistan, and even warfare in 1947, and it is still not close to resolution.

It would seem that self-determination and independence are not for everyone, and seldom for the citizens of sovereign States. States in our time are quite durable, no matter how small, poor, or weak they may be. We have seen since World War II only one established independent State that has actually gone out of business. Zanzibar became independent of Britain on 10 December 1963. The Sultan was overthrown a month later and on 26 April 1964 the new government took the country into a federation with Tanganyika. The new State ultimately became the United Republic of Tanzania.

How can we account for this durability of States in our time and is it permanent? Perhaps we can find some clues to the present and the future of the State as a human institution by examining its evolution.

Chapter 5

THE EMERGENCE
OF STATES

The State is such a familiar phenomenon that we tend to forget that only recently has all of the Earth's land surface except Antarctica come under the jurisdiction and control of States. Today there are roughly 170 States in the world, with another dozen or so colonies and other dependent territories likely to become independent in the next decade. In the past century and a half, the Europeanization of the world has spread not only revolutionary techniques in agricultural, industrial, financial, medical, scientific, and other technical areas, but also European philosophies and techniques of political organization. But in many non-European areas of the world—in Middle and South America, in black Africa, in North Africa and Southwest Asia, and in East and Southeast Asia—States had been developing for thousands of years. Some of these ancient States collapsed almost immediately in the face of the European onslaught, but others, such as Ashanti in West Africa, held out until early in the twentieth century.

BEGINNINGS

Where lie the roots of the State system? How have States emerged? Are there consistencies and similarities in the patterns of emergence of the ancient, indigenous States of the Americas, Africa, and Asia? These questions arise quite naturally in any discussion of the evolution of the State. But it is not easy to discern all the forces and stimuli that contributed to the development of the first true States. Explanations usually involve factors of soil productivity, opportune location, natural protection against enemies, advantageous climate, and similar assets supposedly possessed by those ancient human agglomerations that managed to forge States out of their habitats. But what is not explained is why, in other places where conditions were the same or even more attractive, States failed to emerge.*

CLAN AND TRIBE

One way to approach the whole question of man's territorial organization is to look for the most uncomplicated example that can be found. Early humans found that the numbers of people who could associate closely were limited by mode of life and environmental circumstance. In Southern Africa live a people, the Bushmen, who face just such limitations. Having been driven into dry, inhospitable country by their enemies, the Bushmen continue to subsist on hunting and collecting. They do so mainly in groups, or *clans,* of about 60 individuals. Life for the Bushmen revolves

*One interesting hypothesis that rejects most traditional theories of State origins and emphasizes the role of warfare over circumscribed environments is found in Robert L. Carneiro's "A Theory of the Origin of the State," *Science, 169,* 3947 (August 21, 1970), 733–738.

around the tenuous supply of the water hole, which must draw the animals the people trap and kill while providing them with this essence of life in the Kalahari Desert and Steppe. But,

Although Bushmen are a roaming people and therefore seem to be homeless and vague about their country, each group of them has a very specific territory which that group alone may use, and they respect their boundaries rigidly. Each group also knows its own territory well; although it may be several hundred square miles in area, the people who live there know every bush and stone, every convolution of the ground.*

Undoubtedly there is much in the Bushmen's existence that mirrors the condition of humans at a very early stage in history. What is remarkable is these people's strong attachment to their territory and their sense of its finiteness. These same characteristics also have been associated with a people whose mode of life resembles that of the Bushmen, but in an entirely different part of the world: the Australian aboriginal population. When the Europeans first entered Australia in the eighteenth century, there were perhaps somewhat over three hundred thousand black Australians. This number declined until the 1930s, when about fifty thousand remained. Since then, some growth has again been recorded.

Like the Bushmen, the Australian aborigines were not organized to resist a challenge from more advanced competitors, and except in their reserve areas, nothing remains today of such political institutions as existed when the Europeans arrived. Had Australia experienced the evolution of strong and well-organized tribal States, some influence would no doubt have been exerted in the sequence of events that led to the formation of the modern Australian federation. What might have happened is well illustrated in Africa, where the Bushmen suffered a fate similar to that of their Australian contemporaries,

*Elizabeth Marshall Thomas, *The Harmless People,* New York, Knopf, 1959, p. 10.

but many of the more advanced tribal States in other parts of that continent resisted the European advance for some time. Indeed, while these States were eventually overthrown, their political and economic organization not only withstood the European impact, but to some extent channeled it. And after the termination of the colonial period, many characteristics of the precolonial African tribal States were reflected in the emergent modern black African States.

The fate that befell the Bushmen in southern Africa and the black inhabitants of Australia presumably also overcame other peoples living under similar circumstances elsewhere; they were challenged and ousted from much of their habitat by more numerous, more highly organized competitors. In the case of the Bushmen, the Hottentots posed this challenge initially.

The pastoral Hottentots, who later were themselves to fall victim to the onslaught of still more powerful African peoples, were members of a more complex form of politicoterritorial organization, namely, the *tribe.* It is difficult to define this concept comprehensively; there are very large tribes and very small ones. But comparing it to the Bushmen's clans, several important differences emerge. The power of the tribal leader is greater. Personally given laws emerge. Territorial limits are even more jealously guarded. Hundreds of thousands of people may pay allegiance to the tribal chief or king. And it is especially important to note that the concept and reality of the *central place* exist at this level. Thus the tribal habitat will center on a headquarters (possibly with several subsidiary, minor foci)—a seat of government in the real sense.

Such a central place may be positioned according to some of the same rules that governed the beginnings of some towns that grew into capital cities of world importance: a defensible site, a place at the center of an area rich in resources, a location especially favorable as a market, a point of historical or religious signifi-

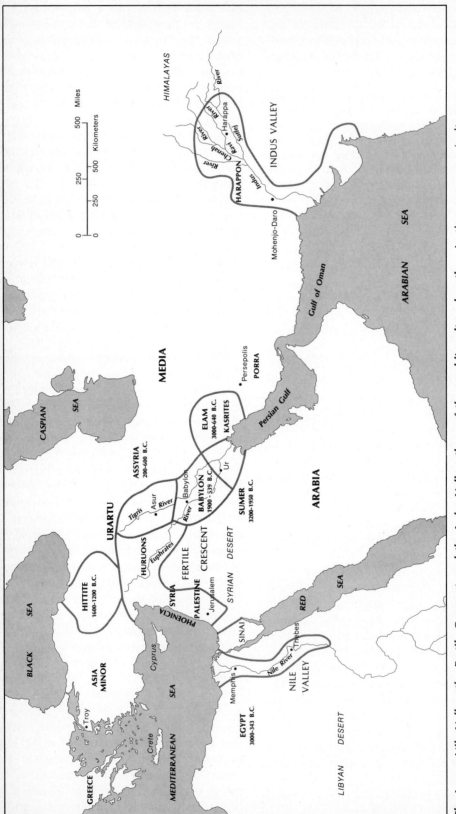

The Lower Nile Valley, the Fertile Crescent, and the Indus Valley, three of the world's culture hearths. In these areas agriculture, cities, organized religion, States, and other attributes of civilization first appeared. These areas are still scenes of discord and even bitter warfare among competing cultures and nationalities.

cance. Long before the coming of whites in Africa and many other parts of the world, important capital cities had developed in tribally organized areas. Many a European traveler described with awe the characteristics of those early cities.

The Central Place as Stimulus

Undoubtedly, many of the headquarters of indigenous tribal States were villages and no more—and they remained so throughout their existence. But certain places were endowed with particular advantages, and they grew into real urban centers. As such, they achieved significance far beyond the limits of the tribal territory in which they were located and functioned as foci for interregional trade. This was the case, for example, with towns located in the "savanna belt" of West Africa. As the strength and wealth of the Arab world to the north increased and the demand for goods there grew, these towns found themselves between one of the major sources of supply—the forested regions of coastal West Africa—and this north African region of demand. Soon the markets of these savanna towns were among the busiest in the world as people and goods moved northward and items of trade came from the north in return. In response, the towns grew into cities, complete with markets, suburbs, religious shrines, places of learning, governmental buildings, and so forth.

In turn, these oversized cities now played a role in transforming tribally organized territory into something more complex than that. Historical geographers do not hesitate when they point to West Africa between A.D. 300 and about 1600; here were true States in an African tradition, with political influences from the Nile valley, religious influences from Mecca, but with a population welded from the characteristic heterogeneity of tribal Africa.

Elsewhere in the precolonial world, large cities similarly passed a certain

threshold of self-perpetuation and became, in effect, the pulse of the indigenous State. In Aztec Middle America, it was Tenochtitlán. It was Cuzco in the Inca empire, Babylon in Mesopotamia, Mohenjo Daro in the Indus valley, Anyang in China, and many others. Some of these cities possessed attributes that permitted them to sustain the impact of the transition to the era of European domination with undiminished strength. Thus Tenochtitlán, though largely destroyed in the conflict with whites, was rebuilt and is today Mexico City, capital of the successor State. Cuzco still remains a town of some importance. Others, however, have failed; there is little left of Timbuktu's old splendor.

ANCIENT STATES
The Middle East

In the Middle East, for thousands of years progress was made that overshadowed all else on the globe. In the Fertile Crescent, people learned to domesticate certain plants. By 5000 B.C., the small farming villages on the lower slopes of the hills were showing the effects of improving farming techniques and higher yields; they were growing larger than they had ever been. Irrigation, metal working, writing, planning, organization—innovations multiplied in numerous spheres and were transmitted to the Nile region, to Mesopotamia, and to the lowland of the Indus River. By 3000 B.C., urban life had become a reality in the Middle East, and unprecedented political consolidation was taking place.

Despite the fact that Mesopotamia probably was ahead of Egypt and the Indus region in the earliest times, ancient Egypt was destined to have the most lasting impact on the politicogeographical world. The political philosophies and practices of Pharaonic Egypt were diffused to West Africa's savanna kingdoms and to many of black Africa's tribal States; to this day there are tribal ceremonies as far away as southern Africa that bear the unmistakable imprint of Egyptian origins.

What were the particular attributes of ancient Egypt, other than its pivotal location, felicitous protection, and productive capacity? The State's longevity is remarkable. Historians often identify the period of "true" empire to have extended from 2650 to 1100 B.C., but there were many centuries of consolidation prior to 2650 and, although subjected to invasion and defeats, the State hardly ceased to exist during the last millenium B.C. At a time when territorial organization was just beginning (in a world context), Egypt rose and survived longer than any other political entity ever. In the process the State changed from a theocratic to a militaristic one, and, in common with some other States we know, it developed an incredibly complex bureaucracy. The peasants, workers, and slaves all fell victim to a most ruthless use of labor. When the power of the State finally began to fall, and conquering, alien enemies rode in triumph through the streets, they were welcomed by the masses as liberators, as were the Persians in 525 B.C.

Nevertheless, the roots of "western" civilization lie in Egypt and Mesopotamia, and among those are the roots of statecraft. The productive capacity of Mesopotamia and Egypt generated the first substantial, urban-based wealthy classes that could afford the luxury of leisure, the arts, and philosophy. What the ancient Egyptians thought and wrote about in politics was not lost on their successors.

East Asia

Middle Eastern contributions may have reached China as early as 2000 B.C., perhaps even before that. From Mesopotamia and the Indus area, it is thought, innovations in agriculture, techniques of irrigation, metal working, and even writing were diffused to the valley of the Hwang He (Yellow River), in the vicinity of its confluence with the Wei River. Whatever the sources of ancient China's stimuli, there is a record of Chinese civilization around five thousand years in length, and from 2000 B.C. on, that record is continuous, quite reliable, and positive proof of China's cultural individuality.

Out of those beginnings in northern China, in the lowland of the Hwang-Wei Rivers, rose a series of dynasties. The oldest of these about which there is substantial knowledge is the Shang (Yin) Dynasty, which lasted from about 1900 B.C. until 1050 B.C. Centered on the ancient capital of Anyang, this dynasty secured the consolidation of China as a true State. The Bronze Age reached China during this period; walled towns developed. Coupled with this were innovations in political organization and control, administration, and the use of military forces. The original core was ready to expand and absorb adjacent frontiers.

This expansion was both easiest and most profitable in a southward direction, and during the Ch'in Dynasty, the lands of the Yangtze Jiang were integrated into the Chinese State. During the Han Dynasty, ancient China experienced a crucial period of growth and further consolidation. Between 202 B.C. and A.D. 220, the Han rulers brought unity and stability to their State, and they also organized the extension of the Chinese sphere of influence to include Korea, Manchuria, and Mongolia as well as distant Xinjiang and Annam.

The Han period was a critical time in the life of the Chinese State. Not only was there military might and great territorial expansion, but the country's internal order changed decisively. The old feudal system of land ownership broke down and individual property was recognized. Internal circulation was unprecedentedly effective. Towns grew into cities; internal trade intensified. The silk trade grew into China's first regular external commerce. Along the silk route across Asia came ideas, innovations—China was being transformed. To this day the people of China, recognizing the germinal character

Angkor.   One of the world's eight great culture hearths developed in Southeast Asia more than a thousand years ago. It was centered on Angkor, near Siem Reap just north of Tonle Sap (Great Lake) in what is now Kampuchea. The site was first settled about 819 A.D. and soon became the focus of a large agricultural region with a very sophisticated irrigation system. The magnificent temple dedicated to Vishnu (called Angkor Wat) was built early in the twelfth century. For some 600 years Angkor was a powerful State but was challenged by Siam late in the fourteenth century. It was abandoned about 1431 A.D. and was soon covered by jungle. The ruins were discovered in 1861. Restoration began in the 1920s but was halted by fighting during the Vietnam War in the early 1970s.

of the Han period, call themselves the People of Han.

Like Egypt, Mesopotamia, the Indus area, and so many other great ancient States and cultures, China faced a conquering enemy. But there was a difference. In A.D. 1280, the Mongols captured control of China and made the State a part of a vast empire that stretched all the way across Asia to Eastern Europe. It was not the Mongols who imposed a permanent new order upon China, however, but the Chinese whose culture was imprinted upon the Mongols. The Mongol period ended in 1368, and history recognizes the Mongol conquest as only an interlude in the continuity of Chinese life. Egypt and Mesopotamia fell, but China managed to shake off its severest challenge. And so today, with China still focused on its ancient northern heartland, it is the only major ancient State whose lineage traces back—spatially as well as temporally—directly to its origins.

West Africa

The savannas of West Africa lie between the desert to the north and the tropical rainforest to the south. For many centuries the peoples of the desert have traded goods with the people of the forests. West Africa's economic mainstays were gold and salt, and the West African savanna peoples found themselves positioned very advantageously between the traders. From the Nile Valley these fortunate savanna dwellers received some political ideas about divine kingship, about the power inherent in a monopoly over the use of iron, and about the use of military forces to exact tribute and maintain control. And so the cities of West Africa grew into great market centers and generated wide realms of control. Here was a form of the *city-state*. But West Africa was also a place of invention and innovation. There are anthropologists who argue that West Africa, along the Southwest Asia, South-

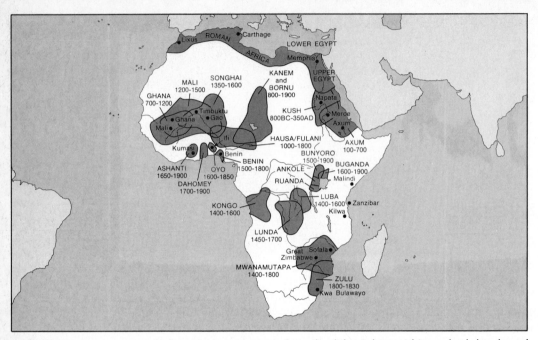

Early African States.   Long before Europeans arrived south of the Sahara, Africans had developed sophisticated and durable States, complete with armies, schools, royal courts, and even tax collectors. Most striking were the succession of States in the Sudan-Sahel region of West Africa, between the desert and the forests. Many of the others were also outstanding.

east Asia, and Middle America, was one of the four major agricultural complexes of the ancient world, where crops were domesticated and farming techniques invented and refined.

In West Africa a series of States of impressive strength and durability arose. The oldest of these about which much is known was ancient Ghana, located in parts of present-day Mali and Mauritania. For perhaps more than 1000 years (the first millennium A.D.), ancient Ghana's rulers managed to weld many diverse groups of people into a stable State. Ghana eventually fell to its northern competitors, and it was succeeded in the savanna belt by the State of Mali, centered on Timbuktu on the Niger River. Timbuktu in its heyday was a university city, a place of learning of international significance.

Mali, too, eventually collapsed. Its initial successor was the State of Songhai, slightly to the east; still later, the center of power shifted farther east again, to the region that is today northern Nigeria and western Chad. But by then, the European

impact was being felt along the West African coast, and the era of the slave trade was soon to upset the patterns of development of nearly two millennia in this part of the world.

Middle and South America

In the Americas prior to the European invasion, there were also large, stable, and well-organized States. Toward the end of the fifteenth century, various sectors of the American Indian realm were at different stages of politicoterritorial organization. In the southern part of South America lived peoples who subsisted by hunting and gathering. This economy prevailed also in parts of what is today Canada and the western half of the United States (though not the Southwest). In the Amazon Basin rainforests, in southeastern North America, and elsewhere in the hemisphere some agriculture accompanied hunting and gathering, but the practice of agriculture was not always accompanied by more advanced organization.

Ancient Zimbabwe. While Mali, Ghana, and Songhai flourished in West Africa, the Shona peoples of South Central Africa between the Zambezi and Limpopo Rivers also had a number of States. Of their remnants, those of Zimbabwe, near Fort Victoria, Zimbabwe are the most imposing. Apparently the site was occupied by at least three successive peoples. This "Temple" or "Elliptical Building" is located on the plain below the "Acropolis" atop a granite hill. It may date from the twelfth century. (Courtesy of the Zimbabwe Tourist Office)

In other parts of the Americas, tribal organization was achieved by many Indian peoples, who established permanent villages, practiced more intensive agriculture, and developed considerable technological skill. The idea of a form of federation also came to the leaders of these peoples, and many multicommunity tribal States developed through voluntary alliance and through conquest, although such alliances were often short-lived.* North of the Rio Grande, only the Iroquois achieved a confederacy prior to the European invasion. Nevertheless, these tribal political entities represent a major advance in the quest for the State.

*It is important to distinguish between indigenously conceived "federal" associations and those that developed after the European impact was felt. The second type were the direct result of common enmity toward the European invader and conceived of as a better means of defense.

Inevitably, the increasing complexity of the society produced clashes of interest between various groups; those who verbalized the positions of the opposing groups were often specialists in the field of politics. Thus there arose, in these societies, a number of political theorists who were able to devote their lives to the direction of political affairs in the growing State.

Aztec State and Inca Empire

The culmination of indigenous political evolution in the Americas, so closely bound up with the development of organized agriculture, was achieved in Middle America and in the Central Andes. Although the Mayan civilization, which flourished in Yucatan and adjacent areas between A.D. 300 and 900, was a forerunner to the significant events later to take place in what is today Mexico, important differences existed between the Mayan societies and those of the Toltecs and Aztecs. The Mayan societies were theocratic, controlled by priest-rulers; Toltec and Aztec society was militaristic and structurally much more complex. The main centers of Mayan culture lay in the forested sections of Yucatan, to the north of the mountain slopes of Guatemala and El Salvador. The central places, however, were not central places in the usual sense. They were ceremonial centers and served as markets, but did not contain large permanent populations.

Ultimately, one ceremonial site, Mayapan, did become a true central place, although after the decline of Mayan power and organization had begun. Mayapan was a walled city, the location of a despotic authority, the home of thousands of residents, and the heart of all Yucatan. But by this time the area that had been unified by Mayan civilization was breaking into warring tribal States, with a corresponding decline in the arts and other fields of achievement.

Elsewhere in Middle America, the strands of progress remained intact, ultimately to lead to the formation of one of the most impressive States to develop anywhere in the world at any time prior to the rise of modern Europe. In fact, the Aztec State was achieved by a relatively small number of Indians who were acquainted with Toltec technology, scattered by the disastrous breakup of Toltec society, and finally regrouped in a very fortuitous location. The choice of headquarters proved a permanent one: the site of the Aztec capital, Tenochtitlán, also was to become the capital of an empire, the seat of government for New Spain, and finally the focus for the modern State of Mexico.

A highly complex bureaucracy developed in the capital, and there were systems of tax collection, law enforcement, mail delivery, rapid messenger communication, and a provincial administration

with governors and district commissioners. In the Valley of Mexico, the Aztecs had found the pivotal geographic feature of Middle America; agricultural productivity was high, and technology, including irrigation practices, had been developing for centuries. Population was relatively dense; the region at that time perhaps contained over 2 million inhabitants. Central places grew rapidly and to sizes that were remarkable for any part of the world: Tenochtitlán had well over a hundred thousand inhabitants, as did a number of other centers of trade and administration.

Later, though, the empire fell suddenly under the onslaught of the Spanish invasion, but the advantages of the Valley of Mexico remained—and played their part in the momentous events that have marked the evolution of the modern State in Middle America.

The American region marked by the culmination of indigenous politicogeographic organization is the Central Andes. Here, over a distance of 2500 miles from Ecuador to central Chile, the Inca Empire evolved. The Inca Empire was itself preceded by earlier civilizations, notably that centered on Tiahuanaco in the area of Lake Titicaca. Better integrated, more stable, less violently militaristic, and generally more benevolent than the Aztec sphere of influence, the Inca domain in many ways was the only real empire to develop in the Americas, Aztec achievements notwithstanding. The Inca Empire was centered on Cuzco, a city located 11,000 feet above sea level, 200 miles from the ocean, and with a population of over a quarter of a million. Here the Inca rulers built the mighty fortress of Sacsahuaman, the Temple of the Sun, and other huge megalithic structures.

Cuzco and its surrounding region formed the core area for a vast State with far-flung possessions and an intricate organization that may with justification be called its major achievement. At the height of its power the empire contained perhaps as many as 11 million subjects, each of whom had an assigned task of sup-

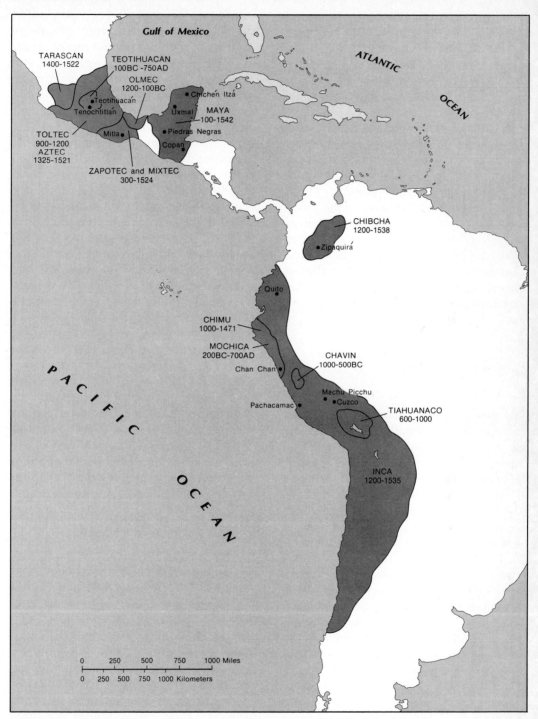

Early States of the Americas. While the Americas south of the Tropic of Cancer were dominated for some 500 years until the Spanish Conquest by the Toltec, Aztec (or Mexica), Maya, and Inca empires, many other people had established States in the region. The Chibchas alone had five States, two of them well organized and quite prosperous for the period; they won the admiration of the first Spanish arrivals who saw them in full flower. Other cultures had stone cities, elaborate legal and religious systems, and extensive international commerce.

plying certain tax or tribute to the central authority, the Inca. In view of the contrasts in terrain within the empire, its stability is admirable, and it could not have been accomplished without a system of central places and efficient communication lines.

A system of roads connected the central places, and along these roads were a number of supply stations at regular intervals, staffed by professional messengers; in relay fashion, an order issued in Cuzco would reach outlying districts in the shortest possible time. When European visitors first arrived in this area, they wondered how such broad and sturdy roads could have been constructed in such difficult terrain. In this respect, as in others, the Inca Empire resembled the Roman Empire.

THE STATE IN EUROPE

The sequence of events that led to the emergence of the European State system began in pre-Christian times. A major factor in the transmission of early stimuli from the Middle East to southeast Europe lay in the growing need for raw materials in the advanced societies of the Nile Valley and the lowland of the Tigris and Euphrates rivers. These societies wanted metals, textiles, and oils, and their search for these and other commodities brought contact with southeast Europe. Greek traders were active in Egyptian towns, Phoenician ships linked Aegean and Mediterranean shores. The traffic was by no means confined to copper and tin; the political ideas of Middle Eastern States filtered into Europe as the trade grew.

Ancient Greece

During the first millennium B.C., the Greeks created what in many respects is the foundation of European civilization. The Hellenic peoples occupied much of Southeast Europe, but they had to adjust to the landscapes of the region, which are characteristically rugged, peninsular, and insular. Hellenic settlements were founded in river valleys and on estuaries, on the narrow littorals of peninsulas, and on islands. They were in contact with one another, but mostly by sea; water often provided a better avenue than the land. Thus the Hellenic civilization was extensive—but it was not contiguous.

This situation produced another version of the city-state. Individuality and independence were strong features of the ancient Greeks, and in their comparative isolation, they experimented with government and politics. At times several of the city-states would join together in a *league* to promote common objectives, but the participants would never give up their autonomy altogether. And just as the Greek city-states competed with each other individually, so there were rival leagues; Athens, Sparta, and Thebes each headed powerful and competing leagues.

During Greek times, political philosophy and public administration became sciences, pursued by the greatest of practitioners. Plato (428 to 347 B.C.) and Aristotle (384 to 322 B.C.) are two of the most famous philosophers of this period, but there were a host of other contributors to the greatness of ancient Greece.

The Roman Empire

The Romans succeeded where the Greeks had failed. Theirs was not a nation dispersed through a string of city-states, but a contiguous and unified *empire* in the true sense of the word, strongly centralized in one unrivaled city, Rome. But Rome itself had not achieved supremacy overnight. Around 400 B.C. (about the time of Greece's greatest glory), Italy was a territory of many tribes and nations. Well-fortified hilltop towns dotted the area; these were the centers of power and protection for local groups. As in Greece, but on a smaller scale, towns entered leagues of mutual defense. As time went on places whose natural protections were less than adequate were abandoned as their inhabitants moved to larger and safer centers. Gradually, the regional influence of Rome increased. Her great achievements were based on her successful absorption of immigrants, the stability of her political institutions, and an effective campaign of Roman colonization in outlying districts. The great empire that was to come had its local predecessor: by the middle of the third century B.C., there was a true Roman Federation extending from near the Po Valley in the north to the Greek-occupied south.

Having achieved unification on its peninsula, Rome now was a major force in the entire Mediterranean area. The defeat of Carthage removed the major obstacle to westward expansion; to the east the Greeks were forced to recognize Roman superiority. But Greek culture had a major impact on the Roman civilization, and Greek identity survived despite the strength of the Roman invasion. It is more accurate to speak of a Roman–Hellenic civilization than of a Roman civilization alone, for the Romans provided the vehicle to disseminate Greek cultural achievements throughout Mediterranean and Western Europe.

Greek culture was a major component of Roman civilization, but the Romans made their own essential contributions. The Greeks never achieved national organization on the scale accomplished by the Romans. In such fields as land communications, military organization, law, and public administration, the Romans made unprecedented progress. Comparative stability and peace marked their vast realm, and for centuries, these conditions promoted social and economic advances.

Only in Asia was there contact with a State of any significance. Thus the empire could organize internally without interference. But its internal diversity was such that isolation of this kind did not lead to stagnation. The Roman Empire was the first truly interregional political unit in Europe. In many ways the Roman Empire was centuries ahead of its time. It was also the first true empire by modern criteria, with outlying colonies and diverse racial groups under its control.

Whatever the impetus for its development, the Roman Empire stood out from all its predecessors and contemporaries, especially in the area of organization. Almost everything else—its power, its size, its longevity, its internal variety—was a function of the Romans' dedication to organize their land and possessions. One aspect of this organization was the effort to build a network of communications. The concept was totally new, and the scale of its execution was grand. Naturally, the network was not dense by modern standards, but it was carefully planned to meet the needs of the State.

Although highly centralized, the Roman Empire's political structure contained elements of the federal as well as the unitary State of today. Peace—and a certain amount of autonomy—was granted to areas west of the Rhine and east of the lower Danube. A true body politic evolved and permitted the disparate peoples of the empire to develop their areas and their abilities in peace.

Feudal Europe

Feudalism. Europe during the period from about the middle of the eighth century to the end of the twelfth was dominated by feudal institutions. Roman judicial processes and other institutions that protected individual rights had weakened. The new kings ruled by private law, personal power, influence, and wealth and through the allegiance of counts, barons, dukes, and other representatives. Although some kings were successful in forging and maintaining States of considerable size, the framework of government was not sufficiently strong, and there was not enough contact and communication with the country to withstand disorder and disorganization. Although he managed to hold his dominions together for a long time, even Charlemagne was compelled to grant increasing powers to local lords. In this way, feudalism served the kings. To secure their allegiance, feudal rights were granted to counts and dukes in outlying parts of the State.*

When a large political unit disintegrated through the death of a king, a weak succession, or an invasion, the feudal counts and barons were quick to assert complete hegemony over their local domains, including all land and people within them. Thus, during the feudal period much of Europe was organized into dozens of jigsawlike feudal units, each with its nobility and vassals and its more or less tenuous connections with higher authority. The connectivity that Rome had brought to Europe was now at a minimum, and reforms were needed to rekindle economic as well as political progress. In England the Norman invasion (1066) destroyed the Anglo-Saxon nobility and replaced it with one drawn from the new immigrants. William the Conqueror strengthened the feudal organization of England to such a degree that he became the most powerful ruler in Western Europe; at the same time he introduced changes in the system of governmental organization that were to outlast the feudal period.

Despite oppression and excesses perpetrated on the population under the divine rights of kings, the roots of modern European nationalism lie in this period. There was a slow but noticeable improvement in circulation, especially in Western Europe, and a feeling of belonging together within a State framework developed among the peoples. Eventually, they began to perceive certain physiographic features of Europe—a river or mountain range—as "their" country's borders. Because the kings had amalgamated diverse groups of people into their empires, the peoples within those borders often were quite varied in terms of language and religion. But if the State or king provided a

*These old units can still be seen in the locations and names of the political subdivisions of certain countries of the Old World, such as the *county,* the *duchy,* and the *march,* or *mark.*

strong common interest, a nation was in the making.

THE DEVELOPMENT OF NATIONALISM

The emergence of modern Europe may be said to date from the second half of the fifteenth century. Western Europe's monarchies, as we have seen, began to represent something more than mere authority; increasingly, they became centers of an emerging national consciousness and pride. At the same time the trend toward fragmentation, which had dominated the feudal period, was reversed as various monarchies, through marriage, alliance, or both, combined to promote territorial unity. Feudal privileges were being recaptured by the central authority, and there was progress in the parliamentary representation of the general population. Renewed interest was shown in Greek and Roman philosophies of government and administration. Europe was ready for change.

Change did come to Europe, and in many forms. The emerging States engaged in intense commercial competition, and *mercantilism* was viewed as the correct economic policy to serve the general interest. The search for precious metals (the standard of wealth) sent the ships of several countries traversing the oceans to lands that lay open for discovery—and appropriation. Kings and nobles sent many of those ships on their way, but in the growing cities, an influential new class was on the rise: the merchants. Before long they were demanding a greater voice in the politics of their countries.

A landmark in the evolution of the State was the Peace of Westphalia, which in 1648 brought to an end the Thirty Years War between Protestants and Catholics. It was a complex settlement among a number of civil and ecclesiastical authorities, but by reducing the power of the Holy Roman Empire and strengthening the emerging States, it made the territorial State, rather than the individual sovereign, the corner-stone of our modern political system. This began a radical reduction in the number of States in Europe. Before Westphalia there were around 900 German States, for example. The settlement reduced them to 355. Napoleon I eliminated more than 200 of these, and by the time the Germanic Confederation was formed in 1815, only 36 were left to join it.

The unification process continued, with the survivors growing stronger, until one German Empire finally emerged in 1871 with the unification of the remaining 24 German States. Similar processes were going on elsewhere in Europe, with Italy (1870) and Germany being the last to consolidate into national States based on common culture, primarily language. In the course of consolidation, people transferred their loyalties from the sovereign to the State, a State they often helped to create by participation in a revolution that swept away the old systems of tribalism, feudalism, and absolutism. This revolution was the development of nationalism.

Chapter 6

MODERN THEORIES
ABOUT STATES

An academic discipline is often judged not so much by its empirical studies, its discoveries, and its inventory of "facts," but by its theories. In Part One we reviewed some of the theories in the body of political geographic thought. Now we turn our attention to some specific theories about the central object of study by political geographers: the State. We will not attempt to discuss or even mention all the theories that have been propounded about the State by political scientists, historians, anthropologists, political geographers, and others, but will only select a few for discussion.

No responsible scholar would suggest that any theory describes exactly the real world of territorial political behavior. Neither do any combination of them nor all of them together provide answers to our innumerable questions about the State and State behavior. They are, after all, more like impressionistic paintings than like photographs. This is not to imply that their only value is to provide the student with material for momentary diversion, coffee-shop debates, or mental gymnastics. These theories have been formulated carefully and scientifically by responsible, experienced, and idealistic individuals who are groping, like the rest of us, toward definitive explanations and solutions we know we may never reach.

Some theories have been rendered obsolete by new information, changing circumstances, or new patterns of thought.

Others may not have survived the frequent questioning, analysis, and testing that any good theory provokes. And many others are still undergoing this process and their validity is uncertain. Nevertheless, they all provide compact descriptions, clues to explanations, and tools for more and better work. They should inspire us to challenge them, extract their valid and useful elements, apply them to real and important problems, and use them as the nuclei of new and better theories.

ENVIRONMENTAL DETERMINISM

One of the most important books ever published was *On the Origin of Species* by Charles Darwin, which appeared in 1859. It set off a revolution in many branches of science and led to the popularization of science. Out of the ferment of thought, discussion, and speculation generated by Darwin came attempts to apply "science"—particularly concepts of natural selection—to many areas of human endeavor and study. Henry Adams and Frederick Jackson Turner in the field of history, for example, Sir Henry Maine in political science, Thorstein Veblen in economics, Oliver Wendell Holmes in law, Lewis Henry Morgan in anthropology, and William Graham Sumner in sociology were only a few of the people who applied Darwin's concepts to human society. This application of Darwinian ideas about the natural world to society is loosely called Social Darwinism. One outgrowth of Social Darwinism in geography is the organic State theory, whose earliest exponent was Friedrich Ratzel. Because of its complexity and importance, we reserve our discussion of this theory for Part Four, on geopolitics. Another derivative of Social Darwinism was the concept of *environmental determinism*.

We have all been exposed at one time or another to such notions as: "Because Japan is very mountainous and has little good agricultural land, its people were forced to turn to the sea for a living." Or, "Because Britain had many good natural harbors and fine oak forests, she was able to build a navy that could control the seas." Or, "The searing deserts and rainsoaked forests to the south and the cold, dreary lands to the north keep their inhabitants in a state of savagery, or at best barbarism; civilization can develop only in the equable climate of the Mediterranean Basin, as in Greece." This concept that one or another element of the physical environment determines the type and level of civilization a society can attain is called environmental determinism.

Many variations of this theory, from extreme determinism to more modest formulations, were expounded by eminent geographers from the 1880s to the 1960s. Among them were the German Friedrich Ratzel, the Americans Ellen Churchill Semple and Ellsworth Huntington, and the British–Australian–Canadian Griffith Taylor. By the late 1930s, the theory was coming under increasing attack; by the late 1950s, "determinist" had become almost a pejorative term. The inadequacy of the theory was revealed in the light of newer concepts backed by more experience and solid evidence. Preston James, for example, demonstrated the importance of culture in determining how a particular people will use a particular environment, contrasting the very different societies that have developed in the very similar physical environments of

Iowa and Uruguay. He summed up the importance of culture by stressing the importance of the "attitudes, objectives, and technical abilities" of a people. Derwent Whittlesey contributed a temporal perspective, demonstrating how, as in California, different cultural groups who occupy a particular region in sequence over time frequently utilize the physical environment in very different ways. Newer studies of the influence of people on their environments and of environmental perception demonstrate even more emphatically that neither individuals, societies, nor States are controlled by any element of their physical environment.

Nevertheless, we cannot ignore the environment altogether. We have not yet learned to control nature, nor should we. The physical environment still offers opportunities to, and imposes limitations on, the people resident in every State. The French geographers Vidal de la Blache and Jean Brunhes, among others, developed concepts of probabilism and possibilism, which recognized these opportunities and limitations as alternatives to determinism. Neither people nor States can do as they please regardless of the environment. It seems highly unlikely, for example, that Bhutan will ever be a great naval power or Tonga a great industrial power. We examine the significance of the physical environmnent again in our discussion of the power inventory; for the present we can say that the physical environment is important in the development of States, though not necessarily determinative.*

---

*Another manifestation of Social Darwinism was the development in 1939 of a cycle theory of the development of States by Samuel van Valkenburg. It was based on the cyclic theory of landscape evolution propounded in 1899 by another American, William Morris Davis, himself strongly influenced by Darwin. Van Valkenburg theorized that States develop through identifiable stages of youth, adolescence, maturity, and old age, each with its own characteristics of State behavior. He also allowed for the possibility of rejuvenation of a State after it has begun to decline and for interruptions of the cycle at any time, bringing the State back to a former stage.

THE FUNCTIONAL APPROACH

A very different view of the State was taken by Richard Hartshorne in his Presidential Address delivered before the Association of American Geographers in 1950. It was titled "The Functional Approach in Political Geography" and was an elaboration of a theme he had introduced a decade earlier in an article on *"raison d'être"* and maturity of States. In his address he proposed that the study of the functioning of the State is the central task of political geography, that political geographers should view the State (and other politically organized areas) in terms of structure and functions. The State is a politically organized space that functions effectively. How does it succeed? By overcoming the *centrifugal* forces—forces that tend to break a State apart—with the cohesion provided by prevalent *centripetal* forces, which bind the State together. Centrifugal forces exist in every State; in some States they are so powerful that they disrupt the

Richard Hartshorne, 1899–. (Courtesy of Geoffrey J. Martin)

State system completely. In Nigeria the Biafran secession movement of the late 1960s was barely overcome; in the early 1970s, the fate of a united Pakistan was sealed and the State broke up. These are highly publicized instances, but even older, more tightly knit States are not without centrifugal elements. Regional differences, ethnic pluralism, religious divisions, and other factors can afflict the State in this way. The centripetal, binding forces must prevail if the State is to have a *raison d'être*, a reason for existing. Hartshorne suggests that "the greatest single weakness in our thinking in political geography" has been our preoccupation with the disruptive, divisive forces affecting politically organized areas, without considering the forces that manage to hold the State together. This set of centripetal forces is the *functioning* State, and, Hartshorne argued a third of a century ago, this process should be studied.

Hartshorne's discourse touched on several additional topics, including the concepts of nation and core area, the internal and external relations of States—all within the context of the functioning of the political unit involved. It still warrants a careful reading today.

PATTERNS OF INTEGRATION AND DISINTEGRATION

Karl Deutsch has contributed much to our understanding of the formation and growth of states; his analyses combine geographic and behavioral approaches. In a classic 1953 article, he summarized his findings of uniformities in the development of nations, or "recurrent patterns of integration."* They include:

1. The shift to exchange economies from subsistence agriculture.
2. Appearance of core areas.
3. Growth of towns.

*K.W. Deutsch, "The Growth of Nations: Some Recurrent Patterns of Political and Social Integration," *World Politics,* 5, 2 (January 1953), 168–195.

4. Development of basic communication grids.
5. Concentration of capital.
6. Growth of individual self-awareness and of group interests.
7. Awakening of ethnic awareness.
8. Merging of this ethnic awareness with political compulsion and sometimes the social stratification of society.

As a nation develops and forms a political system in this manner, it begins to grow and integrate its territory in a number of ways. It perpetuates itself through the generations, it increases the social and political divisions of labor, it accumulates more national symbols, and it acquires the ability to utilize its resources for its own development. Thus, the nation-state "represents a more effective organization than the supra-national but largely layer-cake society or the feudal or tribal localisms that preceded it." By leading to great differences in living standards among States, however, this process tends to turn nationalism to imperialism. But since the process that led one State to become dominant over others is also functioning in the subordinate States, they tend to grow strong enough to weaken or destroy the dominant State. The article ends with the observation that our period is unique in the history of the world, for everywhere there is growth and nowhere a decline to compensate for growth elsewhere.

THE UNIFIED FIELD THEORY

Stephen B. Jones in 1954 integrated some of the ideas of Whittlesey, Hartshorne, and Jean Gottmann, added some of his own, and organized the whole into a coherent explanation of how States develop. He published his concept in an important paper entitled "A Unified Field Theory of Political Geography," in which he viewed movement as a process involving the flow of ideas as well as goods and other tangibles, and in which movement becomes linked with several other processes. The unified field theory holds that there are, in the process of spatial-political organization, at least five clearly recognizable "hubs" of activity, each related to and connected with the other. The first of these is the generation of an *idea*.

A political idea—say, an ideology—may well have spatial ramifications. Colonialism is such an idea, so was geopolitics in interwar Germany and Zionism today. Ideas may lead to *decisions* to occupy territory, to announce claims, to make promises. All this can generate *movement*, of people, goods, ideas, money, and more. While the process proceeds, the "links" in the chain interact with each other. Movement can influence the decision and alter it; the original idea can be changed, too. The whole development takes place in a *field* that, like the movement phase of the model, has tangible (spatial) as well as intangible characteristics. And at the end of the "chain" lies the politically organized *area*.

The unified field theory came to be known as the idea-area chain, suggesting a one-way sequence of development from political idea to politically organized area. But Jones himself cautioned against misinterpretation by emphasizing the underlying principle of two-way interaction; the links in the chain have reverse as well as forward effect.

THE TERRITORIAL STATE

We have already stressed in Chapter Four that a State is different from a nation in that it has territory; without territory a State cannot exist. John H. Herz in 1957 made this fact the central theme in his article "Rise and Demise of the Territorial State." In fact, it is his basic thesis that the territorial State assumed its modern importance only after military technology, particularly gunpowder and large mercenary armies, had rendered obsolete the castles and walled cities of feudal rulers. Security—the primary consideration in the formation of political regions—dictated the expansion of the defensive perimeter to the boundaries of the emerging territorial State. The rise of nationalism personalized the new political units as self-determining, national groups and thus entitled to the same protection as a sovereign. Even the system of collective security (or defensive alliances among sovereign States) has not diminished the significance of the territorial State, but rather is an attempt to maintain and reinforce the "impermeability" of each individual State.

Just as artillery made stone walls obsolete, Herz argues, recent developments in airborne propaganda and thermonuclear weapons have enabled belligerents "to overleap or by-pass the traditional hardshell defense of States." Even industrialization has weakened the territorial State by making it dependent on foreign sources of commodities and foreign markets for its manufactured goods. Therefore, the fundamental need of humans for protection can no longer be served by the territorial State, and it is bound to vanish. Rationally, its replacement would have to be some form of "universalism" that would afford protection to all humanity conceiving of itself as a unit.

Events of the next decade stimulated Herz to rethink both his thesis and his conclusion. In his 1968 article "The Territorial State Revisited—Reflections on the Future of the Nation-State," he explained how some of his ideas had changed. He maintained that his analysis both of "classical" territoriality and of the factors threatening its survival was still valid, but he was no longer certain that "universalism" would replace it. Indeed, area or territory seem to be assuming a new importance to States; there is a trend toward a "new territoriality." He pointed to the survival of Israel and Vietnam, both of which in 1967 faced extinction by States possessing nuclear weapons. Decolonization, then in its most frenetic period, meant that all remaining colonies were bound to be added to a global "mosaic of nation-states." The threat of nuclear (or

thermonuclear) annihilation has stabilized the world through nuclear stalemate rather than forcing States to seek new means of protection, and nationalism "of the self-determining and self-limiting variety (in contrast to expansionist imperialism)" will strengthen the foundations of the individual States. He elaborated on the themes of legitimacy, foreign intervention, and nonalignment and detected diminished antagonisms among States.

In his conclusion Herz rejected the notion of instinctive human territoriality, but he reiterated the centrality of the need for protection, for security against deprivation of scarce resources as well as physical attack. This protection can be provided at least in part by States, but only if the world becomes "modernized" through science and technology and thus freed from scarcity. The "new-old nation-state" may be "the polity of the last decades of this century," but only if four conditions are met. First is "the spread of political, economic, and attitudinal modernity" throughout the world. Territorial disputes must be settled in such a manner that nationalities, or peoples, are satisfied. Systems based on world-revolutionary doctrines must be deradicalized and States must not interfere in the internal affairs of others. Finally, international violence must be reserved only for self-defense in cases of direct attack or invasion and for resistance of the population of an occupied area against the aggressor. Under these conditions the "neo-territorial world of nations" can not only provide the essentials of group identity, protection, and welfare, but also preserve the diversity of life and culture, traditions and civilizations that are threatened by our accelerating rush to "the technological conformity of a synthetic planetary environment."

POLITICAL SYSTEMS ANALYSIS

Early in 1971, an article appeared whose authors, Saul Cohen and L.D. Rosenthal,

sought to build further on the methodological heritage of Whittlesey, Hartshorne, Gottmann, and Jones.* "What is
needed," say the authors, "is a methodology to link effectively political process
and its spatial attributes—a methodology
that, having identified a specific political
process, can pinpoint spatially significant
phenomena which relate to this process,
can observe the impact of these phenomena within some control context, and can
connect one process or parts thereof to
another." Thus, they argue here that political geographers should focus on political processes in their research and on the
spatial expression of these processes.

Their point of departure for analysis of
political process is the *political system,*
"a set of related political objects (parts) and
their attributes (properties) in an environment. . . . political system refers to a functional entity composed of interacting, interdependent parts." It "can be viewed
as the end product of the processes by
which man organizes himself politically in
his particular social and physical environment and in response to outside political
systems with their unique environments."
Other elements in their analysis are the
locational perspective and the *open/
closed political system.* We discussed aspects of locational perspective in Chapter
Three; the degree of openness of a political system is almost as important in understanding decision making within it.
Cohen and Rosenthal consider also the
societal forces and legal systems as especially useful in analyzing the broader
relationship between political process and
spatial attributes.

They construct a model that attempts
"to show man in his political role in society and the relation of that role (as expressed in political ideology, political
structure, and political transaction) to the
land (places, areas, and the general landscape), and to indicate the consequences
of that relation in the formation of the political system."

The authors conclude their paper with
a case study of Venezuela, where two
dominant forces—nationalism and social
democracy—provide vantage points for
the study of the impact of process on geographical space. In the context of laws relating to petroleum exploitation and
immigration, aspects of Venezuela's spatial organization are examined. The conclusion is that the Venezuelan case
confirms the law-landscape thread as the
key to the analysis of the political system.
For example, "the hierarchical role of
administrative centers and the political
significance of place are essential
characteristics in the understanding of
place. Political conflict, as expressed
through rivalry between state capitals
(Ciudad Bolivar) and new development
centers (Ciudad Guayana) over where to
build bridges, to locate factories, or to
build up the strength of local political parties, exemplifies the operations of the political process in space."

The six theories we have presented here
were developed during more than a century and differ widely in their approaches.
As we proceed it would be well to bear in
mind the essentials of these theories and
recall them as we discuss many political
aspects of our world. Perhaps some of
them can help us understand this amaz-
ingly complex world. One of its most complicated elements is territory, to which we
now turn our attention.

*S.B. Cohen and L.D. Rosenthal, "A Geographical
Model of Political Systems Analysis," *Geographical
Review,* 61, 1 (January 1971), 5–31.

Chapter 7

THE TERRITORY OF THE STATE

All modern theories about States agree on at least one thing: a State must have territory. Neither in law, custom, nor current practice, however, are there any guidelines about the territorial characteristics necessary either for formal recognition of a State or for its survival. In this chapter we examine some essential characteristics of a State's territory—acquisition, size, and shape. In subsequent chapters we consider other characteristics.

ACQUISITION OF TERRITORY

A State must have territory, but only a State can acquire territory under international law (with a few minor exceptions). This would seem to be a proverbial vicious circle. In fact, however, international law was and is being created by States, so there can be no formal rules for the acquisition of territory until States make the rules. Furthermore, States came into existence gradually, as we have seen, through the actions of peoples or sovereigns. Therefore, we begin with a survey of how existing States add to their national territory. In Part Five we discuss how new States come into existence out of the territory of old ones.

Occupation

Originally, States became identifiable as such largely by common consent or tacit agreement. European discovery of new lands in the Americas and Africa, however, presented the problem of the legal grounds on which a European State could take possession of land elsewhere. At first, discovery alone was given some status as a basis for a claim, but it was frequently challenged and seldom sustained. By the eighteenth century, discovery alone was no longer adequate; it had to be followed by effective occupation. There have been a great many disputes over definitions of "effective occupation" and its importance vis-à-vis other claims to territory, and some of these disputes survive today. But since there is probably no undiscovered land remaining in the world and little unclaimed land, this basis for claims is of historical and legal interest only.

Prescription

If an area claimed by a State is occupied by another for many years without serious objection by the original claimant, the title, whether clear and recognized or not, may be considered abandoned and may pass to the occupying State. The rules for this are similar to those for occupation. The rise of nationalism, however, has virtually eliminated prescription as a means of transferring territory, except for a few small and remote islands. Some contemporary territorial disputes, however, still involve surviving claims to prescriptive rights.

Conquest and Annexation

Historically, territory has changed hands through conquest at least as often as by any other means. Conquest alone, however, does not confer title. Conquerors must take steps to annex the new territory and incorporate it into their own, extending their laws over it, giving it representation in the national legislatures, appointing administrators, or otherwise making the annexation effective. Generally, this annexation is recognized in a peace treaty or other legal instrument of cession. This method of acquiring territory is no longer considered acceptable in polite society, however, and is rarely used. Territory is still being conquered, of course, but seldom annexed. Even when it is, the annexation is not normally recognized by the international community. A recent example is the portion of Palestine conquered by Transjordan in 1948. Its annexation in 1950 was formally recognized only by Pakistan and the United Kingdom, and even Jordan's Arab allies still refuse to accept it.

There are cases of assimilation of a territory under coercion short of conquest. Korea was annexed by Japan in 1910, Austria by Germany in 1938, and the Baltic States by the Soviet Union in 1940, all without war and without effective opposition from elsewhere. More recently, Western Sahara (formerly Spanish Sahara or Río Muñi) was divided between Morocco and Mauritania with the passive agreement of Spain, but the annexations have not yet been recognized by the international community, and guerrilla warfare continues there.

Voluntary Cession

Formerly, it was quite common for territory (with its inhabitants) to pass from one country to another simply by agreement, with or without cash or other compensation involved. Much of the United States was acquired in this way, for example. Today it is rare. Remote islands are still being transferred, as when Christmas Island in the Indian Ocean was transferred from Britain to Australia in 1950. Minor boundary adjustments, generally involving exchanges of territory, have been quite common in Europe since World War II. Some disputed or loosely held areas are being divided by agreement involving transfers of territory, as in the Jordan–Saudi Arabia settlement of 1965. Generally, however, States just do not like to give up territory, no matter how remote, sparsely settled, or useless it may seem to be.

Accretion

Accretion is the addition of land to a State by natural processes. Most commonly, this results from a gradual shift in the bed of a river that has been adopted as an international boundary. If the river changes course suddenly, as the result of a flood or earthquake, for example, the process is called *avulsion,* and the boundary customarily remains in place. Land is also accreted in deltas, along emerging coastlines, as islands built up by rivers and ocean currents, and so on. The amount of territory involved is generally not very great, but sometimes bitter disputes break out over the ownership of accreted lands.

Acquisition of Rights

Often the use of territory is granted by one State to another without title or sovereignty actually changing hands. Such transfers of rights take the form of leases and servitudes. Russia, Germany, the United Kingdom, and France all leased territories from China in the late nineteenth century, for example, and the Russian and German leases eventually passed to Japan without even the assent of China. The United Kingdom still leases from China the New Territories, adjacent to her crown colony of Hong Kong. The United States in 1903 acquired from Panama rights to the "use, occupation, and control" of a zone for the construction, operation, and de-

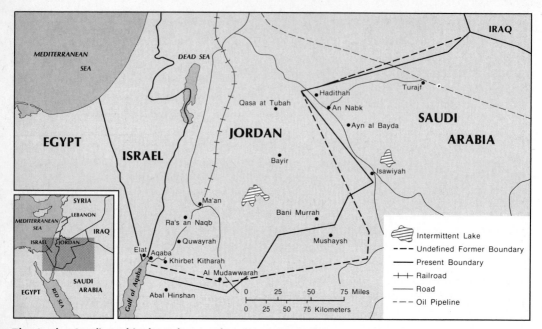

The Jordan-Saudi Arabia boundary settlement, 1965. Like many boundaries through remote, sparsely populated frontiers, the one between Jordan and Saudi Arabia remained undefined for generations. When the two countries did define and delimit their boundary, the result was an exchange of territory claimed or administered provisionally. Perhaps the major effect of this settlement was the lengthening of Jordan's extremely short coastline on the Gulf of Aqaba.

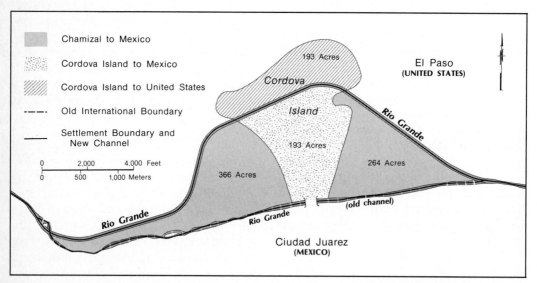

Settlement of the Chamizal dispute, 1963. The Rio Grande is a classic example of the inadequacy of rivers as political boundaries. This stream meanders over much of its length, constantly shifting channels and rearranging the land by accretion and avulsion. Meander cutoffs in the Rio Grande Valley are called *bancos*. There were several hundred of them, most disputed between Mexico and the United States, until they were eliminated by agreement between 1905 and 1970. One, however, known as El Chamizal, resisted settlement because of its location in a densely populated area. The eventual agreement provided for partition of the disputed tract and construction of a new, permanent concrete channel for the river, through which the international boundary now runs.

fense of an interoceanic canal, which she is now gradually giving up. Similarly, the Soviet Union in 1962 granted Finland a 50-year lease over a 4-km strip of land along the Saimaa ship canal, permitting Finland to renovate, improve, and operate the canal built in 1856 to connect the Saimaa Lakes with the Gulf of Finland at Viipuri (now Vyborg).* And many military bases around the world have since 1940 been leased from host countries, primarily by the United States but also by the Soviet Union (Porkkala, Finland and Saseno, Albania, for example) and other countries.

A *servitude* in international law is roughly comparable to an easement in domestic law. Generally, it is a restriction on the sovereignty of a State over its own territory in the form of an obligation to permit a certain use to be made of it by another State or States. There are several forms of servitudes, ranging from rights to free navigation on some international rivers and in the territorial waters of States to special servitudes concerning land territory. These were formerly quite common, but since the consolidation of the nation-state system after the Congress of Vienna in 1815, they have become rare. Rights of way are still in use. Panama had a right of way across the former Canal Zone, for example, and Peru has the right to enter Chilean territory at will to maintain or enlarge the Mauri and Uchusuma irrigation canals, parts of which lie in territory acquired from Peru by Chile in the War of the Pacific. Finally, Norway is required by a 1920 treaty to permit other countries to engage in commercial activities in Svalbard (Spitsbergen) under their own laws.† There are also negative servitudes, the most important of which are demilitarized zones. Even since World War II, many of these have been estab-

lished. Among them have been a number of border areas in southern Europe, some Mediterranean islands, Trieste, places along the 1949 armistice lines around Israel, and strips along both sides of the military demarcation lines separating North and South Vietnam and North and South Korea.

Here is a fertile field of investigation for political geographers. There has not yet been a good geographic study of leased territories or demilitarized zones and other servitudes. Perhaps more information about them would indicate their practicality in resolving many contemporary boundary and territorial disputes.

SIZE

We all know that States vary greatly in size. They range from a half dozen giant States to half a hundred that are very small indeed. But does it really matter how big a country is? The simple answer is that the advantages and disadvantages attributable to size alone seem to be distributed quite randomly over large and small alike. A large country may not necessarily be endowed with resources commensurate with its size, and many of those it has may remain untapped because of the difficulty and expense of utilizing them. It may be easier and cheaper for a small State to import its primary requirements than for a large State to develop its own. The location, physiography, and shape of a State often enhance or diminish the value of large or small size. Defensive depth may be nullified by difficulties of administration and circulation. Population may be large or small, evenly or unevenly distributed, ethnically homogeneous or variegated regardless of the measurements of the territory. This is not to imply that size is irrelevant; it is sufficiently important to warrant further discussion.

A very large State that is sparsely populated may experience internal division, especially if the areas intervening between the populated regions are both

*In 1971, the United States gave up its 1914 lease of the Corn Islands in the Caribbean, though Nicaragua had been administering them with the acquiescence of the United States.

†Only the Soviet Union has so far taken very much advantage of this, however, primarily to mine coal there.

Helgoland. This small island strategically located in the North Sea (1 mile [1.6 km] long, one-third mile [.5 km] wide), 64 kilometers north of the mouth of the Elbe River, was first settled in ancient times by the Frisians. The island was occupied by the Netherlands from 1402 to 1714, then by Denmark and the United Kingdom before Germany acquired Helgoland in 1890. It was heavily fortified by the German empire, which made it a major naval base. After WWI the fortifications were demolished, and the island was formally demilitarized. Hitler fortified it again and after WWII his submarine base, air base, and fortifications were destroyed by the British. The island is again formally demilitarized and its approximately 2400 citizens live primarily by fishing and tourism. (Courtesy of the German Information Center)

difficult to cross and unproductive. Australia's central desert, the Soviet Union's vast eastern domain, and the Canadian Shield all exemplify the barrier effect of vastness, although in each greater political unity exists than in many smaller States that do not have size problems. Nevertheless, most very large States attempt to diminish the "empty" aspect of their sparsely populated regions by encouraging settlement in those areas or by practicing population policies aimed at rapid growth as well as encouraging migration. In Sudan the problem of size is directly related to the forces of fragmentation that have confronted that country almost constantly since independence. The Sudan is so large that it extends from Arab Africa into black Africa. Thus it contains an Arab population in the north, focusing on the capital of Khartoum, while the southern provinces are occupied almost exclusively by peoples who are more closely related to black Africa—racially, culturally, historically. As it happens, the two "heartlands" of the State are separated by desert and swamp, so that communication has always presented problems. The young State has had to expend much of its energy in controlling divisive forces, with distance complicating virtually every aspect of the matter.

The size of a State is related in many ways to its effective national territory, or *ecumene.* Many of the States that evolved in various parts of the world ultimately broke up because their frontiers extended too far outward to be sufficiently integrated with the central area of the State. Continued growth meant growing strength—up to a certain point, after which it meant increasing vulnerability.

This was one of the reasons for the collapse of the Aztec Empire, ancient Ghana, and the Roman Empire. It also has been a major factor in the breakup of more recent colonial empires, and in such States as India and Sudan, it is a critical matter today.

On the other hand, it is obvious that size can present advantages. If some attention is also paid to location (relative location, with reference to environmental regions, mineralized belts, trade routes), a generalization regarding size might even be possible. After all, the United States fits comfortably within the Sahara, and its size there would be meaningless. But the United States lies in middle latitudes, in a world zone of many transitions (in terms of soils and climate, to name two), and fronting two oceans. Depending on location, then, size and environmental diversification are indeed related.

One way to consider this would be as follows. The total land area of the world is limited, and it contains the bulk of the resources on which progress is based. A State that has a larger area than another obviously has a chance to find a greater percentage of such resources within its borders. But these known resources themselves are not evenly distributed; they are scattered in patches across the globe. They are concentrated in certain areas, apparently absent in others. The soils of the lowland tropics are often incapable of sustaining sedentary agriculture; those of the highest latitudes are too shallow, acidic, or frozen. The climates of the polar regions are not conducive to agriculture; those of the low-latitude regions are often excessively dry or excessively moist. A series of maps of mineral resources indicate the remarkable concentrations of significant deposits in rather well-defined belts. Taking these world distributions into consideration when evaluating the effects of size on the wealth and self-sufficiency of the State, we see why some of the largest States are comparatively poor, while others are incomparably rich.

Generally, States exceeding 2.5 million sq. km (1 million square miles) are described as very large, while those under 25,000 sq. km (10,000 square miles) are referred to as very small. Small States range from 25,000 to 150,000 square kilometers (60,000 square miles), and medium size States from 150,000 to 350,000 square kilometers (135,000 square miles). Those over 350,000 but under 2.5 million square kilometers are referred to as large. Thus the following would apply:

Very small	Burundi	Lebanon
Small	Netherlands	Liberia
Medium	United Kingdom	Poland
Large	France	Mexico
Very large	USSR	Canada

One of the remarkable aspects of the group of very large States is the clustering of several of these States around the 8-million-square kilometer mark:

	sq. km	sq. mi.
Canada	9,974,382	3,851,113
China	9,758,475	3,767,751
U.S.A.	9,363,394	3,615,210
Brazil	8,511,631	3,286,344
Australia	7,704,165	2,974,581

At the lower end of the size range, there is another cluster of very small States:

Liechtenstein	160	62
San Marino	62	24
Tuvalu	26	10
Nauru	21	8
Monaco	243 hectares	(600 acres)

With such an enormous range in the territorial size of States, it is useful to have some terms to identify States within certain size categories. For example, the very smallest States, such as Liechtenstein and Monaco are referred to as *microstates*. States somewhat larger than these smallest units, such as Brunei (5765 square kilometers, 2226 square miles), The Gambia (11,292 square kilometers, 4360 square miles), and Cyprus (9251 square kilometers, 3572 square miles) often are called *ministates*. But neither term has any precise definition.

Since the mid-1960s, small States and territories are no longer mere curiosities or subjects for trivia quizzes (in which geographers and stamp collectors have a natural advantage) but have become a major international concern. As more and more small colonies became independent and applied for membership in the United Nations, it became evident that before long they would constitute a significant proportion of the international community. Even the superpowers, despite their public protestations of support for the principles of self-determination and independence, quietly expressed doubts about the desirability of having a large number of UN members representing collectively only a tiny fraction of the world's land area, population, wealth, and military strength casting votes that they probably would not be able to control. Many other questions were raised about the viability of such small States or their potential for sparking conflicts far out of proportion to their size. Numerous studies were produced by the United Nations, governments, and scholars who examined all the questions in great detail.

The result of all the studies, conclusions, and recommendations was that there has been no change at all. The trend toward miniaturization in the international community continues unabated, with three more small countries becoming independent in 1978 alone, three in 1979, five so far in the 1980s, and more on the way.* Nearly all of them have joined or will join the United Nations. But there should be no more concern about small States than there was not long ago about large ones. Size is no indicator of wisdom, talent, or virtue among States any more than among individual human beings. Nor are small States, old or new, necessarily beggars at the tables of the large States; some, such as Liechtenstein, Singapore and Brunei, are, in fact, quite prosperous.

There is also nothing immutable about the size of States. States throughout history, as we have seen, expand and contract, appear and disappear. While Ratzel

and his followers envisioned something inexorable about State growth and modern territorialists foster belief in a biological urge of each society, no matter how small, to have its own sovereignty, the fact remains that the State is only one of a number of institutions created by people to serve their needs. There seems to be, in fact, a tendency in our time toward equilibrium in the size of States. Even since the consolidation of the modern State system, we have seen countries shrink as well as grow. Thailand, Pakistan, Bolivia, and Germany (even East and West Germany together), for example, are smaller than they used to be. There are also separatist movements in many countries that might lead to fragmentation and even more midget actors on the world stage. At the same time movements emerge for closer association among small States, new and old, leading toward varying degrees of economic and eventual political integration into larger units. The evidence belies all theories about the size of States: large or small size is neither good nor bad; what matters is how well the area meets the needs of the people of a State and how they react to that size.

SHAPE

Size is only one of the morphological characteristics of a State that influence its functioning and its international behavior. Another is shape. Again, a quick glance at a map reveals countries of widely varying shapes, some comfortingly geometric (roughly) and some disturbingly erratic. The reactions of the observer to the shapes on the map, however, are not nearly as significant as the effects of those shapes on the people who live in them and on their neighbors. Since no two States have the same shape, it may be helpful to classify them into a few categories and discuss each in turn.

An *elongated* or *attenuated* State may, on the basis of the Chilean example, be defined as one that is at least six times as long as its average width. Thus, Norway,

*See Chapter Twenty-one for more details.

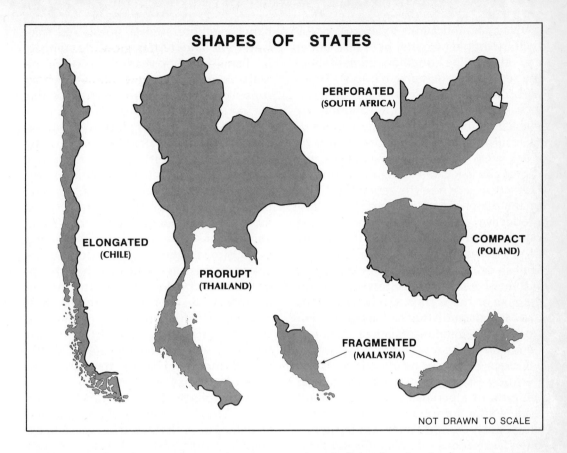

SHAPES OF STATES

PERFORATED
(SOUTH AFRICA)

ELONGATED
(CHILE)

COMPACT
(POLAND)

PRORUPT
(THAILAND)

FRAGMENTED
(MALAYSIA)

NOT DRAWN TO SCALE

Sweden, Togo, The Gambia, Italy, Panama, and Malawi are among the States in this category. Depending to some extent on the State's location with reference to the world's cultural areas, elongation may involve internal division. The north-south division of Italy is one that permeates life in that country and is related to the different exposures of the two regions to European mainstreams of change. The Norwegian administration of the Laplanders is a regional matter, involving the northern parts of that State. Furthermore, the physiographic contrasts within the elongated state may accentuate other divisions. Chile, for example, possesses at least three distinct environmental regions. The central region is Mediterranean in nature, the south is under Marine West Coast conditions, and the north is desert. The country's headquarters lie around the capital city, in the central (Mediterranean) part, and the effects of distance and remoteness are very evident on the Peru-

vian–Bolivian borders in the north and on Tierra del Fuego in the south. On the other hand, the internal diversification of the State resulting from its straddling of several environmental or cultural zones may be advantageous. Again, as in the case of size, much depends on location. Physiographically, location also plays its role. Chile lies astride the Tropic of Capricorn; Norway is bisected by the Arctic Circle. The resulting differences are obvious.

Many States appear to lie spread about their central area and are nearly round or rectangular in shape. Again, some generalizations are possible. Theoretically, since all points of the boundary of such a *compact* State lie at about the same distance from the geometrical center of the State, there are many advantages. First, the boundary is the shortest possible in view of the area enclosed. Second, since there are no peninsulas, islands, or other protruding parts, the establishment of effective communications to all parts of the country should be easier here than under any other shape conditions (unless there are severe physiographic barriers). Third, and consequent to the second point, effective control is theoretically more easily maintained here than in other countries. Of course, additional factors should be considered. In many compact States, the capital city is located along the periphery rather than at or near the geometric center of the State. Often the area of greatest productive capacity lies in one of the quadrants of the State rather than at the center. Finally, a compact State, no less than an elongated one, may be located in a zone of transition—cultural or physiographic, or both. Thus, internal divisions may still occur. Uruguay, Belgium, Poland, Sudan, and Kampuchea (formerly Cambodia) are a few compact States.

Certain States are nearly compact, but they possess an extension of territory in the form of a peninsula or a "corridor," leading away from the main body of territory. Such *prorupt* States and territories often face serious internal difficulties, for the proruption frequently is either the most important part of the political entity or is a distant problem of administration. Perhaps the best example is that of Zaïre, which consists of a huge, compact area with two proruptions, both of which are vital to the country and are in many ways her most important areas. The capital city itself, and the administrative headquarters, lie on the western proruption, which also forms a corridor to the ocean via the Zaïre port of Matadi. The most important area of revenue production, on the other hand, is Shaba Province (formerly Katanga), itself a proruption in the far southeast. Separating the two areas lies the vast Congo Basin, in many ways more a liability than an asset to the State. Another African example is Namibia (formerly Southwest Africa), whose proruption extends to the Zambezi River. Long a useless area, this corridor (the Caprivi Strip) has recently become of great strategic importance and South Africa has established a major mil-

itary base there. In Asia, Burma and Thailand have large territories, fairly compact in shape, but they share a section of the Malayan Peninsula along a boundary that runs almost through the middle of this narrow strip of land.

Other States consist of two or more individual parts, separated by land or by international water, and are therefore *fragmented*. Such fragmentation brings with it obvious consequences. Contact between the various population sectors is more difficult than in a contiguous State, and the sense of unity so necessary in the forging of a nation may be slow to develop. Because any State must have a capital, it will be located in one of the fragments, the choice of which may itself be a source of friction. Governmental control can be rendered ineffective by distance, as was proved repeatedly by the case of Indonesia, where the Java-based government had great difficulty putting down revolts in Sumatra, Celebes, and other islands. In contrast, the fragmented State may lie entirely on land. Pakistan exemplified the continental type. Her two "wings," the West and the East, were united by a common religious faith but divided by numerous cultural contrasts. As happens frequently with fragmented States, one part of the State (in this case the East), felt itself the victim of political discrimination. Charges of "domestic colonialism" abounded, and in the end the State broke up as East Pakistan—now Bangladesh—fought for independence, aided by neighboring India. A third type of fragmented State is that which lies partly on one or more islands. Malaysia consists of a mainland section on the Malayan Peninsula, a host of smaller intervening islands, and a sizable portion of the large Indonesian island of Borneo. Technically, Italy, with its island territories of Sicily and Sardinia, belongs in this category, as do many other States with island possessions.

Finally, there are a few States that completely enclose other States. Such States are *perforated*, and it is impossible to

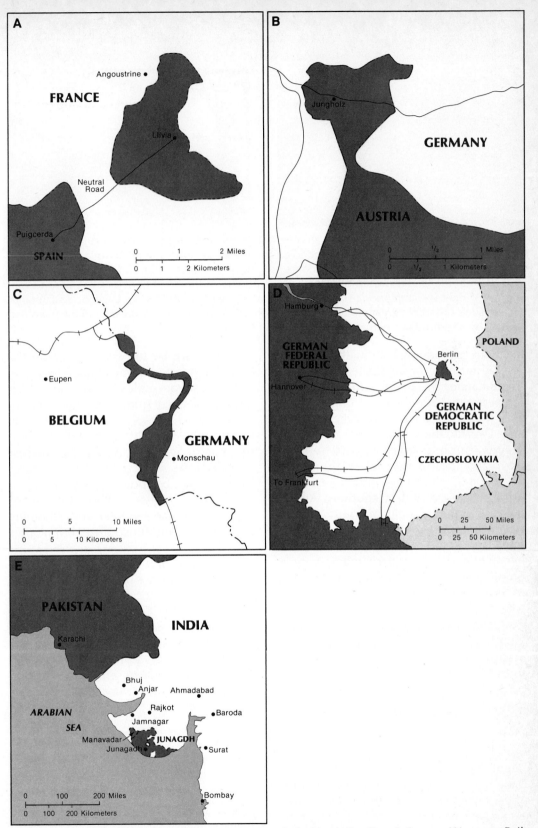

Exclaves. A—Normal: Llivia, Spain. B—Pene: Jungholz, Tyrol. C—Quasi: Raeren–Weywertz Railway Zones, Germany. D—Temporary: West Berlin. (All after G.W.S. Robinson) E—Coastal: Junagadh and Manavadar, Pakistan.

Exclaves. At the left is Point Roberts, an American territory at the tip of a peninsula jutting south of the 49° border with Canada, surrounded on three sides by the Strait of Georgia and Boundary Bay. It is a popular resort area for residents of Vancouver, some 35 km (21 miles) to the north, and much of the land is owned by Canadians. (Courtesy of the Point Roberts Chamber of Commerce) At right is Campione, an Italian exclave on the east shore of Lake Lugano, Switzerland. Its gambling casino is a great attraction and its only "industry." It is separated from the main part of Italy by a high ridge, seen at the right. (Martin Glassner)

reach the perforating State without crossing the territory or air space of the perforated State. This means, obviously, that the perforated State is in a strong position with reference to the land-locked perforator. Usually, these terms refer to tiny enclaves, such as San Marino (which perforates Italy, hence Italy is the perforated State), which have little if any political significance. But in southern Africa, the case of the Republic of South Africa is in a different class. South Africa is perforated by Lesotho, a State of 30,344 square kilometers (11,716 square miles) with a population of about 1 million. Politically, its black nationalism contrasts sharply with South Africa's white rule. In a size category shared with such States as Belgium and Costa Rica, Lesotho is a substantial impediment to the territorial integrity of the South African Republic.

Exclave and Enclave

One or two additional aspects of territorial morphology are relevant. Certain States might appear to be compact on a small-scale map. Close scrutiny of the boundaries, however, reveals that these States also have small pockets of land lying outside the main territory, as islands within the territory of neighboring States. These tiny areas are far too small to render the State fragmented. They may be less than a dozen square miles in area, and their populations may number a few hundred. Nevertheless, these *exclaves* are of some importance in political geography, for they may depend for their survival on their connections with the "homeland." Hence, their boundaries may be under great stress, and from the study of exclaves may come a greater understanding of the nature and functions of boundaries elsewhere.

Exclaves are not always small, and neither are they always unimportant in terms of area or population. West Berlin, as an exclave of West Germany, is one of Europe's most important urban centers and is considered a vital part of the German Federal Republic.* Curiously, West Berlin has its own exclaves embedded within East Germany. All are very small, however, and only one, the hamlet of Steinstücken

*Sometimes the terms *exclave* and *enclave* are confused. The correct usage depends on the point of reference. When an outlying, surrounded area is considered in connection with the homeland of which it forms a part, it is described as an exclave. In this sense, West Berlin is an exclave of West Germany. On the other hand, a State that has such a small territory within its borders (though not necessarily surrounding it) would view the entry as an enclave. West Berlin, then, is an enclave in East Germany.

(population 190 in 1970), is permanently inhabited.

Exclaves can be classified according to their degree of separation from the "homeland."* *Normal* exclaves are parts of certain States completely and effectively surrounded by the territory of other States. Such normal exclaves are usually small. Those in Europe are all under 26 square kilometers (10 square miles) in area and have populations under 2000. There are small Belgian exclaves within the Netherlands, and there is Dutch territory within Belgium. *Pene*-exclaves are "parts of the territory of one country that can be approached conveniently—in particular by wheeled traffic—only through the territory of another country." In other words, these are proruptions, barely connected to the main territory of the State, with the connecting links so narrow or difficult that the only transport lines lie through neighboring territory. *Quasi*-exclaves are those that are technically separated from the motherland, but in reality they are so completely connected with it that they do not function as exclaves. *Virtual* exclaves are areas treated as the exclaves of a country of which they are not legally an integral part. Finally, *temporary* exclaves result from the fragmentation of a State through an armistice; occupation zones or demilitarized areas may create temporary exclaves. West Berlin is such a temporary exclave.

Size and shape are two important characteristics of a State. We turn our attention now to the limits that enclose them.

*The material in this paragraph is taken from G.W.S. Robinson, "Exclaves," *Annals* of the Association of American Geographers, *49*, 3, Part I (September 1959), 283–295.

Chapter 8

FRONTIERS
AND BOUNDARIES

Many studies in political geography have dealt with frontiers and boundaries. Boundaries, on the map and on the ground, mark the limit of the State's jurisdiction and sovereignty. Along boundary lines States make physical contact with their neighbors. Boundaries have frequently been a source of friction between States, and the areas through which they lie are often profoundly affected by their presence. In many ways boundaries are the most obvious politicogeographical features that exist, for we are constantly reminded of them—when we travel, when we read a newspaper map or an atlas, and in many other ways.

Discussions of boundary problems sometimes treat the word "frontier" as though it were synonymous with "boundary." We read of the boundary between Germany and France in one paragraph, and of the French–German frontier in the next. Both references are to the same phenomenon, represented on the map by a line separating these two States. But are both terms correctly applied? Perhaps the answer can be arrived at through a consideration of their respective functions.

THE FRONTIER

We have already described briefly the expansion of States from their heartlands or core areas. Such States as the Aztecs' Mexican Empire or the Empire of the Incas were able to expand into areas that were unable to resist their power. At times in history, several States grew to local power and prominence simultaneously, but never made effective contact. Separating them were natural impediments to communication: lakes, swamps, dense forests, mountain ranges. These States, with very few exceptions, possessed no boundaries in the modern sense of the word, but they were nevertheless separated from their neighbors. Whatever the separating agent—it might have been sheer distance—it functioned effectively to prevent contact.

The modern map showing all States bounded by thin lines that can be precisely represented on maps is, in the politicogeographical world, a very new phenomenon. Although some of the old States, like the Roman Empire, attempted to establish real boundaries by building stone lines across the (in this case, British) countryside and natural features, such as rivers, served as trespass lines, the present, almost total framework of boundaries is a recent development. Maps representing the situation one, two, three, or more centuries ago show vast areas that are either unclaimed, unsurveyed, or merely spheres of influence. And our political evolution has been going on for several thousand years, not just hundreds.

Thus the States and embryonic States of the past were separated, not by lines, but by areas. Still, they were separated, and they were either not in contact or only sporadically and ineffectively so. This intervening area functioned to prevent contact. Today's boundaries do not prevent contact; along them States *make* physical contact!

This, then, is perhaps the best way to view the frontier: as a politicogeographical area lying beyond the integrated region of the political unit and into which expansion could take place. It is the same principle we use when we speak of the "frontiers of science": obviously we mean an ill-defined outer belt, vague and unknown, but into which we are penetrating. So it was with the geographical frontier. In Southern Africa, the European-settled core area forming after 1652 around Cape Town grew stronger, unaware of the powerful Zulu State that was developing in Natal. The frontier functioned to separate these two political units, until penetration of the intervening belt began in the 1830s. Then, finally, a confrontation occurred, and the frontier was replaced by uneasy (and, as it turned out, impermanent) truce lines.

Thus we see the frontier as an area, but it was not always an area separating two or more States. Sometimes it was invaded by a State without any serious obstacles until its "natural" limits were reached. So it was with the westward expansion of the American State and the invasion of the Australian "frontier." But although that invasion may have led to minor hostilities only, the function of the frontier remained the same. The partition of Antarctica is a modern example. Little is yet known about the resources of this continent, but a number of States have staked claims to most of it.

Through frontiers, boundaries were often drawn. Expanding States or spheres met; sometimes they fought over the area involved, and sometimes they settled the disputed area by boundary treaty. The colonial invasion of Africa is replete with examples of boundary treaties. Thus States made contact, and the boundary just established began to function. The frontier, in the original sense of the word, is no more, for the space into which the States had been expanding has been fully occupied. Hence the boundary, a manifestation of integration, is inner oriented. (The sea, however, is still a political frontier and coastal States are now expanding into it. More on this later.)

Why are the terms "boundary" and "frontier" sometimes still used interchangeably? Actually, the use of these terms reflects a reality in political geography—namely, that it is often still possible to recognize frontier characteristics in an area where a boundary does exist. Among the functions of boundaries, as we shall see, is that of division, separation— not physically, but in other ways, such as in economic terms. On either side of an international boundary, therefore, there may be a discernible zone suffering from the interruptive effects of the boundary. This zone, obviously, has spatial characteristics and, hence, may be a frontier.

Boundaries often run through relatively empty, sparsely populated land. In view of the processes of boundary establishment, this is not surprising. Least friction was encountered if, in a disputed area, the boundary was located through useless land. Hence, such a boundary may not be marked on the ground, not even by a line of stones. This is the case along sections of the Portuguese–Spanish border, so that one cannot, off the beaten path, be sure exactly where one is—in Portugal or in Spain. In the absence of a clearly marked boundary, in empty, barren territory, it is not surprising that the word "frontier" comes to mind, the theoretical existence of a boundary notwithstanding.

There are other kinds of frontiers besides natural or politicogeographical ones. Despite the expansion of *homo sapiens* into nearly every nook of the earth in which the species can survive, there are still settlement frontiers. These areas of relatively sparse population but relatively abundant resources capable of supporting larger populations are found primarily today in the more extreme climatic regions of the world: the high latitudes, the humid tropics, and the arid zone. Many States located in these regions, such as the Soviet Union, Colombia, and Egypt, are actively trying to settle and develop these frontiers because they are viewed as real, though still potential, economic assets. Other States are encouraging their nationals to settle in frontier areas near their boundaries, even if they are not very inviting physically. This is largely a defensive measure, to discourage or render more costly invasion or even peaceful encroachment by their neighbors. The Soviet Union, Bolivia, and Israel are among the countries applying this policy adopted in the ancient empires more than two thousand years ago.

Frontiers are frequently zones of transition, and few transitions are more significant on the map or on the landscape than those between one culture and another. Cultural frontiers are often traversed by political boundaries that sometimes obscure the transitional nature of the region. People living along an international boundary tend to be more concerned with their local affairs, which in many cases straddle the boundary, than with affairs in their own capitals, which may be very distant in many ways. Links with neighbors across the border may be stronger than those with compatriots some distance away from the border. The trend, however, since the rise of nationalism and the solidification of States, has been toward hardening of boundaries and complete integration of a country's nationals right up to the borders. Nevertheless, in regions in which boundaries were drawn without regard to ethnic considerations and in regions of considerable international migration, cultures frequently overlap the borders and fade gradually with increasing distance from them. Often as one approaches, passes through, and leaves such a cultural frontier, the observable changes are so gradual it is difficult to know where the international boundary is unless it is clearly marked. Nevertheless, it is there and it affects the lives of everyone on both sides of it.

THE BOUNDARY

Boundaries appear on maps as thin lines marking the limit of State sovereignty. In fact, a boundary is not a line but a plane, a vertical plane that cuts through the airspace, the soil, and the subsoil of adjacent States. This plane appears on the surface of the earth as a line because it intersects the surface and is marked where it does so. But boundaries can be effective underground, where they mark the limit of adjacent States' mining operations in an ore deposit they may share, and they can be effective above the ground, for most countries jealously guard their airspace.

Boundary Making

The ideal sequence of events in the establishment of a boundary is as follows. The first stage involves the description of the boundary and the terrain through which it runs. This description identifies, as exactly as possible, the location of the boundary being established. Reference may be made to hilltops, crestlines, rivers, and even to cultural features, such as farm fences and roads. The more detailed the description, the less likely it is that subsequent friction will occur. As will be seen later, even the most prominent physical features in the landscape have given rise to serious disputes when used as political boundary lines. This first stage, represented by the language found in many treaties, is referred to as the *definition* of the boundary.

When the treaty makers have completed their definition of the boundary in question, their work is placed before cartographers who, using large-scale maps and air photographs, plot the boundary as exactly as possible. The period of time separating this stage of *delimitation* from the initial stage of definition may amount to decades; for example, several African States whose boundaries were defined toward the end of the last century are only now in the process of exactly delimiting their borders. Some are discovering that

MAJOR CHANGES IN
BULGARIA'S BORDERS: 1878-1947

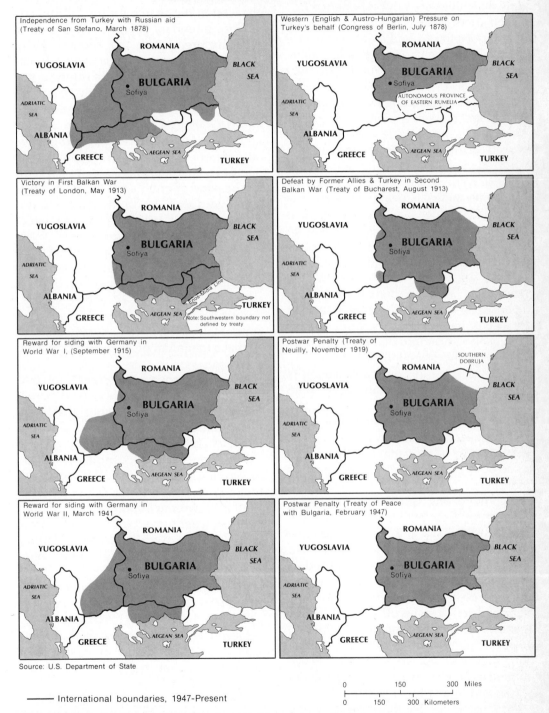

Independence from Turkey with Russian aid
(Treaty of San Stefano, March 1878)

Western (English & Austro-Hungarian) Pressure on
Turkey's behalf (Congress of Berlin, July 1878)

Victory in First Balkan War
(Treaty of London, May 1913)

Defeat by Former Allies & Turkey in Second
Balkan War (Treaty of Bucharest, August 1913)

Reward for siding with Germany in
World War I, (September 1915)

Postwar Penalty (Treaty of
Neuilly, November 1919)

Reward for siding with Germany in
World War II, March 1941

Postwar Penalty (Treaty of Peace
with Bulgaria, February 1947)

Source: U.S. Department of State

———— International boundaries, 1947-Present

The instability of national boundaries through history is particularly evident in the world's shatter belts. A classic example is Bulgaria. Since the formation of the United Nations, however, with its emphasis on the "sovereignty and territorial integrity" of States, boundary changes have been much less frequent.

the original work of definition was done rather crudely and imprecisely, so that many difficulties are now experienced.

Then there is the task of marking the boundary on the ground. For this purpose both the actual treaty and the cartographic material are employed. Boundary *demarcation*, as this process is called, has by no means taken place along every boundary defined and delimited; only a minority of the world's boundaries are actually marked on the surface. When they are demarcated, a wide variety of methods may be employed. A mere line of poles or stones may suffice. Cement markers may be set up, so that from any one of them the adjacent ones on either side will be visible. Fences have been built in certain delicate areas where exact demarcation is required, and on rare occasions walls have been built. Boundary demarcation is an expensive process, and when States do not face problems along their boundaries that absolutely require demarcation, they often delay this stage indefinitely.

The final stage in boundary making is *administration;* that is, establishing some regular procedure for maintaining the boundary markers, settling minor local disputes over the boundary and its effects, regulating the use of water and waterways in the border area, and attending to other "housekeeping" matters. Sometimes special commissions, such as the International Joint Commission of the United States and Canada and the International Boundary Commission of the United States and Mexico, will be established to perform these functions. Usually, the tasks are assigned to an office in the foreign ministry or some other government agency. For many boundary segments, however, there is no regular administration; special *ad hoc* commissions are appointed as needed or a particular problem is assigned to an official to handle.

We must stress that this is an ideal pattern. It appears in practice most commonly in Europe and North America, rather regularly in Latin America, less commonly in Asia, and rarely in Africa.

Criteria for Boundaries

Political geographers, among others, have engaged in the search for the "ideal" criteria for boundary definition, in hopes of reducing international tensions created by boundary disputes. This activity was especially common during the interwar period, and led to intense debate regarding the merits of "artificial" boundaries as opposed to "natural" (physical) ones. In fact, no discussion of criteria is possible without reference to functions. A boundary is one of the parts of the State system, and by speaking of criteria for their establishment, we speak also of the effect the use of these criteria will have. And by effect we mean function. An example will clarify this. Some political geographers have felt that ethnic criteria may be the most appropriate for the definition of international boundaries. In other words, boundaries should be drawn so as to separate peoples who are culturally uniform so that a minimum of stress will be placed on them.

Of course, world population is too heterogeneous and interdigitated to permit the definition of boundaries that completely and exactly separate peoples of different character. There are minorities in nearly every State. Occasionally, instead of altering boundaries, States have attempted to exchange such groups: after World War I some hundreds of thousands of Turks were moved into Anatolia, from where Greeks were repatriated. Among the solutions proposed for the dilemma of Cyprus have been repatriation of Turkish Cypriots to Turkey and partition of the island into Greek and Turkish States. The partition has, in fact, already been achieved by a haphazard exchange of populations since the Turkish invasion in 1974, but it has not noticeably reduced Greek–Turkish friction.*

The criterion of language might also be proposed as a basis for boundary defini-

*In November 1983, the Turkish Cypriots established "The Turkish Republic of Northern Cyprus" with the support of Turkey. It has been recognized by no one else and may be only a bargaining ploy.

tion. But a map of the world's languages shows a patchwork of great complexity that would immeasurably compound the boundary framework existing today. Many States are multilingual and would be fragmented in any such effort; furthermore, it is impossible to reconcile the language criterion at all times with that of "race." This brings up a question that kept political geographers at odds for some time— namely, whether a boundary should be a barrier or a bond between adjacent States. If a boundary separates people who speak different languages, they are not likely to understand each other well, with the result that relations may remain potentially hostile across their mutual boundary. On the other hand, a boundary running through a region of linguistic homogeneity would ensure that people on either side would have, at least, a language in common, and as a result could communicate more easily. This common language across the border would then act as a bond between the two States involved. The point can also be put differently, in terms of physical criteria: a mountain range performs the barrier function, with divisive consequences; a river, with its two banks close together, would form a bond.

One of those criteria that do not coincide with language or race is religion. Peoples of widely varied races and tongues have accepted the same faith, and peoples speaking the same language have adopted different religions. Nevertheless, in areas where religion has been a strong source of internal friction, it has been a major basis for boundary definition. A good example is the partition of the Indian subcontinent into (mainly Hindu) India and (mainly Muslim) Pakistan. The latter State, as a result, became a fragmented State and has shown the weakness of religion as a centripetal force.

Inspection of the world framework of boundaries will indicate that many political boundaries lie along prominent physical features in the landscape. Such boundaries, based on physical features,

have become known as "physiographic political boundaries," a term not to be confused with the physiographic boundary used in physical geography. In political geography a physiographic boundary refers to any prominent physical feature paralleled by a political boundary: a river, mountain range, or escarpment. These would seem to be especially acceptable criteria since pronounced physical features often also separate culturally different areas. But in practice such boundaries have also produced major problems. No State corresponds exactly to a physiographic province, and very homogeneous physiographic regions are sometimes divided by boundaries based on some insignificant feature, such as a small stream or a low divide. In the early days of boundary establishment, physiographic features were useful because they were generally known and could be recognized as the trespass line. But this function has ceased now, and the most divisive and obvious physical features have created major difficulties between States.

Any mountain range has recognizable crestlines, but they rarely coincide with the watershed. As a result, water flowing from one side of the State boundary (which coincides with a physiographic, but not with a hydrographic feature) feeds the streams of the State on the other side of the boundary. If the State that possesses the source areas decides to dam those waters, it may impede the water supply of the other State and friction may result. There are many other examples of problems arising along international boundaries in mountains—transhumance, the use of passes, the need for tunnels.

Rivers, too, seem obvious and useful boundary features, but again many problems attend their use as boundaries. Use of the water by the riparian States is one major issue for debate.* Furthermore, riv-

*Riparians are those States through or along which a river flows. The term is also used as a synonym for the adjective "littoral" to indicate a State that abuts a lake or the sea.

ers tend to shift their course, producing new circumstances that require a redefinition of the boundary, which may lead to friction. In addition, rivers have breadth as well as length and depth, and countless disputes have erupted over whether the boundary should be along the left bank, the right bank, the *thalweg* (main navigable channel), the middle, or somewhere else.

Thus physical features, which seem almost to be nature's own provision for boundary establishment, are at times as problematic as boundaries based on the other criteria enumerated. We might also observe that arguments on behalf of "natural" boundaries are usually advanced by representatives or academics of States that wish to expand their territories. Can anyone imagine a State offering to move its boundary *back* to a "natural" line deep inside its own territory?

Since *all* political boundaries are by definition artificial, a *morphological* classification of them is not particularly useful.

It hardly matters when one is at a border and looking across or flying over whether it is geometric or not. How the border functions, how it is reflected on the landscape, and how the people on both sides feel about it are much more important. Thus, other classifications have been devised that can be quite useful in analyzing boundaries and boundary problems.

A *functional* classification might reflect, for example, whether the boundary was originally or is still designed primarily for defensive purposes, as a separator of cultures, according to economic factors, simply for legal or administrative purposes (as in Spanish America and French Africa), or on ideological bases (communist or noncommunist areas, Catholic or Protestant areas, black or white-controlled areas in Africa). A *genetic* classification would be based on when the boundary was laid out. A *pioneer* boundary is drawn through essentially unoccupied territory. Or a boundary may be *antecedent* to intensive settlement and land use, *subsequent* to

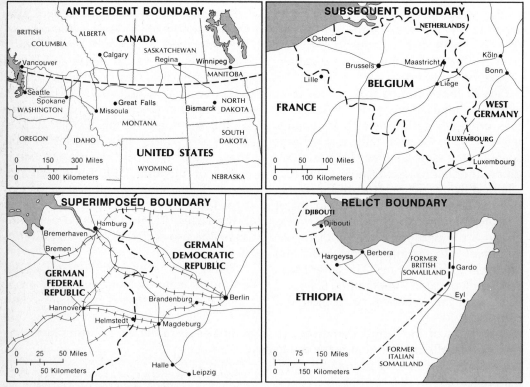

Genetic classification of boundaries.

the establishment of people with different cultures in the region and taking them into account, *superimposed* on an existing cultural pattern, or *relict,* a former boundary that no longer functions as such.

A *legal* classification could consider those boundaries that are settled and recognized in international law; those recognized only by the adjacent countries and some others; *de facto* boundaries, which are disputed by one adjacent State; and *fictitious* boundaries, which exist on maps but not in the real world, usually relict boundaries or the limits dreamed of by separatist or irredentist groups. While all of these concepts are useful for analysis and understanding of boundaries, we must remember that no type of boundary is necessarily better than any other. The best boundary may be the one that performs the fewest functions; certainly it is the one falling between good neighbors.

Functions

The functions of boundaries change over time. This is perhaps best illustrated through a consideration of the role of the boundary as a line of defense. Until quite recently, it was conceivable for a State to attempt to fortify its boundary to such an extent that it would be invincible. French hopes until 1940 were pinned on the Maginot Line, a line of fortifications constructed along its northeastern boundary. The idea is as old as the Chinese Wall, and the principle is the same. Plateaus with sheer escarpments have afforded protection to societies that used these natural barriers to their advantage and considered the scarps to mark the limit of their domain; the histories of Lesotho and Ethiopia illustrate the case.

But advancing technology has diminished the importance of the defensive function of boundaries, and States no longer rely on fortified borders for their security. In some parts of the world, where guerrilla activities short of open warfare occur, a river or mountain range may still present strategic advantages. To the major powers of the world, however, and those States possessing modern military equipment, the naturally or artificially fortified boundary is no longer an asset. On the other hand, this should not suggest that the boundary as a mark of territorial inviolability has thereby also vanished. Many States are presently demarcating their boundaries without intending or attempting to fortify them. Rather, the aim is to mark the limit of State sovereignty, which may have become necessary as a result of emerging friction.*

The boundary, then, is one of the interacting parts of the State system. It has an impact on many forms of organization within the State and, in turn, on the way in which the State's various arms of organization affect the nature of the boundary. The first case is exemplified by the commercial function of the boundary. The government can erect tariff walls against outside competition for its market and thus assist internal industries. These industries may prosper and owe their prosperity to the protection thus afforded. However, the price differential on either side of the boundary will affect the location of outlets for the products of the various industries affected by tariff provisions, and while the industry may prosper, the area under the shadow of the boundary may not and smuggling will be encouraged.

The second case can best be envisaged in terms of contrast. Some boundaries, because of the close and positive rela-

*Nevertheless, boundary walls and fences continue to be built. The most famous, of course, is "the Iron Curtain," purely metaphoric in fact but symbolized on the borders of East Germany and Czechoslovakia with the West by fences, minefields, guardtowers, and so on. The similar installations along Israel's borders with Lebanon and Syria, however, perform a very different function: to keep armies and terrorists out, not to keep citizens in. Morocco completed in 1985 a 4000-kilometer (1550-mile) wall of stone and sand in Western Sahara to protect the most valuable portion of the territory from Algerian-backed rebels seeking independence. Thailand has built an antismuggler and antiterrorist fence along the Malaysian border. South Africa has erected an electrified fence along parts of its border with Zimbabwe. And there are others.

Two contrasting boundaries. At the top is a fine example of a relict boundary, the one formerly separating the French and Spanish protectorates in Morocco. This impressive ruin is on the main highway near the Atlantic coast. (Martin Glassner) On the bottom is a segment of the Berlin Wall, perhaps the best example of a superimposed boundary today. The photograph shows an area near Bernauer-Strasse away from the central part of Berlin. (Courtesy of the German Information Center)

Two very different border towns. At the top is Jogbani, which is on the border between the Indian state of Bihar and the southeastern corner of Nepal, just south of Biratnagar. Nepal and India have an open border, and there are no restrictions on the movement of their citizens across it. There is a tiny customs-immigration post on the Indian side to service the occasional foreign traveler; on the Nepali side, beyond the striped pole, there is no post at all. At the bottom is Le Perthus, a Catalan town straddling the boundary between France and Spain in the Pyrenees south of Perpignan, France. This is the main highway between the French and Spanish Rivieras, and the town is generally crowded with tourists. A small stone pillar to the left of the car marks the actual boundary. The customs and immigration posts for each country are some distance down the mountain slopes where there is some level land. (Martin Glassner)

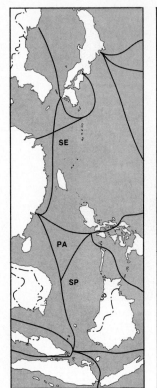

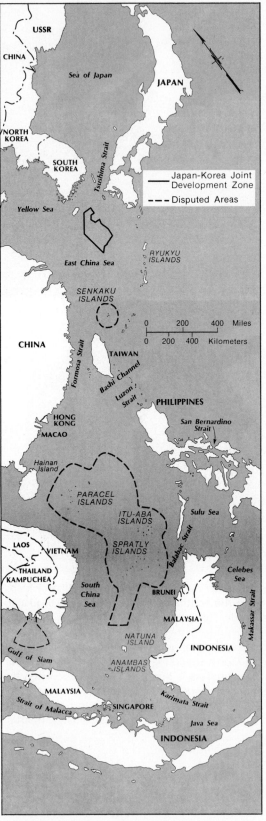

tionship between the societies they separate, do not mark great changes in modes of government, prosperity, or other phenomena. The U.S.–Canada border is an excellent example. Some other borders, on the contrary, mark lines across which contrasts are severe and prominent. The societies they separate may have different institutions, different levels of development, and different ideological ties. Such a boundary may be very interruptive in every way. The boundary between East Germany and West Germany functions in this manner.

The boundary, of course, also has a legal function. National law prevails to this line. Taxes must be paid to the government by anyone legally subject to taxation, whether the person resides 1 or 100 miles from the border. Even though residents living within sight of the border may have closer linguistic, historical, and religious ties with the people on the other side, they are subject to the regulations prevailing on their side of the boundary. Furthermore, the government is capable of controlling emigration and immigration at points along the border.

A number of contemporary trends are combining to reduce the significance of boundaries, even as many countries are energetically seeking to fix and demarcate theirs. For one thing, as peoples are brought closer together by vastly better communications and transportation, they are more and more being forced or enticed to look beyond their own national boundaries and become more familiar with the people out there. As trade, travel, and tourism increase, boundary formalities are being recognized as unnecessary and very harmful impediments and are being removed. Many groups of States around the world are integrating their economies in varying degrees, and fundamental to all integration is free movement across borders. Boundaries are utterly meaningless to missiles and television broadcasts alike; not only terror but understanding can readily pass through a political boundary, no matter how well defended. Boundaries are becoming more permeable and less hostile. This is not to say, however, that boundaries are disappearing or even that States no longer care much about them; indeed, quite the contrary is true. There are still a great many boundary and territorial disputes around the world. Some of them are dormant, some very much alive, and some of uncertain status. All, however, are interesting and deserve study.

BOUNDARY AND TERRITORIAL DISPUTES

It is impossible to count precisely all contemporary boundary and territorial disputes. Certainly the number is near one hundred, far more if we count every dis-

---

Disputed islands in the East and South China Seas. As competition grows for marine resources, ever smaller islands are becoming objects of dispute. The seas off East and Southeast Asia provide many examples. The Senkaku islands are currently controlled by Japan but also claimed by China and Taiwan; the Paracels are occupied by China, which displaced the Vietnamese claimants in a brief battle in 1974; various Spratly islets are controlled by Taiwan, Vietnam, and the Philippines, all of which have claims on all 200 of them, as does China. All these islets, reefs, rocks, and cays are uninhabited except, in some cases, where they have military garrisons. Vietnam and Kampuchea have not agreed on their maritime boundary or on ownership of the islands in the disputed area. The value of these islands lies in the fisheries nearby, the potentially large petroleum deposits in their continental shelves, and, as shown in the outset map, their proximity to some of the world's most strategic sea lanes. Japan and Korea, unable to agree on their southern continental shelf boundary, established the disputed area as a zone to be developed jointly for at least 50 years. China has also claimed a portion of this shelf.

puted island separately. And these are only land areas; maritime boundary disputes constitute a huge category of their own, though closely related to disputes over land, and we discuss them later. All we can do now is to survey a few outstanding problems to illustrate the point that there is still a great deal of work to be done by political geographers—and students of political geography—on boundary and territorial disputes.

As we turn more and more to the sea for its resources as well as its transport and strategic value and as States expand their national jurisdiction farther and farther out into the sea frontier, more and more islands are being disputed, often ones scarcely known until recently and which may have no intrinsic value. What makes them worth disputing is the 200-nautical-mile exclusive economic zone to which each island is entitled under the evolving Law of the Sea. The United States alone disputed 18 islands in the Pacific with the United Kingdom and 7 with New Zealand. While these disputes have been

resolved as the United States relinquished its claims, other island disputes are getting more intense.

These disputes involve a number of islands in the far Western Pacific. In the South China Sea, China and Vietnam are vying for possession of the Paracel and Spratly islands, uninhabitable coral islets and reefs that lie in an area of potentially considerable submarine petroleum deposits. Similarly, in the East China Sea, Japan, Taiwan, and China are disputing ownership of the Senkaku Islands. A different type of island dispute is represented by the Falkland Islands, administered since the eighteenth century by the United Kingdom (with an interval of abandonment) but claimed by Argentina. This is more of a traditional territorial dispute, similar to those between Guatemala and Belize or between Guyana and Venezuela, which claims about a third of that former British colony. In each case the problems are complex, resulting from competing claims among colonial powers to little-known territories based on con-

Traditional territorial disputes in northern South America.   All three of these areas are in dispute because of varying interpretations of boundary definitions and conflicting maps made at a time when the region was largely unexplored and scarcely known to Spain, France, the Netherlands, and the United Kingdom. At the moment all these disputes are quiet, but none has been resolved.

Propaganda sign at La Quiaca, Argentina on the Bolivian border. "The Malvinas (Falklands) are Argentine." The people are Bolivians crossing the border to buy cheap food in Argentina to take back to La Paz for sale. (Martin Glassner)

flicting principles of international law and disputed facts. In 1982, the Falklands dispute broke into open warfare as Argentina occupied them briefly before being displaced by a sizable military force sent from Britain. There are still no signs of a peaceful resolution of this problem.

There are other boundary and territorial disputes in Latin America, but considering the casual way those boundaries were determined in the first place, it's remarkable more don't occur. This is in part attributable to the adoption by the leaders of the struggle for independence from Spain early in the nineteenth century of the ancient principle of Roman law, *uti possidetis juris*. This means that whoever possesses a thing has a right to it. The Latin American leaders called their version "The Uti Possidetis of 1810," meaning that the boundaries of the new States that would emerge there would be the administrative limits in effect when the revolutions broke out in 1810. The Organization of African Unity adopted the same principle in 1963 in an effort to prevent innumerable conflicts over imposed colonial

boundaries of the emerging States of Africa. As in Latin America, the effort has not been wholly successful: witness the wars between Algeria and Morocco and between Somalia and Ethiopia over desert areas. Nevertheless, the policy did prevent major wars that might have spread from the Nigerian civil war and the repeated attempts by Shaba to secede from Zaïre. It is likely that more dormant or latent boundary and territorial disputes will develop in Africa, but they stand a good chance of being resolved peacefully. There is very strong resistance to a "Balkanization of Africa."

While Spain has been trying to wrest Gibraltar from the United Kingdom, Morocco would like to retrieve from Spain five presidencies, small islands and enclaves along the Mediterranean coast, that Spain has held as part of her sovereign territory for centuries. This poses a problem for Spain: Does she believe in self-determination? Or decolonization? In one case but not the other? No resolution is in sight. To complicate matters further, an irredentist group in Portugal has been claiming Olivenza, a 1500-square-kilometer (600-square-mile) wedge of territory just south of the Spanish town of Badajoz, which was part of Portugal from 1228 to 1801. But the Portuguese government could hardly press the claim while stoutly refusing to give up her colonial possessions in Africa and Asia. Boundary disputes are seldom conducted in a vacuum; they are generally linked with many other factors in each of the disputing countries and even considerations halfway around the world.

The irredentist claims of China to huge areas of the Soviet Union have been well publicized, as have the several wars fought between India and Pakistan over Kashmir and the Rann of Kutch. Unlike the examples given earlier, these cases (and a number of others in Asia) involve areas of considerable size, population, natural resources, and/or strategic importance. They could well lead to more serious fighting before the disputes are settled. In Asia

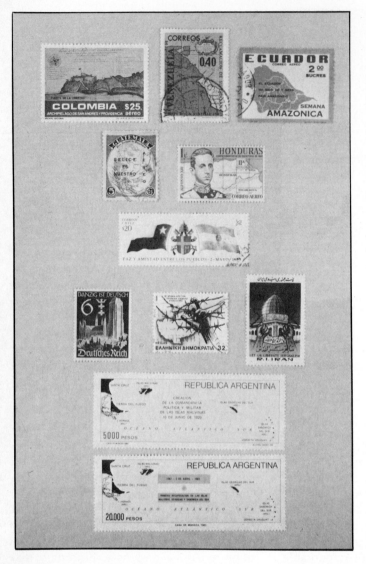

Boundary and territorial disputes on stamps. Top row, left to right: Colombia and Nicaragua both claim San Andrés and Providencia Islands in the Caribbean Sea; one of a series of maps on stamps supporting Venezuela's claim to two-thirds of Guyana; Ecuador's claim to territory held by Peru in the Amazon Basin. Second row: "Belize is Ours," on a Guatemalan map of the country; Honduran stamp honoring settlement of a territorial dispute with Nicaragua. Third row: Chilean stamp honoring settlement of Beagle Channel dispute with Argentina through papal arbitration. Fourth row: "Danzig is German," on a German stamp issued after the Nazi conquest of Poland and Danzig in 1939; Greek stamp decrying Turkish occupation of much of Cyprus; Iranian stamp demanding return of sovereignty over Jerusalem to an unspecified "us." Bottom: Stamps honoring the brief Argentine occupation of the Falkland Islands in 1982. (Martin Glassner)

there is no regional organization comparable to the Organization of American States or the Organization of African Unity to moderate and help resolve disputes in the region. And no *uti possidetis* principle operates there either.

That the principle does not always work even in the region that first adopted it in modern times is illustrated by the chronic boundary disputes between Argentina and Chile. Their mainland boundary in the Andes was settled only one segment at a time in a series of agreements extending over more than a century, the most recent being La Palena in 1966. Despite the inscription on the giant statue in Uspallata

Pass, erected in 1902 to commemorate settlement of an earlier boundary dispute, the two countries are anything but friendly. The inscription reads, "These mountains will crumble into dust before the peoples of Argentina and Chile break the peace sworn at the feet of Christ the Redeemer." Yet the two countries came perilously close to war in 1978 over their boundary in the Beagle Channel at the southern tip of the continent, and the dispute was not resolved until 1985, when they finally accepted an arbitral decision of Pope John Paul II. Despite many ethnic and historical similarities and links, Chileans and Argentines are not particularly friendly toward one another. Speaking of the Christ of the Andes, the Argentines will say that he faces Argentina with his hand raised in benediction over them because he favors them over the Chileans, on whom he has turned his back. The Chileans will respond that Christ is only facing the Argentines because he has to watch them carefully; he doesn't dare turn his back to them!

It may be true, as Robert Frost said, that "Good fences make good neighbors." It is much more likely, though, that it takes good neighbors to make good fences.

Chapter 9

CORE AREAS AND CAPITALS

We noted earlier that the original States of Europe, Asia, Africa, and Latin America developed around core areas of relatively dense population, generally focused on at least one city, served by a good circulation system, and supported by a firm agricultural base. During the past few hundred years, however, this pattern has been joined by others and there are examples of core areas that do not fit into any particular pattern. Every adequately functioning State system, however, has a nucleus, a central, essential, enduring heart. It was probably the German geographer Ratzel who first tried to define this reality in politicogeographical terms. States, he said, tended to begin as "territorial cells," which would then become larger through the addition of land and people, and eventually evolve into States or even empires. Whittlesey elaborated on this theme, emphasizing the role of the core as the area in which, or about which, a State originates. Normally the capital city is situated in this core area. These generalizations are all incontestable; problems arise, however, in applying them to the real world of the twentieth century.

For one thing, there is the problem of quantifying the definition. How dense should the population be? Or the transportation network? What is the relative importance of agriculture, industry, ethnic homogeneity in delimiting the core area? It would seem that it is both impossible and unnecessary to quantify the def- inition, for in fact the criteria would necessarily differ from one country to another and even in different periods of history. Thus, when we study core areas, we must study among other things the level of political, social, and economic development of each State system individually rather than try to establish universally applicable criteria. As we do, we find different types of core areas.

TYPES OF CORE AREAS

Core areas may be classified in several ways. The Canadian geographer Andrew Burghardt identifies three types based on historical development. The case in which a small territory grows into a larger State, perhaps over a period of centuries, as described by Ratzel and Whittlesey, he calls a *nuclear core*. In some States the *original core* was always the area of greatest political and possibly economic importance within an already larger framework. Finally, the *contemporary* core is presently the area within the State of greatest economic and/or political importance, although it may have superceded one or more earlier cores. Pounds and Ball analyzed European core areas in some detail and concluded that they can be classified on the basis of function. Thus they recognized States with *distinct* core areas, such as France, Czechoslovakia, and Russia; those with *peripheral* core areas, including Yugoslavia and Portugal; and

CORE AREAS AND CAPITALS

those *without* distinct core areas, such as Albania and Belgium. Another possible classification is based on spatial considerations. In France and South Africa, for example, the core area is *centrally* located. In Brazil, Argentina, and many other States, the core area is *marginal* in the national territory. It is even noted that in some States that have experienced territorial division or shifting boundaries, the core area may currently be *external*.

Spatial considerations immediately lead to another problematic characteristic of States and core areas; certain States possess more than one focus. Thus, we might recognize *multicore, single-core,* and *no-core* States. Nigeria, for example, has three core areas: one in the southwestern part of the country, one in the southeast, and a third in the north. Ecuador may be said to have two core areas: one centered on Guayaquil on the coast and another focusing on Quito in the highland interior. Thailand has a single core area; Mauritania and Chad might well classify as no-core units.

In addition, some States have quite distinct core areas *and* incipient core areas emerging elsewhere. Are such States multicore States or are they of a different variety? Consider, for example, the United States. Although there might be argument about the exact boundary of the core area of this country, there is little doubt that it lies in the northeastern corner. But core area characteristics are developing to the west of this core, notably on the west coast. The urban-industrial-population agglomeration in the west is large enough to constitute a core area in almost any State but the United States, where it is still overshadowed by the northeastern core region. What is the solution? It is probably one of *scale*, the same solution we alluded to when we suggested that national rather than universal criteria ought to be employed in determining core areas. Thus we suggest that the major core area, that of the northeast, be designated as the *primary* core, while the western (and other emerging areas) be identified as *second-*

ary core areas. Burghardt suggests that this criterion of scale can be carried farther, that there are *continental* and *world* core areas as well. In a sense, the United States–Canadian core in eastern North America is such a continental core area, and in Europe a developing contential core area can be recognized as well.

CORE AREAS AROUND THE WORLD

The types of core areas just described can be illustrated by a survey of different parts of the world. In Europe most core areas are of the type described by Ratzel, Whittlesey, and Pounds, perhaps because they drew their conclusions primarily from analyses of those very core areas. Most of them—nuclear, original, or contemporary—include one or more urban centers. The Paris Basin is the core area of France, and Paris is the focus of the Paris Basin. Normally in Europe, these cities are national capitals as well as the largest cities. In the United States and Canada, the core area is located in the eastern portions of the countries. In each case the core area contains roughly half the total population of the country and nearly three quarters of the industrial employment. It is also the cultural and political focus of the State, the area in which the State idea originated, from which the westward movement began, where the capital cities emerged. Both countries are developing subsidiary core areas to the west, but the eastern core area, essentially one core area shared by two States, remains unrivaled.

In Latin America most States similarly have distinct core areas. This results largely from the original settlement pattern. When Europeans first arrived, they found dense clusters of Indian population in highland basins from Mexico to Chile. Many of these clusters, or nodes, of settlement became centers for Spanish administration and economic activity and eventually became the core areas of States that emerged from the revolutions of the period 1810 to 1825. Two other types of settlements were founded by the Span-

Core areas of Europe.   (After Pounds and Ball)

iards and Portuguese: mining centers and
seaports. Each became the focus of a lively
and productive region, but many declined
or disappeared entirely when, because of
exhaustion of minerals, changing trade
patterns, or other factors, they lost their
raison d'être. Some, though, survived to
become nuclei of core areas and later of
States; Panamá, Tegucigalpa, and Asun-
ción are examples. Chile represents still a
different pattern, evident primarily in the
South Cone of South America. Here the

core area is the Central Valley, which contains neither a dense Indian population
nor a mining center nor a seaport. It is
almost entirely an agriculture-based core.

African core areas, in a number of the
emergent States of that continent, are still
developing. Without doubt the most significant, in terms of productivity, degree
of development, variety of activities, and
intensity of urbanization, among other
factors, is the core of South Africa. This
consists of an east–west axis that extends

from the coal-mining town of Springs in the east to the gold-mining town of Klerksdorp in the west, along the Witwatersrand, centered on Johannesburg. There is also a north–south axis to this core area and a number of secondary core areas elsewhere in the country, notably at Cape Town. Most of the countries south of the Sahara have single cores, and most of the capitals are located in these cores. An important exception is Zambia, whose core area is the Copperbelt along the Zaïre border. Most of the West African coastal States are dominated by their capital cities, which are nearly all important seaports. This is no accident, as the ports were developed and used by Europeans for penetration into the interior and remained their primary links with home. Along the North African coast, the core area of Egypt is one of the best defined in the world. Westward, the core areas that have developed are on or near the coast.

A number of African countries possess more than one core area, and since several of these States are also among the most important and populous of the continent, they deserve special mention. Zaïre has two major core areas, which we have already described. Tanzania has no traditional core, but Dar es Salaam, the capital and largest seaport, constitutes a relatively new one. The area round Arusha, near the Kenya border, has developed spectacularly in recent years and, together with Moshi and the port of Tanga, constitutes a second core. A third is on the shores of Lake Victoria, centered on the port of Mwanza. Smaller ones are found in the southeast and southwest. In Ethiopia, the area around the capital, Addis Ababa, is clearly the dominant region of the country, but the area of Asmara and Massawa in rebellious Eritrea provides strong competition. A third core is developing around Diredawa.

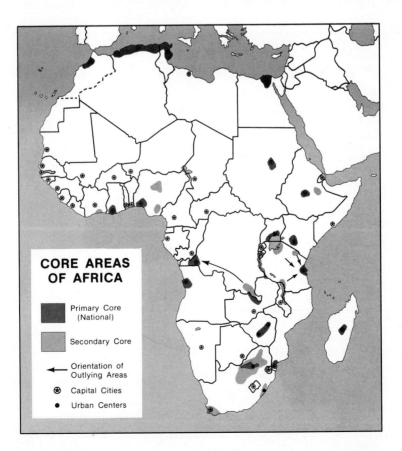

CORE AREAS OF AFRICA

■ Primary Core (National)

▨ Secondary Core

← Orientation of Outlying Areas

⊛ Capital Cities

● Urban Centers

A number of States in Africa are so sparsely populated or such recent creations that they have no true cores at all. Most of them are in former French West and Equatorial Africa. In each case a core is developing around the capital, a process quite the reverse of that observed in Europe.

The core of the Soviet Union is a rather large area centered on Moscow. It is difficult to say whether this core is simply the heart of a larger but more diffuse core that includes Leningrad, the Ukraine, and the Volga–Urals region or whether each of these constitutes a separate secondary core. In any case the areas around Novosibirsk and Irkutsk constitute secondary cores, but they have become so only in the past half century.

Core areas in Asia show a great deal of variety. China's core, like those of Mexico, Egypt, Peru, and Iraq, is one of the world's great culture hearths. Here in the northeast, in the North China Plain, are

such cities as Shanghai, Nanjing, Jinan, Tianjin, and Beijing as well as most of the country's manufacturing and much of its productive agricultural land. India, on the other hand, has several cores. For a long time the major one has extended from Delhi along the Ganges to the Bangladesh border and from there south to Calcutta. Other cores focus on Madras and Bombay, and there may be still others. Japan is unusual in that its core area, stretching from just north of Tokyo along the north shore of The Inland Sea to Kitakyushu on Kyushu Island, occupies a considerable percentage of the national territory. Tokyo–Yokohama dominates the region— and the country—more completely than any other large-country capital region in Asia.

Vietnam, since reunification in 1975, again has two distinct cores; the lower reaches of the Red River in the north and the Mekong in the south, dominated, respectively, by Hanoi and Ho Chi Minh City

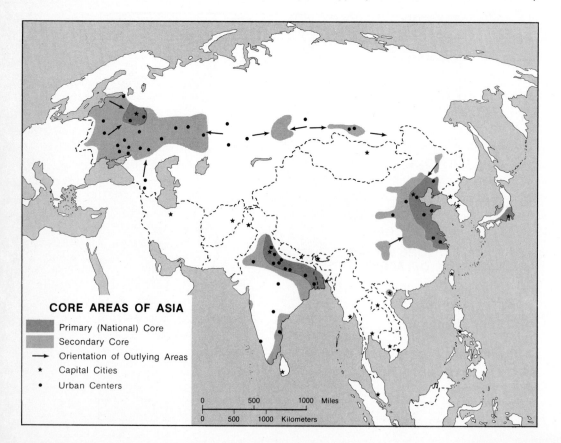

CORE AREAS OF ASIA

■ Primary (National) Core
■ Secondary Core
→ Orientation of Outlying Areas
★ Capital Cities
• Urban Centers

0   500   1000   Miles
0   500   1000   Kilometers

(formerly Saigon). Should Korea be re-unified, she too would have two distinct cores, since Pyongyang is unlikely to lose completely the importance it has attained as the capital and focal point of the Democratic People's Republic of Korea. Most of the other countries of mainland Asia also have a major core area and one or more secondary ones. All the island and archipelago States—and Taiwan—have their core areas on the most productive and densely populated island.

The Australian core area is located in the southeastern part of the country, with the two ports of Melbourne and Sydney forming the focal points of the region. Beyond its margins are several nuclei that are developing into subsidiary areas. Each is the area of a state capital and each is located on tidewater, a reflection of the pattern of population distribution of the country as a whole. New Zealand's less well-defined core area, around Auckland on North Island, is similarly on the periphery of the State.

SIGNIFICANCE OF THE CORE AREA

In considering the relevance and importance of the core areas, a number of questions arise. Only a few core areas of the States of the world have been briefly outlined here, but it is clear that core areas must be viewed in two ways: first on a world scale and second on a State scale. We have seen that we cannot apply the same criteria to define the core areas of the United States and Niger. But we can use the same criteria in determining the core areas of the United States, Argentina, the Soviet Union, China, the United Kingdom, Japan, and Germany. Comparisons among these States, therefore, are possible. When considering the core areas of individual States, our attention is drawn to the contrasts between conditions within the core and those in the remaining parts of the country. Here we must vary our judgments, although the factors of significance remain the same (population

density, productivity, communications, for example).

The most important question regarding core areas, and the most difficult one to answer, is whether the core functions satisfactorily as the binding agent for the State. This depends on a number of factors, including the adequacy of the capital city, its position with reference to the population distribution within the State, its ethnic content and the values held, the efficacy of various core-centered organizations in areas outside the core, and many more.

The core area may also be viewed as performing two major functions within a State, both related to scale. First, it may be considered as the nucleus of a State's *ecumene*. Ecumene derives from the Greek word *oikoumene*, which originally meant the inhabited world. Now we define it in several different ways. It can mean that part of the total territory of a State that actually contributes to its economic viability, in which the inhabitants participate in the national economic system, in which the economic system functions effectively. It can be defined politically also; it would be that part of a State's territory in which the government functions effectively, in which the people participate somehow in the political life of the country (if there is any). Preston James summed up the concept by defining the ecumene as the *effective* national territory, as distinguished from the *total* territory. In many countries even today, there are significant areas outside the effective national territory in which the government does not exercise effective control, and/or where the people are oriented toward a neighboring country rather than toward their own capital. Among these countries are Colombia, Brazil, Chad, Ethiopia, Burma, and Papua New Guinea. We mention this factor again when we discuss the power inventory.

We have observed that a country's capital is frequently located in its core area. Even if the capital is outside the core, a city within the core probably was the capital at some time in the past. Generally the country's largest city is located in the core, whether it is the capital or not. We may then view the core area as the hinterland of the capital, largest city, or cluster of large cities. This interpretation assigns to the core the dynamic role of supporting a significant urban area instead of simply the passive role of being the area in which certain things happen or have happened in the past. Good core area studies testing this hypothesis would be most welcome.

CAPITALS

Many States, as we have seen, originally grew around urban centers that possessed nodality and attained strength and permanence. Many were market centers for large tributary areas; others were fortifications to which the population retreated every night after farming the surrounding lands. As the influence of those cities expanded, far-flung territories came under the control of the political authority located there. Rivalries between "city-states" occurred, in which the largest and best-organized center had the greatest chance of survival and of absorbing its competitors.

Capitals have often evolved as centers of trade and government because of their fortuitous situation. But not all capitals can claim such a past. In simpler as well as complex societies of history, rulers and their entourages moved about within their realm, not by preference, but because supplies could be stored for them at various places visited at regular intervals; it was simpler to take the rulers to their food than food to the rulers. A ruler may have expressed a liking for one of his temporary abodes over others, and ultimately arrangements were made to permit the Chief of State to occupy such a place permanently.

Whatever the origins of the capital, it is destined to embody and exemplify the nature of the core area of the State and be a reflection of its wealth, organization, and power. And indeed, it sometimes is more than that. Some States have poured money into their capital cities to create there an image of the State as it will be in the future, a goal for the people's aspirations, and a source of national pride. Anyone who has seen the shining capitals of some Latin American countries and the abject poverty of areas within sight of the rooftops of the skyscrapers will attest to this phenomenon. Addis Ababa, capital of Ethiopia, has been described as "a mask, behind which the rest of the country is hidden."[*]

Certain States have invested heavily in the creation of wholly new capitals that are constructed with specific aims. Brasilia was built at tremendous expense in large part to draw the Brazilian nation's attention toward the interior. Here the functions of the capital, long performed by Rio de Janeiro, were taken out of the core area and placed elsewhere. Other motives underlie Pakistan's decision to replace Karachi as its capital with Islamabad, in what may still be described as a northern frontier. Islamabad lies close to areas disputed with India and Afghanistan, and thus the government will be in a critical position to determine policy.

Emergent, formerly colonial States have at times decided to replace their European-developed capital with another town possessing either traditional importance or a more fortuitous location from the point of view of the new government. Hence, Malawi decided in 1964 to move its capital from the town of Zomba to more centrally located Lilongwe; Tanzania in 1973 announced it would move its capital from coastal Dar es Salaam to more central Dodoma; Nigeria is planning a new capital district near Abuja; Côte d'Ivoire moved its capital from Abidjan to Yamoussoukro (the president's home village); and even Argentina plans to replace Buenos Aires as the country's capital with the small Patagonian town of Viedma.

[*]D. Mathew, *Ethiopia*, London; Longmans, Green, 1947, p. 5.

FUNCTIONS AND TYPES

Capital cities perform certain distinct functions. Some are obvious: this is the place for legislative gatherings and the residence of the Chief of State. Here is a prime place for the State's reception of external influences, for embassies and international trade organization offices are located here and the turnover of foreign visitors is likely to be greater than anywhere else. In most States the capital city is also the most "cosmopolitan" city of the country.

Capital cities must also act as binding agents, for example in a federation such as Brazil or Nigeria. In a federal State of great internal diversity, the capital city may be the only place to which all the people can look for guidance. Often, in such a State, a territory is separated from the rest of the country and made into "federal territory," so that none of the other entities in the federation can claim bias in the location of the national capital. In the United States, the District of Columbia serves this function.

Capital cities must also be a source of power and authority, either to ensure control over outlying and loosely tied districts of the State or to defend the State against undesirable external influences. This function is diminishing, but the capital is most frequently located in the economic heart of the country, from which much of the image of strength of the State emanates.

Functions of capital cities have changed much over time. London provides a good example. After the Roman conquest in the first century A.D., it was the trade center of the Roman Province of Britain, but not the capital, which was located about 20 miles away at Verulamium. Under Norman influence it became the capital, and after the conquest of Wales and Scotland those territories fell under its jurisdiction also. The city's salutary location became an even greater asset during the years of increasing overseas trade and empire building. Thus, once a local center, then a regional capital, London became the capital of a State and then the headquarters of an empire.

A planned capital: Brasilia, Brazil. It is centrally located, replacing a peripheral capital established in colonial times. The city is located in open, level land and has considerable open space between modernistic buildings, most uncharacteristic of other cities in Latin America. Still not very popular with officials, Brasilia is achieving one objective by attracting some pioneer settlers to the areas around it. (Martin Glassner)

Types

There are several distinct types of capitals, so that classification may be possible. But caution is needed here. Some geographers, for example, have argued that there are "natural" (that is, *evolved*) capitals and "artificial" capitals. This categorization would suggest that certain capitals have emerged and developed as the State system grew increasingly complex, whereas others have been simply the result of arbitrary decision. O.H.K. Spate attacked this whole concept, arguing that any decision leading to the establishment of an "artificial" capital is itself the result of pressures created within and by the system.

It was perfectly "natural," for example, for Belize in 1970 to inaugurate a new, planned capital in almost the exact center of the country, leaving Belize City to continue serving as the colony's chief port. Like Brasilia, Islamabad, and other new capitals, Belmopan was created for a variety of reasons. The trigger for the move was Hurricane Hattie in 1961, which destroyed 80 percent of Belize City, but the city of more than forty thousand people was already in a deplorable condition, with no feasible means of improving it. Establishment of a second major urban center and the need for development of the interior were also important motives. There was nothing arbitrary about the move.

The Primate City

Another approach might be based on Jefferson's "The Law of the Primate City": "A country's leading city is always disproportionately large and exceptionally expressive of national capacity and feeling."* In many countries the largest city is the capital, but this need not be so. In States that have selected new capital sites, for example, such as Pakistan, the primate city did not lose its primacy when it lost its

*Mark Jefferson, "The Law of the Primate City," *Geographical Review*, 29, 2 (1939), p. 226.

major political function. Certain federal States, such as Canada, Australia, and Nigeria, similarly have cities larger and culturally more expressive of their nations than their capitals. In some other States, such as Spain and Ecuador, no city, not even the capital, dominates the country sufficiently to be designated as primate. While the status of national capital automatically confers political power on a city, other cities, by virtue of their size, wealth, and concentration of influential people, can also be politically powerful.

A MORPHOLOGICAL APPROACH

Perhaps the most productive approach is a morphological one. Here we view capital cities in relation to their position with reference to the State territory and the core area of the State. This results in three classes of capital cities:

1. **Permanent Capitals.** Permanent capitals might also be called historic capitals; they have functioned as the leading economic and cultural center for their State over a period of several centuries. Obvious examples are Rome, London, Paris, and Athens, the leading cities in their respective States for many centuries and through numerous stages of history. The Japanese capital, Tokyo, however, is not in this category. It has been the Japanese headquarters for just over a century—the century that spanned the emergence of the modern nation-state of Japan. A whole new cycle began in the life of the Japanese State in the late 1860s, and with it the old capital of Kyoto gave up its primacy. Tokyo symbolized the new Japan, and although the city seems destined to become the permanent Japanese capital, it does not yet rank with Rome and Paris (or Beijing) as a headquarters of continuity.

2. **Introduced Capitals.** Tokyo, in fact, was *introduced* to become the focal point for Japan when the revolutionary

events referred to as the Meiji Restoration occurred. Recent history has seen similar choices made in other countries, but while Tokyo (then called Edo or Eastern City) was already a substantial urban center, other capitals were created, literally, from scratch. They replaced older capitals in order to perform new functions, functions perhaps in addition to those normally expected of the seat of government. Several examples of this type have already been mentioned.

Introduced capitals have also come about by less lofty actions. Intense interstate rivalries among Australia's individual states made it impossible to select one of that country's several large cities as the permanent national capital, and a compromise had to be reached. That compromise was the new capital of Canberra, built on federal territory carved out of the State of New South Wales.

One other factor must be taken into account in connection with introduced capitals. Despite the general absence of planning for a time when the colonial city in Africa would serve as a national capital, the vast majority of the emerging States have retained the former European headquarters as the national capital. Some of these capitals are really quite embryonic, such as Mbabane (Swaziland), Libreville (Gabon), and Kigali (Rwanda), but others have been converted into modern symbols of the nation's aspirations. In the most unusual case, when the British protectorate of Bechuanaland attained independence, it had— incredibly—no capital, since the territory had for 80 years been administered from a town called Mafeking, located *outside* its boundaries. Thus Gaberone was chosen as a new, internal (and introduced) capital of Botswana and construction of the necessary facilities begun.

3. **Divided Capitals.** In certain States the functions of government are not concentrated in one city, but divided among two or even more. Such a situation suggests—and often reflects— compromise rather than convenience. In the Netherlands (a kingdom) the parliament sits in The Hague (the legislative capital), but the royal palace is in Amsterdam (the "official" capital). In Bolivia, intense rivalry between the cities of La Paz and Sucre produced the arrangements existing today whereby the two cities share the functions of government. In South Africa, following a war between Boer and Briton, a Union was established in which the Boer capital, Pretoria, retained administrative functions, while the British headquarters, Cape Town, became the legislative headquarters. As a further compromise, the judiciary functions in Bloemfontein, capital of one of the old Boer republics that fought in the Boer War.

THE SEARCH FOR CONCEPTUAL UTILITY

The capital city is the focus of the State system, the node, often, of the core area. Political geographers have described many capital cities, they have made attempts at classification, and they have explained why certain cities became capitals while others did not. But as in the case of the core-area concept, it has not been easy to define exactly *how* the capital functions in the integration of the total State system. Among the first geographers to consider this whole subject was Vaughan Cornish. More than a half century ago, he published a book entitled *The Great Capitals,* and it is still worth a careful reading. He discusses the origin and evolution of capital cities in many parts of the world, and suggests that

they fall into three categories of *Natural Storehouses, Crossways,* and *Strongholds.* The first is the original and fundamental character, the second and third dependent upon it, for easy movement and natural barriers are only important if there be inducement to develop or obstruct traffic. It is therefore the world's Storehouses of natural wealth which sooner or later determine the importance of Crossways and Strongholds.*

Following an elaboration of this theme, Cornish adds still another category: the *foreword capital,* one whose strategic position vis-à-vis frontier regions or potential enemies gives it special importance. Some governments today use capital cities for purposes over and above those having to do with administration. As national foci they enjoy the attention of the nation, an attention that can be manipulated through the manipulation of the capital itself. The frontierlike position of Islamabad, Pakistan's capital, underscores that State's determination to confirm its presence in nearby, still-contested areas.

Spate, too, struggles with the problem of defining the functional qualities of the capital in the State system. He proposes the idea of the *headlink* function, whereby the capital is viewed as the primary link in a chain representing the organizing direction of the State. He then analyzes the cases of Belgrade and Prague, the capitals of Yugoslavia and Czechoslovakia, two States which, at the time his article was written, were still quite young. Belgrade became a "head-link" capital for at least two resons: its preemptive position (it was already a regional capital) and the role played by Serbia, the "active organizing principle in the creation of the new state." In Czechoslovakia, Prague was the regional center of Bohemia (one of the regions of the new State) and, as Spate points out, the city was "the very flower of Slav civilization, enriched by western contacts." The choice of Prague was inescapable. Although "the geographical center of Czechoslovakia would have been somewhere near Brno . . . there could be only one political center, and that was eccentric Prague; centrally focal to Bohemia, to Czechoslovakia as a

whole it was the link with western cultural and social influences, the type of a 'head-link' capital."

The short-lived Federation of the West Indies presents a fascinating study of both the importance and the difficulty of proper selection of a national capital. The federation broke apart before it could become independent for many reasons. There was little enthusiasm for it in the larger territories, Jamaica and Trinidad and Tobago, and a State idea never really developed. The 10 participants disagreed on many vital matters that had to be settled before a State could be created. Among them were representation in the federal parliament, control over economic development, freedom of movement among the islands, and—most emotional of all—the location of the federal capital.

What we note about the West Indies applies to much of the politicogeographical world, although the parochialism and individualism of the several competitors for the capital's functions were made worse by the insular character of the federal State that was being created. The three major aspirants were Jamaica, Trinidad, and Barbados, and for a while there was the possibility of a small-island compromise: St. George's, Grenada. Both Barbados and Trinidad rejected this proposal, however, and the compromise was negated. That negation was confirmed by a special commission that argued that certain minimal service and living conditions were to be met by the town or city chosen as capital. This eliminated all but three of the proposed Federation's cities: Bridgetown, Barbados; Port of Spain, Trinidad; and Kingston, Jamaica. The commission chose Barbados.

All this emanated from a London—and still colonial—commission, of course. But in the West Indies, it was received with anger almost everywhere, except Barbados. When, finally, the Standing Federation Committee met to consider the Barbados proposal, it rejected it outright. In the subsequent balloting, Jamaica was first eliminated, and in the second ballot

---

*V. Cornish, *The Great Capitals,* London: Methuen, and New York: Doran, 1923, p. vii. Italics added.

Trinidad was selected by 11 votes to 5 over Barbados.

Finally, the parties were able to agree on a site: Chaguaramas, near Port of Spain. It was far from centrally located, but it did have a number of advantages, chief among which may have been that the parties were too exhausted and dispirited after years of haggling and bitterness to object to it. There was just one small problem with the site chosen, however: it happened to be the location of a U.S. naval base!

The federal government had to send a delegation to negotiate with the United States for the land for their new capital. By the time the Americans finally agreed to relinquish a portion of the base area, it was too late; Jamaica was voting to leave the federation, signaling its inevitable and ignominious demise. The Federation of the West Indies, born with much fanfare in 1958, died almost unnoticed in 1963— and it never even had a capital.

The West Indies case is certainly an extreme, at least in modern times, but it could be repeated elsewhere as small countries gradually move toward closer association and perhaps eventual political union. And such problems will surface not only among the newer countries of the world. We need only recall the intense competition in the 1960s and 1970s among Luxembourg, Strasbourg, and Brussels for selection as the headquarters of the merged European Communities (the European Coal and Steel Community, Economic Community, and Atomic Energy Community) to realize that these problems can arise anywhere. The economic, political, cultural, strategic and—perhaps most important—psychological and emotional importance of a national capital cannot be overstressed.

We have discussed some of the features of capitals of both federal and unitary States; now we will see what these classes of States are like.

Chapter 10

UNITARY, FEDERAL, AND REGIONAL STATES

Political geographers are naturally interested in the ways in which States are organized in spatial-political terms. Such organization is the result of lengthy processes of experimentation and modification, processes that have not ceased. State systems are continually being altered, sometimes through deliberation and consultation and at other times because the system cannot withstand certain centrifugal pressures or forces. We are all aware that the federal State that is the United States of America is not the United States of 100 or even 25 years ago. For the France of today, Napoleon laid certain foundations, but the France inherited by Charles de Gaulle was very different. The political system—and the politicoterritorial framework—of these States had in some manner failed, and a new order was necessary.

THE UNITARY STATE

The word "unitary" derives from the Latin *unitas* (unity) that, in turn, comes from *unus* (one). It thus emphasizes the *one*ness of the State and implies a high degree of internal homogeneity and cohesiveness. The term "federal," on the other hand, has its origins in the Latin *foederis*, meaning league. Its implication is one of alliance, contract, and coexistence of the State's internal, diverse regions and peoples. The unitary State, therefore, theoretically has one strong focus, and its internal differences are few; the federal

State may have a number of separate foci, and its component areas may have quite different characteristics.

In practice, the two terms refer to the functions of the central authority of the State. All States are divided for political purposes into administrative units. Each of these units has a local government to deal with local matters. In the case of the unitary State, the central authority controls these local governments and determines how much power they will have. The central authority may, under certain circumstances, temporarily take over the functions of a local government. It can impose its decisions on all local governments, regardless of their unpopularity in certain parts of the country. In a national emergency, the central authority can assume greater powers to meet the crisis, while in times of stability, it may grant increased responsibilities to local governments.

From these statements, it should not be difficult to formulate a model unitary State. In the first place such a State should not be in the "large" or "very large" categories of State territory. The larger a State, the more likely it is to straddle more than one cultural region, and the more varied will be the historical mainstreams to have touched its peoples. The larger a State, the greater may be the physiographic impediments to effective communications and transportation. Anything that would intensify the centrifugal forces present in any State reduces the efficacy of a single, central authority.

Second, the ideal unitary State is compact in shape. A fragmented or prorupt territory may present obstacles to unity and cohesion and require a measure of autonomy for various individual regions. Racial, religious, and linguistic diversity are likely in a fragmented State, and only a federal political framework may function to the satisfaction of the majority of the people.

Third, the unitary State should be relatively densely populated and effectively inhabited. There should be no vast territory with separate concentrations of population interspersed with empty and unproductive areas. This leads to isolation and regionalism.

Finally, the unitary State should have only one core area. Multicore States reflect strong regionalism, an undesirable condition in unitary States. Theoretically, the most suitable location for the single core area of the unitary State is central to its compact territory. This brings all peripheral areas within the shortest distance of the capital city and makes the presence of the core area and capital strongly felt in all parts of the State. A single urban center that is disproportionately large and influential in the affairs of the State, where the central authority resides, and where national feeling is strongest, obviously constitutes a binding agent and a focus not only for the core area, but for the State as a whole.

Evolution and Present Distribution

Few of the unitary States in existence today conform to the model just described. Several examples that show a close approximation to the ideal occur in Western Europe. The old European States fostered a strong central authority, and it is here that the unitary State as it is known today emerged. But this system of politicoterritorial organization has been copied in many other parts of the world—and the copies are often far removed from the original. Nevertheless, in 1988, there were fewer than two dozen federal States and over a hundred unitary States on the world map.

France is often cited as the best example of the unitary State. Though large by European standards (213,000 square miles), this country, apart from Corsica, is compact in shape, has a core area with a lengthy history, and at its heart possesses a capital city of undoubted eminence as well as a large, politically conscious population with much historical momentum and strong traditions. Modern France was forged, in effect, by Napoleon, who swept away the old system of loosely tied divisions (Artois, Piccardy, etc.) and replaced them with 90 separate "departments," based on rough equality of size. Each of these departments possessed the same relationship to the central political authority as did the next, and each sent representatives to Paris. Napoleon also developed an entirely new system of communications, focusing very strongly on Paris, to act as a unifying agent. Until the days of Napoleon, allegiances in France had been to individual divisions rather than to France, despite the forces of revolution and the overthrow of the monarchy. Today the French nation-state is close to reality.

Many of the formerly colonial territories of the world have adopted a unitary form of organization, especially those in which the indigenous population has taken control. In those areas where Europeans remain dominant (Australia, Canada, Brazil) the federal arrangement has sometimes been employed. In Latin America, including the Caribbean, all countries except Venezuela, Brazil, Mexico, and Argentina function as unitary States. In Africa and the Middle East, all Arab countries are unitary States and the majority of the black African States have also chosen this form of organization. Africa south of the Sahara affords some excellent examples of recent experimentation with European concepts of government. The former French terri-

Administrative districts of France. Since Napoleon's reforms, France has been divided into *départements*, the first order civil divisions. At present there are 96 *départements*, each headed by a prefect representing the government. All these *départements* have been regrouped for economic and administrative purposes into 22 regions, each headed by a regional prefect who coordinates activities in the *départements* composing his region. Many of the regions' names are the names of pre-revolutionary provinces.

tories have become unitary States, but some of the former British dependencies have adopted a federal system. British-influenced unitary States in Africa include Kenya, Tanzania, Sierra Leone, and Ghana. In Asia, only India and Malaysia are not unitary States. Again, the European contribution can be recognized: the fragments of former French Indo-China have become unitary States, as have Indonesia and the Philippines.

Viewing these unitary States in relation to the model described previously, it is clear that comparatively few approach the stated ideal. A number are territorially fragmented, including Japan, the Philippines, Indonesia, and New Zealand. Several are in size categories that would appear to contradict the required internal homogeneity and unity, such as China and Sudan.

Another group of unitary States possess more than one core area, although it may be argued that the core areas of several of these States are yet too distinct and undeveloped to be so labeled. Nevertheless,

Ecuador and Colombia in South America, Tanzania and Chad in Africa, and Spain and Sweden in Europe display multicore characteristics that reflect centrifugal forces within these unitary States.

Types of Unitary States

Although unitary States are based on the premise of national unity and considerable homogeneity, their high degree of political centralization often belies their internal cultural variety. The imposition of the will of a minority on the majority of the people within a State may produce a unitary State that might function better as a federal State. In certain former colonies, indigenous, Europeanized elites have taken control and have administered the country in much the same way as did the colonizers themselves. Latin America has shown that such political arrangements can be very durable. In these unitary States, the response to heterogeneity and regional problems often is an even greater degree of concentration of power in the central government. The flexibility of the federal arrangement would, theoretically at least, result in just the opposite action.

Whatever the mode of government in the unitary State—whether a monarchy, a dictatorship, or a democracy—certain adjustments are made in the politicoterritorial system that reflect the internal conditions of the State. As we pointed out earlier, some unitary States have emerged simply because of the almost total assent and satisfaction on the part of the population with regard to the existing authority. On the other hand, certain other unitary States have seen an increase in the power vested in the central authority, whether a person, a group of persons, or the representatives of a minority ethnic group. As we shall see, changes of this kind—whether in the direction of greater centralization of power or the reverse— often reflect geographical conditions within the State. In some emergent States, tribalism remains a force that obstructs the evolution of real nations; in many the

governing authority has taken over more and more of the powers of the local chiefs. The following types may be recognized:

1. **Centralized.** This is the "average" unitary State, true to the basic rule of centralization of governmental authority but without excesses either in the direction of totalitarianism or in the direction of devolution of power. Normally, in such a State, stability has been achieved by virtue of the homogeneity of the population and the binding elements of culture and traditions.

   States in this category usually possess only one core area. They are generally older States, in the mature stage of the Van Valkenburg classification. Most examples are found in Europe, such as Denmark, Sweden, and the Netherlands. These three monarchies, by their very retention of this form of government, reflect the satisfaction of the majority of the population with the *status quo*. In centralized unitary States, the population participates in the government through the democratic election of representatives. No single ethnic minority or political party has sole claim to leadership, which doesn't mean that there is no ethnic diversity within centralized unitary States and no regionalism. Despite its small size, the Netherlands has at times been mildly aware of regionalism in Friesland, with Frisians demanding that their language be taught in Dutch schools if Frisians must learn Dutch. The overriding factors of proximity, interdigitation, interdependence, and historical association produce the centripetal forces that bind the State together.

   Other centralized unitary States are Ireland, Finland, and France. The case of France once again provides useful insights, for the shift toward greater power for the president that accompanied the demise of the Fourth and the advent of the Fifth Republic was related to problems in the Empire, first in Indo-China, then in Algeria, and barely warded off in black Africa. Elsewhere, Japan and New Zealand, and perhaps Uruguay, fit this centralized category.

2. **Highly Centralized.** In this type of unitary State, internal diversity or dissension, ethnic heterogeneity, tribalism or regionalism that threaten to disrupt the State system are countered by tight and omnipotent control. The leader or leaders often are the representatives of a minority group within the country or of the only political party that may operate within the State. Three major groups of States may be recognized within this category: unitary States within the communist sphere, one-party States in Africa and other parts of the decolonized world, and dictatorships elsewhere.

Although the Soviet Union, Czechoslovakia, and Yugoslavia are federal States, the majority of the communist States are unitary and highly centralized. Their geographical bases for unitary organization are reinforced by the character of communist control. Single-core, relatively densely populated, compact States, such as Poland, Hungary, Romania, and Bulgaria, would appear suited to the unitary system under any circumstances, but the nature of communist control over practically every aspect of State organization differentiates this group from those described as centralized. Despite recent evidence of amelioration in some communist States, virtual dictatorships still prevail in others, for example in Albania. The largest unitary State functioning today, China, fits this category of highly centralized State.

States displaying these characteristics are, however, not confined to the communist world. Paraguay and Chile in Latin America, the Central African Republic and Libya in Africa, and Saudi Arabia and Indonesia in Asia are some current noncommunist, highly centralized unitary States, and it is not difficult to list others. Some of these States have recently emerged from several generations of colonial rule, and although some were endowed with a multiparty system, many have become what is described as "one-party democracies."

In other emergent States, the shift toward highly centralized control on the part of a single party, a small group of individuals, or a single individual has simply involved the transfer of power from the European colonists to a local, acculturated, and privileged elite. This has occurred in some of the French-influenced States in Africa and Asia, and the example of South Vietnam before its merger with North Vietnam requires no elaboration.

THE FEDERAL STATE

In the unitary State, the central government exercises its power equally over all parts of the State. The federal framework, on the other hand, permits a central government to represent the various entities within the State where they have common interests—such as defense, foreign affairs, and communications—yet allows these various entities to retain their own identities and to have their own laws, policies, and customs in certain fields. Thus, each entity (such as a state, province, or region) has its own capital city, its own governor or premier, and its own internal budget. And each entity is, in turn, represented in the federal capital, so that it has a voice in matters concerning the entire federation.

History has shown that federal States, like unitary States, evolve and change over time. In North America, when the thirteen states found themselves with more common interests to unite them than conflicting ideas to divide them, a federation emerged that was, in many ways, a far cry from the United States of today. The number of functions of government has grown tremendously, and the federal government must involve itself more and more in the affairs of the states, affairs that at one time would have been considered domestic to these states. Thus, just as some unitary States show signs of adjustment in the direction of decentralization, many federations are shifting toward greater centralization of authority. Some federal States are so highly centralized that they function, in effect, as unitary States.

Theoretically, the federal framework is especially suitable for States in the large and very large categories. Poor communications and ineffective occupation of large areas within the State still affect, for example, Brazil and Zaïre. These impediments to contact and control might disrupt a unitary State, whereas a federal framework can withstand such centrifugal forces. In terms of shape it is obvious that fragmented States and States with pronounced and important proruptions may be best served by a federal system. An elongated State possessing more than one core area also might turn to a federal arrangement.

Federal States can adjust to the presence of more than one core area (or a number of subsidiary cores) more easily than unitary States. Several contemporary federal States are multicore States, and in the case of Canada, Nigeria, Pakistan, and Australia, it is probable that only a federal constitution could have bound the diverse regions together. Thus, the federal State often has a number of individual population agglomerations separated by large areas that are sparsely populated and relatively unproductive. Canada and Australia, two of the largest federations in terms of area and two of the smallest in terms of population, illustrate this principle.

Ideally, the government of the federal State functions in a capital city located in an area of federal territory set off within the State for the specific purpose of administration. This forestalls any friction over the choice of an existent major city as the capital and prevents regional fa-

voritism from occurring, as it has in some federal States. In Nigeria, an existing city was separated from the Western Region (one of the political divisions of the State) and made the capital, for two reasons: first, it was a long-term colonial capital and housed most government records and existing facilities, and second, it happens to be the leading port of the country, thus guaranteeing the land-locked north an exit through a federal rather than a regional port.

The federal arrangement is also a political solution for those territories occupied by peoples of widely different ethnic origins, languages, religions, or cultures. This is especially true in cases where these differences have regional expression, where various peoples see individual parts of their country as a homeland. The promise of an autonomous Macedonian Republic, part of a Balkan Federation, rallied Macedonians around Tito's war effort. Although the Macedonian Republic (which was to include parts of Greece and Bulgaria as well as Yugoslavia) did not materialize, Macedonians today have a small national territory within the Yugoslav federal State. For them, it has been the only alternative to absorption, and it constitutes a closer approximation to their original goal than they have enjoyed in centuries. In unitary Bulgaria, such status is impossible.

By their flexibility, federal frameworks have been able to accommodate expanding territories. Provisions are often made for areas not yet incorporated. The United States, India, Mexico, and Brazil have promoted several of their territories into the ranks of states, equal to the others already in these federations.

Another advantage offered by the federal arrangement is its encouragement of individual and local enterprise. Economic development in the United States took place as fast as it did largely for this reason; the westward push of Brazil, as exemplified by the relocation of the capital, is an effort to stimulate a similar chain of events in its hinterlands. Of course, this accommodation of individualism has not prevented the emergence of conflicts of interest. Some, like the Civil War in the United States, helped bring about revisions in the federal framework; others resulted in the collapse of the State itself.

Types of Federal States

"A federation," wrote K.W. Robinson in an article about Australian federalism, "is the most geographically expressive of all political systems. It is based on the existence of regional differences, and recognizes the claims of the component areas to perpetuate their individual characters. . . . Federation does not create unity out of diversity; rather, it enables the two to coexist."* In the creation of the State system, internal variety and diversity can be treated in two ways. One way, as we noted earlier in this chapter, is to render the State the great equalizer, the eliminator of internal differences, even to the point of making certain expressions of regional individuality illegal. The other way permits the perpetuation of internal contrasts and provides the possibility of coexistence within one State of peoples with varied backgrounds and interests. Thus, the advantages of participation in the State must, for all these peoples, outweigh the liabilities that inevitably are involved. To put it another way, the centripetal forces must outweigh the centrifugal forces present in any federal State. Clearly, most federal States *born* of one set of circumstances are now *sustained* by conditions that are quite different.

There are several different *types* of federal States. In some present-day federal States, such as Australia and Argentina, internal variety and diversity seem so insignificant (compared to that existing in other countries) that a unitary arrangement might be just as effective. Other federal States, including Canada and Nigeria, incorporate such diversity that a certain

amount of give and take was, and remains, essential for the well-being of the State. In still other federal States, the geographical obstacles to any unitary system rendered a federal arrangement imperative. It is important to realize that these types are not mutually exclusive, for the elements of two or more kinds of federalism may be recognized in some federal States today. Furthermore, a federal State established for one set of reasons may continue to exist today for an entirely different set. The case of South Africa exemplifies both instances: this federal State originally resulted from an armistice agreement between two warring factions, and the major element in its creation was *compromise*. But today, the erstwhile enemies have grown together, and they face a totally new threat in the advance of black African nationalism. Hence *mutual interest* is by far the stronger centripetal force in the State at the present time.

Another category of federations that must be considered is those that have *failed*. The world is littered with the wreckage of federations that have been proposed but never consummated, that have been created only to fragment relatively quickly, and that have survived but only after conversion into unitary States. Most resulted from the breakup of empires, and federation was seen as a way of managing, if not solving, many of the different problems engendered by decolonization.

In the early 1950s, there was much talk of a united, independent Maghreb, to be composed of Morocco, Tangier, Algeria, Tunisia, and perhaps Libya. Organizations were formed to support the idea and at least the intellectuals in each country favored it. Despite the many real differences among them, all except Libya had a great deal in common. France granted independence to Morocco and Tunisia in 1956, and Tangier and the Spanish protectorates soon joined Morocco. But France fought a long and bloody war to keep Algeria French. By the time Algeria was granted independence in 1962, the other two po-

*K.W. Robinson, "Sixty Years of Federation in Australia," *Geographical Review*, 51, 1 (January 1961), p. 2.

tential partners had gone their separate ways, and the chance for federation (if indeed there ever was one) was lost. A similar situation developed in East Africa. The British territories of Kenya, Uganda, and Tanganyika had been cooperating in increasingly closer association since the 1920s, and the movement toward formal federation after independence (possibly to include Zanzibar) was growing stronger. But there was no great enthusiasm for federation among the ordinary people and little among the black African leaders. Moreover, while Tanganyika received independence in 1961 and Uganda in 1962, Kenya's was delayed, at least in part by the Mau Mau rebellion, and she did not become a State until December 1963. By that time it was certainly too late.

Early in the nineteenth century, when most of the Spanish colonies in the Americas won their independence, two federations emerged. Within a short time, however, both broke up. Central America split into Guatemala, Honduras, El Salvador, Nicaragua, and Costa Rica; Gran Colombia fragmented into Venezuela, Colombia, and Ecuador.

A century and a half later, another decolonization period spawned a number of federations. The British promoted the federation as, it was thought, an honorable and practical device for cutting loose a large number of small dependent territories while giving them a chance to survive in the highly competitive world of sovereign States. They tried the device in the West Indies, Central Africa, South Arabia, and Malaysia. The first three were failures and the last only a limited success. The West Indies and Central African federations broke up before independence, and each unit went its own way, most toward independence. A civil war broke out in Aden and South Arabia before the British could escape, and the winning faction almost immediately formed a unitary State centered in Aden. Malaysia remains a federation (with a federation as its major constituent), but without Singapore, which

Twilight of the Federation of the West Indies. Fourteen months after this photograph was taken at the Queen's Birthday Parade at Up Park Camp in Kingston, Jamaica, the colony was an independent State and the Federation was dead. The flags, from top to bottom, are the Union Jack—the official flag of Jamaica as well as of the UK at the time; the Royal Standard, flown only when the sovereign is present, on this occasion in the person of the Governor-General; and the flag of the Federation, rarely seen anywhere in the Federation except on public occasions. (Martin Glassner)

was expelled in 1965 after only eight years of togetherness.

But federations failed elsewhere besides in the British Empire. French Soudan and Senegal joined in January 1959 to form the federation of Mali, became independent in June 1960, and split just two months later into Senegal and Mali. In a move unconnected with decolonization, Egypt and Syria joined in 1958 to form the United Arab Republic, but after an unhappy marriage were divorced in 1961. A bewildering number of other federations involving various combinations of Libya, Egypt, Sudan, Iraq, Syria, and Jordan have been proposed, announced, or attempted—but none has ever become effective.

Then there are the countries that have been federations in recent years but have become unitary States. Eritrea was federated with Ethiopia by the United Nations in 1952, but just a decade later lost its autonomy. French Cameroon became independent in 1960 and was joined in 1961 by the southern portion of British Cameroon. The Federal Republic of Cameroon, after a referendum in 1972, became the United Republic of Cameroon. Uganda at independence was a federation of four kingdoms and one territory, plus another territory and 10 administrative districts (with the King of Buganda as president of the country). Only five years later, however, the kingdoms were abolished, and the country became a unitary State. Indonesia's federal period was the shortest of all, lasting less than eight months after independence and ending in August 1950.

How can we explain this dreary succession of failures? Isn't the federation "the most geographically expressive of all political systems," as Robinson, Dikshit, and others have proclaimed? If so, why are there only about a dozen true federations in the whole world and why are even some of these dubiously federal? There are many answers, of course, at least as many answers as failures. Because a federation is a compromise form of territorial organization, it requires very special conditions for initiation and a great deal of hard work and dedication for survival. A common factor in all the failures in the first two categories—those that were never born and those that broke up—was the lack of a *raison d'être*, or a State idea. Typically, the federation was justified in terms of certain practical advantages that could be quickly realized. In such a case, if the advantages do materialize, the federation may no longer be needed; if they do not, the federation is exposed as bankrupt. Without a real will for a federation on the part of the people, failure is inevitable. Since both a balance of forces within a federation and its practical, day-to-day

management are so difficult, the tendency toward centralization is strong indeed, as evidenced by the third category of failure. Federation in the classic sense may not survive even in those countries where it is best developed.

REGIONAL STATES

In territorial organization, the same force we have observed in the size of States may well be operating: a tendency toward equilibrium. As federal States become more centralized and unitary States grant more autonomy to regions within them, we face increasing difficulty in applying the old labels to new situations. A solution to the dilemma, and a possible label for States approaching a midway area between federalism and unitarianism, is suggested by the Spanish scholar Juan Ferrando. This is the term *regional State*.*

Into this category we may place those unitary States in which considerable autonomy has been granted to regions within them, generally regions of ethnic distinctiveness or remoteness from the core area. All the really good examples at the moment are in Europe. Perhaps that is not so strange, however, for both unitary and federal States, after all, began here.

The United Kingdom is perhaps a good example of a regional State. Though now experiencing a period of direct rule, Northern Ireland long enjoyed a large measure of home rule. Scotland has had some autonomy and is gradually receiving more, and even Wales may receive some before long as the "devolution" of power from the center to the peripheries continues.† In Italy, Sardinia, Sicily, Trentino-

*Ferrando Badia, Juan, *El Estado Unitario, El Federal y El Estado Regional*, Madrid, 1978.

†Despite the rejection by the people of Scotland and Wales of proposals for their own assemblies that would give them a measure of home rule, in referenda on 1 March 1979, it still seems likely that the United Kingdom, while remaining a unitary State technically, will continue to devolve some responsibilities to these regions, especially to Scotland.

Administrative districts of Italy. Like France, Italy has conceded to regional feelings to some extent by grouping her 94 provinces into 20 regions with some autonomy. Those labeled with capital letters were the original five regions and still exercise somewhat greater authority than the other 15, which have been functioning only since 1970.

Alto Adige, Valle d'Aosta, and Friuli-Venezia Giulia have for some time been functioning under special statutes, while another 15 regions elected their first regional councils (parliaments) in 1970. In Finland, the Aaland Islands constitute an autonomous county, demilitarized and maintaining their own traditions. The Faeroe Islands and Greenland are largely—but not completely—independent of Denmark. Spain granted regional autonomy to Cataluña in 1977 and to three

of the four Basque provinces in 1978, completing the process in 1983 by granting autonomy to Extremadura, Castilla-León, the Balearic Islands, and Madrid. The Azores were granted autonomy by Portugal in 1975. In 1970, Belgium adopted a form of "federalization without federalism" by creating four linguistic regions (Dutch [or Flemish], French, German, and Brussels [French and Flemish]) that have autonomy in many cultural matters.

A special case that might fit into this

category is Cyprus, in which a *de facto* Turkish federated State has been established by the Turkish Cypriots in the northern part of the island, occupied by troops from Turkey in 1974. Other candidates for consideration as regional States are Burma, Iraq, and China, all nominally unitary States with regions exercising uncertain degrees of self-government. Perhaps the Netherlands Antilles' status would allow the Netherlands to be included. France may move in this direction by giving some autonomy to Brittany, Corsica, and the overseas departments.

Other regional States are those with federal constitutions in which federalism either was never very real, has gradually given way to centralization, or alternates with unitarianism in law or practice. Here we might consider the Union of Soviet Socialist Republics, whose constitution provides that each constituent republic may conduct its own foreign relations and may secede at will. One can hardly get more federalist than that! In practice, however, no republic has conducted its own foreign affairs, not even the Ukraine and Byelorussia, which are members of the United Nations. And no republic would dare to try to secede, not even one in which local nationalism is very strong. The State is, in fact, highly centralized and kept under control by the national army, the communist party, and the practice of settling Great Russians in all nationality areas, frequently giving them the best jobs. South Africa has also become virtually a unitary State, even though it is supposedly a federation. The "Bantustans" created by the Nationalist government present something of a dilemma. Even though four of them have been declared independent (Transkei, Bophuthatswana, Venda and Ciskei), not one country has recognized this 'independence" and the United Nations has denounced their creation.

Mexico is essentially a "one party democracy" in which the party and the central government manipulate state elections and in which from time to time the president will oust an elected state governor and appoint his own person to bring the deviant state back into line. Tanzania has been whittling away at the autonomy of Zanzibar almost since federation, and the trend is apparently toward total incorporation. In Nigeria, Yugoslavia, and India, there are different kinds of limitations on federalism in practice that raise some questions about how they should actually be classified. We could provide other examples, but the point is clear.

This notion of a regional State is quite new and still untested. Even the countries we have listed and the categories in which we have placed them are offered tentatively. There is both ample scope and real need for investigations into the utility of home rule or autonomy as a device for governing areas inhabited by minority ethnic groups or those separated from the core area of a country or those with greatly different economic, physiographic, or political conditions. Even the suggestion offered here of a tendency toward equilibrium is worth testing. Perhaps it offers hope for resolution of some problems of government in this restless world.

Chapter 11

ANOMALOUS POLITICAL UNITS

Although the European State system has spread around the world during the past several hundred years, resulting in the creation of over 170 States, by no means has all the earth's land area been organized in such a neat and seemingly definitive manner. We have already discussed leased areas as an example of territory legally under the sovereignty of one State but functioning as part of another, and we will later discuss territories possessing little or no sovereignty but rather functioning merely as colonies of a State. Here we are concerned with types of territorial organization that fall between these two extremes—or outside them altogether. Some of these anomalies are remnants of the period before the organization of nation-states; some are simply *ad hoc* arrangements resulting from war or decolonization, many of which seem to be taking on the appearance of permanence. The following review, by no means a complete inventory of such anomalies, attempts to indicate some alternatives to either complete sovereignty or none whatever.

MILITARY OCCUPATION

We have already discussed how territory is acquired through conquest and annexation and through assimilation involving coercion short of overt violence. States also acquire control over territory, though not sovereignty, through conquest without annexation. We need not go back further than World War II to find many examples. Even since that disastrous war, portions of sovereign States have been occupied by invading forces during smaller wars. Examples of such areas that have taken on a degree of semipermanence are northern Cyprus, occupied by Turkey since 1974, and "the West Bank," taken by Israel from Jordan in 1967. In each case, there is no doubt about sovereignty, but the territory is controlled by a State other than the sovereign.

TERRITORIES OF INTERMEDIATE STATUS

Partition of a dependent territory has been used on occasion as a method of dealing with otherwise intractable problems resulting from cultural heterogeneity in the territory. Examples are Ireland in 1922, Palestine in the same year and again in 1948, and India in 1947. (Partition has also been suggested for Cyprus and Lebanon.) In each case, the status of the separated pieces is quite clear; each is or is part of a sovereign State (except for the West Bank and Gaza). In other cases, however, the status is not so clear. Since 1945, Korea, Germany, Vietnam, China, and Kashmir have been divided into two or more segments by the fortunes of war. Although the circumstances in each case are different, only Vietnam has been reunited and the rest are likely to remain divided for some time. Gradually, the new situations are being formalized and frozen as more and more States recognize the *de facto* independence of East Germany and North Korea as well as their noncommunist counterparts, transfer recognition from Taipei to Beijing and accept the incorporation into India and Pakistan of the portions of the former State of Jammu and Kashmir they have occupied.

In this intermediate status also are such *quasi-states* as Andorra, Vatican City, Taiwan, and the South African Bantustans of Transkei, Bophuthatswana, Venda, and Ciskei. Andorra is governed by foreign co-princes, the president of France and the bishop of Urgel in Spain, and they conduct its foreign affairs, defense, and judicial system. But it is not a colony, has had its own international personality since 1278, and is a member of UNESCO. The State of Vatican City is the tiny remnant of the former Papal States, annexed by Italy in 1870. After that act, the number of States having diplomatic relations with the Holy See fell to 4, but now there are nearly 90. It is still questionable, however, whether a territory of 44 hectares (109 acres), whose residents are all citizens of other States, can properly be called a "State" under our definition. The South African government has been pursuing a policy of "separate development" of the "races" in the country, as it has defined these terms itself. The logical outcome of this policy is the subtraction of portions of the national territory and their conversion into independent States. The government claims to be doing this, having created 4 of these "States" already and with more promised. But since they were not created by the peoples of the territories and have very greatly circumscribed "sovereignty," few in South Africa and fewer still outside the country take them seriously. There is a chance, though, that one or more may survive and achieve statehood and legitimacy.

Another group of countries in this intermediate status are a number of colonies, protectorates, and other units of present or former colonial empires that have been or have recently become autonomous, or semi-independent. They include such territories as the West Indies Associated States, Puerto Rico, Niue and the Cook Islands, the Faeroe Islands, the Northern Marianas and a number of others. In each case, while a metropolitan country handles its foreign relations, currency, and other matters, it enjoys such a substantial degree of self-government that it cannot be called a colony without some qualification. Some, in fact, are teetering on the brink of independence, but others, particularly the smaller ones, may opt to continue indefinitely their semi-independent status.*

TERRITORIES OF UNCERTAIN STATUS

The decolonization process has not been easy. Not all former colonies, protectorates, mandates, and trusteeships have become independent States or legally parts of independent States. The portions of the former Palestine mandate occupied by Transjordan and Egypt, for example, are now occupied by Israel but are in political and legal limbo. Jordan annexed a portion of Palestine in 1950 and thus may have some legal claim to it, but she is unlikely to have it returned as an integral part of the State. The Gaza Strip is legally still part of Palestine since it was never annexed by Egypt or Israel and has no government of its own. Israel, for example, still administers Gaza's 3-mile territorial sea, even though Israel has a 6-mile territorial sea and Egypt's is 12 miles broad. Another example is Western Sahara, formerly Spanish Sahara. Spain wanted to withdraw from the territory in 1975, but it

*Undoubtedly, the most curious anomaly is the Sovereign Military Order of Malta, commonly known as the Knights of Malta. The Order was founded in 1099, was a power in the Mediterranean for seven hundred years, but now is sovereign over only an estate of 1.2 hectares (3 acres) in Rome. Nevertheless, it operates hospitals and leper asylums around the world; issues its own coins, stamps, and passports; and has diplomatic relations with some 45 countries, including Spain, Italy, Brazil, Austria, Togo, and Mauritania.

Frente POLISARIO supporters in Western Sahara. These people in Ausred, which is located in the southern part of the former Spanish Sahara, are demonstrating their support for the Algerian-backed rebels against the Moroccan occupation of their country. (Courtesy of the United Nations. Photo by Y. Nagata)

was claimed by Morocco, Mauritania, and Algeria, and no one knew what the people of the territory, mostly nomads, wanted. As Spain withdrew, Morocco and Mauritania divided it between them and later Mauritania withdrew. This action has been protested by Algeria and continues to be a lively debate issue in the United Nations. It seems unlikely that the Western Sahara will be resurrected in any form under any name, but one can never be certain about such things.

Another region of uncertain status is Antarctica. Part of the continent has never been formally claimed by any State, and the United States and the Soviet Union (the countries most active in Antarctic research and exploration) have never made any territorial claims. Most of the continent and its outlying islands have, though, been claimed by several countries. In the area of the Antarctic Peninsula (also called Graham Land and the Palmer Peninsula), the claims of Chile, Argentina, and the United Kingdom overlap. All claims, however, are frozen by the 1959 Antarctic Treaty. (No pun intended; there's just no better way of describing the situation.) Some countries assert that their claims are part of the national territory, others that they are dependent territories. But none is recognized by the United States or the

Soviet Union and each claim, as well as the unclaimed area, must be considered as legally *res nullius,* or belonging to no one. Antarctica is discussed in detail in Chapter Thirty-One.

INSURGENT STATES AND NONSTATE NATIONS

The current turmoil of decolonization and intensifying nationalism among peoples in multinational or plural States has spawned a large number of revolutions, rebellions, guerrilla wars, terrorist activities, and other acts of violence by a wide assortment of "national liberation movements" and of various political groups that have no particular national distinctiveness. These activities, in turn, have spawned a large number of studies. Some of the basic work has been done by the geographer Robert McColl and the political scientist Judy Bertelsen, on insurgent States and nonstate nations, respectively. Since neither term is widely accepted, we may take some liberties with them (perhaps doing them an injustice) and consider them as separate stages of the same process. We might consider a *nonstate nation* as being a people living as a minority in one or more States who want a State of their own in territory currently included in one or

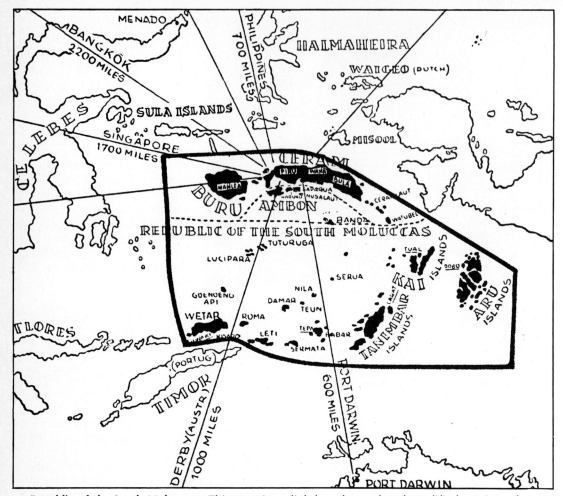

Republic of the South Moluccas. This map is a slightly enhanced and modified version of one found in the pamphlet *The Forgotten War: An Appeal from the Republic of the South Moluccas* (New York: Information Office of South Moluccas Republic). This is a good example of a "nonstate nation." The "republic" does not exist; its advocates live mostly in the Netherlands, control no territory in Indonesia, and try to win sympathy for their cause chiefly through propaganda and terrorism.

more States. It may engage in a variety of activities to win its point, ranging from polite petitions through intense propaganda and economic sanctions to widespread terrorism. It does not, however, control effectively any significant territory within the area it claims as a national homeland. It may, in exceptional cases, perform certain activities normally performed by States, but only with the permission of, and under the rules laid down by, the government. An *insurgent state* might develop within a country if the rebellious group is actually able to win and

retain control of a territorial base of operations. Within this territory, the insurgents operate a state-within-a-State in which only they and not the regularly constituted authorities perform all governmental functions.

According to these criteria, we might consider among the nonstate nations of the postwar era such groups as the Catholics of Ulster, the Greek Cypriots demanding *enosis,* the Jewish community in Palestine that performed some civil functions through the Jewish Agency, the South Moluccans in The Netherlands demanding independence from Indonesia, the Palestine Liberation Organization demanding an independent Palestinian Arab State, and other nationalist groups that use both peaceful and violent tactics but without a territorial base of their own.

Other groups in this same period, however, have had territorial bases in which they have established something resembling a State, what McColl has labeled an "insurgent state." This has occurred in Greece, China, South Vietnam, the Philippines, Malaya, Indonesia, Cuba, Portuguese Guinea, Angola, Mozambique, Algeria, Nigeria, Eritrea, Thailand, Burma, Iraq, and elsewhere. In most cases a nationalist group was fighting for independence from colonial rulers; in some cases national minorities were fighting for autonomy within or secession from an independent State; and in others the movement was political, generally opposing a corrupt or tyrannical government or favoring establishment of a communist government. Regardless of motive, however, the patterns of operation have been similar. McColl and others have tried to apply the same analysis to "urban guerrillas" and even the university campus, but with less justification and less success. The basic theses of the nonstate nation and insurgent state concepts will undoubtedly be tested repeatedly in the coming years, for we are not yet free of our "times of troubles."

BINATIONAL TERRITORIES

A number of small areas around the world are administered by two States, neither of which has exclusive sovereignty over them. Although they fall into two different classes of entities, we may group them together as binational territories. They are *condominiums* and *neutral zones*.

Undoubtedly, the largest and most important condominium of modern times was the Anglo–Egyptian Sudan. This vast territory, the largest in Africa, was governed jointly by Britain and Egypt from 1899 through 1955. Britain was definitely the senior partner and the Sudan was generally accepted as being part of the British Empire, but in various ways Egypt did participate in the country's operation until she abrogated the condominium agreements in 1951. Sudan became independent on 1 January 1956.

The last remaining condominium of any importance was New Hebrides, a group of 12 main islands and nearly 70 smaller ones in the South Pacific, administered jointly by France and Britain since 1906. Its government was known as the Joint Administration, of which the joint and equal heads were British and French officials. Some services were provided by the British and French National Services and some by the Joint Service. This arrangement ended with the granting of independence to the territory in 1980, when it became Vanuatu. Another surviving condominium is also in the South Pacific, uninhabited Canton and Enderbury islands, claimed by both the United States and the United Kingdom and administered jointly by them since 1938. Andorra, as mentioned earlier, is technically governed by Spanish and French coprinces, but in reality is virtually independent. Finally, embedded within and adjacent to the United Arab Emirates (UAE) are a number of enclaves administered jointly by Oman and Ajman, by Oman and Sharjah, and by Sharjah and Fujairah. Oman is not a member of the federation.

Neutral zones also used to be rather common around the world in frontier regions or separating contending States. Today there are few left. The largest and most important, the one between Kuwait and Saudi Arabia, was partitioned between them in 1969. Iraq and Saudi Arabia jointly administer a similar neutral zone that lies between them. They agreed in 1975 to divide the zone, but as yet the agreement has not been implemented. The British colony of Gibraltar, at the tip of a peninsula jutting south from Spain, is separated from Spain by a mile-wide neutral zone called La Linea running across the peninsula. Within the UAE is another

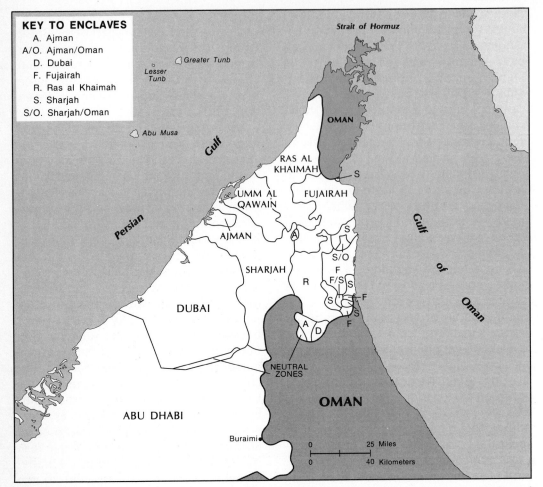

KEY TO ENCLAVES
- A. Ajman
- A/O. Ajman/Oman
- D. Dubai
- F. Fujairah
- R. Ras al Khaimah
- S. Sharjah
- S/O. Sharjah/Oman

The United Arab Emirates. In this one small peninsula of Southwest Asia, we see a number of politicoterritorial phenomena very common in Europe in the seventeenth century but rare anywhere today. Oman is a fragmented State and the federation itself contains not only seven members, but also a neutral zone, three condominiums, and 12 enclaves. The Buraimi Oasis was long fought over by Oman (formerly Muscat and Oman), Abu Dhabi, and Saudi Arabia. The Tunb Islands and Abu Musa in the Persian Gulf were occupied and annexed by Iran when the British withdrew in late 1971, despite claims to them by several of the emirates (formerly the Trucial States).

neutral zone, 1 kilometer (.6 mile) wide and 18 kilometers (11.2 miles) long, between Dubai and Abu Dhabi. A map of the UAE, in fact, resembles a map of seventeenth-century Europe or nineteenth-century India, with one large component (Abu Dhabi), 6 more semiautonomous states linked together with it in a loose confederation (Ajman, Dubai, Fujairah, Ras al Khaima, Sharjah, and Umm al Qawain), and with 12 enclaves besides the neutral zone and 3 condominiums already mentioned. And the UAE itself is the sur-

vivor of the former Gulf Federation, which from 1968 until 1971 included Bahrain and Qatar. Like the West Indies Federation, it could not agree on terms for unity before independence and all three became independent separately in 1971.

INTERNATIONAL TERRITORIES

Finally, a number of territories have been placed under international supervision and/or control in the twentieth century. By far the largest and most important class in this category is composed of the mandated territories of the League of Nations and the trust territories of the United Nations. We discuss *mandates* and *trusteeships* in more detail in Chapter Twenty-One; here we mention the only remaining examples—which happen also to be the most unusual. German South-West Africa was assigned by the League of Nations to the Union of South Africa after World War I. Under the terms of the mandate agreement, South Africa was to guide the territory for the benefit of its people. After World War II, when the United Nations succeeded the League of Nations, all the mandates were either terminated (those in the Middle East) or converted into UN trusteeships—except South-West Africa. South Africa instead extended its apartheid system and other laws to the territory and moved toward its annexation. After futile efforts to convince South Africa to

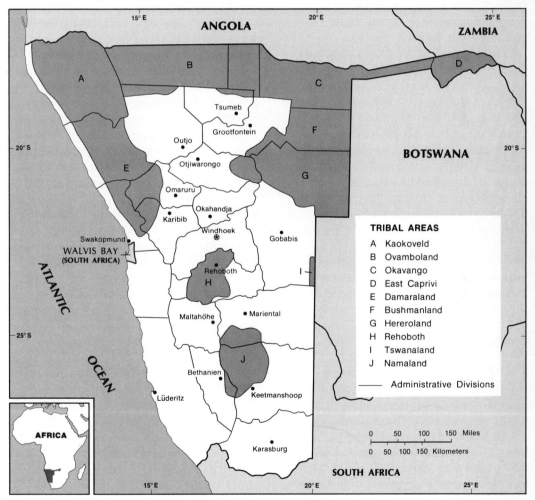

Namibia. The shaded areas, generally the least economically attractive in the territory, have been set aside by South Africa as "homelands" for the native tribes. The Rehoboth Basters are people of mixed descent who claim territory and special rights to protect their interests. The Caprivi Strip, in the far northeast, is evidence of Germany's drive toward her East African territory in the late nineteenth century. Walvis Bay, which has a splendid natural harbor, has been administered as an integral part of South Africa since 1922. It is the site of a major South African naval base. Namibians and the United Nations, however, claim that it should be part of Namibia since it will be essential to its economy and security after independence.

fulfill its mandatory obligations, sign a trusteeship agreement, or grant the territory independence, the United Nations voted in 1966 to assume control of the territory itself. It established a council to administer it, but South Africa would not recognize the council or permit it to enter the territory (now called Namibia). Instead, South Africa proceeded with plans to grant autonomy to Bantustans it had created within the territory and otherwise to modify but strengthen its own position there. South Africa is now taking steps to grant independence to the territory—on South Africa's terms.

Quite a different situation prevails in the Pacific Ocean. There, Japan was forced after World War II to relinquish her mandate over the former German colonies of the Marshall, Caroline, and Mariana islands. All these islands (except Guam, which has been a U.S. colony since 1898) were grouped together into the Trust Territory of the Pacific Islands, with the United States as Trustee.* Since it was designated a strategic trust, with fortifications permitted, it was placed under the supervision of the Security Council rather than the General Assembly—a unique situation. While the United Nations continues to supervise this last remaining trust territory, the Northern Marianas were separated from it in 1976 after a plebiscite in which the people voted to become a commonwealth of the United States, much like Puerto Rico. The rest of the territory is gradually being granted more self-government as the Federated States of Micronesia, the Marshall Islands, and Palau.

The United Nations has also had experience with *direct administration* of territory. Its own headquarters in New York is international territory, owned and administered by the United Nations under terms of an agreement with the United States. More important as a precedent, however, was the United Nations Temporary Executive Authority. Between 10 October 1962

and 1 May 1963, this agency directly administered the western half of New Guinea. It served as a temporal rather than a spatial buffer, separating the departing Dutch and the arriving Indonesians, as control was passed from one to the other after a long and bitter struggle. It performed all the functions of a government, except for conducting foreign relations. Should the United Nations actually assume control of Namibia, even on an interim basis, it would have a useful and successful precedent to draw on.

Another class of international territory is the *free city*. Three times in this century disputes over ownership of strategically located cities have resulted in the establishment of separate regimes in these cities and their immediate hinterlands, creating in effect modern city-states, but ones lacking sovereignty.* The first was Danzig, governed by a largely German administration from its creation in 1919 to its reincorporation into Germany after Hitler's invasion of Poland in 1939. Tangier was governed by an international administration, largely French and British, from 1923 until it was incorporated into Morocco in 1956. The Free Territory of Trieste was a little different for several reasons: (1) it was a much larger area than either Danzig or Tangier; (2) it was only a temporary arrangement pending settlement of the dispute over its disposition between Italy and Yugoslavia; and (3) the territory was governed by military authorities of the United States and Britain in Zone A (Trieste city and a strip of territory to the northwest) and of Yugoslavia in Zone B (the remainder of the territory to the east and south). It was partitioned in 1954, with Italy acquiring Zone A plus about 5 square miles and Yugoslavia receiving the rest. None of these international city administrations was notably successful. If experience is any guide, such an admin-

*A proposal to make Fiume (now Rijeka, Yugoslavia) a free state run by an Italo-Fiuman-Yugoslav consortium was approved by the city's electorate in 1921 but was never implemented.

*This territory is also known, somewhat inaccurately, as Micronesia.

Anomalous political units on stamps. The top four were produced by South Africa for the four "independent" Bantustans. The middle stamps were issued by the Isle of Man and by one of the Channel Islands, neither colonies nor parts of the United Kingdom, but dependencies of the Crown. In the next row are stamps issued by the short-lived Republic of Biafra during the Nigerian civil war and by Taiwan, which still claims to be part of the "Republic of China." Last is a Chilean stamp honoring the Knights of Malta, the world's smallest "State." (Martin Glassner)

istration for Jerusalem, recommended by the United Nations in 1947 but rejected by the Palestinian Arabs and the Arab States at the time, would be no more successful. This proposal is still raised in varying forms from time to time but is unlikely to be adopted.

The territories described in this chapter are anomalies because the accelerating spread of the modern State system concomitant with the disintegration of colonial empires has swept away many of the world's great variety of territorial organizations. The pressure is great, perhaps inexorable, to eliminate the last of them and incorporate every last square centimeter of land into some State. It is doubtful whether this is the wisest course. Perhaps a condominium arrangement similar to that of Andorra would be the best solution for the West Bank, for example, with Israel and Jordan jointly handling its foreign relations, currency, and so on. A similar arrangement could be worked out between Israel and Egypt for the Gaza Strip and between Morocco and Algeria for Western Sahara. The United Nations could administer either directly or through trusteeships many of the remaining nonself-governing territories in the world. Other territories might be willing to settle for permanent autonomous or protectorate status under a State rather than press for sovereignty. These suggestions are unlikely to be adopted in the near future, but we can envisage a time when once again the nation-state will not be considered the ideal form of territorial organization and divergent forms will not only be tolerated, but welcomed.

Chapter 12

POWER ANALYSIS

A whole generation of political geographers, political scientists, military intelligence specialists, foreign ministry officials, and others have devoted countless hours and immeasurable energy to the definition, quantification, and analysis of the "power" of many if not all States of the world. Not only has it been a popular exercise, even for graduate students, but it has been accepted as a required one in the field of political geography. Indeed, some definitions of "State" even include "power" as one of the requirements. There might have been some value in the ranking of the dozen or so most important States in the world prior to about 1950, and it is still undoubtedly essential for military and civilian leaders of a State to have a reasonably accurate picture of the strengths and weaknesses of potential enemies and allies, but the exercise known as the power analysis or power inventory, as generally practiced, seems of little value beyond the classroom. So many things have changed since World War II, and the pace of change is accelerating so rapidly, that we find it difficult to keep up with the currently significant elements of power in the first place and then to quantify them, assign them realistic values, apply them to individual countries, and rank the countries.

It is quite obvious to anyone who reads even a hometown newspaper that there are two superpowers in the world, with the United States somewhat more powerful than the USSR. Below them are perhaps six or eight countries that could bring some power to bear outside their own immediate regions on a sustained basis. Below this group are the remaining 160 or so countries whose ranking would be anyone's guess. No computer could possibly produce a ranking acceptable to the majority of observers as both accurate and useful. Even if it were possible to do so, the ranking would quickly go out of date and have to be recalculated.

For these reasons our discussion of the topic is confined to a survey of the indicators that should be included in any power inventory and to some comments on the difficulty of analyzing them in any meaningful way.* The following listing is by no means definitive; it is suggestive only. The organization of the factors is partly arbitrary and they are not necessarily listed in order of importance. Finally, we must stress that many factors overlap and most are interrelated, if not interdependent.

TERRITORY

We have already discussed the importance of size and shape of a State and con-

*Readers who are interested in quantification of these indicators are referred to two basic tools: The United Nations *Statistical Yearbook;* and the *World Handbook of Political and Social Indicators,* 3rd ed., 2 vols., Charles Lewis Taylor and David A. Judice, New Haven, CT, and London: Yale University Press, 1983.

cluded that there is no ideal size or shape. Nevertheless, in measuring the power of an individual State, these factors can be very significant. All other things being equal, reasonably large size affords space for maneuver, defense in depth, variety of resources, and other advantages; a small ratio of perimeter to area makes internal communications and external defense easier than a fragmented or prorupt shape. But a compact shape can leave valuable resources nearby but still outside a State's boundaries and a large territory may be vulnerable to attack from different directions.

"Strategic" or "favorable" location is often cited as an asset to a State, and is sometimes cited by determinists as explaining a particular country's success. We discuss this in more detail later in our chapters on geopolitics, but for now we can point out that since technology, economies, politics, and strategies are constantly changing, so are the values of particular locations. It does make a difference to a State whether it has one seacoast or two or none. Buffer States can benefit from the intermediary role in trade, transportation, and communications, but are also subject to buffeting from both sides. States located at the western edge of the Pacific Ocean, the eastern edge, or the middle clearly derive both advantages and handicaps from these locations. Who ever heard of Diego Garcia 30 years ago? Wasn't Korea "The Hermit Kingdom" in the past century? Wasn't Valparaíso a major port with a strategic location before the Panama Canal was opened? When the Suez Canal was closed from 1967 to 1974, eastern and southern Africa assumed a new importance. But who now knows or cares much about Chateau Thierry, Tarawa, Kasserine Pass, Galipoli, Pork Chop Hill, or the A Shau Valley—battlefields where hundreds of thousands died not long ago because of their "strategic importance"? Who can predict which State

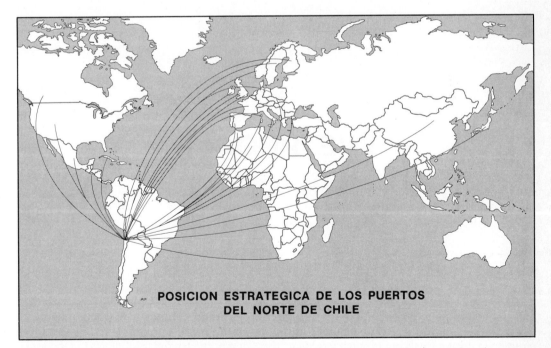

POSICION ESTRATEGICA DE LOS PUERTOS DEL NORTE DE CHILE

"Strategic Position of the Ports of Northern Chile." This map appears in the forward to the report *Antofagasta and the North of Chile, Bridge of Exit and Entry of the Central West South American Hinterland, in the Sphere of Action of the G.E.I.C.O.S.* (Interregional Business Group of the South American Central West, a private organization of Chilean, Argentine, Paraguayan and Bolivian business people trying to develop a new transit corridor through Antofagasta.)

TRADITIONAL SOCIO-POLITICAL STRUCTURE IN LATIN AMERICA

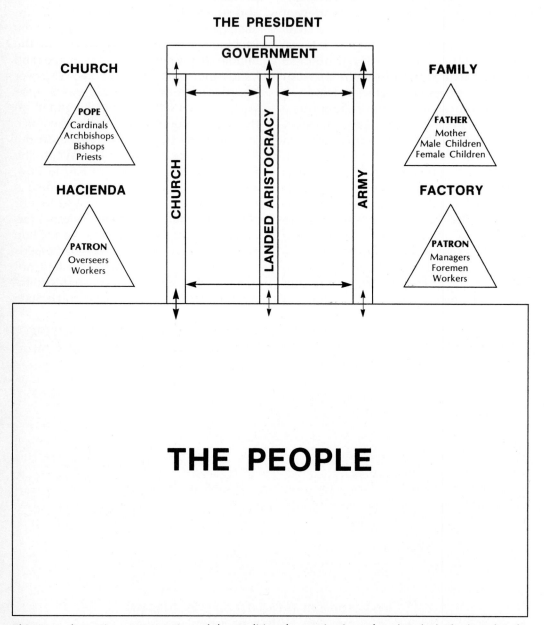

This is a schematic representation of the traditional organization of society in Latin America. In some countries of the region it still exists in very much this form; in others it has disappeared entirely; in most it exists but is much modified. The Church, the landed aristocracy and the army form an interlocking triumvirate controlling the entire structure. They exploit and repress the masses of the people while supporting and providing leadership for the government. At the apex of this pyramidal structure is the President (or a junta), a symbol of authority whether exercising real power or not. The size of the arrows represents the relative strength of the reciprocal links among the various sectors of society. The relatively rigid pyramidal structure is the product of will occupy a "strategic position" tomorrow?

The "natural resources" of a State are also considered valuable and quantifiable, so they tend to rank high in everyone's power inventory. But what is a "natural resource"? Thirty years ago titanium was used primarily as an ingredient in some paints; today mine waste piles are being searched for this metal that is essential for the skins of high-performance supersonic aircraft. Wars were fought over natural nitrates only a century ago, yet few people today even know what they are. Natural resources would have to include soils, water, forests, and many other elements of the physical environment that are difficult to evaluate. Where in the State's territory are the resources located—out near a hostile border, deep in the inaccessible interior, in the frigid and empty northland? Which resources are currently being utilized, which are known but inaccessible, which suspected but not proven, which too costly to process? What about resources that are needed but totally absent in the national territory? Can they be obtained from friends? At what cost? Even in times of stress? What is the value of resources in the colonies or of investments in the resources of other States? How do we factor in the roles of substitutes, synthetics, conservation, reuse? How do we evaluate resources in a country that squanders them on cosmetics and pet food while another uses them for tools and weapons?

POPULATION

In any power analysis it is quite reasonable to assume that a country with a large population has a distinct advantage over a country with a small population—all other things being equal. But the fact that all other things are not equal makes it difficult to evaluate population as a power factor. Although sheer numbers are important to some extent in every country, their importance may be enhanced or diminished by such factors as their distribution over the national territory and demographic trends. Qualitative factors may be even more important than quantitative ones. The age/sex ratio, for example, affects the pool of potential military personnel and the proportions of producers and consumers in the society. Immigration and emigration, education levels, health factors, the distribution of wealth and degree of poverty—all must be considered and evaluated.

Cultural and "racial" homogeneity or heterogeneity may strengthen or weaken a society depending on the peoples involved, their attitudes toward one another, the role of government in fostering integration or segregation, whether the various groups complement or compete with one another. Social organization can be very important, including such things as the degree of socioeconomic stratification and social mobility in the society, the land tenure systems and distribution of good agricultural land, and the status and role of elites. Religious and cultural values are intangibles that can bear heavily on a State's ability to mobilize its strength to pursue some policy. The mobilization might be ideologically motivated: patriotism, communism, democracy, or some other cause might be considered worth working or dying for—or might not.

GOVERNMENT

The political system of a country invariably figures in its power, but invariably it is also difficult to quantify. Does the government effectively control the whole country; is

---

similar structures in all other important aspects of life, symbolized by the triangles in the upper corners. Even the major pre-Columbian civilizations of the region, of which many of the people are heirs, were hierarchical. Such a structure is inherently conservative, resistant to change. It has hindered economic, social, and political progress in the region and where it has not been displaced or greatly modified it must be considered a negative factor in any power analysis.

the ecumene coextensive with the national territory? Is the government popular, likely to be supported and rallied around in times of stress? Or is there widespread opposition either to the government in power or to the whole political system? Are the officials trained and experienced, efficient and honest, or is the government weakened by corruption, nepotism, bloated payrolls, and inefficiency? Is the government capable of providing wise, effective, and inspiring leadership?

ECONOMY

It is self-evident that wealth in our world is a form of power or can be translated readily into power, so wealthy countries, like populous countries or those richly endowed by nature, have an automatic advantage over those less fortunate. But wealth, even material wealth, consists of more than just "natural resources" or even GNP (Gross National Product). If a State is to be powerful, it must have productive capacity, not only to meet its needs, but to supply surpluses to sell abroad in peacetime and to meet expanded needs in wartime. Science and technology must be well developed, and research must receive adequate investment and other support. Industry must have adequate supplies of efficient, skilled, and dedicated managers and workers. Capital must move readily from where it is to where it is needed, and agriculture must get a fair share. Most societies rest on an agricultural base (no matter how marginal) and in times of stress a country may be cut off from agricultural imports completely. Foreign trade is both a strength and a weakness, and many countries have adopted autarkic policies in order to capitalize on its strength and avoid its weakness. Autarky has no place in the twentieth century, however; all countries are and must be interdependent if they are to prosper.

Banking and insurance systems are fundamental to any modern economy, yet they are seldom included in power inventories. How can we measure their quality or value in a power system? The question of quality runs through any useful economic analysis, for the quality of goods produced and services performed may be more important than the quantity.

CIRCULATION

Transportation and communications—the movement of goods, people, and ideas—are fundamental to any modern society and to any modern State. Without a good circulation system to tie together all sections of the country, all elements of the population, all units of government, all sectors of the economy, and all of them with one another, a State would simply disintegrate. It would project not power but weakness. It must have not only an adequate physical infrastructure, but also efficient management and workers, adequate capitalization, and proper maintenance. This includes all forms of transportation plus radio, telephone, telegraph, and postal systems; a free and vigorous press; and other means of communication. These can be misused, of course, to persuade as well as inform, for mind control as well as mind enrichment. People must be wary of this, especially in times of stress, when they tend to relax their defenses against internal threats in order to concentrate on external ones.

MILITARY STRENGTH

All these factors are elements of, or can be translated into, military strength. Every conventional power inventory counts every piece of military hardware and every person in military uniform it can identify in a State and assigns values to the totals. But is this enough? Can we say that a country with a big, well-equipped army is more powerful than one with a small, poorly equipped one? Of course not. In order to evaluate the effectiveness of a military force, we must reckon with many intangibles. These would include the education

POWER FACTORS AND COMPONENT FORCES

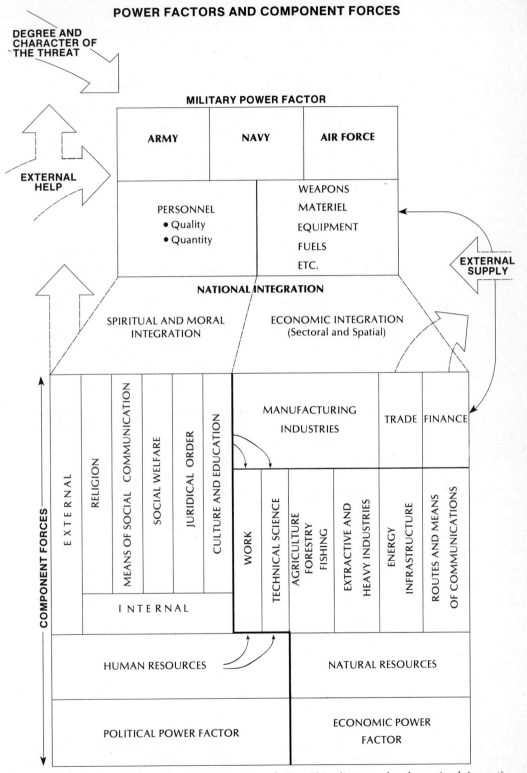

DEGREE AND CHARACTER OF THE THREAT

EXTERNAL HELP

MILITARY POWER FACTOR

ARMY	NAVY	AIR FORCE

PERSONNEL
• Quality
• Quantity

WEAPONS
MATERIEL
EQUIPMENT
FUELS
ETC.

EXTERNAL SUPPLY

NATIONAL INTEGRATION

SPIRITUAL AND MORAL INTEGRATION

ECONOMIC INTEGRATION
(Sectoral and Spatial)

COMPONENT FORCES

EXTERNAL

RELIGION

MEANS OF SOCIAL COMMUNICATION

SOCIAL WELFARE

JURIDICAL ORDER

CULTURE AND EDUCATION

MANUFACTURING INDUSTRIES

TRADE FINANCE

WORK

TECHNICAL SCIENCE

AGRICULTURE FORESTRY FISHING

EXTRACTIVE AND HEAVY INDUSTRIES

ENERGY INFRASTRUCTURE

ROUTES AND MEANS OF COMMUNICATIONS

INTERNAL

HUMAN RESOURCES

NATURAL RESOURCES

POLITICAL POWER FACTOR

ECONOMIC POWER FACTOR

A Latin American conception of national power analysis. This diagram, by the retired Argentine general Juan E. Guglialmelli, represents one way of organizing one's thinking about national power. Note that these factors are not weighted in any way. (Translated by Martin Glassner) and skill of the troops; the quality of their leadership; their motivation, morale, loyalty; their combat experience; the logistical system; and even whether they are up-to-date in military doctrine, strategy, and tactics. (It is no good to be, as was said of the French army, always superbly ready to fight the last war.) The country must have an effective intelligence service and an efficient system for mobilizing reserves.

A country's military strength can be enhanced by having distant bases in strategic places—unless the cost of supplying and protecting them nullifies their value. Its strength can also be enhanced through collective security, having allies pledged to aid one another ("all for one and one for all"), if they all have something tangible to contribute and will remain steadfast in the face of adversity or temptation. How can we measure the potential reliability of an ally under a variety of real or hypothetical circumstances?

FOREIGN RELATIONS

Today no State can remain isolated and be either strong or prosperous. States that have chosen the route of virtual isolation, such as Albania and Burma most recently, must rely on their own human and natural resources. This may be good for the soul, but not for the belly or the mind. Today every State, every people, is somehow dependent on every other one. Even those that have chosen not to join military alliances must still be involved with others. Broadly speaking, it seems that those countries that participate most actively in world affairs—political, economic, and cultural—have the best opportunities to maximize their assets and overcome their handicaps and thus increase their national "power." International trade, cultural links (even with the former "mother country"), participation in transfers of development funds, participation in the work of the United Nations and other world organizations and of regional organizations, all elevate a country, make it a more valued member of the international community, and perhaps render it less likely to be attacked.

ANALYSIS

What conclusion can we draw? Only that "power" is exceedingly difficult to define and to measure. After considering the superpowers and perhaps three or four other countries, it becomes almost impossible to evaluate power. Some recent examples may illustrate the point. How do we rank Israel, for example? A country of only 4 million people in a land the size of Vermont with few natural resources and surrounded by hostile forces with overwhelming strength, she has nevertheless won five wars in less than 30 years. Her contributions to the world in science, medicine, scholarship, music, and other fields would be impressive in a country many times her size. Her air and shipping lines, banks, development aid programs, educational and cultural exchanges, importers and exporters operate around the world, and she has even been able to project her military power as far away as Uganda, Iraq, and Tunisia.

And how do we rank Cuba, still poor but with a sizable army, parts of which are operating across Africa and in the Arabian peninsula? Or Vietnam, poorer still but strong enough to conquer Kampuchea (Cambodia) and challenge China? Or India, which, though still poor, has exploded a nuclear device and has kept her army very busy indeed since 1947? Pakistan was unquestionably weakened by her civil war and the breakaway of East Bengal in 1971 to become the new State of Bangladesh, but by how much? Are Pakistan and Bangladesh together equal to the Pakistan of 1970? Was Nigeria weakened or strengthened by the unsuccessful secession effort of Biafra? Iran spent billions of dollars on the most modern conventional weapons available, perhaps in an attempt to recreate the empire of Darius, but none of that helped the Shah to retain power; indeed, it may have cost him his throne.

Then there are the States with negligible military power but considerable economic power. Japan, of course, is the prime example of an economic superpower. But she has serious weaknesses in her social and industrial structures and is utterly dependent on the importation of commodities (foods, fuels, and raw materials) and the export of manufactured goods for her economic health—and perhaps her survival. Is she a giant with feet of clay? Even Switzerland is something of an economic power, especially if we factor in her banking system. OPEC (the Organization of Petroleum Exporting Countries) has proven the efficacy of concerted action by States that individually have little power. In 1973 to 1974, this cartel, including such countries as Gabon, Ecuador, and the United Arab Emirates, was able to shake the economies and threaten the safety of the United States and Western Europe by withholding exports of oil for a time and quadrupling its price when exports resumed. How then do we measure the "power" of Gabon, Ecuador, and the United Arab Emirates?

Perhaps it would be best to cease measuring and ranking "power" and try to measure influence, the ability of a State to influence the behavior of other States. This might be less precise and less geographic, but it would be more rewarding. Nevertheless, the exercise would still have to be repeated periodically to take into account new and altered circumstances. Influence, like power, is constantly changing and quite unpredictable.

REFERENCES FOR PART TWO

Books and Monographs

A

Ackerman, Edward A., "Population and National Resources," in P.M. Hauser and O.D. Duncan (eds.), *The Study of Population*. Chicago: Univ. of Chicago Press, 1959, 621–648.

Acquah, Ioné, *Accra Survey: A Social Survey of the Capital of Ghana*. London: Univ. of London Press, 1958.

Adami, Vittorio, *National Frontiers in Relation to International Law* (1919), trans. T.T. Behrens. London: Oxford Univ. Press, 1927.

Agarwal, H.O., *Kashmir Problem*. Allahabad, India: Kitab Mahal Private, 1979.

Aitken, H.G.J. (ed.), *The State and Economic Growth*. New York: Social Science Research Council, 1959.

Ake, Claude, *A Theory of Political Integration*. Homewood, Il: Dorsey Press, 1967.

Albino, Oliver, *The Sudan: A Southern Viewpoint*. New York: Oxford Univ. Press, 1970.

Alexander, David G. and others, *Atlantic Canada and Federation*. Toronto: Univ. of Toronto Press, 1983.

Arkadie, Brian Van, *Benefits and Burdens: A Report on the West Bank and Gaza Strip Economies Since 1967*. New York: Carnegie Endowment for International Peace, 1977.

Arnade, Charles W., *The Emergence of the Republic of Bolivia*. Gainesville: Univ. of Florida Press, 1957.

Asiwaju, A.I. (ed.), *Partitioned Africans; Ethnic Relations Across Africa's International Boundaries 1884–1984*. New York: St. Martin's Press, 1985.

Attalides, Michael A., *Cyprus*. New York: St. Martin's Press, 1979.

Awa, Eme O., *Federal Government in Nigeria*. Berkeley: Univ. of California Press, 1964.

B

Banks, A. Leslie (ed.), *The Development of Tropical and Subtropical Countries, with Particular Reference to Africa*. London: Arnold, 1954.

Barros, James, *The Aland Islands Question; Its Settlement by the League of Nations*. New Haven, CT: Yale Univ. Press, 1968.

Bavkis, Herman and William Chandler (eds.), *Federalism and the Role of the State*. Toronto: Univ. of Toronto Press, 1987.

Becker, William H. and Samuel F. Wells, Jr., (eds.), *Economics and World Power: An Assessment of American Diplomacy Since 1789*. New York: Columbia Univ. Press, 1984.

Beckinsale, Robert P., "Rivers as Political Boundaries," in *Water, Earth and Man*. London, 1969.

Benvenisti, Meron, *The West Bank Data Project: A Survey of Israel's Policies*. Lanham, MD: Univ. Press of America, 1984.

Bertelsen, Judy S. (ed.), *The Non-State Nation in International Politics*. New York: Praeger, 1977.

Best, Harry, *The Soviet State and Its Inception*. New York: Philosophical Library, 1950.

Betts, R.R. (ed.), *Central and South East Europe, 1945–1948*. London: Royal Institute of International Affairs, 1950.

Bew, Paul and others, *The State in Northern Ireland 1921–72*. New York: St. Martin's Press, 1980.

Bianchi, William J., *Belize: The Controversy Between Guatemala and Great Britain over the Territory of British Honduras in Central America*. New York: Las Américas, 1959.

Birch, A.H., *Political Integration and Disintegration in the British Isles*. London: Allen & Unwin, 1977.

Birdwood, Christopher B.B., *Two Nations and Kashmir*. London: Hale, 1956.

Bloomfield, Lincoln M., *The British Honduras–Guatemala Dispute*. Toronto: Carswell, 1953.

Bloomfield, Lincoln M. and Gerald F. Fitzgerald, *Boundary Waters Problems of Canada and the United States*. Toronto: Carswell, 1958.

Boal, Frederick W. and J. Neville H. Douglas (eds.), *Integration and Division: Geographical Perspectives on the Northern Ireland Problem*. London and New York: Academic Press, 1982.

Boggs, S. Whittemore, *International Boundaries: A Study of Boundary Functions and Problems*. New York: Columbia Univ. Press, 1940.

Bowman, Isaiah, *The Pioneer Fringe*. New York: American Geographical Society, 1931.

Bradford, S., *Spain in the World*. Princeton, NJ: Van Nostrand, 1962.

Brandt, Karl, *Europe, the Emerging Third Power: Phenomenon and Portent*. New York: City News Publishing Co., 1958.

Braveboy-Wagner, Jacqueline Anne, *The Venezuela–Guyana Border Dispute; Britain's Colonial Legacy in Latin America*. Boulder, CO: Westview, 1984.

Brawer, Moshe, "Geographical Factors in Boundary Delimitation in the Sinai Peninsula," in *Essays in Political Geography*. Calcutta: Indian National Council of Geography, 39–45.

Brecher, Michael, *The Struggle for Kashmir*. New York: Oxford Univ. Press, 1953.

Breuilly, John, *Nationalism and the State*. New York: St. Martin's Press, 1982.

Briand, Michael (ed.), *A Way Out: Federalist Options for South Africa*. San Francisco: ICS Press, 1987.

Brigham, Albert Perry, *Geographic Influences in American History*. Boston: Ginn, 1903.

Broek, Jan O.M., *The Problem of Natural Frontiers*. Berkeley: Univ. of California Press, 1940.

Brown, Peter G. and Henry Shue (eds.), *Boundaries: National Autonomy and Its Limits*. Totowa, NJ: Rowman & Littlefield, 1981.

Brownlie, Ian, *African Boundaries: A Legal and Diplomatic Encyclopedia*. London: Hurst; Berkeley and Los Angeles: Univ. of California Press, 1979.

Brunnschweiler, Dieter, *The Llanos Frontier of Colombia*. East Lansing; Michigan State Univ., Latin American Studies Center, 1972.

Bryce, J., *The Holy Roman Empire*. New York: Macmillan, 1919.

Bueler, William M., *U.S. China Policy and the Problem of Taiwan*. Boulder: Colorado Associated Univ. Press, 1971.

Bull, Vivian A., *The West Bank–Is It Viable?* Lexington, MA: Heath, 1975.

Bunday, Laszlo, *Dismembered Hungary*. London: Grant Richards, 1923.

Burns, R.M., and others, *Political and Administrative Federalism*. Canberra: Australian National Univ. Press, 1976.

Burrows, Bernard and others (eds.), *Federal Solutions to European Issues*. New York: St. Martin's Press, 1978.

Butler, Jeffrey and others, *The Black Homelands of South Africa*. Berkeley and Los Angeles: Univ. of California Press, 1977.

C

Calvert, Peter, *The Falklands Crisis: The Rights and Wrongs*. London: Frances Pinter, 1982.

Canaway, A.P., *The Failure of Federalism in Australia*. London: Oxford Univ. Press, 1930.

Carney, John and others (eds.), *Regions in Crisis; New Perspectives in Europe Regional Theory*. London: Croom Helm, 1982.

Carter, F.W., *Dubrovnik (Ragusa): A Classic City-State*. New York and London: Seminar Press, 1972.

Catudal, Honoré M., *The Enclave Problem of Western Europe*. University: Univ. of Alabama Press, 1979.

Chadwick, H.M., *The Nationalities of Europe and the Growth of National Ideologies*. Totowa, NJ: Rowman & Littlefield, 1973.

Chang, Luke T., *China's Boundary Treaties and Frontier Disputes*. New York, London, Rome: Oceana, 1982.

Chapin, Miriam, *Contemporary Canada*. New York: Oxford Univ. Press, 1959.

Chay, John, and Thomas E. Ross (eds.), *Buffer States in World Politics*. Boulder, CO: Westview, 1986.

Clark, Gordon L. and Michael Dear, *State Apparatus*. Boston: Allen & Unwin, 1984.

Coedes, G., *The Making of Southeast Asia*. Berkeley: Univ. of California Press, 1983.

Cohen, Saul B., *The Geopolitics of Israel's Border Question*. Boulder, CO: Westview, 1987.

Coker, F.W., *Organismic Theories of the State*. New York: Columbia Univ. Press, 1910.

Collingwood, R.G. and J.N.L. Myers, *Roman Britain and the English Settlements*. Oxford and New York: Oxford Univ. Press, 1957.

Cottam, Richard W., *Nationalism in Iran*. Pittsburgh: Univ. of Pittsburgh Press, 1979.

Coulborn, Rushton, *The Origin of Civilized Societies*. Princeton, NJ: Princeton Univ. Press, 1959.

Crawford, James, *The Creation of States in International Law*. Oxford: Oxford Univ. Press, Clarendon, 1979.

Crepeau, P.A. and C.B. Macpherson (eds.), *The Future of Canadian Federalism*. Toronto: Univ. of Toronto Press, 1965.

Cressey, George B., *Soviet Potentials; A Geographic Appraisal*. Syracuse, NY: Syracuse Univ. Press, 1962.

Cukwurah, A.O., *The Settlement of Boundary Disputes in International Law*. Manchester, Eng.: Manchester Univ. Press, 1967.

Curran, Joseph M., *The Birth of the Irish Free State*. University: Univ. of Alabama Press, 1979.

Currie, David P., *Federalism in the New Nations*. Chicago: Univ. of Chicago Press, 1964.

Curzon, Lord, *Frontiers*. London: Oxford Univ. Press, 1908.

D

Day, Alan J. (ed.), *Border and Territorial Disputes*. Detroit: Gale Research Co., 1982.

Dennis, William Jefferson, *Documentary History of the Tacna–Arica Dispute*. Port Washington, NY: Kennikat Press, 1971.

Devine, D.J., "Status of Rhodesia in International Law," *Acta Juridica (Cape Town), 1967.*

Dickinson, Robert E., *City, Region and Regionalism*. New York: Oxford Univ. Press, 1947.

————, *The West European City*. London: Routledge & Kegan Paul, 1951.

Dikshit, Ramesh D., *The Political Geography of Federalism*. New York: Wiley, 1975; Delhi: Macmillan Co. of India, 1975.

Doolin, Dennis J., *Territorial Claims in the Sino-Soviet Conflict; Documents and Analysis*. Stanford, CA: Hoover Institution Press, 1965.

Duchacek, Ivo D., *Comparative Federalism: The Territorial Dimension of Politics*. Lanham, MD: Univ. Press of America, 1987.

Dulles, Foster Rhea, *America's Rise to Power 1898–1954*. New York: Harper & Row, 1954.

Duncan, Otis Dudley and others, *Metropolis and Region*. Baltimore: Johns Hopkins Press for Resources for the Future, 1960.

Dyson, K.H.F., *The State Tradition in Western Europe: A Study of an Idea and Institution*. Oxford: Martin Robertson, 1980.

E

East, W. Gordon, *An Historical Geography of Europe*. London: Methuen, 1935.

————, *The Political Division of Europe*. London: Univ. of London Press, 1948.

East, W. Gordon and O.H.K. Spate, *The Changing Map of Asia*. London, 1961.

East, W. Gordon, "The Concept and Political Status of the Shatter Zone," in *Geographic Essays on Eastern Europe,* Pounds, Norman J.G., (ed.). Bloomington: Indiana Univ. Press, 1961, reprinted, Westport, CT: Greenwood, 1972, 1–27.

Eckes, Alfred E., Jr., *The United States and the Global Struggle for Minerals*. Austin: Univ. of Texas Press, 1979.

Elazar, Daniel J., *Judea, Samaria, and Gaza: Views on the Present and Future*. Lanham, MD: Univ. Press of America, 1982.

————, *American Federalism: A View from the States*. New York: Harper & Row, 1984.

————, *Federalism and Political Integration*. Ramat Gan, Israel: Turtledove; Lanham, MD: Univ. Press of America, 1985.

Elegant, Robert S., *The Dragon's Seed; Peking and the Overseas Chinese*. New York: St. Martin's Press, 1959.

Emeny, Brooks, *Mainsprings of World Politics*. New York: Foreign Policy Association, 1956.

Epstein, David G., *Brasilia: Plan and Reality*. Berkeley: Univ. of California Press, 1973.

Etzioni, Amitai, *Political Unification; A Comparative Study of Leaders and Forces*. New York: Holt, Rinehart & Winston, 1985.

F

Falklands; Whose Crisis? London: Latin America Bureau Special Brief, 1982.

Farah, Tawfic E. (ed.), *Pan-Arabism and Arab Nationalism*. Boulder, CO: Westview, 1987.

Fawcett, Charles Bungay, *Frontiers: A Study in Political Geography*. London and Oxford: Oxford Univ. Press, 1918.

Febvre, L.P.V. and Lionel Bataillon, *A Geographical Introduction to History*. New York: Knopf, 1925.

Ferrer Vieyra, Enrique, *An Annotated Legal Chronology of the Malvinas (Falkland) Islands Controversy*. Córdoba, Arg: 1985.

Ferris, W.H., *The Power Capability of Nation-States*. Lexington, MA: Lexington Books, 1973.

Fishel, Wesley R., *The End of Extraterritoriality in China*. Berkeley: Univ. of California Press, 1952.

Fitzmaurice, John, *Quebec and Canada*. New York: St. Martin's Press, 1985.

Franck, Thomas Meral, *Why Federations Fail*. New York: New York Univ. Press, 1968.

Franklin, H., *Unholy Wedlock: The Failure of the Central African Federation*. London: Allen & Unwin, 1964.

Fraser, T.G., *Partition in Ireland, India and Palestine; Theory and Practice*. New York: St. Martin's Press, 1984.

G

Gamba, Virginia, *The Falklands/Malvinas War:* London: Allen & Unwin, 1986.

Gerson, Allan, *Israel, the West Bank and International Law*. Totowa, NJ: Frank Cass, 1978.

Goad, H.E., *The Making of the Corporate State*. London: Christophers, 1932.

Goebel, Julius, *The Struggle for the Falkland Islands*. New Haven, CT: Yale Univ. Press, 1982.

Golay, J.F., *The Founding of the Federal Re-*

public of Germany. Chicago: Univ. of Chicago Press, 1958.

Gottmann, Jean, *Megalopolis*. New York: Twentieth Century Fund, 1961.

——, *The Significance of Territory*. Charlottesville: Univ. Press of Virginia, 1973.

Grant, George, *Lament for a Nation; The Defeat of Canadian Nationalism*. Toronto: McClelland & Stewart, 1965.

H

Hadawi, Sami, *Palestine Partitioned, 1947–1958*. New York: Arab Information Center, 1959.

Hamel, J.A. Van, *Danzig and the Polish Problem*. New York: Carnegie Endowment for International Peace, 1933.

Hanna, Willard A., *The Separation of Singapore from Malaysia*. American Universities Field Staff Reports, Southeast Asia Series, Vol. 13, No. 21. 1965.

——, *The Republic of the South Moluccas*. American Universities Field Staff Reports, Southeast Asia Series, Vol. 23, No. 2. 1975.

Hardy, W.G., *From Sea unto Sea: Canada 1850 to 1910; The Road to Nationhood*. Garden City, NY: Doubleday, 1960.

Harrison, David, *The White Tribe of Africa: South Africa in Perspective*. Berkeley: Univ. of California Press, 1982.

Harrison, John A., *Japan's Northern Frontier; A Preliminary Study in Colonization and Expansion with Special Reference to Relations of Japan and Russia*. Gainesville: Univ. of Florida Press, 1953.

Hawkins, Robert B., Jr., (ed.), *American Federalism: A New Partnership for the Republic*. San Francisco: ICS Press, 1982.

Hayes, Carlton J.H., *Nationalism: A Religion*. New York: Macmillan, 1960.

Hertslet, Sir Edward, *The Map of Europe by Treaty 1875–1891*. London: Butterworths, 1891.

——, *The Map of Africa by Treaty*. London: Frank Cass, 1967.

Hertz, F., *Nationality in History and Politics*. London: Kegan Paul, 1944.

Heslinga, M.W., *The Irish Border as a Cultural Divide: A Contribution to the Study of Regionalism in the British Isles*. New York: Humanities Press, 1963.

Hill, Norman, *Claims to Territory in International Law and Relations*. New York: Oxford Univ. Press, 1945.

Hinsley, F.H., *Power and the Pursuit of Peace*. London: Cambridge Univ. Press, 1963.

Hitti, Philip K., *Capital Cities of Arab Islam*. Minneapolis: Univ. of Minnesota Press, 1973.

Hodgkins, Jordan A., *Soviet Power: Energy Resources, Production and Potentials*. Englewood Cliffs, NJ: Prentice-Hall, 1961.

Hof, Frederic C., *Galilee Divided: The Israel–Lebanon Frontier, 1916–1984*. Boulder, CO: Westview, 1985.

Hoffman, George W., *The Balkans in Transition*. Princeton, NJ: Van Nostrand, 1963.

Hoffman, Fritz L., and Olga Mingo Hoffman, *Sovereignty in Dispute; The Falklands/Malvinas, 1493–1982*. Boulder, CO: Westview, 1984.

Holdich, Thomas H., *Political Frontiers and Boundary Making*. London: Macmillan, 1916.

Hutchison, Bruce, *The Struggle for the Border*. New York: Longmans, Green, 1955.

I

Ireland, Gordon, *Boundaries, Possessions and Conflicts in South America*. Cambridge: Harvard Univ. Press, 1938; reprinted, New York: Octagon, 1971.

——, *Boundaries, Possessions, and Conflicts in Central and North America and the Caribbean*. Cambridge: Harvard Univ. Press, 1941; reprinted, New York: Octagon, 1971.

Isard, Walter, *Location and Space Economy*. New York: Technology Press of MIT and Wiley, 1956.

J

James, Alan, *Sovereign Statehood: The Basis of International Society*. London, Boston, and Sydney: Allen & Unwin, 1986.

Jeffries, William W. (ed.), *Geography and National Power*, 4th ed. New York: Arco, 1971.

Jennings, R.Y., *The Acquisition of Territory in International Law*. Manchester, Eng.: Manchester Univ. Press, 1963.

Jesman, C., *The Ethiopian Paradox*. London: Oxford Univ. Press, 1963.

Johnston, Ray E. (ed.), *The Politics of Division, Partition and Unification*. New York: Praeger, 1976.

Johnston, Ronald J., *Geography and the State*. New York: St. Martin's Press, 1983.

Jones, E.L., *The European Miracle*. New York: Cambridge Univ. Press, 1987.

Jones, Stephen B., *Boundary Making*. Washington: Carnegie Endowment for International Peace, 1945.

——, *Theoretical Studies of National Power*. New Haven, CT: Yale Univ. Press, 1955.

K

Kalia, Ravi, *Chandigarh*. Carbondale: Southern Illinois Univ. Press, 1987.

Karnes, Thomas L., *The Failure of Union: Central America, 1824–1960*. Chapel Hill: Univ. of North Carolina Press, 1961.

Kedourie, Elie, *Nationalism*. London: Hutchinson, 1960.

Kelly, J.B., *Eastern Arabian Frontiers*. London: Faber & Faber, 1964.

Kendall, Frances and Leon Louw, *After Apartheid: The Solution for South Africa*. San Francisco: ICS Press, 1987.

Kiang, Ying Cheng, *China's Boundaries*. Lincolnwood, IL: Institute of China Studies, 1984.

Kibulya, H.M., and B.W. Langlands, *The Political Geography of the Uganda–Congo Boundary*. Kampala, Uganda: Makerere Univ. College, 1967.

Kieffer, John E., *Realities of World Power*. New York: McKay, 1952.

Kimmerling, B., *Zionism and Territory—The Socio-territorial Dimension of Zionist Politics*. Berkeley, CA: Institute of International Studies, 1983.

Kimmich, Christoph M., *The Free City; Danzig and German Policy, 1919–1934*. New Haven, CT: Yale Univ. Press, 1968.

Kliot, Nurit, and Stanley Waterman (eds.), *Pluralism and Political Geography*. New York: St. Martin's Press, 1983.

Knight, David B., *A Capital for Canada: Conflict and Compromise in the 19th Century*. Research Paper No. 182. Chicago: Univ. of Chicago, Dept. of Geography, 1977.

——— (ed.), *Our Geographic Mosaic*. Ottawa: Carleton Univ. Press, 1985.

Knorr, Klaus, *The World Potential of Nations*. Princeton, NJ: Princeton Univ. Press, 1956.

Kohn, Hans, *The Idea of Nationalism*. New York: Macmillan, 1944.

Kostelski, Z., *The Yugoslavs: The History of the Yugoslavs and Their States to the Creation of Yugoslavia*. New York: Philosophical Library, 1962.

Kratochwil, Friedrich and others, *Peace and Disputed Sovereignty: Reflections on Conflict over Territory*. Lanham, MD: Univ. Press of America, 1985.

Krenz, Frank E., *International Enclaves and Rights of Passage*. Geneva and Paris: Droz & Minard, 1961.

Krosby, H. Peter, *Finland, Germany and the Soviet Union, 1940–1941; The Petsamo Dispute*. Madison: Univ. of Wisconsin Press, 1968.

Kuhn, Delia and Ferdinand Kuhn, *Borderlands*. New York: Knopf, 1962.

L

Lamb, Alistair, *The China–India Border*. London: Chatham House, 1964.

———, *The McMahon Line*. London: Chatham House, 1966.

Lamborn, Alan C. and Stephen P. Mumme, *Statecraft, Domestic Politics, and Foreign Policy Making*. Boulder, CO: Westview, 1987.

Larus, J., *Comparative World Politics*. Belmont, CA: Wadsworth, 1964.

Laswell, Harold D., and Abraham Kaplan, *Power and Society: A Framework for Political Inquiry*. New Haven, CT: Yale Univ. Press, 1950.

Latouche, Daniel, *Canada and Quebec, Past and Future: An Essay*. Toronto: Univ. of Toronto Press, 1987.

Lattimore, Owen, *Inner Frontiers of Asia*. American Geographical Society, Research Series, No. 21. New York, 1940.

———, *Inner Asian Frontiers of China*. Boston: Beacon Press, 1962.

———, *Studies in Frontier History: Collected Papers 1928–1958*. Paris: Mouton, 1962.

Leith, C.K., *Pivot of Asia: Sinkiang and the Inner Asian Frontier of China and Russia*. Boston: Little, Brown, 1950.

Lesch, Ann Mosley, *Arab Politics in Palestine, 1917–1939: The Frustration of a National Movement*. Ithaca, NY: Cornell Univ. Press, 1979.

Leslie, Peter, *Federal State, National Economy*. Toronto: Univ. of Toronto Press, 1987.

Lessing, O.E. (ed.), *Minorities and Boundaries*. New York: Van Riemsdyck, 1931.

Levie, Howard S., *The Status of Gibraltar*. Boulder, CO: Westview, 1983.

Leys, Colin, and Peter Robson, *Federation in East Africa: Opportunities and Problems*. Nairobi, Kenya: Oxford Univ. Press, 1965.

Li, Victor H. (ed.), *The Future of Taiwan: A Difference of Opinion*. Armonk, NY: Sharpe, 1980.

Lowenthal, David (ed.), *The West Indies Federation: A Perspective on a New Nation*. New York: Columbia Univ. Press, 1961.

Luttwak, Edward N., *Strategic Power: Military Capabilities and Political Unity*. Beverly Hills, CA: Sage, 1976.

M

Maas, Arthur (ed.), *Area and Power; A Theory of Local Government*. Glencoe, IL: Free Press, 1959.

MacMahon, A. (ed.), *Federalism, Mature and Emergent*. Garden City, NY: Doubleday, 1955.

Magocsi, Paul Robert, *Ukraine: A Historical Atlas*. Toronto: Univ. of Toronto Press, 1985.

Mallison, W. Thomas and Sally V. Mallison, *The Palestine Problem in International Law and World Order*. White Plains, NY: Longman, 1986.

McColl, Robert W., *The University as a Mini-State. . . .* Occasional Paper No. 5. Lawrence: Univ. of Kansas, Dept. of Geography, December 1970.

McWhinney, Edward, *Comparative Federalism, States' Rights and National Power*. Toronto: Univ. of Toronto Press, 1962.

Miller, J.M., *Lake Europa: A New Capital for the United Europe*. New York: Books International, 1963.

Milton, J. *Ethnic Conflict in the Western World*. Ithaca, NY: Cornell Univ. Press, 1977.

Missakian, J., *A Searchlight on the Armenian Question (1878–1950)*. Boston: Hairenik, 1950.

Moodie, A.E., *The Italo–Yugoslav Boundary; A Study in Political Geography*. London: George Philip, 1945.

Morley, Sylvanus Griswold, *The Ancient Maya*. Stanford, CA: Stanford Univ. Press, 1946.

Morton, W.L., *The Canadian Identity*. Madison: Univ. of Wisconsin Press, 1961.

Moseley, George V.H., III, *The Consolidation of the South China Frontier*. Berkeley: Univ. of California Press, 1973.

Murdoch, Richard K., *The Georgia–Florida Frontier, 1793–1796; Spanish Reaction to French Intrigue and American Design*. Berkeley: Univ. of California Press, 1951.

N

Nakleh, Emile A., *The West Bank and Gaza: Toward the Making of a Palestinian State*. Washington: American Enterprise Institute for Public Policy Research, 1979.

———, *A Palestinian Agenda for the West Bank and Gaza*. Lanham, MD: Univ. Press of America, 1980.

Newbigin, Marion I., *Geographical Aspects of the Balkan Problems in Their Relation to the Great European War*. New York: Putnam, 1915.

Niebuhr, R., *The Structure of Nations and Empires*. New York: Scribner's, 1959.

Nijim, Basheer K. (ed.), *Toward the De-Arabization of Palestine/Israel 1945–1977*. Dubuque, IA: Kendall/Hunt, 1984.

Nisan, M., *Israel and the Territories*. Ramat Gan, Israel: Turtledove, 1978.

Norrie, K., and others, *Federalism and Economic Union in Canada*. Toronto: Univ. of Toronto Press, 1987.

Novak, Bogdan C., *Trieste 1941–1954; The Ethnic, Political, and Ideological Struggle*. Chicago: Univ. of Chicago Press, 1970.

Nuseibeh, H.Z., *The Ideas of Arab Nationalism*. Ithaca, NY: Cornell Univ. Press, 1956.

O

Oduho, J. and William Deng, *The Problem of the Southern Sudan*. London: Oxford Univ. Press, 1963.

O'Halloran, Clare, *Partition and the Limits of Irish Nationalism*. Atlantic Highlands, NJ: Humanities Press International, 1987.

Ojeda, Mario, *Mexico: The Northern Border as a National Concern*. El Paso: Univ. of Texas, 1983.

Orico, Osvaldo, *Brazil's Capital: Brasilia*. New York: Brazilian Government Trade Bureau, 1958.

P

Pankhurst, E. Sylvia and Richard K. Pankhurst, *Ethiopia and Eritrea; The Last Phase of the Reunion Struggle, 1941–1952*. Woodford Green, Essex, Eng.: Lalibela House, 1953.

Parker, W.H., *The Superpowers: The United States and the Soviet Union Compared*. London: Macmillan, 1972.

Penlington, Norman, *The Alaska Boundary Dispute: A Critical Re-Appraisal*. Toronto: McGraw-Hill, 1972.

Perl, Raphael, *The Falkland Islands Dispute in International Law and Politics*. London, Rome and New York: Oceana, 1983.

Plischke, Elmer, *A Geographical Study of the Dutch–German Border*. Münster in Westfalen: Selbstverlag der Geographischen Kommission, 1958.

———, *Microstates in World Affairs: Policy Problems and Options*. Washington: American Enterprise Institute for Public Policy Research., 1977.

Pounds, Norman J.G., *Divided Germany and Berlin*. Princeton, NJ: Van Nostrand, 1962.

Prescott, John Robert Victor, *Boundaries and*

Frontiers. London: Croom Helm; Totowa, NJ: Rowman & Littlefield, 1978.

———, *Political Frontiers and Boundaries.* Winchester, MA: Allen & Unwin, 1987.

Prescott, John Robert Victor and others, *Frontiers of Asia and Southeast Asia.* Carlton, Victoria: Melbourne Univ. Press, 1977.

Problem Regions of Europe. Series of monographs, D.I. Seagill (ed.). London: Oxford Univ. Press.

Q

Quandt, William B. and others, *The Politics of Palestinian Nationalism.* Berkeley: Univ. of California Press, 1973.

R

Reese, David, *The Soviet Seizure of the Kuriles.* Praeger, 1985.

Reeves, T. Zane, *The U.S.–Mexico Border Commissions: An Overview and Agenda for Further Research.* El Paso: Univ. of Texas, 1984.

Richardson, J.E., *Patterns of Australian Federalism.* Canberra: Australian National Univ. Press, 1973.

Robinson, E.A.G. (ed.), *Economic Consequences of the Size of Nations.* London: Macmillan, 1960.

Robson, C.B., *Berlin: Pivot of German Destiny.* Chapel Hill: Univ. of North Carolina Press, 1960.

Roys, Ralph B., *The Political Geography of the Yucatan Maya.* Washington: Carnegie Institution of Washington, Publication No. 613, 1957.

Rubin, Neville, *Cameroun: An African Federation.* Amsterdam, Neth.: Rudolf Muller, 1971.

Russell, Frank M., *The Saar, Battleground and Pawn.* Stanford, CA: Stanford Univ. Press, 1951.

S

Sarkar, B.K., *The Politics of Boundaries and Tendencies in International Relations.* Calcutta: N.M. Ray Chowdhury, 1938.

Sathyamurthy, T.V., *Nationalism in the Contemporary World: Political and Sociological Perspectives.* London: Frances Pinter, 1982.

Sawer, Geoffrey, *Federation: An Australian Jubilee Study.* Melbourne, 1952.

Schiller, A. Arthur, *The Formation of Federal Indonesia, 1945–1949.* The Hague: W. Van Hoeve, 1955.

Schwartz, Mildred A., *Politics and Territory: The Sociology of Regional Persistence in Canada.* Montreal and London: McGill–Queen's Univ. Press, 1974.

Shafer, Boyd C., *Faces of Nationalism: New Realities and Old Myths.* New York: Harcourt Brace Jovanovich, 1972.

Shambi, Hisham B., *Nationalism and Revolution in the Arab World.* Princeton, NJ: Van Nostrand, 1966.

Shamgar, Meir (ed.), *Military Government in the Territories Administered by Israel 1967–1980: The Legal Aspects,* Vol. 1, Jerusalem: Harry Sacher Institute for Legislative Research and Comparative Law, Hebrew University of Jerusalem, 1982.

Shevtsov, V.S., *National Sovereignty and the Soviet State.* Moscow: Progress Publishers, 1974.

Singh, Elen C., *The Spitsbergen (Svalbard) Question: United States Foreign Policy, 1907–1935.* Oslo: Univ. of Oslo Press, 1980.

Smith, Anthony D., *State and Nation in the Third World.* New York: St. Martin's Press, 1983.

———, *The Ethnic Origins of Nations.* New York: Basil Blackwell, 1987.

Smith, B.C., *Decentralization; The Territorial Dimension of the State.* Boston: Allen & Unwin, 1985.

Smith, David M., *Apartheid in South Africa.* New York: Cambridge Univ. Press, 1985.

Solch, J., *Natural Boundaries.* Innsbruck, Aus.: Innsbruck Univ. Press, 1924.

Southall, Roger, *South Africa's Transkei; The Political Economy of an "Independent" Bantustan.* New York; Monthly Review Press, 1983.

Spencer, Robert and others (eds.), *The International Joint Commission Seventy Years On.* Toronto: Univ. of Toronto, Centre for International Studies.

Sprout, Harold H. and Margaret Sprout (eds.), *Foundations of National Power.* Princeton, NJ: Princeton Univ. Press, 1945.

Stacey, C.P., *The Undefended Border: The Myth and the Reality.* Ottawa: Canadian Historical Association, 1953.

Stoessinger, John G., *The Might of Nations.* New York: Random House, 1963.

Stokes, William S., *Honduras; An Area Study in Government.* Madison: Univ. of Wisconsin Press, 1950.

Stokke, Olav, *Integration and Disintegration: The Case of the Nigerian Federation up to June 1967.* Uppsala, Swed.: Scandinavian Institute of African Studies, 1970.

Strassoldo, Raimondo, *Boundaries and Regions: Explorations in the Growth and Peace Potential of the Peripheries.* Trieste, Italy: Lint, 1973.

Stultz, Newell M., *Transkei's Half Loaf; Race Separatism in South Africa.* New Haven, CT: Yale Univ. Press, 1979.

Symmons-Symonolewicz, Konstantin, *Modern Nationalism: Towards a Consensus in Theory.* New York: Polish Institute of Arts and Sciences in America, 1968.

Symonds, Richard, *The Making of Pakistan.* London: Faber & Faber, 1950.

Szaz, Zoltan Michael, *Germany's Eastern Frontiers: The Problem of the Oder–Neisse Line.* Chicago: Regnery, 1960.

T

Thom, Derrick J., *The Niger–Nigeria Boundary 1890–1906: A Study of Ethnic Frontiers and a Colonial Boundary.* Papers in International Studies, Africa Series No. 23. Athens: Ohio Univ. Center for International Studies, 1975.

Thompson, J.E.S., *The Rise and Fall of Maya Civilization.* Norman: Univ. of Oklahoma Press, 1954.

Tibi, Bassam, *Arab Nationalism.* New York: St. Martin's Press, 1981.

Tilly, C. (ed.), *The Formation of Nation States in Western Europe.* Princeton, NJ: Princeton Univ. Press, 1975.

Timm, C.A., *The International Boundary Commission; United States and Mexico.* Austin: Univ. of Texas Press, 1941.

Tivey, L. (ed.), *The Nation-State.* Oxford: Martin Robertson, 1981.

Touval, Saadia, *Somali Nationalism.* Cambridge: Harvard Univ. Press, 1963.

———, *The Boundary Politics of Independent Africa.* Cambridge: Harvard Univ. Press, 1972.

Turner, Frederick Jackson, *The Frontier in American History.* New York: Holt, 1921.

U

U.S. Department of State. Office of The Geographer. *International Boundary Studies.* 1961ff.

V

Vaillant, G.C., *The Aztecs of Mexico.* New York: Doubleday, 1950.

van Walt van Pragg, Michael C., *The Status of Tibet.* Boulder, CO: Westview, 1987.

W

Whalley, John, *Regional Aspects of Confederation.* Toronto: Univ. of Toronto Press, 1987.

Wheare, Kenneth C., *Federal Government.* London and New York: Oxford Univ. Press, 1947.

Wheatcroft, A.J.M., *The World Atlas of Revolutions.* New York: Simon & Schuster, 1983.

White, H.L. (ed.), *Canberra, a Nation's Capital.* Sydney: Angus & Robertson, 1954.

Widstrand, Carl Gosta (ed.), *African Boundary Problems.* Uppsala, Swed.: Scandinavian Institute of African Studies, 1969.

Wilson, C.M. and L.M. Hanks, *The Burma–Thailand Frontier over Sixteen Decades.* Athens: Ohio Univ. Press, 1984.

Wilson, Joe F., *The United States, Chile and Peru in the Tacna and Arica Plebiscite.* Lanham, MD: Univ. Press of America, 1979.

Wiskemann, Elizabeth, *Germany's Eastern Neighbours: Problems Relating to the Oder–Neisse Line and the Czech Frontier Regions.* London: Oxford Univ. Press, 1956.

Wittfogel, Karl A., *Oriental Despotism; A Comparative Study of Total Power.* New Haven, CT: Yale Univ. Press, 1957.

Wood, Bryce, *Aggression and History; The Case of Ecuador and Peru.* Ann Arbor, MI: Univ. Microfilms, 1978.

Periodicals

A

Ahmad, Nafis, "The Indo–Pakistan Boundary Disputes Tribunal, 1949–1950," *Geographical Review,* 43, 3 (July 1953), 329–337.

———, "The Evolution of the Boundaries of East Pakistan," *Oriental Geography,* 2, 2 (July 1958), 97–106.

———, "China's Himalayan Frontiers: Pakistan's Attitude," *International Affairs,* 38, 4 (1962), 478.

Ake, Claude, "Political Integration and Political Stability: A Hypothesis," *World Politics,* 19, 3 (1967), April 486–499.

Alcock, N.Z., and A.G. Newcombe, "The Perception of National Power," *Journal of Conflict Resolution*, 14, 3 (September 1970), 335–344.

Alexander, Lewis M., "Recent Changes in the Benelux–German Boundary," *Geographical Review*, 43, 1 (January 1953), 69–76.

Alexander, Lewis M., "The Arab–Israeli Boundary Problem," *World Politics*, 6, 3 (April 1954), 322–337.

Allen, W.E.D., "New Political Boundaries in the Caucasus," *Geographical Journal*, 69, 5 (May 1927), 430–440.

Anderson, Albin T., "The Soviets and Northern Europe," *World Politics*, 4, 4 (July 1952), 468–487.

Anthony, John Duke, "The United Arab Emirates," *Middle East Journal*, 26, 3 (1972), 271–287.

Apthorpe, Raymond, "The Introduction of Bureaucracy into African Politics," *Journal of African Administration*, 12, 3 (July 1960), 125–134.

Archibald, Charles H., "The Failure of the West Indies Federation," *World Today*, 18, 6 (June 1962), 233–242.

Arciszewski, Franciszek, "Some Remarks About the Strategical Significance of the New and Old Soviet–Polish Border," *Polish Review*, 1, 2–3 (1956), 89–96.

Armstrong, Hamilton Fish, "Where India Faces China," *Foreign Affairs*, 37, 4 (July 1959), 617–625.

Arora, R.S., "The Sino–Indian Border Dispute: A Legal Approach," *Public Law*, (1963), 172.

Atwood, Wallace W., "The Increasing Significance of Geographic Conditions in the Growth of Nation–States," *Annals, AAG*, 25, 1 (March 1935), 1–16.

Augelli, John P., "Brasilia: The Emergence of a National Capital," *Journal of Geography*, 62, 6 (September 1963), 241–252.

———, "The Nationalization of Frontiers: The Dominican Borderlands Under Trujillo," (abstract) *Annals, AAG*, 57, 1 (March 1967), 166.

Austin, D.G., "The Uncertain Frontier: Ghana–Togo," *Journal of Modern African Studies*, 1 (1963), 139–146.

B

Barlow, I. Max, "Political Geography and Canada's National Unity Problem," *Journal of Geography*, 79, 7 (December 1980), 259–263.

Barton, William, "Pakistan's Claim to Kashmir," *Foreign Affairs*, 28, 2 (January 1950), 299–308.

Beloff, Max, "The Federal Solution in Its Application to Europe, Asia and Africa," *Political Studies*, 1, (1953), 118.

Bennett, George, "The Eastern Boundary of Uganda in 1902," *Uganda Journal*, 23, 1 (March 1959), 69–72.

Berry, Brian J.L., "By What Categories May a State Be Characterized?" *Economic Development and Cultural Change*, 15 (1966), 93.

Best, Alan C.G., "Gaberone: Problems and Prospects of a New Capital," *Geographical Review*, 60, 1 (January 1970), 1–14.

Bird, J., "The Foundation of Australian Seaport Capitals," *Economic Geography*, 43, 3 (October 1965), 283–289.

Bishop, W.W., Jr., "Case Concerning Sovereignty over Certain Frontier Lands, 1959," *American Journal of International Law*, 53 (1959), 937.

Blood, Hilary, "Federation in the Caribbean," *Corona*, 8, 5 (May 1956), 166–169.

———, "The West Indian Federation," *Journal of the Royal Society of Arts*, 105, 5009 (August 2, 1957), 746–757.

Boateng, E.A., "The Growth and Functions of Accra," *Bulletin, Ghana Geographical Association*, 4, 1 (January 1959), 9–13.

Bose, N.K., "Bengal Partition and After," *Calcutta Geographical Review*, 9, 1–4 (1949), 14–22.

Bouchez, Leo J., "The Fixing of Boundaries in International Boundary Rivers," *International and Comparative Law Quarterly*, 12, 31 (July 1963), 789–817.

Brigham, Albert Perry, "Principles in the Determination of Boundaries," *Geographical Review*, 7 (1919), 201–219.

Broek, Jan O.M. and M.A. Junis, "Geography and Nationalism," *Geographical Review*, 35, (1945), 301–311.

Brown, David J.L., "The Ethiopia–Somaliland Frontier Dispute," *International and Comparative Law Quarterly*, 5 (1956), 245–264.

———, "Recent Developments in the Ethiopia–Somaliland Frontier Dispute," *International and Comparative Law Quarterly*, 10, 1 (January 1961), 167–178.

Brown, P.M., "Costa Rica–Nicaragua," *American Journal of International Law*, 11 (1917), 156–160.

Brown, R.N. Rudmose, "Spitsbergen Terra Nullius," *Geographical Review*, 7 (1919), 311–321.

Burghardt, Andrew F., "The Bases of Territorial Claims," *Geographical Review*, 63, 2 (April 1973), 225–245.

———, "Nation, State and Territorial Unity: A Trans-Outaouais View," *Cahiers de géographie du Québec*, 24 (1980), 123–134.

Burns, Alan, "Towards a Caribbean Federation," *Foreign Affairs*, 34, 1 (October 1955), 128–140.

Butland, Gilbert J., "Frontiers of Settlement in South America," *Revista Geográfica*, 65 (1966), 93–108.

Byrnes, Andrew and Hilary Charlesworth, "Federalism and the International Legal Order: Recent Developments in Australia," *American Journal of International Law*, 79, 3 (July 1985), 622–640.

C

Capital Cities. Special Issue of *Ekistics*, 50, 299 (March–April 1983).

Catudal, Honoré M., "Steinstucken: The Politics of a Berlin Enclave," *World Affairs*, 134, 1 (Summer 1971), 51–62.

Chang, S., "Peking: The Growing Metropolis of Communist China," *Geographical Review*, 55, 3 (July 1965), 313–327.

Chao, K.T., "Legal Nature of International Boundaries," *Chinese Yearbook of International Law and Affairs*, 5 (1985), 29–89.

Child, Clifton J., "The Venezuela–British Guiana Boundary Arbitration of 1899," *American Journal of International Law*, 44, 4 (October 1950), 682–693.

Church, R.J. Harrison, "New Franco–Italian Frontier," *Geographical Journal*, 111 (1948), 143, 293–294.

Clarke, Colin, "Political Fragmentation in the Caribbean: The Case of Anguilla," *Canadian Geographer*, 15 (1971), 13–29.

Classen, H. George, "Keepers of the Boundary," *Canadian Geographical Journal*, 65, 4 (1962), 122–129.

Clegern, W.N., "New Light on the Belize Dispute," *American Journal of International Law*, 52 (1958), 280–297.

Clifford, E.H.M., "The British Somaliland–Ethiopia Boundary," *Geographical Journal*, 87 (1936), 289–307.

———, "Boundary Commissions," *Royal Engineers Journal*, 51 (1937), 369.

Close, C.F., "The Western Frontier of the Sudan," *Geographical Journal*, 66 (1925), 349.

Clough, Ralph N., "Taiwan's International Status," *Chinese Yearbook of International Law and Affairs*, 1 (1981), 17–34.

Cohen, Saul B., "Jerusalem: A Geopolitical Imperative," *Midstream*, (May 1975), 1–6.

Cole, K.C., "Theory of the State as a Sovereign Juristic Person," *American Political Science Review*, 42 (1948), 16–31.

Connell, John, "The India–China Frontier Dispute," *Royal Canadian Asian Journal*, 47, 3–4 (1960), 270–285.

Cornwall, J.H.M., "The Russo–Turkish Boundary and Territory of Nakhchivan," *Geographical Journal*, 61 (1923), 445.

Cowan, L. Gray, "Federation for Nigeria," *International Journal*, 10, 1 (1954), 51–60.

Cox, R.E., "Notes on the Anglo–Liberian Frontier," *Geographical Journal*, 24 (1904), 427.

Craig, John Keith, "Vienna: A Geographical Analysis," *Tydskrif vir Aardrykskunde—Journal of Geography*, 1, 6 (April 1960), 10–21.

Cree, D., "Yugoslav–Hungarian Boundary Commission," *Geographical Journal*, 25 (1925), 89–112.

Crone, G.R., "The Turkish–Iranian Boundary," *Geographical Journal*, 91 (1938), 57–59; 92 (1938), 149–150.

Cumpstom, J.H.L., "The Story of the Boundaries," *Walkabout*, 17, 4 (April 1951), 18–20.

D

Dale, Edmund H., "The West Indies: A Federation in Search of a Capital," *Canadian Geographer*, 5, 2 (Summer 1961), 44–52.

———, "The State-Idea: Missing Prop of the West Indies Federation," *Scottish Geographical Magazine*, 78, 3 (December 1962), 166–176.

Darby, H.C., "The Medieval Sea-State," *Scottish Geographical Magazine*, 47 (1932), 136–149.

Dear, Michael and G. Clark, "The State and Geographic Process: A Critical Review," *Environmental Planning A*, 10 (1978), 173–183.

Dennis, William C., "The Venezuela–British Guiana Boundary Arbitration," *American Journal of International Law*, 44 (1950), 720–727.

Deutsch, Herman J., "The Evolution of the International Boundary in the Inland Empire of the Pacific Northwest," *Pacific Northwest Quarterly*, 51, 2 (1960), 63–79.

———, "A Contemporary Report on the 49° Boundary Survey, Pacific Northwest," *Pacific Northwest Quarterly*, 53, 1 (January 1962), 17–33.

Deutsch, Karl W., "Germany's Frontiers," *World Affairs*, n.s. 1, 3 (July 1947), 262–277.

Dickinson, Robert E., "Germany's Frontiers," *World Affairs*, 1, 3 (1947), 262–277.

Dikshit, Ramesh D., "Geography and Federalism," *Annals, AAG*, 61, 1 (March 1971), 97–115.

————, "The Failure of Federalism in Central Africa; a Politico-Geographical Postmortem," *Professional Geographer*, 23, 3 (July 1971), 224–228.

Dillman, C. Daniel, "Recent Developments in Mexico's National Border Program," *Professional Geographer*, 22, 5 (September 1970), 243–247.

Dodge, Stanley D., "The Finnish–Russian Boundary North of 68°," *Geographical Journal*, 72, 3 (September 1928), 297–298.

Dore, Isaak I., "Recognition of Rhodesia and Traditional International Law: Some Conceptual Problems," *Vanderbilt Journal of Transnational Law*, 13, 1 (Winter 1980), 25–41.

Dugdale, John, "Can Federation Work in Rhodesia?" *New Commonwealth*, 33, 2 (1957), 61–62.

E

East, W. Gordon, "The New Frontiers of the Soviet Union," *Foreign Affairs*, 29, 4 (July 1951), 591–607.

Elliott, G.S., "The Anglo–French Niger–Chad Boundary Commission," *Geographical Journal*, 24 (1904), 505.

Eyre, John D., "Japanese–Soviet Territorial Issues in the Southern Kurile Islands," *Professional Geographer*, 20, 1 (January 1968), 11–15.

F

Farran, C. d'Olivier, "International Enclaves and the Question of State Servitude," *International and Comparative Law Quarterly*, 4, 2 (April 1955), 294–307.

Fawcett, Charles Bungay, "Some Geographical Factors in the Growth of the State," *Scottish Geographical Magazine* (1922), 221–232.

Feer, Mark C., "India's Himalayan Frontier," *Far Eastern Survey*, 22, 11 (October 1953), 137–141.

Fenwick, C.G., "The Honduras–Nicaragua Boundary Dispute," *American Journal of International Law*, 51, 4 (October 1957) 761–765.

Fifield, Russell, "The Postwar World Map: New States and Boundary Changes," *American Political Science Review*, 42, 3 (June 1948), 533–541.

Finkelstein, Lawrence S., "The Indonesian Federal Problem," *Pacific Affairs*, 24, 3 (September 1951), 284–295.

Fischer, Eric, "On Boundaries," *World Politics*, 1, 2 (January 1949), 196–222.

Fisher, F.C., "Arbitration of the Guatemalan–Honduran Boundary Dispute," *American Journal of International Law*, 27, (1933), 403–427.

Fisher, Margaret W. and Leo E. Rose, "Ladakh and the Sino–Indian Border Crisis," *Asian Survey*, 2, 8 (October 1962), 27–37.

Floyd, Barry N., "Pre-European Political Patterns in Sub-Saharan Africa," *Bulletin, Ghana Geographical Association*, 8, 2 (1963), 3–11.

Flugal, Raymond R., "The Palestine Problem: A Brief Geographical, Historical, and Political Evaluation," *Social Studies*, 48, 2 (February 1957), 43–51.

Foulkes, C.H., "The Anglo–French Boundary Commission, Niger to Chad," *Royal Engineers Journal*, 73, 4 (December 1959), 429–437.

Franck, Dorothea Seelye, "Pakhtunistan Dispute—Disposition of a Tribal Land," *Middle East Journal*, 6, 1 (1952), 49–68.

Freeman, T.W. and Mary M. MacDonald, "The Arctic Corridor of Finland," *Scottish Geographical Magazine*, 54, 4 (1938), 219–230.

Freshfield, Douglas W., "Ruwenzori and the Frontier of Uganda," *Geographical Journal*, 28, 5 (November 1906), 481–486.

————, "The Southern Frontiers of Austria," *Geographical Journal*, 46, 4 (December 1915), 413–435.

G

Gangal, S.C., "An Approach to Indian Federalism," *Political Science Quarterly*, 77, 2 (June 1962), 248–253.

Garner, J.W., "The Doctrine of the Thalweg as a Rule of International Law," *American Journal of International Law*, 29 (1935), 309–310.

Geographic Notes. Washington: U.S. Dept. of State. Issued irregularly by The Geographer.

Geographical Digest. London: George Philip. Published annually from 1963 to 1984, *The New Geographical Digest*. London: George Philip. Published irregularly beginning 1986.

German, F. Clifford, "A Tentative Evaluation of World Power," *Journal of Conflict Resolution*, 4, 1 (March 1960), 138–144.

Gibson, James R., "Russia on the Pacific: The Role of the Amur," *Canadian Geographer*, 12, 1 (1968), 15–27.

Gilbert, E.W., "Practical Regionalism in England and Wales," *Geographical Journal*, 94, 1 (July 1939), 29–44.

GilFilland, S. Columb, "European Political Boundaries," *Political Science Quarterly*, 39, 3 (September 1924), 458–484.

Giles, F.L., "Boundary Work in the Balkans, *Geographical Journal*, 75 (1930), 300–312.

Gillin, John and K.H. Silvert, "Ambiguities in Guatemala," *Foreign Affairs*, 34, 3 (April 1956), 469–482.

Glassner, Martin Ira, "The Bedouin of Southern Sinai Under Israel Administration," *Geographical Review*, 64, 1 (January 1974), 31–60.

Goetzmann, W.H., "The United States–Mexican Boundary Survey, 1848–1853," *Southwestern Historical Quarterly*, 62, 2 (October 1958), 164–190.

Greenidge, C.W.W., "The British Caribbean Federation," *World Affairs*, 4, 5 (July 1950), 321–334.

Grey, Arthur L., "The Thirty-Eighth Parallel," *Foreign Affairs*, 29, 3 (April 1951), 482–487.

Grossman, David, "The Process of Frontier Settlement: The Case of Nikeland, Nigeria," *Geografiska Annaler*, 53B (1971), 107–128.

Grossman, David and Zeev Safrai, "Satellite Settlements in Western Samaria," *Geographical Review*, 70, 4 (October 1980), 446–461.

———, "The Expansion of the Settlement Frontier of Hebron's Western and Southern Fringes," *Geographical Research Forum*, 5 (1982), 57–73.

H

Haas, M., "Paradigms of Political Integration and Unification," *Journal of Peace Research*, 21, (1984), 47–60.

Hall, Arthur R., "Boundary Problems in Cartography," *Surveying and Mapping*, 12, 2 (April–June 1952), 138–141.

Hall, H.D., "Zones of the International Frontier," *Geographical Review*, 38, 4 (October 1948), 615–625.

———, "The International Frontier," *American Journal of International Law*, 42 (1948), 42–65.

Hamdan, G., "The Political Map of the New Africa," *Geographical Review*, 53, 3 (July 1963), 418–439.

———, "Capitals of the New Africa," *Economic Geography*, 40 (July 1964), 239–253.

Hartshorne, Richard, "Geographic and Political Boundaries in Upper Silesia," *Annals, AAG*, 23, 4 (December 1933) 195–228.

———, "Suggestions on the Terminology of Political Boundaries," *Annals, AAG*, 26, 1 (March 1936), 56–57.

———, "The Concept of *Raison d'Être* and Maturity of States," *Annals, AAG*, 30, 1 (1940), 59–60.

———, "The Franco–German Boundary of 1871," *World Politics*, 2, 2 (January 1950), 209–250.

———, "The Functional Approach in Political Geography," *Annals, AAG*, 40, 2 (June 1950), 95–130.

Harvey, David, "The Marxian Theory of the State," *Antipode*, 8 (1976), 80–89.

Hasson, S. and N. Gosenfeld, "Israeli Frontier Settlements, a Cross-temporal Analysis," *Geoforum*, 11 (1980), 315–322.

Haupert, John, "Some Aspects of Boundaries and Frontiers of Israel," *California Council of Geography Teachers*, 4, 3 (June 1957), 206.

———, "Political Geography of the Israeli–Syrian Boundary Dispute, 1949–1967," *Professional Geographer*, 21, 3 (May 1969), 163–171.

———, "Jerusalem: Aspects of Reunification and Integration," *Professional Geographer*, 23, 4 (October 1971), 312–319.

Hay, W.R., "Demarcation of the Indo–Afghan Boundary in the Vicinity of Arandu," *Geographical Journal*, 82, 4 (October 1933), 351–354.

Hay, Sir Rupert, "The Persian Gulf States and Their Boundary Problems," *Geographical Journal*, 120, 4 (December 1954), 433–443.

Held, Colbert C., "The New Saarland," *Geographical Review*, 41, 4 (October 1951), 590–605.

Helin, Ronald A., "Finland Regains an Outlet to the Sea: the Saimaa Canal," *Geographical Review*, 58, 2 (April 1968), 167–194.

Hensinkweld, Harriet M., "Separatist Tendencies in the Yucatan Peninsula," *Professional Geographer*, 19, 5 (September 1967), 258–260.

Herman, Theodore, "Group Values Toward the National Space: The Case of China," *Geographical Review*, 49, 2 (April 1959), 164–182.

Herz, John H., "The Rise and Demise of the Territorial State," *World Politics*, 9, 4 (July 1957), 473–493.

———, "The Territorial State Revisited; Reflections on the Future of the Nation-State," *Polity*, 1, 1 (Fall 1968), 11–34.

Hill, James E., Jr., "El Chamizal: A Century-Old Boundary Dispute," *Geographical Review*, 55, 4 (October 1965), 510–522.

Hinks, Arthur R., "Boundary Delimitations in the Treaty of Versailles," *Geographical Journal*, 54, 2 (August 1919), 103–112.

———, "The Boundaries of Czecho-Slovakia," *Geographical Journal*, 54, 3 (September 1919), 185–187.

———, "The Progress of Boundary Delimitation in Europe," *Geographical Journal*, 54, 6 (December 1919), 363–366.

———, "The New Boundaries of Bulgaria," *Geographical Journal*, 55, 2 (February 1920), 127–138.

———, "The Slesvig Plebiscite and the Danish–German Boundary," *Geographical Journal*, 56, 6 (December 1920), 484–491.

———, "The Belgian–German Boundary Demarcation," *Geographical Journal*, 57, 1 (January 1921), 43–50.

———, "Notes on the Technique of Boundary Delimitation," *Geographical Journal*, 58, 6 (December 1921), 417–443.

Hodgkiss, A.G. and Robert W. Steel, "The Shatter-belt in Relation to the East–West Conflict," *Journal of Geography*, 51, 7 (October 1952), 265–275.

Hoffman, George W., "Boundary Problems in Europe," *Annals, AAG*, 44, 1 (March 1954), 102–106.

———, "The Changing Face of Africa," *Geography*, 46, Pt. 2, No. 211 (April 1961), 156–160.

Holford, William, "The Future of Canberra," *Town Planning Review*, 29, 3 (October 1958), 139–162.

Holdich, Thomas H., "The Perso-Baluch Boundary," *Geographical Journal*, 9, 3 (March 1897), 416–422.

———, "Political Boundaries," *Scottish Geographical Magazine*, 32 (1916), 497.

———, "Geographical Problems in Boundary Making," *Geographical Journal*, 47, 6 (June 1916), 421–439.

———, "The Geographical Results of the Peru–Bolivia Boundary Commission," *Geographical Journal*, 48, 2 (February 1916), 95–115.

Horvath, R.J., "The Wandering Capitals of Ethiopia," *Journal of African History*, 10, 2 (1969), 205–219.

House, John W., "The Franco–Italian Boundary in the Alpes Maritimes," *Transactions and Papers, Institute of British Geographers*, Publ. No. 26 (1959), 107–131.

Hudson, Manley O., "Irish Boundary Question," *American Journal of International Law*, 19 (1925), 150.

Humphreys, R.A., "The Anglo–Guatemalan Dispute," *International Affairs*, 24, 3 (July 1948), 387–404.

Huttenback, R.A., "A Historical Note on the Sino–Indian Dispute over the Aksai Chin," *China Quarterly*, 18 (April–June 1964), 201–207.

Hyde, Charles Cheney "Notes on Rivers as Boundaries," *American Journal of International Law*, 6 (1912), 901–909.

———, "Maps as Evidence in International Boundary Disputes," *American Journal of International Law*, 27, 2 (April 1933), 311–316.

———, "Looking Towards the Arbitration of the Dispute over the Chaco Boreal," *American Journal of International Law*, 28 (1934), 718–723.

I

Inlow, E. Burke, "The McMahon Line," *Journal of Geography*, 63, 6 (September 1964), 261–272.

Innis, H.A., "Canadian Frontiers of Settlement: A Review." *Geographical Review*, 25, 1 (January 1935), 92–106.

Innis, H.A. and Jan O. Broek, "Geography and Nationalism: A Discussion," *Geographical Review*, 35, 2 (1945), 301–311.

J

James, Preston E., "Forces for Unity and Disunity in Brazil," *Journal of Geography*, 38 (1939), 260–266.

James, Preston E. and Speridião Faissol, "The Problem of Brazil's Capital City," *Geographical Review*, 46, 3 (July 1956), 301–317.

Jefferson, Mark, "The Problem of the Ecumene: The Case of Canada," *Geografiska Annaler*, 16 (1934), 146–158.

John, I.G., "France, Germany and the Saar," *World Affairs*, 4, 3 (July 1950), 277–293.

Johnson, Douglas Wilson, "The Role of Political Boundaries," *Geographical Review*, 4, (1917), 208–213.

Johnson, R.E., "Partition as a Political Instrument." *Journal of International Affairs*, 27, 2 (1973), 159–174.

Johnston, Ronald J., "Marxist Political Economy; The State and Political Geography," *Progress in Human Geography*, 8 (1984), 473–492.

Jones, Stephen B., "The Forty-Ninth Parallel in the Great Plains: The Historical Geography of a Boundary," *Journal of Geography*, 31, 9 (December 1932), 357–368.

———, "The Cordilleran Section of the Canada–United States Borderland," *Geographical Journal*, 89, 5 (May 1937), 439–450.

———, "The Description of International Boundaries," *Annals, AAG*, 33 (1943), 99–117.

———, "A Unified Field Theory of Political Geography," *Annals, AAG*, 44, 2 (June 1954), 111–123.

———, "The Power Inventory and National

Strategy," *World Politics,* 6, 4 (July 1954), 421–452.

———, "Boundary Concepts in the Setting of Place and Time," *Annals, AAG,* 49, 3, Pt. I (June 1959), 241–255.

Jost, Isabelle, "Territorial Evolution of Canada," *Canadian Geographical Journal,* 75 (1967), 134–141.

K

Kapil, R.L., "On the Conflict Potential of Inherited Boundaries in Africa," *World Politics,* 18, 4 (July 1966), 659–673.

———, "Political Boundaries and Territorial Instability," *International Review of History and Political Science,* 5 (1968), 46–78.

Karan, Pradyumna P., "The Fringes and Frontiers of India and Pakistan," *Geography,* 4, 1 (May 1951), 6–14.

———, "Indo-Pakistan Boundaries: Their Fixation, Functions and Problems," *Indian Geographical Journal,* 28, 1–2 (January–June 1953), 19–23.

———, "The India–China Boundary Dispute." *Journal of Geography,* 59 (1960), 16–21.

———, "Dividing the Waters: A Problem in Political Geography," *Professional Geographer,* 13, 1 (January 1961), 6–10.

———, "The Sino–Soviet Boundary Dispute," *Journal of Geography,* (1964), 216–222.

———, "The India–Pakistan Enclave Problem," *Professional Geographer,* 18, 1 (January 1966), 23–25.

———, "The Indochinese Boundary Dispute," *Journal of Geography.* 59, 1 (January 1969), 16–21.

Kearns, Kevin C., "Belmopan: Perspective on a New Capital," *Geographical Review,* 63, 2 (April 1973), 147–169.

Kimmerling, B., "Change and Continuity in Zionist Territorial Orientation and Politics," *Comparative Politics.* 14 (1982), 191–210.

King, L.N., "The Work of the Jubaland Boundary Commission," *Geographical Journal,* 72, 5 (November 1928), 420–434.

Kingsbury, Robert C., "The Changing Map of Africa," *Journal of Geography,* 59, 5 (May 1960), 220–224.

Kirchheimer, Otto, "The Decline of Intra-State Federalism in Western Europe," *World Politics,* 3, 3 (April 1951), 281–298.

Kirk, W., "The Sino–Indian Frontier Dispute: A Geographical Review," *Scottish Geographical Magazine,* 76, 1 (April 1960), 3–13.

———, "The Inner Asian Frontier of India," *Transactions and Papers, Institute of British Geographers,* Publ. No. 31 (December 1962), 131–168.

Klieman, Aaron S., "The Politics of Partition," *Journal of International Affairs,* 18 (1964), 161–162.

Knight, David B., "Gaberones: A Viable Proposition," *Professional Geographer,* 17, 6 (November 1965), 38–39.

———, "Impress of Authority and Ideology on Landscape: A Review of Some Unanswered Questions," *Tijdschrift voor Economische en Sociale Geografie,* 63 (1971), 383–387.

———, "Regionalisms and Nationalisms and the Canadian State," *Journal of Geography,* 83 (1984), 212–220.

Knorr, Klaus, "The Concept of Economic Potential for War," *World Politics,* 10, 1 (October 1957), 49–62.

Kozicki, Richard J., "The Sino–Burmese Frontier Problem," *Far Eastern Survey,* 26, 3 (March 1957), 33–38.

Krengel, Rolf, "Soviet, American and West German Basic Industries, a Comparison," *Soviet Studies,* 12, 2 (October 1960), 113–125.

Kristof, Ladis K.D., "The Nature of Frontiers and Boundaries," *Annals, AAG,* 49, 3 Pt. I (June 1959), 269–282.

———, "The State Idea, the National Idea and the Image of the Fatherland," *Orbis,* 11 (1967), 238–255.

Kung, J.L., "Guatemala Versus Great Britain: In re Belice," *American Journal of International Law,* 40 (1946), 383–390.

Kureshy, Kahil Ullah, "Choice of Pakistan Capital: A Politico-Geographical Analysis," *Pakistan Geographical Review,* 5, 1 (1950), 13–25.

———, "The National Frontier of Pakistan," *Pakistan Geographical Review,* 7, 1 (1952), 35–52.

Kusielewicz, Eugene, "New Light on the Curzon Line," *Polish Review,* 1, 2–3 (1956), 82–88.

L

Lamb, Alastair, "The Indo–Tibetan Border," *Australian Journal of Politics and History,* 6, 1 (May 1960), 28–40.

Laski, Harold, "The Obsolescence of Federalism," *New Republic,* 98 (1939), 367–369.

Lattimore, Owen, "The New Political Geography of Inner Asia," *Geographical Journal,* 119, 1 (March 1953), 17–32.

Lauterpacht, E., "River Boundaries: Legal Aspects of the Shatt-Al-Arab Frontier," *Inter-*

national and Comparative Law Quarterly, 9 (1960), 206–236.

Laws, J.B., "A Minor Adjustment in the Boundary Between Tanganyika and Rwanda," *Geographical Review*, 80 (1932), 244–247.

Lewis, I.M., "The Problems of the Northern Frontier District of Kenya," *Race*, 5, 1 (1963), 48–60.

Linge, G.J.R., "Canberra After Fifty Years," *Geographical Review*, 51, 4 (October 1961), 467–486.

Livingstone, William S., "A Note on the Nature of Federalism," *Political Science Quarterly*, 67, 1 (March 1952), 81–95.

Lloyd, H.I., "The Geography of the Mosul Boundary," *Geographical Journal*, 68, 2 (August 1926), 104–117.

Lloyd, Trevor, "The Norwegian–Soviet Boundary; a Study in Political Geography," *Norsk Geografisk Tidsskrift*, 15, 5–6 (1956), 187–242.

Lord, Robert Howard, "The Russo–Polish Boundary Problem," *Procedings, Massachusetts Historical Society, October 1944–May 1947*, 68 (1949), 407–423.

Lowenthal, David, "The West Indies Chooses a Capital," *Geographical Review*, 48, 3 (July 1958), 336–364.

Lydolph, Paul E., "The New Map of the Soviet Union," *Professional Geographer*, 10, 4 (July 1958), 13–17.

M

Macartney, C.A., "The Slovak–Hungarian Frontier," *Geographical Magazine*, 20, 8 (December 1947), 293–295.

MacKintosh, John P., "Federalism in Nigeria," *Political Studies*, 10, 3 (October 1962), 223–247.

MacKirdy, K.A., "Conflict of Loyalties: The Problem of Assimilating the Far West into the Canadian and Australian Federations," *Canadian Historical Review*, 32, 4 (December 1951), 337–355.

———, "Geography and Federalism in Australia and Canada," *Australian Geographer*, 6, 2 (March 1953), 38–47.

Mandel, R., "Roots of the Modern Interstate Border Dispute," *Journal of Conflict Resolution*, 24 (1980), 427–454.

Mariam, Mesfin Wolde, "The Background of the Ethio–Somalian Boundary Dispute," *Journal of Modern African Studies*, (1964), 189–219.

Markham, Clements R., "The Putumayu and the Question of Boundaries between Peru and Colombia," *Geographical Journal*, 41, 2 (February 1913), 145–147.

Maron, S., "The Problem of East Pakistan," *Pacific Affairs*, 28, 2 (1955), 132–144.

Mason, Philip, "Partnership in Central Africa," *International Affairs*, 33, 2 (April 1957), 154–164; 33, 3 (July 1957), 310–318.

Matthews, Robert O., "Interstate Conflicts in Africa: A Review." *International Organization*, 24, 2 (Spring 1970), 335–360.

Mayfield, Robert C., "A Geographical Study of the Kashmir Issue," *Geographical Review*, 45, 2 (April 1955), 181–196.

McCune, Shannon, "Physical Bases for Korean Boundaries," *Far Eastern Quarterly*, (1946), 272–288.

———, "The Thirty-eighth Parallel in Korea," *World Politics*, 1, 2 (January 1949), 223–232.

McDonald, James R., "Europe's Restless Regions," *Focus*, 28, 5 (May–June 1978).

McKay, J. Ross, "The Interactance Hypothesis and Boundaries in Canada: A Preliminary Study," *Canadian Geographer*, 11 (1958), 1–8.

McKee, J.O., "The Rio Grande: The Political Geography of a River Boundary," *Southern Quarterly*, 4, 1 (1965), 29–40.

———, "An Application of the Jones Field Theory to Rhodesia and the Concept of the Historical Time Period," *Virginia Geographer*, 6, 1 (Spring–Summer 1971), 11–14.

McMahon, A. Henry, "The Southern Borderlands of Afghanistan," *Geographical Journal*, 9, 3 (March 1897), 393–415.

———, "International Boundaries," *Journal of the Royal Society of Arts*, 84 (1935), 2–16.

McManis, Donald R., "The Core of Italy: The Case for Lombardy–Piedmont," *Professional Geographer*, 19, 5 (September 1967), 251–257.

Meadows, Martin, "The Philippine Claim to North Borneo," *Political Science Quarterly*, 77, 3 (September 1962), 321–335.

Melamid, Alexander, "The Economic Geography of Neutral Territories," *Geographical Review*, 45, 3 (July 1955), 359–374.

———, "Partitioning Cyprus; a Class Exercise in Applied Political Geography," *Journal of Geography*, 59, 3 (March 1960), 118–122.

———, "Political Boundaries and Nomadic Grazing," *Geographical Review*, 55, 2 (April 1965), 287–290.

Melbourne, W.H., "Exploration and Survey on the Bolivia–Peru Border," *Geographical Journal*, 126, 4 (December 1960), 455–458.

Merrill, Gordon, "The West Indies—the Newest Federation of Commonwealth," *Canadian Geographical Journal*, 56, 2 (February 1958), 60–69.

Mikesell, Marvin W., "The Myth of the Nation State," *Journal of Geography*, 82 (1983), 257–260.

Minghi, Julian Vincent, "Point Roberts, Washington: The Problem of an American Exclave," *Yearbook, Association of Pacific Coast Geographers*, 24 (1962), 29–34.

———, "Boundary Studies and National Prejudice," *Professional Geographer*, 15, 1 (January 1963), 4–8.

———, "Boundary Studies in Political Geography," *Annals, AAG*, 53, 3 (September 1963), 407–428.

———, "The Pacific Coast Section of the Canada–United States Boundary," *Scottish Geographical Magazine*, 70 (1964), 41–42.

Misra, S.D., "The Sino–Indian Dispute: A Geographical Analysis," *Indian Geographical Journal*, 34, 3–4 (1959), 59–64.

Modelski, George, "The Long Cycle of Global Politics and the Nation-State," *Comparative Studies in Society and History*, 20 (1978), 214–235.

Monroe, Elizabeth, "The Arab–Israel Frontier," *International Affairs*, 29, 4 (October 1953), 439–448.

Moodie, A.E., "The Italo–Yugoslav Boundary," *Geographical Journal*, 101, 2 (February 1943), 49–65.

———, "States and Boundaries in the Danubian Lands," *Slavonic and East European Review*, 26, 67 (April 1948), 422–437.

———, "The Cast Iron Curtain," *World Affairs*, 4, 3 (July 1950), 294–305.

———, "Some New Boundary Problems in the Julian March," *Transactions and Papers, Institute of British Geographers*, Publ. No. 16 (1950), 83–93.

Morris, Michael A., "The 1984 Argentine–Chilean Pact of Peace and Friendship," *Oceanus*, 28, 2 (1985), 93–96.

Morse, S.J., "National Identity from a Social Psychological Perspective: Two Brazilian Case Studies," *Canadian Review of Studies in Nationalism*, 4 (1976), 52–76.

Morse, S.J. and others, "National Identity in a 'Multi-nation' State: A Comparison of Afrikaners and English-speaking South Africans," *Canadian Review of Studies in Nationalism*, 4 (1977), 225–246.

———, "National Identity from a Social Psychological Perspective: A Study of University Students in Saskatchewan," *Canadian Review of Studies in Nationalism*, 7 (1980), 299–312.

Morton, W.L., "The Geographical Circumstances of Confederation," *Canadian Geographer*, (1965), 74–87.

Mosely, Philip E., "The Occupation of Germany; New Light on How the Zones Were Drawn," *Foreign Affairs*, 28, 4 (July 1950), 580–604.

Mukerji, A.B., "Kashmir: A Study in Political Geography," *Geographical Review India*, 17, 1 (March 1955), 19–32; 18, 1 (1956), 15–29.

Munger, Edwin S., "Boundaries and African Nationalism," *California Geographer* (1963).

Murphey, Rhoads, "New Capitals of Asia," *Economic Development and Cultural Change*, 5, 3 (April 1957), 216–243.

Myhre, Jeffrey D., "Title to the Falkland–Malvinas Islands Under International Law," *Millennium: Journal of International Studies*, 12, 1 (Spring 1983), 25–38.

N

Nadan, Ram, "Jammu and Kashmir," *Focus*, 13, 1 (September 1962).

National and International Boundaries, Thesaurus Acroasium (Thessaloniki), 14 (1985).

Neumann, Franz L., "Approaches to the Study of Political Power," *Political Science Quarterly*, 65, 2 (June 1950), 161–180.

Newman, David, "Gush Emunim and Settlement in the West Bank," *Bulletin, British Society for Middle Eastern Studies*, 8 (1981), 33–38.

———, "The Evolution of a Political Landscape: Geographical and Territorial Implications of Jewish Colonization in the West Bank," *Middle Eastern Studies*, 21, 2 (April 1985), 192–205.

———, "Functional Change and the Settlement Structure in Israel: A Study of Political Control, Response and Adaptation," *Journal of Rural Studies*, 2, 2 (1986), 127–137.

Nugent, W.V., "Geographical Results of the Nigeria–Kamerun Boundary Demarcation Commission of 1912–13," *Geographical Journal*, 43, 6 (June 1914), 630–651.

O

On the State. Special issue of *International Social Science Journal*, 32, 4 (1980).

Organski, A.F.K., "Population and Politics in Europe: Demographic Factors Help Shape the Relative Power of the Communist and

Non-Communist Blocs," *Science*, 133, 3467 (June 9, 1961), 1803–1807.

Orridge, Andrew W., and Colin H. Williams, "Autonomous Nationalism," *Political Geography Quarterly*, 1, 1 (January 1982), 19–39.

P

Pachai, B. "The Story of Malawi's Capitals; Old and New: 1891–1969," *Society of Malawi Journal*, 24, 1 (January 1971), 35–56.

Parker, R.S., "Australian Federation: The Influence of Economic Interests and Political Pressures," *Historical Studies of Australia and New Zealand*, 4, 13 (1949), 1–24.

Patten, Simon N., "Unnatural Boundaries of European States," *Survey*, 34 (1915), 24–32.

Paullin, C.O., "The Early Choice of the 49th Parallel as a Boundary Line," *Canadian Historical Review*, 4 (1923), 127–131.

Payne, R.H., "Divided Tribes: A Discussion of African Boundary Problems," *New York University Journal of International Law and Politics*, 2, 2 (Winter 1969), 243–266.

Peake, E.R.L., "Northern Rhodesia—Belgian Congo Boundary," *Geographical Journal*, 83, 4 (April 1934), 263–280.

Pearcy, G. Etzel, "Boundary Types," *Journal of Geography*, 64, 7 (1965), 300–303.

———, "Boundary Functions," *Journal of Geography*, 64, 8 (1965), 346–349.

Pitts, Forrest R., "The 'Logic' of the Seventeenth Parallel as a Boundary in Indochina," *Yearbook, Association of Pacific Coast Geographers*, 18 (1956), 42–56.

Platt, Robert S., "Brazilian Capitals and Frontiers," *Journal of Geography*, 53, 9 (December 1954), 369–374; 54, 1 (January 1955), 5–17.

———, "The Saarland, an International Borderland," *Erdkunde*, 15, 1 (1961), 54–68.

Pluralism and Federalism. Special issue of *International Political Science Review*, 5, 4 (1984).

Potter, Pitman B., "Stephen B. Jones: Boundary-Making," *American Journal of International Law*, 34 (1945), 859.

Potts, Deborah, "Capital Relocation in Africa: The Case of Lilongwe in Malawi," *Geographical Journal*, 151, 2 (July 1985), 182–196.

Pounds, Norman J.G., "The Origin of the Idea of Natural Frontiers in France," *Annals, AAG*, 41, 2 (June 1951), 146–157.

———, "France and 'Les Limites Naturelles' from the Seventeenth to the Twentieth Centuries," *Annals, AAG*, 44, 1 (March 1954), 51–62.

———, "History and Geography: A Perspective of Partition," *Journal of International Affairs*, 18, 2 (1964), 161–172.

Pounds, Norman J.G. and Sue Simons Ball, "Core Areas and the Development of the European States System," *Annals, AAG* 54, 1 (March 1964), 24–40.

Pratt, R.C., "The Future of Federalism in British Africa," *Queens Quarterly*, 67 (1960), 188–200.

Prescott, John Robert Victor, "The Geographical Basis of Nigerian Federation," *Nigerian Geographical Journal*, 2, 1 (June 1958), 1–15.

———, "Geographical Problems Associated with the Delimitation and Demarcation of the Nigeria–Kamerun Boundary, 1885–1916," *Research Notes*, Ibadan Univ. College, Dept. of Geography, No. 12 (February 1959), 1–14.

———, "The Evolution of Nigeria's Boundaries," *Nigerian Geographical Journal*, 2, 2 (March 1959), 80–104.

———, "Africa's Major Boundary Problems," *Australian Geographer*, 9, 1 (March 1963), 3–12.

Procházka, Theodore, "The Delimitation of Czechoslovak–German Frontiers After Munich," *Journal of Central European Affairs*, 21, 2 (July 1961), 200–218.

Proctor, Jesse Harris, "Britain's Pro-Federation Policy in the Caribbean; an Inquiry into Motivation," *Canadian Journal of Economic and Political Sciences*, 22, 3 (August 1956), 319–331.

———, "The Effort to Federate East Africa: A Post-Mortem," *Political Quarterly*, 37, 1 (January–March 1966), 46–69.

Q

Qureshi, S.M.M., "Pakhtunistan: The Frontier Dispute Between Afghanistan and Pakistan," *Pacific Affairs*, 39, 1–2 (Spring–Summer 1966), 99–114.

R

Rajabov, S.A., "Geographical Factors and Certain Problems of Federalism in the USSR," *International Social Science Journal*, 30, 1 (1978), 88–97.

Rao, K. Krishna, "The Sino–Indian Boundary Question and International Law," *International and Comparative Law Quarterly*, 11, 2 (April 1962), 375–415.

Ratzel, Friedrich, "The Territorial Growth of States." *Scottish Geographical Magazine*, 12 (1896), 351–361.

Raup, Philip M., "The Agricultural Significance of German Boundary Problems," *Land Economics*, 26, 2 (May 1950), 101–114.

Ravenstein, E.G., "The Anglo–French Boundaries in West Africa," *Geographical Journal,* 12, 1 (July 1898), 73–75.

Rawlings, E.H., "The India–China Border," *Asian Review,* n.s., 58, 213 (January 1962), 21–26.

Reeves, Jesse Siddall, "International Boundaries," *American Journal of International Law,* 38 (1944), 533–545.

Reilly, Sir Bernard and J.C. Morgan, "South Arabian Frontiers," *Corona,* 10, 3 (March 1958), 88–91.

Reinhardt, G.F., "Rectification of the Rio Grande in the El Paso–Juarez Valley," *American Journal of International Law,* 31 (1937), 44–45.

Renner, George T., "Arizona's Lost Seaport," *Journal of Geography,* 61, 2 (1962), 57–59.

Reyner, A.S., "The Case of an Indeterminate Boundary: Algeria–Morocco," *Duquesne University Institute of African Affairs,* No. 18 (1964), 1–7.

Reynolds, David R. and Michael L. McNulty, "On the Analysis of Political Boundaries as Barriers: A Perceptual Approach," *East Lakes Geographer,* 4 (1968), 21–38.

Richards, J. Howard, "Changing Canadian Frontiers," *Canadian Geographer,* 5, 4 (Winter 1961), 23–29.

Robinson, G.W.S., "West Berlin: The Geography of an Exclave," *Geographical Review,* 43, 4 (October 1953), 540–557.

Rogge, John R., "The Balkanization of Nigeria's Federal System: A Case Study of the Political Geography of Africa," *Journal of Geography,* 76, 4 (April–May 1977), 135–140.

Rose, Archibald, "Chinese Frontiers of India," *Geographical Journal,* 39, 3 (March 1912), 193–223.

Rosen, S., "The Occupied Territories," *Jerusalem Quarterly.* 5 (1977), 125–144.

Rubin, Alfred P., "The Sino–Indian Border Disputes," *International and Comparative Law Quarterly,* 9, 1 (January 1960), 96–125.

Ruddock, G., "Capital of East Pakistan," *Pakistan Quarterly,* 7, 1 (Spring 1954), 49–58.

Rumble, Gary A., "Federalism, External Affairs and Treaties: Recent Developments in Australia," *Case Western Reserve Journal of International Law,* 17, 1 (Winter 1985). 1–42.

Rupen, Robert A., "Mongolian Nationalism," *Royal Central Asian Journal,* 58, 2 (1958), 157–178.

Ryder, C.H.D., "The Demarcation of the Turco–Persian Boundary in 1913–1914," *Geographical Journal.* 66, 3 (September 1925), 227–242.

S

Sabbagh, M. Ernst, "Some Geographical Characteristics of a Plural Society: Apartheid in South Africa," *Geographical Review,* 58, 1 (January 1968), 1–28.

Sage, Walter N., "British Columbia and Confederation," *British Columbia Quarterly,* 15, 1–2 (January–April 1951), 71–84.

Salisbury, Howard G., III, "The State Within a State: Some Comparisons Between the Urban Ghetto and the Insurgent State," *Professional Geographer,* 23, 2 (April 1971), 105–112.

————, "The Israeli–Syrian Demilitarized Zone: An Example of Unresolved Conflict," *Journal of Geography,* 71, 2 (February 1972), 109–116.

Sandford, K.S., "Libyan Frontiers," *Geographical Journal,* 96, 6 (December 1940), 377–388.

————, "Western Frontiers of Libya," *Geographical Journal,* 99, 1 (January 1942), 29–40.

Sanguin, André-Louis, "Territorial Aspects of Federalism: A Geography Yet to Be Made," *Scottish Geographical Magazine,* 99, 2 (September 1983), 66–75.

Savelle, Max, "The Forty-ninth Degree of North Latitude as an International Boundary, 1719: The Origin of an Idea," *Canadian Historical Review,* 38 (1957), 183–201.

Schoenrich, Otto, "The Venezuela–British Guiana Boundary Dispute," *American Journal of International Law,* 43, 3 (July 1949), 523–530.

Scholler, Peter, "The Division of Germany: Based on Historical Geography," *Erdkunde,* 19 (1965), 161–164.

Scott, J.B., "Swiss Decision in the Boundary Dispute Between Colombia and Venezuela," *American Journal of International Law,* 16 (1922), 428.

Semple, Ellen Churchill, "Geographical Boundaries," *Bulletin, American Geographical Society,* 39 (1907), 385–397, 449–463.

Shafer, Boyd C., "If We Only Knew More About Nationalism," *Canadian Review of Studies in Nationalism,* 7 (1980), 197–218.

Shaudys, Vincent K., "Geographic Consequences of Establishing Sovereign Political Units," *Professional Geographer,* 14, 2 (March 1962), 16–20.

Shaw, W.B.K., "International Boundaries of Libya," *Geographical Journal,* 85, 1 (January 1935), 50–53.

Shelvankar, K.S., "China's Himalayan Frontiers: India's Attitude," *International Affairs,* 38, 4 (1962), 472.

Shute, J., "Czecho-Slovakia's Territorial and Population Changes," *Economic Geography*, 24, 1 (January 1948), 35–44.

Simey, T.S., "A New Capital for the British West Indies," *Town Planning Review*, 28, 1 (April 1957), 63–70.

Singh, Ujagir, "New Delhi—Its Site and Situation," *National Geographic Journal of India*, 5, 3 (September 1959), 113–120.

Smith, C.G., "Arab Nationalism, a Study in Political Geography," *Geography*, 43, 202 (1958), 229–242.

Smith, H. Carrington, "On the Frontier of British Guiana and Brazil," *Geographical Journal*, 92, 1 (July 1938), 40–54.

Smith, S.G., "The Boundaries and Population Problems of Israel," *Geography*, 37, 177 (July 1952), 152–165.

Snyder, D.E., "Alternate Perspectives on Brasilia," *Economic Geography*, 40, 1 (January 1964), 34–45.

Solomon, R.L., "Boundary Concepts and Practices in South East Asia," *World Politics*, 23, 1 (October 1970), 1–23.

Spain, James W., "Pakistan's North West Frontier," *Middle East Journal*, 8, 1 (Winter 1954), 27–40.

Spate, O.H.K., "Factors in the Development of Capital Cities," *Geographical Review*, 32 (1942), 622–631.

———, "The Partition of the Punjab and of Bengal," *Geographical Journal*, 110, 4–6 (October–December 1947), 201–222.

———, "The Partition of India and the Prospects for Pakistan," *Geographical Review*, 38, 1 (January 1948), 5–29.

———, "Two Federal Capitals: New Delhi and Canberra," *Geography Outlook*, 1, 1 (January 1956), 1–8.

Stacey, C.P., "The Myth of the Unguarded Frontier, 1815–1871," *American Historical Review*, 56 (October 1950), 1–18.

Starr, H. and B.A. Most, "Contagion and Border Effects on Contemporary African Conflict," *Comparative Political Studies*, 16 (1983), 92–117.

Symmons-Symonolewicz, Konstantin, "Sociology and Typologies of Nationalism," *Canadian Review of Studies in Nationalism*, 9 (1982), 15–22.

T

Tamaskar, B.G., "The Concept of Boundaries in India Through the Ages," *Indian Political Science Review*, 12, 1 (January 1978), 59–72.

Tarlton, Charles, "Symmetry and Asymmetry as Elements of Federalism," *Journal of Politics*, 27, 4 (1965).

Taylor, Griffith, "Frontiers of Settlement in Australia," *Geographical Review*, 16, 1 (January 1926), 1–25.

Tayyeb, Ali, "A Note on the Political Geography of the India–China Border," *Canadian Geographer*, 4, 16 (July 1960), 22–26.

Teal, John J., Jr., "Europe's Northernmost Frontier," *Foreign Affairs*, 29, 2 (January 1951), 263–275.

Thomas, H.B., "The Kagera Triangle and Kagera Salient," *Uganda Journal*, 23, 1 (March 1959), 73–78.

Toppin, H.S., "The Diplomatic History of the Peru–Bolivia Boundary," *Geographical Journal*, 47, 2 (February 1916), 81–95.

Tregonning, K.G., "The Claim for North Borneo by the Philippines," *Australian Outlook*, 16 (December 1962), 283–291.

Tregonning, K.W., "The Partition of Brunei," *Journal of Tropical Geography*, 11 (1958), 84–89.

Tuckermann, W., "Territorial Reorganization in European Russia: A Note on the Political Map." *Geographical Review*, 14 (1924), 270–274.

Turnock, David, "Bucharest: The Selection and Development of the Romanian Capital," *Scottish Geographical Magazine*, 86, 1 (April 1970), 53–68.

U

Unstead, J.F., "The Belt of Political Change in Europe," *Scottish Geographical Magazine*, 39 (1923), 183–192.

V

Valkenier, Elizabeth K., "Eastern European Federation; a Study in the Conflicting National Aims and Plans of the Exile Groups," *Journal of Central European Affairs*, 14, 4 (January 1955), 354–370.

Venkatachar, C.S., "The Unity of India (A Study in Geopolitics)." *Indian Yearbook of International Affairs*, 3 (1954), 9–20.

Visher, Stephen S., "What Sort of International Boundary Is Best?" *Journal of Geography*, 31 (1932), 288–296.

———, "Territorial Expansion: A Study in Political Geography." *Scientific Monthly*, 40 (1935), 440–449.

———, "Where Should the International Boundaries Be?" *Social Science*, 14 (1939), 55–58.

Volacic, M., "The Curzon Line and Territorial Changes in Eastern Europe," *Bielorussian Review*, 2 (1956), 37–72.

W

Waddell, D.A.G., "Developments in the Belize Question 1946–1960," *American Journal of International Law*, 55, 2 (April 1961), 459–469.

Waldock, C.H.M., "Disputed Sovereignty in the Falkland Island Dependencies," *British Yearbook of International Law*, 25 (1948), 311–353.

Walker, P.C. Gordon, "The Future of City and Island States," *New Commonwealth*, 29, 8 (1955), 369–371.

Ward, Michael, "The Northern Greek Frontier," *Geographical Magazine*, 21, 9 (January 1949), 329–335.

Weigend, Guido G., "Effects of Boundary Changes in the South Tyrol," *Geographical Review*, 40, 3 (July 1950), 364–375.

Weissberg, Guenter, "Maps as Evidence in International Boundary Disputes: A Reappraisal," *American Journal of International Law*, 57 (1963), 781–803.

Welensky, Roy, "Toward Federation in Central Africa," *Foreign Affairs*, 31, 1 (October 1952), 142–149.

Whebell, C.F.J., "Models of Political Territory," *Proceedings, AAG*, 2 (1970), 152–156.

Whittam, Daphne E., "The Sino–Burmese Boundary Treaty," *Pacific Affairs*, 34, 2 (1961), 174–183.

Whittlesey, Derwent, "Trans-Pyrenean Spain: The Val d'Aran," *Scottish Geographical Magazine*, 49 (1933), 217–228.

———, "The Impress of Effective Central Authority upon the Landscape," *Annals, AAG*, 25, 2 (June 1935), 85–97.

Wiley, S.C., "Kashmir," *Canadian Geographical Journal*, 62, 1 (January 1961), 22–31.

Williams, Colin H. and A.D. Smith, "Conceived in Bondage—Called into Liberty: Reflections on Nationalism," *Progress in Human Geography*, 9, 3 (September 1985), 331–355.

Wilson, Curtis M., "The Geographical Basis of National Power," *Ohio Journal of Science*, 50, 1 (January 1950), 33–44.

Winch, Michel, "Western Germany's Capital," *Geographical Magazine*, 31, 1 (May 1958), 32–41.

Wiskemann, Elizabeth, "The Saar Moves Towards Germany," *Foreign Affairs*, 34, 2 (January 1956), 287–296.

———, "Berlin Between East and West," *World Today*, 16, 11 (November 1960), 463–472.

Woolsey, L.H., "The Bolivia–Paraguay Dispute (1928)," *American Journal of International Law*, 23 (1929), 110–112.

———, "The Tacna–Arica Settlement," *American Journal of International Law*, 23 (1929), 605–610.

———, "The Bolivia–Paraguay Dispute (1929)," *American Journal of International Law*, 24 (1930), 122–126, 573–577.

———, "The Chaco Dispute (1931)," *American Journal of International Law*, 26 (1932), 796–801.

———, "The Leticia Dispute Between Colombia and Peru (1932)," *American Journal of International Law*, 27 (1933), 317–324, 525–527.

———, "The Chaco Dispute (1933)," *American Journal of International Law*, 28 (1934), 724–729.

———, "The Leticia Dispute Between Colombia and Peru (1934)," *American Journal of International Law*, 29 (1935), 94–99.

———, "The Ecuador–Peru Boundary Controversy," *American Journal of International Law*, 31 (1937), 97–100.

———, "The Settlement of the Chaco Dispute," *American Journal of International Law*, 33 (1939), 126–129.

———, "The Polish Boundary Question," *American Journal of International Law*, 38 (1944), 441–448.

Wright, J.K., "Sections and National Growth," *Geographical Review*, 22 (1932), 353–360.

Wright, L.A., "A Study of the Conflict Between the Republics of Peru and Ecuador," *Geographical Journal*, 98, 5–6 (November–December 1941), 253–272.

Wurtel, D., "Okinawa: Irredenta on the Pacific," *Pacific Affairs*, 35, 4 (1962), 353–374.

Y

Yarham, E.R., "Mystery Island of the South Atlantic," *Contemporary Review*, 212, 1229 (June 1968), 307–311.

Z

Zartman, L.W., "The Sahara: Bridge or Barrier?" *International Conciliation*, No. 541 (January 1963), 3–62.

———, "The Politics of Boundaries in North and West Africa," *Journal of Modern African Studies*, 3 (1965), 155–173.

Zarur, J., "Berlin Between East and West," *World Today*, 16, 11 (November 1960), 463–472.

Part Three

POLITICAL GEOGRAPHY WITHIN THE STATE

Chapter 13

FIRST ORDER
CIVIL DIVISIONS

In Part Two we described many characteristics of States, including origin, growth, morphology, organization, and power. One characteristic, however, deserves special attention: the internal administrative organization of the State. All States, even the smallest, have internal divisions for administrative purposes.* The largest general purpose administrative or governmental units within a State are called first order civil divisions. In most States, these have subdivisions within them, ranging downward in some cases through second, third, and fourth order civil divisions. Local names for all these divisions vary considerably and can be very confusing if we try to generalize too much about them. The commune, for example, is a first order civil division in Liechtenstein but fourth order in France. The important thing for us, however, is not their names but their functions, and we consider organization and functions in this chapter and the next.

By definition, the first order civil divisions in federal States have more authority and usually more functions than those in unitary States. In many States, federal and unitary, the capital city is located in a separate district in which its functions are performed by the national government and the municipality. In a few States, notably the United States, the first order civil divisions may form regional consultative groups and even formal regional authorities to perform specific functions. All are governmental in character, whether or not they have independent taxing power (probably the single most important measure of "sovereignty" at the sub-State level), and even if they are simply administrative arms of the central government. In addition, many States have established special purpose districts, which we discuss in Chapter Fifteen.

UNITARY STATES

In unitary States, any civil divisions are created by the central government. The central government, whether by constitutional or other procedures, determines their number, boundaries, authority, and operations. Generally, the central government finances them, largely or completely, and appoints their chief officials. In many unitary States, even the first order civil divisions have no legislative or judicial functions and little decision-making power. Thus they function as administrative districts, designed to make the work of the central government easier or more effective. When conditions in the country change, or the needs of the government, or even the philosophy of the party or faction in power, then modifications are made in the civil divisions.

*San Marino, for example, has 9 castellos, Burundi has 8 provinces, The Gambia has 5 divisions, Andorra has 6 parroquias (parishes), Liechtenstein has 11 Gemeinden (communes), and Nauru has 14 districts. Even the British colony of Hong Kong has 2 departments divided into 17 districts, and the tiny British Caribbean colony of Montserrat has 3 parishes.

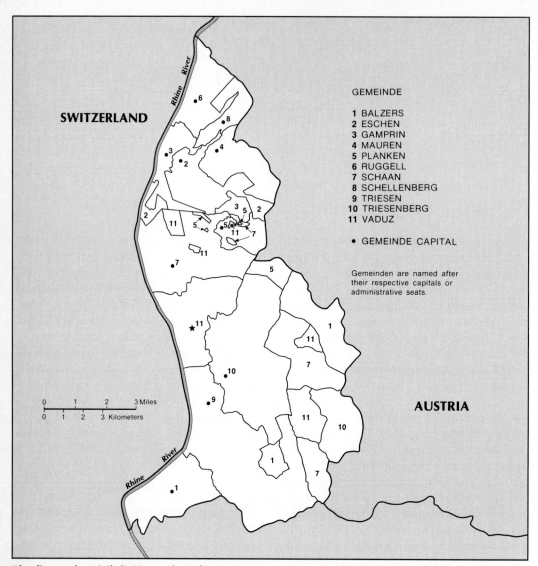

The first order civil divisions of Liechtenstein.   Here we see an example of fragmentation of administrative districts within a very small country. Note the numerous exclaves and the oddly shaped *gemeinden*. Something other than administrative convenience must account for them.

We are not saying that civil divisions in unitary States have no personalities of their own or that in practice they are mere appendages of the central government. This is largely true with the lower order civil divisions but does not necessarily describe those of the first order, particularly in old countries or those in which the civil divisions are based on linguistic, tribal, ecclesiastical, historical, or other cultural associations. In these cases the central government must exercise great care in making changes, otherwise the people involved

may become angered and even rebellious.

Central governments also face the classic dilemma of the number of levels of civil divisions: If there are too many levels, power tends to gravitate toward the lower levels since the central government finds it difficult to keep track of who is doing what at which level; on the other hand, if there are too few levels, the central government can be overwhelmed with the detail passed upward from a large number of units at each level. In some cases the government decides to create districts on the basis of some cultural, economic, or physiographic criteria so as to aid rational planning and encourage the cooperation of the people involved. In other cases it decides to make the divisions arbitrary, deliberately breaking up and rearranging traditional or "natural" units so as to weaken or destroy traditional loyalties and foster loyalty to the State.

There is nothing necessarily undemocratic about unitary government or even about repeated reshuffling of civil divisions. In both unitary and federal democratic States, these changes are usually made with the feelings and needs of the people concerned in mind. In those States that are not democratic, whether unitary or federal, the process of creating, changing, or abolishing civil divisions is much easier than in their democratic counterparts. A few examples will help to illustrate the general observations made thus far.

The United Kingdom

The United Kingdom of Great Britain and Northern Ireland is composed of three ancient kingdoms—England, Scotland, and Wales—and part of a fourth—Northern Ireland. Over a period of several centuries they were welded together into a single unitary State headquartered in London. (The nearby Isle of Man and the Channel Islands are dependencies of the Crown, having a great deal of local autonomy, but they are not part of the United Kingdom.) Each of the four constituent units of the UK is divided into civil divisions that, despite reorganization in 1974 and 1975, retain many characteristics of, and links with, political units of ancient times, at least as far back as the Saxon period. Numerous studies, both official and academic, had shown that the traditional hierarchy of governmental units was anachronistic— long on sentiment, short on efficiency. Repeated tinkering with the system met considerable opposition from traditionalists, could only compromise on piecemeal changes, and generally resulted in more

complexity but no greater efficiency. Finally, Parliament passed the Local Government Acts of 1972 (for England, Wales, and Northern Ireland) and 1973 (for Scotland).

The general effect of the reorganization has been a simplification of the administrative system. Prior to reorganization there had been 48 counties and 78 county boroughs in England (the latter chiefly cities that had been detached from their counties after 1888 and given equal status); since 1 April 1974 there are only 47 counties and 6 metropolitan counties. In Wales 13 counties and 4 county boroughs have been reduced to 8 counties. In each case the lower order civil divisions were also simplified. The county and metropolitan boroughs and the rural and urban districts were abolished. The counties are divided into districts and these into parishes. For Scotland, which still retains some autonomy from its period as a kingdom, special legislation has often been passed by Parliament taking into account the special conditions there. Thus, the former 33 counties and 4 counties of a city were reduced on 16 May 1975 to 9 regions and 3 island areas (Shetland, Orkney, Western Isles). The lower order civil divisions were also simplified. Northern Ireland has only one level of local administration; now there are 26 districts instead of the former 6 counties and 2 county boroughs.

As might be expected, this reorganization was the result of many compromises made in many quarters since the initiation of the work of the Royal Commission on Local Government in England (Radcliffe–Maud Commission) in 1966. It is therefore wholly satisfactory to no one and has already been criticized on several grounds, primarily for failure to include cities and their hinterlands uniformly in the same units. Nevertheless, it seems likely that there will be a reasonable trial period before another reform of local government is undertaken.

Meanwhile, attention in the UK has turned to the possible devolution of power from London to Scotland and Wales and

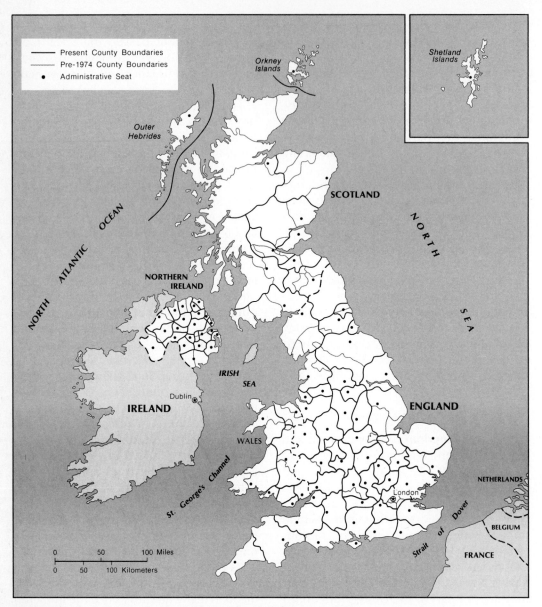

United Kingdom: Reorganization of first order civil divisions, 1974.

the restoration of home rule to Northern Ireland when security and social conditions there permit it. The differing types of civil divisions in each of the four provinces, however, contribute to the rationale for considering the United Kingdom to be already a regional rather than a unitary State.

France

One of the many changes wrought in France by the Revolution of 1789 was the reorganization of her administrative structure. The major objective was to create a highly centralized State to which all citizens would transfer their allegiance from the provinces, remnants of medieval and earlier independent States. One device for achieving this centralization was the fragmentation of the historic provinces into 83 *départements*, which have since been increased to 96. (In addition, there are now 5 overseas *départements:* French Guiana, Guadeloupe, Martinique, Reunion, and St. Pierre and Miquelon.) Rather than retaining any of the traditional names, the revolutionary government assigned to the new units names of physical features in the area. The old provinces did not die, however. The *département* boundaries often follow old provincial boundaries, people still identify with the provinces, and the 22 new regions of France (groupings of *départements* for economic and administrative purposes) generally bear the names of old provinces, such as Bretagne, Picardie, Lorraine, Alsace, Aquitaine, Languedoc, and Auvergne.

The *départements* are divided into *arrondissements*, these into *cantons*, and these into *communes*. All, however, serve as administrative arms of the central government in Paris. The creation of regions in the early 1970s was in part a response to the evident need for larger units for planning and carrying out economic development and environmental management programs, and in part a concession to the durable loyalties of the French people to their traditional provinces. We may see more changes in France in the near future, perhaps the secession of the overseas *départements*, the abolition of the largely nonfunctional *arrondissements* and *cantons*, and the consolidation of many mainland *départements*.

China

China is the oldest State in the world, with a continuous history of some five thousand years. During that time her territory has expanded and contracted; she has been split, invaded, and occupied; her rulers have frequently changed her administrative structure. Throughout, however, she has remained either in theory or practice a unitary State. During times of weakness and disorder at the center, peripheral portions of the Empire detached and maintained themselves as autonomous States, often still recognizing the nominal suzerainty of the emperor. Even within what used to be called China Proper, local warlords often controlled large regions with little deference paid to the central government. This was the situation that prevailed in China during the waning days of the Manchu (Ch'ing) Dynasty and throughout the Nationalist period (1911–1949). When the communists took over the mainland, they did as many emperors had done before them: they established a strong government over nearly all of China.

After the new government had brought under its control all of what had nominally been China at the end of the Manchu Dynasty (except for Mongolia and Tannu Tuva), it began a series of reorganizations of its territory. At times a number of provinces were grouped together into large economic and political units that were subsequently broken up. Today the first order civil divisions consist of 21 provinces (22 if Taiwan is included as the government insists), 5 autonomous regions, and the centrally administered metropolitan areas of Beijing and Shanghai. The autonomous regions are all on the peripheries and are the homes of most of the non-Han citizens of China. They are Nei Mongol, Xinjiang Uygur, Xizang (Tibet), Ningxia Huizu, and in the south Guangxi Zhuangzu, formerly a province. These autonomous regions constitute 42 percent of the territory of China but contain only some 6 percent of the population. Even some of the lower order civil divisions are based on ethnic groups, but in no case does their autonomy extend to more than cultural affairs; China's government is highly centralized. It uses a variety of technical and administrative devices to maintain control, one of which is the commune, the smallest minor civil division.

Nepal

Nepal was united into a kingdom in the latter half of the eighteenth century by King Prithvi Narayan Shah of Gorkha out of a large conglomeration of petty kingdoms. At first the kingdoms retained varying degrees of local autonomy within the unified State, depending on how vigorously they had resisted Prithvi. After they consolidated power, however, the Shah dynasty began reducing the prerogatives of the

The first order civil divisions of China.

local rajas and centralizing the administration. By 1950, the powers of the local rulers had been reduced to limited judicial functions, revenue collection, and the right to exact forced labor. By 1961, even these had been abolished. All this time local administration had been developing in those areas controlled by the king. By 1950 there were 32 districts headed by chief administrative officers and three units in the Kathmandu Valley headed by magis- trates, all appointed by the central government. The districts were reorganized partially in 1956 and completely in 1961. The 1961 reorganization divided the country into 14 zones, most of which were arranged vertically, including the lowland Terai in the south, the central belt of "hills," and the Himalayas in the north.

The major objectives of this arrangement were stated to be administrative uniformity and better implementation of de-

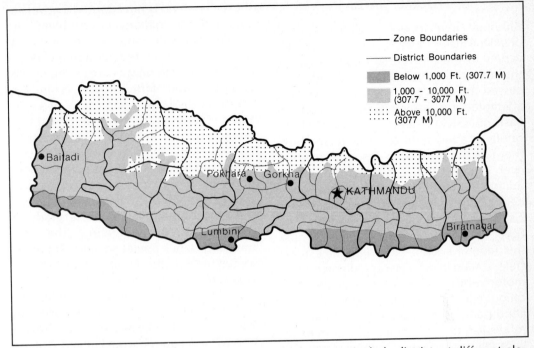

The districts and zones of Nepal. Note how most of the zones include districts at different elevations, thus bringing together for planning purposes and administration not only a variety of natural resources, but also different ethnic groups, while separating related ethnic groups that tend to occupy the same types of physical regions.

velopment projects. Unstated, but very likely of considerable importance, was the objective of breaking with the past and orienting the diverse peoples of Nepal (some 30 ethnic groups) toward the national development program. The zones are subdivided into 75 districts. Within the districts, villages or groups of villages with a population of two thousand or more are organized into village assemblies, while a town of ten thousand or more may have a town council. These lower order divisions send representatives to the district assemblies, which in turn send members to the national legislature. This system took some time to put completely into effect, but it functions now as an administrative system of the king, in whom the real power resides.

Norway

While Norway is a unitary State, it is also an old one and a democratic one. Its ad-

ministrative structure is similar to that of the United Kingdom in some ways, since similar forces are at work in both countries. The first order civil division is the county, of which there are 19. The counties (except for Oslo) are subdivided into 47 urban and 407 rural municipalities, down from a total of 680 in 1950. The counties are primarily responsible for social services, utilities, fire fighting, education, health care, and roads, and they play a large role in development planning. Norway devolved more responsibility from the state to the county and from the county to the municipality. This is not a shift toward either federalism or regionalism, however, but simply a reordering of administrative tasks. The central government retains the real authority and assigns tasks to the other levels of government as it deems appropriate according to a general formula. The central government continues to supervise both the counties and the municipalities.

FEDERAL STATES

Although there are only a relative handful of federal States today, they exhibit as much variety in their territorial organization as the unitary States. We have already discussed the nature of federalism and have attempted to classify federal States on the basis of certain criteria. Now we can look briefly at a few federal States to see how they function. No paper description, however, can give an accurate picture of federalism in action. One of the most bewildering aspects of the United States to a visitor or an immigrant from Sweden or Egypt or Honduras, for example, is just this federal system—the difference from state to state in marriage and divorce laws, motor vehicle codes, professional certification, educational systems, elections, and a thousand other aspects of everyday life. All we can do here is describe certain salient features of a few federal States and try to derive some patterns from them.

The Union of Soviet Socialist Republics

The Soviet Union is not only the largest country in the world, but the one with the most complex territorial organization.

The complexity is due not so much to its size as to its ethnic diversity and centrally planned economy. The Tsarist Empire had a relatively simple administrative structure because it made few concessions to its non-Russian subjects and because, while it was autocratic and highly centralized, it engaged in little detailed economic planning. When the Bolsheviks came to power in 1917, their first task was to win the support of the masses, including if possible, the non-Russian masses. In the turmoil of the Revolution, World War I, the Civil War, and the War with Poland, much of the empire in Europe fell away—Finland, the Baltic States, Poland, and the Transcaucasian Federated Republics. Some areas were lost to neighboring countries: Kars and Ardahan to Turkey, Bessarabia to Romania, Southern Sakhalin and the Kurils to Japan. In the face of all

these disasters, Lenin adopted a novel approach to the "nationalities problem": he would grant autonomy to most of the nationalities within a federal framework.

The Soviet Union was organized in 1922 as a federation of the Ukraine, White Russia (Byelorussia) and, largest of all and itself a federation, the Russian Soviet Federated Socialist Republic. The first order civil divisions, the union republics, were organized on an ethnic basis, the first experiment of its kind, at least in modern times. Through persuasion and coercion, the outlying nationalities were gradually brought into the federation. The Turkmen and Uzbek Soviet Socialist Republics joined in 1925; the Tadjik SSR in 1929; Kazakhstan, Armenia, Azerbaijan, and Georgia in 1936, and Moldavia (Bessarabia), Estonia, Latvia, and Lithuania in 1940. The Karelo-Finnish SSR was created in 1940 out of territory acquired from Finland, but lost its union republic status in 1956. The 15 union republics today have, according to the Soviet constitution, a great deal of autonomy; as pointed out earlier, however, this autonomy is, in fact, fictitious. Each republic is closely controlled by the central government in Moscow. The republics are also coequal in theory; in practice, however, the Russian republic is far more equal than all the others.

The lower order civil divisions are organized in two parallel tiers, one based on nationality, the other on essentially economic criteria. In the first tier are the autonomous Soviet socialist republics (20 of them, of which 16 are within the RSFSR), then the autonomous *oblasts* (regions) and autonomous *okrugs* (areas). All of the autonomous *okrugs* and most of the autonomous *oblasts* are within the RSFSR. The second tier consists of *oblasts* and *krays* (territories), about equal to ASSR's, with the *krays* differing from the *oblasts* only in that they have autonomous *okrugs* within them. The *rayon* is a still lower order civil division, and the municipality is below that.

All these units vary greatly in size and population, and all are subject to change

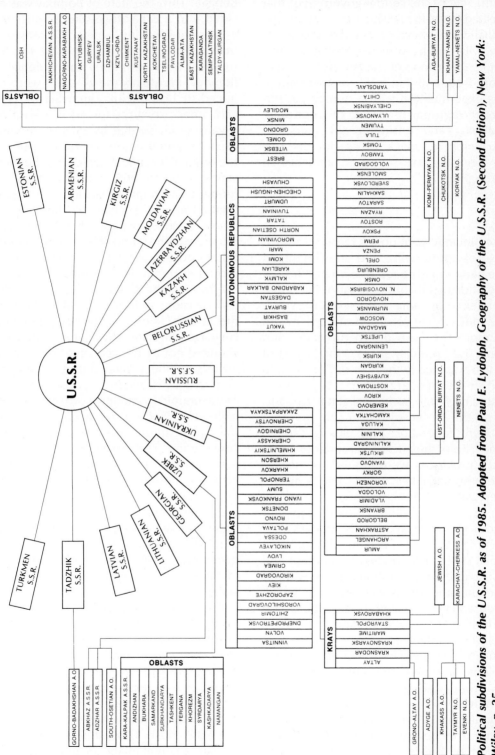

Political subdivisions of the U.S.S.R. as of 1985. Adopted from Paul E. Lydolph, Geography of the U.S.S.R. (Second Edition), New York: Wiley, p. 25.

at the will of the central government that created them. In practice, however, the nationality areas have remained relatively stable, with only a few up- or down-gradings, while the economic areas are changed more frequently. In fact, the government has experimented with a number of different regional planning or economic groupings and considered still others, but the union republics, except for the Karelo-Finnish, have remained undisturbed throughout.

India

The Indian federation has required a great deal of compromise and adjustment on the part of many people. With partition of the Indian subcontinent and the independence of India and Pakistan in 1947, the Indian State consisted of a patchwork of more than two dozen provinces and some 562 "princely states," the largest of which was nearly the size of France, and the smallest no larger than a large American city. These traditional states had survived the period of British rule, but in a modern federation they clearly had no place. Thus the internal boundaries of India's provinces were altered to include these principalities, while some "unions" were created out of a large number of these political entities. One of the greater subdivisions thus created—Saurashtra—included more than two hundred of these small states. The process of partition and consolidation did not take place without friction, and there were riots in which thousands were killed. Finally, the Union of India emerged with 27 states and a number of federal territories covering the capital city, several islands, recently acquired colonial enclaves, and some of the northern frontier areas.

Almost as soon as this framework was established, pressures on the government came from several of the states for a revision of boundaries according to language divisions. More than two hundred languages and dialects are spoken in India, including Hindi, Urdu, Gujarati, and Bengali. Thus the state of Bombay was divided into a Gujarati-speaking north and a Marathi-speaking south (Gujarat and Maharashtra, respectively), a move that led to street demonstrations and violence in Bombay. Eventually the 27 states were reduced to 14, but riots, guerrilla warfare, and other pressures forced renewed reorganization, so that by 1974 there were 21 states and 9 union territories, but not the original 9, some of which had become states. Then in 1975, India annexed the former protectorate of Sikkim, to bring the total number of states to 22. In 1987, the union territories of Arunachal Pradesh and Mizoram in the far northeast were granted statehood, as was Goa, but not the 3 other former Portuguese territories in India.

Although the basis for the first order civil divisions in India is similar to that in the Soviet Union, in this case generally language, the states have been created and changed not by the acts of a few political leaders in the capital, but through the democratic process in response to the will of the people (sometimes expressed quite violently). The central government has this power under Article 3 of the Indian constitution. The process has never been easy since any state created on linguistic grounds is bound to include speakers of other languages and exclude some speakers of the dominant language. In addition, India has many tribal groups and a number of religious minorities whose interests are sometimes given inadequate attention in the process of creating or abolishing states.

The most recent partition of a state, for example (other than the states created out of former districts of Assam), occurred in 1966 when Punjab (which had already been partitioned with Pakistan in 1947) was reduced by the creation of the Hindi-speaking state of Haryana and the transfer of other areas to the union territory of Himachal Pradesh. This left the remainder of Punjab largely Punjabi-speaking, but also largely Sikh rather than Hindu. A problem arose in this partition that is rather interesting. The beautiful new planned city of

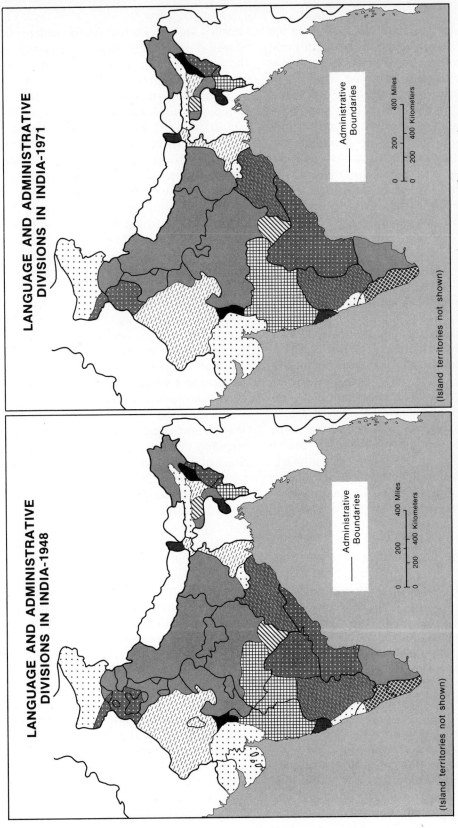

LANGUAGE AND ADMINISTRATIVE DIVISIONS IN INDIA—1971

(Island territories not shown)

Administrative Boundaries

0   200   400 Miles
0   200   400 Kilometers

LANGUAGE AND ADMINISTRATIVE DIVISIONS IN INDIA—1948

(Island territories not shown)

Administrative Boundaries

0   200   400 Miles
0   200   400 Kilometers

Note how the current state and territorial boundaries nearly coincide with the distribution of the major languages. This experiment in ethnic federalism will be interesting to watch over the next generation or two. Since 1971 a number of territories have become states and several other changes in politicoterritorial patterns have taken place, but few of them have been based on language. The "linguapolitical" situation appears stable for the present.

Chandigarh, capital of Punjab, was on the linguistic boundary, and the two states vied for possession of it as their capital. Eventually, they agreed to share it, so today one city serves as the capital of two states.

While this division of India on a linguistic basis is still experimental, it does appear to be working. Its flexibility has enabled the country to avoid—or postpone— many problems, and there do not appear to be significant centrifugal forces operating within the country today, except for the Sikh extremists.

Nigeria

Nigeria constitutes another compromise federation. Its initial federal framework, developed during discussions between the (British) colonial rulers and Nigerian political leaders, was put into effect before independence, in 1954. Nigeria has a population approximating 100 million and three distinct and separate core areas. These core areas form the foci for three of the dominant peoples in the country: the Yoruba in the southwest, the Ibo in the southeast, and the Hausa in the north. When the State achieved independence in 1960, it consisted of three major federal regions and a federal district at Lagos, the federal capital. Because Nigeria contains literally hundreds of tribal peoples and more than two hundred languages are spoken today, the measure of compromise can be understood.

Some Nigerian politicians immediately warned against the inadequacies of the system, insisting that the State would not survive unless a greater number of regions were established to satisfy the demands of minority peoples. In 1963 the decision was reached to establish a Midwest Region, carved out of territory of the Western Region and populated dominantly by the Edo people. But the Northern Region, by far the most populous unit within the State, was in a position to dominate the State. Counterbalancing this was the

north's land-locked situation, which rendered it dependent on southern ports for its external trade. Exacerbating the situation was the great contrast between Nigeria's northern and southern regions. Indeed, the north is a world apart from the south, with the strength of its Islamic traditions deriving in large measure from British methods of indirect colonial rule. When independence was under discussion, there were appeals from the north that sovereignty be delayed—views that appeared treasonable to the independence-oriented south. Northern fears were aggravated by the prevalence of southerners (especially Ibos) in key positions and jobs in the north.

Following independence, outbreaks of violence occurred against Ibo people living in the north, and the survivors returned to the south. In May 1967, in an attempt to placate further many ethnic groups and avoid civil war, the government reorganized the country into 12 states. The Ibo-controlled government of the Eastern region rejected the plan and seceded, declaring itself the sovereign State of Biafra. This touched off a civil war which ended in the surrender of the Ibos and the dissolution of Biafra on 12 January 1970. In 1976 the country was again reorganized, this time into 19 states and a Federal Capital Territory to which the capital eventually is to be moved.

Nigeria is an example of a country that inherited a federal system upon attaining independence, but one clearly not suited to its needs. Tribal unrest, internal frictions and jealousies, economic dislocations, repeated violent changes of government, and experiments with territorial organization have kept the country from achieving what its size, population, location, and resources (including petroleum) had promised for it at independence.

Few generalizations can usefully be made about first order civil divisions in federal States and examinations of them in such true federations as Canada, Australia, Switzerland, Germany (Federal Re-

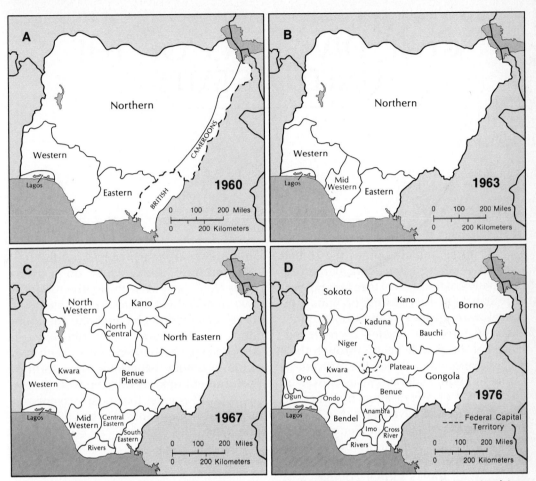

Nigeria: changes in first order civil divisions. Map A In 1954 the territory was reorganized into three regions; the federal district of Lagos; and the Southern Cameroons, part of a British trust territory administered through Nigeria. This was the situation at independence in 1960. *Map B* In a United Nations-sponsored plebiscite in 1961, the Northern Cameroons voted to join Nigeria and Southern Cameroons voted to join French Cameroun to create the Federal Republic of Cameroun. In 1963, the Western Region was partitioned to create a Midwestern Region. *Map C* In May 1967 the four regions were reorganized into 12 states, but this move led to the secession of the Eastern Region (which called itself Biafra) and to civil war. *Map D* Biafra was defeated in 1970 but ethnic rivalries continued, so the 12 states were redivided into 19 in 1976 and a new federal capital territory was designed at Abuja, in a sparsely populated, centrally located area not dominated by any major ethnic group.

public), and the United States will reveal as much diversity as appears in the three nominal federations just described. Since the United States was the first modern State to be organized as a federation, however, and is still one of the most successful, we devote the entire next chapter to a review of its civil divisions.

Chapter 14

CIVIL DIVISIONS OF THE UNITED STATES

The territorial organization of the United States of America may be considered representative of the class of true federations. The State and its constituent units (the states) were created by the Constitution, which in turn, was created by representatives of the people acting through their state governments under the Articles of Confederation. Federation was chosen as the method of organization not because of any theoretical advantages, but simply as a compromise, an expedient dictated by the objective conditions of the time and the perceptions of the decision makers. Unity of the 13 former colonies was clearly necessary for survival. Confederation had proved unworkable, and a unitary State was both impossible and undesirable; therefore the only way to achieve unity without destroying diversity and enforcing uniformity was through federalism.

The framers of the Constitution drew ideas from British and French political philosophers, from the practice of the Iroquois Confederacy, from their own experience under the Articles of Confederation, and from the conditions in their individual states. The result was "a bundle of compromises" so complex and so subtle that generations of scholars, officials, and judges have made careers of analyzing and interpreting it. It has, moreover, been so transformed by amendment, interpretation, and practice that the federal system of today works quite differently from that envisaged by the founders two hundred years ago. Yet its elements have remained unchanged and the founders would have little difficulty in recognizing the descendent of their creation, a unique experiment in a form of government unknown in the world at the time. Small wonder that most foreigners—and many Americans—are bewildered by the system!

THE STATES

The first order civil division of the United States is the state. There are presently 50 states plus the semiautonomous federal district in which the capital is located, the District of Columbia. Thirty-seven states have joined the union so far "on an equal footing with the original states in all respects whatever," in the words of the Northwest Ordinance of 1787. This act, one of the last of the Confederation, proved to be among the sturdiest foundations of the federal system, a device for expansion of the national territory without creating a permanent colonial empire. It was similar to the Iroquois system, under which the Tuscorora nation was first admitted to the Confederacy about 1714 as a kind of junior partner (much like the later American territories) and then, about 1789, as a full member with an equal vote.*

The Constitution granted a number of executive, legislative, and judicial powers (called the enumerated powers) to the national government; others are implied in the wording of the Constitution, and still others, particularly in the area of foreign affairs, are inherent in the existence of a national government. The Tenth Amendment provides that "The powers not delegated to the United States by the Constitution, nor prohibited by it to the States, are reserved to the States respectively, or to the people."† The states are thus invested with some measure of sovereignty; within their sphere, they are free to do as they please, provided they do not violate the federal consitution, federal laws, or treaties.

The Constitution provides also for an elaborate system of checks and balances regulating not only the separation of powers of the branches of the federal government, but also the division of powers between the federal government and the state governments. The federal judiciary acts as the major interpreter of all these provisions and leavens the whole system. The system is enormously complex, firm yet flexible, and constantly changing. In our brief survey, we can only scan some aspects of the system that are particularly geographic.

Since the states are in some measure sovereign, they have exclusive jurisdiction over their territory, except for federal land.

Thus, even before the Constitution was adopted, the United States could not govern the newly acquired lands west of the Appalachians that were claimed by seven states until these states had ceded the lands to the federal government (1781–1802). In the subsequent territorial acquisitions west of the Mississippi, all the land became part of the public domain except for the State of Texas.*

There have been scores of boundary disputes, not only between the United States and her neighbors, but also among the several states. This is because the territory of the United States was acquired over a long period of time through royal grants, purchases, conquest, annexation, and other means; because the states have much sovereignty in territorial matters; and because much of the land over which they claimed sovereignty was sparsely populated and scarcely known, and surveying methods were primitive, especially under difficult conditions. Most of these disputes have been settled by negotiation, arbitration or Supreme Court decisions, but some persist today. Most involve rivers whose channels are meandering; some involve maritime boundaries.

In a unitary State, such problems either would not arise at all or would be very quickly resolved by a decision made in the national capital. In a federal state, however, these problems can result in protracted, costly, and bitter struggles.† Geographers have been employed as experts in a number of these interstate boundary disputes and there is every like-

---

*The Land Ordinance of 1785 had provided for the surveying of the Old Northwest and the laying out of townships 6 miles square (93 sq. km) each consisting of 36 sections of 640 acres (2.59 sq. km) with one section reserved for education. This township and range system was later extended to much of the western land acquired by the United States and was another instrument for the peaceful and orderly integration of new territory. It is also vividly impressed on the map and on the landscape even today in the form of political boundaries, property lines, and road and field patterns.

†In Canada the reverse is true; powers not specifically delegated to the provinces are reserved for the central government.

*As an independent republic before voluntarily joining the United States as a state without first being a territory, Texas was able to negotiate unusual terms for admission, including retention of her public lands, her 9-mile territorial sea, her navy, her land survey system, and many other features that still distinguish her from other states.

†The United States is not alone in this; even in Brazil, where the states have less sovereignty, a boundary dispute between Espirito Santo and Minas Gerais was not settled until 1963, when the two governors signed a treaty incorporating the suggestions of a joint commission.

lihood that more will be so employed in the future.

As with boundaries, so with state capitals; the states are responsible for their locations. We have already discussed the importance of national capitals. State capitals, while generally not arousing the emotions of national capitals, may have much greater importance in the everyday life of the average citizen of a federal State—and in the economy and politics of the individual state. Consequently, state capital locations have often engendered much negotiation, propaganda, and even improper conduct as rival factions sought to locate the capital here or there.

In some cases existing cities competed for the prize; in others a new central location was desired for the convenience of the citizens or to thwart the designs of existing claimants; in still others the location was based on existing or projected population concentration, transportation facilities or economic activity—or simply on the contemporary balance of political forces in the state. Decisions have been made by state legislatures, by special commissions, and by popular vote. Some states have changed their capitals one or more times. Pennsylvania, for example, moved its capital from Philadelphia to Harrisburg in 1812; Ohio's first capital was Chillicothe, then Zanesville for two years, then back to Chillicothe, and finally Columbus in 1816; Michigan followed a similar pattern; Illinois has had three capitals, Oklahoma two, and North Carolina eight. Connecticut even had two capitals at the same time, New Haven and Hartford, until 1873.

Most of these changes took place during the early years of statehood, since the capital, like the boundaries, tends to become fixed with time. Alaska, however, which became a state in 1959, decided in a referendum in 1976 to move its capital from Juneau in the far south to a more central site near Willow, north of Anchorage. Thereafter doubts arose about whether the site was far enough north of Anchorage, if the original cost estimates were realistic, and whether the target date of 1982 for transfer could be met.*

Over the years many proposals have been made for changes in state boundaries, frequently by disgruntled citizens of a peripheral area of a state demanding that the area be detached and added to an adjoining state or made into a new state. These secession movements were successful a number of times in the nineteenth century. Vermont, Maine, Kentucky, Tennessee, and West Virginia were all parts of, or claimed by, other states before they entered the union. After becoming states, Maryland and Virginia ceded land to the federal government for the creation of the new federal capital district, Missouri added territory in its northwest corner in 1836, Nebraska received small tracts from Dakota Territory in 1870 and 1882 and exchanged some small tracts with Iowa in 1943, New York annexed a part of Massachusetts in 1853, Pennsylvania bought an area in its northwest corner from the federal government in 1792, West Virginia annexed two counties of Virginia in 1866, and a number of states have gained or lost small bits of territory after the settlement of boundary disputes.

Even in recent years there have been proposals, generally short-lived, to transfer southeast Oregon to Nevada; to attach the part of Inyo County, California, lying east of Death Valley National Monument, to Nevada; to make a separate state of the Upper Peninsula of Michigan; and in 1977, to detach Nantucket and Martha's Vineyard islands from Massachusetts (whereupon Governor Ella Grasso of Connecticut graciously offered to take them in!). And as regularly as the summer fogs of San Francisco or the Santa Ana winds of Los Angeles, someone proposes the partition of California into northern and southern

*In 1978 the voters rejected a proposed bond issue to pay for the move but approved an initiative requiring voter approval of all "bondable costs" before any funds are spent on the move. In 1982, the voters apparently killed the move completely, at least for the foreseeable future.

states. No secession movement has been successful, however, since the exceptional period of the Civil War, and it seems unlikely that any state boundaries will be substantially changed in the future.* Even imaginative proposals to reorganize the country into more rational first order civil divisions arouse little enthusiasm in a conservative population satisfied with the present arrangement.†

INTERSTATE RELATIONS

One virtue of the American federal system is flexibility. Without built-in flexibility, it is doubtful if the system could have survived the changes of the past two hundred years, changes that would have been almost inconceivable to the system's creators. Enormous demographic and territorial growth; the development of party politics; new technology; involvement in world wars and world affairs generally; immigration of masses of people of almost every existing race and culture; the greatest migration in history, from the countryside to the city; great increases in crime and poverty; environmental degradation; the shift of the population to the south and west—all and more have created problems too big for individual states to handle and frequently also outside the province of the federal government. The states have therefore been forced to cooperate with one another, formally and informally, on a bilateral, regional, and national basis, despite their historic propensity to compete instead. At times this cooperation has had to be encouraged by the federal government.

The states are bound together by an intricate network of conferences, information exchanges, informal agreements, and formal compacts. Many agreements grow out of governors' conferences and the Council of State Governments. The National Governors' Conference meets annually; there are also conferences of Democratic and Republican governors. Most important in practice, however, are the regional conferences of governors, often held several times a year in the South, New England, Middle Atlantic states, the Midwest, and the West.* Other state officials and legislators meet from time to time with their out-of-state counterparts. Both the conferences and the agreements they spawn sometimes have permanent offices and staffs. Boston, for example, contains several regional offices, including those of the New England Governors' Conference and the New England Board of Higher Education. The most formal agreement among states is the interstate compact. Some compacts require the consent of Congress before they can become effective. There are currently more than 160 of these compacts in operation. Some of interest to geographers include the Atlantic States Marine Fisheries Commission, the Southern States Forest Fire Compact, the Ohio River Valley Water Sanitation Commission, and the Colorado River Compact.

The Appalachian Regional Commission was established by the Congress in 1965 to plan and coordinate economic development programs in the 13-state area on a joint federal–state basis. Similar commissions for regional economic development in 11 other regions were established between 1966 and 1979 under Title V legislation, and 7 commissions were established during the same period for river basin planning under Title II. There are also 10 Federal Regional Councils, created

---

*In 1973, for example, G. Etzel Pearcy proposed a 38-state arrangement and in his 1974 book *Geography and Politics in America* Stanley Brunn suggested 16 states. In both cases all the proposed states would have completely new boundaries.

†In 1986, several predominantly black sections of Boston threatened to secede and form a new city called Mandela. The proposal was rejected in a referendum.

*In addition, the New England governors meet annually with the premiers of the eastern provinces of Canada. Out of these meetings have come agreements on energy and joint tourism promotion.

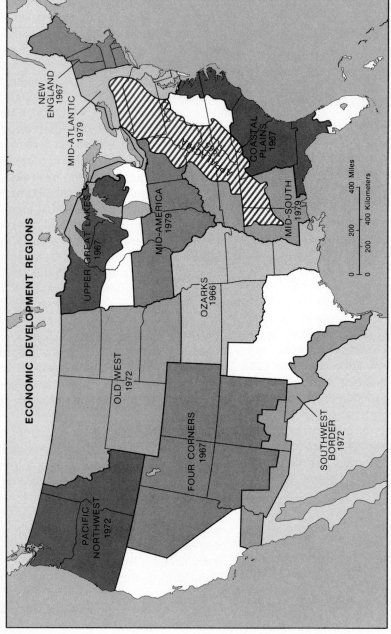

ECONOMIC DEVELOPMENT REGIONS

NEW ENGLAND 1967

MID-ATLANTIC 1979

COASTAL PLAINS 1967

MID-SOUTH 1979

UPPER GREAT LAKES 1967

MID-AMERICA 1979

OZARKS 1966

OLD WEST 1972

FOUR CORNERS 1967

SOUTHWEST BORDER 1972

PACIFIC NORTHWEST 1972

0   200   400 Miles

0   200   400 Kilometers

The Appalachian Regional Commission was established by federal law in 1965 as a mechanism for economic development planning in a chronically depressed area. Gradually such commissions were established in regions covering most of the United States, primarily in response to increasing pressure from state governors. In January 1979 the president announced the creation of three new commissions (dated 1979 on the map) and four specific federal actions to support the work of all the commissions. All of the commissions, except for the Appalachian, were abolished by the Reagan administration during the 1980s.

174

in 1969 by the president to bring together the regional directors of the major federal grant-in-aid agencies in each region. None of these interstate and regional organizations has received sufficient attention from geographers; even the information about the better-known ones is scattered through political science literature. In this book, all we can do is examine the principal features of three of the best-known interstate and regional agencies.

The Port Authority of New York and New Jersey (originally the Port of New York Authority) was organized in 1921 through an interstate compact that supplemented an 1834 harbor treaty between the two states. It operates under the direction of six unsalaried commissioners from each state and its programs are subject to the veto of either state legislature. It has the power to purchase, construct, lease, and operate any terminal or transportation facility within

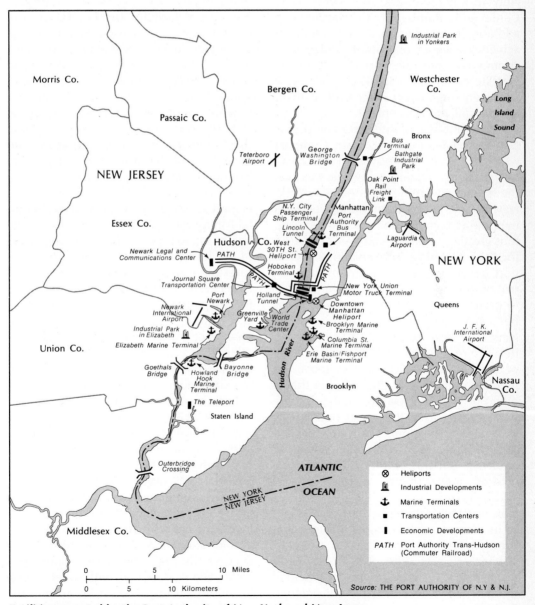

Facilities operated by the Port Authority of New York and New Jersey.

the port district, but it may not levy taxes and it receives no governmental financial support. It does, however, have the power of eminent domain. At present it owns and operates four major commercial airports, four bridges, two vehicular tunnels under the Hudson River, one motor truck terminal, two heliports, two interstate bus terminals, Port Newark, the Hoboken and Elizabeth Marine Terminals, most of the docks in New York City, a grain terminal in Brooklyn, a commuter railroad between New York and Newark, the World Trade Center and a number of other facilities. In addition, it engages in a wide range of activities designed to promote and protect commerce using the Port of New York. Despite criticism for its fiscal policies, it has been remarkably successful in managing a difficult assignment.

A different type of interstate compact is the 1961 Delaware River Basin Compact which created the *Delaware River Basin Commission*. Its unique feature was that the federal government joined New York, New Jersey, Pennsylvania, and Delaware in both the compact and the commission, thus creating a unified federal–state agency for "the planning, conservation, utilization, development, management and control of the water resources of the basin." In order to carry out this mandate, it has been given powers similar to those of the New York Port Authority. It was formed because of a series of disputes over uses of the Delaware River water, most immediately the fear of Philadelphia that it might be hurt by New York City's extraction of water from the upper river for its own needs. The commission now has authority to deal with additional aspects of river use, such as pollution control, watershed management, recreation, and hydroelectric power. While the commission was still forming in 1961, it was put to its first test—a prolonged drought that brought interstate relations to a water-sharing crisis by 1965. The commission managed to work out a solution that helped the states survive the drought until it ended in 1967. For some years after, the com-

mission largely confined itself to technical matters: gathering data, making analyses, investigating complaints, warning of pollution spills and impending floods, and similar management functions. More recently, it has become active in such areas as wetlands conservation, preparation of an annual Water Resources Program, studies of waste treatment and water salinity, and preparation of recreation maps.

The Tennessee Valley Authority is yet another type of regional organization, this one a wholly federal corporation. Its genesis was in facilities the federal government had built at Muscle Shoals on the Tennessee River in Alabama to manufacture explosives during World War I and fertilizer from the nitrogen after the war. There was considerable disagreement, however, about what to do with the facility after the war. The debate dragged on in the Congress until finally the vision and dynamism of Senator George Norris of Nebraska prevailed in the early days of the Roosevelt administration. In May 1933 Congress passed a comprehensive act establishing the TVA "to improve the navigability and to provide for the flood control of the Tennessee River; to provide for reforestation and the proper use of marginal lands in the Tennessee Valley; to provide for the national defense. . . ." Many specific powers were enumerated in the act, enabling the TVA to be instrumental in the rebirth of a large region of the United States that had been chronically depressed and poverty-ridden since the Civil War.

TVA operates in a 106,000-square-kilometer (41,000-square-mile) basin covering parts of seven states, sells electric power in an area more than twice this size, and has influence throughout the United States. It has pioneered in the comprehensive, integrated development of an entire river drainage basin, and, despite incessant criticism from special interests and ideologues, has been a conspicuous success. Its organization and operations have been reproduced, with suitable local variations, in many countries around the world—but not in the United States. De-

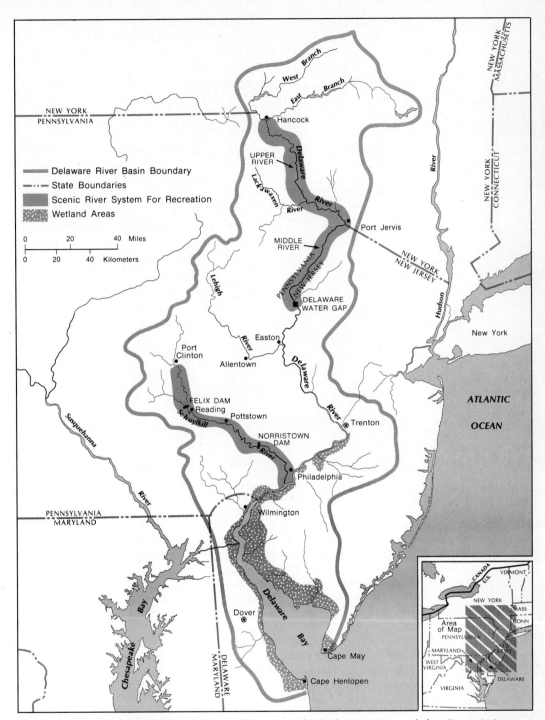

The Delaware River Basin Commission. The major physical operations of the Commission are shown on this map. In 1978 the federal government designated both the Upper and Middle River stretches of the Delaware as part of the National Wild and Scenic Rivers system. This apparently has prevented construction of the long-authorized, but most unpopular, Tocks Island Dam just upstream from the Delaware Water Gap; the reservoir behind the dam would have flooded the valley nearly as far upstream as Port Jervis. The Commission's offices are in West Trenton, N.J.

The Tennessee Valley Authority.   The area bounded by the solid line is the area within which TVA
is authorized to sell electric power. (Note that it does not coincide with the river basin itself.)
Of the power-generating plants shown, seven (including those under construction) are nuclear
and 11 burn fossil fuels, evidence of the inability of the region's abundant hydroelectricity to
meet the needs of its people and their rapidly developing economy.

spite its remarkable achievements, it is
likely to remain singular, largely because
of the preemption of its work by other
federal agencies in other river basins. Yet
it demonstrated the value of unified regional development and the lessons
learned here have been applied elsewhere.

These interstate relations and many others like them, combined with imaginative
and ever-changing relations between the

states and the central government, are
the kind of "creative federalism" found
wanting in Latin America, Africa, and
most of Asia. All successful federations
must make adaptations of this kind. In the
Commonwealth of Australia, for example, we find a similar pattern, developed
in response to comparable conditions on
the basis of a broadly similar, though
much newer, federal structure (dating
from 1901).

LOWER ORDER CIVIL DIVISIONS

All the states of the United States except Alaska are subdivided into counties and these into minor civil divisions.* These lower order civil divisions are "creatures of the states" in which they are located and for this reason they vary considerably. The older states have tended to retain the civil divisions they inherited from colonial times, while in general, the pattern for the newer states was set while they were territories. In the early settlement period, the county was an important governmental unit, less so in New England than elsewhere. Fascinating stories abound in the literature of the formation, division, combination, and abolition of counties, especially in the West. Representation was by county in some state legislatures, and that factor alone made them important. Land values, accessibility to government services, law enforcement, and regional pride all contributed to their importance and generated many fierce political battles. The location of the county seat aroused much emotion; the stories of our county seats alone could fill several volumes.

Today there are over 3000 counties in the United States, ranging in number from 3 in Delaware to 254 in Texas; and in area from Arlington, Virginia (24 square miles)† to San Bernadino County, California which, with over 52,000 square kilometers (20,000 square miles), is bigger than the four smallest states plus the District of Columbia. Twenty-three counties have over 1 million people, while 24 have fewer than 1000. Virginia had 38 independent cities in 1970, which in terms of their charters were similar to counties, but much smaller in area.

The functions of the county depend entirely on those assigned to it by the state. In general, counties have been concerned primarily with law enforcement, the administration of justice, building and maintaining highways, supervising county property, assessing and collecting taxes, recording vital statistics, recording legal documents, directing poor relief (now usually called welfare), and conducting elections. Other functions were performed by some counties, others only during certain periods. By the middle of the twentieth century, however, many functions had passed upward to the states or downward to the mushrooming cities. As their functions shrank, so did their staffs and their budgets. Many of them became hollow shells, little more than lines on the map, and there was much talk of abolishing them altogether. Two states did, in effect, abolish their counties, but the others retained them, often only for sentimental or political reasons. The series of Supreme Court decisions in the postwar period that required representation in state legislatures by population rather than by county further weakened the county.

At the same time, however, the same societal changes that forced interstate cooperation—population growth, urban sprawl, rise in crime rates, pollution, demands of citizens for more amenities, and so on—also forced many states to assign new functions to their counties. While rural areas continue to lose population, the suburbs are still growing, and it is suburban counties and those containing smaller urban areas that have seen the greatest revival in the past three decades. Many suburban counties, in fact, have become virtually states themselves in terms of both functions and organization. They now provide and maintain recreation areas; deal with water supply, flood control, and sewage disposal; provide library, airport, bus, police, fire, and health and hospital services, among other things. They have become big business, with more than 2 million employees spending over $50 billion in 1977. Many counties have outgrown the old unsalaried board of three or four supervisors; they now have full-time, well-paid chief executives and leg-

---

*Connecticut and Rhode Island have deprived their counties of all governmental functions; they survive as statistical and judicial units only. In Louisiana the counties are called parishes.

†Arlington County is a Washington suburb that has no incorporated places within it; it functions much like a city.

islatures, and employ computers and full-time planners and managers. Although a case can still be made for the abolition of counties in rural and large urban areas, it is either too soon or too late to consider elimination of the county entirely as a governmental unit. It appears to be here to stay.

In New England the original unit of settlement was the *town* rather than the county, and it performed many of the functions performed by counties elsewhere. As population grew, the towns tended to be divided instead of having their powers increased. As they became more urbanized, they tended to lose most of their functions to the municipalities. Under the same pressures, the famed New England town meeting has tended to disappear, just as its Swiss counterpart has, to be replaced by more orthodox forms of government. The town system functioned well for a century and a half. It spread, in fact, southward through the Middle Atlantic states, then westward through the Midwest and even to parts of Washington State, until today there are nearly 17,000. Outside New England, New York, and Wisconsin, however, it is called a *township*. Three factors probably explain the confinement of this tier of government to this region: the plantation system in the South, which like *latifundia* everywhere, involved a large measure of political as well as economic and social function; the sparseness of population in the West, which made the county much more significant than it had been in the Northeast; and the migrations of New Englanders westward through the northern tier of states, carrying the town concept in their cultural baggage. Today, in the 19 states in which it exists, the town or township *is* losing its *raison d'être*, is atrophying, and may eventually disappear.

The smallest general purpose administrative unit in the United States is the *municipality*. Today there are over 19,000 of them, and they range in population from a few hundred people to over 7 million in New York City, and in area from a few

hectares to the 2142 square kilometers (827 square miles) of Jacksonville, Florida (nearly the size of Rhode Island!).* Since municipalities are creatures of the state and since the country's population is rapidly urbanizing, the trend is toward the creation of ever more municipalities. This is in sharp contrast to the other units of government, whose numbers have remained essentially stable in the twentieth century. There is almost as much variety in the organization and functions of municipalities as there is in their size, and there is not necessarily any correlation among size, organization, and functions. State law (and in some cases the state constitution) determines all these factors, except for population, though some states specify minimum and/or maximum populations for different types of incorporation. In California, for example, a community may incorporate as a "general law" city with as few as five hundred people, or as a "charter" city with a minimum of thirty-five hundred residents—and these are the only types of municipalities. In Connecticut there are only towns (townships) and cities. New Jersey, however, charters municipalities as cities, towns, boroughs, and villages, all without regard to size; unincorporated areas are governed by the townships and, as in California, by the county.

Functions are also assigned in some states without regard to the size of the municipality, so that in a small one a big-city function may simply be performed by fewer people on a smaller scale for fewer residents than in a big city. In other states, however, formulas determine which classes of municipalities provide which services. Similarly with revenues: the state determines which municipalities may raise money in which ways, and what kind and what level of financial support the state will provide. All of this variety has the advantage of flexibility, of course, but it is

---

*Three municipalities in Alaska (Sitka, Juneau, and Anchorage) have areas exceeding 1700 square miles, but these are special cases. Another special case is the City and County of Honolulu, which includes the entire island of Oahu.

increasingly being considered a nuisance at best and frequently a serious impediment to good government at the local level. The relations among municipalities (particularly adjacent ones) and among municipalities, townships, counties, and states are becoming so complex that many states have embarked on reform programs—or at least are discussing them. At the risk of overgeneralization, we can say that the overall trend at present is for functions to be passed upward through the hierarchy of civil divisions and for financial aid to be passed downward. Both movements include the federal government. There is considerable experimentation with methods of dealing with the manifold problems of urban areas, many involving the political organization of space.

DEALING WITH URBAN PROBLEMS

One way to deal with urban problems is to avoid them altogether by building and governing *new towns* with the best available methods. The new-town movement began in Britain and Scandinavia just after World War II and has spread over much of the world. In the United States the first two, and perhaps still the best known, are Reston, Virginia and Columbia, Maryland. Since then, 15 others were started with federal assistance (until it was discontinued in 1975) in Minnesota, Maryland, Illinois, Texas, Arkansas, New York, North Carolina, South Carolina, Georgia, and Ohio. The success of these communities is debatable (probably Columbia is the most successful so far), but even if it were possible to avoid problems in new towns, they cannot help solve problems in old ones.

Another approach is to *restrict the growth of communities* deliberately. This has long been done through zoning—for instance, requiring minimum building lot sizes of 1 acre or more. Since the Supreme Court has outlawed such "exclusionary zoning," in some places used to keep out "undesirables," other devices have been tried. These include a simple ban on more peo-

ple moving in through a refusal to register new deeds or issue building permits. Another, more constructive, approach is to purchase vacant land around the community to maintain as publicly-owned greenbelts.

The most common technique historically has been *annexation* of surrounding unincorporated land into the municipality. Some cities, including Albuquerque and Los Angeles, have sprawled out over enormous areas in this way. In some cases annexation solved some major problems, but in other cases it simply postponed the day of reckoning and, when it came, the problems were bigger because the cities were bigger. More important, though, annexation is of little or no value to cities hemmed in by physiographic features or by other incorporated places. Many unincorporated communities, in fact, have incorporated at considerable expense solely to avoid being "swallowed up" by a neighboring city. Nevertheless, annexations continue at the rate of about 5000 per year, most commonly by cities in the 20,000 to 250,000 population range.

Other methods of dealing with urban problems include authorization by states of *extraterritorial powers* for municipalities (i.e., they may engage in certain activities outside their corporate limits) and *transfer of functions* to county government. There are also *councils of governments* or other voluntary associations of small and large governments within a metropolitan area, most of which are devoted to research, communications among the governments, coordination, and recommendation, but some have modest governmental powers as well; *special purpose districts* (discussed in more detail in the next chapter); and *intergovernmental contractual arrangements*, most popular in California but adopted nationwide, under which a county will perform certain functions for a city on a fee basis or a city will provide services for its suburbs or even one suburb will provide another with services, all on a contract basis.

None of the methods mentioned so far

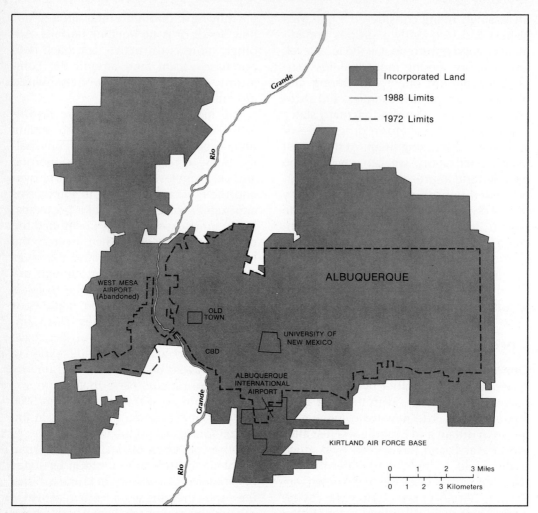

Albuquerque, New Mexico: A city that has grown by annexation.   Settled in 1706 at a ford on the Rio Grande, the town began expanding eastward when a railroad was laid down a mile to the east of the center (now called Old Town) in 1880. Since then its expansion has been primarily eastward and most recently northward and northwestward. Further expansion is limited by the Sandia Mountains to the east and northeast, the Air Force base and Sandia Scientific Laboratories to the south, and Indian lands to the southwest. Additional annexations are planned in the west and northwest.

involves disturbing existing governmental structures. While some are imaginative, all are moderate in their approach to metropolitan area problems. Many observers feel, however, that the governmental structure itself is one of the major causes of contemporary urban problems and that the best way to resolve the problems is to reorganize urban area government in some way. If we consider all the special purpose districts, including school districts, as gov-

ernmental units, there are about eighty thousand governments in the United States, with about fourteen hundred in the New York metropolitan area alone. (The total has been reduced from over ninety thousand in 1962, largely because of school district consolidation.) There is no question that this fragmentation has caused overlapping, duplication, "empire building," jurisdictional disputes, petty jealousies, waste, corruption, dangerous gaps,

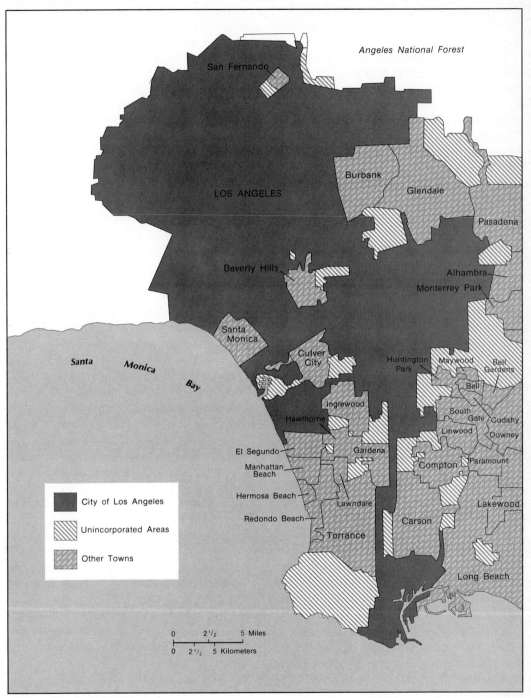

Angeles National Forest

San Fernando

Burbank

Glendale

LOS ANGELES

Pasadena

Beverly Hills

Alhambra
Monterrey Park

Santa
Monica

Culver
City

Huntington
Park

Maywood

Bell
Gardens

Santa Monica Bay

Bell

Inglewood

South
Gate

Cudahy

Hawthorne

Linwood

Downey

El Segundo

Gardena

Compton

Paramount

Manhattan
Beach

Hermosa Beach

Lawndale

Redondo Beach

Carson

Lakewood

Torrance

Long Beach

City of Los Angeles

Unincorporated Areas

Other Towns

0   2½   5 Miles
0   2½   5 Kilometers

Los Angeles: A patchwork city.   Founded in 1781 and incorporated in 1850, the town did not begin to grow rapidly until connected by rail with San Francisco in 1876 and the east in 1885. Another milestone was the opening of the "Free Harbor" of San Pedro to the south in 1914 and its connection with Los Angeles through the "shoestring strip," annexed in 1906. Expansion since World War I has been rapid, but numerous communities have resisted annexation and survive as independent municipalities surrounded by Los Angeles. The county of Los Angeles controls the unincorporated areas, which creates many problems, including some related to law enforcement.

and a host of other ills. Four approaches to reform of this patchwork arrangement have been tried.

City–county separation was tried in Baltimore, San Francisco, and St. Louis in the nineteenth century and in Denver in 1912, for example. In some cases, however, the experiment was abandoned after a while. Today only in Virginia are cities commonly separated from their counties, when they reach a population of ten thousand. Separation (except in Virginia) would very likely cause more problems than it would solve today, and it is not seriously considered.

Consolidation of similar government units is more attractive to many people, but it also arouses much more opposition, most of which tends to be on emotional rather than practical grounds. The opposition has been overcome and some consolidation effected in a few places. (Very few counties have merged, mostly before World War II.) In 1958 Warwick and Newport News, Virginia consolidated; in 1961 the town of Winchester and the city of Winsted, Connecticut were merged and Tampa and Port Tampa, Florida consolidated; and in New Mexico in 1970 Las Vegas city and Las Vegas town merged. One of the largest consolidations in the United States involved North Sacramento and Sacramento, California, which in 1965 formed a united city of over a quarter of a million people. Nevertheless, consolidation efforts have failed more often (as in Pittsburgh, Cleveland, and St. Louis) than succeeded recently, and it does not appear that the success rate will increase notably in the near future.*

City-county merger, on the other hand, has met with somewhat more acceptance. Usually it involves one major city in a rapidly urbanizing county in a moderate-sized metropolitan area. Normally the smaller municipalities have the option of joining or remaining outside the merged unit. Consolidation by referendum, however,

*During the period 1970 to 1975 there were only 24 mergers of this type in the whole country, nearly all of them involving very small communities.

is largely a postwar phenomenon. Mergers in the nineteenth century and the first decades of the twentieth were generally accomplished by state legislation. The first was New Orleans–Orleans Parish in 1805 and others included Honolulu, Boston, New York, Philadelphia, and San Francisco. Since World War II the only city–county mergers attempted or accomplished solely by legislation have been Indianapolis–Marion County in 1969, Las Vegas–Clark County in 1975 and several in Virginia.

Even those accomplished by referendum have been relatively few. Although scores of such proposals have been made by reform-minded citizens all over the country, only 40 different communities have gotten beyond the proposal stage and actually held referenda. Of the 56 consolidation referenda held between 1947 and 1976, only 13 were approved by the voters, and only two communities approved after having rejected consolidation at least once. Two—Tampa, Florida and Augusta, Georgia—have rejected it three successive times. Most successful consolidation attempts have occurred in the South.* The distribution of blacks and whites between the central city and the surrounding county has certainly been an issue in these cases, as in other cases outside the South, but its effect is unpredictable: an equal number of consolidation efforts in the South have failed.† Consolidation efforts will undoubtedly continue around the country, since consolidation does result in more efficient government at lower cost. In order to be successful, however, these efforts will have to include carefully planned and executed campaigns to educate the

*These include Nashville–Davidson County, Tennessee (1962); Virginia Beach–Princess Ann County, Virginia (1962); Jacksonville–Duval County, Florida (1967); Columbus–Muscogee County, Georgia (1970); and Lexington–Fayette County, Kentucky (1974).

†Of the 23 cases so far in which some form of city–county merger has taken place, 8 involve state capitals: Baton Rouge, Boston, Carson City, Denver, Honolulu, Indianapolis, Nashville, and Juneau, which merged with Greater Juneau Borough in 1970. Richmond is an independent city.

voters on the advantages, counter the often emotional appeals of the opposition, and turn out the vote.

Finally, governmental reform in urbanized areas is being tried through what has been called a "two-tier" approach: *metropolitan government*. This is a kind of federalism at the local level. Under this system, the entire metropolitan area, city and suburbs, is placed under a single government that performs a variety of area-wide functions, while the preexisting municipalities retain their identity and responsibility for some local functions. The prototype for this system is the Municipality of Metropolitan Toronto, created by the Ontario government in 1954. Subsequent provincial action in 1967 consolidated Metro Toronto's 13-member government into six units. Metro Toronto was the result of a petition to the Ontario government by the city of Toronto for consolidation and a petition from the suburb of Mimico for some type of areawide

administration for the provision of urban services. Public hearings were held, studies were conducted, recommendations made and debated, and finally the legislature acted. Despite the usual federal problems of distribution of powers and representation on the governing council and complaints that the area covered is not large enough, the experiment has been quite successful and the residents are generally enthusiastic about it.

The system has also been adopted in Edmonton, Alberta and Winnipeg, Manitoba, as well as by a number of smaller communities in Ontario. In the United States this approach has so far been adopted only in Miami-Dade County, Florida, in 1957. It differs from the Toronto model, however, because in Ontario the county was abolished, whereas in Florida home rule was granted in 1956 to the county, which then adopted a new charter reorganizing county government and transferring many city functions to the

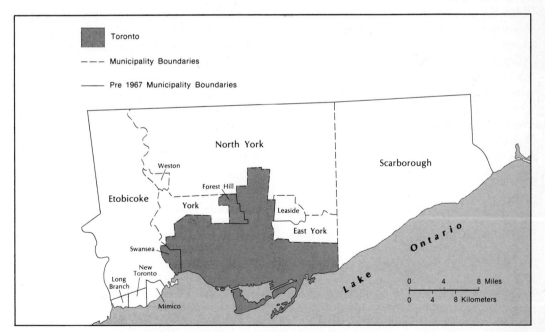

The Municipality of Metropolitan Toronto. Created in 1954 with the city of Toronto as its core, Metro Toronto was reorganized in 1967 to bring its component units from 13 to six. The entire area within the outer boundary is governed as one unit, but the Townships of Etobicoke, York, North York, East York and Scarborough have some degree of local autonomy. This is the prototype for metropolitan government elsewhere.

county. It was not a consolidation, however, for the 26 municipalities in the county still exist and retain jurisdiction over certain local matters. Nevertheless, the county's authority is very broad and it has been expanding its jurisdiction. The success of this experiment is by no means certain; it has barely survived a variety of crises and may have to be revised again in the near future.

Variations on metro government are being applied in Alaska and the Twin Cities. The Alaska legislature provided in 1961 for the creation of boroughs (essentially regional governments incorporating at least one metropolitan area) on a local option basis and in 1963 boroughs were made mandatory under certain conditions. Two classes of boroughs were established, based not on population but on powers, and the borough chose its class. By 1964, there were one first-class borough (Greater Juneau) and eight second-class boroughs. The remainder of the state constitutes a single unorganized borough. All functions normally performed by lower order civil divisions are divided between the borough and municipal governments. In the Minneapolis–St. Paul area of Minnesota, the Metropolitan Council, created by the state legislature in 1967, was given jurisdiction over areawide matters, as in Toronto and Dade County, but it was strictly a policy-making body and had no executive powers. It relied on the preexisting governments and agencies to implement its policies. Since establishment, however, it has been granted taxing power and the power to execute its proposals in matters of zoning, solid waste disposal, pollution, and noise abatement. It may be granted additional powers as experience accrues.

With so many obvious advantages to be gained from metro government, why has it not been adopted more widely, especially in the United States? Suburbs often resist metro government for the same reasons they resist annexation and consolidation. But recently, as this resistance has begun to weaken, new opposition has arisen from within the central cities themselves. Because of changing demographic patterns (primarily the exodus of whites to the suburbs and the inflow of blacks into the cities) and sweeping political and social changes, blacks have at last attained some measure of political power in many cities, and they are unwilling to have this power diluted or possibly obliterated in a governmental unit which would include a number of white-dominated suburbs.*

Metropolitan reorganization is another field in which geographers can make major contributions, but as yet few have done so. It is enormously complex and calls for a thoroughly interdisciplinary approach. One of the chief complicating factors is the multiplicity of special purpose districts, found throughout the country but concentrated in urban areas. This subject is covered in our next chapter.

---

*Another reason for the greater success of metro government in Canada than in the United States is that Canadian provinces have greater power to create such units than American states do.

Chapter 15

SPECIAL PURPOSE DISTRICTS

As numerous as the general purpose governmental units are in the United States, they are more than equaled by special purpose districts. While other countries have them also, and for similar reasons, nowhere else are they so prolific and so varied. They have become so important partly because of the nature of the American federal system; the high mobility of Americans (changing residence on the average every five years); the emphasis on private enterprise and individual decision making regarding land use; the almost total lack of planning in most American cities until recently; and the postwar suburbanization or urban sprawl that has so transformed the American landscape. Historically, the special district (aside from the school district) was created in rural areas to administer drainage, irrigation, water conservation, flood control, fire protection, soil conservation, hydroelectric power, river transportation, and similar functions. Since World War II, however, with cities and suburbs exploding (spatially), the special district has experienced a phenomenal rate of growth within or overlapping metropolitan areas, especially those included in MSAs.* Just in the decade 1952–1962, for example, their numbers increased by nearly 50 percent, and the growth continues.

The traditional special districts continue to function in metropolitan areas, though often their organization and services are adjusted to meet changing needs, and new ones are being created to meet increased demands for their services. Newer types of districts, however, associated with largely urban needs, are being formed in even greater numbers. These include, as a representative sample, transportation (airport, bus, rail, seaport, bridge, highway, tunnel), law enforcement, judicial, utilities, fire protection, housing and urban renewal, air and water pollution control, solid waste disposal, library, sewer, park, and hospital districts. They are usually operated by authorities, commissions, or boards. Some districts perform multiple but related functions. A popular kind of district is a planning district or region, which may cover a very extensive area around a city or in a depressed rural area of a state. Some districts that straddle state boundaries were discussed in the previous chapter. The only essential difference between them and intrastate districts is that interstate districts must conform to federal law.

---

*A metropolitan statistical area (MSA) is defined as "an integrated economic and social unit with a recognized large population nucleus." The boundaries are normally county boundaries and thus often enclose rural as well as urban areas. Two important MSAs that have no central city as such are located in suburban Long Island, New York, and Orange County, California. Besides the basic MSA, of which there were 261 in 1987, the Bureau of the Census has identified 71 larger Primary MSAs and 20 still-larger Consolidated MSAs.

How can we account for the popularity of special purpose districts? Why can local government units (counties, townships, municipalities) not perform these functions? At least five sets of circumstances related to local governments seem to encourage the creation of special districts.

1. They frequently cover inadequate areas and/or lack the resources to do the job.
2. Their expansion is restricted by statutory and even constitutional provisions.
3. Their quality of service or performance record may be poor.
4. Special interest groups want independent units established that will obtain the best possible results from their point of view.
5. The existing units themselves support the creation of new units to handle functions they cannot handle as well.

The districts themselves can operate within areas delimited for their specific purposes, disregarding existing political boundaries. They can hire experts to concentrate on particular problems; they are generally nonpolitical and do not threaten existing political interests; and, since most have independent power to tax, borrow, and levy service charges, they can be financed in a more efficient and flexible manner than frequently is possible under the numerous restrictions and limitations imposed on local government units. There is not necessarily a conflict either between a special and a general government unit. In Orange County, California, for example, the county serves as the tax-collection agent for a number of special districts; its regular tax bill includes an itemized list of these special taxes.

Special districts are not without their problems and their detractors, however. After a decision has been made to establish such a district (which itself may have been a complex process), a major problem and one of special interest to geographers is how to draw its boundaries.

Management and staffing methods are sometimes questionable; responsibility to the voters and taxpayers is sometimes obscured or forgotten; single purpose sometimes becomes single minded, with the district pursuing its own plans without balancing them against other needs; a multiplicity of special districts overlapping one another and the local government units leads to jurisdictional disputes and duplication of effort, and individual citizens are likely to be ignorant, confused, or dismayed (or all three) about the dozen or more governmental agencies that tax them and regulate their activities. These and other problems have led to the demands for reform of local government organization described in the last chapter. They have not, however, noticeably slowed the proliferation of special purpose districts. Instead, these are likely to continue increasing in the future—although once again the greater growth will probably be in rural rather than metropolitan areas.

While political scientists, historians, and economists have studied various kinds of special purpose districts, geographers have tended to ignore them. Here we single out three of the most interesting, if atypical, types of districts for special attention: school districts, public lands, and Indian reservations.

SCHOOL DISTRICTS

Education has had a high priority in American society since colonial times, and the United States was one of the first countries to encourage publicly financed and operated educational systems. Only two years after independence was secured by a peace treaty with Britain, the Land Ordinance of 1785 required that one section of each township in the new lands of the West was to be reserved for public education. The land grant college system, established by the Morrill Act in the midst of another national crisis in 1862, was also a demonstration of this commitment. Education itself, however, was a matter left

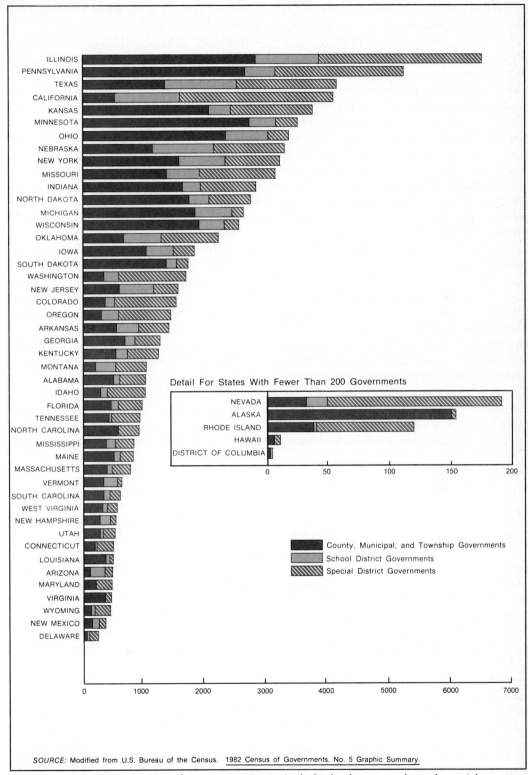

Local governments in the United States. Note particularly the large number of special purpose districts in many states.

by the Constitution to the states. Although the federal government has come to the aid of education since the 1950s by supplying financial and technical assistance and setting some standards and guidelines, the states still control public education. In only five states (Alaska, Hawaii, Maryland, North Carolina, and Virginia) and the District of Columbia are all local school systems dependent on some local governmental unit; in Tennessee and five New England states most school systems are dependent in this way. Elsewhere all public school systems are operated through independent school districts, in order, the theory goes, to keep them free from local politics.*

In practice, the local school districts, dependent or independent, have been the locale of some momentous political and social battles of modern times: church–state separation, division of powers between state and federal governments, local control versus state control of education, racial segregation and desegregation, censorship, sex education, Darwinian concepts of evolution—all have been fought out in both the educational and political systems. Geographers, however, are concerned primarily with the spatial aspects of education, and at the local level, these focus on the independent school district.

In the 1930s, there were more than 120,000 school districts in the United States. By 1982 there were fewer than 15,000, and the consolidation movement was still strong. Despite the cost reductions resulting from this consolidation, education still accounts for nearly half the expenditures of local government and 60 percent of all state aid to local governmental units. Whether the schools are referred to as consolidated, unified, central, or regional, they still cost money, and

school financing is currently the major issue in education.

Typically, school districts are endowed by state governments with considerable administrative, curricular, and fiscal independence. Most are administered by unsalaried, elected boards that hire professional superintendents and other administrators and are, in turn, supervised by state boards or departments of education. Many are large, wealthy, and dynamic units that completely overshadow some of the local general purpose governments. Many, at the other extreme, particularly in the Midwest, are "nonoperating" districts that own no property and provide no instruction but only furnish tuition and transportation for resident students attending schools in other districts. Most of these result from consolidation.

Consolidation has become not only desirable, but both essential and feasible as the country and society have changed. It became increasingly difficult to attract and retain good teachers and administrators for rural districts, and the growing complexity and mobility of society increased the need for expansion, improvement, and standardization of curricula. The most important development impelling school district consolidation, however, has been the great increases in the costs of construction, maintenance, instruction, and administration. Most of the districts abolished were unable to finance needed expansion and improvements or even their current operating budgets. In addition, there are economies of scale to be realized; it costs less to operate 1 high school for one thousand students than 5 for two hundred each. Consolidation makes feasible the purchase and operation of expensive equipment for transportation, science instruction, athletics, dramatics, vocational training, and so on. Adult education, programs for the handicapped and bedridden, joint programs with universities, exchange programs with other districts and even other countries, sabbaticals for teachers, and many other val-

*A unique bistate school district is that encompassing Norwich, Vermont, and Dresden (Hanover), New Hampshire, which took an Act of Congress to make possible. The community of interest focused on Dartmouth College and Mary Hitchcock Hospital in Hanover.

uable contributions to quality education for a democratic society become feasible in larger districts.

Two major problems arise with consolidation, however: how to delineate the new districts and how to work out relations between the new districts and other units of government. When consolidation is proposed, frequently by the state, a number of questions arise. Among them, the most important from our perspective, are: Which districts should be included in the plan? How should the boundaries be drawn? Where should the new central schools be located? What procedures and routes should be arranged for transportation? What would the social and economic effects be on both the communities losing schools and those gaining them?

Opposition to consolidation comes from those who philosophically oppose surrendering "local control" of schools, even though local control does not necessarily mean good education; from rural families who do not want their children infected with an urban viewpoint or "radical" ideas; from parents who do not want their children mixing with others of different race, religion, or social background; from local school board members and administrators who fear losing prestige and power, if not jobs; and from a variety of other groups with other objections. These objections have slowed but not halted consolidation.

School districts can, in fact, get too big, so big that economies of scale pass the point of diminishing returns; that parents and taxpayers do lose contact with their educational system; that the new education establishment becomes a powerful political force itself, sometimes challenging political officials and institutions over many issues; that community identities and values are sacrificed to the cause of efficiency. For these reasons, New York City decentralized its school system in 1969, creating 31 (later 32) neighborhood school districts that have their own school boards and considerable autonomy. The central board retained control over high schools and certain financing and contractual matters. It is too early to reach a definitive conclusion on the effectiveness of this decentralization, but to date its results have not been particularly encouraging. It has not noticeably increased local participation in decision making, produced innovative programs, or saved money. Perhaps more than two decades is needed to produce these results.

Consolidation of districts, especially in metropolitan areas, often brings the new districts into conflict with established general or special purpose districts. Problems arise regarding police and fire protection, traffic control and pedestrian safety, use of library and recreation facilities, adult education and vocational training, and health services. These problems have to be resolved by formal or informal arrangements among the various jurisdictions, or neither the educational process, the children, the parents, or the taxpayers will realize the benefits of either consolidation or local control of education through school districts.

PUBLIC LANDS

Another type of special purpose district is not normally considered as such but should be mentioned here. This is land in the public domain that is managed under special provisions of law for particular purposes. All land acquired by the United States from other countries and from individual states since independence (except for properties privately owned under laws of the former sovereigns) automatically entered the public domain. This totaled 728 million hectares (1.8 billion acres) between 1781 and 1867 (exclusive of Pacific and Caribbean territories), of which 445 million hectares (1.1 billion acres) had been disposed of by 1982. The remaining public domain plus another 27 million hectares (68 million acres) acquired since then (usually by purchase or gift from citizens) is managed by a number of governmental agencies, most of them within the U.S. Departments of Interior,

Table 15–1   U.S. Federal Public Lands (in million acres)*

	1975	1980	1983	1983–84	
Total	761	720	732.0	Bureau of Land Management	337.0
Department of the Interior	538	502	506.6	National Park Service	75.0
Department of Agriculture	188	184	192.5	Bureau of Indian Affairs	53.0
Department of Defense	31	30	29.1	Tribally owned	42.4
Tennessee Valley Authority			1.0	Trust allotted	10.2
Department of Energy			0.6	U.S. government owned	0.4

*These figures have been compiled from various government sources and are not always consistent. They should, therefore, be considered approximations.

Agriculture, and Defense, and the Nuclear Regulatory Commission. In 1984, 32.2 percent of all the land in the United States was owned by the federal government (298 million hectares, 736 million acres) and 6.8 percent (63 million hectares, 155 million acres) by state and local governments.

These lands are scattered widely over the United States, but nearly 94 percent of all federally-owned land is in the West; 64 percent of the land in the 13 Western states is federally owned, including 85 percent of Nevada and over 86 percent of Alaska. The 1971 Native Claims Settlement Act granted outright title to 40 million acres of federal land in Alaska to native Indians, Aleuts, and Eskimos and also authorized the Secretary of the Interior to withdraw from the public domain 32 million hectares (80 million acres) for possible inclusion in national parks, national forests, wildlife refuges, and wild and scenic rivers. The Statehood Act of 1959 had allocated another 42 million hectares (104 million acres) to the new state. These three allocations total in area nearly the

size of Texas, and Texas-sized controversies over their precise distribution and use are still raging.

These public lands represent zoning on a state and national scale since their uses are controlled by statute and regulation. Even land once owned by the federal government but granted or sold cheaply to homesteaders, railroads, irrigators, and states, amounting to over 400 million hectares (1 billion acres), carried with it requirements for, or limitations on, its use. Through a variety of programs, the federal government encourages the states to support soil, water, forest, and wildlife conservation, in part by restrictions on land use, required reclamation of strip-mined land, and other sound management practices.

Federal land ownership and management, particularly the environmental and natural resource regulation involved, have long been resented in the West. This smoldering resentment broke into the open in 1979 when the State of Nevada, 86-percent owned by the federal government, filed a lawsuit claiming the land within the state

Table 15–2 Percentage of Land Owned by Federal Government in the United States (in million acres)

States	Federally Owned	Total in States	Percentage Federally Owned
Alaska	321.5	365.5	88.0%
Western	358.3	752.9	47.6
South Atlantic, South Central, DC	29.6	561.2	5.3
Northeastern, North Central	21.9	587.6	3.7
Hawaii	0.7	4.1	17.1
TOTAL	732.0	2,271.3	32.2

Source: U.S. General Services Administration. *Summary Report of Real Property Owned by the United States Throughout the World as of September 30, 1983.* July 1984, p. 20.

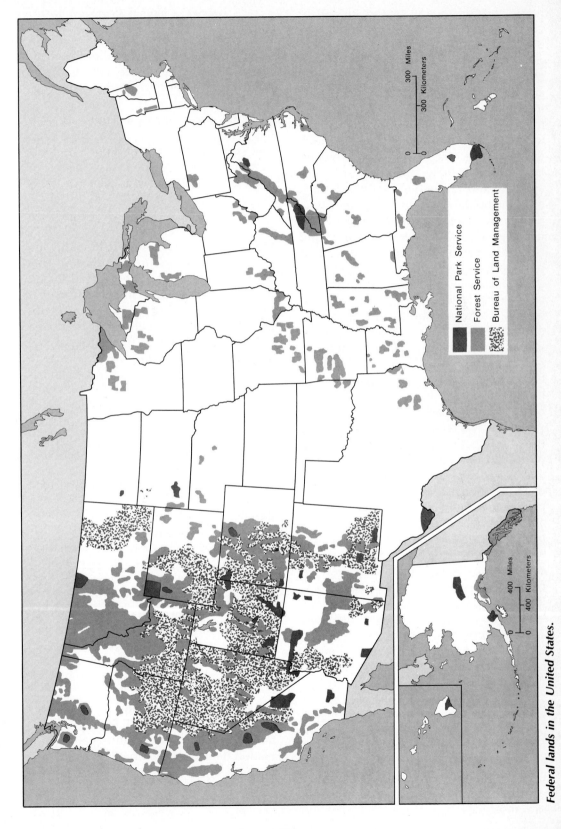

National Park Service

Forest Service

Bureau of Land Management

300 Miles

300 Kilometers

400 Miles

400 Kilometers

Federal lands in the United States.

under the jurisdiction of the Bureau of Land Management. This triggered "The Sagebrush Rebellion", soon joined by six other western states. In 1981, the new Reagan administration gave considerable support to the rebellion, but only by seeking to loosen the regulations, not by giving away the land. But support in the West faded when the costs and benefits of acquiring huge tracts of federally-owned land were carefully calculated. The Reagan administration, meanwhile, did sell off some public land, but only a small fraction of it; relaxed enforcement of land-use regulations; accelerated timber and minerals leasing; froze acquisition of parkland and wildlife refuges; slowed down wilderness designations; and encouraged private rather then public conservation efforts. Short-term economic considerations took precedence over long-term environmental ones. Thus, the federal government defused the rebellion by making many concessions; Nevada lost its lawsuit in 1981 and the Sagebrush Rebellion wound down.

Though the rebellion ended, the controversy lingers on. It flared again in 1986 when the Bureau of Land Management and the Forest Service proposed a trade of 14 million hectares (35 million acres) of lands under their jurisdiction in the West, and again in 1987 over addition of public lands to the National Wilderness Preservation System. Ranchers, miners, foresters, and conservationists are likely to continue jousting over public lands for a long time to come.

Even in a huge, rich, and relatively sparsely populated country, such as the United States land for grazing, mining, and forestry is highly prized, no less now than in pioneer days. And the controversies over land use in the West are as bitter as those over land ownership or over land use in the crowded Northeast. These continuing controversies underscore the absence of, and need for, a national land-use policy, one that will help us make wisest use of one of our most precious natural resources.

INDIAN RESERVATIONS

Another special type of district, very different from the ones described so far, is the reservation set aside for aboriginal Americans. Other countries, such as Canada, Chile, Brazil, Dominica, and Australia, have set aside lands for their indigenous peoples in which they exercise some degree of autonomy as specified by statute. Nowhere else, however, is the aboriginal reserve as numerous, varied, and complex as in the United States.

According to the U.S. Department of Commerce figures for the end of 1972, there are 259 Indian reservations in the United States, excluding Alaska. This number alone may be very misleading and requires some explanation. In the first place, the very word "reservation" is not clearly defined.* In general, it is an area of land belonging to an Indian tribe, some of which may be held in trust by a state or the federal government. An examination of a reservation, however, may reveal much land within it that is owned by or leased to whites, owned by individual Indians either in trust or in fee simple, and owned or controlled by a state, county, or municipality. Furthermore, Indians and whites both may own land adjacent to the reservation, much of it formerly reservation land but left outside when the reservation was reduced in size.

Of the 259 reservations, 26 are state reservations, primarily in the East; the remainder are federal reservations, mostly in the West. Many of them are very small. Of the 76 federal reservations in California, for example, 17 of them are smaller than 40 hectares (100 acres), including three of 2.2, .5, and .37 hectares (5.5, 1.32, and 0.92 acres). One state reservation in Connecticut is 0.1 hectare (0.26 acre). The largest, on the other hand, is the Navajo reservation, with nearly 5.66 million hectares (14 million acres). Land quality, population density, accessibility, mineral and

*See Imre Sutton, "Sovereign States and the Changing Definition of the Indian Reservation," *Geographical Review*, 66, 3 (July 1976), 281–295.

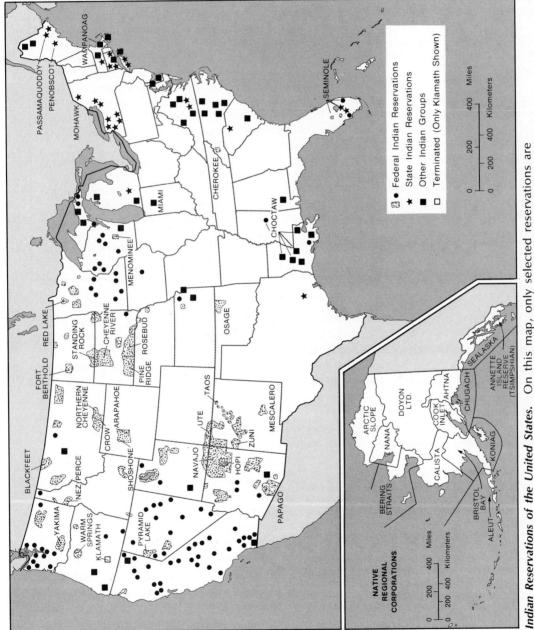

Indian Reservations of the United States. On this map, only selected reservations are named for reference. Only a representative sample of the 76 rancherías and other Indian lands in California are shown. Comparison of this map with a map of distribution of Indians in the United States will show marked discrepencies, among which are concentrations of Indians in Los Angeles, Chicago, and New York.

SOME UNUSUAL INDIAN RESERVATIONS

There are some interesting anomalies with regard to reservations. Until recently, there have been 13 reservations in Alaska for Eskimos, Aleuts, and two different Indian groups. Under the Alaska Native Claims Settlement Act of 1971, all outstanding native claims against the government for lost land are to be settled by allowing newly formed Native Village Corporations (205 of them) and the new Regional Corporations (11) to select allocations of public land totaling 38 million acres. The village corporations will take the place of 12 of the reservations. The remaining one, the Annette Island Reserve of the Tsimshian Tribe in the southeast, is exempted from the provisions·of the Act and will continue in its present status.

In Oklahoma, there are 32 tribes and tribal groups who own 43,500 hectares (107,500 acres) tribally and 607,000 hectares (1.5 million acres) in allotments, yet the state has no reservations as such.* Many tribes are organized under state or federal laws or regulations and most members live on or near their "reservations," which are federal trust areas. Most of them have assimilated into the white society.

The Red Lake Reservation of the Chippewa Tribe in Minnesota is an autonomous unit in the traditional tribal homeland that was never formally ceded to the United States. It is thus exempt from all state jurisdiction that may apply to other reservations in Minnesota, though it is under federal jurisdiction. The Warm Springs Reservation in Oregon (Warm Springs, Northern Paiute, and Wasco Confederated Tribes) enjoys a somewhat similar status.

*There is a continuing dispute over the Osage Reservation. The federal government claims it still exists, but the state denies it.

forest resources, and other characteristics are similarly varied. Generally, however, reservations are located in areas bypassed or unwanted by white settlers and land speculators. White society, however, has reached many of them and engulfed some, generating many problems.*

The Indian reservation is a distinctive kind of territory. Though located usually within the boundaries of a single state, often within a county, it is not part of the state or any lower order civil division. Over the past two centuries government policy toward Indians has fluctuated from extermination to paternalism to termination. This indecision, combined with different conditions in various parts of the country, different Indian cultures and reactions to white intrusion, and changing circumstances in American society generally, accounts for the great variety we find today in Indian reservations and their relations with other units of government and with individual Indians.

As American citizens, Indians are free to come and go as they please and there is considerable movement to and from the reservations. The trend for the past quarter century, though, has been for increasing numbers to move away from the reservation more or less permanently, aided by government relocation programs. Today only about half of the 800,000 to 1 million Indians (outside

*The term "tribe" as applied to Indians also requires some explanation. In many cases it was simply a term of convenience applied by whites to loosely associated Indians who shared some cultural traits. Some of these groups were nations, as we have defined the term earlier. Some were simply hunting bands or independent villages of much larger cultural groups. Some were amalgamations or confederations of tribes. Some were branches or offshoots of other tribes. Most of the Indian groups in the West were not organized tribally, but generally by clan (lineage) or village. Today the term is a formal one applied to those Indian groups that organized as such under the Indian Reorganization Act of 1934 or other legislation. A number of Indian groups in the East, still lacking official tribal status, are at a disadvantage in applying for federal or state Indian benefits and in pressing land and other claims, whereas others have secured the coveted designation and are entitled to some federal services.

Choctaw Reservation, This consists of over 17,000 acres of ancestral lands in East-central Mississippi. Tribal headquarters is in Pearl River, Mashoba County. Most of the tribe was removed to Oklahoma in the 19th century and over 90,000 of them live there today. Only about 30,000 live on this reservation. Most of the tribal income here derives from forestry, with additional funds coming both from agriculture and an arts and crafts shop in nearby Philadelphia, Mississippi. The industrial park is new and not fully developed. Some reservations are wealthier than this one, with income from oil wells, forest products, manufacturing, and other activities. Many reservations have virtually no resources and must resort to raising funds from bingo and other forms of gambling, and the sale of tax-free cigarettes and other items to tourists. (Courtesy of the Choctaw Community News. Photo by Julie Kelsey.)

Alaska) are living on reservations. Off the reservation they have the same status as other citizens and are subject to the same laws. On the reservation the situation is more complex. In general, reservation Indians are subject only to federal and tribal law, and the jurisdiction of each is fairly well defined. In practice, however, the federal government has gradually whittled away the immunities granted to Indians by constitutional, treaty, and statutory provisions, particularly in regard to land tenure. Indian lands were opened to white homesteaders, for example, although tribes retained their legal status. Land was allotted to individual Indians, eroding the political power of many tribes and allegedly increasing state authority over the privately owned land—and the landown-

ers. The most serious erosion of tribal authority has occurred since 1953. In that year the federal government adopted a policy of working toward termination of federal responsibility for the tribes. It also passed P.L. 280, which permitted the states to assume civil and criminal jurisdiction over the federal reservations if their legislation was acceptable to the federal government.

Between 1954 and 1960, there were 61 tribes, groups, communities, allotments, and *rancherías* (small reservations in California) terminated by the withdrawal of federal services and protection. The largest tribes choosing termination were the Klamath in Oregon and the Menominee in Wisconsin (whose reservation became a county). The termination policy was just short of disastrous and has been abandoned.* P.L. 280 continues in force, though much modified by the Civil Rights Act of 1968 and numerous judicial decisions. Under it Alaska, California, Iowa, Kansas, and Minnesota assumed jurisdiction over the reservations within their boundaries. Wisconsin, Florida, Idaho, Nevada, and Washington assumed jurisdiction in whole or in part by state legislation under the authority of the federal law. Other states have taken the opportunity to extend certain kinds of jurisdiction over reservations. All of this has led to great confusion, much squabbling over jurisdiction, and many court battles. Some issues involved are land tenure, housing codes, environmental protection (including air and water pollution control, forest-fire prevention, restrictions on traditional fishing), and most important, resource exploitation and commercial and industrial development on Indian reservations. Some tribes have set up large-scale gambling on their reservations to attract outside money and this, too, has become controversial.

Some future damage to Indian reservations was prevented by a provision in the 1968 Civil Rights Act that requires tribal consent before a state may assume civil and criminal jurisdiction over reservations, but the current picture is still very much a muddle. Federal, tribal, and state (including county and municipal) jurisdiction coexist (and sometimes overlap) in different strengths in different states and even within states. Current laws conflict with old treaties. Different philosophies of land ownership and land use still clash. It will be a long time before all the problems are resolved.

The Indian reservation is, in Sutton's words (pp. 292–293), "a unique and semi-autonomous enclave. . . . There is no obvious counterpart in our political system: a legally sanctioned island that only indi-

*Federal trusteeship was restored to the Menominees in 1973 and their reservation was reestablished, but a great deal of damage had already been done to tribal resources and institutions.

genous Americans can possess." The most significant element of this uniqueness is tribal autonomy within the reservation. But the dimensions and even the basis for this autonomy are questionable. The Supreme Court has ruled that Indian sovereignty is nonterritorial; that is, it is limited to functions within the reservation instead of applying to the territory of the reservation itself. The reservation is not a state, and its boundaries are more permeable than those of the weakest state. It is autonomous but not sovereign. It bears many similarities to a developing country (as we will see later) as it begins to assume control over its own economic development.

It does not fit into the hierarchy of civil

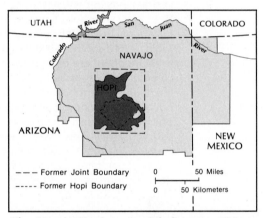

The Navajo-Hopi territorial dispute. This is similar in many ways to international boundary and territorial disputes. The Hopi Reservation in Arizona is an enclave within the Navajo Reservation. For more than a century the pastoral Navajos have been occupying Hopi land without Hopi consent and the agricultural Hopis have been trying to oust the Navajos. In 1943, the Bureau of Indian Affairs approved the Navajo request for "joint use" of three-quarters of the Hopi Reservation and this was confirmed by a 1963 court decision. But the Navajos prevented Hopi use of the land and the dispute went on. In 1976 the Hopis sued the Navajos in court. In 1977 a compromise was reached: the disputed land would be partitioned and the new boundaries would be fenced. The lightly shaded area is the new Hopi reservation. A decade later, however, the compromise had still not been implemented as Navajos living on Hopi land refused to be relocated.

divisions—state, county, township, municipality—yet it performs many of the same functions and overlaps all of them. It is not a special purpose district because it was not created to serve a public purpose as an arm of the state. Yet all the jurisdictional problems we discussed earlier in this chapter exist in reservation areas. They are governed in part by treaties that have little or no standing in international law and cannot be amended as international treaties can. They have some of the boundary and territorial problems of States, yet seek redress in federal court. Indians are American citizens and under federal authority wherever they are in the United States; they may also be tribal citizens but are under tribal authority only when on the reservation and are subject to state law when off it. As the King of Siam says in *The King and I*, "It's a puzzlement!"

Chapter 16

THE GEOGRAPHY
OF ELECTIONS

In 1913, *André Siegfreid* published a seminal study of elections in western France and their relationship to geographic and socioeconomic factors. In 1949, he produced another landmark study of elections in the *Département* of Ardèche, on the west bank of the river Rhône, from 1871 to 1940. During the intervening period, not only did Siegfried develop the technique of comparing maps of electoral results with maps of geographic and other factors, but other French scholars, including geographers, did similar work. So did a few people elsewhere, mostly sociologists and political scientists. In 1949 also, the noted American political scientist *V.O. Key, Jr.*, published his classic study of voting in the American South. After that more Americans became involved in the study of elections. It was not until the late 1960s, however, after the advent of the so-called "quantitative revolution," that American, British, and other political geographers became seriously interested in the subject. Since then we have been inundated with studies of all kinds on elections. The flood has become so great, in fact, that, on the basis of sheer magnitude alone (if for no other reason), electoral geography has come to be recognized as a subdivision of political geography.

Examining the literature on voting produced since 1913 in North America, Western Europe, Australia, New Zealand, India, Israel, and the very few other places where it has attracted professional interest, we find some good studies on referenda and plebiscites; voting in the United Nations, the European Communities, and other intergovernmental organizations; and voting in national legislatures. Overwhelmingly, however, attention has been fixed on contested elections for seats in legislatures at the national and subnational level, and elections for president in States with a presidential system. We may wonder whether this fixation is based on the true significance of these elections or on the ready availability of enormous quantities of statistical material and the machines to play with them. Certainly, playing with election returns is a marvelous exercise for graduate students; whether professional political geographers can make a really useful contribution doing so, however, is not yet entirely clear. In any case some interesting patterns and theories are beginning to emerge about voting at the national and subnational level, and for this reason electoral geography is being presented here along with subdivisions of the State.

THEMES IN ELECTORAL GEOGRAPHY

Peter Taylor and *Ronald Johnston*, in their numerous publications on the subject, have identified "three main foci of geographical interest in electoral studies." First is *the geography of voting*. Generally, studies in this genre try to explain the patterns of voting after a particular election or group of elections. The emphasis is on statistical methods and, while the work follows the tradition established by Siegfried, maps have been largely replaced by statistics and formulas in illustrating the results.* It is questionable whether this represents a true advance toward a clearer understanding of elections and their results. Are numbers really better than maps?

The second major theme, or focus, is *the geographic influences on voting.* There are four aspects of voting that can be explained in part by examining the geographic background to an election. They are voting on issues, voting for candidates, the effects of election campaigns and—most geographic of all—"the neighborhood effect." This is the relationship between election results and the hometowns or home districts of the candidates. A study of either of these two main themes requires an understanding of the electoral system in use at the time. There are many such systems, and many combinations and variations. To name only the major ones: proportional representation, winner-take-all, single-member constituency, multiple-member constituency, majority, plurality, weighted plurality, representation by party, representation by socioeconomic group, etc. Then there are the methods of selecting candidates: primary elections, party conventions, nomination by party chiefs, selection by traditional social/political leaders, and so on. The more one delves into the intricacies of the political process, the farther one goes from geography and the less geography has to contribute to an understanding of it.

The theme in electoral geography that is undoubtedly most geographic is that of *the geography of representation.* In those countries in which there are elections to the legislature (or legislatures if the civil divisions also have elected legislatures) based on constituencies or districts, the number of districts and their boundaries can have profound influences on the composition of the legislature independently of the actual total votes for candidates and/or parties. This often, even in the most "democratic" countries, results in a disparity between the number of votes won by a party and the number of seats it wins in the legislature. This electoral bias is most evident in those countries, such as the United States, United Kingdom, Canada, and New Zealand, in which the plurality-majority system is used. Here the winner of a seat is the candidate with a plurality of votes, even if it is not close to a majority, and the party that organizes the legislature (and in a parliamentary system forms the government) is the one that can command a majority of seats. Because of their importance and their geographic nature, we turn our spotlight now on electoral districts, using the United States to illustrate our main points. It should be borne in mind, however, that similar situations exist in other countries. Recent studies of electoral districts in the United Kingdom, Norway, France, and Australia show this graphically.

ELECTORAL DISTRICTS

Most Americans live within a number of electoral districts, typically congressional districts and separate districts for the two houses of the state legislature. Many, especially in metropolitan areas, live within additional districts (wards, zones) for local or metropolitan legislative bodies. Thus, not only does the individual citizen cast votes in as many as six or seven districts within a few seconds on election day, but the thousands of electoral districts generate enough statistics to enable us to make some useful generalizations. Each district is in itself a study, for its size, shape, population numbers and characteristics, land use, and so on. Combining districts in various ways reveals patterns and raises questions only hinted at by the individual district. Studies of the same information for a number of successive elec-

*In Taylor and Johnston's 1979 book, *Geography of Elections,* for example, of 466 pages of text, only 22 are devoted to maps, many of poor quality.

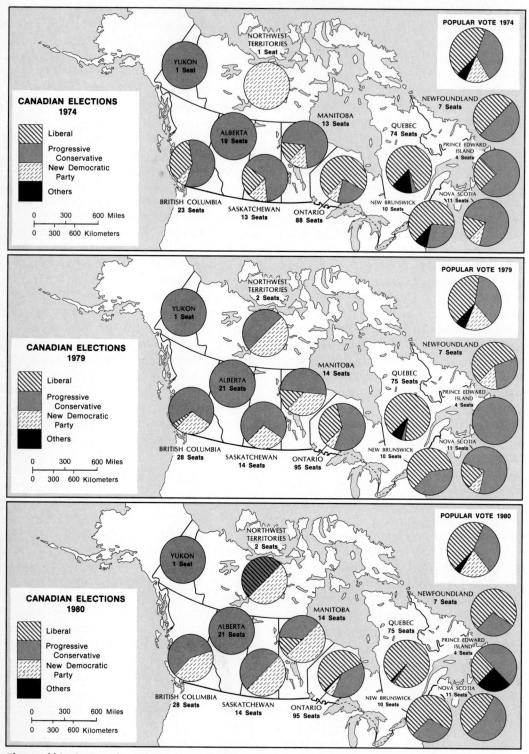

Electoral bias in Canada. The disparity between votes received by the various parties in successive elections and the seats won in the national parliament is clearly evident in these maps. This is a common problem in many countries that do not have proportional representation. However, PR frequently generates other and more serious problems.

tions or selected elections over a longer period of time reveal patterns of change and raise additional questions. Studies of this type must be based, however, on a clear understanding both of the voting system used and method of application of the system. In the United States this is relatively simple, but in other countries it can trap the unwary. A basic principle in studying election data is that the most meaningful patterns are revealed by the smallest possible voting units. This means manipulating more figures and makes mapping more difficult, but it is frequently worth the extra trouble.

The aspect of the electoral process in the United States that most readily lends itself to geographical analysis is the voting district. Because the number of districts in legislative elections is limited by the number of seats to be filled, while the electorate itself changes (sometimes dramatically) in both numbers and distribution, the size (population) and shape of the districts can be important influences on the outcome of the elections.

The federal Constitution mandates that a decennial census of the population be conducted to determine representation in the House of Representatives. Since the number of members of the House is fixed by law (currently 435), increases in population (even through the admission of new states) cannot be accommodated by simply adding more House seats. States that are growing relatively slowly or are actually losing population must therefore surrender seats to the more rapidly growing states. Each census, therefore, results in a *reapportionment* of the 435 seats among the several states. In the states that neither gain nor lose seats, the congressional districts normally remain unchanged. In the rest, *redistricting* is necessary; that is, the district boundaries must be redrawn to allow for the additional seats or the loss of seats. Since this process has been mandated by the Constitution and enforced by the courts, it has generally been accomplished on schedule.

This regularity does not necessarily apply to state legislative districts, however, which in the first instance are governed by state constitutions and statutes. By 1960, a number of states had not redistricted in half a century or more, sometimes in defiance of their own constitutions and statutes. Idaho had not redistricted since 1911, for example, Louisiana since 1912, Tennessee since 1901. Others had not redistricted since 1931 (South Dakota, Colorado, Rhode Island, Connecticut, Georgia, and Alabama). In all of them (and others) major demographic changes had taken place, so that the districts were grossly malapportioned. To cite only a few examples, lower house districts ranged in population from 236,000 to 635,000 in Alabama, 257,000 to 410,000 in Idaho, 319,000 to 690,000 in Connecticut and 216,000 to 952,000 in Texas.

The effects of this malapportionment were many. For one thing, rural districts were heavily overrepresented while rapidly growing urban districts could scarcely be heard in the legislatures. The work load for the legislators varied widely, conservatives were perpetuated in office, committee chairmanships and other positions based on seniority were held by conservative rural representatives, and, most glaring of all, one voter's vote was worth two or three times another's.

It was this last point that formed the basis for the landmark Supreme Court decision of 1962 in the case of *Baker* v. *Carr*. The Court ruled that the apportionment of the Tennessee General Assembly by means of its 1901 statute debased the votes of the plaintiffs and denied them equal protection of the laws under the Fourteenth Amendment. Subsequent decisions in 1964 ruled that (in *Reynolds* v. *Sims*) state senates as well as lower houses had to be apportioned on the basis of population, not counties or other units ("Legislators represent people, not trees or acres."), and that (*Wesberry* v. *Sanders*) in congressional districts as well, as nearly as practicable, one person's vote in an election is to be worth as much as another's. These and similar decisions precipitated a wave of reapportionments throughout the country based on the 1960 and 1970 censuses. They did not, however, solve all the problems connected with apportionment and districting.

For one thing, the reapportionment came too late to help many urban areas, for the great rural-to-urban migration had already largely given way to an urban-to-suburban migration, and many of the seats lost by rural voters went straight to the suburbs, bypassing the cities entirely. Second, a number of states either did not redistrict at all or did so in such a manner that there was still a fairly wide discrepancy between the districts with the largest and smallest populations. Third, while the Court established broad guidelines regarding the shape of districts, saying only that they should be contiguous and compact, ample scope remained for the state legislatures in drawing new boundaries to continue the time-dishonored practice of

The original gerrymander. The term (though not the practice) originated in 1812 when Governor Elbridge Gerry of Massachusetts signed into law an oddly shaped district in Essex County. Painter Gilbert Stuart saw a map of the new district and penciled in a head, wings, and claws, saying, "That will do for a salamander!" Editor Benjamin Russell replied, "Better say a Gerrymander."

gerrymandering. Subsequent Court decisions and federal and state legislation have mitigated some of these problems, but gerrymandering survives.

Gerrymandering is a device to give an advantage to a particular party or group by drawing district boundaries in advantageous shapes. Blacks and whites, Republicans and Democrats, rural and urban and suburban voters have all been helped or hurt by this tactic in nearly every state for nearly two centuries. There are several types of gerrymandering. One of them—simply failing to redistrict as population changes—is no longer very common, but three others are. Perhaps the most common is the practice of drawing the boundaries in such a way that one group (e.g., Chicanos, Republicans, blue-collar workers) is concentrated in the fewest possible districts so that they win overwhelmingly there but their influence is not felt in other districts where they might have had a chance of winning. This is called the *excess vote* technique. Its counterpart is the *wasted vote* technique. Here the lines are drawn so as to break up a concentration of voters (e.g., urbanites, Democrats, blacks) into a number of districts so that their votes are wasted through dispersion and they are unable to elect anyone. A third method is to draw circuitous boundaries enclosing grotesque shapes to enclose pockets of strength of the group in power or to avoid areas of weakness. This is the *stacked* type of gerrymander. Regardless of the type, the practice does infringe the "one person one vote" concept and is inherently undemocratic. It is also exceedingly difficult to purge from the political scene.

REFORMING ELECTORAL DISTRICTS

Nearly everyone can agree in theory that gerrymandering is "wrong." It makes one person's vote (for a candidate or in the legislature) worth more than another, and this is quite undemocratic, though in some cases it might be more efficient and practical than strict equality of voting

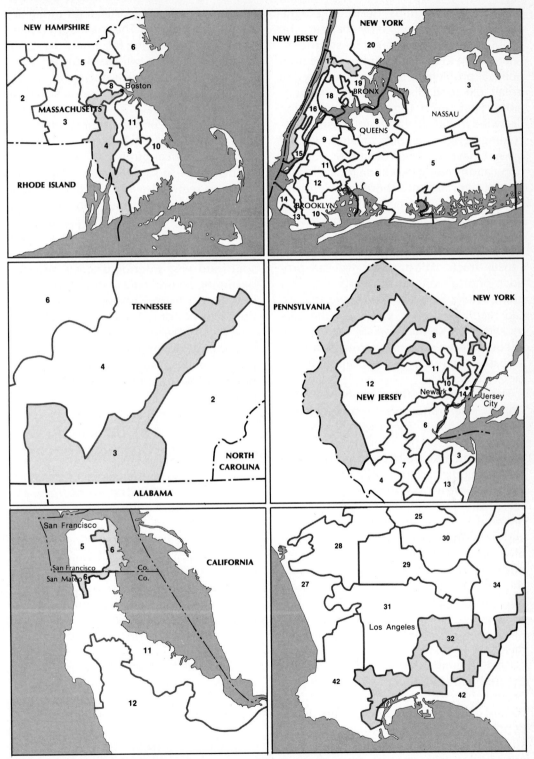

Contemporary gerrymandered congressional districts. Redistricting after the 1980 census left a number of congressional districts throughout the conterminus United States in what appear to be flagrant violations of the concept of compact districts. Here are a few examples.

strength. If reform is desired, geographers can contribute to it by analyzing the problem and offering some solutions. There seem to be, for example, three nonpolitical problems that lead to drawing of unfair electoral districts, even inadvertently. First, even though we are being drowned in statistics and our census is becoming ever more sophisticated, we still do not have sufficient accurate data on population numbers, characteristics, and distribution. Second, there is little or no correlation between census units and traditional political precincts, though results are tabulated by precincts within districts, while seats are allocated on the basis of census tracts. Third, while we can produce photographs of the earth from orbiting satellites so large and sharp that we can read the license plates on the cars parked in the Kremlin, we still do not have adequate large-scale maps suitable for use as base maps in drawing boundaries of electoral districts.

The solutions to these problems seem simple: use computers to compile adequate data bases, including all relevant population factors; use computers to generate adequate base maps; and have the census people and local politicians get together to create statistical units that will serve the needs of both. Ignoring the practical obstacles to each of these steps and assuming that they have been accomplished, we can then proceed to utilize our computers once again either to adjust the existing districts by shifting their boundaries slightly until all districts have roughly the same population or create entirely new idealized districts and then make the necessary minor adjustments.

Even assuming that we have overcome the problems inherent in all of these procedures, we will still be faced with difficulties. If, for example, some independent, apolitical body has produced strictly neutral district boundaries, the new districts will inevitably favor the majority party in whatever area (such as a state or county) is being redistricted. An alternate technique is to have the redistricting done by a bipartisan (or multipartisan) commission. The inevitable horse-trading and jockeying for power that goes on in such bodies will inevitably produce a number of safe seats for both (or all) parties, so that the parties are satisfied but the voters will really have little influence on the outcome of the elections. In either case, democracy will be sacrificed on the altar of statistical purity.

This is not to say, however, that an attempt to rectify the most egregious cases of gerrymandering should not be made. Considering the dramatic population shifts now occurring in the United States—cities to rural areas, northeast to south and west, interior to coasts, suburbs becoming central cities—the 1990 census is likely to force considerable reapportionment of congressional and state legislative districts. This will be a fine opportunity to improve, if not idealize, the districting, and geographers, if allowed to do so, can assist greatly in the process.

THEMES AND THEORIES IN ELECTORAL GEOGRAPHY

Besides the three themes just discussed, others have attracted some interest among political geographers in recent years, but they really deserve more attention than they have been getting. Perhaps when the fascination with numbers and computers has waned, scholars will again turn their attention to serious questions not amenable to quantification. Some of these themes are: political parties themselves and their bases of support; representation and voting patterns in legislatures; the relationships among voting, geography, and power (or influence); and the role of ideology in producing electoral results and the geographic variations in this role.

Recently, a few political geographers have begun to go beyond merely counting votes to try to develop a geography of elections that will help to explain more general political patterns. This includes studying the relationships between geography and elections in (1) the process of forming a government; (2) the functioning of a State; and (3) the world economy. The last has probably been best developed so far. In the rather elaborate body of theory emerging from its study, mostly with a Marxist perspective, political parties are viewed as a part of the overall political development of the world economy. While much of this theory is quite abstruse and perhaps even farfetched, we must remember that this whole field is quite new and give credit to those who are trying to learn something useful from huge masses of raw election statistics.

One of the basic problems in trying to detect and explain worldwide patterns in election statistics is that the statistics in many places are still not adequate nor are the maps necessary for plotting them nor are electoral systems or statistics compatible from one country to another. Furthermore, elections themselves are important in choosing governments in relatively few countries in the world; perhaps two dozen countries have governments freely elected according to democratic, constitutional procedures on a more or less regular basis. Most of these countries are rich, industrialized, heavily engaged in international trade, members of the OECD (discussed in Chapter Twenty-seven). There are a few others: Israel, India, Costa Rica, most of the former British colonies in the Caribbean and the Pacific, for example. But elsewhere governments are chosen through heredity, councils of chiefs, selection by the Party, military control, or some other method. Elections, if held at all, have largely symbolic value, and their statistics are meaningless and useless for analytical purposes.

Nevertheless, the geography of elections will continue to have only limited and local value until more cross-cultural studies are included, until techniques are more standardized, and until there are more free elections around the world to begin with. Geographers can help to bring about all three conditions. If they do not, electoral geography will continue to be little more than an intellectual exercise of little practical value, and the really useful electoral studies will continue to be done by sociologists and political scientists. One step toward accomplishing this goal would be a return to mapping election results. Anyone can play with computers and statistics, but only geographers can map the data in such a way as to make them meaningful to ordinary mortals.

REFERENCES FOR PART THREE

Books and Monographs

A

Abedin, Najmul, *Local Administration and Politics in Modernizing Societies: Bangladesh and Pakistan*. New York: Oxford Univ. Press, 1973.

Advisory Committee on Intergovernmental Relations, *Multistate Regionalism*. Washington, 1972 (updated to 1978).

Agnew, John A., *Place and Politics*. Winchester, MA: Allen & Unwin, 1987.

Archer, J. Clark and Fred M. Shelley, *American Electoral Mosaics*. Washington: Association of American Geographers, 1986.

Arian, A., *The Elections of Israel—1969*. Jerusalem: Jerusalem Academic Press, 1972.

———, *The Elections of Israel 1973*. Jerusalem: Jerusalem Academic Press, 1974.

Arnold, Millard W. and others, *Zimbabwe: Report on the 1985 General Elections*. Washington: International Human Rights Law Group, 1986.

Arnold, R, *Alaska Native Land Claims*. Anchorage: Alaska Native Foundation, 1976.

Aucoin, Peter, *Regional Responsiveness and the National Administrative State*. Toronto: Univ. of Toronto Press, 1987.

B

Barlow, I. Max, *Spatial Dimensions of Urban Government*. Chichester, Sussex, Eng. and elsewhere: Wiley, 1981.

Barnett, M.R. (ed.), *Electoral Politics in the Indian States*. Delhi: Monohar, 1975.

Batteau, Allen (ed.), *Appalachia and America: Autonomy and Regional Dependence*. Lexington: Univ. Press of Kentucky, 1983.

Beaglehole, John Holt, *The District: A Study in Decentralization in West Malaysia*. New York: Oxford Univ. Press, 1976.

Berkley, George E. and Douglas M. Fox, *80,000 Governments: The Politics of Subnational America*. Boston: Allyn & Bacon, 1978.

Berry, M.C., *The Alaska Pipeline: The Politics of Oil and Native Land Claims*. Bloomington: Indiana Univ. Press, 1975.

Bogdanor, V., *Multi-Party Politics and the Constitution*. Cambridge: At the University Press, 1983.

Bogdanor, V. and D.E. Butler (eds.), *Democracy and Elections*. Cambridge: At the University Press, 1983.

Bushnell, E., *Impact of Reapportionment on the Thirteen Western States*. Salt Lake City: Univ. of Utah Press, 1970.

Busteed, Mervyn A., *Geography and Voting Behavior*. London: Oxford Univ. Press, 1975.

C

Cain, Bruce E., *The Reapportionment Puzzle*. Berkeley: Univ. of California Press, 1984.

Campbell, P., *French Electoral Systems and Elections Since 1789*. London: Faber & Faber, 1958.

Case, D.S., *The Special Relationship of Alaska Natives to the Federal Government*. Anchorage: Alaska Native Foundation, 1978.

Caspi, D. and others (eds.), *The Roots of Begin's Success*. Beckenham, Eng.: Croom Helm, 1984.

Cease, Ronald C. and Jerome R. Saroff, *The Metropolitan Experiment in Alaska*. New York: Praeger, 1968.

Cohen, Felix S., *Federal Indian Law*, rev. ed., Charlottesville, VA: Michie Bobbs-Merrill, 1982.

Confederation of American Indians, *The Indian Reservations; A State and Federal Handbook*. Jefferson, NC, and London: McFarland, 1986.

Congressional Quarterly, *Representation and Apportionment*. Washington, 1966.

———, *State Politics and Redistricting*, Parts I and II. Washington, 1982.

———, *Congressional Districts in the 1980's*. Washington, 1983.

Council of Energy Resource Tribes, *The Control and Reclamation of Surface Mining on Indian Lands*. Washington: U.S. Dep. of the Interior, Office of Surface Mining, 1979.

Cox, Kevin R., *Conflict, Power and Politics in the City: A Geographic View*. New York: McGraw-Hill, 1973.

Crewe, Ivor and David Denver, *Electoral Change in Western Democracies*. New York: St. Martin's Press, 1985.

Cullingworth, J. Barry, *Urban and Regional*

Planning in Canada. New Brunswick, NJ: Transaction Books, 1987.

D

Derthick, Martha, *Between State and Nation; Regional Organizations of the United States.* Washington: Brookings Institution, 1974.

De Vorsey, Louis, Jr., *The Georgia–South Carolina Boundary; A Problem in Historical Geography.* Athens: Univ. of Georgia Press, 1982.

Dickinson, Robert E., *City Region and Regionalism.* New York: Oxford Univ. Press, 1947.

Dixon, R.G., *Democratic Representation: Reapportionment in Law and Politics.* London and New York: Oxford Univ. Press, 1968.

Dumars, C.T., and others, *Pueblo Indian Water Rights: Struggle for a Precious Resource.* Tucson: Univ. of Arizona Press, 1984.

Duncan, Otis Dudley, *Metropolis and Region.* Baltimore: Johns Hopkins Univ. Press, 1960.

Dye, Thomas R., *Politics in States and Communities,* 3rd ed. Englewood Cliffs, NJ: Prentice-Hall, 1977.

E

Ennis, P.H., "The Contextual Dimension in Voting," in W.N. McPhee and W.A. Glaser, *Public Opinion and Congressional Elections.* Glencoe, IL: Free Press, 1962.

F

Farley, Lawrence T., *Plebiscites and Sovereignty.* Boulder, CO: Westview, 1986.

Foss, Phillip O. (ed.), *Federal Lands Policy.* Westport, CT: Greenwood Press, 1987.

Freeman, T. Walter, *Geography and Regional Administration.* London: Hutchinson, 1968.

G

Getches, David H. and others, *Federal Indian Law; Cases and Materials.* St. Paul, MN: West, 1979.

Ginsberg, Benjamin and Alan Stone (eds.), *Do Elections Matter?* Armonk, NY: Sharpe, 1986.

Goldman, R.M. (ed.), *Transnational Parties: Organizing the World's Precincts.* Lanham, MD: Univ. Press of America, 1983.

Goodall, Leonard E. and Donald P. Sprengel, *The American Metropolis,* 2nd ed. Columbus, OH: Merrill, 1975.

Goodey, B.R., *The Geography of Elections: An Introductory Bibliography.* Grand Forks: Univ. of North Dakota Press, 1968.

Graves, W. Brooke, *American Intergovernmental Relations.* New York: Scribner's, 1964.

Grofman, Bernard and others (eds.), *Representation and Redistricting Issues.* Lexington, MA: Lexington Books, 1982.

Gudgin, Graham and Peter J. Taylor, *Seats, Votes and the Spatial Organization of Elections.* London: Pion, 1979.

Gunlicks, Arthur B., *Local Government in the German Federal System.* Durham, NC: Duke Univ. Press, 1986.

H

Hardy, LeRoy and others (eds.), *Reapportionment Politics: The History of Redistricting in the 50 States.* Beverly Hills, CA: Sage, 1981.

Harrigan, John J. and William C. Johnson, *Governing the Twin Cities Region: The Metropolitan Council in Comparative Perspective.* Minneapolis: Univ. of Minnesota Press, 1978.

Hartman, H., *Political Parties in India.* New Delhi: Meenakshi Prakashow, 1980.

Heard, Kenneth A., *General Elections in South Africa, 1943–1970.* London: Oxford Univ. Press, 1974.

Hoffmann, Phillip, *School District Reorganization in Upstate New York; Case Studies from Homer and Truxton.* Discussion Paper No. 29. Syracuse, NY: Syracuse Univ. Dept. of Geography, 1977.

Howitt, Arnold, *Managing Federalism: Studies in Intergovernmental Relations.* Congressional Quarterly, CQ Press, 1984.

I

Indian Water Policy in a Changing Environment: American Indian Lawyer Training Program. Washington: Island Press, 1982.

J

Johnston, Ronald J., *People, Places and Votes.* Armidale, Austral.: Univ. of New England Press, 1977.

———, *Political, Electoral and Spatial Systems.* Oxford: Oxford Univ. Press Clarendon, 1979.

K

Kammer, Jerry, *The Second Long Walk; The Navajo–Hopi Land Dispute.* Albuquerque: Univ. of New Mexico Press, 1980.

Kaplan, Harold, *Urban Political Systems: A Functional Analysis of Metro Toronto.* New York: Columbia Univ. Press, 1967.

Kleppner, Paul, *Continuity and Change in Electoral Politics, 1893–1928.* Westport, CT: Greenwood Press, 1987.

Knight, J., and N. Baxter-Moore, *Republic of Ireland: The General Elections of 1969 and 1973.* London: Arthur MacDougall Fund, 1973.

Kraemer, Joel (ed.), *Jerusalem; Problems and Prospects.* Praeger, 1980.

Kugisaki, C. *Reapportionment in Hawaii*. Honolulu: Hawaiian Legislative Reference Bureau, 1978.

L

Leach, Richard H., *Interstate Relations in Australia*. Lexington: Univ. Press of Kentucky, 1965.

Litchfield, F.H., *Voting Behavior in a Metropolitan Area*. Ann Arbor: Univ. of Michigan Press, 1941.

Lodge, Juliet, *Direct Elections to the European Parliament 1984*. New York: St. Martin's Press, 1986.

M

Martin, Roscoe C. and others, *River Basin Administration and the Delaware*. Syracuse, NY: Syracuse Univ. Press, 1960.

Massam, Bryan H., *The Spatial Structure of Administrative Systems*. AAG Resource Paper No. 12. Washington, 1972.

Matthews, O.P., *Water Resources, Geography and Law*. Washington: Association of American Geographers, 1984.

Miller, W.L., *Electoral Dynamics*. New York: St. Martin's Press, 1978.

Morrill, Richard L., *Political Districting and Geographic Theory*. Washington: Association of American Geographers, 1981.

N

Neils, Elaine M., *Reservation to City*. Research Paper No. 131. Chicago: Univ. of Chicago, Dep. of Geography, 1971.

Nice, David C., *Federalism; The Politics of Intergovernmental Relations*. New York: St. Martin's Press, 1987.

Nicholson, Norman L., *The Boundaries of Canada, Its Provinces and Territories*. Ottawa: Dept. of Mines and Technical Surveys, Geographical Branch, 1954.

O

Otis, D.S., *The Dawes Act and the Allotment of Indian Lands*. Norman: Univ. of Oklahoma Press, 1973.

P

Page, Edward C. and Michael J. Goldsmith (eds.), *Central and Local Government Relations*. Newbury Park, CA: Sage, 1987.

Palmer, N.D., *Elections and Political Development: The South Asian Experience*. Durham, NC: Duke Univ. Press, 1975.

Parman, Donald L., *The Navajos and the New Deal*. New Haven, CT and London: Yale Univ. Press, 1976.

Peroff, Nicholas C., *Menominee Drums: Tribal Termination and Restoration, 1954–1974*. Norman: Univ. of Oklahoma Press, 1982.

Picard, Louis A. and Raphael Zariski (eds.), *Subnational Politics in the 1980's*. New York: Praeger, 1986.

Porter, Frank, III, *Strategies for Survival: American Indians in Eastern United States*. Westport, CT: Greenwood Press, 1986.

Prucha, Francis Paul, *The Great Father: The United States Government and the American Indians*. Lincoln: Univ. of Nebraska Press, 1986.

R

Reed, Steven R., *Japanese Prefectures and Policymaking*. Pittsburgh: Univ. of Pittsburgh Press, 1986.

Reynolds, David R. and J. Clark Archer, *An Inquiry into the Spatial Basis of Electoral Geography*. Discussion Paper No. 11. Iowa City: Univ. of Iowa, Dep. of Geography, 1969.

Rokkan, Stein, *Citizens, Elections, Parties*. New York: McKay, 1970.

Rubenstein, James M., *The French New Towns*. Baltimore: Johns Hopkins Univ. Press, 1978.

S

Schusky, Ernest L., *The Forgotten Sioux: An Ethnohistory of the Lower Brule Reservation*. Chicago: Nelson Hall, 1975.

Simeon, Richard, *Intergovernmental Relations*. Toronto: Univ. of Toronto Press, 1987.

Singh, V.B., *Elections in India*, 2nd ed. Newbury Park, CA: Sage, 1987.

Singh, V.B. and Shankar Bose, *State Elections in India; Volume I: The North (Part 1)*. Newbury Park, CA: Sage, 1987.

Smallwood, Frank, *Greater London, the Politics of Metropolitan Reform*. Indianapolis: Bobbs–Merrill, 1965.

Smiley, Donald V. and Ronald L. Watts, *Intrastate Federalism in Canada*. Toronto: Univ. of Toronto Press, 1987.

Smith B.C., *Decentralizations*. Winchester, MA: Allen & Unwin, 1985.

Smith, G. Hubert, *Like-a-Fishhook Village and Fort Berthold, Garrison Reservoir, North Dakota*. Anthropological Paper No. 2. Washington: U.S. Dept. of the Interior, National Park Service, 1972.

Stewart, John, *The New Management of Local Government*. London: Allen & Unwin, 1986.

Sutton, Imre (ed.), *Irredeemable America; the*

Indians' Estate and Land Claims. Albuquerque: Univ. of New Mexico Press, 1985.

T

Taylor, Peter J. and Ronald J. Johnston, *Geography of Elections.* London: Croom Helm, 1979.

Taylor, Theodore W., *The States and Their Indian Citizens.* Washington: U.S. Dept. of the Interior, Bureau of Indian Affairs, 1972.

———, *American Indian Policy.* Mt. Airy, MD: Lomond, 1983.

Teaford, Jon C., *City and Suburb: The Political Fragmentation of Metropolitan America, 1850–1970.* Baltimore and London: John Hopkins Univ. Press, 1979.

Tyler, S. Lyman, *A History of Indian Policy.* Washington: U.S. Dept. of the Interior, Bureau of Indian Affairs, 1973.

U

U.S. Department of Commerce, *Federal and State Indian Reservations and Indian Trust Areas.* Washington, 1974.

———, Bureau of the Census, *County and City Data Book.*

———, *Congressional District Atlas.* Washington, biennial.

———, *Boundary and Annexation Survey 1970–1975.* Washington, 1978.

W

Weiner, Myron and Ergun Özbudun (eds.), *Competitive Elections in Developing Countries.* Durham, NC: Duke Univ. Press, 1987.

Wikstrom, Nelson, *Councils of Governments: A Study of Political Incrementalism.* Chicago: Nelson Hall, 1976.

Wilkinson, Charles F., *American Indians, Time and the Law.* New Haven, CT: Yale Univ. Press, 1986.

Williams, P.M., *French Politics and Elections 1931.* Cambridge: At the University Press, 1970.

Wollmington, E.R., *A Spatial Approach to the Measurement of Support for the Separatist Movement in Northern New South Wales.* Armidale Austral.: Univ. of New England Press, 1966.

Wood, Robert, *1400 Governments: The Political Economy of the New York Metropolitan Region.* Cambridge: Harvard Univ. Press, 1961.

Worcester, Donald E. (ed.), *Forked Tongues and Broken Treaties.* Caldwell, ID: Caxton, 1975.

Wynant, Lillian K., *Westward in Eden; The Public Lands and the Conservation Movement.* Berkeley, Los Angeles, and London: Univ. of California Press, 1982.

Studies of National Elections and Referenda Published by the American Enterprise Institute, Lanham, MD

Country	Type	Date	Editor	Publication Date
Australia	national elections	1975	Howard R. Penniman	1977
Australia	national elections	1977	Howard R. Penniman	1980
Australia	national elections	1980 & 1983	Howard R. Penniman	1983
Canada	general election	1974	Howard R. Penniman	1975
Canada	general election	1979 & 1980	Howard R. Penniman	1981
France	presidential election	1974	Howard R. Penniman	1975
France	national assembly elections	1978	Howard R. Penniman	1980
Greece	national elections	1974 & 1976	Howard R. Penniman	1981
India	parliamentary elections	1977	Myron Weiner	1978
India	parliamentary elections	1980	Myron Weiner	1983
Ireland	Dáil elections	1977	Howard R. Penniman	1978
Israel	Knesset elections	1977	Howard R. Penniman	1978
Italy	parliamentary elections	1976	Howard R. Penniman	1977
Italy	parliamentary elections	1979	Howard R. Penniman	1981
Japan	House of Councillors	1974	Michael K. Blaker	1976
Japan	House of Representatives	1976	Herbert Passin	1979
	House of Councillors	1977		
Micronesia	plebiscites	1983	Howard R. Penniman & Austin Ranney	1985
New Zealand	general election	1978	Howard R. Penniman	1980
Scandinavia	parliamentary elections	1973	Karl H. Cerny	1977
Spain	national elections	1977, 1979, 1982	Howard R. Penniman & Eusebio M. Mujal-Lesn	1985
Switzerland	national elections	1979	Howard R. Penniman	1984
UK	parliamentary elections	1974	Howard R. Penniman	1975
UK	Referendum on European Communities	1975	Anthony King	1977
UK	general election	1979	Howard R. Penniman	1981
UK	general election	1983	Austin Ranney	1985
Venezuela	national elections	1978	Howard R. Penniman	1980
West Germany	Bundestag election	1976	Karl H. Cerny	1978

Periodicals

A

Adams, B., "A Model Reapportionment Process: The Continuing Quest for 'Fair and Effective' Respresentation," *Harvard Journal of Legislation,* 14, (1977), 825–904.

Adejuyibge, O. "Ife–Ijesa Boundary Problems," *Nigerian Geographical Journal,* 13 (1970), 23–28.

Agnew, John A., "Place Anyone? A Comment on the McAllister and Johnston Papers," *Political Geography Quarterly,* 6, 1 (January 1987), 39–40.

Ahmad, Hamiduddin, "Election Data Analysis as a Tool of Research in Political Geography," *Pakistan Geographical Review,* 21 (1966), 34–40.

Alford, R.R., "The Role of Social Class in American Voting Behavior," *Western Political Quarterly,* 16, 1 (March 1963), 80–86.

Andrew, B.H., "Some Queries Concerning the Texas–Louisiana Sabine Boundary," *Southwestern History Quarterly,* 53 (1949), 1–18.

Appel, J.S., "A Note Concerning Apportionment by Computer," *American Behavioral Scientist,* 7 (1965), 36.

Archer, J. Clark, "Some Geographical Aspects of the American Presidential Election of 1980," *Political Geography Quarterly,* 1, 2 (April 1982), 123–135.

Archer, J. Clark and others, "Counties, States, Sections, and Parties in the 1984 Presidential Election," *Professional Geographer,* 37, 3 (August 1985), 279–287.

Arora, Satish Kumar, "The Reorganization of Indian States," *Far Eastern Survey,* 25, 2 (February 1956), 27–30.

B

Backstrom, Charles H., "The Practice and Effect of Redistricting," *Political Geography Quarterly,* 1, 4 (October 1982), 351–359.

Beatty, F.W., "Ontario–Manitoba Boundary," *Annual Report of the Association of Ontario Land Survey,* (1949), 138–139.

Billington, Monroe, "The Red River Boundary Controversy," *Southwestern Historical Quarterly,* 62, 3 (January 1959), 356–363.

Birdsall, Stephen S. "Preliminary Analysis of the 1968 Wallace Vote in the Southeast," *Southeastern Geographer,* 9, 2 (November 1969), 55–66.

Boddy, Martin, "Central-local Government Relations Theory and Practice," *Political Geography Quarterly,* 2, 2 (April 1983), 119–138.

Bodman, Andrew R., "Regional Trends in Electoral Support in Britain, 1950–1983," *Professional Geographer,* 37, 3 (August 1985), 288–295.

———, "Habeas Corpus: A Reply to Johnston and Taylor," *Professional Geographer,* 39, 1 (February 1987), 64–65.

Bondi, Liz, "School Closures and Local Politics: The Negotiation of Primary School Rationalization in Manchester," *Political Geography Quarterly,* 6, 3 (July 1987), 203–224.

Bowden, J.J., "The Texas–New Mexico Boundary Dispute Along the Rio Grande," *Southwestern Historical Quarterly,* 63 (1959), 221–237.

Bowman, Isaiah, "An American Boundary Dispute: Decision of the United States with Respect to the Texas–Oklahoma Boundary," *Geographical Review,* 13 (1923), 161–189.

Brightman, George, F. "The Boundaries of Utah," *Economic Geography,* 16, 1, (January 1940), 87–95.

Brunk, Gregory G., "On Estimating Distance-Determined Voting Functions," *Political Geography Quarterly,* 4, 4 (October 1985), 55–65.

Brunn, Stanley D. "The Defeat of a Youngstown School Levy: A Study in Urban-Political Geography," *Southeastern Geographer,* 9, 2 (November 1969), 67–79.

Brunn, Stanley, D. and others, "Some Spatial Considerations of the Flint Open Housing Referendum," *Proceedings, AAG,* 1 (1969), 26–32.

Bunge, William, "Gerrymandering, Geography, and Grouping," *Geographical Review,* 56, 2 (April 1966), 256–263.

Burghardt, Andrew, "Regions of Party Support in Burgenland (Austria)," *Canadian Geographer,* 7 (1963), 91–98.

———, "The Bases of Support for Political Parties in Burgenland," *Annals, AAG,* 54, 3 (September 1964), 372–390.

———, "The Proposed Regional Government for the Hamilton Area: An Analysis of City and Suburban Positions," *Communication au XXIIe congrès international de géographie* (Section VII). Montreal, 1972.

———, "The Redesigning of Scotland's Administrative Areas: An Outsider's View," *Scottish Geographical Magazine*, 98, 3 (December 1982), 130–142.

Bushman, D.O. and W. R. Stanley, "State Senate Reapportionment in the Southeast," *Annals, AAG*, 61, 4 (December 1971), 654–670.

Busteed, M.A., "The Belfast Region: Local Government in Need of Change," *Public Affairs*, 1, 3 (1968), 12–13.

———, "Reform of Local Government Structure in Northern Ireland," *Irish Geography*, 6 (1971).

C

Carpenter, W.C., "The Red River Boundary Dispute (Oklahoma–Texas)," *American Journal of International Law*, 19 (1925), 517–529.

Chandrasekhar, S., "The New Map of India," *Population Review*, 1, 1 (January 1957), 32–36.

Chapman, B.B., "The Claims of Texas to Greer County," *Southwestern Historical Quarterly*, 53 (1949), 401–424.

Charney, Jonathan I., "The Delimitation of Lateral Seaward Boundaries Between States in a Domestic Context," *American Journal of International Law*, 75, 1 (January 1981), 28–68.

Chatterje, S.P., "The Changing Map of India," *Geographical Review of India*, 19, 2 (June 1957), 1–5.

Christopher, A.J., "Parliamentary Delimitation in South Africa, 1910–1980," *Political Geography Quarterly*, 2, 3 (July 1983), 205–217.

Comeaux, Malcolm L., "Attempts to Establish and Change a Western Boundary," *Annals, AAG*, 72, 2 (June 1982), 254–271.

Conradt, D.P., "Electoral law Politics in West Germany," *Political Studies*, 18, (1970), 341–356.

Cope, C.R., "Regionalization and the Electoral Districting Problem" *Area*, 3 (1971), 190–195.

Cox, Kevin Robert, "Suburbia and Voting Behavior in the London Metropolitan Area," *Annals, AAG*, 58, 1 (March 1968), 111–127.

———, "The Voting Decision in a Spatial Context," *Progress in Geography*, 1 (1969), 81–117.

———, "The Voting Decision in Intra-Urban Space," *Proceedings, AAG*, 1 (1969), 43–47.

———, "The 'Individual,' the 'Social' and Reconceptualizing Contextual Effects," *Political Geography Quarterly*, 6, 1 (January 1987), 41–43.

Criddle, B., "Distorted Representation in France," *Parliamentary Affairs*, 26 (1975), 154–179.

Crisler, Robert M., "Republican Areas in Missouri," *Missouri Historical Review*, 42 (1948), 299–310.

———, "Voting Habits in the United States," *Geographical Review*, 62, 2 (April 1952), 300–301.

Cushing, Sumner W., "The Boundaries of the New England States," *Annals, AAG*, 10 (1920), 17–40.

D

Das Gupta, Sivaprasad, "The Changing Map of India," *Geographical Review of India*, 22, 3 (September 1960), 23–33; 22, 4 (December 1960), 13–32.

Davies, Richard B. and Robert Crouchley, "The Determinants of Party Loyalty: A Disaggregate Analysis of Panel Data from the 1974 and 1979 Elections in England," *Political Geography Quarterly*, 4, 4 (October 1985), 307–320.

Dawson, Andrew H., "Local Government and the Idea of the Region: A Comment on the Present Situation in Scotland," *Scottish Geographical Magazine*, 100, 2 (September 1984), 113–122.

Day, Winifred M., "Relative Permanence of Former Boundaries in India," *Scottish Geographical Magazine*, 65, 3 (December 1949), 113–122.

Dean, V.K., "Geographical Aspects of the Newfoundland Referendum," *Annals, AAG*, 29, 1 (March 1949), 70–77.

Deutsch, H.J., "The Evolution of State and Territorial Boundaries in the Inland Empire of the Pacific Northwest," *Pacific Northwestern Quarterly*, 51 (1960), 115–131.

Douglas, J.N.H. and R.D. Osborne, "Northern Ireland's Increased Representation in the Westminster Parliament," *Irish Geography*, 14 (1981), 37–40.

Downing, Bruce and others, "The Decline of Party Voting: A Geographical Analysis of the 1978 Massachusetts Election," *Professional Geographer*, 32, 4 (November 1980), 454–461.

Dykstra, T.L. and R.G. Tronside, "The Effects of the Division of the City of Lloydminster by the Alberta–Saskatchewan Inter-Provincial Boundary," *Cahiers de géographie de Québec*, 16, 38 (1972), 263–283.

E

Edgerton, Robert B., "Menominee Termination: Observations on the End of a Tribe," *Human Organization*, 21, 1 (Spring 1962), 10–16.

Edwards, K.C., "The New Towns of Britain," *Geography*, 49 (1964), 272–285.

Electoral Studies. London: Butterworths. Published three times a year since 1981.

F

Foladare, I.S., "The Effect of Neighbourhood on Voting Behavior," *Political Science Quarterly*, 16 (1968), 266–272.

G

Ganong, W.I., "A Monograph on the Evolution of the Boundaries of New Brunswick," *Transactions, Royal Society of Canada*, 7, 2 (1901), 139–149.

Gilbert, E.W., "Practical Regionalism in England and Wales," *Geographical Journal*, 94, 1 (July 1939), 29–44.

———, "The Boundaries of Local Government Areas," *Geographical Journal*, 111, 4–6 (April-June 1948), 172–206.

Glanz, O., "The Negro Voter in Northern Industrial Cities," *Western Political Quarterly*, 13 (1960), 1006–1007.

Glassner, Martin Ira, "Drawing Boundaries of Planning Regions: A Political Geographer's Contribution," *ITCC Review*. 1, 3 (July 1972), 35–40.

———, "The New Mandan Migrations: From Hunting Expeditions to Relocation," *Journal of the West*, 13, 2 (April 1974), 59–74.

Goldberg, A.L., "The Statistics of Malapportionment," *Yale Law Journal*, 72 (1962), 90–101.

Gosnell, Harold F., "The Negro Vote in Northern Cities," *National Municipal Review*, 30 (1941), 264.

Gradus, Yehuda, "The Role of Politics in Regional Inequality—the Israeli Case," *Annals, AAG*, 73, 3 (September 1983), 388–403.

Grofman, Bernard, "Reformers, Politicians, and the Courts: A Preliminary Look at US Redistricting," *Political Geography Quarterly*, 1, 4 (October 1982), 303–316.

Gudgin, Graham and Peter J. Taylor, "Electoral Bias and the Distribution of Party Voters," *Transactions, Institute of British Geographers*, 63 (1974), 53–74.

H

Hajdú, Zoltan, "Political Geography Around the World VII: Administrative Geography and Reforms of the Administrative Areas in Hungary," *Political Geography Quarterly*, 6, 3 (July 1987), 269–278.

Hale, Richard W., Jr., "The Forgotten Maine Boundary Commission," *Proceedings, Massachusetts Historical Society*, (October 1953–May 1957), 71 (1959), 147–155.

Hanson, G.H., "The Geographic Factor and Its Influence in Utah Administrative Units," *Yearbook, Association of Pacific Coast Geographers* (1937), 3–8.

Hare, F. Kenneth, "Regionalism: A Development in Political Geography," *Public Affairs* (1946), 34–39.

———, "Regionalism and Administration: North American Experiments," *Canadian Journal of Economic and Political Sciences*, 15, 3 (August 1949), 344–352.

———, "The Labrador Frontier," *Geographical Review*, 42, 3 (July 1952), 405–424.

Harris, C. Alexander, "The Labrador Boundary," *Contemporary Review*, 131 (1927), 415–421.

Hart, Paxton, "The Making of Menominee County," *Wisconsin Magazine of History*, 43 (Spring 1960), 181–184.

Helin, Ronald A., "The Volatile Administrative Map of Rumania," *Annals, AAG*, 57, 3 (September 1967), 481–502.

Hellevik, O. and N.P. Gleditsch, "The Common Market Decision in Norway: A Clash Between Direct and Indirect Democracy," *Scandinavian Political Studies*, 8 (1973), 227–235.

Hirsch, W.Z., "Local Versus Area Wide Urban Government Services," *National Tax Journal*, 17, 4 (1964), 331–339.

Hoffmeister, Harold, "The Consolidated Ute Indian Reservation," *Geographical Review*, 35 (1945), 601–623.

Holmes, J.M., "Regional Boundaries in the Murray Valley," *Australian Geographer*, 1 (1944), 197–203.

Hoppe, E.O., "Australian Capitals," *Canadian Geographical Journal*, 44, 3 (March 1952), 97–107.

I

Ikporukpo, C.O., "Politics and Regional Policies: The Issue of State Creation in Nigeria," *Political Geography Quarterly*, 5, 2 (April 1986), 127–139.

Ireland, Willard E., "The Evolution of the Boundaries of British Columbia," *British Columbia Historical Quarterly*, 3 (1939), 263–282.

J

Jaensch, D., "A Functional Gerrymander—South Australia, 1944–1970," *Australian Quarterly*, 42 (1970), 96–101.

James, J.R. and others, "Local Government Reform in England: A Symposium," *Geographical Journal*, 136, 1 (March 1970), 1–23.

Jeans, D.N., "Territorial Divisions and the Location of Towns in New South Wales, 1826–1842," *Australian Geographer*, 10 (1967), 243–255.

Johnston, Ronald J., "Local Effects in Voting at a Local Election," *Annals, AAG*, 64, 3 (September 1974), 418–430.

———, "Spatial Structure, Plurality Systems, and Electoral Bias," *Canadian Geographer*, 20, 3 (Fall 1976), 310–328.

———, "The Compatability of Spatial Structure and Electoral Reform: Observations on the Electoral Geography of Wales," *Cambria*, 4 (1977), 125–151.

Johnston, Ronald J. and J. Ballantine, "Geography and the Electoral System," *Canadian Journal of Political Science*, 10 (1977), 857–866.

Johnston, Ronald J. and C. Hughes, "Constituency Delimitation and the Unintentional Gerrymander in Brisbane," *Australian Geographical Studies*, 16 (1978), 79–110.

Johnston, Ronald J., "Class, Conflict and Electoral Geography," *Antipode*, 11, 3 (1979), 36–43.

———, "Xenophobia and Referenda: An Example of the Exploratory Use of Ecological Regression," *L'Espace géographique*, 9 (1980), 73–80.

———, "Changing Voter Preferences, Uniform Electoral Swing, and the Geography of Voting in New Zealand," *New Zealand Geographer*, 37 (1981), 13–19.

———, "Short-term Electoral Change in England: Estimates of Its Spatial Variation," *Political Geography Quarterly*, 1, 1 (January 1982), 41–55.

———, "Redistricting by Independent Commissions: A Perspective from Britain," *Annals, AAG*, 72, 4 (December 1982), 457–470.

Johnston, Ronald J. and others, "The Changing Electoral Geography of the Netherlands: 1946–1981," *Tijdschrift voor Economische en Sociale Geografie*, 74 (1983), 185–195.

———, "The Neighborhood Effect Won't Go Away: Observations on the Electoral Geography of England in the Light of Dunleavy's Critique", *Geoforum*, 14, (1983), 161–168.

———, "Proportional Representation and Fair Representation in the European Parliament Area," *Area*, 15 (1983), 347–355.

———, "The Geography of the Working Class and the Geography of the Labour Vote in England 1983: A Prefatory Note to a Research Agenda," *Political Geography Quarterly*, 6, 1 (January 1987), 7–16.

———, "What Price Place," *Political Geography Quarterly*, 6, 1 (January 1987), 51–52.

———, Review Essay: "Can We Leave Electoral Reform to Politicians? (House of Commons, Second Report from the Home Affairs Committee Session, 1986–87, Redistribution of Seats; Report of the Royal Commission on the Electoral System: Towards a Better Democracy.) *Political Geography Quarterly*, 6, 3, (July 1987), 279–282.

Jones, C.D., "The Impact of Local Elections Systems on Black Political Representation," *Urban Affairs Quarterly*, 11 (1976), 745–754.

Jones, E., "Problems of Partition and Segregation in Northern Ireland," *Journal of Conflict Resolution*, 4, 1 (1960), 96–105.

Jones, Stephen B., "The Forty-ninth Parallel in the Great Plains: The Historical Geography of a Boundary," *Journal of Geography*, 31, 9 (1932), 357–368.

———, "Intra-State Boundaries in Oregon," *Commonwealth Review*, 16 (1934), 105–126.

K

Kasperson, Roger E., "Toward a Geography of Urban Politics: Chicago, A Case Study," *Economic Geography*, 41, 2 (1965), 95–107.

———, "On Suburbia and Voting Behavior," *Annals, AAG*, 59, 2 (June 1969), 405–410.

Kirk-Greene, A.H.M., and M.J. Campbell, "The Capitals of Northern Nigeria," *Nigeria*, No. 54 (1957), 243–272.

Klimasewski, Ted, "Analysis of Spatial Voting Patterns: An Approach in Political Socialization," *Journal of Geography*, 72, 3 (March 1973), 26–32.

Knowles, R.D., "Malapportionment in Norway's Parliamentary Elections Since 1921," *Norsk Geografisk Tidsskrift*, 35 (1981), 147–159.

Kollmorgen, W., "Political Regionalism in the United States: Fact or Myth," *Social Forces*, 15 (1936), 111–122.

Krebheil, Edward, "Geographic Influences in

British Elections," *Geographical Review*, 2, 6 (1916), 419–432.

L

Langlois, C., "The State of Local Government in Greater Montreal," *Canadian Geographer*, 8, 3 (1964), 160–162.

Laponce, J.A., "Assessing the Neighbor Effect on the Vote of Francophone Minorities in Canada," *Political Geography Quarterly*, 6, 1 (January 1987), 77–87.

Leonski, Zbigniew, "Aspects of Territorial Subdivisions in European Socialist States," *International Social Science Journal*, 30, 1 (1978), 47–56.

Lewis, Pierce F., "Impact of Negro Migration on the Electoral Geography of Flint, Michigan, 1932–1962," *Annals, AAG*, 55, 1 (March 1965), 1–25.

Logan, W.S., "The Changing Landscape Significance of the Victoria–South Australia Boundary," *Annals, AAG*, 58, 1 (March 1968), 128–154.

Long, J.A., "Maldistribution in Western Provincial Legislatures," *Canadian Journal of Political Science*, 2 (1969), 345–355.

M

Mackay, J. Ross, "The Interactance Hypothesis and Boundaries in Canada: A Preliminary Study," *Canadian Geographer*, 11 (1958), 1–8.

Maddick, Henry, "Decentralization in the Sudan," *Journal of Local Administration Overseas*, 1, 2 (1962), 75–83.

Marando, V.C., "Inter-local Cooperation in Metropolitan Areas: Detroit," *Urban Affairs Quarterly*, 4 (1968), 185–200.

Martin, Lawrence, "The Michigan–Wisconsin Boundary Case in the Supreme Court of the United States, 1923–26" *Annals, AAG*, 20, 3 (September 1930), 105–163.

———, "The Second Wisconsin–Michigan Boundary Case in the Supreme Court of the United States, 1930–1936," *Annals, AAG*, 28, 2 (June 1938), 77–126.

Massam, Bryan H. and Andrew Burghardt, "The Administrative Subdivision of Southern Ontario: An Attempt at Evaluation," *Canadian Geographer*, 12, 3 (1968), 125–134.

———, "A Test of a Model of Administrative Areas," *Geographical Analysis*, 3, 4 (1971), 402–406.

Massam, Bryan H. and M.F. Goodchild, "Temporal Trends in the Spatial Organization of a Service Agency," *Canadian Geographer*, 15, 3 (1971), 193–206.

———, "Evolution of Local Government Units (Toronto)," Montreal, 1972, *Communication au XXIIe Congrès International de Géographie*, Section VII.

McAllister, Ian, "Social Context, Turnout and the Vote: Australian and British Comparisons," *Political Geography Quarterly*, 6, 1 (January 1987), 17–30.

———, "Comment on Johnston," *Political Geography Quarterly*, 6, 1 (January 1987), 45–49.

McColl, Robert W., "Development of Supra-provincial Administrative Regions in Communist China 1949–1960," *Pacific Viewpoint*, 4 (1963), 53–64.

McDonald, G.T., "Rural Representation in Queensland State Parliament," *Australian Geographical Studies*, 14 (1976), 33–42.

McDonough, P., "Repression and Representation in Brazil," *Comparative Politics*, 14 (1982), 73–99.

McGee, T.G., "The Malayan Elections of 1959: A Study in Electoral Geography," *Journal of Tropical Geography*, 16 (1962), 72–99.

———, "The Malayan Parliamentary Elections 1964," *Pacific Viewpoint*, 6 (1965), 96–101.

McGrath, Patrick, "The Labrador Boundary Decision," *Geographical Review*, 17 (1927), 632–660.

McPhail, I.R., "The Vote for Mayor of Los Angeles in 1969," *Annals, AAG*, 61, 4 (December 1971), 744–758.

Mellor, Roy, "Trouble with the Regions: Planning Problems in Russia," *Scottish Geographical Magazine*, 75, 1 (April 1959), 44–47.

Mintz, Eric, "Election Campaign Tours in Canada," *Political Geography Quarterly*, 4, 1 (January 1985), 47–54.

Morrill, Richard L., "Ideal and Reality in Reapportionment," *Annals, AAG*, 63, 4 (December 1973), 463–477.

———, "Redistricting Standards and Strategies After 20 Years," *Political Geography Quarterly*, 1, 4 (October 1982), 361–370.

———, "Redistricting, Region and Representation." *Political Geography Quarterly*, 6, 3 (July 1987), 241–260.

Morrison, Joel A., "The Evolution of the Territorial-Administrative System of the USSR," *American Quarterly on the Soviet Union*, 1 (1938), 25–46.

N

Nelson, Howard J., "The Vernon Area of California, A Study of the Political Factor in Urban Geography," *Annals, AAG,* 42, 2 (June 1952), 177–191.

Nicholson, Norman L., "Some Aspects of the Political Geography of the District of Keewatin," *Canadian Geographer,* 1, 3 (1953), 73–83.

——, "Boundary Adjustments in the Gulf of Saint Lawrence Region," *Newfoundland Quarterly,* 53 (1954), 13–17.

O

Ogier, J.C.H., "The Victorian State Boundary," *Victorian Geographical Journal,* 23 (1905), 78–106.

Oliver, Robert, "Legal Status of the American Indian Tribes," *Oregon Law Review,* 38 (1959), 193–245.

O'Loughlin, John, "The Identification and Evaluation of Racial Gerrymandering," *Annals, AAG,* 72, 2 (June 1982), 165–184.

O'Loughlin, John and Anne Marie Taylor, "Choices in Redistricting and Electoral Outcomes: The Case of Mobile, Alabama," *Political Geographical Quarterly,* 1, 4 (October 1982), 317–339.

Orr, D.M., Jr., "The Persistence of the Gerrymander in North Carolina Congressional Redistricting," *Southeastern Geographer,* 9, 2 (November 1969), 39–54.

Osei-Kwame, Peter and Peter J. Taylor, "A Politics of Failure: The Political Geography of Ghanaian Elections, 1954–1979," *Annals, AAG,* 74, 4 (December 1984), 574–589.

P

Panikar, K.M., "Indian States Reorganization," *Asian Review,* n.s., 52, 192 (October 1956), 247–258.

Park, Richard L., "East Bengal: Pakistan's Troubled Province," *Far Eastern Survey,* 23, 5 (1954), 70–74.

Parker, A.J., "The 'Friends and Neighbors' Voting Effect in the Galway West Constituency," *Political Geography Quarterly,* 1, 3 (July 1982), 243–262.

——, "Localism and Bailiwicks: The Galway West Constituency in the 1977 General Election," *Proceedings, Royal Irish Academy,* 83c (1983), 17–36.

Peake, H.J.E., "Geographical Aspects of Administrative Areas," *Geography,* 15, 7 (1930), 531–546.

Peake, Linda J., Review Essay: "How Sarlvik and Crewe Fail to Explain the Conservative Victory of 1979 and Electoral Trends in the 1970's" *Political Geography Quarterly,* 3, 2 (April 1984), 161–167.

Pedersen, M.H., "Preferential Voting in Denmark: The Voters' Influence on the Election of Folketing Candidates," *Scandinavian Political Studies,* 1 (1966), 167–187.

Phelps, Glenn, "Representation Without Taxation: Citizenship and Suffrage in Indian Country," *American Indian Quarterly,* (Spring 1985).

Philbrick, Allen K., "Principles of Areal Functional Organization in Regional Human Geography," *Economic Geography,* 33, 4 (October 1957), 299–336.

Pirie, Gordon H. and others, "Covert Power in South Africa: The Geography of the Afrikaner Broederbond," *Area,* 12 (1980), 97–104.

Prescott, John Robert Victor, "The Function and Methods of Electoral Geography," *Annals, AAG,* 49, 3 (September 1959), 296–304.

——, "Nigeria's Regional Boundary Problems, *Geographical Review,* 49, 4 (October 1959), 485–505.

——, "The Evolution of Nigeria's Boundaries," *Nigerian Geographical Journal,* 2, 2 (1959), 80–104.

——, "The Evolution of the Anglo–French Inter-Cameroons Boundary," *Nigerian Geographical Journal,* 5, 2 (December 1962), 103–120.

R

Rantala, O., "The Political Regions of Finland," *Scandinavian Political Studies,* 2 (1967), 117–142.

Reeves, "Vermont Versus New Hampshire," *American Journal of International Law,* 27 (1933), 506–508.

Richards, J. Howard, "Changing Canadian Frontiers," *Canadian Geographer,* 5, 4 (1961), 23–29.

——, "Provincialism, Regionalism and Federalism as Seen in Joint Resource Development Programmes," *Canadian Geographer,* 9, 4 (1965), 205–215.

Richards, P.G., "Local Government Reform: Smaller Towns and the Countryside," *Urban Studies,* 2 (1965), 147–162.

Roberts, M.C. and K.W. Rumage, "Spatial Variations in Urban Left-wing Voting in the London Metropolitan Area," *Annals, AAG,* 55, 2 (March 1965), 161–178.

Robinson, K.W., "The Political Influence in Australian Geography," *Pacific Viewpoint,* 3, 2 (1962).

Rose, A.J., "The Border Between Queensland and New South Wales; a Study of Political

Geography in a Federal Union," *Australian Geographer*, 6, 4 (January 1955), 3–18.

Rowley, C., "Electoral Behavior and Electoral Behaviour: A Note on Certain Recent Developments in Electoral Geography," *Professional Geographer*, 21, 6 (November 1969), 398–400.

Rumley, Dennis, "Structural Effects in Different Contexts," *Political Geography Quarterly*, 6, 1 (January 1987), 31–37.

Ryan, Joe, "Compared to Other Nations," *American Indian Journal of the Institute for the Development of Indian Law*, 3, 8 (1977), 2–13.

Rydon, J., "Malapportionment Australian Style," *Politics*, 3 (1968), 133–147.

———, "Compulsory and Differential: The Distinctive Features of Australia's Voting Methods," *Journal of Commonwealth Political Studies*, 6 (1968), 183–201.

S

Sauer, Carl O., "Geography and the Gerrymander," *American Political Science Review*, 12 (1918), 403–426.

Savage, Mike, "Understanding Political Alignments in Contemporary Britain: Do Localities Matter?" *Political Greography Quarterly*, 6, 1 (January 1987), 53–76.

Shelley, Fred M., "A Constitutional Choice Approach to Electoral District Boundary Delineation," *Political Geography Quarterly*, 1, 4 (October 1982), 341–350.

Stanislawski, D., "Tarascan Political Geography," *American Anthropologist*, 49 (1947), 46–55.

Swauger, John, "Measuring Regional Agreement in Senate Voting, 1977," *Professional Geographer*, 32, 4 (November 1980), 446–453.

T

Thomas, Benjamin E., "Boundaries and Internal Problems of Idaho," *Geographical Review*, 39, 1 (January 1949), 99–109.

———, "Demarcation of the Boundaries of Idaho," *Pacific Northwest Quarterly*, 40 (1949), 23–34.

———, "The California–Nevada Boundary," *Annals, AAG*, 42, 1 (March 1952), 51–68.

Tinkler, I., "Malayan Elections: Electoral Pattern for Plural Societies?" *Western Political Quarterly*, 9, (1956), 258–282.

Tucker, Harvey, J., "State Legislative Apportionment: Legal Principles in Empirical Perspective," *Political Geography Quarterly*, 4, 1 (January 1985), 19–28.

U

Ullman, Edward L., "Political Geography in the Pacific Northwest," *Scottish Geographical Magazine*, 54 (1938), 236–239.

———, "The Eastern Rhode Island–Massachusetts Boundary Zone," *Geographical Review*, 29, 2 (1939), 291–302.

V

Valen, H. and Stein Rokkan, "The Norwegian Program for Electoral Research," *Scandinavian Political Studies*, 2 (1967), 294–305.

Vogeler, Ingolf, and Terry Simmons, "Settlement Morphography of South Dakota Reservations," *Yearbook, Association of Pacific Coast Geographers*, 37 (1975), 91–108.

W

Wake, William H., "State Reorganization in India: A Centrifugal or Centripetal Force for the Future," *California Geographer* (1963), 35–47.

Warntz, W., "A Methodological Consideration of Some Geographical Aspects of the Newfoundland Referendum on Confederation with Canada, 1948," *Canadian Geographer*, 1, 6 (1955), 39–49.

Waterman, Stanley, "The Dilemma of Electoral Districting in Israel," *Tijdschrift voor Economische en Sociale Geografie*. 71, (1980), 88–97.

Whebell, C.F.J., "Core Areas in Intrastate Political Organisation," *Canadian Geographer*, 12, 2 (1968), 99–112.

Windmiller, Marshall, "The Politics of States Reorganization in India: The Case of Bombay," *Far Eastern Survey*, 25, 9 (September 1956), 131–143.

Wolfe, James H., "International Law and the Regime of the Sea in Mississippi's Coastal Zone," *Mississippi College Law Review*, 2, 3 (June 1981), 239–264.

Y

Yonekura, J., "Historical Development of the Political-Administrative Divisions of Japan," *Abstract of Papers, Rio de Janeiro, 18th International Geographical Congress, 1956.*

Z

Zamora, Stephen, "Voting in International Economic Organizations," *American Journal of International Law*, 74, 3 (July 1980), 566–608.

Zarur, J., "The New Brazilian Territories," *Geographical Review*, 34, 1 (January 1944), 142–144.

Part Four

GEOPOLITICS

Chapter 17

DEVELOPMENT THROUGH WORLD WAR II

Now that we have examined the State in some detail, we turn our attention to the activities of a State beyond its boundaries. In Part One we discussed the nature of political geography and some of its aspects. Unfortunately, many people, including some geographers, confuse political geography with geopolitics. Geopolitics, however, is only one of the subjects studied by political geographers. It is concerned basically with the application of geographic information and geographic perspectives to the development of a State's foreign policies. It has been called, with some justification, "applied political geography." The basic philosophy of a State's foreign policy and its method of utilizing geography as a tool for executing it, however, make all the difference. Geopolitics was given a proverbial black eye in the interwar period by German geographers who distorted some of the basic ideas of geopolitics into the pseudoscience of "*Geopolitik*," the chauvinist, aggressive, and antidemocratic version of geopolitics that led to so much physical and intellectual destruction before and during World War II.

Geopolitics developed toward the end of the nineteenth century as new developments in science and technology led people to take a broader view of the world than they had previously. The consolidation of the modern State system with the unification of Germany and Italy, the apogee of European imperialist expansion,

the appearance of Japan and the United States as new imperialist powers on the fringes of Europe's sphere of interest, rapid population growth and pressures on resources, and differential development all took place in this period and contributed to the new perspectives of scholars and policymakers. Out of this ferment of new thinking (at least new in modern times) came two streams of thought that were geopolitical in nature. One of them emerged from the Social Darwinism fashionable in the period; this was the *organic State theory*. The other was based more on geographic facts and the policies that should be influenced by them; this is often called *geostrategy*. Neither term is entirely satisfactory, as we shall see, but they can help us separate in our minds two very different sets of ideas that were developing simultaneously.

ORGANIC STATE THEORY

Friedrich Ratzel (1844–1904) was a distinguished and prolific scholar, a geographer trained originally in biology, chemistry, and other sciences. He was greatly influenced both by Darwin's discoveries and by Social Darwinism. In his writings, particularly his classic *Politische Geographie* (1896), he used similies and metaphors from biology in his analysis of political science and geography, comparing the State with an organism. We can summarize

Ratzel's theory of an organic State in this manner:

The State is land, with man on the land, linked by the State idea and conforming to natural laws, with development tied to the natural environment. Therefore, for example, States, like plants and people, do not do well in desert or polar regions. States need food in the form of *Lebensraum* (living space) and resources, and they constantly compete for them. States, like organisms, must grow or die. They live through stages of youth, maturity, and old age, with possible rejuvenation. The vitality of a State can generally be gauged by its size at a given time. In 1896, Ratzel produced what he called the seven laws of State growth. They were:

1. The space of States grows with the expansion of the population having the same culture.

2. Territorial growth follows other aspects of development.

3. A State grows by absorbing smaller units.

4. The frontier is the peripheral organ of the State that reflects the strength and growth of the State; hence, it is not permanent.

5. States in the course of their growth seek to absorb politically valuable territory.

6. The impetus for growth comes to a primitive State from a more highly developed civilization.

7. The trend toward territorial growth is contagious and increases in the process of transmission.

There is a great deal more to the theory, but even from this sample we can detect that Ratzel had a very deterministic view of the world. Nevertheless, he was a careful scientist, emphasized that his description was only based on an analogy to an organism, and did consider the interrelationships between people and their environment in both directions. He took the position of the detached observer, making no policy recommendations. His American disciple, Ellen Churchill Semple, wrote in the same spirit in her work. Not all of Ratzel's followers were so careful, however, and ignored his cautions.

Like Ratzel, *Rudolf Kjellén* (1864–1922) was a university professor, Ratzel at Leipzig, Kjellén at Uppsala. Kjellén, however, was a political scientist and a member of the Swedish parliament. He was a Germanophile, impressed with the new work in natural science and especially imbued with Ratzel's work in political geography. Unlike Ratzel, however, Kjellén was not a careful scientist. Instead, he took Ratzel's analogies literally and insisted flatly that the State *is* an organism. He even titled his most important book *Staten som Lifsform* (*The State as an Organism*, 1916). Here he presented his theory that the State is composed of five organs:

1. *Kratopolitik*—government structure.
2. *Demopolitik*—population structure.
3. *Sociopolitik*—social structure.
4. *Oekopolitik*—economic structure.
5. *Geopolitik*—physical structure.

Kjellén introduced aspects of the quality of the population, the nation whose aggregate constitutes the body of the State. In addition to moral capacities, there is the will, the cumulative psychological force of the State. The great power of the State is a dynamic, psychological concept. Kjellén saw the State in a condition of constant competition with other States; larger ones would extend their power over smaller ones, and ultimately the world would have only a few very large and extremely powerful States. He envisioned in Europe a superstate controlled by Germany.

This book, in which Kjellén originated the terms *Geopolitik* and *Autarky*, was translated into German in 1917, when the war was already going badly for Germany. At the end of the war, it was seized upon by some German political scientists, geographers, and nationalists who used it to lend the authority of the new evolutionary natural science to the older German political philosophy. It became a tool for rebuilding Germany into a world power and was subsequently used in the same way by Italy and Japan.

GEOSTRATEGY

Meanwhile, other scholars were focusing not on the State, but on the world and trying to find patterns in State development and behavior. They took a global view of geopolitical affairs and actually recommended policies, or strategies, to be followed by their governments. The first was *Alfred Thayer Mahan* (1840–1914). Mahan was a naval historian who eventually reached the rank of admiral in the U.S. Navy. He was a prolific writer, producing some 20 books altogether, among them *The Life of Nelson* (1897). His most influential books, however, were *The Influence of Sea Power upon History, 1660–1783* (1890) and *The Influence of Sea Power upon History; the French Revolution and Empire, 1793–1812* (1892). In his books he argued that control of the sea lanes to protect commerce and wage economic warfare was very important to a State. He therefore advocated a big navy. But there were six fundamental factors that affected the development and maintenance of sea power:

1. *Geographical position* (location). Whether a State possesses coasts on a sea or ocean (or perhaps more than one), whether these waters are interconnected; whether it also has vulnerable, exposed land boundaries; whether it can maintain overseas strategic bases and command important trade routes.
2. *Physical "conformation" of the State* (the nature of its coasts). Whether the coastline of a State possesses natural harbors, estuaries, inlets, and outlets; an absence of harbors will prevent a people from having its own sea trade, shipping, or navy; the importance of navigable rivers to internal trade but

their danger as avenues of penetration by enemies.
3. *Extent of territory* (length of the coastline). The ease with which a coast can be defended.
4. *Population numbers*. A State with a large population will be more capable of building and maintaining a merchant marine and navy than a State with a small population.
5. *National character*. Aptitude for commercial pursuits; sea power is "really based upon a peaceful and extensive commerce."
6. *Governmental character*. Whether government policy is taking advantage of the opportunities afforded by the environment and population to promote sea power.

Mahan, too, refers (in terms 5 and 6) to the question of the national "will," in terms reminiscent of Kjellén's. Indeed, Mahan's writing in places is quite similar to that of the German geopoliticians; he "was characteristically disposed to view politics in terms of force, whose efficacy was qualified only by inexorable natural and moral laws . . . only the fittest nations could successfully sustain themselves in the constant grappling. Fitness was measured in terms of national strength. This national strength was characteristically viewed by Mahan in military terms. It depended, in turn, on the moral and martial fiber of the population."[*]

And Mahan was a practical man. Like the German geopoliticians, he had prescriptions for United States foreign policy, and his advice did not go unheeded. He advocated that the United States should occupy the Hawaiian Islands, take control of the Caribbean, and build a canal to link the Atlantic and Pacific oceans. President Theodore Roosevelt's administration used several of Mahan's suggestions as the basis of its foreign policy. He was even more influential in Germany, Britain, and Japan.

[*]Charles D. Tarlton, "The Styles of American International Thought: Mahan, Bryan, and Lippmann," *World Politics, 17*, 4 (July 1965), pgs. 585, 586.

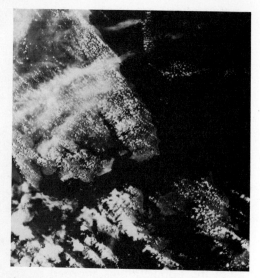

Singapore, a classic geopolitical outpost.
Founded in 1819 as a trading post on an island
off the tip of the Malay Peninsula, Singapore
evolved during the heyday of the British Empire as a major naval base and ship-servicing
center near the strategic Straits of Malacca, to
the left in the photo, along the lines advocated
by Admiral Mahan. The island became independent in 1965 and has since become not only
the world's fourth largest seaport, but also a
major industrial and banking center. (NASA)

Of greater interest to the political geographer are the glimpses of Mahan's world
view from his later book, *The Problem of
Asia.* In this work Mahan recognizes a core
area in Asia and Russia's domination of it;
he anticipates a struggle between Russian
land power and British sea power. Not surprisingly, he presumed that British sea
power would be able to contain Russian
expansionism. He also predicted that the
containment of Russia and the control of
China would become the joint concern of
the United States, Great Britain, Germany,
and Japan. (It is well to remember that he
was writing *before* the turn of the present
century.) Thus, he proposed that Russia
should be provided warm-water ports in
China by guaranteeing it the use of those
exits. By this means, he argued, the Russian push toward the sea, felt in Europe
and the Near East, might be alleviated. It
almost seems that Mahan forgot where

the Russian concentration of population
and productive capacity were located—
the factor of distance alone negated his
idea.

While Mahan emphasized sea power,
Sir Halford J. Mackinder (1861–1947), much
younger than Mahan, felt that the great
age of naval warfare was over; that changing technology, especially the railroad,
had altered the relationship between sea
power and land power. Still, his approach
to global strategy was similar to that of
Mahan, but with a different emphasis and
different forecasts. As Professor of Geography at Oxford and Director of the
London School of Economics, Mackinder
helped raise geography to a high level in
England. He was also a member of Parliament from 1910 to 1922 and Chairman of
the Imperial Shipping Committee from
1920 to 1945.

In 1904, Mackinder read a paper at the
Royal Geographical Society in London entitled "The Geographical Pivot of History." This was a true milestone in the
geopolitical debate of that period; in fact,
the contents of that article (afterward pub-

Sir Halford J. Mackinder, 1861–1947. (Courtesy
of Brian Blouet)

lished in the *Geographical Journal*) are still a subject of discussion and evaluation today, three generations later.

It is easy to regard Mackinder's paper as sophisticated speculation and to suggest that it has little value as a contribution to political geography (other than as a manifestation of "proper" geopolitical analysis). That Mackinder was closer to the truth than he himself may have realized is sometimes dismissed as accident. But it is worth remembering the main elements of the world situation when he produced his remarkable piece. Russia was losing a disastrous war with Japan, Germany was still a youthful, organizing State. Yet Mackinder in 1904 expressed the view that there was a Eurasian core area that, protected by inaccessibility from naval power, could shelter a land power that might come to dominate the world from its continental fortress. This Eurasian core area Mackinder called the Pivot Area in his original paper, but later he renamed it the Heartland. As Mackinder suggested, the Heartland's rivers drain into the Arctic, distances to warm-water oceans are huge, and only the Baltic and Black seas could form avenues for sea power penetration, but these are easily defended.

Mackinder reasoned that this Eurasian Pivot Area would become the source of a great power that would dominate the Far East, southern Asia, and Europe—most of what he called the "World-Island," which he conceptualized as consisting of Eurasia and Africa. He presumed that the Pivot Area contained a substantial resource base capable of sustaining a power of world significance. The key, he argued, lay in Eastern Europe, the "open door" to the pivotal Heartland. Thus he formulated his famous hypothesis:

Who rules East Europe commands the Heartland
Who rules the Heartland commands the World-Island
Who rules the World-Island commands the World*

Democratic Ideals and Reality, 1919.

Because Germany occupied the west of Mackinder's hypothetical Heartland, and Russia the east, he thought that effective control of this region for the purposes he had identified could be achieved only by the alliance of two or more States.

In 1924, Mackinder propounded a little-known counterhypothesis: the potentialities of the Heartland could be balanced in the future by Western Europe and North America which (as Mackinder read the lessons of World War I) "constitute for many purposes a single community of nations." He called the North Atlantic "the Midland Ocean," in the midst of the area from the Volga to the Rockies, which he called "the main geographical habitat of Western civilization."*

In 1943, in the midst of World War II, Mackinder blended all these ideas and modified them in an article titled "The Round World and the Winning of the Peace."† He moved the Heartland east of the Yenesei River and renamed it Lenaland. It would oppose the Midland Basin (the North Atlantic and bordering lands separated by Central Europe and surrounded by deserts and the Arctic). He felt that the Heartland contained soils and minerals equal to those of North America, but that the two regions would combine against Germany.

Mackinder made a significant contribution to our perspectives of the world, and in a broad sense his assumptions about the Heartland were later substantiated. However, there were three major weaknesses in his work. First, he did not give enough weight to the growing power of North America; second, he failed to explain the seeming contradiction between his thesis of the power of the possessor of the Heartland and the relative weakness of Russia/USSR until World War II; and, third, he failed to take into account the growing and very obvious importance of air power and other tech-

The Nations of the Modern World, vol. 2, London, George Philip & Son, 1924, p. 251
†*Foreign Affairs*, 21 (July 1943), 595–605.

nological developments. Like Mahan, he oversimplified history and leaned too far in the direction of determinism.

Mackinder had many critics. Prominent among them was *Nicholas John Spykman* (1893–1943). Born in Amsterdam, he studied at Berkeley, became a professor of International Relations at Yale in 1920, and became a US citizen in 1928. In his work he emphasized the power relations among States and the impact of geography on politics, but he rejected the German school of *Geopolitik*. In two books, *America's Strategy in World Politics* (1942) and *The Geography of the Peace* (1944), Spykman pointed out two of the basic weaknesses of Mackinder's theories. First, he felt Mackinder overemphasized the power potential of the Heartland; its importance was in fact reduced by the major problem of internal transportation and by access through the barriers that surrounded it. Second, history involving the Heartland was never a matter of simple sea power–land power opposition. Instead, Spykman felt, the real power potential of Eurasia lay in what Mackinder had called the "Inner or Marginal Crescent," and what Spykman called the Rimland. This area is vulnerable to both land and sea power and hence must operate in both modes. Historically, alliances have always been made among Rimland powers or between Heartland and Rimland powers. Spykman, therefore, proposed his own dictum:

Who controls the Rimland rules Eurasia;
Who rules Eurasia controls the destinies of the world.*

Spykman advocated that the Allies should base their postwar policy on preventing any consolidation of the Rimland. While there is no evidence that George Kennan (who proposed the "containment" policy of the Cold War era) ever read Spykman, this policy became fundamental in the anticommunist position of the Western powers because of the

*The Geography of the Peace, New York: Harcourt, Brace, 1944, p. 43.

change in thinking represented by Spykman. It was still, however, basically a nineteenth century view of the world.

GEOPOLITIK

While the two streams of thought in geopolitics were developing, a new school of geopoliticians was forming in Germany, chief of whom was *Karl Haushofer* (1869–1946). A career officer in the Bavarian army, Haushofer served in Japan (1908–1910) and rose to the rank of major general in the army general staff, serving through World War I. Before the war he took a Ph.D. in Geography at the University of Munich. He was embittered by Germany's defeat in the war, blaming it in part on the incompetence of her generals, but also on a too-early start of the war and the lack of links between the State's leaders and the land. He was convinced that Germany should have won the war and wanted somehow to avenge the defeat. He had been impressed by what he had seen of the power and expansionist ambitions of Japan, and also by the power of Britain. He therefore devoted himself to the study of geopolitics.

Haushofer began lecturing at the University of Munich in 1919 and became a professor of geopolitics after the Nazis came to power in 1933. He gathered around him a group of disciples, including journalists who helped spread his ideas. In 1924, he founded the *Zeitschrift für Geopolitik*, a monthly journal in which he and his colleagues propagated their new "science" whose name he had drawn from Kjellén. They also produced a flood of books and other publications, including maps, which they used as a major weapon. They remained active through World War II, working to justify Germany's drive for conquest by "science." There was, in truth, little that was scientific in *Geopolitik*.

Haushofer and his group blended together the organic State theory of Ratzel, its refinements and elaborations by Kjellén, and the geostrategic principles of Ma-

A German propaganda map in the Haushofer tradition.   This map, by Friedrich
Lange, appeared in *Volksdeutsche Kartenskizzen,* published in Berlin in 1936.
"This drawing was done in anger," reads the caption, and goes on to de-
nounce German students who only visit places in "the beleaguered East"
rather than settling there "because struggle, real patriotic struggle of the
people" is to be found there, "where to be called a German demands cour-
age, has disadvantages, and requires pledging of one's whole personality,
the risking of one's economic position, and even the sacrifice of personal

han and Mackinder, added a heavy dose of German chauvinism, willful ambiguity, and mysticism, and created a case for a German policy of expansionism. They used not only maps, but slogans and pictographs to influence people. The notion of *Lebensraum* was drawn from Ratzel but distorted to justify as "natural" Germany's growth at the expense of her less virile neighbors, Poland and Czechoslovakia. Kjellén's advocacy of autarky was revived and elaborated, with emphasis on the political character of the German quest for economic self-sufficiency, again at the expense of other countries in southeastern Europe, the Middle East, and elsewhere. Economic policy was, in fact, used as a political weapon by Nazi Germany. Influenced by the Heartland concept, the German geopoliticians advocated an alliance with the Soviet Union and apparently were dismayed when Hitler invaded that vast country. Not only did they develop new versions of geostrategy, all for the benefit of Germany, but they also attempted to create geo-medicine, geo-psychology, geo-economics, and geo-jurisprudence, not as true sciences or legitimate branches of geography, but as weapons for German conquest. There was much, much more, of course, but this sample gives something of the flavor of *Geopolitik*.

How influential was *Geopolitik*? There is still no consensus on this question, but the evidence seems to indicate that its influence was great in some areas, nil in others. Within Germany, it was pervasive throughout the educational system during the Nazi period, and it helped condition the intelligentsia to a policy of aggression. Its emphasis on planning, especially with regard to natural resources, did influence in part the direction of Nazi expansion. There is no evidence, however, that Hitler, his top advisors, or his generals were notably influenced by *Geopolitik*, even though Rudolf Hess was a disciple of Haushofer. Outside Germany, *Geopolitik* found fertile ground only in Japan, a country with which Haushofer had special ties and about which he wrote six books. Elsewhere, it was greeted with opposition or indifference. A few of its ideas have survived and have been widely adopted: the extensive use of maps to convey ideas, for example, and the need for governments to have a store of accurate and up-to-date information about the earth; concepts of propaganda and psychological warfare, of total war and of the importance of air power. But *Geopolitik* as such died with the collapse of Nazi Germany in 1945—and none need mourn its passing. Geopolitics, however, lives on to add spice and color to political geography.

---

freedom. This kind of struggle is going on today on a broad front behind the borders, where German men and women cling to the soil which is their fate in a spirit of steadfastness, spite and family sense. It takes place all over the East, to the north and south of Czechoslovakia, which brags about being the Slavic fist in Middle Europe. But, to be sure, only he can keep a border who has the right border spirit and is willing to put himself along the border." Such maps helped prepare the way for the German *Drang nach Osten*, (March to the East), which triggered World War II. (Courtesy of E. J. Huddy, British Library)

Chapter 18

GEOPOLITICS SINCE WORLD WAR II

Otto Maull, of Haushofer's *Institut für Geopolitik*, once claimed that "*Geopolitik* is concerned with the spatial *requirements* of a state while political geography examines only its space conditions." In one sense, this is quite true. Every State has needs, just as every society and every person does, and there is certainly nothing wrong in studying these needs. *Geopolitik*, however, so warped and twisted this perfectly legitimate study that its odium spread over all of geopolitics and even over political geography. In the wreckage of *Geopolitik* lies the organic State theory, but the obvious value of geostrategy has enabled this aspect of geopolitics to survive the denunciations of the immediate postwar period. Indeed, geopolitical views of the world, of regions, and of States are still vigorously expressed in many quarters. The postwar literature in geopolitics, in fact, is abundant. In this survey we only mention some of the highlights.

GEOSTRATEGY

Advocacy of air transport and air warfare is not new. It began with the Wright brothers, continued through Lindbergh and Billy Mitchell, Glenn Curtis, Eddie Rickenbacker, and many others around the world. One strategist, however, was the first to advocate forcefully a geopolitical view of the world based on air power. He was *Major Alexander P. de Seversky* (1894–1974). Born in Russia, he served in the Russian navy during World War I and became a naval aviator. He lost a leg on a bombing mission but continued fighting and became Russia's leading naval ace. In 1918, he was sent on a naval mission to the United States, to whom he offered his services after Russia dropped out of the war. He served in various capacities, became an American citizen in 1927, invented the world's first fully automatic bombsight, founded an aircraft company which became Republic Aviation Corporation, and continued to advance aviation through his aircraft design, innovative combat strategy and tactics, speed records, production of aircraft, and work in civilian aviation. He was both a practical engineer and businessman, and an imaginative thinker. Among his many writings, two books had enormous influence on government officials as well as the general public. The first was *Victory Through Air Power* (1942). In it he reviewed the course of the war to that point, declared "the twilight of sea power," deplored the insufficient attention to air warfare being paid by the Allies (especially the United States), and advocated a totally new strategy and organization for victory through air power. The book reads well even today.

After World War II had ended and the "Cold War" had begun, de Seversky published another innovative book, *Air Power: Key to Survival* (1950). Here he re-

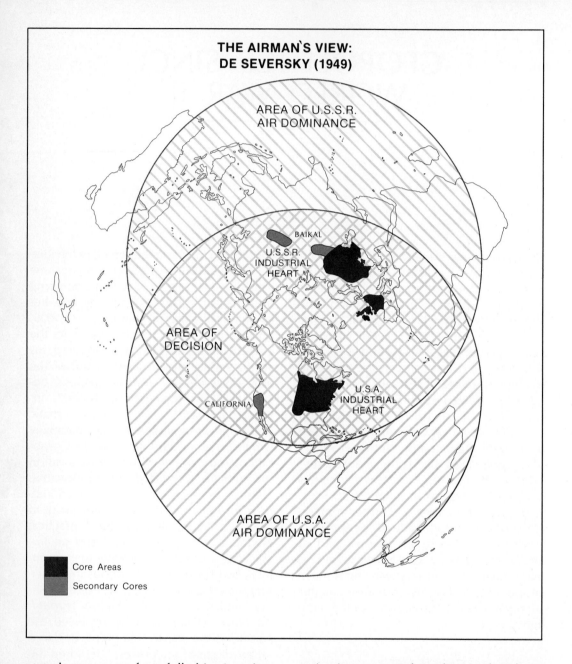

**THE AIRMAN'S VIEW:
DE SEVERSKY (1949)**

AREA OF U.S.S.R.
AIR DOMINANCE

BAIKAL

U.S.S.R.
INDUSTRIAL
HEART

AREA OF
DECISION

CALIFORNIA

U.S.A.
INDUSTRIAL
HEART

AREA OF U.S.A.
AIR DOMINANCE

■ Core Areas

▨ Secondary Cores

stated even more forcefully his view that land and sea power had been subordinated to air power. He urged the development of massive air superiority for the United States. He advocated defense of the Western Hemisphere, avoidance of small wars as a useless sapping of American strength, abandonment of overseas bases as costly luxuries. For the first time in such geopolitical writings, he used a map drawn on an azimuthal equidistant projection centered on the North Pole to show clearly how close the United States and the Soviet Union really were. It also showed the vast areas of air dominance by both countries and how these areas overlapped over the North Polar region, in what he called the "Area of Decision." As a result of this "new" view of the world, the United States and Canada erected at great expense three lines of radar stations and air bases stretching across Alaska and

Canada for the defense of North America against attack from the USSR by the shortest routes—over the North Pole. It is easy to criticize this view in an era of intercontinental ballistic missiles and space travel, but at the time it performed a most useful function by tearing us away from our Mercator-view of the world, by developing an interim defense system, and by emphasizing defense instead of expansion as the prime goal of geostrategy. The "fortress America" concept was never accepted by policymakers, however, and the United States has become deeply involved in political affairs around the world.

The more traditional world view of Mackinder and Spykman, however, has not yet been supplanted. The American geographer *Donald W. Meinig* and the British geographer *David J.M. Hooson* developed modifications of the Heartland and Rimland concepts in 1956 and 1964, respectively. Meinig, in his "Heartland and Rimland in Eurasian History," suggests that there is more variety and flexibility within the Rimland than Spykman

had recognized and proposed a "continental" Rimland composed of States that were oriented "inward" (toward the Heartland) and a "maritime" Rimland of "outward"-oriented States. Since this situation is dynamic, not static, he also proposed a category of neutral or transitional States. Hooson redefined the Heartland in terms of such criteria as the rate of population growth and rate of urbanization, location of accessible resources, specialization in both agriculture and industry, ethnic factors, and historical association and cohesiveness; in short, the scale of contribution to the Soviet economy of various parts of the Heartland. He concluded that the Heartland now should be considered as a core around Moscow and the Black Sea, and a significant Volga–Baikal Region. It is this eastern zone that Hooson views as critical, considering that what happens there might well be decisive in the future of the Soviet Union as a world power. Although these two modifications are useful, they do not represent any breakthroughs in geostrategic thinking.

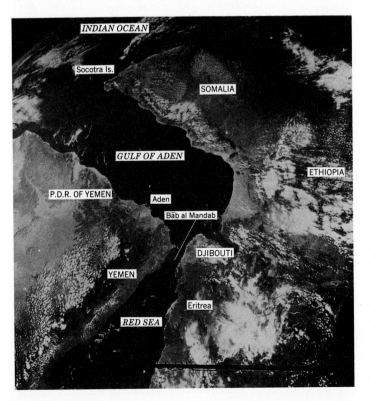

The Strait of Bab el Mandeb. One of the most strategic waterways in the world, this strait connects the Red Sea (and beyond it the Suez Canal and the Mediterranean) with the Gulf of Aden and the Indian Ocean. At the top of the photo (southeast of the strait) is the Horn of Africa; surrounding the strait are Aden and Djibouti, formerly major British and French naval bases, and the warwracked Yemen and Eritrea. (NASA)

THE DEVELOPMENT OF GEOPOLITICS

THE ORGANIC STATE

GEOSTRATEGY

FRIEDRICH RATZEL
1844-1904
"The Laws of the Spatial Growth of States" (1895)
Political Geography (1896)

Admiral ALFRED THAYER MAHAN
1840-1914
*The Influence of Sea Power Upon
History 1660-1783* (1890)

RUDOLF KJELLEN
1864-1922
The State as an Organism (1916)

Sir HALFORD J. MACKINDER
1861-1947
"The Geographical Pivot of History" (1904)
Democratic Ideals and Reality (1919)

General KARL HAUSHOFER
1869-1946
Journal of Geopolitics (1924-1968)

NICHOLAS JOHN SPYKMAN
1893-1943
America's Strategy in World Politics (1942)
The Geography of the Peace (1944)

ALEXANDER P. DE SEVERSKY
1894-1974
Victory Through Air Power (1942)
Air Power: Key to Survival (1950)

DONALD W. MEINIG
1924-
"Heartland and Rimland in Eurasian
History" (1956)

SAUL B. COHEN
1928-
*Geography and Politics in a World
Divided* (1963)

DAVID J. M. HOOSON
1926-
A New Soviet Heartland? (1964)

Other observers have taken fresh views of world relationships and have presented innovative concepts.

One innovator is the American geographer *Saul B. Cohen*. In his *Geography and Politics in a World Divided* he considers the entire world as being divided into "geostrategic regions." Like de Seversky, he takes the Americas into account as well as new technology and aims at global equilibrium. "The major premise of the work is that the dynamic balance that characterizes relations among States and larger regions is inherent in the ecology of the global political system. This world is organized politically in rational, not random fashion."* He rejects the notion, popular in the immediate postwar period, that spheres of influence are obsolete, even reprehensible. He insists, in fact, that "spheres of influence are essential to the preservation of national and regional expression."† His geostrategic regions are essentially the spheres of influence of the United States, Maritime Europe, the Soviet Union, and China.

He still preserves the concept of maritime and continental powers, but expands it to include in the "Trade Dependent Maritime World" all of the Americas, all Western Europe, all Africa except for the northeastern corner, and all of offshore Asia and Oceania; and in the "Eurasian Continental Power" all the Soviet Union and Eastern Europe, and Eastern and Inner Asia. He still does not take into account, however, the fact that the Soviet Union has become a leading maritime power, with naval, merchant, fishing, and ocean-

*Preface to the second edition, New York: Oxford, 1973, p. vi.
†*Ibid.*, p. viii.

ographic fleets ranking among the world's largest.

South Asia he classifies as an Independent Area, and he identifies the Middle East and Southeast Asia as shatter belts. The term "shatter belt" or "shatter zone" has customarily been applied to Central and Eastern Europe, a region of chronic instability in which States appear, disappear, and reappear with frequently changing names and boundaries. Cohen has omitted reference to this original shatter belt in deference to its partition in the 1940s between Soviet and American power systems. His new ones are quite justified in view of their chronic instability, but not the omission of Central and Eastern Europe. The current stability of this region may be deceiving since ancient rivalries and animosities have been only temporarily suppressed by communist togetherness. Even in the former French Indo-China, where all three States (Laos, Kampuchea, and Vietnam) are under communist rule, all three are fighting rebels or one another. Shatter belts do not disappear very readily.

Cohen has a great deal more to say, including recommendations for American policymakers, but a major thesis bears some special mention. He advocates the maintenance of the unity of Europe and the Maghreb by "subtle economic and political persuasion" rather than by force. This emphasis on cooperation, on economic power, on persuasion and propaganda is a realistic summation of actual trends in the postwar period, trends which go on, as it were, under an umbrella (protective or threatening) of nuclear missiles and space satellites.

These newer trends, however, were

The two streams of thought that comprise the theoretical foundations of geopolitics are symbolized in this diagram by the leading figures at each stage of development of the field. The organic theory of the State and geostrategy were blended by Haushofer, who added a heavy dose of German chauvinism and liberal portions of racism and militarism to create what he called *Geopolitik*. After the demise of the Third Reich in 1945, *Geopolitik* was no longer seriously advocated publicly; though Haushofer's *Zeitschrift* resumed publication for several years after the war, it was quite different from the pre-1945 version. With *Geopolitik* died the organic State theory, but geostrategy survived and flourishes today.

viewed differently by *Lin Piao*, late Defense Minister of China. In 1965, he expounded a theory of world revolution that viewed the world as similar to a city and the surrounding countryside. The rich, industrialized, largely Western countries represent the city and the poor, agricultural countries, largely former colonies of the Western countries, represent the countryside. The poorer areas will gradually be converted to communism and, using tactics similar to those described by Cohen, but supplemented by guerrilla warfare, will confront and eventually overwhelm the cities. These are the tactics the Chinese Communists used so successfully against the Nationalists and the Japanese during the 1930s and 1940s, and are described by McColl in his studies of "insurgent states." Applying them on a world scale, however, seems a bit unrealistic, if only because of the diversity within the developing world and the industrial world's great strength. Insurgencies, in fact, have not all been successful, and even China has gone over to fighting conventional wars when it seems appropriate.

Although all these geostrategic views have serious flaws, they do have the virtue of analyzing the world as a whole, rather than as scores of discrete political units. In view of the increasing interdependence of the world, we need a good deal more of this holistic thinking, not directed toward formulation of strategies for confrontation, but toward strategies of cooperation; not a geostrategy of war, but a geostrategy of peace.

GEOPOLITICS ON A SMALLER SCALE

Geographic, political, and historical literature are replete with discussions of buffer States and buffer zones, areas of weakness that serve to separate areas of strength so as to reduce the chances of conflict between them. Classic buffer States include Bolivia, Paraguay, and Uruguay, which almost completely separate Brazil from Chile and Argentina; Poland and the Balkans separating Germany and Russia/USSR; Afghanistan lying between Russia/USSR and British India; the Himalayan kingdoms of Nepal, Sikkim, and Bhutan guarding the major mountain passes between China and India; Mongolia between Russia/USSR and China; and Laos and perhaps Thailand separating British and French power in Asia. These and others have largely, if not completely, lost their buffer functions as a result of rising nationalism and the strengthening of the buffers themselves, changing technology, the role of ideologies other than nationalism, and the general tendency throughout the world to eliminate frontier areas in favor of clearly defined and delimited boundaries.

Even the kind of buffer zone that became prominent after World War II—the ideological buffer—has tended to disintegrate. On the grand scale, the new States that emerged from the dissolving empires combined with a number of older ones, including most of those listed as traditional buffers, to form a neutralist bloc, aligned with neither the Soviet nor the American blocs. This channeled some of the energies of the superpowers into attempts to win friends and influence in the neutralist countries, perhaps contributing to a reduction of tensions between them and avoidance of World War III. Another type of buffer zone developed across southern Africa as black Africans to the north attained independence between 1957 (Ghana) and 1963 (Kenya), while the white minority regimes in Southern Rhodesia and South Africa/South-West Africa retained tight control.

Both buffers have virtually disappeared. In Africa, the tide of "Africanization" has swept southward and even penetrated the bastion of white supremacy with the granting of independence to the former British High Commission territories of Bechuanaland (now Botswana), Basutoland (now Lesotho), and Swaziland, and the advent of black majority rule in Zimbabwe. The "nonaligned" bloc has also lost much of its value because many of its members have become aligned (though perhaps not admitting it publicly), as new centers of communist ideology have developed to challenge the original one in Moscow, and as relations have improved between the United States and the major communist States, the USSR and China.

We do not claim that buffer zones no longer exist or that no new ones will be created. After all, no intercontinental missiles or weapons from space have been fired at anyone so far, so conventional warfare—military, economic, and ideological—is still important. The Soviet Union, for example, set about after World War II to surround herself with concentric rings of defense against the kinds of invasion she has suffered so often in the past. The innermost ring consisted of territory actually annexed from other countries: Karelia, the Baltic States, eastern Poland, the Carpatho–Ukraine, Bessarabia, Tannu Tuva, southern Sakhalin, and the Kuril Islands. These acquisitions enabled her to regain most of the Czarist Em-

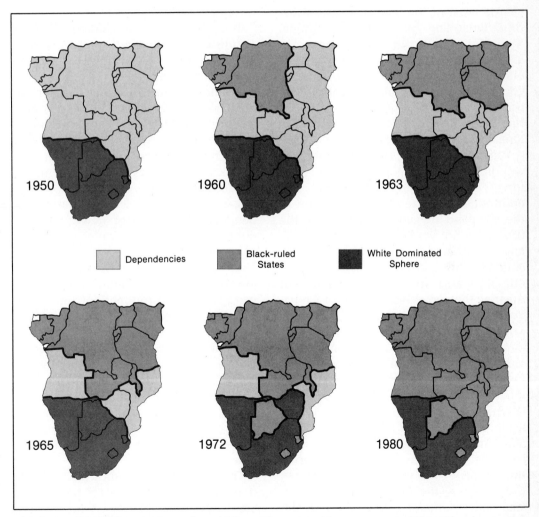

The decay of the buffer zone in southern Africa.   The zone of European and white settler domination affording the Republic of South Africa and its mandated territory of Namibia (Southwest Africa) some degree of protection from native African nationalism to the north steadily shrank during the past quarter century, as more and more colonies achieved independence.

pire, take most of the bases Hitler had used for invasion of her territory, create defense in depth for vital and exposed Leningrad, lengthen her coastline, attain common borders with Norway and Hungary, reach the Danube and obtain a seat on the Danube River Commission, and virtually enclose the Sea of Okhotsk, all most useful geopolitical objectives, the value of which she has demonstrated repeatedly in the past forty years.

The second ring was composed of States closely aligned with her; some, in fact, were virtually colonies. These included most of the States of Eastern Europe (including East Germany), China, Mongolia, and North Korea. Somewhat less reliable but still useful were States either neutral by treaty or neutralist by conviction (or necessity): Finland, Austria, Turkey, Afghanistan. The gaps in this ring were filled by Soviet troops in Iran, Xinjiang, and Manchuria.

Very soon, though, beginning with the withdrawal (under UN pressure) of Soviet forces from northern Iran in 1947, this enormous and extensive buffer zone in the "Rimland" began to crumble and the process is not yet complete. The Soviets have withdrawn from Xinjiang and Manchuria, from Porkkala in Finland and Saseno in Albania, from Vienna and eastern Austria; they have lost control of China, Yugoslavia, Albania, and perhaps North Korea; Romania has developed an independent foreign policy and Poland has attained a large measure of independence. One is tempted to comment on the disintegration of the Soviet empire by exfoliation, reaching into the Soviet Union itself, with revolts in the Ukraine, Lithuania, Armenia and elsewhere. Nevertheless, she has penetrated and overleaped these rings to win important positions in Cuba, Africa, the Middle East, Afghanistan, and Southeast Asia. But out in the distant, unfamiliar world she has had about as many failures and retreats as she has had successes and advances. Only the new and powerful Soviet navy, ranging all the oceans of the world, is undefeated—

and untested. There is still plenty of scope for geopolitical analysis here.

SOME TENTATIVE GEOPOLITICAL CONCEPTS

Because of lack of space, we can only suggest here three geopolitical concepts that could use extensive and careful study by political geographers. Two have been dominating the strategic thinking of the military and political leaders of the United States since about 1950, whereas the third is nonideological and based on our own observations.

The dominant and guiding vision of American policymakers since very soon after World War II has been that of a bipolar world: the Soviet Union and her "satellites" versus the United States and her "friends and allies." There has been a recognition that not all countries in the world could in any manner be crammed into one or another of these neat categories of "bad guys" and "good guys." Indeed, the "Non-Aligned Movement" began as early as 1955, at a meeting in Bandung, Indonesia of newly independent countries, some colonies about to become independent, and assorted other countries that did not want to be swallowed up by either "camp." Their leaders were Sukarno of Indonesia, Nehru of India, Tito of Yugoslavia, Nasser of Egypt, and (a little later) Nkrumah of Ghana. In recognition of her split from the Soviet Union, China was (after considerable debate), invited to attend. Their acceleration of neutralism in the Cold War marked the origin of the term "Third World," a term that has since been given so many meanings that it has become virtually meaningless.

The reaction of the United States to Bandung was epitomized by U.S. Secretary of State John Foster Dulles, who declared flatly, "Neutralism is immoral." With the modest exception of Yugoslavia, it was assumed that any country with a centrally planned economy was a Soviet satellite or, in another then-current phrase, a "captive nation." It was many years before policymakers accepted the Sino–Soviet split as genuine, deep, and abiding and recognized China as a separate force in the world. Meanwhile, led by the Italian Communist Party leader Palmiro Togliatti, what may have been a reasonably solid bloc of communist parties in and out of power in Europe began fragmenting. Uprisings against communist rule in East Germany, Poland, and Hungary, although suppressed by the power of the Soviet Union, hastened this fragmentation. The Asian communist parties lined up with the Soviet Union or China—or split between them. So did parties elsewhere in the world, and some went off in a third direction, or a fourth. Meanwhile, Western Europe, largely through the instrument of the European Communities (discussed in Chapter Twenty-five) was developing into another "pole," and so, in a sense, was Japan. Yet the United States continued to behave as if the world were still bipolar (if, indeed, it ever was). Though it waned somewhat during the 1970s, the myth of a bipolar world gained new life and strength during the Reagan administration, in which some unreconstructed "cold warriors" gained positions of considerable influence and the geopolitical saber rattling began again in earnest.

The second concept derives directly from the first; this is the "domino theory." Like "containment," "massive retaliation," "the balance of terror" and other Cold War doctrines, this one assumed that the Soviets (and all communists and most socialists everywhere) were, and are, unqualifiedly evil, that they were bent on world domination, that they were fiendishly clever and that any small victory by them would automatically lead to many more. The "domino theory" originated, apparently, with U.S. Admiral Arthur Radford, who in 1953 urged a carrier-based nuclear strike against the Viet Minh in Vietnam to relieve their pressure on the French at Dien Bien Phu on the theory that a Viet Minh victory there would set off a chain reaction of countries going com-

munist "like a row of falling dominoes." The theory came into prominence during the Vietnam War in the 1960s. The argument went that the United States had to fight and win in Vietnam, for if South Vietnam "went Communist," then automatically, like falling dominoes, Cambodia, Laos, Thailand, Burma, and perhaps India would also "go Communist." In the other direction, the Philippines, Taiwan, Hawaii, and even mainland United States were similarly threatened.

Although the theory is based on false assumptions about the nature of local nationalist movements, despite the lack of evidence of any grand Soviet scheme to knock over "a row of dominoes," despite vastly different conditions in countries "threatened" by communism, despite the huge problems of distance and logistics from one "domino" to another, despite the obvious fact that many countries actually bordering on the Soviet Union, China, and Cuba have not "fallen," despite, that is, the lack of any credible evidence whatever of its validity, it is still guiding much of American foreign policy today. It has led the United States to unleash military power against Libya, Grenada, and Nicaragua; to support "destabilizing" movements in Chile, Iran, Angola, and many other countries; and to threaten military action in the Persian Gulf, Central America, and elsewhere. This appears to be a revival of the Cold War strategy of fighting "brushfire wars" in the Rimland, in traditional or emerging shatter belts. It may be only coincidental, but these wars have all been fought against countries that were small, poor, and weak. Some of the "non-aligned" States may be wondering whence cometh the threat of falling dominoes.

The third concept that needs more geographic analysis has been almost entirely ignored in the geographical literature. This is the notion of changing orientations. Three examples of changing orientation may be given briefly. First is the case of Zambia. Prior to independence, Zambia, as Northern Rhodesia, was a

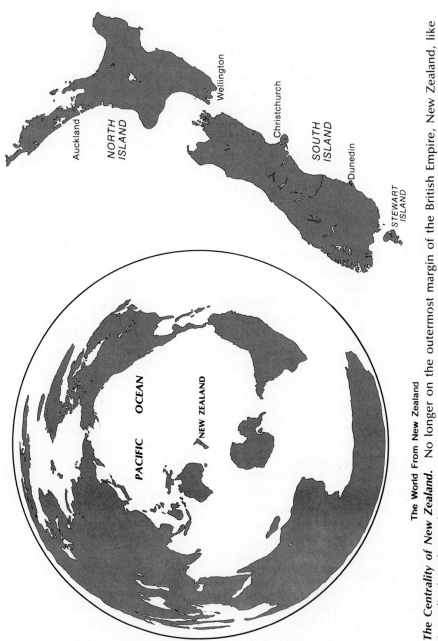

The World From New Zealand

The Centrality of New Zealand. No longer on the outermost margin of the British Empire, New Zealand, like Australia, has changed its orientation since World War II and now takes a broader view of its place in the world, even developing an independent foreign policy. This new world view is reflected in this official government map.

member of the Central African Federation. She remained very much a central African country until after Southern Rhodesia declared her independence in November 1965, whereupon Zambia cut her ties to the south and began strengthening those to the north and east. Zambia has thus become the southernmost State in East Africa.

A less dramatic but nonetheless important change in orientation has taken place in the Commonwealth Caribbean. Until World War II this region was oriented almost exclusively toward the United Kingdom and to a lesser extent to the rest of the Commonwealth and Western Europe. During the war these ties were severely frayed and new links were developed with North America. By 1965, it could be fairly said that these new ties were far more important than the traditional ones. Since then, we can detect a second shift in orientation, gradual and still inchoate, but unmistakable: a turning away from North America toward Latin America and, to a very much lesser extent, Africa. A third example is that of Australia. Until the Second World War, Australia had the same orientations as the Commonwealth Caribbean: toward Britain first, and less importantly toward the rest of the Commonwealth. During the war and for about two decades thereafter, ties with Britain weakened and Australia began looking northeastward, toward the United States, while still maintaining close relations with New Zealand. More recently, although not abandoning any of these important relationships, Australia has become continually more closely tied to Japan while strengthening her relations with nearby Pacific Ocean States. Now, for the first time, Australia is looking at the Indian Ocean not as a vast void to be traversed enroute to or from the British Isles, but as an arena for Australian interest and activity.

What other countries around the world are similarly shifting their orientations? Why? What effects will these shifts have on the US–USSR rivalry? What effects will they have on international trade? On regionalism? Geographers should investigate these and many other questions so we can better understand the dynamic nature of history—and of geopolitics.

THE CURRENT STATUS OF GEOPOLITICS

Like political geography itself, geopolitics has experienced a recent revival. In the past decade, it has become acceptable—even respectable—to write on geopolitical theories without being branded a chauvinist or a warmonger. In the United States and Britain especially, geopolitics is again attracting attention and some creativity is being shown. Some of the credit must be given to Henry Kissinger, National Security Advisor and Secretary of State under Presidents Richard Nixon and Gerald Ford, but geopolitics was reviving independently in Italy, Germany, and France at the same time. New texts specifically on geopolitics appeared in 1985 (G. Parker) and 1986 (O'Sullivan). The new academic work in geopolitics (as distinguished from the Kissinger-type governmental theses) is distinctly less bellicose and imperialistic than the older work was. We may be witnessing (and perhaps participating in) the development of a new kind of geopolitics: a geopolitics of peace. We examine this notion in the next chapter, but first we should point out that while geopolitics was languishing in most of the world, it was flourishing in Latin America.

All over Latin America, but especially in the Cono Sur, the Southern Cone of South America, geopolitics is not only a major professional field, but it is of great popular interest as well. While in North America and Western Europe books on the subject were produced chiefly by military academies and organizations and geopolitical articles appeared almost exclusively in military journals, in Brazil, Chile, Argentina, Uruguay, Bolivia, Peru, and to a lesser extent Ecuador, there has been a steady and voluminous outpouring of writings on the subject by diplomats, professors, and government officials as well as by military men. Daily newpapers contain long, weighty geopolitical analyses; geopolitical books and journals are readily available in public libraries and on newsstands and in bookstores. Radio and television carry geopolitical themes into the homes even of illiterates.

Most of the themes focus on South America and nearby areas, of course. Among the most prominent are: the South Atlantic as a zone of potential conflict between the United States and the Soviet Union; Brazilian expansionism (real and imagined); the strategic value of the straits at the southern tip of South America; the triangular relationships of Brazil–Peru–Chile or Brazil–Argentina–Chile or Peru–Chile–Argentina; potential conflict over the western Amazon Basin; the integration of the "American Antarctic" into South America; and the equivalence of sea and land in calculating national territory and power. The writings are often amply illustrated with maps and diagrams replete with triangles, arrows, arcs, stars, and other figures that illustrate areas of action, directions of movement, and relationships of all kinds.

It is difficult to say how much policy is influenced by this kind of geopolitical thinking, or whether, in fact, geopolitics is simply used to justify policy. In South America there is a constant rotation of people (almost entirely men) among academia, the military, the diplomatic service, and the government, and many, many people have served in all four sectors.* Even after retirement, they continue to write and to be read and quoted. And not all of them are parochial. Some write on geopolitics in other parts of the world, and the German geopoliticians of the Hitler era are still quoted approvingly.

We can hope that with the revival of geopolitics in North America and Western Europe, some of the better Latin American geopolitical literature will become available to us in translation. It might be useful to us and it would certainly be interesting. Meanwhile, we can thank such people as Jack Child, Howard Pittman, and Philip Kelly for bringing it to our attention.

*The present dictator of Chile, General Augusto Pinochet Ugarte, has a doctorate in Geography and has published a widely known book on geopolitics.

Chapter 19

THE GEOGRAPHY OF WAR AND PEACE

In our last chapter, we discussed the revival of geopolitics as an academic field and some contemporary geopolitical concepts. We have also witnessed in the past decade the growth of what might be considered an even more fundamental study: the relationship between geography and war and peace. Geographers have long contributed to the study of warfare itself, and there is a substantial literature in the field. But the study of the geographic factors that can lead to war, or a geographic perspective on the causes of war, is relatively new. Geopolitics, it is often said, is nothing but a justification for imperialism, and imperialism usually leads to war. We discuss imperialism in some detail in the next chapter and it will be seen that there is some truth in the accusation. But it is an oversimplification. Geopolitics does not always lead to imperialism and imperialism is not the only cause of war. In an introductory text, we cannot analyze this observation in great detail. We shall, therefore, simply survey the subject, focus on a few topics of special interest to geographers, look at four major world regions to see how some principles apply there, and exhort students to become involved in the search for, and application of, a geopolitics of peace.

William Bunge, an American geographer living in Canada (as well as many other scientists), has pointed out in great detail the effects on our species and on our entire planet of the uncontrolled explosion of even one thermonuclear bomb—not to mention a full-scale nuclear war. It should be clear to every thinking person on earth by now that a nuclear war cannot be won and must never be fought. Yet all the major world powers have developed thermonuclear weapons, continue to test new ones, and persist in producing them at a furious pace. The colossal expenditure of natural resources, productive capacity, time, talent, energy, and capital on preparation for nuclear war is almost unimaginable and—to a rational being—inexcusable. Yet it is only a part of the military expenditures made by countries large and small, rich and poor, old and new all over the world since World War II was supposed to have taught us—again—the folly and futility of war. A few statistics gathered from various sources will help to illustrate the point.

1. During the period 1945–1989, the United States alone will have spent approximately *ten trillion dollars* ($10,000,000,000,000), in 1987 dollars, on military weapons, equipment, and personnel—more than a third of it in the eight years of the Reagan administrations alone.

2. In 1985, all the countries of the world together spent nearly two-thirds of a *trillion* ($663,120,000,000), in 1980 dollars, for military purposes, and this fig-

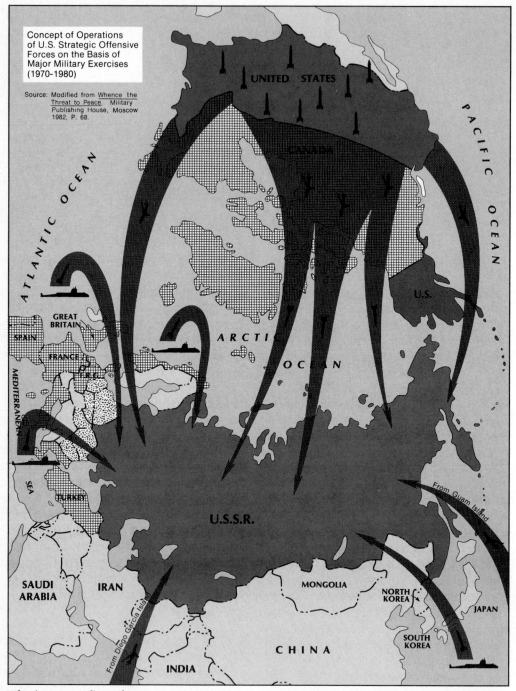

Concept of Operations
of U.S. Strategic Offensive
Forces on the Basis of
Major Military Exercises
(1970-1980)

Source: Modified from Whence the
Threat to Peace. Military
Publishing House, Moscow
1982, P. 68.

Who is surrounding whom? In Chapter Three we pointed out that direction is essentially a matter of perception. We used as an example the perception of the American people that they were surrounded by a chain of their own military bases protecting them from attack by the Soviet Union. Here is a more recent Soviet map showing some of those same bases, supplemented by nuclear-armed submarines, as seen from the Soviet Union. These very different perceptions of the same basic facts help to create a climate that can lead to war. By presenting spatial facts objectively and analytically, political geographers can help to create a climate that will lead to peace.

ure has been steadily rising with no sign that it will diminish in the foreseeable future.

3. In 1986, all the countries of the world together spent an average of about $30,000 to support each soldier, but only about $455 to educate each school-age child.

4. In a single *hour*, the world spent *twice* as much on military materiel as it cost the United Nations to stop a locust plague in Africa in 1986, saving enough grain to feed 1.2 million people.

5. The superpowers are not entirely to blame for these staggering expenditures, nor even the rich countries together. Military expenditures are increasing fastest in those countries least able to afford them, 800 percent between 1960 and 1986, even adjusting for inflation; between 1974 and 1985, this amounted to $250,000,000 worth of arms imported by poor countries from rich ones!

In Chapter Twenty-eight, we discuss the arms trade in a different context, but it is clear enough even from these few figures that huge sums are being diverted from productive use and squandered on a vast array of lethal weapons and their support systems. Geographers are not the only ones who can think of better ways to use our finite resources. But the scale of military expenditures is only one aspect of the problem. Another, even more geographic, is their distribution. As indicated in Point 5, the poor countries are spending not only an increasing proportion of all expenditures on arms, but also an increasing proportion of their limited resources as well, often going into debt to do so. In 1985, they accounted for over 17 percent of total world military expenditures; if China is counted in, the total rises to 21.5 percent. Moreover, it is not the United States, the Soviet Union, and the other major world powers alone that have chosen to fight their wars on the soil of developing countries; since 1945 a total of

119 wars have been fought there, most of them between developing countries themselves. No major part of Latin America, Africa, or Asia has been free of armed conflict since 1945. Furthermore, in addition to the extraordinarily wide dispersal of sophisticated weapons, the range of ballistic missiles has increased to the point that there is no place at all that they cannot reach. And numerous spying satellites revolving around the earth assure that very little of consequence on the planet goes undetected. From surveillance or attack, there is no longer any safe place on earth, no place to hide.

NUCLEAR-FREE ZONES AND ZONES OF PEACE

The realization of this fact has led many States to propose that certain areas of the earth be declared off limits. The first such proposal in modern times came in 1957, when Adam Rapacki, Foreign Minister of Poland, proposed a zone in central Europe, on both sides of "the Iron Curtain", that would be freed and kept free of all nuclear weapons as a contribution to the reduction of Cold War tensions and of the possibility of the outbreak of a hot war. The United States and her allies rejected the Rapacki Plan and since then, with one exception, the United States has rejected every proposal for a nuclear-free zone in the inhabited world. Typically, these proposals include the prohibition of the manufacture, testing, acquisition, storage, installation, or use of any nuclear weapon of any kind within the region and provide for on-site inspection and verification to ensure that the ban is strictly honored.

The first such nuclear-free zone was established in Latin America by the 1967 Treaty of Tlatelolco. It was not the product of abstract theorizing or a naive desire to change the world, but of a very strong and widespread perception of the imminent danger of the introduction of nuclear weapons into the region by an outside power. When Algeria became independent of France in 1962, after a protracted and bitter war, France was forced to remove her nuclear weapon-testing facilities from the Algerian Sahara. She then began building a new facility in French Guiana. She claimed it was to be only a scientific research base from which space vehicles would be launched, but the Latin Americans (and others) had reason to believe otherwise. After Latin America was declared a nuclear-free zone, with inspection and verification to be conducted by the International Atomic Energy Agency, the Kourou station became a joint U.S.–France rocket research base and the French established their new nuclear weapons testing facilities on Moruroa Atoll in French Polynesia.

Since then, other nuclear-free zones have been proposed in various parts of the world: the Balkans, the Adriatic, and the Mediterranean; Africa; Northern Europe; the Middle East and South Asia among others. None has actually been legally established by treaty, however, and it is known or presumed that nuclear weapons have been produced in, or introduced into, all of these regions. Four areas have, however, been declared nuclear-free zones to date: Antarctica (1959), outer space (1967), the seabed (1970), and the South Pacific (1985–1986). The first three are discussed in Part Seven and the last later in this chapter.

Even more sweeping, and more idealistic, are the proposals that have been made over the years for zones of peace. The first of any consequence in modern times was the Declaration of South-East Asia as a Zone of Peace, Freedom, and Neutrality adopted by the Association of Southeast Asian Nations (ASEAN) in November 1971—in the midst of the Vietnam War. It referred specifically to the Treaty of Tlatelolco but aimed at "the neutralization of South-East Asia". The UN General Assembly in November 1986 adopted the Declaration of a Zone of Peace and Cooperation of the South Atlantic. It is much broader and more detailed than the

ASEAN declaration, but nowhere mentions "neutralization" of the region. It is still unclear what significance either of these declarations will have.

Meanwhile, the UN General Assembly adopted in December 1971 the Declaration of the Indian Ocean as a Zone of Peace, discussed in more detail below, and King Birendra of Nepal in February 1975 proposed that his country be singled out as a Zone of Peace. Both of these proposals have been pursued with varying degrees of vigor for more than a decade, yet neither has actually been implemented. The Nepal proposal is especially interesting since, unlike the "neutral" statuses of Finland, Austria, and Switzerland, it does not derive from wars with neighbors and, unlike the "neutralist" statuses of the "non-aligned" countries, it has nothing to do with the rivalry between the United States and the Soviet Union—at least not directly. Instead, it concerns a country which has long been at peace with both of her neighbors, is on good terms with both of them at present, and is not threatened by anyone else.

It is unclear what value all of these proposals for, and declarations of, zones of peace and nuclear-free zones may have in the worldwide search for peace. Perhaps their greatest values lie in the deterrent effect they might have on extraregional powers that might be tempted to intervene in the regions, and in the catalytic effect they might have on other countries and regions seeking some permanent relief from the chaos, misery, and bloodshed they have experienced since World War II.

RESOURCE WARS?

For millennia States have fought over territory with the resources and population it contains. Even since the introduction of ideology (going back to the Arab conquests of the seventh century, if not before, and including the Crusades, Napoleon's campaigns, and the activities of the Comintern)* and of nuclear weapons and "star wars," it is still true that most wars are fought and are likely to continue to be fought to gain and hold territory. While irredentism is still important and there could conceivably be more wars of independence, the danger of wars over resources continues unabated. This was one of the major objectives of the three UN conferences on the Law of the Sea (described in Chapters Twenty-nine and Thirty): to devise a public order of marine space and resources that would reduce, if not eliminate, conflicts over them. To some extent this is also an objective of the "confidence-building measures" currently being pursued within and outside the United Nations and of the work that organization has been doing to foster peaceful decolonization, liberalized world trade, economic integration, and the removal of Antarctica and outer space from territorial competition—all discussed later.

History records numerous conflicts over minerals, including hydrocarbons. In the next chapter, for example, we point out that the quest for assured supplies of commodities, especially of fuels and raw materials, was one of the impulses leading to European conquest and colonization of much of the Southern Hemisphere, while greed for precious metals was probably the most important motivation in the earlier Spanish conquest of most of Latin America. During the world wars of this century, minerals were prime targets of military campaigns and diplomatic intrigues. After the First World War, the U.S. War Department drew up a list of 28 materials, mostly minerals, that had presented supply problems during the war, and the concept of "strategic minerals" became important in the interwar period.

*Comintern. The Third (Communist) International, an association of communist parties around the world, organized and controlled by the Soviet Union, whose purpose was to protect the Soviet Union against "capitalist encirclement" and, to a lesser extent, to spread communist ideology as widely as possible.

Definitions of the term vary considerably and frequently change, but generally speaking, they include three factors.

1. The mineral is essential for defense.
2. Its supplies are found largely or entirely outside the country.
3. During wartime strict conservation and control of it are necessary.

The gargantuan consumption of minerals during and since the Second World War enhanced fears of exhaustion of some minerals, and the chronic instability of the postwar world enhanced fears of interruption or interdiction of mineral supplies destined for the rich, industrialized countries. The politically motivated, Arab-led embargo of petroleum shipments by members of the Organization of the Petroleum Exporting Countries (OPEC) to most Western countries and the subsequent quadrupling of the prices of petroleum in 1973–1974 nearly resulted in panic in the importing countries (of which, it should be noted, the most seriously affected were not the rich, long-industrialized countries that could well afford the higher prices, but the newly industrializing countries of Latin America, Africa, and Asia that could not). The movement for a New International Economic Order (discussed in Part Nine), which also began in 1973, included a demand that all "natural resources" in developing countries be under the ownership and control of their own governments, and this heightened the concern of the Western industrialized countries still further.

They took a number of measures to ensure their supply of "strategic minerals." They stepped up the search for new sources of supply, not only in the traditional land areas, but in the earth's frontier regions as well: Antarctica, the seabed and outer space (see Part Seven for details). They devised new methods of retaining or even enhancing control over supplies and prices; accelerated the search for substitutes for some of the

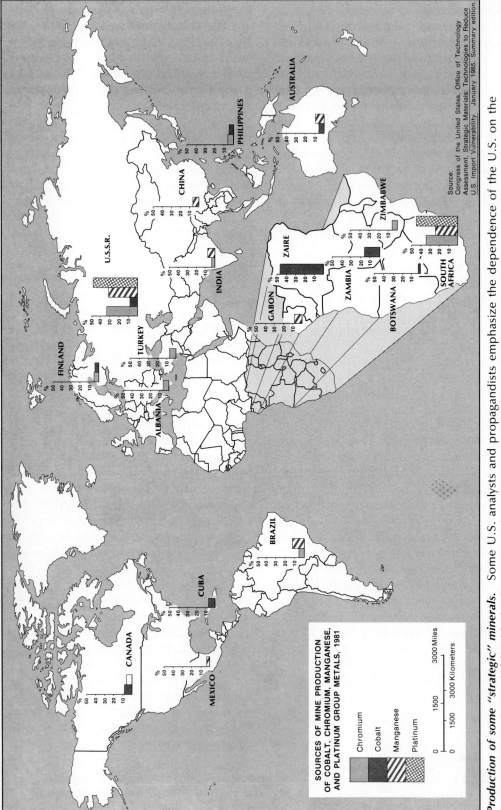

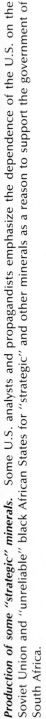

**SOURCES OF MINE PRODUCTION
OF COBALT, CHROMIUM, MANGANESE,
AND PLATINUM GROUP METALS, 1981**

Chromium
Cobalt
Manganese
Platinum

Source:
Congress of the United States, Office of Technology
Assessment. Strategic Materials: Technologies to Reduce
U.S. Import Vulnerability. January 1985. Summary edition.

Production of some "strategic" minerals. Some U.S. analysts and propagandists emphasize the dependence of the U.S. on the
Soviet Union and "unreliable" black African States for "strategic" and other minerals as a reason to support the government of
South Africa.

248

"strategic minerals"; intensified conservation, recycling, and stockpiling—and lent crucial support to South Africa.

Despite its reprehensible sociopolitical system, South Africa has claimed—and received—support from the Western democracies on three distinctly geopolitical grounds:

1. She is both a bulwark against communism, which has taken over much of the continent to the north, and a staunch ally against Soviet expansionism.

2. She is the guardian of the vital Cape Route, the route around the Cape of Good Hope followed by most supertankers that bring petroleum from the Persian Gulf to Western Europe—and by many other vessels as well.

3. She is the noncommunist world's major supplier of "strategic minerals," producing in 1984, for example, 40 percent of its manganese, 47 percent of its chrome, and about 80 percent of its platinum-group metals.

It will be very difficult for the West to abandon South Africa or insist that she abandon mineral-rich Namibia, regardless of apartheid, because of these factors, of which the last is by far the most important.

Does this mean that the need for minerals is the most important element in determining the foreign policies of industrialized countries? Or that they—or anyone else—will again go to war over minerals? It's possible, but not likely. For one thing, there are now so many factors contributing to a breakdown in world order that minerals have simply declined in relative importance. So has the equation between mineral possession and military potential; a review of Chapter Twelve on power analysis should make this clear. Third, while instability in the world has grown, so have the mechanisms for preventing and containing armed conflicts of all kinds; despite the scores of conflicts since 1945, none of them was over minerals and the world system has not broken

down and dissolved into another world war. Finally, the world is now so utterly interdependent, so laced together with innumerable linkages, and so influenced by many countervailing forces that there are few if any insoluble problems and few if any that can be isolated from all others. This is not to say that no country will ever again go to war in order to assure its supply of a particular mineral or minerals, only that if a country needs minerals and does not want to go to war, there are many ways of satisfying its needs peacefully. If it wants to go to war, a need for minerals can no longer serve as an excuse. Food or water, perhaps, but not minerals.

THE MIDDLE EAST

In order to illustrate some of the themes presented in this chapter, and in the book as a whole, we will consider very briefly four case studies, regions where issues of war and peace, of interdependence and interrelationships are most apparent, leaving aside the Soviet–American, East–West rivalry. We discuss them in descending order of level of conflict, beginning, of course, with the Middle East, the most tempestuous region of the world since 1945.

To understand the Middle East, one must begin at the beginning, with its location. The terms "Middle East" and "Near East" have been defined and used variously for centuries. Currently, definitions include many combinations of States, ranging from Morocco through Afghanistan and from Turkey through Somalia. All agree, however, that the heart of the Middle East is that area encompassed in the irredentist term "Greater Syria": Syria, Lebanon, Israel, Jordan, and Iraq, along with Egypt and the Arabian peninsula. The Levant, those countries fronting on the eastern shore of the Mediterranean (Syria, Lebanon, Israel) is the heart of the heart, the great land bridge connecting Asia, Africa, and Europe, the western curve of the ancient Fertile Crescent, highway of armies,

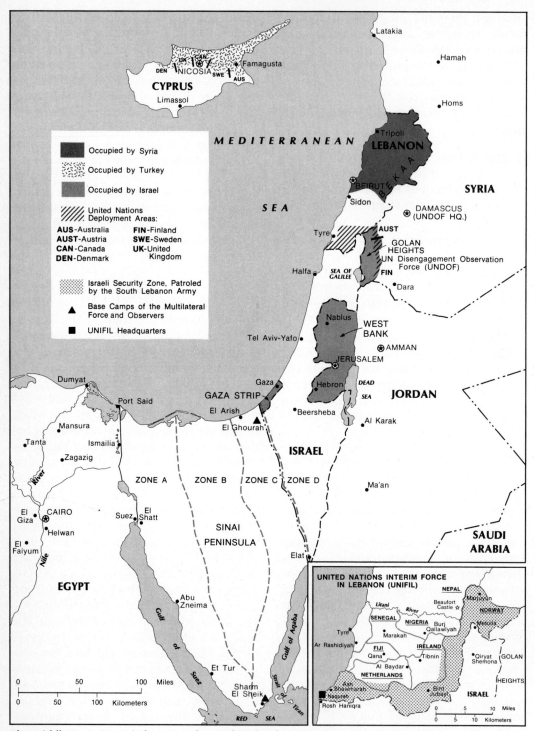

The Middle East: Occupied areas and peacekeeping forces. UN troops were withdrawn from Sinai in mid-1979 while Israel was gradually returning the peninsula to Egypt. They were replaced by the Multilateral Force and Observers, primarily US troops, who continue to monitor the relevant terms of the Israel–Egypt peace treaty. Within Zones A-C decreasing numbers of Egyptian troops and decreasingly powerful weapons are permitted. In Zone D only lightly-armed Israeli police are permitted. UN troops remain in Cyprus, southern Lebanon and the Golan Heights of Syria, in all cases separating hostile forces.

merchants, preachers, pilgrims, refugees, students, scholars, and countless others since the beginning of recorded history and probably earlier. Here is an equable climate, water, good soils, some good natural harbors, varied terrain with a fairly high proportion of level land, and even a little oil.

Here, over the centuries, have gathered a bewildering variety of peoples, probably the greatest variety per unit of land of any region of the world. Here there have been wars and lesser conflicts for millennia; here there are wars and lesser conflicts still. The most prominent by far has been the conflict between Israel and her Arab neighbors, not over the territory occupied by Israel, but over the very existence of the State. Yet even the most ardent Arab nationalist would admit that the region knew little peace before the creation of Israel in 1948 or even before the first Zionist settlements in Palestine in the 1890s, and, upon serious consideration, would also admit that would Israel vanish instantly, tonight, there would still be conflicts and even warfare throughout the region for generations to come. The tensions between Christian and Muslim, Arab and Persian, Turk and Arab, Libya and Egypt, Hashemites and other royal houses, Bedouin and farmer, socialist and capitalist, French and British, old and new, oil rich and oil poor, water rich and water poor, Sunni and Shia, Kurds and their neighbors, monarchist and republican, democrat and dictator, revolutionary and traditionalist, Maronite and Druze, smuggler and policeman, foreigner and native are all independent of Israel.

Iraq and Iran have since 1980 been fighting the region's bloodiest war in centuries, a frightful drain of lives and resources that neither party can afford. What are the stakes? Their mutual boundary in the Shatt-al-Arab, control of the head of the Persian Gulf, the Arab residents of western Iran, internal political power in each country, the Shia majority in Iraq ruled by Sunnis, rival revolutionary ideologies—all these and more. Yet the war has linkages

to other conflicts in the region. Both sides have been fighting the Kurds on and off for generations, as have from time to time the Turks. Neither they nor the Soviets or Syrians, also hosts to Kurdish settlements for millennia, want the Kurds to have their own State or even autonomy within the existing States. Iran is supported by Syria and Libya, who also support terrorist groups operating throughout the region and beyond, but Iran does not support the expansionist goals of either ally. Libya at various times has invaded Egypt, "federated" with Egypt, occupied northern Chad, and threatened Sudan and Tunisia; Syria has at various times fought with Iraq and Jordan, allied herself with Jordan and Egypt, and occupied most of Lebanon.

We could go on and on with these linkages, exploring how Saudi Arabia and the other conservative monarchies of the Arabian peninsula are involved, and the two Yemens, and Sudan, and Pakistan, and Greece—and, of course, Britain, France, the Soviet Union, the United States, and many other countries. But the point has been made: the Middle East is such a convoluted region with so many complex, interlocking and ancient rivalries that few theories about peace and war can emerge from it except in tatters. It will take heroic efforts by many people over a long period of time to bring peace to this dynamic but troubled region.

THE CARIBBEAN BASIN

The Caribbean is a much simpler region to understand, though not without its own complexities. Though for more than three centuries it was the scene of almost continual jockeying and even warfare among Britain, France, Spain, and the Netherlands for control of the region's resources and trade routes, the United States has for more than a century and a half considered it an American lake and resisted any "foreign" intrusion into it. In fact, the parallels between the Caribbean and Eastern Europe since World War II are very close indeed. Just as Yugoslavia broke away from

Soviet hegemony in 1948, Cuba broke away from American hegemony in 1959. Thereafter, neither country has hesitated to use its clandestine forces and even its regular troops to maintain its control over its "sphere of influence": the Soviet Union in East Germany, Poland, Hungary and Czechoslovakia; the United States in Guatemala, the Dominican Republic, Grenada, and Nicaragua—in addition to the bungled invasion of Cuba in 1960.

In Chapter Three, we pointed out how decision makers are often influenced by their mental maps, their perceptions of the political world. It would be hard to find a better example of this than a statement made by General David C. Jones, Chairman of the American Joint Chiefs of Staff (the highest-ranking uniformed officer in all the armed services), to the Chicago Bar Association in March 1981. In defending the use of American military advisors to help the conservative government of El Salvador suppress a leftist-led rebellion, he said that the United States must end "erosion in our hemisphere" and "get the Cubans not to play around in our back yard." He went on to invoke the domino theory again, "[If left alone,] El Salvador would probably go under Cuban influence—then you'd see another country and another country." Now, even a child can look at a map of the Caribbean Basin and see that El Salvador, like most other countries, is certainly "in the back yard" of several countries, but the United States is just as certainly not one of them.

The United States does, indeed, have legitimate interests in the region: the Panama Canal, drug trafficking from South America to the United States, petroleum flows from Mexico and Venezuela, heavy private American investment and substantial markets in the region, numerous American citizens residing there, considerable migration to the United States from the region, and various military and civilian installations. It is questionable, however, whether these interests are entirely compatible with the legitimate interests of the peoples of the region or whether they confer on the United States a right to attempt to control all governments and economies around the perimeter of the Caribbean Sea and the Gulf of Mexico.

But the United States is not the only important player on the Caribbean stage; nor is it responsible for all of the region's problems; nor have all of its activities been malevolent, or even consistent; nor is the situation in the region static. As in the Middle East, if the most powerful and controversial country were to be magically removed altogether, the Caribbean would still have very serious problems, though on a smaller scale and a lower level of intensity. There is in the Caribbean much more of a sense of collegiality, if not of true unity, than in the Middle East, a longer history of serious joint efforts to relieve fundamental economic and social problems and, of course, far more democratic governments. While there have been clashes around the basin, notably between Haiti and the Dominican Republic and between El Salvador and Honduras, they have been contained in duration, scope, and intensity and have never been a threat to the region.

The various subregional intergovernmental organizations, such as the Central American Common Market, the Caribbean Community (CARICOM), and the Organization of Eastern Caribbean States, while not precisely shining successes, are nevertheless indications of a genuine desire to work together as much as possible. The introduction by the United States of its Caribbean Basin Initiative in early 1982, while clearly a reaction to the accession to power of the Sandinista government in Nicaragua and not entirely altruistic, did, like its more ambitious predecessor the Alliance for Progress, demonstrate some genuine American interest in assisting the countries of the region to develop economically and socially. To date it has not been notably successful and may, like the alliance, vanish soon with scarcely a trace left behind, but it is arguably a very great improvement over (though not yet a substitute for) military intervention. All things considered, there is a better chance for a durable peace reasonably soon in the Caribbean than in the Middle East.

THE PACIFIC BASIN/PACIFIC RIM

For many years we were accustomed to thinking of the Pacific Basin as an entity, largely because its size, nearly symmetrical shape, and sprinkling of Polynesian islands seemed to endow it with a kind of unity, if not uniformity, much like the Caribbean Basin. Yet, on closer inspection, we noted far more differences than similarities from one part of the ocean to another, and little unity of any kind. Now, for some inexplicable reason, "Pacific Rim" seems to be the more favored term, perhaps because of the rising importance of the newly industrialized countries (South Korea, Taiwan, Hong Kong, Singapore) on the western edge of the basin or perhaps because of the strong and tightening links among most of the States around the periphery of the basin.

Japan has been the dynamo driving this growing unity of the region, forging economic and cultural ties with many of the countries around the rim, largely because of her constant need for more sources of commodities and more markets for her manufactured goods and services. The ANZUS (Australia–New Zealand–United States) alliance continues to be viable, despite New Zealand's refusal to permit American (or any other) nuclear-powered or nuclear-armed vessels to enter her waters. Australia and Japan have formed an especially close relationship, based largely on symbiotic economies. Transpacific trade has become so important and is growing so rapidly that it may soon rival transatlantic trade in volume and importance. ASEAN and the South Pacific Commission have at last achieved significance and are bringing tangible benefits to their peoples, while tying them ever more closely together. The U.S.–Canadian relationship continues strong, despite familial squabbles from time to time. The spectacular growth of California combined with recent heavy immigration into the state from most countries in and around the basin have reinforced California's traditional westward orientation and the state, in turn, seems to be pulling the whole country along in a shift from an Atlantic to a Pacific emphasis. The same is true to a lesser extent of Washington, Oregon, and British Columbia. Australia and New Zealand have forged "closer economic relations," virtually a common market, and retain strong ties with their former colonies and mandates in the South Pacific. And the Pacific Economic Cooperation Council—a nongovernmental group from around the Pacific—meets biennially to discuss ways of cooperating more effectively.

These and many other developments are tending to validate the vision of Professor Kiyashi Kojima (Hitotsubashi University of Japan) and Hiroshi Kurimoto (UN Economic Commission for Asia and the Far East), who in 1965 first proposed the creation of a Pacific Free Trade Area. Yet the forces operating in the region are not all centripetal. The Soviet Union not only maintains a formidable Pacific fleet, with a major base at Cam Ranh Bay in Vietnam, but has begun to negotiate aid and fishing agreements with several South Pacific island States. China and Taiwan are still competing for support around the world. Libya has begun to catalyze and support revolutionary movements in a largely peaceful South Pacific. Australia, as already noted, has begun to take the Indian Ocean seriously, planning in 1987 to shift half of her navy from Sydney to Perth within the next decade. Melanesian nationalists have become more united, vocal and aggressive in asserting their rights against France, Indonesia, and Australia. The Philippines is still unstable. Palau continues to reject a "Compact of Free Association" with the United States because it does not want nuclear weapons on its territory.

There are still enormous disparities in economic well-being and political development within the region, and incredible cultural diversity. Transportation and communications are still poor in the South Pacific. The United States refuses to sign the South Pacific Nuclear-Free Zone Agreement, a strong force for unity in the region. Political tensions and continuing warfare on the Asian mainland detract from unity efforts. The Soviet Union, the United States, Canada, Japan, and China, the most powerful countries around the Pacific Rim, all have vital interests outside the region. And the Latin American countries of the Pacific Rim have yet to develop the same intensity of interest in closer relations as their counterparts to the north and west.

Nevertheless, there seems to be evolving some form of closer association in a region which, though heavily armed and beset by rivalries and competition, does have commonalities and an incipient community of interest. During the past two hundred years we have seen the center of gravity in the world shift from the Mediterranean to the North Atlantic. It may now be shifting toward the Pacific. We can hope that this shift will be more peaceful than the last.

THE INDIAN OCEAN

While the Indian Ocean has been crisscrossed by vessels of every description for millennia, it has remained the least-known of the three largest oceans. Because no great powers have developed along its shores and no major industrial countries emerged on its periphery, it has never become a major battleground except, at times, at its outer margins. Two World Wars and the Cold War left it tranquil and of little interest to the world at large. With growing interest in the sea after World War II and growing dependence of the industrialized countries on Persian Gulf petroleum, interest in the Indian Ocean began to awaken. This interest was reinforced by the beginning of the decolonization process on three sides of the ocean. Still, largely under the informal but effective supervision of the Royal Navy and other British forces, it remained outside the sphere of big-power rivalries.

A landmark in its history was the International Indian Ocean Expedition of 1964, sponsored by the International Oceanographic Commission (an agency of UNESCO), the first large-scale, concerted effort to obtain basic scientific information about the ocean. Another was the British decision in 1968 to withdraw from all its remaining important positions "east of Suez." Since then Mauritius, Bahrein, Qatar, the United Arab Emirates, Mozambique, the Comoros, the Seychelles, and Djibouti have joined the ranks of independent States in and around the ocean. In 1965, the British created the British Indian Ocean Territory out of a number of uninhabited and sparsely inhabited islands in midocean, of which they still retain the Chagos Archipelago. Australia and France also retain some small islands in the eastern and western sectors, respectively. Otherwise, only small or midsize countries occupy the land in and around the Indian Ocean.

The withdrawal of the British created a classic power vacuum in a huge and increasingly strategic region. The United States and the Soviet Union began sending naval units into the ocean on show-the-flag missions. India and Iran began building navies that could operate outside their coastal waters. Australia turned hesitantly from the Pacific and began debating building a navy base at Cockburn Sound in Western Australia. Intensive jockeying began for control of the Bab el Mandeb, the southern entrance to the Red Sea, with Israel more than just an interested bystander. Japan grew nervous about her vital oil-supply route from the Persian Gulf. China began making friends in East Africa. The Soviet Union for a time had the use of shore facilities in Somalia for her navy and offered to lease a former Royal Air Force base in the Maldives. Iran helped Oman suppress a rebellion in Dhofar province. The closure of the Suez Canal from 1967 to 1974 diverted considerable maritime traffic around the east

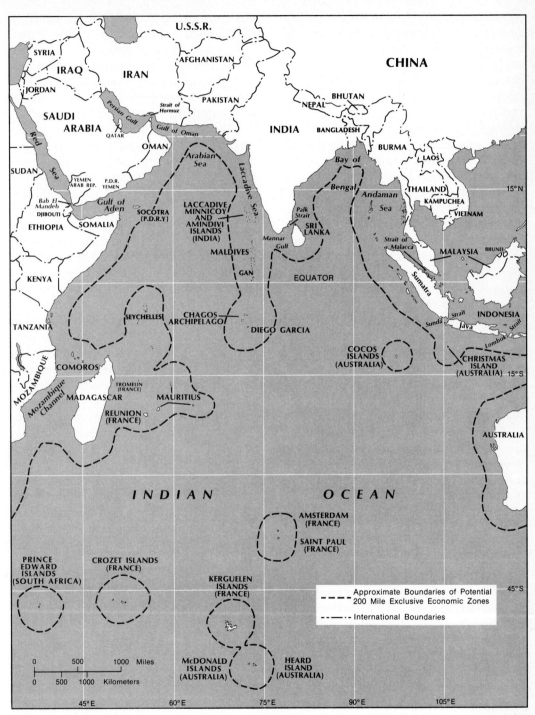

The Indian Ocean: Testing ground for a new geopolitics of peace.   Note the large proportion of the ocean that would come under national jurisdiction if 200-mile exclusive economic zones are claimed by all of the States in and around the basin. (See Chapter 30.) Note also the formerly remote, but now "strategic," location of Diego Garcia island and the critically important straits of Bab el Mandeb, Hormuz, and Malacca around the northern arc of the ocean. This northern arc is a part of Spykman's Rimland.

coast of Africa. The United States and Britain set up joint communications and refueling stations on Aldabra Island in the Seychelles and Diego Garcia in the Chagos Archipelago, almost precisely in mid-ocean. Iran occupied the disputed Tunb and Abu Musa Islands near the Strait of Hormuz in 1971. The United States in 1976 began upgrading its austere facility on Diego Garcia into a full-fledged carrier task force support base.

While all this maneuvering was going on, nearly every country bordering the Indian Ocean experienced internal unrest, up to and including actual warfare, and a number engaged in localized international wars. Clearly, the Indian Ocean had become "destabilized." The situation has been watched carefully at the United Nations and around the world. There is ample evidence that no one wants the Indian Ocean to become a new arena for big-power rivalry, but the middle-rank powers of the littoral—India, South Africa, Australia—all have their own interests to protect. The war with Iraq and internal problems have temporarily diverted the attention of Iran from the great sea to the south, but she'll be back in due course. Indonesia, Thailand, Pakistan, Saudi Arabia, and Kenya may also, in time, begin to project themselves into the Indian Ocean, complicating what is at present a relatively simple situation.

To prevent these rivalries from developing into conflicts and to keep the region peaceful, three approaches are being pursued simultaneously. First, bilateral negotiations between the United States and the Soviet Union have been conducted irregularly since 1977. Second, there is an incipient sense of regional community developing, exemplified by the Indian Ocean Marine Affairs Co-operation, which held its first conference in Colombo, Sri Lanka, in July 1985; and in the north, the South Asian Association for Regional Cooperation, which held its first formal meeting, also in Colombo, in August 1983. Third, negotiations are taking place within the United Nations, which in 1971 passed a Declaration of the Indian Ocean as a Zone of Peace. The UN continues trying to implement this declaration through a variety of devices. One of them was a July 1979 "meeting of littoral and hinterland States" that would lead toward a general conference on the Indian Ocean. But the *ad hoc* committee established to prepare for the conference had not been able to complete its work by the end of 1986, and the conference continues to recede into the future.

Will these moves work? Will the reluctance of the superpowers to make major military commitments in the Indian Ocean and the opposition of the littoral States to such commitments sustain the present "balance of no power" situation? Can the Indian Ocean remain a power vacuum in a tumultuous world? Is a new kind of geostrategy at work here, a geostrategy of peace? It's much too early to tell, but certainly what we are witnessing is an advance over the ethnocentric strategies we have discussed in the two preceding chapters, strategies based on power. And it is better than the evolving and growing—and related—conflicts between rich and poor and between energy-rich and energy-poor. Perhaps there is a sign of hope here in the Indian Ocean that sanity can prevail even where vital interests of many countries, large and small, are at stake.

WAR AND PEACE AND GEOGRAPHY

Having introduced the subject of war and peace, considered the possibility of resource wars, and glanced at four important geopolitical regions, we can now review some of the major themes in the contemporary literature on the subject. Few of these writings, it should be noted, are by geographers, a deficiency that should be rectified in the near future. We *broadly* follow here the rough outline of an analysis by John O'Loughlin and Herman van der Wusten, American and Dutch geographers, respectively, in the December 1986 issue of *Progress in Human Geography*.

1. *Geography of War.* Broadly speaking, there have been three major approaches to the geography of war. Traditionally, international relations theory has been characterized by emphasis on power relationships among States. Foreign policies of individual States, sometimes analyzed in their historical, cultural, geographic, and ideological contexts, determine the outcomes of given situations and the long-term trends in world affairs. Typical of this *traditionalist* view is the vision of the world as split between East and West (socialist and capitalist economic systems) and between North and South (rich and industrialized, poor and agricultural), yielding a quadripartite division of the world, with competition among the four the dominant theme. We have already seen, though, that while such competition surely does exist, it is seldom clear-cut and seldom decisive.

Behavioralism has dominated international relations research for the past 20 years or so, with its quantitative techniques and emphasis on the relationship between a State's domestic qualities and its foreign policy. Here, geographic factors, such as location, population, and alliances, join with economic attributes, political orientation, and arms races as predictors of conflict. Deriving from this approach are some of the newer geopolitical theories, notably that of Saul Cohen, described in the preceding chapter. This approach is also incomplete because it does not adequately consider the historical evolution of States and tends to concentrate too much on particular features of the political world while neglecting others.

More recently, *structuralist* perspectives have attracted considerable attention. Otherwise known as world-systems theory, these perspectives derive largely from the work of Wallerstein and Modelski. They take a global and historical view of the world rather than one centered on States and the contemporary scene in order to explain how the world works today. While they differ in some respects, they both emphasize the links between the dynamic global economic and political systems, the cycles of international hegemony, and the occurrence of conflict. Wallerstein's view tends to be rather Marxian and includes such features as the core-periphery model of the world and both the class struggle between proletariat and bourgeoisie and the political struggle between different bourgeois. Modelski's contribution has been the long-cycle model of successive long periods of world dominance by single States.

2. *Geography of Peace.* For lack of a better practical definition, peace can best be defined as the absence of armed conflict. This does not preclude competition, even rivalry, but it does preclude violence. The pattern of peace studies so far has generally tended to follow the same pattern as war studies. The relatively few scholars active in the field concentrate on analysis of policies that result in the avoidance of violent conflict and hence the maintenance of peace. Generally speaking, for both States and groups of States, there are two ways to avoid war: dissociation and association.

Dissociation means essentially separation of States from other States that might engage in violence. Geographical isolation helps, of course, but (as we have seen and shall see again later) is no guarantee of peace and in any case does not result from government policy. Neutrality, nonalignment, deterrence, partition, peace-keeping forces, and political/economic isolation are policies designed to prevent conflict by reducing contact. Because of the many features of the contemporary world already discussed in this book and to be discussed later (e.g., long-range missiles, economic interdependence, global environmental problems) these policies are seldom entirely successful.

Association is based on interaction, sharing, or cooperation. This means essentially the kinds of relationships covered in our Part Six: international law, economic integration, regional and worldwide functional cooperation, efforts toward intercultural understanding, and so on. But proximity alone, even common membership in organizations, does not assure peace.

There must be serious and sustained efforts to resolve differences peacefully or proximity may simply make war more tempting and more likely.

In summary, then, we can say that geographers can and should contribute to the achievement and maintenance of world peace, in both their personal and professional activities. One way to do this is through the various peace academies being opened around the world. Another is through the University for Peace, created by the United Nations in 1981, but operating in Costa Rica outside the UN framework, to promote peace through research and education. Another is to get involved in some of the activities described in the next few chapters.

REFERENCES FOR PART FOUR

Books and Monographs

A

Abdulghani, Jasim, *Iraq and Iran; the Years of Crisis*. Johns Hopkins Univ. Press, 1985.

Abrahamsson, Bernhard J. and Joseph L. Steckler, *Strategic Aspects of Seaborne Oil*. Beverly Hills, CA: Sage, 1973.

Adie, W.A.C., *Oil, Politics and Seapower: The Indian Ocean Vortex*. New York: Crane, Russak, 1975.

Alford, Jonathan (ed.), *Sea Power and Influence: Old Issues and New Challenges*. Montclair, NJ: Allanheld, Osmun, 1980.

Allen, Charles D., Jr., *The Uses of Navies in Peacetime*. Lanham, MD: Univ. Press of America, 1980.

Allen, Philip M., *Security and Nationalism in the Indian Ocean*. Boulder, CO: Westview, 1987.

Amin, Sayed Hassan, *International and Legal Problems of the Gulf*. Wisbech, Cambridgeshire, Eng.: Middle East and North African Studies Press, 1981.

———, *Political and Strategic Issues in the Persian–Arabian Gulf*. Glasgow: Royston, 1984.

Amirsadeghi, Hossein (ed.), *The Security of the Persian Gulf*. New York: St. Martin's Press, 1981.

Anderson, Thomas P., *The War of the Dispossessed; Honduras and El Salvador, 1969*. Lincoln: Univ. of Nebraska Press, 1981.

———, *Geopolitics of the Caribbean: Mini-states in a Wider World*. New York: Praeger, 1984.

Angell, Norman, *Raw Materials, Population Pressure and War*. New York: National Peace Conference, 1936.

Anthony, John Duke, *The Middle East: Politics and Development*. Lanham, MD: Univ. Press of America, 1975.

Ashby, Timothy, *The Bear in the Back Yard; Moscow's Caribbean Strategy*. Lexington, MA: Lexington Books, 1987.

Asprey, R.B., *War in the Shadows: The Guerrilla in History*. Garden City, NY: Doubleday, 1975.

Ausland, John, *Nordic Security and the Great Powers*. Boulder, CO: Westview, 1986.

B

Bagley, Bruce M. (ed.), *Contadora and the Diplomacy of Peace in Central America*, 2 Vols. Boulder, CO: Westview, 1987.

Bailey, Sydney D., *The Making of Resolution 242*. Dordrecht, Boston, Lancaster: Nijhoff, 1985.

Bakhash, Shaul B., *The Politics of Oil and Revolution in Iran*. Washington: Brookings Institution, 1982.

Banse, Ewald, *Germany Prepares for War*. New York: Harcourt, Brace, 1934.

Barley, Norman A., *Latin America: Politics and Hemisphere Security*. New York: Praeger, 1965.

Barnett, A. Doak, *Communist China and Asia: Challenge to American Policy*. New York: Harper & Row, 1961.

Barton, John H., *The Politics of Peace: An Evaluation of Arms Control*. Stanford: Stanford Univ. Press, 1981.

Bayne, E.A., "Chiarascuro on the Horn," *American Universities Field Staff Report, Northeast African Studies*, 15, 1 (October 1968), 1–11.

Beard, Charles A. and George Radin, *The Balkan Pivot: Yugoslavia*. New York: Macmillan, 1929.

Beling, Willard A., *Middle East Peace Plans*. New York: St. Martin's Press, 1987.

Bell, J. Bowyer, *The Horn of Africa; Strategic Magnet in the Seventies*. New York: Crane, Russak, 1974.

Ben-Dor, Gabriel and David B. Dewitt (eds.), *Conflict Management in the Middle East*. Lexington, MA: Lexington Books, 1987.

Bennet, James T. and Walter E. Williams, *Strategic Minerals: The Economic Impact of Supply Disruptions*. Washington: Heritage Foundation, 1981.

Ben-Rafael, Eliezer, *Israel–Palestine*. Westport, CT: Greenwood Press, 1987.

Bergesen, Helge Ole and others, *Soviet Oil and Security in the Barents Sea*. New York: St. Martin's Press, 1987.

Betts, Richard K., *Nuclear Blackmail and Nuclear Balance*. Washington: Brookings Institution, 1987.

Blechman, Barry M. and Stephen S. Kaplan, *Force Without War: U.S. Armed Forces as a Political Instrument*. Washington: Brookings Institution, 1978.

Blouet, Brian, *Sir Halford Mackinder, 1861–1947: Some New Perspectives*. Research Paper No.

13. Oxford: Oxford Univ. School of Geography, May 1975.

Booth, Kenneth, *Law, Force and Diplomacy at Sea*. London, Boston, and Sydney: Allen & Unwin, 1985.

Booth, R., *The Armed Forces of African States*. Toronto: Institute of Strategic Studies, Canadian Institute of International Affairs, 1970.

Boulding, Kenneth E., *Conflict and Defense: A General Theory*. New York: Harper & Row, 1963.

———, "Toward a Theory of Peace," in R. Fisher (ed.), *International Conflict and Behavioral Science*. New York: Basic Books, 1964, 70–87.

———, *Stable Peace*. Austin: Univ. of Texas Press, 1978.

Bowett, Derek W., *Self-Defence in International Law*. Manchester, Eng.: Manchester Univ. Press, 1958.

Bowman, Larry W. and Ian Clark (eds.), *The Indian Ocean in Global Politics*. Boulder, CO: Westview, 1981.

Brandt, Karl, "Foodstuffs and Raw Materials as Elements of National Power," in Hans Speier and Alfred Kahler (eds.), *War in Our Time*. New York: Norton, 1939, 105–131.

Braun, Aurel, *The Middle East in Global Strategy*. Boulder, CO: Westview, 1987.

Braun, D., *The Indian Ocean*. New York: St. Martin's Press, 1983.

Braveboy-Wagner, Jacqueline, *The Caribbean in World Affairs*. Boulder, CO: Westview, 1987.

Brebner, John Bartlett, *The North American Triangle; the Interplay of Canada, the United States and Great Britain*. New Haven, CT: Yale Univ. Press, 1954.

Broek, Jan O.M., "The German School of Geopolitics," in Russell H. Fitzgibbon (ed.), *Global Politics*. Berkeley: Univ. of California Press, 1944, 167–177.

Brookfield, Harold C., *The Pacific in Transition: Geographical Perspectives on Adaptation and Change*. Toronto: Macmillan, 1974.

Brown, James and William P. Snyder (eds.), *The Regionalization of Warfare*. New Brunswick, NJ: Transaction Books, 1984.

Bruce, Robert H., *Perspectives on International Relations in the Indian Ocean Region*. Hong Kong: Asian Research Service, n.d.

Bruton, Henry J., *The Promise of Peace: Economic Cooperation Between Egypt and Israel*. Washington: Brookings Institution, 1981.

Bunge, William, *The Nuclear War Atlas*. New York: Basil Blackwell, 1987.

Burrell, R.M., *The Persian Gulf*. Beverly Hills, CA, and London: Sage, 1972.

———, *Politics, Oil and the Western Mediterranean*. Lanham, MD: Univ. Press of America, 1987.

Burt, Richard (ed.), *Arms Control and Defense Postures in the 1980's*. Boulder, CO: Westview; London: Croom Helm, 1982.

Buss, Claude A. (ed.), *National Security Interests in the Pacific Basin*. Stanford, CA: Hoover Institution Press, 1985.

Buss, R., *Wary Partners: The Soviet Union and Arab Socialism*. Toronto: Institute of Strategic Studies, Canadian Institute of International Affairs, 1970.

Butterfield, H., *International Conflict in the Twentieth Century*. London: Routledge & Kegan Paul, 1960.

Butts, Kent Hughes, and Paul R. Thomas, *The Geopolitics of Southern Africa*. Boulder, CO: Westview, 1986.

Byers, R.B. (ed.), *The Denuclearisation of the Oceans*. New York: St. Martin's Press, 1986.

Bywater, Hector Charles, *Sea Power in the Pacific: A Study of the American-Japanese Naval Problem*. Boston: Houghton Mifflin, 1921.

———, *Navies and Nations*. Boston: Houghton Mifflin, 1927.

C

Cable, James, *Gunboat Diplomacy, 1919–1979: Political Applications of Limited Naval Force*. New York: St. Martin's Press, 1986.

Caldwell, C., *Air Power and Total War*. New York: Coward-McCann, 1943.

Cameron, Eugene N. (ed.), *The Mineral Position of the United States, 1975–2000*. Madison: Univ. of Wisconsin Press, 1973.

Carroll, Berenice A. and others, *Peace and War: A Guide to Bibliographies*. Santa Barbara, CA, and Oxford: ABC–Clio, 1983.

Castle, Emery and others, *U.S. Interests and Global Natural Resources; Energy, Minerals, Food*. Johns Hopkins Univ. Press, 1983.

Chamberlain, James, *Air Age Geography and Society*. Philadelphia: Lippincott, 1945.

Chandra, Satish (ed.), *The Indian Ocean*. Newbury Park, CA: Sage, 1987.

Chayes, Abram, *The Cuban Missile Crisis*. New York: Oxford Univ. Press, 1974.

Child, Jack, "The American Southern Cone: Geopolitics and Conflict," in Barry Lentnek (ed.) *Contemporary Issues in Latin America*, Muncie, IN: Conference of Latin Americanist Geographers, 1983, 200–213.

————, *Geopolitics and Conflict in South America: Quarrels Among Neighbors.* New York: Praeger, 1985.

————, *Conflict in Latin America.* New York: St. Martin's Press, 1986.

Clarke, Duncan L., *Politics of Arms Control. The Role and Effectiveness of the U.S. Arms Control and Disarmament Agency.* London: Collier Macmillan; New York: Free Press, 1979.

Cohen, Saul B., "The Emergence of a New Second Order of Powers in the International System," in C. Enloe and O. Marwa, *Nuclear Proliferation and the Near Nuclear Countries.* Cambridge, MA: Ballinger, 1976.

Coll, Alberto R. and Anthony C. Arend (eds.), *The Falkland War: Lessons for Strategy, Diplomacy and International Law.* Boston: Allen & Unwin, 1985.

Collins, Joseph J., *The Soviet Invasion of Afghanistan.* Lexington, MA: Lexington Books, 1985.

Conference on Middle Eastern Affairs, *The Arabian Peninsula, Iran and the Gulf States, New Wealth, New Power: A Summary Record of the 27th Annual Conference of the Middle East Institute, Washington, DC, September 28–29, 1973.* Washington: 1974.

Cottrell, Alvin J. and R.M. Burrell (eds.), *The Indian Ocean: Its Political, Economic and Military Importance.* New York: Praeger, 1972.

————, *Politics, Oil and the Western Mediterranean.* Beverly Hills, CA: Sage, 1973.

Cottrell, Alvin J. and Thomas H. Moorer, *U.S. Overseas Bases: Problems of Projecting American Military Power Abroad.* Washington Papers, Vol. 5, No. 47. Lanham, MD: Univ. Press of America, 1977.

Cottrell, Alvin J. and Frank Bray, *Military Forces in the Persian Gulf.* Washington Papers, Vol. 6, No. 60. Lanham, MD: Univ. Press of America, 1978.

Cox, C. and R. Scruton, *Peace Studies: A Critical Survey.* London: Institute for European Defence and Strategic Studies, 1984.

Cox, Kevin R. and others (eds.), *Locational Approaches to Power and Conflict.* New York: Wiley, 1974.

Cranwell, J.P., *The Destiny of Sea Power.* New York: Norton, 1941.

Crassweller, R.D., *The Caribbean Community: Changing Societies and U.S. Policy.* New York: Praeger, 1972.

Crawford, *Raw Materials and Pacific Economic Integration.* Vancouver: Univ. of British Columbia Press, 1986.

Curtis, Michael (ed.), *The Middle East.* New Brunswick, NJ: Transaction Books, 1986.

D

Dankewych, M., *Siberia in Soviet Power Politics: Economic, Strategic, and Geopolitical Factors.* Washington: Free Siberia Society, 1970.

Dawisha, Adeed and Karen Dawisha, *The Soviet Union in the Middle East: Policies and Perspectives.* New York: Holmes & Meier, 1982.

Day, Arthur R., *East Bank/West Bank,* New York: Council on Foreign Relations, 1986.

Deeb, Mary-Jane, *Libya's Foreign Policy in North Africa.* Boulder, CO: Westview, 1987.

Demangeon, A., *America and the Race for World Dominion.* New York: Doubleday, Page, 1921.

De Mille, John B., *Strategic Minerals.* New York: McGraw-Hill, 1947.

De Seversky, Alexander P., *Victory Through Air Power.* New York: Simon & Schuster, 1942.

————, *Air Power: Key to Survival.* New York: Simon & Schuster, 1950.

————, *America: Too Young to Die.* New York: McGraw-Hill, 1961.

Deudney, David, *Whole Earth Security: A Geopolitics of Peace.* Washington: Worldwatch Institute, 1983.

Deutsch, Karl W., *Political Community and the North Atlantic Area.* Princeton, NJ: Princeton Univ. Press, 1957.

Dickinson, Robert E., *The German Lebensraum.* London: Routledge & Kegan Paul; New York: Penguin, 1943.

Dominguez, Jorge I., *U.S. Interests and Policies in the Caribbean and Central America.* Lanham, MD: Univ. Press of America, 1982.

Dorpalen, A., *The World of General Haushofer.* New York: Holt, Rinehart, 1942.

Douhet, Guilio, *The Command of the Air.* New York: Coward-McCann, 1942.

Dowdy, William L. and Russell B. Trood (eds.), *The Indian Ocean; Perspectives on a Strategic Arena.* Durham, NC: Duke Univ. Press, 1985.

Drysdale, Peter and Kiyoshi Kojima (eds.), *Australia–Japan Economic Relations in the International Context: Recent Experience and the Prospects Ahead.* Acton, Australian Capital Territory: Australia–Japan Economic Relations Research Project, 1978.

E

Eckel, Edmund C., *Coal, Iron and War*. New York: Holt, 1920.

Eckes, Alfred, Jr., *The United States and the Global Struggle for Minerals*. Austin: Univ. of Texas Press, 1979.

El Azhary, M.S. (ed.), *The Iran–Iraq War*. New York: St. Martin's Press, 1984.

Elazar, Daniel J., *The Camp David Framework for Peace: A Shift Toward Shared Rule*. Lanham, MD: Univ. Press of America, 1979.

———, *Self Rule/Shared Rule: Federal Solutions to the Middle East Conflict*. Lanham, MD: Univ. Press of America, 1985.

Eliot, George Fielding, *Bombs Bursting in Air: The Influence of Air Power on International Relations*. New York: Reynal & Hitchcock, 1939.

Ellison, Herbert J. (ed.), *Japan and the Pacific Quadrille*. Boulder, CO: Westview, 1987.

Emeny, Brooks, *The Strategy of Raw Materials*. New York: Macmillan, 1934.

Epstein, Joseph M., *Strategy Force Planning: The Case of the Persian Gulf*. Washington: Brookings Institution, 1987.

Erisman, H. Michael, *The Caribbean Challenge; U.S. Policy in a Volatile Region*. Boulder, CO: Westview, 1984.

Evron, Y., *The Middle East: Nations, Superpowers and Wars*. London, 1978.

F

Fairgrieve, James, *Geography and World Power*, 8th ed. New York: Dutton, 1941.

Farer, Tom J., *War Clouds on the Horn of Africa; The Widening Storm*, 2nd ed. New York: Carnegie Endowment for International Peace, 1979.

Farid, Abdel Majid (ed.), *The Red Sea*. New York: St. Martin's Press, 1984.

Fauriol, Georges A., *Foreign Policy Behavior of Caribbean States: Guyana, Haiti and Jamaica*. Lanham, MD: Univ. Press of America, 1984.

Fawcett, J.E.S. and Audrey Parry, *Law and International Resource Conflict*. Oxford: Oxford Univ. Press, Clarendon; New York: Oxford Univ. Press, 1981.

Fernandez, Damian J., *Cuba's Foreign Policy in the Middle East*. Boulder, CO: Westview, 1987.

Fielden, D.G., *The Political Geography of the Red Sea Region*. Occasional Publications (New Series) No. 13. Durham, Eng.: Univ. of Durham, Dept. of Geography, Eng., 1978.

Fifield, Russell H. and G. Etzel Pearcy, *Geopolitics in Principle and Practice*. Boston: Ginn, 1944.

Fischer, Dietrich, *Preventing War in the Nuclear Age*. Totowa, NJ: Rowman & Allanheld, 1984.

Fischman, Leonard L., *World Mineral Trends and United States Supply Problems*. Washington: Resources for the Future, 1980.

Forsberg, Randall (ed.), *World Weapon Database*. Lexington, MA: Lexington Books, 1987.

Franda, Marcus F., *The Indian Ocean: A Delhi Perspective*. American Universities Field Staff Reports, Southeast Asia Series, Vol. 19, No. 1. 1975.

Freedman, Lawrence, *The Evolution of Nuclear Strategy*. New York: St. Martin's Press, 1981.

———, *Atlas of Global Strategy*. London: Macmillan, 1985.

Freedman, Robert O., *The Middle East Since Camp David*. Boulder, CO: Westview, 1984.

G

Gaddis, J.L., *Strategies of Containment: A Critical Approach to Postwar American National Security Policy*. New York: Oxford Univ. Press, 1982.

Galtung, J. (ed.), *Essays in Peace Research*. Copenhagen: Chr. Ejlers, vol. I, 1975; vol. II, 1976.

George, James L. (ed.), *Problems of Sea Power as We Approach the Twenty-first Century*. Washington: American Enterprise Institute for Public Policy Research, 1978.

Gilpin, R., *War and Change in the International System*. New York: Cambridge Univ. Press, 1981.

Glassner, Martin Ira (ed.) *Global Resources: Challenges of Interdependence*, New York: Praeger, 1983.

Golovin, N. and A.D. Bubnov, *The Problem of the Pacific in the 20th Century*. New York: Scribner's, 1922.

Gordon Lincoln, *Eroding Empire: Western Relations with Eastern Europe*. Washington: Brookings Institution, 1987.

Gorshkov, Sergei G., *The Sea Power of the State*. Elmsford, New York: Pergamon, 1979.

Gottmann, Jean (ed.), *Centre and Periphery*. Beverly Hills, CA: Sage, 1980.

Graham, Norman A. and Keith L. Edwards, *The Caribbean Basin to the Year 2000: Demographic, Economic, and Resource Use Trends in Seventeen Countries: A Compendium of Statistics and Projections*. Boulder, CO: Westview, 1984.

Gray, Colin S., *The Geopolitics of the Nuclear Era: Heartland, Rimlands and the Technological Revolution*. New York: Crane, Russak for the National Strategy Information Center, 1977.

———, *Maritime Strategy, Geopolitics, and the Defense of the West*. New York: National Strategy Information Center, 1986.

Greene, James R. and Brent Scowcroft (eds.), *Western Interests and U.S. Policy Options in the Caribbean Basin*. Boston: Oelgeschlager, Gunn & Hain, 1984.

Grenfell, R., *Sea Power*. New York: Doubleday, Doran, 1941.

Griffith, William E., *The World and the Great-Power Triangles*. Cambridge: MIT Press, 1975.

The Gulf: Implications of British Withdrawal. Washington: Center for Strategic and International Studies, 1969.

Gürkan, Ihsan, *NATO, Turkey and the Southern Flank: A Midwestern Perspective*. New York: National Strategy Information Center, 1980.

Gyorgy, Andrew, *Geopolitics: The New German Science*. Berkeley: Univ. of California Press, 1944.

H

Hamilton, Nora and others, *Crisis in Central America*. Boulder, CO: Westview, 1987.

Harden, Sheila, *Small Is Dangerous: Micro States in a Macro World*. London: Frances Pinter, 1985.

Harkavy, Robert E. and Stephanie G. Neuman (eds.), *The Lessons of Recent Wars in the Third World*. Lexington, MA: Lexington Books, 1986.

Harris, N., *Of Bread and Guns: The World Economy in Crisis*. Harmondsworth, Middlesex, Eng.: Penguin, 1983.

Harrison, Selig S., *China, Oil, and Asia: Conflict Ahead?* New York: Columbia Univ. Press, 1977.

Haward, H.N., *Turkey, the Straits, and United States Policy*. Baltimore: Johns Hopkins Univ. Press, 1974.

Hearnshaw, F.J.C., *Sea-Power and Empire*. London: Harrap, 1940.

Helms, Christine Moss, *Iraq: Eastern Flank of the Arab World*. Washington: Brookings Institution, 1984.

Hershwy, Burnet, *The Air Future: A Primer of Aero Politics*. New York: Duell, Sloan & Pearce, 1943.

Hessell, Mary Stanley and others, *Strategic Materials in Hemispheric Defense*. New York: Hastings House, 1942.

Hinton, Harold C., *The China Sea: The Amer-*

ican Stake in its Future. New York: National Strategy Information Center, 1980.

Hooson, David J.M., *A New Soviet Heartland?* New York: Van Nostrand, 1964.

Howe, Jonathan Trumbull, *Multicrises: Sea Power and Global Politics in the Missile Age*. Cambridge: MIT Press, 1971.

Hull, Galen Spencer, *Pawns on a Chessboard: The Resource War in Southern Africa*. Washington: Univ. Press of America, 1981.

I

Ismael, Tareq Y., *Iraq and Iran; Roots of Conflict*. Syracuse, NY: Syracuse Univ. Press, 1982.

J

Jabber, Paul, *Not by War Alone*. Berkeley: Univ. of California Press, 1980.

Jawatkar, K.S., *Diego Garcia in International Diplomacy*. Bombay: Popular, 1983.

Jervell, Sverre and Kare Kyblom, *The Military Buildup in the High North: American and Nordic Perspectives*. Lanham, MD: Univ. Press of America, 1986.

Johnston, Ronald J. and Peter J. Taylor (eds.), *A World Crisis: Geographical Perspectives*. Oxford: Basil Blackwell, 1985.

Jones, Rodney W., *Nuclear Proliferation: Islam, the Bomb, and South Asia*. Washington Papers, Vol. 9, No. 2. Lanham, MD: Univ. Press of America, 1981.

Jones, Stephen B., *Australia and New Zealand and the Security of the Pacific*. New Haven, CT: Yale Institute of International Studies, 1944.

Jordan, Amos A. and Robert A. Kilmarx, *Strategic Mineral Dependence: The Stockpile Dilemma*. Beverly Hills, CA: Sage, 1979.

Joshua, Wynfred and Walter F. Hahn, *Nuclear Politics: America, France, and Britain*. Beverly Hills, CA: Sage, 1973.

K

Kantorowicz, H.U., *The Spirit of British Policy and the Myth of the Encirclement of Germany*. New York: Oxford Univ. Press, 1932.

Kaplan, Stephen S., *Diplomacy of Power: Soviet Armed Forces as a Political Instrument*. Washington: Brookings Institution, 1981.

Karsh, Efraim, *The Cautious Bear*. Boulder, CO: Westview, 1986.

Kato, I. and others, *Environmental Law and Policy in the Pacific Basin*. New York: Columbia Univ. Press, 1981.

Kaufman, William W., *A Thoroughly Efficient Navy*. Washington: Brookings Institution, 1987.

Keith, Ronald C. (ed.), *Energy, Security and Economic Development in East Asia.* New York: St. Martin's Press, 1987.

Kemp, Geoffrey and others, *The Superpowers in a Multinuclear World.* Lexington, MA: Lexington Books, 1974.

Kerner, R.J., *The Urge to the Sea: The Course of Russian History.* Berkeley: Univ. of California Press, 1942.

Kidron, Michael, *The War Atlas: Armed Conflict–Armed Peace.* New York: Simon & Schuster, 1983.

Kihl, Young Whan and Lawrence E. Grinter, *Asian–Pacific Security.* London: Frances Pinter, 1986.

Kilmarx, Robert A., *Toward a Coherent U.S. Policy on Strategic Minerals: A Guide to the Facts.* CSIS Significant Issues Series, Vol. 4, No. 3, Lanham, MD: Univ. Press of America, 1982.

Kipnis, Kenneth and Diana T. Meyers (eds.), *Political Realism and International Morality.* Boulder, CO: Westview, 1987.

Kirk, William, *Geographical Pivots of History,* Leicester, Leicestershire, Eng.: Leicester Univ. Press, 1965.

Kissinger, Henry A., *Nuclear Weapons and Foreign Policy.* New York: Harper, 1957.

Klieman, Aaron S., *Israel, Jordan, Palestine: The Search for a Durable Peace.* Washington Papers, Vol. 9, No. 83. Lanham, MD: Univ. Press of America, 1981.

Kraus, Willy and Wilfried Lütkenhorst, *The Economic Development of the Pacific Basin.* New York: St. Martin's Press, 1987.

Krause, Lawrence B. and Sueo Sekiguchi, *Economic Interaction in the Pacific Basin.* Washington: Brookings Institution, 1980.

L

La Feber, Walter, *Inevitable Revolutions; The United States in Central America.* New York and London: Norton, 1984.

Lakoff, Sanford and Randy Willoughby (eds.), *Strategic Defense and the Western Alliance.* Lexington, MA: Lexington Books, 1987.

Lall, Arthur S. (ed.), *Multilateral Negotiation and Mediation: Instruments and Methods.* Elmsford, NY: Pergamon, 1985.

Laqueur, Walter and Robert Hunter (eds.), *European Peace Movements and the Future of the Alliance.* New Brunswick, NJ: Transaction Books, 1985.

Larson, David L. (ed.), *The "Cuban Crisis" of 1962: Selected Documents, Chronology and Bibliography.* Lanham, MD, New York, and London: Univ. Press of America, 1986.

Larson, Joyce E. (ed.), *New Foundations for Asian and Pacific Security.* New Brunswick, NJ: Transaction Books, 1980.

Leith, Charles Kenneth and others, *World Minerals and World Peace.* Washington: Brookings Institution, 1943.

Lenszowski, George, *Russia and the West in Iran 1918–1948.* Ithaca, NY: Cornell Univ. Press, 1949.

Levie, Howard S., *The Code of International Armed Conflict,* 2 Vols. New York and elsewhere: Oceana, 1986.

Levran, Aharon and Zeev Eytan, *The Middle East Military Balance—1986.* Boulder, CO: Westview, 1987.

Liddell Hart, B.H., *Axis Plans in the Mediterranean: An Analysis of German Geopolitical Ideas on Italy, France, Balearic Islands, Gibraltar, Catalonia and Spain.* London: London General Press, 1939.

Lim Joo Jock, *Geo-strategy and the South China Sea Basin.* Singapore: Singapore Univ. Press, 1979.

———, *Territorial Power Domains, Southeast Asia, and China.* Singapore: Institute of Southeast Asian Studies, 1984.

Lincoln, G.A., *Economics of National Security.* Englewood Cliffs, NJ: Prentice-Hall, 1954.

Lloyd, Jones C., *Caribbean Interests of the United States.* New York: Appleton, 1916.

Long, David E., *Confrontation and Cooperation in the Gulf.* Washington: Middle East Institute, 1974.

Looney, Robert E., *The Political Economy of Latin Defense Expenditures.* Lexington, MA: Lexington Books, 1986.

Lowe, J.Y., *Geopolitics and War; Mackinder's Philosophy of Power.* Washington: Univ. Press of America, 1981.

Lowenthal, Abraham, F., *The Dominican Intervention.* Cambridge: Harvard Univ. Press, 1972.

Luttwak, Edward N., *The Strategic Balance.* New York: Library Press, 1972.

———, *The Political Use of Sea Power.* Baltimore: Johns Hopkins Univ. Press, 1974.

——— and Robert G. Weinland, *Sea Power in the Mediterranean: Political Utility and Military Constraints.* Washington Papers, Vol. 6, No. 61. Lanham, MD: Univ. Press of America, 1979.

Lyon, Thoburn C., *Air Geography: A Global View.* New York: Van Nostrand, 1951.

M

Mackinder, Halford T., *Britain and the British Seas*. Oxford: Clarendon; New York: Appleton, 1902.

———, *Democratic Ideals and Reality; A Study in the Politics of Reconstruction*. London: Constable; New York: Holt, 1919.

Mahan, Alfred Thayer, *The Influence of Sea Power upon History, 1660–1783*. Boston: Little, Brown, 1890.

———, *The Influence of Sea Power upon History; the French Revolution and Empire, 1793–1812*. Boston: Little, Brown, 1892.

———, *The Interest of America in Sea Power, Present and Future*. Boston: Little, Brown, 1898.

———, *The Problem of Asia and Its Effect upon International Politics*. Boston: Little, Brown, 1900.

———, *Letters and Papers of Alfred Thayer Mahan*, Annapolis, MD: Naval Institute Press, 1975.

Majeed, Akhtar (ed.), *Indian Ocean—Conflict and Regional Cooperation*. New Delhi: ABC, 1986.

Malin, James C., "The Contriving Brain as the Pivot of History. Sea, Landmass and Air Power: Some Bearings of Cultural Technology upon the Geography of International Relations," in George LaVerne Anderson (ed.), *Issues and Conflicts: Studies in Twentieth Century Diplomacy*, Lawrence: Univ. of Kansas Press, 1959.

Marsden, Ralph W. (ed.), *Politics, Minerals, and Survival*. Madison: Univ. of Wisconsin Press, n.d.

Marshall-Cornwall, J.H., *Geographic Disarmament: A Study of Regional Demilitarization*. London, 1935.

Martz, John D., *Central America: The Crisis and the Challenge*. Chapel Hill: Univ. of NC Press, 1959.

Martz, Mary Jeanne Reid, *The Central American Soccer War; Historical Patterns and Internal Dynamics of OAS Settlement Procedures*. Athens: Ohio Univ. Center for International Studies, 1978.

Matheson, Neil, *The "Rules of the Game" of Superpower Military Intervention in the Third World 1975–1980*. Lanham, MD: Univ. Press of America, 1982.

Mattern, Johannes, *Geopolitik—Doctrine of National Self Sufficiency and Empire*. Baltimore: Johns Hopkins Univ. Studies in Historical and Political Science, 1942.

May, Ronald James and Hank Nelson, *Melanesia: Beyond Diversity*, 2 vols. Canberra: Australian National Univ. Press, Research School of Pacific Studies, 1982.

McColl, Robert W., *A Geographic Look at Guerrilla Wars and Urban Riots*. Occasional Paper No. 2, Lawrence: Univ. of Kansas, Dept. of Geography, n.d.

McGwire, Michael, *Military Objectives in Soviet Foreign Policy*. Washington: Brookings Institution, 1987.

McLuhan, Marshall, and Quentin Fiore, *War and Peace in the Global Village*. New York: McGraw-Hill, 1968.

McMillen, Donald Hugh (ed.), *Asian Perspectives on International Security*. New York: St. Martin's Press, 1984.

McNaugher, Thomas L., *Arms and Oil: U.S. Military Strategy and the Persian Gulf*. Washington: Brookings Institution, 1985.

Mehta, J.S., *Third World Militarization: A Challenge to Third World Diplomacy*. Austin: LBJ School of Public Affairs, Univ. of Texas, 1985.

Meyer, H.C., *Mitteleuropa in German Thought and Action: 1815–1945*. The Hague: Nijhoff, 1955.

Mikdashi, Fuhayr, *The International Politics of Natural Resources*. Ithaca, NY: Cornell Univ. Press, 1976.

Millar, T.B., *The Indian and Pacific Oceans: Some Strategic Considerations*. Toronto: Institute of International Affairs, 1969.

———, *International Security in the Southeast Asian and Southwest Pacific Region*. St. Lucia, London, and New York: Univ. of Queensland Press, 1983.

Misra, R.N., *Indian Ocean and India's Security*. Delhi: Mittal, 1986.

Mitchell, William, *Winged Defense: The Development and Possibilities of Modern Air Power—Economic and Military*. New York: Putnam, 1925.

Moodie, Michael, *Sovereignty, Security and Arms*. Lanham, MD: Univ. Press of America, 1987.

Moodie, Michael and Alvin J. Cottrell, *Geopolitics and Maritime Power*. Beverly Hills, CA: Sage, 1981.

Morenus, R., *Dewline: Distant Early Warning, the Miracle of America's Pivot Line of Defense*. New York: Rand McNally, 1957.

Morley, James W. (ed.), *Security Interdependence in the Asia Pacific Region*. Lexington, MA: Lexington Books, 1986.

Morris, Michael A., and Victor Millán (eds.),

Controlling Latin American Conflicts: Ten Approaches. Boulder, CO: Westview, 1983.

Mottale, Morris Mehrdad, *The Arms Buildup in the Persian Gulf.* Lanham, MD: Univ. Press of America, 1986.

Mouzon, Olin T., *International Resources and National Policy.* New York: Harper & Row, 1959.

Mushkat, Marion (ed.), *Violence and Peace Building in the Middle East: Proceedings of a Symposium Held by the Israeli Institute for the Study of International Affairs.* Munich: K.G. Saur Verlag, 1981.

———, *The Third World and Peace.* New York: St. Martin's Press, 1983.

N

Nathan, K.S. and M. Pathmanathan (eds.), *Trilaterlism in Asia.* Honolulu: Univ. of Hawaii Press, 1986.

Nemetz, Peter H. (ed.), *The Pacific Rim; Investment, Development and Trade.* Vancouver: Univ. of British Columbia Press, 1987.

Newcombe, Hanna, *Design for a Better World.* Lanham, MD: Univ. Press of America, 1983.

Nugent, Jeffrey, and others, *Bahrain and the Gulf; Past Perspectives and Alternate Futures.* New York: St. Martin's Press, 1985.

O

Oborne, Michael West, and Nicholas Fourt, *Pacific Basin Economic Cooperation.* Paris: OECD Development Centre, 1983.

O'Connell, D.P., *The Influence of Law on Seapower.* Manchester, Eng.: Manchester Univ. Press; Annapolis: Naval Institute Press, 1975.

Odell, Peter R., *Oil and World Power.* New York: Penguin, 1979.

Ogunbadejo, Oye, *The International Politics of Africa's Strategic Minerals.* London: Frances Pinter, 1985.

Olson, William J. (ed.), *U.S. Strategic Interests in the Gulf Region.* Boulder, CO: Westview, 1987.

O'Sullivan, Patrick, *Geopolitics.* Beckenham, Eng.: Croom Helm; New York: St. Martin's 1986.

O'Sullivan, Patrick and Jesse W. Miller, *The Geography of Warfare.* London: Croom Helm, 1983.

P

Pacific–Asian Security Policies. Honolulu: Univ. of Hawaii Press, 1983. A private seminar of the Pacific Forum, October 1982.

Pacific Regional Security: The 1985 Pacific Symposium. Washington: National Defense Univ. Press, 1987.

Page, Stanley W., *Geopolitics of Leninism.* New York: Columbia Univ. Press, 1982.

Page, Stephen, *The USSR and Arabia: The Development of Soviet Policies and Attitudes Towards the Countries of the Arabian Peninsula.* London: Central Asian Research Centre, 1971.

Parker, Geoffrey, *Western Geopolitical Thought in the Twentieth Century.* New York: St. Martin's Press; London: Croom Helm, 1985.

Parker, Richard, *North Africa; Regional Tensions and Strategic Concerns.* Praeger, 1984.

Parker, W.H., *Mackinder: Geography as an Aid to Statecraft.* Oxford: Oxford Univ. Press, Clarendon, 1982.

Payne, Anthony, *The International Crisis in the Caribbean.* Baltimore: Johns Hopkins Univ. Press, 1984.

Payne, Anthony and others, *Grenada.* New York: St. Martin's Press, 1986.

Peltier, Louis C. and G. Etzel Pearcy, *Military Geography.* Princeton, NJ: Van Nostrand, 1966.

Pepper, David and Alan Jenkins (eds.), *The Geography of Peace and War.* New York: Basil Blackwell, 1986.

Pierre, Andrew J., *Third World Instability: Central America as a European–American Issue.* New York: Council on Foreign Relations, 1984.

Pirages, Dennis, *Global Ecopolitics, the New Context for International Relations.* North Scituate, MA: Duxbury, 1978.

Pittman, Howard T., "Geopolitics and Foreign Policy in Argentina, Brazil and Chile," in Elizabeth G. Ferris and Jennie K. Lincoln, *Latin American Foreign Policies: Global and Regional Dimensions.* Boulder, CO: Westview, 1981.

Pogany, István S., *The Security Council and the Arab-Israeli Conflict.* New York: St. Martin's Press, 1984.

Possony, Stefan T., *Strategic Air Power: The Pattern of Dynamic Security.* Washington: Infantry Journal Press, 1949.

Potter, E.B. (ed.), *The United States and World Sea Power.* Englewood Cliffs, NJ: Prentice-Hall, 1955.

Price, A. Grenfell, *The Geopolitical Transformation of the Pacific and Its Present Significance.* Reprint from Royal Geographic Society, South Australian Branch, Session 1950–1951, Vol. 52, 1951.

Price, David Lynn, *Oil and Middle East Security.* Washington Papers, Vol. 4, No. 41. Lanham, MD: Univ. Press of America, 1976.

Pridham, B.R. (ed.), *The Arab Gulf and the West.* New York: St. Martin's Press, 1985.

Q

Quandt, William B., *Camp David: Peacemaking and Politics.* Washington: Brookings Institution, 1986.

Quester, George H., *Sea Power in the 1970's: A Symposium Sponsored by Cornell University Program on Peace Studies.* New York: Quellen, 1975.

Quddus, Syed Abdul, *Afghanistan and Pakistan: A Geopolitical Study.* Lahore, Pakistan: Feroz, 1982.

R

Rais, Rasul B., *The Indian Ocean and the Superpowers.* Totowa, NJ: Barnes & Noble, 1987.

Ramazani, Rouhollah K., *The Persian Gulf: Iran's Role.* Charlottesville: Univ. of Virginia Press, 1972.

Raymond, Gregory A., *Conflict Resolution and the Structure of the State System. An Analysis of Arbitrative Settlements.* Montclair, NJ: Allanheld, Osmun, 1980.

Razis, Vincent Victor, *Swords or Ploughshares: South Africa and Political Change.* Athens: Ohio Univ. Press, 1980.

Rohwer, Jurgen, *Superpower Confrontation on the Seas: Naval Development and Strategy Since 1945.* Washington Papers, Vol. 3, No. 26. Lanham, MD: Univ. Press of America, 1975.

Rona, Thomas, *Our Changing Geo-Political Premises.* New Brunswick, NJ: Transaction Books, 1982.

Rosen, S., *Military Geography and the Military Balance in the Arab–Israel Conflict.* Jerusalem: Hebrew Univ. Press, 1977.

Rotfeld, Adam Daniel (ed.), *From Helsinki to Madrid; Conference on Security and Cooperation in Europe; Documents 1973–1983.* Warsaw: Polish Institute of International Affairs, 1985.

Rubinstein, Alvin Z., *The Arab–Israeli Conflict Perspective.* New York: Praeger, 1984.

Russet, Bruce M. (ed.), *Peace, War and Numbers.* Beverly Hills, CA: Sage, 1972.

Rustow, Dankwart A., *Turkey: America's Forgotten Ally.* 1987.

S

Safran, Nadav, *From War to War.* New York: Bobbs-Merrill, Pegasus, 1969.

Saivetz, Carol R., *The Soviet Union and the Gulf in the 1980's.* Boulder, CO: Westview, 1987.

Sandler, Shmuel and Hillel Frisch, *Israel, the Palestinians and the West Bank: A Study in Intercommunal Conflict.* Lexington, MA: Lexington Books, 1984.

Schelling, T., *Arms and Influence.* New Haven, CT: Yale Univ. Press, 1965.

Schilling, Warner R. and others, *American Arms and a Changing Europe: Dilemmas of Deterrence and Disarmament.* New York: Columbia Univ. Press, 1973.

Schnitzer, Ewald W., *German Geopolitics Revised: A Survey of Geopolitical Writing in Germany Today.* Santa Monica, CA: RAND Corp., 1954.

Schoenhals, Kai P. and Richard A. Melanson, *Revolution and Intervention in Grenada.* Boulder, CO: Westview, 1985.

Schulz, Donald E. and Douglas H. Graham, *Revolution and Counterrevolution in Central America and the Caribbean.* Boulder, CO: Westview, 1984.

Seilliere, E., *The German Doctrine of Conquest: A French View.* Dublin and London: Maunsel, 1914.

Sen, D. *Basic Principles of Geopolitics and History.* Delhi: Concept, 1975.

Sharp, G., *The Politics of Non-Violent Action.* Boston: Porter Sargent, 1973.

Shimkin, D.M., *Minerals: A Key to Soviet Power.* Cambridge: Harvard Univ. Press, 1953.

Singh, K. Rajendra, *Politics of the Indian Ocean.* Delhi: Thomson Press (India), 1974.

Sivard, Ruth Leger, *World Military and Social Expenditures.* Washington: World Priorities, annual.

Small, M. and J.D. Singer, *Resort to Arms: International and Civil Wars, 1816–1980.* Beverly Hills, CA: Sage, 1982.

Smoke, Richard and Willis Harman, *Paths to Peace.* Boulder, CO: Westview, 1987.

Spaight, James M., *The Beginnings of Organised Air Power.* London: Longmans, Green, 1927.

———, *Air Power in the Next War.* London: Bles, 1938.

Spiegel, Steven L., and others (eds.), *The Soviet American Competition in the Middle East.* Lexington, MA: Lexington Books, 1987.

Sprout, Harold H., *Towards a Politics of the Planet Earth.* New York: Van Nostrand, Reinhold, 1973.

Sprout, Harold and Margaret Sprout, *Toward*

a New Order of Sea Power. Princeton, NJ: Princeton Univ. Press, 1940.

Spykman, Nicholas John, *America's Strategy in World Politics.* 1942; reprinted: Hamden, CT: Shoe String Press, Archon Books, 1970.

Stephan, John J. and V.P. Chichkanov (eds.), *Soviet–American Horizons on the Pacific.* Honolulu: Univ. of Hawaii Press, 1986.

Stephenson, Carolyn M., *Alternative Methods for International Security.* Lanham, MD: Univ. Press of America, 1982.

Stern, Eliahu and Yehuda Hayuth, "Developmental Effects of Geopolitically Located Ports," in Brian Stewart Hoyle and David Hilling (eds.), *Seaport Systems and Spatial Change.* New York and elsewhere: Wiley, 1984, 239–255.

Stockman-Shomron, Israel (ed.), *Israel, the Middle East and the Great Powers.* New Brunswick, NJ: Transaction Books, 1985.

Stokesbury, James L., *Navy and Empire.* New York: Morrow, 1983.

Strausz-Hupe, Robert, *Geopolitics.* New York: Putnam, 1942.

———, *The Balance of Tomorrow.* New York: Putnam, 1945.

———, *Protracted Strategy.* New York: Harper & Row, 1963.

Suter, Keith, *An International Law of Guerrilla Warfare; The Global Politics of Law-Making.* New York: St. Martin's Press, 1984.

Sutlov, Alexander, *Minerals in World Affairs.* Salt Lake City: Univ. of Utah Printing Services, 1973.

Szuprowicz, Bohdan O., *Strategic Materials Geopolitics.* New York: Wiley–Interscience, 1981.

T

Tahir-Kheli, Shirin (ed.), *U.S. Strategic Interests in Southwest Asia.* New York: Praeger, 1982.

Tahtinen, Dale R. *Arms in the Indian Ocean: Interests and Challenges.* Washington: American Enterprise Institute for Public Policy Research, 1977.

Tardanico, Richard, *Crises in the Caribbean Basin.* Newbury Park, CA: Sage, 1987.

Taylor, E.G.R., *Geography of an Air Age.* London: Royal Institute of International Affairs, 1945.

Taylor, G.E., *America in the New Pacific.* New York: Macmillan, 1942.

Taylor, Griffith, *Canada's Role in Geopolitics.* Toronto: Ryerson, 1942.

———, *Our Evolving Civilization: An Introduction to Geopacifics. Geographical Aspects of*

the Path Toward World Peace. London and Toronto: Oxford Univ. Press, 1946.

———, "Geopolitics and Geopacifics," in Griffith Taylor (ed.), *Geography in the Twentieth Century.* New York: Philosophical Library; London: Methuen; 1951, 587–608.

Thompson, W.R., ed. *Contending Approaches to World System Analysis.* Beverly Hills: Sage, 1983.

Thorsson, Inga, *In Pursuit of Disarmament; Conversion from Military to Civil Production in Sweden.* Stockholm, 1984.

Till, Geoffrey, *Maritime Strategy and the Nuclear Age,* 2nd ed. New York: St. Martin's Press, 1984.

Tripp, Charles (ed.), *Regional Security in the Middle East.* New York: St. Martin's Press, 1984.

U

UNESCO Yearbook on Peace and Conflict Studies. Paris: UNESCO, published annually.

The United Nations Disarmament Yearbook. New York: United Nations, published annually.

V

Valenta, Jiri and Herbert Ellison, *Grenada and Soviet/Cuban Policy.* Boulder, CO: Westview, 1986.

van Doorn, J. (ed.), *Armed Forces and Society.* The Hague and Paris: Mouton, 1968.

Vasey, Lloyd R. (ed.), *Pacific Asia and U.S. Politics.* Honolulu: Univ. of Hawaii Press, 1978.

Väyrynen, Raimo and others (eds.), *The Quest for Peace.* Newbury Park, CA: Sage, 1987.

Vedovato, Claudio, *Politics, Foreign Trade and Economic Development.* New York: St. Martin's Press, 1986.

Voss, Carl Hermann, *The Palestine Problem Today: Israel and Its Neighbors.* Boston: Beacon Press, 1953.

W

Walters, Robert E., *Sea Power and the Nuclear Fallacy: A Reevaluation of Global Strategy.* Holmes & Meier, 1980 (reprint of 1974 ed.).

Wallerstein, Immanuel, *Modern World System* (2 vols.), New York: Academic Press, 1974 and 1980.

———, *The Politics of the World Economy.* London: Cambridge, 1984.

Weigert, Hans W., *Generals and Geographers: The Twilight of Geopolitics.* New York: Oxford Univ. Press, 1942.

Wesson, Robert G., *The Russian Dilemma: A*

Political and Geopolitical View. New Brunswick, NJ: Rutgers Univ. Press, 1974.

West, Philip and Frans A.M. Alting von Geusau, *The Pacific Rim and the Western World*. Boulder, CO: Westview, 1987.

Westing, Arthur H., *Warfare in a Fragile World: Military Impact on the Human Environment*. London: Taylor & Francis; New York: Crane, Russak, 1980.

Whittlesey, Derwent, *German Strategy of World Conquest*. New York: Holt, Rinehart, 1942.

Wiens, Herold J., *Pacific Island Bastions of the United States*. Princeton, NJ: Van Nostrand, 1962.

Wildavsky, Aaron (ed.), *Beyond Containment: Alternative American Politics Toward the Soviet Union*. San Francisco: ICS Press, 1983.

Woolsey, R. James (ed.), *Nuclear Arms: Ethics, Strategy, Politics*. San Francisco: ICS Press, 1984.

Wu, Yuan-li, *Raw Material Supply in a Multipolar World*. New York: Crane, Russak, 1973.

Z

Zoller, Elizabeth, *Peacetime Unilateral Remedies*. Ardsley-on-Hudson, NY: Transnational, 1984.

Zoppo, C.E., and C. Zorgbibe (eds.), *On Geopolitics; Classical and Nuclear*. Dordrecht, Neth.: Nijhoff, 1985.

Periodicals

A

The Aftermath of Nuclear War, Special issue of *Ambio*, 11, 2–3 (1982).

Alexander, Lewis N., "The New Geopolitics: A Critique," *Journal of Conflict Resolution*, 5 (1961), 407–410.

B

Bagley, H.W., *Seapower and Western Security: The Next Decade*. Adelphi Papers, No. 139. London: International Institute for Strategic Studies, 1977.

Baldwin, Hanson W., "The Soviet Navy," *Foreign Affairs*, 33, 4 (July 1955), 587–604.

Balfour-Paul, H.G., "Recent Developments in the Persian Gulf," *Royal Central Asian Journal*, 56 (February 1969), 12–19.

Barnett, Harold J., "The Changing Relation of Natural Resources to National Security," *Economic Geography*, 34, 3 (July 1958), 189–201.

Behre, Charles H., Jr., "Mineral Economics and World Politics," *Geographical Review*, 30, 4 (October 1940), 676–678.

Berg, P. and S. Lodgaard, "Disengagement Zones: A Step Towards Meaningful Defense?" *Journal of Peace Research*, 20, 1 (1983), 5–15.

Berry, John A., "Oil and Soviet Policy in the Middle East," *Middle East Journal*. 26, 2 (Spring 1972), 149–160.

Bidwell, P.W., "Self-containment and Hemispheric Defense," *Annals, American Academy of Political and Social Science*, 218 (November 1941), 175–185.

Bizzilli, P., "Geopolitical Conditions of the Evolution of Russian Nationality," *Journal of Modern History*, 2 (1930), 27–36.

Blechman, B. and M.R. Moore, "A Nuclear Weapon-Free Zone in Europe," *Scientific American*, 248 (1983), 29–35.

Blouet, Brian W., "The Maritime Origins of Mackinder's Heartland Thesis," *Great Plains–Rocky Mountain Geographical Journal*, (1973), 6–11.

———, "Halford Mackinder's Heartland Thesis," *Great Plains–Rocky Mountain Geographical Journal*, 5 (1976), 2–6.

Bowman, Isaiah, "The Military Geography of the Atacama," *Educational Bi-monthly*, 6 (1911), 1–21.

———, "Political Geography of Power," *Geographical Review*, 32 (1942), 349–352.

———, "The Strategy of Territorial Decisions," *Foreign Affairs*, 24, 2 (January 1946), 177–194.

Boyle, Francis A. and others, Letter: "International Lawlessness in Grenada," *American Journal of International Law*, 78, 1 (January 1984), 172–175.

Brewer, William D., "Yesterday and Tomorrow in the Persian Gulf," *Middle East Journal*, 23 (September 1969), 149–158.

Brush, John E., "Peace Research and Geography," *Professional Geographer*, 16 (1964), 49.

Bryan, G.S., "Geography and the Defense of

the Caribbean and the Panama Canal," *Annals, AAG*, 31, 2 (June 1941), 83–94.

Bull, Hedley, "Kissinger: The Primacy of Geopolitics," *International Affairs*, 56, 3 (1980), 484–487.

Bulletin of the Atomic Scientists. Chicago: Educational Foundation for Nuclear Scientists. Published 10 times per year.

Bundy, William P., "Elements of Power," *Foreign Affairs*, 56, 1 (October 1977), 1–26.

Bunge, William, "The Geography of Human Survival," *Annals, AAG*, 63, 3 (September 1973), 275–295.

C

Cahnman, Werner J., "Concepts of Geopolitics," *American Sociological Review*, 8, 1 (1943), 55–59.

———, "The Concept of Raum and the Theory of Regionalism," *American Sociological Review*, 9, 5 (1944), 455–462.

———, "Frontiers Between East and West in Europe," *Geographical Review*, 39, 4 (October 1949), 605–642.

Calder, Kent E., "The Emerging Politics of the Trans-Pacific Economy," *World Policy Journal*, 2, 4 (Fall 1985), 593–623.

Cameron, Alan W., "The Strategic Significance of the Pacific Islands: A New Debate Begins," *Orbis*, 19 (Fall 1975), 1012–1036.

Chatterjee, M.N., "New Ideas for Old: Socio-Politics vs. Geopolitics," *Antioch Review*, 7, 1 (1947), 64–72.

Chaudhuri, M., "Geopolitical Aspect of World Politics," *Calcutta Geographical Review*, 2 (1949), 48–54.

Child, Jack, "Geopolitical Thinking in Latin America," *Latin American Research Review*, 14, 2 (1979), 89–111.

Chubb, Basil, "Geopolitics," *Irish Geographer*, 5, 1 (1954), 15–25.

Clokie, H.McD., "Geopolitics—New Super Science or Old Art?" *Canadian Journal of Economic and Political Sciences*, 215 (1941), 66–73.

Cohen, Saul B., "A New Map of Global Geopolitical Equilibrium: A Developmental Approach," *Political Geography Quarterly*, 1, 3 (July 1982), 223–241.

Cooper, J.C., "Air Power and The Coming Peace Theatre," *Foreign Affairs*, 24, 3 (1946), 441–452.

Cottrell, Alvin J., "Iran, the Arabs and the Persian Gulf," *Orbis*, 17 (Fall 1973), 978–988.

Crone, Gerald R., "A German View of Geopolitics," *Geographical Journal*, 111, 1–3 (1948), 104–108.

Current Research on Peace and Violence, Tampere, Finland-Tampere Peace Research Institute. Published quarterly.

D

Darling, F.C., "Geopolitics of American Foreign Policy," *United Asia*, 15, 5 (1963), 370–376.

Davies, W.W., "Geopolitics and the United States," *Contemporary Review*, 172 (August 1947), 87–91.

Dayan, Moshe, "Israel's Border and Security Problems," *Foreign Affairs*, 33, 2 (January 1955), 250–267.

Debre, Michel, "France's Global Strategy," *Foreign Affairs*, 49, 3 (April 1971), 395–406.

Delcoigne, G., "An Overview of Nuclear Weapon-Free Zones," *International Atomic Energy Bulletin*, 24 (1982), 50–55.

De Seversky, Alexander P., "The Twilight of Seapower," *American Mercury*, 52 (June 1941), 647–658.

———, "Air Power Ends Isolation," *Atlantic Monthly*, 168 (October 1941), 407–416.

Deutsch, Karl W. and David J. Singer, "Multipolar Power Systems and International Stability," *World Politics*, 16, 3 (1964), 390–406.

Diehl, P.F., "Arms Race and Escalation: A Closer Look," *Journal of Peace Research*, 20 (1983), 205–212.

Dietrich, S.De R., "Some Geopolitical Aspects of the Indo-Pak. Sub-Continent," *United Asia*, 15, 3 (1963), 225–233.

Doran, Ronald J., "Changing Political Views," *Journal of Geography*, 58, 1 (January 1959), 32–33.

Dow, Roger, "A Geopolitical Study of Russia and the United States," *Russian Review*, 1, 1 (November 1941), 6–19.

Dowdy, William L. and Russell B. Trood, "The Indian Ocean: An Emerging Geostrategic Region," *International Journal*, 38, 3 (1983), 432–458.

Drambyants, G., "The Persian Gulf: Time of Changes," *International Affairs*, 6 (June 1968), 34–39.

———, "The Persian Gulf: Twixt the Past and the Future," *International Affairs*, 10 (October 1970), 66–72.

Dryer, Charles R., "Mackinder's 'World Island' and Its American 'Satellite'," *Geographical Review*, 9, 1 (March 1920), 205–207.

Dugan, A.B., "Mackinder and His Critics Reconsidered," *Journal of Politics*, 24 (1962), 241–257.

Dunn, John M., "American Dependence on Materials Imports: The World-Wide Resource Base," *Journal of Conflict Resolution*, 4, 1 (March 1960), 106–122.

E

Earle, Edward Mead, "Power Politics and American World Policy," *Political Science Quarterly*, 58, 1 (March 1943), 94–106.

Earney, Fillmore C.F., "The Geopolitics of Minerals," *Focus*, 31, 5 (May–June 1981), 1–16.

Emme, Eugene M., "Some Fallacies Concerning Air Power," *Annals, American Academy of Political and Social Science*, 299 (May 1955), 12–24.

Ewell, J., "The Development of Venezuelan Geopolitical Analysis Since World War II," *Journal of Interamerican Studies and World Affairs*, 24 (1982), 295–320.

F

Faber, J. and others, "Diffusion of War: Some Theoretical Considerations and Empirical Evidence," *Journal of Peace Research*, 22 (1985), 31–46.

Feis, Herman, "Raw Materials and Foreign Policy," *Foreign Affairs*, 16, 4 (1938), 574–586.

Field, Michael, "The Gulf," *World Survey*, 37 (January 1972), entire issue.

Fifield, Russell H., "Geopolitics After Hitler," *Social Studies*, 35, 6 (1944), 243–246.

Fisher, Charles A., "The Expansion of Japan: A Study in Oriental Geopolitics," *Geographical Journal*, 115, 1–3 (January–March 1950), 1–19; 115, 4–6 (April–June 1950), 179–193.

———, "Containing China? I The Antecedents of Containment," *Geographical Journal*, 136, 4 (December 1970), 534–556. II "Concepts and Applications of Containment," 137, 3 (September 1971, 281–310.

Flanders, D.P., "Geopolitics and American Post-War Policy," *Political Science Quarterly*, 60 (1945), 578–585.

Furniss, Jr. Edgar S., "The Contribution of Nicholas John Spykman to the Study of International Politics," *World Politics*, 4, 3 (April 1952), 382–401.

G

Galtung, Johan, "The Geopolitics of War: Total War and Geostrategy," *Journal of Politics*, 5, 4 (1943), 347–362.

———, "The Geography of Conflict," *Journal of Conflict Resolution*, 4 (1960), 1–161.

———, "Violence, Peace and Research," *Journal of Peace Research*, 6 (1969), 167–191.

———, "Social Cosmology and the Concept of Peace," *Journal of Peace Research*, 18 (1981), 183–200.

———, "Twenty-five Years of Peace Research: Ten Challenges and Some Responses," *Journal of Peace Research*, 22 (1985), 141–158.

Gilbert, E.W. and W.H. Parker, "Mackinder's 'Democratic Ideals and Reality' After 50 Years," *Geographical Journal*, 135, 2 (June 1969) 228–231.

Gimlin, H., "Global Waterways: Access and Control," *Editorial Research Reports*, 2 (1967), 823–842.

Gorman, Stephen M., "Geopolitics and Peruvian Foreign Policy," *Inter-American Economic Affairs*, 36, 2 (Autumn 1982), 65–88.

Gottmann, Jean, "The Background of Geopolitics," *Military Affairs*, 6 (1942), 197–206.

———, "French Geography in Wartime," *Geographical Review*, 36 (1946), 80–91.

———, "Geography and International Relations," *World Politics*, 3, 2 (1950), 153–173.

Greenleaf, W.H., "Imperialism and Geopolitics," *World Affairs*, 1, 2 (1947), 181–188.

Gustavson, C.G., "Historical Maps and Seapower," *Journal of Geography*, 55, 8 (November 1946), 317–321.

Gyorgy, Andrew, "The Application of German Geopolitics: Geo-Science," *American Political Science Review*, 37 (1943), 677–686.

H

Hall, A.R., "Mackinder and the Course of Events," *Annals, AAG*, 45, 2 (June 1955), 109–126.

Harding, Sir John, "The Cyprus Problem in Relation to the Middle East," *International Affairs*, 34, 3 (July 1958), 291–296.

Harkema, Roelof C., "Zambia's Changing Pattern of External Trade," *Journal of Geography*, 71, 1 (January 1972), 19–27.

Hayes, John D., "Peripheral Strategy—Mahan's Doctrine Today," *Proceedings, United States Naval Institute*, (November 1953).

Henrikson, Alan K., "The Moralist as Geopolitician," *Fletcher Forum*: 51, 7 (October 1952), 391–414.

Herrick, Francis H., "World Island and Heartland—The Strategical Theories of Sir Halford John Mackinder," *South Atlantic Quarterly*, 43 (June 1944), 248–255.

Hessler, W.H., "A Geopolitics for America," *Proceedings, United States Naval Institute*, 70, 3 (1944), 245–253.

Hoffman, George W., "The Shatter-Belt in Relation to the East–West Conflict," *Journal of Geography*, 51 (1952), 265–275.

———, "Nineteenth-Century Roots of American World Power Relations," *Political Geography Quarterly*, 1, 3 (July 1982), 279–292.

Hoivik, T., "The Demography of Structural Violence," *Journal of Peace Research*, 14 (1977), 59–74.

Holden, David, "The Persian Gulf: After the British Raj," *Foreign Affairs*, 49, 4 (July 1971), 721–735.

Hoopes, Townsend, "Overseas Bases in American Strategy," *Foreign Affairs*, 37, 1 (October 1958), 69–82.

Hooson, David J.M., "The Middle Volga—An Emerging Focal Region in the Soviet Union," *Geographical Journal*, 126, 2 (June 1960), 180–190.

———, "A New Soviet Heartland?" *Geographical Journal*, 128, 1 (March 1962), 19–29.

House, John W., "War, Peace and Conflict Resolution: Towards an Indian Ocean Model," *Transactions, Institute of British Geographers ns*, 9 (1984), 3–21.

Howard, H.N., "The United States and the Problems of the Straits," *Middle East Journal*, 1, 1 (1947), 59–72.

Hughes, C.E., "Observation of the Monroe Doctrine," *American Journal of International Law*, 17 (1923), 611–628.

Hurewitz, J.C., "Unity and Disunity in the Middle East," *International Conciliation*, No. 481 (May 1952), 197–260.

———, "The Persian Gulf; British Withdrawal and Western Security," *Annals, American Academy of Political and Social Sciences*, 401 (May 1972), 106–115.

Huszar, G.B. de, "Use of Satellite Outposts by the U.S.S.R.," *Annals, American Academy of Political and Social Science*, 271 (1950), 157–164.

Huth, Paul and Bruce M. Russett, "What Makes Deterrence Work? Cases from 1900 to 1980" *World Politics*, 36, 4 (July 1984), 496–526.

J

Jackson, W.A. Douglas, "Mackinder and the Communist Orbit," *Canadian Geographer*, 6, 1 (Spring 1962), 12–21.

Jahn, E., "Peace Research and Politics Within the Field of Social Demands," *Journal of Peace Research*, 20 (1983), 253–259.

Jay, Peter, "Regionalism as Geopolitics," *Foreign Affairs*, 58, 3 (Spring 1980), 485–514.

Johnson, A.H., "Mahan on Sea Power," *English Historical Review*, 45, 4 (1955), 492–508.

Jones, B. and K. Mehnert, "Hawaii and the Pacific: A Survey of Political Geography," *Geographical Review*, 30 (1940), 358–375.

Jones, Stephen B., "Views of the Political World," *Geographical Review*, 45, 3 (July 1955), 309–326.

———, "Global Strategic Views," *Geographical Review*, 45, 4 (October 1955), 492–508.

Journal of Conflict Resolution. New Haven, CT: Yale Univ. Published quarterly.

Journal of Peace Research. Oslo: International Peace Research Institute. Published quarterly.

Joyner, Christopher C., "Reflections on the Lawfulness of Invasion," *American Journal of International Law*, 78, 1 (January 1984), 131–144.

K

Kane, F.X., "Space Age Geopolitics," *Orbis*, 14 (1971), 911–933.

Karan, Pradyumna P., "India's Role in Geopolitics," *India Quarterly*, 9, 2 (April–June 1953), 160–169.

Kelly, John B., "The British Position in the Persian Gulf," *World Today*, 20 (July 1964), 238–249.

Kelly, Philip L., "Escalation of Regional Conflict: Testing the Shatterbelt Concept," *Political Geography Quarterly*, 5, 2 (April 1986), 161–180.

Kemp, A., "Image of the Peace Field: An International Survey," *Journal of Peace Research*, 22 (1985), 129–140.

Kendi, I., "Twenty-five Years of Local Wars," *Journal of Peace Research*, 7 (1971), 5–22.

Kennan, George (writing under pseudonym "X"), "The Sources of Soviet Conduct," *Foreign Affairs*, 25 (1947), 566–582.

Kennedy, Edward M., "The Persian Gulf: Arms Race or Arms Control?" *Foreign Affairs*, 54 (October 1975), 14–35.

Khan, Fazlur, R., "The Geographical Basis of Pakistan's Foreign Policy," *Pakistan Geographical Review*, 7, 2 (1952).

Khilnani, N.M., "Bahrain Islands: Their Economic and Geopolitical Importance to South-West Asia," *United Asia*, 24 (July–August 1972), 233–236.

Kirk, W., "Indian Ocean Community," *Scottish Geographical Magazine*, 67 (1951), 161–177.

Kish, George, "Political Geography into Geopolitics: Recent Trends in Germany," *Geo-*

graphical Review, 32, 4 (October 1942), 632–645.

Klare, M. and others, "The Global Reach of the Superpowers," South Magazine, (1983), 9–17.

Knight, David B., "Changing Orientations: Elements of New Zealand's Political Geography," Geographical Bulletin, 1 (1970) 21–30.

Koch, Howard E. and others, "Some Theoretical Notes on Geography and International Conflict," Journal of Conflict Resolution, 4, 1 (1960), 4–14.

Kohler, G., "War, the Nation-State Paradigm and the Imperialism Paradigm: British War Involvements," Peace Research, 7 (1975), 31–41.

Kohler, H., and N. Alcock, "An Empirical Table of Structural Violence," Journal of Peace Research, 13 (1976), 343–356.

Krishan, Radha, "Geopolitics of Petroleum," Indian Geographer, 6, 1 (August 1961), 90–110.

Kristof, Ladis K.D., "The Origins and Evolution of Geopolitics," Journal of Conflict Resolution, 4, 1 (March 1960), 15–51.

———, "The Geopolitical Image of the Fatherland: The Case of Russia," Western Political Quarterly, 20, 4 (1967), 941–954.

Krushchev, Nikita S., "On Peaceful Coexistence," Foreign Affairs, 38, 1 (October 1959), 1–18.

Kruszewski, Charles, "Germany's Lebensraum," American Political Science Review, 34, 5 (October 1940), 964–975.

———, "The Pivot of History," Foreign Affairs, 32, 3 (April 1954), 388–401.

Kulski, W.W., "The Problem of the Heartland of Europe," Journal of Central European Studies, 7, 3 (1947), 255–261.

L

Lacoste, Y., "An Illustration of Geographical Warfare: Bombing of the Dikes of the Red River," Antipode, 5 (1973) 1–13.

La Feber, Walter, "A Note on the 'Mercantilistic Imperialism' of Alfred Thayer Mahan," Mississippi Valley Historical Review, 68, 4 (1962), 674–685.

Landheer, B., "Interstate Competition and Survival Potential," Journal of Conflict Resolution, 3 (1959), 162–171.

Larson, David L., "Security, Disarmament and the Law of the Sea," Marine Policy, 3, 1 (January 1979), 40–58.

Lattimore, Owen, "Yunnan, Pivot of Southeast Asia," Foreign Affairs, 21, 3 (1943), 476–493.

Livermore, Shaw, "International Control of Raw Materials," Annals, American Academy of Political and Social Science, 281 (1951), 157–165.

Lorwin, L.L., "Geo-Economics Versus Geo-Politics: A Basis for United Nations Policy," World Economics, 1, 1–2 (1943), 10–28.

Lottich, K., "Geopolitics of the Philippine Islands," United Asia, 15, 3 (1963), 256–262.

Luce, William, "Britain in the Persian Gulf," Round Table, No. 227 (July 1967), 277–283.

———, "A Naval Force for the Gulf," Round Table, No. 236 (1969), 347–356.

M

Mackinder, Halford J., "The Physical Basis of Political Geography," Scottish Geographical Magazine, 6, 2 (February 1890), 78–84.

———, "The Geographical Pivot of History," Geographical Journal, 23, 4 (April 1904), 421–444.

Mahan, Alfred Thayer, "The United States Looking Outward," Atlantic Monthly, 66 (1890), 816–824.

———, "Hawaii and Our Future Sea Power," Forum, 15 (March 1893), 1–11.

———, "The Future in Relation to American Naval Power," Harper's Magazine, 91 (1895), 767–775.

———, "The Problem of Asia," Harper's Magazine, 100 (1900), 536–547.

———, "The Panama Canal and Sea Power in the Pacific," Century Magazine, 82 (1911), 240–248.

———, "Strategic Features of the Caribbean Sea and the Gulf of Mexico," Harper's New Monthly Magazine, 95 (1911), 680–691.

———, "Importance of Command of the Sea," Scientific American, 105 (December 9, 1911), 512.

———, "The Panama Canal and the Distribution of the Fleet", North American Review, 200 (1941), 406–417.

———, "The Round World and the Winning of Peace," Foreign Affairs, 21, 4 (July 1943), 595–605.

Malin, James C., "Space and History: Reflections on the Closed-Space Doctrines of Turner and Mackinder and the Challenge of Those Ideas by the Air Age," Agricultural History, 18, Pt. 1 (1944), 65–74.

Mangold, Peter, "Force and Middle East Oil," Round Table, No. 261 (January 1976), 93–101.

Marlowe, John, "Arab–Persian Rivalry in the Persian Gulf," *Royal Central Asian Journal,* 51, (January 1964), 23–31.

Martin, Geoffrey J., "Political Geography and Geopolitics," *Journal of Geography,* 58, 9 (December 1959), 441–444.

Martin, L., "The Geography of the Monroe Doctrine and the Limits of the Western Hemisphere," *Geographical Review,* 30 (1940), 525–528.

McColl, Robert W., "A Political Geography of Revolution: China, Vietnam, and Thailand," *Journal of Conflict Resolution,* 11, 2 (June 1967), 153–167.

———, "The Insurgent State: Territorial Bases of Revolution," *Annals, AAG,* 59, 4 (December 1969), 613–631.

———, "Geographical Themes in Contemporary Asian Revolutions," *Geographical Review,* 65, 3 (July 1975), 301–310.

Meinig, Donald W., "Heartland and Rimland in Eurasian History," *Western Political Quarterly,* 9, 3 (September 1956), 553–569.

Melamid, Alexander, "The Political Geography of the Gulf of Aqaba," *Annals, AAG,* 47, 3 (September 1957), 231–240.

Meyer, Henry Cord, "Mitteleuropa in German Political Geography," *Annals, AAG,* 36, 3 (September 1946), 178–194.

Meyerhoff, Arthur A., "Economic Impact and Geopolitical Implications of Giant Petroleum Fields," *American Scientist,* 64, 5 (May 1976), 530–540.

Midlarsky, Manus I., "Boundary Permeability as a Condition of Political Violence," *Jerusalem Journal of International Relations,* 1, 2 (Winter 1975), 53–69.

Military Balance. London: International Institute for Strategic Studies. Published annually.

Mills, D.R., "The U.S.S.R.: A Re-Appraisal of Mackinder's Heartland Concept," *Scottish Geographical Magazine,* 72, 3 (December 1956), 144–152.

Modelski, G., "The Premise of Geocentric Politics," *World Politics,* 22, 4 (July 1970), 617–635.

———, "The Long Cycle of Global Politics and the Nation-State," *Comparative Studies in Society and History,* 20 (1978), 214–235.

Moll, Kenneth L., "A.T. Mahan, American Historian," *Military Affairs,* 27 (1963), 131–140.

Momsen, Janet Henshall, "Caribbean Conflict: Cold War in the Sun," *Political Geography Quarterly,* 3, 2 (April 1984), 145–151.

Moore, John Norton, "Grenada and the International Double Standard," *American Journal of International Law,* 78, 1 (January 1984), 145–168.

———, "The Secret War in Central America and the Future of World Order," *American Journal of International Law,* 80, 1 (January 1986), 43–127.

Morris, Michael A., "Maritime Geopolitics in Latin America," *Political Geography Quarterly,* 5, 1 (January 1986), 43–55.

Morrison, Joel A., "Russia and Warm Water: A Fallacious Generalization and Its Consequences," *Proceedings, United States Naval Institute,* 78, 11 (1952), 1169–1179.

Most, Benjamin A. and Harvey Starr, "Diffusion, Reinforcement, Geopolitics, and the Spread of War," *American Political Science Review,* 74, 4 (December 1980), 932–946.

———, "Conceptualizing War: Consequences for Theory and Research," *Journal of Conflict Resolution,* 27 (1983), 137–159.

Murphey, Rhoads, "China and the Dominoes," *Asian Survey,* 6 (1966), 510–515.

N

Nathan, Robert S., "Geopolitics and Pacific Strategy," *Pacific Affairs,* 15 (1942), 154–163.

Nijim, Bashir K., "Israel and the Potential for Conflict," *Professional Geographer,* 21, 5 (September 1969), 319–323.

O

Odingo, R.S. "Geopolitical Problems of East Africa," *East Africa Journal,* 2, 9 (February 1966), 17–24.

O'Keefe, Phil and others, "A Sad Note for Grenada," *Political Geography Quarterly,* 3, 2 (April 1984), 152–159.

O'Loughlin, John and Herman van der Wusten, "Geography, War and Peace: Notes for a Contribution to a Revived Political Geography," *Progress in Human Geography,* 10, 4 (December 1986), 484–510.

Openshaw, S. and P. Steadman, "On the Geography of a Worst Case Nuclear Attack on the Population of Britain," *Political Geography Quarterly,* 1, 3 (July 1982), 263–278.

O'Sullivan, Patrick, "Antidomino," *Political Geography Quarterly,* 1, 1 (January 1982), 57–64.

———, "A Geographical Analysis of Guerrilla Warfare," *Political Geography Quarterly,* 2, 2 (April 1983), 139–150.

O'Tuathail, Geróid, "Political Geography of Contemporary Events VIII: The Language and

Nature of the 'New Geopolitics'—the Case of U.S.–Salvador Relations," *Political Geography Quarterly*, 5, 1 (January 1986), 73–85.

Owens, R.P., "The British Withdrawal from the Persian Gulf," *World Today*, 28 (February 1972), 75–81.

P

Peace and Disarmament; Academic Studies. Moscow: Scientific Research Council on Peace and Disarmament. Published annually.

Peace Research. Oakville, Ontario: Canadian Peace Research Institute. Published quarterly.

Peace Research Reviews. Dundas, Ontario: Peace Research Institute. Published quarterly.

Pedone, R.J., "The Geopolitics of Cambodia," *United Asia*, 15, 3 (1963), 251–255.

Pepper, David and Alan Jenkins, "A Call to Arms: Geography and Peace Studies," *Area*, 14 (1983), 202–208.

———, "Reversing the Nuclear Arms Race: Geopolitical Bases for Pessimism," *Professional Geographer*, 36, 4 (November 1984), 419–427.

Petrie, C., "The Strategic Concept of Modern Diplomacy," *Quarterly Review*, (1952), 289–301.

Polisk, A.N., "Geopolitics of the Middle East," *Middle Eastern Affairs*, 4, 8–9 (August–September 1953), 271–277.

Posony, Stefan T. and Leslie Rosenzweig, "The Geography of the Air," *Annals, American Academy of Political and Social Sciences*, 299 (May 1955), 1–11.

Pounds, Norman J.G., "The Political Geography of the Straits of Gibraltar," *Journal of Geography*, 51, 4 (1952), 165–170.

Powers, R.J., "Containment: From Greece to Vietnam—and Back?" *Western Political Quarterly*, 22, 4 (1969), 846–861.

R

Randolph, R. Sean, "Pacific Overtures," *Foreign Policy*, 57 (Winter 1984–85), 128–142.

Redler, R., "How the Geopolitician Looks at Our World," *Canadian Geographical Journal*, 29, (July 1944), 4–9.

Reinhardt, George C., "Air Power Needs Its Mahan," *Proceedings, United States Naval Institute, (April 1952)*.

Resources Policy. London: Butterworths. Published quarterly since 1974.

Roett, Riordan, "Brazil Ascendant: International Relations and Geopolitics in Late 20th Century," *Journal of International Affairs*, 29 (1975), 139–154.

Rose, A.J., "Strategic Geography and the Northern Approaches," *Australian Outlook*, 13, 4 (December 1959), 304–314.

Roucek, Joseph S., "The Geopolitics of Danubia," *American Journal of Economics and Sociology*, 5 (1946), 211–230.

———, "Geopolitics of Poland," *American Journal of Economics and Sociology*, 7, 4 (1948), 421–427.

———, "The Geopolitics of the Baltic States," *American Journal of Economics and Sociology*, 8, 2 (January 1949), 171–175.

———, "The Geopolitical Implications of the Macedonian Problem," *World Affairs Interpreter*, 21, 1 (1950), 95–107.

———, "Geopolitical Trends in Central-Eastern Europe," *Annals, American Academy of Political and Social Science*, 271 (1950), 11–19.

———, "Moscow's European Satellites," *Annals, American Academy of Political and Social Science*, 271 (September 1950), 1–253.

———, "The Geopolitics of the U.S.S.R.," *American Journal of Economics and Sociology*, 10, 1 (October 1950), 17–26; 10, 2 (January 1951), 153–159.

———, "The Geopolitics of Antarctica and the Falkland Islands," *World Affairs Interpreter*, 22, 1 (1951), 44–56.

———, "The Geopolitics of the Aleutians," *Journal of Geography*, 40, 1 (January 1951), 24–29.

———, "The Geopolitics of Greenland," *Journal of Geography*, 50, 6 (September 1951), 239–246.

———, "Some Geographic Factors in the Strategy of the Kuriles," *Journal of Geography*, 51 (1952), 297–301.

———, "The Geopolitics of the Adriatic," *American Journal of Economics and Sociology*, 11, 2 (January 1952), 171–178.

———, "The Geopolitics of Tibet," *Social Education*, 15, 4 (May 1952), 217–218, 299.

———, "The Geopolitics of Albania," *World Affairs Interpreter*, 23, 1 (1952), 83–96.

———, "Geopolitics and Air Power," *Air University Quarterly*, 6 (1952), 52–73.

———, "Notes on the Geopolitics of Singapore," *Journal of Geography*, 7 (February 1953), 78–81.

———, "The Geopolitics of the Mediterranean," *American Journal of Economics and*

Sociology, 12, 4 (July 1953), 346–354; 13, 1 (October 1953), 71–86.

——, "The Geopolitics of Pakistan," *Social Studies,* 44, 7 (November 1953), 254–258.

——, "The Geopolitics of the Philippine Islands," *World Affairs Interpreter,* 25, 1 (1954), 79–90.

——, "The Geopolitics of Thailand," *Social Studies,* 45, 2 (February 1954), 57–63.

——, "The Geopolitics of Spain," *Social Studies,* 46, 3 (March 1955), 89–93.

——, "The Geopolitics of the United States," *American Journal of Economics and Sociology,* 14, 2 (January 1955), 185–192; 14, 3 (April 1955), 287–303.

——, "The Geopolitics of Yugoslavia," *Social Studies,* 47, 1 (January 1956), 26–29.

——, "The Geopolitics of Afghanistan," *Social Studies,* 48, 4 (1957), 127–129.

——, "The Development of Political Geography and Geopolitics in the United States," *Australian Journal of Politics and History,* 3, 2 (1958), 204–217.

——, "Tibet and Its Geopolitical Aspects," *United Asia,* 12, 4 (1960), 377–380.

——, "Cuba in Its Geopolitical Setting," *Contemporary Review,* 198 (1960), 663–667.

——, "Vietnam in Geopolitics," *Contemporary Review,* 200 (1961), 420–423.

——, "The Geopolitics of the Congo," *United Asia,* 14, 1 (January 1962), 81–85.

——, "Thailand's Geopolitics," *United Asia,* 14 (1962), 381–387.

——, "Albania in Geopolitics," *Contemporary Review,* 201 (1962), 17–20.

——, "Yemen in Geopolitics," *Contemporary Review,* 202 (1962), 310–317.

——, "Nigerian Geo-Politics," *United Asia,* 14, 3 (March 1962), 182–185.

——, "The Geopolitics of Mongolia," *United Asia,* 15, 3 (1963), 240–244.

——, "The Geopolitics of Asia," *United Asia,* 15, 4 (1963), 290–296.

——, "The Geopolitics of Japan," *United Asia,* 15, 4 (1963), 297–301.

——, "The Geopolitics of Korea," *United Asia,* 15, 4 (1963), 304.

——, "The Geopolitics of Formosa (Taiwan)," *United Asia,* 15, 4 (1963), 305–308.

——, "Venezuela in Geopolitics," *Contemporary Review,* 203, Pt. I (February 1963), 84–87; 203, Pt. II, (March 1963), 126–132.

——, "Haiti in Geopolitics," *Contemporary Review,* 204 (July 1963), 21–29.

——, "Iraq in Geopolitics," *Contemporary Review,* 204 (September 1963), 116–123.

——, "Peru in Geopolitics," *Contemporary Review,* 204 (1963), 310–315; 205 (1964), 24–31.

——, "The Geopolitics of Portuguese Colonialism," *Contemporary Review,* 205 (June 1964), 284–296.

——, "Changing Geopolitical Patterns Along the Persian Gulf," *Il Politico* (Univ. of Pavia), 29, 2 (1964), 440–474.

——, "Ecuador in Geopolitics," *Contemporary Review,* 205 (February 1964), 74–82.

——, "The Geopolitics of South-East Asia," *France–Asia,* 19, 182 (1964), 1077–1094.

——, "Cambodia in Global Geopolitics," *International Review of History and Political Science,* 1, 1 (1964), 13–32.

——, "Portugal in Geopolitics," *Contemporary Review,* 205 (1964), 476–488.

——, "The Pacific in Geopolitics," *Contemporary Review,* 206, 1189 (February 1965), 63–76.

——, "Chile in Geopolitics, *Contemporary Review,* 206 (1965), 127–141.

——, "The Dominican Republic in Geopolitics," *Contemporary Review,* 206 (1965), 289–294; 207 (1965), 12–21.

——, "Geopolitical Trends Behind the Iron Curtain," *Central Europe Journal,* 14, 9 (1966), 267–279.

——, "Geopolitics of Zambia," *New Africa,* 9 (July–August 1967), 9–12.

——, "Africa in Geopolitics," *International Review of History and Political Science,* 4, 2 (1967), 76–96.

——, "The Geopolitics of Hong Kong," *Revue du Sud-Est Asiatique* (Brussels), 2 (1967), 155–189.

——, "Burma in Geopolitics," *Revue du Sud-Est Asiatique et l'Extrême Orient,* 1 (1968), 47–82.

——, "South Africa in Geopolitics," *Africa Quarterly,* 8, 2 (1968), 166–176.

——, "Okinawa in Geopolitics," *Study of Current English,* 24, 11 (1969), 8–18, 24; 12 (1969), 15–21.

——, "Cambodia in Geopolitics," *Revue du Sud-Est Asiatique,* (1970), 197–223.

——, "South-East Asia in Global Geopolitics," *Il Politico* (Univ. of Pavia), 35, 3 (1970), 472–495.

——, "Modern Implications of the Geopolitical Heartland Theory for Eurasia and the Pacific Ocean," *International Behavioral Scientist*, 21 (1970), 19–46.

———, "The Pacific in World Geopolitics," *International Review of History and Political Science*, 8, 3 (1971), 1–30.

———, "The Indian Ocean in Global Geopolitics," *International Review of History and Political Science*, 8, 4 (1971), 57–77.

———, "The Geopolitics of the Adriatic Sea," *Il Politico* (Univ. of Pavia), 36, 3 (1971), 569–594.

———, "Geopolitics of Indonesia," *International Behavioral Scientist*, 2 (1972), 49–72.

RUSI and Brassey's Defence Yearbook. London: Royal United Services Institute. Published annually.

Russett, Bruce M., "The Asia Rimland as a Region for Containing China," *Public Policy*, 16 (1967), 226–249.

S

Saleem Khan, M.A., "Oil Politics in the Persian Gulf Region," *India Quarterly*, 30 (January–March 1974), 25–41.

Schwartz, Benjamin I., "The Maoist Image of World Order," *International Affairs*, 21, 1 (1967), 92–102.

Schweinfurth, U., "The Himalayas—Divider of Landscape, Region of Withdrawal and Zone of Political Tension," *Geographische Zeitschrift*, 53 (1965), 241–260.

Semmel, Bernard, "Sir Halford Mackinder: Theorist of Imperialism," *Canadian Journal of Economic and Political Sciences*, 24, 4 (November 1958), 554–561.

Senghaas, D., "The Cycles of War and Peace," *Bulletin of Peace Proposals*, 14 (1983), 119–134.

Shafer, Boyd C., "Webs of Common Interests: Nationalism, Internationalism and Peace," *Historian*, 26, 3 (1974), 3–32.

Singer, J.D., "Accounting for International War: The State of the Discipline," *Journal of Peace Research*, 18 (1981), 1–18.

Slessor, Sir John, "Air Power and World Strategy," *Foreign Affairs*, 33, 1 (October 1954), 43–53.

Smith, C.G., "Arab Nationalism; a Study in Political Geography," *Geography*, 43 (1958), 229–242.

———, "Israel After the June War," *Geography*, 53 (1968), 315–319.

Smith, G.H., "The Emergence of the Middle East," *Journal of Contemporary History*, 3 (1968), 3–17.

Smolansky, O.M., "Moscow and the Persian Gulf: An Analysis of Soviet Ambitions and Potential," *Orbis*, 14 (Spring 1970), 92–108.

Soffner, Heinz, "Geopolitics in the Far East," *Amerasia*, (1943), 513–521.

Soustelle, Jacques, "France and Europe: A Gaullist View," *Foreign Affairs*, 30 (1952), 545–553.

Sprout, Harold, "Geopolitical Hypotheses in Technological Perspective," *Journal of Conflict Resolution*, 15, 2 (January 1963), 187–212.

Spykman, Nicholas John, "Geography and Foreign Policy," *American Political Science Review*, 32, 1 (February 1938), 28–50; 32, 2 (April 1938), 213–236.

Spykman, Nicholas John and Abbie A. Rollins, "Geographic Objectives in Foreign Policy," *American Political Science Review*, 33, 3 (June 1939), 391–410; 33, 4 (August 1944), 591–614.

———, "Frontiers, Security, and International Relations," *Geographical Review*, 32, 3 (July 1942), 436–447.

Staley, Eugene, "The Myth of Continents," *Foreign Affairs*, 19, 3 (April 1941), 481–494.

Standish, J.F., "British Maritime Policy in the Persian Gulf," *Middle Eastern Studies*, 3 (July 1967), 324–354.

———, "Pursuit of Peace in the Persian Gulf," *World Affairs*, 132 (December 1969), 235–244.

Starr, H. and Benjamin A. Most, "Contagion and Border Effects on Contemporary African Conflict," *Comparative Political Studies*, 16 (1983), 92–117.

Stefansson, Vihjalmur, "The Arctic," *Air Affairs*, 3 (1950), 391–402.

Stolberg, Irving, "Geography and Peace Research," *Professional Geographer*, 17, 4 (July 1965), 9–12.

Stone, Adolf, "Geopolitics as Haushofer Taught It," *Journal of Geography*, 52, 4 (April 1953), 167–171.

Strang, Sir W., "Germany Between East and West," *Foreign Affairs*, 33, 3 (1955), 387–401.

Strategic Analysis. New Delhi: Institute for Defence Studies and Analysis. Published monthly.

Strategic Studies. Islamabad, Pak.: Institute of Strategic Studies.

Sullivan, Robert R., "The Architecture of Western Security in the Persian Gulf," *Orbis*, 14 (Spring 1970), 71–91.

Swann, Robert, "Laos, Pawn in the Cold War," *Geographical Magazine*, 32, 8 (1960), 365–375.

T

Takeuchi, K., "Geopolitics and Geography in Japan Reexamined," *Hitotubashi Journal of Social Studies*, 12 (1980), 14–24.

Tambs, Lewis A., "Latin American Geopolitics. A Basic Bibliography," *Revista Geográfica*, 73 (1970), 71–106.

Taylor, E.G.R., "Geography in War and Peace," *Geographical Review*, 38, 1 (January 1948), 132–141.

Teggart, Frederick J., "Geography as an Aid to Statecraft; an Appreciation of Mackinder's 'Democractic Ideals and Reality,' " *Geographical Review*, 8, 4–5 (October–November 1919), 227–242.

Theberg, James, "Brazil's Future Position in the Hemisphere and in the World," *World Affairs*, 132, 1 (1969), 39–47.

Thermaenius, Edward, "Geopolitics and Political Geography," *Baltic and Sandinavian Countries*, 2 (1938), 165–173.

Thompson, William R. "Uneven Economic Growth, Systemic Challenges, and Global Wars," *International Studies Quarterly*, 27, 3 (September 1983), 341–355.

———, "World Wars, Global Wars and the Cool Hand Luke Syndrome: A Reply to Chase–Dunn and Sokolovsky," *International Studies Quarterly*, 27, 3 (September 1983), 369–374.

Thompson, William R. and G. Zuk, "War, Inflation and Kondratieff's Long Waves," *Journal of Conflict Resolution*, 26 (1982), 621–644.

Tomasic, D., "The Structure of Soviet Power and Expansion," *Annals, American Academy of Political and Social Science*, 271 (1950), 32–42.

Troll, Carl, "Geographical Science in Germany During the Period 1933–1945," *Annals, AAG*, 39, 2 (June 1949), 99–137.

Turner, R., "Technology and Geo-Politics," *Military Affairs*, 7, 1 (1943), 5–15.

U

Unstead, J.F., "H.J. Mackinder and the New Geography," *Geographical Journal*, 113, 1 (January–June 1949), 47–57.

V

Vagts, Alfred, "Geography in War and Geopolitics," *Military Affairs*, 7 (Spring 1943), 79–88.

Vagts, Detlev F., "International Law Under Time Pressure: Grading the Grenada Take-Home Examination," *American Journal of International Law*, 78, 1 (January 1984), 169–172.

van der Wusten, Herman and John O'Loughlin, "Claiming New Territory for a Stable Peace: How Geography Can Contribute," *Professional Geographer*, 38, 1 (February 1986), 18–28.

Van Hollen, Christopher, "The Tilt Policy Revisted: Nixon–Kissinger Geopolitics and South Asia," *Asian Survey*, 20, 4 (April 1980), 339–361.

Van Valkenburg, Samuel, "Pan Europa," *Journal of Geography*, 29, 4 (1930), 132–140.

Väyrynen, R., "Regional Conflict Formations: An Intractable Problem of International Relations," *Journal of Peace Research*, 21 (1984), 337–359.

Vitkovskiy, V., "Political Geography and Geopolitics: A Recurrence of American Geopolitics," *Soviet Geography; Review and Translation*, 22 (1981), 586–593.

W

Waite, G.F. "The Geopolitics of China," *United Asia*, 15, 3 (1963), 235–239.

Walford, R., "Geography and the Future," *Geography*, 69 (1984), 193–208.

Wang, C.C., "The Pan-Asiatic Doctrine of Japan," *Foreign Affairs*, 13, 1 (1934), 59–67.

Watt, D.C., "The Persian Gulf—Cradle of Conflict?" *Problems of Communism*, 21 (May–June 1972), 32–40.

Weigert, Hans Werner, "German Geopolitics; a Workshop for Army Rule," *Harpers*, 183 (November 1941), 586–597.

———, "Haushofer and the Pacific," *Foreign Affairs*, 20, 4 (July 1942), 732–742.

———, "Iceland, Greenland and the United States," *Foreign Affairs*, 23, 1 (October 1944), 112–122.

———, "The Northward Course," *Virginia Quarterly Review*, 20, 1 (1944), 114–122.

———, "Mackinder's Heartland," *American Scholar*, 15, 1 (Winter 1945–46), 43–54.

———, "U.S. Strategic Bases and Collective Security," *Foreign Affairs*, 25, 2 (January 1947), 250–262.

Whebell, C.F.J., "Mackinder's Heartland Theory in Practice Today," *Geographical Magazine*, 42 (1970), 630–636.

Whittington, G., "Some Effects on Zambia of Rhodesian Independence," *Tijdschrift voor Economische en Sociale Geografie*, 58, 2 (April 1967), 103–106.

World Armaments and Disarmament; SIPRI

Yearbook. Stockholm: Stockholm International Peace Research Institute. Published annually.

Wright, Denis, "The Changed Balance of Power in the Persian Gulf," *Asian Affairs*, 60 (October 1973), 255–262.

Y

Yariv, A., "Strategic Depth," *Jerusalem Quarterly*, 17 (1980), 3–12.

Yorke, Valeria, "Security in the Gulf: A Strategy of Pre-emption," *World Today*, 36, 7 (July 1980), 239–250.

Z

Zabur, J., "Geopolitics: The Struggle for Space and Power," *Revista Brasileira de Geografia*, 4, 4 (December 1942), 849–852.

Part Five

IMPERIALISM, COLONIALISM, AND DECOLONIZATION

Chapter 20

COLONIAL EMPIRES

The study of imperialism or colonialism is, in part, the study of power and geopolitics. Expansionism has been a characteristic of many States, and it has always been made possible by superior power and organization. Such expansionist tendencies were exhibited by States long before the most recent wave of colonialism: the Inca Empire grew by colonial acquisitions and the subjugation of outlying areas and peoples; so did the Roman Empire. Many European States in recent centuries sought to control parts of the non-European world because their populations and technology permitted them to do so, and the scale of European colonialism was unprecedented. We are here concerned with colonialism as a phenomenon related to the emergence of the modern nation-state, so that our primary interest is in the European drive for colonial possessions. However, it should be remembered that territorial acquisitiveness is not unique to European States, nor to the States that have existed only during the past two or three centuries.

MODERN IMPERIALISM AND COLONIALISM

We have seen that some terms in political geography are not well defined, and few present as many difficulties as *colonialism*. Cohen has defined colonialism as separate from imperialism: "Colonialism, as a process, involves the settlement from a mother country, generally into empty lands and bringing into these lands the previous culture and organization of the parent society. Imperialism, as distinct from colonialism, refers to rule over indigenous people, transforming their ideas, institutions, and goods."* Frankel has attempted to bring greater clarity to the term colonization by recognizing *primary* colonization as the occupation of the lands and the domination of indigenous peoples, while *secondary* colonization is the acquisition by a colonial power of virtually empty territory.† For our purposes, we may divide the modern colonial empire into three categories. First is the empire that resulted from the overland expansion of a State, conquering or otherwise acquiring territories occupied by peoples of different cultures and in some cases different races. These would include the Russian, Austro-Hungarian, Prussian, Ottoman, and Chinese empires and perhaps others. Second is the empire built up primarily of overseas territories by the States of Europe during and following the Age of Exploration beginning in the late fifteenth century and extending into the nineteenth century. During this period the Netherlands, Portugal, Britain, France, and Spain not only built up enormous empires, but also lost significant portions of them.

*S.B. Cohen, *Geography and Politics in a World Divided*, New York, Random House, 1963, p. 204.
†S.H. Frankel, *The Concept of Colonization*, Oxford University Press, Clarendon, 1949.

After the Napoleonic Wars, a third phase of imperialism began, one based largely, as we have seen, on nationalism, geopolitics, and to a lesser extent, economic and missionary motives. It resulted in the creation of new or expanded continental or overseas empires. Some of the original colonizing powers dropped out of the new competition, simply holding on to what they had; others expanded their holdings, and other countries (mostly newly unified as nation-states or just emerging from colonial status themselves) joined in the competition. This was the heyday of "the white man's burden," "manifest destiny," "the civilizing mission," and similar euphemisms for the political, economic, and cultural domination of weaker peoples. We are mainly concerned with the empires founded, flowering, or terminating between roughly 1815 and 1919, and on the methods employed in ruling these empires. Now that nearly all colonies have received or are scheduled for independence, these policies are of much more than historical interest, for they have had a powerful influence on the process of decolonization and on the behavior of the newly independent States.

MOTIVES AND REWARDS

The geopolitical aspects of colonialism are self-evident. Taking the term at its narrowest meaning (i.e., referring to the German school of *Geopolitik*), the relationships between German fears of "encirclement" and British imperialist activity are clear. Haushofer had personally visited the Eurasian perimeter, ringed by British bases; his doctrines constituted a direct reaction to that situation. Comparisons between land and sea powers, such as those of Mahan and Mackinder, served to emphasize the value of colonial possessions. Britain possessed a worldwide network of bases that more often than not were located within larger colonial entities, and Mahan advocated the acquisition of similar facilities by the United States.

While the British achieved success on all continents, to establish the greatest colonial empire ever to exist, the French attained a vast contiguous colonial territory in Africa and another in Southeast Asia. German colonial activity was in large part an attempt to thwart the designs of Britain and France. In West Africa the Germans succeeded in establishing themselves in Togo, thereby canceling any British plans to link their Gold Coast and Nigerian dependencies. In Kamerun, the Germans pushed far northward and came close to fragmenting the extensive French West African realm. In East Africa, Germany obtained the mainland part of what is today the United Republic of Tanzania, thereby frustrating British plans for a Cape-to-Cairo railway. In South-West Africa, Germany made use of a lapse in British attentiveness to gain a foothold adjacent to the richest part of the continent and came very close to uniting South-West Africa with Tanganyika; a relic of that effort is the Caprivi Strip.

The Treaty of Tordesillas (1494) may have greatly influenced the course of colonial history, for it appeared to divide between Spain and Portugal not only the Americas but all newly discovered and undiscovered lands. It has been postulated that this may be why Portugal, which for decades had almost unobstructed opportunity to extend its sphere of influence in Africa, actually established only a few centers needed to facilitate and protect the Cape route to the Indies. Subsequently, when Britain, France, Germany, and Belgium entered the scene, Portugal could only claim historic rights over areas that by the realities of power fell to other States. Even in South America the treaty did not ward off other invaders; the British, French, and Dutch established colonial domains in the area where, according to the treaty, Portuguese and Spanish colonial realms should have met.

While European expansion overseas was proceeding, Russia was expanding eastward and the United States was expanding westward. Neither stopped at the Pacific Ocean but continued onward, with the

Russians establishing colonies as far from home as San Francisco and the Americans as far away as the Philippines, for essentially the same motives and by use of the same methods as the Europeans.

Coupled with the geostrategic motives for colonization were a number of other inducements. "God, glory, and gold" sustained the Spanish in their penetration of the Americas, implying that they were there not only to gain wealth, but also to make a contribution to the spiritual well-being of their wards. This missionary zeal was not unique to the Spanish in America. The calling to bring the benefits of what Europeans saw as their more advanced civilization and their "true" religion to the backward colonial world constituted another driving force. Indeed, the first penetration was often made by missionaries, who established stations that sometimes became the centers from which colonial control was extended. One example is South-West Africa, which made its first permanent contacts with Europe through several German missionaries. The small German missionary population found itself in trouble during an ensuing intra-African war, appealed for help—and brought German occupation to this area.

Missionary zeal was not confined to missionaries. A relatively small number of European explorers and adventurers, wishing to make a contribution to their fatherlands, had an enormous impact on the process of colonization. Cecil Rhodes was possessed of this spirit, and he almost singlehandedly penetrated the region of the Zambezi River on behalf of the British Crown. He strongly influenced the course of events in South Africa, and the map of Africa eventually carried his name in Northern and Southern Rhodesia. Luderitz, a wealthy German merchant, similarly helped initiate and sustain the German drive in South-West Africa, while Karl Peters did the same in East Africa. De Brazza represented France in the area of the lower Congo River. In nineteenth century Europe, such activities carried the highest honor and approval.

At times the governments of the States of Europe, as a result of the colonial scramble in which they were involved, found themselves confronted with requests for "protection" by peoples facing the threat of penetration by colonists. Such protection, if granted by the government, would prevent the colonists from conquering the indigenous people by force and taking their land. "Protectorates" were occasionally granted, especially if the government of one State received such a request from a people facing colonization by colonists from another State. We might consider this type of colonization as resulting from a particular kind of moral obligation, as distinct from that exhibited by the individual missionaries and explorers.

In many parts of the colonized world, the major motive that drove the colonial power to control was the economic one. For three centuries the Netherlands East Indies produced enormous revenues for the Dutch government from the sale of coffee, rubber, petroleum, and other commodities.

Some other colonies also sustained the home economy for many decades. King Leopold, in his acquisition of the Congo, foresaw great profits to be derived from ivory and rubber. Portugal introduced a system of compulsory cropping in its African domain to support the metropolitan economy. But very often, the vast wealth of the colonial world turned out to be more imagined than real, and some colonies became liabilities rather than assets. Until the ascent to power of the Salazar regime, Portugal was barely able to administer Angola and Mozambique; Salazar's acceptance of the leadership of Portugal was made under condition that he should be empowered to initiate economic reform programs in the colonies aimed at their contributing to Portuguese finances. Mauritania, Chad, Somalia, Surinam, New Guinea, Laos—all produced net economic losses to the colonial power. Then, some colonized territories produced richly for a few decades, but were ultimately exhausted. The quest for gold in the Americas initially met with great success, but the sources were rapidly expended.

VARIETIES OF IMPERIALISM

We have spoken so far of a particular kind of imperialism, one that results in a radical change in the political and military *status quo* by the extension of a State's sovereignty over territory either empty or occupied by peoples of different cultures. This military imperialism may be practiced in adjacent territory, as in the cases of Russia's eastward expansion and the westward expansion of the United States, or it may be practiced in noncontiguous lands, as in the British and French empires. It may or may not involve the settlement of large numbers of colonists from the imperial country. If it does, it can certainly be called colonialism; if not, it may still be called colonialism, depending at least in part on the degree of self-government permitted the acquired territories.

There are, as we have seen, many motives for imperialism/colonialism. Lenin wrote a book titled *Imperialism, the Highest Stage of Capitalism*. However, Lenin seems to have been wrong, for a careful study of modern imperialism shows that it was seldom practiced for exclusively economic reasons (although one of its products was "the colonial economy," which we discuss later). It would be much more accurate to refer to "imperialism, the highest stage of nationalism." Virulent nationalism has long been a driving force behind imperialism. The missionary drive has also been instrumental in extending political power over weaker peoples, to offer them—or thrust on them—the conqueror's culture, or selected parts of it. Nevertheless, the balance sheet of nineteenth-century-style imperialism/colonialism has yet to be computed. But there are other kinds of imperialism as well.

One variety of imperialism that does not depend on military conquest is *economic imperialism*. This is a more subtle and generally less effective process. It involves a number of techniques, all designed to tie one country to another so tightly through economic means that one becomes in practice an economic colony of the other without giving up its formal sovereignty and its trappings. The United States, for example, has long practiced "dollar diplomacy" in much of Latin America; the Central American countries are virtually economic appendages of the United States and others are tied nearly as strongly. The British have been the masters in this field, making London the world center for banking, insurance, shipping, and other "invisible exports" that we discuss further in Chapter Twenty-four. British investments worldwide but especially in the Middle East, Argentina, and Portugal gave the United Kingdom enormous influence even where her military power did not reach. The politics of oil, in the Middle East and elsewhere, was until 1974 an aspect of economic imperialism.

Another variety of imperialism is *cultural imperialism*. It is still more subtle, sometimes less effective, and certainly less geographic than the other forms, so we will not dwell on it but only mention a few examples. Cultural imperialism aims not at controlling territory or economies but minds. Typically, it does not conquer on its own but prepares a target for later penetration by economic, political, and military weapons. Nazi "fifth columns" in Europe and South America between the wars practiced it and achieved notable successes. The Third (Communist) International was even more widely successful with its brand of ideological imperialism, and its work is being carried on still by other groups. Before the Bolshevik Revolution, the Russian Tsars exploited the Orthodox Church as an instrument of their foreign policy and gained preponderance in the Balkans in this way. France deliberately used the more attractive elements of French culture as a device for maintaining control in her colonies but also conducted her cultural imperialism in Eastern Europe, the Eastern Mediterranean, and elsewhere.

The result of all this imperialist activity

was a kind of hierarchy of colonies: political-military, economic, and cultural, but with much overlapping. British money and French culture competed in the Middle East, for example, and in some cases the local peoples were quite willing to accept both. The Ottoman Empire disintegrated in part because of the infiltration of European political ideas, notably nationalism and self-determination. American economic power penetrated the British sphere of the Caribbean so thoroughly that Britain began losing her colonies there economically before she began giving them political independence. Soviet and Japanese ideological warfare were instrumental in stimulating independence movements in many European colonies in Asia. The picture of modern imperialism, then, is far more intricate than many ideologists would have us believe. We now will observe how varied was the pattern even within the formal colonial empires.

COLONIAL POLICIES

With few exceptions (notably the United States, Germany, and Japan), the empires were created over several centuries through different circumstances and with no prior plan for their administration. Thus there was generally some adaptation to local conditions, experimentation, and changes that reflected political, social, and economic changes at home. During the nineteenth century, there was great and continual rivalry among the major European States (and later others as well) for bases, trading posts, colonies, and spheres of interest around the world. As a result of these rivalries and wars, some territories changed hands many times, boundaries and alliances (including alliances with indigenous rulers) shifted kaleidoscopically. With all these changes came varying methods of rule, administrators with different personalities, and varying degrees of profitability of the colonies. Economic conditions changed greatly during the colonial period. The Industrial Revolution,

with its growing capitalist class, its ever-increasing demands for commodities (foods, fuels, and raw materials); its need for external markets to achieve economies of scale; its generation of surplus population in Europe; its provision of superior transport, weapons, and communications to the imperialist governments; contributed mightily to the middle and later stages of nineteenth-century colonialism.

The settlement of Europeans from the metropole in the colonies varied considerably; generally speaking, the colonies with climates similar to those of the home countries and the colonies with the sparsest aboriginal population tended to attract the most European settlers. This pattern was modified by several factors, including imperial policies, the duration of European rule, and the availability of more attractive settlement areas. After their rule was firmly established, the Europeans tended to suppress slavery and local wars and to introduce modern health and welfare systems. This resulted in a "population explosion" that wrought all kinds of changes in the indigenous societies. Most colonizers introduced into their colonies notions of nationalism and self-determination, European style, though seldom with any premeditation or deliberation. European colonial rulers, especially Spain and Portugal, took various measures to encourage or require colonies to trade exclusively with the "mother country" and to suppress local capital formation and investment in industrialization. The responses of the local peoples to all these variations and changes were nearly as varied and changeable. Variation and change were the two constants of colonial rule.

Portugal's vast worldwide empire was greatly reduced in 1822 when a liberal king, Dom João VI, returned home from a long exile in Brazil and left his even more liberal and popular son, Dom Pedro I, to rule there with permission to separate from Portugal if the people wished. They did, and Dom Pedro became emperor of an independent Brazil. That was only a momentary lapse, however, in Portugal's traditional policy of holding on to her colonies at virtually any cost. For nearly a century and a half thereafter, through war and depression and revolution and burgeoning nationalism, Portugal administered her empire with no concessions whatever. Her rule was autocratic, based on mercantilist economic policies, the political policy of transforming the colonies into "overseas provinces" of Portugal, and the social policy of permitting (though not encouraging) a small native elite to become "*assimilados*," Portuguese citizens represented in the Portuguese government. She was steadfast, even obdurate, in resisting pressures for change, supremely confident that her method of carrying out her "civilizing mission" was approved (if not ordained) by heaven itself.

Not even India's forcible takeover of her colonies there (Goa, Damão, Diu, Dadra, and Nagar Haveli) in 1961 shook her resolve to maintain her empire forever.* Beginning in 1961, she was continually fighting nationalist rebels in Angola, Mozambique, and Portuguese Guinea. Finally, a military coup in Portugal carried out in 1974 by professionals dissatisfied with these colonial wars and convinced of their futility set off a chain of events that led to the evaporation of her four hundred-year-old African empire in the incredibly brief period of 14 months, from September 1974 to November 1975.

Portuguese colonial policies were most clearly displayed in her vast African domain. Politically, the "provinces" were administered through the governor general in the capital city. For purposes of administrative control, each territory was subdivided into subprovinces, where decisions made in Lisbon and transmitted through the governor general were implemented. Economically, however, there was a system of producing regions that affected the population much more directly than the internal political subdivi-

sions. These economic units, in many ways, were the *de facto* political subdivisions of the provinces, where the European's control over the daily lives of the indigenous population was most strongly felt. This system of the *circumscription* was modified in the face of international objections, but its impact continues to be felt.

Belgium saw her colonial task as a paternal one. The Belgians never viewed the Congo as an "overseas Belgium," nor did they intend to make Belgians of the Congolese population. Apart from the mandate and trusteeship over Ruanda–Urundi, the Belgians possessed only one dependency: the Congo. Once the personal possession of King Leopold II, the Congo administration was taken over by the Belgian State as recently as 1907, after the reputation of the country had suffered through the exposure of the "Congo atrocities" that took place during the king's administration.

The shape characteristics and the multicore aspects of Zaïre were discussed earlier. These realities faced the Belgian colonist as they do the Zaïre government today. Kinshasa (formerly Leopoldville) lies on the periphery of the vast Congo Basin, and its communications with outlying parts of the country have never been good. The Belgians thus established six major provinces, each with a capital and each extending over a section of the highland rim of the Congo as well as over a part of the interior basin. The capital cities became the headquarters of lieutenant governors, whose tasks included the implementation of Brussels policy as conveyed by the governor general from his Leopoldville office. Each province was, in turn, divided into a number of districts, where lesser officials were in charge.

A glance at even the most general geographic patterns in the Congo reveals the shortcomings, from the Zaïrian viewpoint, of this arrangement. Each province lay astride a major transport route to Leopoldville, but provincial boundaries cut through tribal and linguistic units. Some

---

*Indeed, she did not formally recognize Indian incorporation of her territories until late 1974 and did not formally end Portugal's sovereignty there until 1 January 1975!

Macau, a fragment of empire. The photograph shows auspicious phrases and blessings being put up on a wall, perhaps prophetically, on Street of Hope. Macau consists of a peninsula, three islets, and two larger islands totaling 16 sq km (6 sq mi) at the mouth of the Si Jiang. It was established in 1557—the first European enclave in China—and became renowned as a center of trade, opium traffic, fishing, manufacturing, and gambling. Portugal was finally granted perpetual occupation and government of Macau by China in 1887. Macau was granted broad autonomy in 1976, and is scheduled to be returned to China in 1999, shortly after nearby Hong Kong is returned to China in 1997. The population is approximately 400,000. (Macau Tourist Information Bureau)

(such as the Kivu–Katanga boundary) were geometric; the system obviously was created for administrative convenience rather than the integration of the whole country. This was reflected by the events immediately after independence, when the cities of Stanleyville, Luluabourg, and Elizabethville—all provincial capitals—became centers of control for secessionist leaders who took over from the Europeans and failed to heed Leopoldville's dictates. Thus Belgium, unwittingly, prepared the country for near-feudal conditions by imposing the administrative framework of the six provinces.

Decisions regarding the Congo were made only in Brussels. In the Congo, neither the European representatives of the Belgian government, the settlers, nor any other European group possessed any political rights or any voice in the fate of the territory. Neither, until a few months before independence, did the indigenous inhabitants. Furthermore, the paternalist approach to colonial administration had its effects in economic, educational, and social spheres: much progress was made at elementary levels but very little at higher ones. While the Portuguese opened a narrow avenue for Africans to attain *assimilado* status, theoretically implying full Portuguese citizenship, the Belgian *evolués* could not aspire to Belgian nationality and equality. Even the "evolved" indigenous population was viewed as in need of paternal restrictions.

France once adhered to the concept of a *France d'Outre Mer,* an Overseas France, but that idea was largely abandoned with the demise of the Fourth Republic. If the best one-word definition for Belgian colonial policy is paternalism, the most appropriate term for French colonialism is assimilation. Like the Portuguese, but with more to offer and more success, the French have brought French culture to their overseas domain, where it remains very much in evidence, despite the fact that French control has ceased. Unlike the Belgians, the French desired the quick development of an educated, acculturated elite, which would have French interests at heart and French culture to boast. In Southeast Asia as well as Africa, many modern leaders of the independent "French" emergent States are products of this elite and the system that helped erect it.

French colonial subjects were to be assimilated in the greater French Empire, and they could obtain a voice in the politics of the French realm through representation in Paris. A number of overseas territories of France had such representation, and French dependencies were placed in a hierarchy, with their position in terms of closeness to France determined by their historical associations with the motherland.

The politicoterritorial aspects of French colonialism are perhaps most clearly revealed in Africa, where the bulk of French territory was located. Algeria always occupied a special position because of its large settled French population and its proximity and effective communications with European France. The remainder of French Africa was divided into two major groupings, French West Africa and French Equatorial Africa. Within them a number of separate administrative entities were established, often defined in West Africa by geometric boundaries, each focusing on a single core area and capital. French Equatorial Africa consisted of Moyen Congo, Gabon, Ubangi–Shari, and Chad, with their respective capitals of Brazzaville, Libreville, Bangui, and Fort Lamy. The administrative headquarters of French Equatorial Africa, however, were in Brazzaville, so that this city attained a greater importance than the others and grew to greater size. The four States of former French Equatorial Africa (now Congo, Gabon, the Central African Republic, and Chad) are in the same size category as the provinces of the former Belgian Congo, but the French always saw them as individual and separate entities, despite their association for administrative convenience in a "federation." Major colonial policies were transmitted through Brazzaville, but decisions regarding each individual territory were implemented directly through the local capitals.

French West Africa also had its administrative headquarters, Dakar, today the capital of the Republic of Senegal. Again, each individual territory had its own focus, and some core areas are of considerable significance today. Southeastern Côte d'Ivoire, with its impressive capital, Abidjan, is a core area of the first order in West Africa. Dakar remains French-influenced Africa's most important port, but Conakry and Porto Novo-Cotonou serve Guinea and Dahomey, respectively.

Britain possessed an empire that was larger than any other colonial realm and was confronted by a wider variety of indigenous cultures and peoples than any other colonial power. British colonial policies were often adapted to the requirements of individual dependencies, but British policy, unlike that of Belgium and France, had a basic premise: indirect rule. British colonial policy was always the least centralized among the European colonial powers.

The principle of indirect rule was intended to prevent the destruction of indigenous culture and organization. In such fields as tribal authority, law, and education, the British often recognized local customs and permitted their perpetuation, initially outlawing only those practices that constituted, in British eyes, serious transgressions. Then the people were slowly introduced to the changes British rule inevitably brought. Prior to World War II, the emphasis was especially on slowness; haste in bringing about reforms was seen as an evil and danger. Despite this attitude, however, the British from very early on professed that their ultimate goal was the independence and self-determination of the peoples under their colonial flag. After World War II, the pace of change became faster than anyone had expected. The British response was generally to help dependencies destined for independence develop parliamentary systems of government. Less attention was paid to the principle of indirect rule and more to the development of elected local government.

The variety of British approaches to colonial rule in their numerous colonies also was related to the size and character of the European "settler" population in these dependencies. What was said in the previous paragraphs concerning indirect rule was especially true in territories with a small immigrant European population, little land alienation, and a history of peaceful penetration rather than violence and conquest. In territories where the settler population was large and powerful, the desires of London were often overridden by those of the local whites. Examples include the former colonies of Kenya and Southern Rhodesia. But elsewhere, as in Sudan, Ghana, Sierra Leone, and Uganda, the white colonists presented far fewer obstacles in the path toward independence.

Apart from the white-settler element playing a role in the Crown's colonial administration, the historical factor also influenced the status of a territory within the framework of the empire. Any territory invaded, conquered, and settled by a white immigrant population became a *colony*, implying a considerable amount of self-determination for the settler population. However, those territories whose indigenous leaders had requested and been granted Crown "protection" became *protectorates*. Although the principle of indirect rule, in the white-controlled colonies, was mainly replaced by local European control, it was adhered to quite strictly in any territory that had been granted the status of protectorate. Some-

times adjacent territories possessed a different status: Northern Rhodesia was a protectorate, but Southern Rhodesia, by virtue of conquest from the south, became a colony; Uganda became a protectorate, whereas Kenya was a colony. Sometimes different parts of the same political entity came under different kinds of administrative control: southern Nigeria, in contact with Britain much more effectively than the north, became a colony, whereas northern Nigeria remained under indirect rule. Britain also administered a number of territories in cooperation with other colonial and noncolonial powers. This joint administration produced the *condominium*, discussed in Chapter Eleven.

British colonial rule was a conspicuous failure in Ireland, Southern Rhodesia, and the Middle East, and less than completely successful in South Africa and Burma. Nevertheless, considering the immensity, duration, and diversity of their empire, the British did astonishingly well. Their proverbial sense of fair play combined with their realism and genius for political improvisation helped them administer and withdraw from their empire with aplomb. It is noteworthy that of all the territories of the modern British Empire, only Burma and those in the Middle East have chosen not to join the Commonwealth on attaining independence, and only three Commonwealth members have formally withdrawn (Ireland, Pakistan and South Africa, as she was about to be expelled for her intransigence on the *apartheid* and South-West Africa issues). Perhaps this is the surest evidence of a point made earlier—that colonialism was not all bad.

If the British Empire was acquired "in a fit of absentmindedness," as one observer once remarked, then the empire of the *United States* was acquired enthusiastically but administered absentmindedly. In a real sense the United States acquired two separate empires consecutively: one continental and the other (with one exception) insular. The continental empire was assembled between 1803 and 1867 by peaceful purchase, by negotiation and settlement of territorial disputes with other countries, by voluntary annexation, and by conquest and annexation.* For the most part, it was administered in an orderly fashion according to the Constitution, the Northwest Ordinance, and other authority designed to bring the individual territories into the union as states equal to the original ones. The native peoples were subjugated, herded onto reservations, and largely forgotten. In 1912, the last remaining territories in the conterminous United States, Arizona and New Mexico, became states. Alaska and Hawaii became states only in 1959. Except for the fact that in each territory the dominant population group was white American, at least from the time of its formal organization, this first American empire resembled the other continental empires of its time, but now forms a true nation-state.

The second American empire still exists and is quite different from the first. Perhaps its most distinctive feature is that despite its obvious similarities to all other overseas colonial empires, it is never officially (and rarely unofficially) called an "empire." Nor are the individual units ever referred to as colonies. Instead, the United States uses such euphemistic terms as "insular possessions," "island responsibilities," "dependencies," "territorial areas under U.S. administration" and "territories under U.S. purview," but *never* "colony" or "empire." The reason for all this circumlocution is plain enough: the United States is ambivalent at best and perhaps even embarrassed by having a colonial empire, since she herself had fought to break free from a colonial empire and ever since has been critical of empires. This attitude has produced a desire to ignore the colonies and forget the empire. Few American citizens have been, or are now, aware of the empire; there has never been the kind of emotional attachment to it that characterized the citizens of the European metropoles. Americans never went out to the colonies to settle in any real numbers (except recently to Hawaii); their peoples were of little interest, their status debated largely in Washington, and their existence of little interest except as sources of sugar, tropical fruits, and exotic music. For this reason, perhaps more than for any other, the second American colonial empire has (again with the exception of Hawaii) been administered almost absentmindedly.

Despite some very real contributions to her colonies in the fields of health, welfare, and education, the United States has a rather mixed record as a colonial administrator. Cuba (which she had undertaken not to occupy) was granted independence in 1902—but under a form of American protectorate that lasted until 1933. The Philippines, which had been subjugated only in 1905 after a six-year war against nationalists led by Emilio Aguinaldo, became a commonwealth (self-governing colony) in 1935 and was granted independence in 1946. Puerto Rico was not granted responsible government until 1952, when it, too, became a commonwealth. Until that time, Puerto Rico and all the other American colonies, except for Cuba and the Philippines, had been treated with, in Daniel Moynihan's phrase, "benign neglect."

Various other forms of colonial status were gradually evolved, ranging downward from "incorporated territory," in which the U.S. Constitution was applicable (as in Hawaii after 1900) through "organized unincorporated territories" (Guam and the U.S. Virgin Islands) and "unorganized unincorporated territories" (American Samoa) to simply "island possessions" or some other term (anything but colony!). Most territories were originally under military government (army, navy, or air force) and some still are. Others are or have been under the jurisdiction of the Department of the Interior, as the continental territories had been. Within the departments, responsibility for admin-

---

*The states formed out of lands ceded to the United States by existing states, and the District of Columbia, all colonial in a sense, must be considered in separate categories.

istering the territories has been shifted from one office to another, and there has been reorganization after reorganization. The United States has never had a colonial office or its equivalent, a reflection of the haphazard way the territories were acquired, the indecisive way they have been administered, and the indifference of the American people to them once they were acquired.

Spain and the *Netherlands,* like Portugal, exploited their colonies as much as possible, giving very little in return. After they lost the largest and richest portions of their empires, however, the two countries took very different paths. Spain continued to invest almost nothing in her remaining colonies, moderated her rigid rule somewhat, and awaited further developments. Little by little, under mounting pressures for decolonization, Spain relinquished control in one territory after another, with little planning or preparation. Her once vast worldwide empire has now shrunk to five tiny *plazas de soberanía,* or presidencies, on the north coast of Morocco and in the Mediterranean nearby, and these she has long claimed as integral parts of Spain. Spain was an authoritarian and unimaginative ruler but not always as cruel as her enemies alleged.

The Netherlands had exploited her East Indies in much the same fashion as the Belgians had exploited the Congo. But after the colony won independence as Indonesia, she created a partnership with her remaining colonies in 1954, called the Tripartite Kingdom of The Netherlands. It was much like a federation and Surinam (formerly Dutch Guiana) and the Netherlands Antilles did participate in the affairs of the kingdom, especially its foreign affairs. Surinam, however, moved toward independence and was granted it on request in 1975, leaving the Netherlands Antilles very much a junior partner, riven by internal disputes and uncertain of its future.*

*In 1978 the official spelling was changed to Suriname.

Germany and *Italy* became unified nation-states at about the same time and almost immediately plunged into the race for colonies, the nineteenth-century status symbol of the Great Power. Germany got a head start, picking up a number of Spanish islands in the Pacific (now the Trust Territory of the Pacific Islands), northeast New Guinea and other Pacific islands, and several territories in west, southwest, and east Africa, all through daring, skillful, or lucky competition with the existing Great Powers. Italy, the last European country to embark on an imperialist adventure, is alleged to have dreamed of recreating the Roman Empire (or at least a reasonable imitation of it). But she was too late to get the most desirable colonies in the first instance and too weak to fight any of the colonial powers for what they already had, so she had to be satisfied with the leftover areas of Africa that no one else wanted. Even so, she had a difficult time subduing Libya and was routed by the Ethiopians at the Battle of Adowa in 1896. She finally conquered Ethiopia in 1935–1936 with the aid of aircraft, heavy weapons, and mustard gas, thus completing her African empire. In 1912, she seized the Dodecanese Islands, in 1939 annexed Albania, in which she had exercised great influence since 1926, unsuccessfully invaded Greece in 1940, and occupied part of Yugoslavia during World War II.

Neither Germany nor Italy held her empire very long. Germany lost her colonies during and after World War I; Italy lost hers during and after World War II. Both ruled their colonies autocratically, with little thought of preparing the inhabitants for self-government. Both were exploitative and indifferent to local cultures. More Italians migrated to the colonies as settlers than Germans, but in neither case were the numbers large. Both countries were interested in the colonies primarily for whatever prestige they could bring, but Germany did have geopolitical motives and Italy economic and demographic motives as well.

Japan's expansion from her home islands began with her acquisition of the Kuril Islands from Russia in 1875 and her conquest of Korea, southern Manchuria, the Pescadores, and Taiwan in a war with China in 1894–1895. She then defeated Russia in 1904–1905 and received certain rights in Manchuria and southern Sakhalin. She received a mandate over the former German Pacific islands north of the equator in 1919, occupied Manchuria in 1931, and continued on to occupy much of China and a very considerable portion of Southeast Asia and the Western Pacific. Her motives were almost single-mindedly economic—to gain unrestricted access to the resources of the conquered areas to feed her growing population and new industries. She set up a puppet state in Manchuria (calling it Manchukuo) and elsewhere she used some local administrators to help her rule, but on the whole the empire was centrally—and harshly—administered from Tokyo. Some public works were undertaken, but everything was geared toward integrating the colonies into Japan's economy, so the indigenous peoples benefited little from Japanese rule. She did try to win friends among the local peoples, partly by capitalizing on their anti-European feelings, but made scarcely any headway. Few (except some diehard Japanese traditionalists) were sorry to see Japan lose her empire entirely during and after World War II.

Australia and *New Zealand* became colonial rulers even before they themselves became fully independent. In neither case was this the result of any long-range plan. Instead, territories (mostly islands in the Pacific and Indian oceans) were transferred to them bit by bit over more than half a century by the United Kingdom and the League of Nations. The largest of all these territories was the eastern half of New Guinea, including the Admiralty, Bismarck, northern Solomon, and other islands to the east. This was organized originally as two colonies, German in the north and British in the south. British New Guinea was placed under Australian authority in 1902, and its name was changed to Territory of Papua in 1905. Australia received a League mandate for Northeast New Guinea in 1920. This was converted later into a UN trusteeship and the two territories were combined in 1945–1946 as an administrative union. Papua New Guinea became independent in 1975. The most important of New Zealand's territories was Western Samoa, originally a German colony that New Zealand administered under a mandate and then a trusteeship. It became independent in 1962. Both countries remain responsible for a number of other islands, mostly small in size and population.

All things considered, these two countries have probably been the best of all colonial administrators (along with Norway and Denmark). Despite some lapses, mistakes, and exceptions, they have shown great respect for indigenous traditions and institutions while introducing some of the better features of "Western civilization." They have guided their territories toward self-government and independence without forcing the pace or even determining the islanders' futures for them. There has never been a formal independence movement in any of the territories and no heavy outside pressures for decolonization. The islanders enjoy free access to the metropolitan countries, even as permanent migrants, and most of them are citizens of the administering countries. Niue and Cook Islands (under New Zealand administration) are very nearly independent but are not anxious to take the final step. Governing small, remote, relatively homogeneous, and not particularly valuable islands is certainly not comparable to governing a Congo, an India, an Algeria, or even a Puerto Rico, but the attitudes and policies of Australia and New Zealand have been commendable nonetheless.

This review of colonial policies has been necessarily brief. We have scarcely mentioned the Ottoman Empire, that fascinating enigma that bridged the gap between ancient oriental and modern European empires. Nor have we mentioned the colonial areas and peoples within indepen-

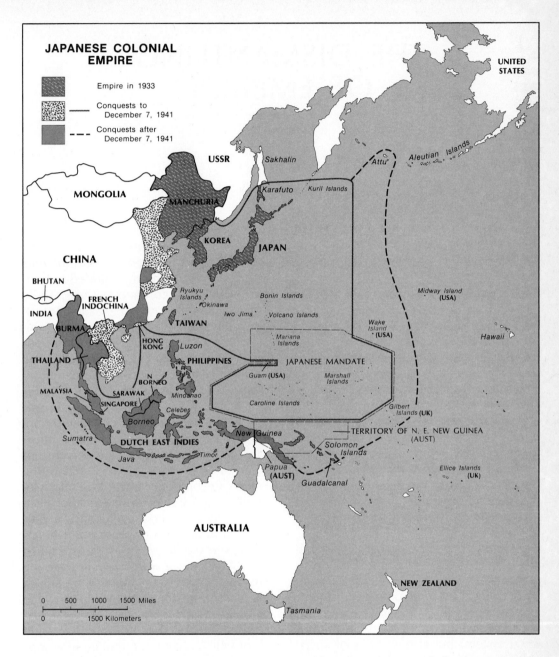

JAPANESE COLONIAL EMPIRE

Empire in 1933

Conquests to
December 7, 1941

Conquests after
December 7, 1941

UNITED STATES

USSR

MONGOLIA

MANCHURIA

Sakhalin

Karafuto   Kuril Islands

Attu°   Aleutian Islands

KOREA   JAPAN

CHINA

BHUTAN

FRENCH
INDOCHINA

INDIA

BURMA

THAILAND

MALAYSIA

SARAWAK
SINGAPORE

N
BORNEO

HONG
KONG   Luzon

TAIWAN

PHILIPPINES

Mindanao

Celebes

Borneo

Ryukyu
Islands
·Okinawa

Iwo Jima

Bonin Islands

·Volcano Islands

Mariana
Islands

Guam (USA)

Caroline Islands

Marshall
Islands

JAPANESE MANDATE

Wake
Island
·(USA)

Midway Island
(USA)

Hawaii

Gilbert
Islands (UK)

Sumatra

DUTCH EAST INDIES

Java

Timor

New Guinea

Papua
(AUST)   Guadalcanal

Solomon
Islands

TERRITORY OF N. E. NEW GUINEA
(AUST)

Ellice Islands
(UK)

AUSTRALIA

NEW ZEALAND

Tasmania

0   500   1000   1500 Miles

0   1500 Kilometers

dent countries (aside from China and the USSR), or the role of private companies in creating and administering colonial empires, or many other features of colonialism essential to a complete understanding of the subject. Nevertheless, we have sketched the major outlines of the system, emphasizing variety and change, good and bad, and the dangers of discussing it in slogans and glib generalizations. The same principles are equally valid for our discussion of the decline and eventual demise of the system, to which we now turn our attention.

Chapter 21

THE DISMANTLING OF EMPIRES

The imposing colonial structure we have just described disintegrated in far less time than it took to erect it—generations rather than centuries. A trickle of new States became a flood in the 1960s, 44 in just that decade and 18 in 1960 alone. Yet nothing in human affairs ever happens suddenly; just as imperialism and colonialism had deep roots, so does decolonization.

The process of decolonization began long before the period of modern colonization had ended—before it had fairly begun, in fact. Indeed, some of the same forces that led to the creation of the modern empires also led to their demise. Here we can only list the more important forces that became prominent, beginning in the mid-nineteenth century, and contributed to this process.

Certainly, one of the most subtle, yet most critical, factors was the steady evolution of democracy in Western Europe, particularly in the United Kingdom and to a lesser extent in France. There were not only significant developments in the political system, but also a series of social and economic reforms, so that social, political, and economic democracy tended to develop together, though at different rates. The new doctrines in the Western world of progress and of evolutionary improvement, one of many offshoots of Social Darwinism, pervaded thinking in many areas of life. The great flowering of the Industrial Revolution generated remarkable developments in science, medicine, and technology that not only spread around the world, but also led to developments in related areas, among them education, transportation, communications, trade unions, Marxism, and a population growth that, combined with land enclosure movements and other agrarian reforms, sent millions of Europeans out to the colonies.

The spectacular success of Japan in industrializing, in defeating a Great Power of Europe (Russia) in a European-style war, and in starting her own colonial empire demonstrated that independence, industrialization, wealth, and power were not reserved only for Europeans. Certainly, the French Revolution was an essential element in the growth both of nationalism and democracy, and it did have a profound influence on the Spanish colonials in the New World, yet it was not until much later that its ideals were transmitted to the French colonies of the modern imperialist age. Of more immediate interest in the colonies of Africa and Asia were the revolutions in Mexico, Russia, and Turkey, in all of which a despotic regime in a poor country was overthrown by (it seemed to them) the downtrodden masses. And the very nationalism that in its most extreme form became imperialism, in more moderate forms spread to the victims of imperialism.

The course of human history is punctuated with momentous events that mark true divides between eras. There have not been many of them and they are not always easy to identify, but certainly World War I is one. It marked the end of the way of life that had evolved in much of the world through the nineteenth century. The frightful carnage of the war was unprecedented; not only were people and cities and forests brought down, but empires as well—Germany, Austria–Hungary, Russia, the Ottoman Empire, gone. Their colonies and minority peoples were transferred to presumably more enlightened guardians or given independence altogether. During the Great War, the horizons of the colonials were expanded as they were called to the colors of the "mother country" and fought in Europe, the Middle East, and elsewhere. European soldiers were sent to the colonies in greater numbers and battles were fought there with modern weapons and new ideas. Trading patterns were disrupted and some colonies began to understand how dependent they were on Europe. And the ferment of the war stimulated new interest in self-determination.

The Great Depression and World War II reinforced all these trends, which had become so clear during World War I, and actually intensified them. Dependence, freedom, vulnerability of Europeans, exploitation, nationalism, progress, equality, all were themes brought home to the peoples of the colonies by two world wars and a worldwide depression. In their wake came improved education and communications, bringing to the remotest nations the revolutionary message that they did not have to live in bondage and misery, that there were alternatives, that they could be free, could control their own destinies, could share in the fruits of the new world of abundance created by industrialization. The economists call this "the international demonstration effect"; more commonly it is called "the revolution of rising expectations."

Two world wars and a worldwide depression also left most of the colonial powers, winners and losers alike, dreadfully weakened and no longer capable of ruling huge empires. The weakness was not only military, but economic and, in a sense, spiritual. It became clear to all but the old-school imperialists that the world had changed, that colonialism, if not imperialism itself, was now outmoded and had to be replaced by a new version of international and interhuman relations.

This is not to imply that all imperialists gathered under the tree of enlightenment and suddenly became humane and democratic. Far from it. Portugal and Belgium, for example, blithely ignored the new world and made no moves toward giving their colonies even limited self-government until it was too late. Immediately after World War II France and the Netherlands fought bitter wars against colonials who did not want them to return to their countries just liberated from the Japanese. And the Soviet Union not only annexed whole States and parts of others, but actually colonized Eastern Europe and parts of Asia for a time. Nevertheless, from World War I on, nineteenth-century colonialism was doomed.

Beginning with the British North America Act of 1867, the United Kingdom adopted the practice of granting gradually increasing degrees of self-government to her colonies. She also made extensive use of the protectorate system, in which she recognized the inherent sovereignty of native rulers but performed certain governmental functions for them. These practices were later adopted by other powers, even by France who established protectorates over Tunisia in 1881 and Morocco in 1912.* There were also many devices used by the imperialists to control foreign territories without incorporating them formally into their colonial systems. They included spheres of influence formally recognized by the other powers; concessions (essentially enclaves) in port cities and coastal areas, chiefly in China; extraterritoriality, usually combined with some other technique, as in Morocco and China;

*Spain also established protectorates over northern and southern zones of Morocco in 1912, but it took her more than 20 years to subdue nationalist guerrillas in the northern Rif Mountains and secure the entire territory.

and "advising" local rulers, as the British did in the Trucial States.

Some of these devices were simply tactics in the game of power politics, of course, and were used where the indigenous cultures were too strong to be subjected to outright colonization; nevertheless, they did represent looser colonial relationships that were relatively easy to modify or abandon as local nationalisms or other pressures began to be felt. The most important and radical departure from colonialism, however, was a product of World War I: the League of Nations mandate.

THE MANDATE SYSTEM OF THE LEAGUE OF NATIONS

After World War I the victorious Allies were tempted to continue the pattern of most previous wars by gaining territory at the expense of the losers. They did agree that Austria–Hungary should be split into independent States and that Germany and Turkey should be stripped of their empires. They stopped short, however, of simply dividing up the German and Ottoman colonies among themselves. Influenced by all the new forces described above, reminded by the Arab subjects of the Sultan about Allied promises of freedom in exchange for aid in the war against Turkey, and badgered by Woodrow Wilson, who in a very real sense was then the spokesman for much of humanity, they agreed on a compromise. They established a system of international supervision of the German and Ottoman colonies, with themselves as the administering powers, but not the sovereigns. The Allies would assign the colonies, determine their boundaries, and define the terms of the mandates under which they were to be governed. The whole system was to be supervised by the League of Nations, specifically its Permanent Mandates Commission.

Article 22 of the League Covenant spelled out the purpose of the mandate system and, in general terms, its procedures. Its tone was moralistic, even paternalistic, declaring that since the peoples of the former colonies in Africa, the Middle East, and the Pacific were "not yet able to stand by themselves under the strenuous conditions of the modern world, there should be applied the principle that the well-being and development of such peoples form a sacred trust of civilization" and that they should therefore be placed under the "tutelage" of "advanced nations," taking into account the different conditions in the various colonies.

Three essential features of this mandate system are important for our purposes. First, the mandated territories were divided into three classes based on their degree of development. Class A mandates were those former Turkish provinces whose independence would be provisionally recognized until they could sustain independence on their own. Iraq and Palestine were assigned to Britain and Syria to France. Britain partitioned Palestine, creating a territory on the eastern side of the Jordan River for Emir Abdullah as a reward for his help in World War I; it was called Transjordan and received independence in 1946. France partitioned Syria, creating Lebanon, essentially as a State in which the Christians could be a majority. Syria and Lebanon became independent in stages between 1941 and 1946. Iraq became independent in 1932.

Most of the German colonies in Africa fell into Class B, in which the mandatory was responsible for the welfare of the people and for their administration. Togoland and the Cameroons (Kamerun) were both divided between Britain and France, Tanganyika was assigned to Britain, and Ruanda–Urundi went to Belgium. Class C mandates were those that could "best be administered under the laws of the mandatory as integral portions of its territories, subject to the safeguards in the interests of the indigenous population" that were specified in the mandates. In this category were South-West Africa; West-

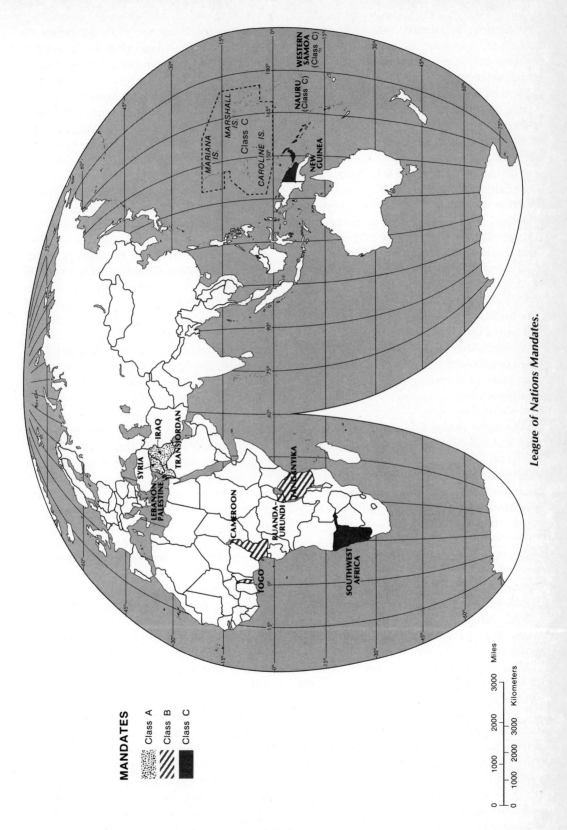

League of Nations Mandates.

ern Samoa; Northeast New Guinea; the Marshall, Caroline, and Mariana islands (except for Guam); and Nauru Island. The mandatory powers were, respectively, the Union of South Africa, New Zealand, Australia, Japan, and Australia, which administered Nauru and its rich phosphate deposits on behalf of Great Britain, New Zealand, and herself.

The second important feature of the mandate system was that it actually laid out the responsibilities of the mandatory powers. These fell into three categories: (1) guarantees of freedom of conscience and religion; (2) prohibition of abuses, such as the traffic in arms and liquor and slavery and slave trading; and (3) prohibition of fortifications, naval and military bases, and the military training of the natives, except for police and home defense duties. These responsibilities to the local people and to the League, and the contemplation in the Covenant of eventual self-government and independence, distinguished a mandate from a protectorate.

The third major feature of the system also was a departure from all previous colonial policies and techniques: the League had an unqualified right to supervise the mandates to be certain that their terms were being carried out faithfully. The principal means for doing this was the annual report required from each mandatory for each mandated territory. The Permanent Mandates Commission could also receive written petitions and memorials from or concerning the mandated territories.

This whole system unquestionably marked an advance in international relations and international organization, no matter how obvious its weaknesses (such as lack of enforcement machinery) or how severe its critics. It also marked the end of nineteenth-century-style imperialism/colonialism. It was no longer fashionable. The mandate system worked as well as could reasonably be expected, considering that the world was plunged into a severe depression shortly after it began functioning, to be followed immediately by a cruel war.

During the 1920s and 1930s, all the forces described were still working. The free trade union movement was burgeoning; unions in many colonies became the nuclei of political parties and independence movements as well as training grounds for their leaders. Nationalism was growing in many colonies, inspired anew by the principle of self-determination that dominated Wilson's Fourteen Points. After the success of the Bolshevik Revolution in Russia, Marxist ideology became more attractive and inspired some colonials around the world to become revolutionaries rather than work for freedom through peaceful means. Mahatma Gandhi and his followers in India provided a different sort of model: nonviolent resistance to colonial rule and patient negotiation for concessions leading to freedom. Education, travel, communications, and the example of the mandates all played their parts. It was World War II, however, that effectively destroyed colonialism.

All the forces that had led to the undermining of the colonial system before and during World War II continued working after the war and many of them seem to have merged into what we might call a new kind of nationalism, one based not on unity of a culture requiring expression in a territory of its own, but founded largely on anticolonialism. Among the many peoples thrown together within a colony, the joint goal of decolonization may have been pursued for different purposes. Certain groups of people found themselves in a position to succeed the colonial power in its control of the territory involved. Often their joint efforts to oust the foreign ruler were followed by a breakdown in cooperation and a struggle for supremacy. The Ganda of Uganda, seeing their intent to dominate an independent State of Uganda thwarted by a British-designed constitution, threatened to secede from the country even before sovereignty was achieved. The apparent centripetal nature of nationalism becomes real only when anticolonialism ceases to have a function in the emergent State.

THE ROLE OF THE UNITED NATIONS

The Charter of the United Nations commits that body, in the Preamble and more specifically in Article 1, to "equal rights and self-determination of peoples." Articles 73 and 74 deal in more detail with the matter under the heading "Declaration Regarding Non-self-governing Territories." They lay out the responsibilities of colonial powers to the subject peoples, including assisting them to develop self-government, and to the international community, specifically the United Nations. The Charter, moreover, in Articles 75–91 provides for an international trusteeship system to replace the League of Nations mandate system.

The trusteeship system has the same general objectives as the mandate system, except that one of its objectives is *specifically* to promote the trust territories' "progressive development towards self-government or independence." There are, however, some very important differences between the two systems. First, the trusteeship system may apply to three categories of territories: "a. territories now held under mandate; b. territories which may be detached from enemy States as a result of the Second World War; and c. territories voluntarily placed under the system by States responsible for their administration." In practice, only the first category (a.) has been really important, since only Italian Somaliland was added to the former mandates that had not been terminated, and subparagraph (c.) has never been utilized—and is unlikely to be. Second, petitioners from the trust territories are permitted to testify in person before the Trusteeship Council or the General Assembly in addition to submitting written petitions. Third, the Charter provides for periodic visits by UN missions to the trust territories in addition to written annual reports from the administering power. Both devices have been extensively used and have contributed greatly to the success of the trusteeship system.

It has been so successful, in fact, that every trust territory has peacefully become an independent State or part of a neighboring independent State except for one. The remaining one, the Trust Territory of the Pacific Islands, was designated a "strategic area" under Articles 82–84 and is thus under the jurisdiction of the Security Council rather than the General Assembly. The Trusteeship Council still performs the actual supervision and reports to the Security Council. The administering power, the United States of America, negotiated with the people of the Trust Territory a series of agreements during the 1970s and early 1980s on their future status. In 1975, a covenant was concluded to create a Commonwealth of the Northern Mariana Islands in Political Union with the United States and under its sovereignty. The Commonwealth became official in November 1986.

Between 1982 and 1986, "Compacts of Free Association" were negotiated between the United States and the Federated States of Micronesia (Caroline Islands), the Republic of the Marshall Islands, and the Republic of Palau. They provide for considerable self-government but not independence. The first two were approved by the people and the Trusteeship Council. The last, with Palau, was repeatedly rejected by the people because of provisions for the storage of nuclear weapons there and visits by nuclear powered and/or armed warships. The Palau constitution was amended and the compact was approved in a referendum in August 1987, but the entire process was still being challenged in mid-1988. Although the United States considers the trusteeship terminated, the agreements have not yet been approved by the Security Council, so it legally remains in effect.

There have been two important exceptions to the otherwise excellent performance of the mandate-trusteeship system. Because of conflicts between Arab and Jewish Palestinians in the remaining portion of the Palestine mandate (west of the Jordan River) over the future of the

A trust territory becomes a republic.   The new flag of the Republic of Nauru is raised on 31 January 1968 by the Head of State in front of the Nauruan Administration Offices. Witnessing the ceremony are representatives of Australia, New Zealand, the United Kingdom (the Trustee Powers) and the United Nations. The tiny phosphate-rich island in the South Pacific has only about 6000 people, of whom nearly half are Chinese, Europeans, and other Pacific Islanders. (UN)

territory, the British were unable or unwilling to replace the mandate with a trusteeship. As the conflict became more intense and the British could offer no satisfactory solution to the problem, they turned it over to the United Nations in 1947 and announced that they would terminate the mandate and leave Palestine in May 1948. In November 1947, the General Assembly voted to partition Palestine again, creating Jewish and Arab States out of it. Jerusalem was to be an international enclave.

The Jews accepted the proposal, although the area proposed was much less than they had hoped for, and they were unhappy about not having Jerusalem, their ancient capital and focus of their religion, included in their new State. The Arabs of Palestine and the Arab States rejected the proposal completely, objecting to any Jewish State in the region whatever its boundaries. Preparations nevertheless went ahead to carry out the UN decision,

at least in the area destined for Jewish control. Communal fighting broke out and the British had a difficult time maintaining a semblance of order until they departed on schedule, even as the armies of six Arab states invaded in a vain attempt to extinguish the new State of Israel. Israel became a member of the United Nations in 1949 but has had to battle challenges to her existence ever since.

The other mandate that was neither terminated with the independence of the territory nor replaced by a trusteeship agreement was that of South Africa over South-West Africa. Indeed, South Africa moved vigorously toward outright annexation of the territory, in violation of the terms of the mandate. The attempt failed, largely because of determined UN opposition and little support from the Western Powers. Nevertheless, she did extend her *apartheid* system to the territory and has governed it as a *de facto* fifth province, with representation (white only) in the South

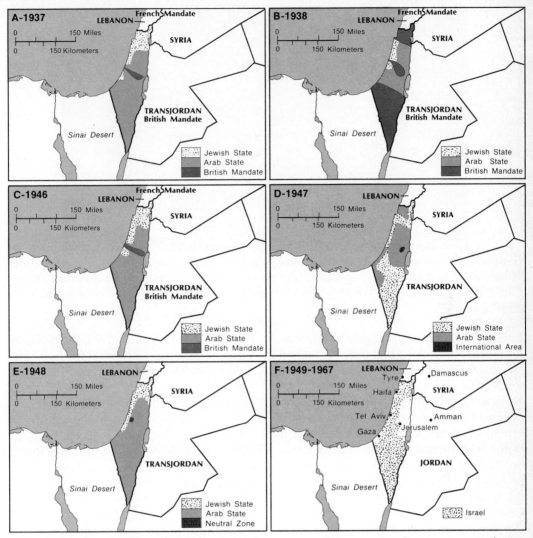

Partition plans for Palestine. After Britain partitioned her Palestine mandate to create a throne for Emir Abdullah, called Transjordan, Jewish and Arab nationalism developed in the smaller western portion. Attempting to accommodate them both while maintaining her own position in the Middle East, Britain began proposing a series of partition plans. After World War II other agencies attempted to mediate with partition plans. None was actually effected.

A. Peel Commission Proposal, 1937.

B. One of two proposals of the Woodhead Commission; the other included a larger Jewish State, 1938.

C. Proposal of the Anglo-American Commission, 1946.

D. Proposal of the United Nations Special Committee on Palestine, 1947. This proposal was voted by the General Assembly and accepted by the Jews of Palestine but rejected by the Arabs.

E. In an attempt to secure agreement of the Arabs to a Jewish State, Count Folke Bernadotte, United Nations mediator, proposed early in 1948 a very much smaller Jewish State. It too was rejected by the Arabs.

F. This is the territory included within the "Green Line," the armistice lines agreed on by Israel and her neighbors after the fighting stopped and they exchanged territory they had occupied. This territory, included within *de facto* boundaries, remained intact until 1967 when during the Six Day War Israel occupied all of Sinai, Gaza, the "West Bank" and the Golan Heights.

Consolidated Diamond Company's mine at Oranjemund, Namibia.   One of the major reasons South Africa refuses to grant independence to Namibia is the substantial revenues South Africa obtains from exploitation of the mineral and fisheries resources of the territory. (UN/Contact/Alon Reininger)

African Parliament. She argued that since the League of Nations had expired, so had the mandate, and that she was therefore relieved of any responsibility to the international community for the territory. The International Court of Justice ruled otherwise in an advisory opinion in 1950. It stated that the mandate was still in force, South Africa was still obligated to abide by it, and South Africa was bound to submit to the supervision and control of the General Assembly, which inherited the supervisory role of the Council of the League. South Africa refused to accept the Court's opinion and continued to oppose any form of UN supervision over the territory's affairs.

The United Nations voted in 1966 to terminate the mandate, in 1967 to establish the United Nations Council for South West Africa to administer the territory until independence, in 1968 to change the name of the territory to Namibia (from the Namib Desert), and continued, through the General Assembly, Security Council, International Court of Justice, and Secretary-General, to try to persuade South Africa to cooperate—all without success. The Court in 1971 rendered another advisory opinion that South African presence in

Namibia was illegal. In 1978, the Security Council adopted a proposal for free elections in Namibia under UN supervision and control. South Africa rejected this proposal also and the stalemate continues.*

These two exceptions to the success of the mandate-trusteeship system resulted from particular local conditions not found in the other territories. As we have already pointed out, however, they were not unique in the colonial world. The Cyprus problem, for example, bears some resemblance to the Palestine problem and still defies solution. Decolonization has certainly not been easy. Considering the number and complexity of the potential problems, however, and the magnitude and historic significance of the dismantling of empires, it is remarkable that it was not much more painful, prolonged, and violent than it has been. Much of the credit for the relative orderliness of the

*A small parcel of territory that has become a major issue is Walvis Bay. It was a British enclave in the German territory, important because it has the best natural harbor in the region. South Africa has a major naval base and other military facilities there and has taken steps to incorporate it completely into her own territory. The United Nations has objected, claiming it is and must remain an integral part of Namibia. The stalemate on this issue continues also.

process must go to the United Nations, even without the trusteeship system.

Its first experience with colonies was the matter of the disposition of the former Italian colonies. The UN decided that Libya should become independent almost immediately, in December 1951; that Italian Somaliland should become independent after a 10-year period of trusteeship with Italy as the administering power; and that Eritrea should be autonomous and federated with Ethiopia.*

Meanwhile, in 1946, the General Assembly had listed those territories that came under the purview of Articles 73 and 74 and had established machinery to carry out their provisions. The major provision was that the administering power was obligated to transmit regularly to the Secretary-General information on the territories. This procedure has been followed rather consistently and the debates on the reports have been the principal instrument for urging the powers to grant greater self-government to their dependent territories. In 1953, the first territory was removed from the list when the United States informed the UN that Puerto Rico had become a commonwealth (*Estado Libre Asociado*) associated with the United States. Greenland was next in 1954, then Surinam and the Netherlands Antilles, Alaska and

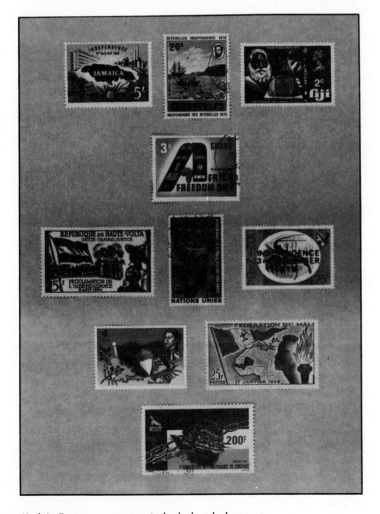

Decolonization on stamps. It is customary for each colony as it becomes independent to issue a set of stamps to commemorate the historic event. Frequently the stamps depict the country's history, culture, and modern characteristics. Three of these stamps are a little unusual. In the second row is one of three stamps issued by Ghana in 1960 to honor Africa Freedom Day; below that is a United Nations stamp calling attention to Namibia; at the bottom is one issued by the Central African Republic to honor the first anniversary (1981) of the independence of Zimbabwe. (Martin Glassner)

*Italy's European conquests had already been returned to their prior sovereigns, Yugoslavia, Albania, and Greece.

Table 21–1 Chronology of Decolonization: Countries Receiving Independence Since 1943

Year	Date	No.	Country	Year	Date	No.	Country
1943	Nov. 22	1	Lebanon		Oct. 9	50	Uganda
1944	Jan. 1	2	Syria	1963	Dec. 12	51	Kenya
	June 17	3	Iceland	1964	July 6	52	Malaŵi
1946	Mar. 22	4	Transjordan		Sept. 21	53	Malta
			(Jordan)		Oct. 24	54	Zambia
	July 4	5	Philippines	1965	Feb. 18	55	The Gambia
1947	Aug. 14	6	Pakistan		July 26	56	Maldives
	Aug. 15	7	India		Aug. 9	57	Singapore
1948	Jan. 4	8	Burma	1966	May 26	58	Guyana
	Feb. 4	9	Sri Lanka		Sept. 30	59	Botswana
	May 15	10	Israel		Oct. 4	60	Lesotho
	Aug. 15	11	Korea		Nov. 30	61	Barbados
1949	Mar. 8	12	Vietnam	1967	Nov. 30	62	Yemen (Aden)
	July 19	13	Laos	1968	Jan. 31	63	Nauru
	Nov. 8	14	Cambodia		Mar. 12	64	Mauritius
			(Kampuchea)		Sept. 6	65	Swaziland
	Dec. 28	15	Indonesia		Oct. 12	66	Equatorial
1951	Dec. 24	16	Libya				Guinea
1956	Jan. 1	17	Sudan	1970	June 4	67	Tonga
	Mar. 2	18	Morocco		Oct. 10	68	Fiji
	Mar. 20	19	Tunisia	1971	Aug. 14	69	Bahrain
1957	Mar. 6	20	Ghana		Sept. 3	70	Qatar
	Aug. 31	21	Malaysia		Dec. 2	71	United Arab
1958	Oct. 2	22	Guinea				Emirates
1960	Jan. 1	23	Cameroon	1972	Apr. 4	72	Bangladesh
	Apr. 27	24	Togo	1973	July 10	73	Bahamas, The
	June 27	25	Madagascar	1974	Feb. 7	74	Grenada
	June 30	26	Congo (Zaïre)		Sept. 10	75	Guinea-Bissau
	July 1	27	Somalia	1975	June 25	76	Mozambique
	Aug. 1	28	Dahomey		July 5	77	Cape Verde
			(Benin)		July 12	78	São Tomé &
	Aug. 3	29	Niger				Príncipe
	Aug. 5	30	Upper Volta		Sept. 16	79	Papua New
			(Burkina				Guinea
			Faso)		Nov. 11	80	Angola
	Aug. 7	31	Ivory Coast		Nov. 25	81	Suriname
			(Côte		Dec. 31	82	Comoros
			d'Ivoire)	1976	June 28	83	Seychelles
	Aug. 11	32	Chad	1977	June 27	84	Djibouti
	Aug. 13	33	Central African	1978	July 7	85	Solomon
			Republic				Islands
	Aug. 15	34	Congo-		Oct. 1	86	Tuvalu
			Brazzaville		Nov. 3	87	Dominica
	Aug. 16	35	Cyprus	1979	Feb. 22	88	Saint Lucia
	Aug. 17	36	Gabon		July 12	89	Kiribati
	Aug. 20	37	Senegal		Oct. 27	90	Saint Vincent &
	Sept. 22	38	Mali				the
	Oct. 1	39	Nigeria				Grenadines
	Nov. 28	40	Mauritania	1980	Apr. 18	91	Zimbabwe
1961	Apr. 27	41	Sierra Leone		July 30	92	Vanuatu
	June 19	42	Kuwait	1981	Sept. 21	93	Belize
	Dec. 9	43	Tanganyika		Nov. 1	94	Antigua &
			(Tanzania)				Barbuda
1962	Jan. 1	44	Western Samoa	1983	Sept. 19	95	Saint
	July 1	45	Burundi				Christopher
	July 1	46	Rwanda				& Nevis
	July 5	47	Algeria	1984	Jan. 1	96	Brunei
	Aug. 6	48	Jamaica				Darusalaam
	Aug. 31	49	Trinidad &				
			Tobago				

Hawaii, and the parade of territories that became fully independent.

During the General Assembly session in 1960, however, this rather deliberate process was drastically altered. Early in the session, 17 new members, mostly Francophone, were admitted, creating a new majority of formerly dependent territories. In December, the Assembly adopted a Declaration on the Granting of Independence to Colonial Countries and Peoples. It proclaimed "the necessity of bringing to a speedy and unconditional end colonialism in all its forms and manifestations." It also linked very firmly self-determination with independence of dependent peoples and added that "the integrity of their national territory shall be respected."

In 1961, the Assembly created a special committee (known as the Committee of 24) to oversee implementation of this declaration. This committee has been extremely active ever since in accelerating decolonization. Among its activities, it has sent missions to visit a number of territories to investigate conditions or to supervise elections. It even sent a mission to the "liberated" area of Portuguese Guinea in 1972 without the permission of the Portuguese authorities, an historic precedent. It has held many of its meetings away from New York, mostly in Africa, as another way of expediting and publicizing its work. In 1964, it began giving attention to the activities of foreign economic interests in the colonial territories; in 1967, to military activities and arrangements in the territories. It works with governments, other UN organs, nongovernmental organizations, and "national liberation movements."

It is questionable whether all this activity has actually contributed substantially to decolonization, but it certainly has generated interest in the subject, not only in the colonies, but in the world at large. Other UN activities, however, have been of a more practical nature. Examples are the United Nations Temporary Executive Authority that governed West New Guinea between the Dutch departure and the Indonesian arrival, the plebiscite conducted in Bahrain (claimed by both Iran and Saudi Arabia) after the British announced their intention to withdraw, and the prevention of South African annexation of South-West Africa. But it is doubtful if the insistence on completely obliterating colonialism "in all its forms and manifestations" is really the wisest course.

As long as only relatively large territories with some reasonable chance of sustaining themselves as independent States were involved, there were few serious problems. Even such small new countries as Western Samoa (1962), Rwanda and Burundi (1962), and Malta (1964) did not cause much alarm. The first three had been trust territories and were (presumably) groomed for independence, and Malta was European and had a fairly sturdy economy. The admission of the Maldives in 1965, however, initiated a debate about the future of small territories that has not yet ended. Despite reservations, the phrase "self-determination *and* independence" has almost replaced both components used separately. Pressure has continued to give independence to the most minute inhabited island. Once a country has become independent, however, "self-determination" is no longer applicable to peoples within it. This raises many questions, some of which we consider in the next chapter. Meanwhile, several of the remaining territories have announced dates for independence or incorporation into another country, leaving about three dozen inhabited dependent territories whose future is still uncertain. Some of them, very likely, will become independent before long.

Table 21–2   Remaining Dependent Territories

Name	Dependency of	Population, Early 1980s
American Samoa	United States	40,000
Anguilla	United Kingdom	6,500
Aruba	Netherlands	67,000
Ashmore & Cartier Islands	Australia	—
Baker, Howland, & Jarvis Islands	United States	—
Bermuda	United Kingdom	63,000
British Indian Ocean Territory	United Kingdom	2,000
British Virgin Islands	United Kingdom	12,034
Canton & Enderbury Islands	United States/United Kingdom	—
Cayman Islands	United Kingdom	18,000
Channel Islands	United Kingdom	130,000
Christmas Island	Australia	3,094
Cook Islands	New Zealand	20,000
Coral Sea Islands	Australia	—
Falkland Islands & Dependencies	United Kingdom	2,000
French Polynesia	France	160,000
Gibraltar	United Kingdom	30,000
Greenland	Denmark	51,000
Guam	United States	120,000
Heard & McDonald Islands	Australia	—
Hong Kong	United Kingdom	5,420,000
Isle of Man	United Kingdom	60,000
Johnston Atoll	United States	300
Kingman Reef	United States	—
Macau	Portugal	390,000
Mayotte	France	53,000
Midway Islands	United States	2,200
Montserrat	United Kingdom	12,160
Namibia	South Africa	1,510,000
Netherlands Antilles	Netherlands	200,000
New Caledonia	France	150,000
Niue	New Zealand	6,000
Norfolk Island	Australia	2,175
Pitcairn Island	United Kingdom	68
Puerto Rico	United States	3,270,000
St. Helena, Ascension, Tristan da Cunha	United Kingdom	6,000
Swan Islands	United States	—
Tokelau	New Zealand	1,552
Trust Territory of the Pacific Islands	United States	116,974
Turks & Caicos Islands	United Kingdom	7,000
U.S. Virgin Islands	United States	95,951
Wake Island	United States	1,600
Wallis & Futuna Islands	France	11,943

N.B. Not included are the French overseas departments, Antarctic claims of various countries, Western Sahara, East Timor, and a number of small islands.

Chapter 22

THE NEW STATES

Between 1943 and 1984, 96 new States joined the ranks of independent actors on the world stage (97 if we include Zanzibar, independent only briefly). The "strenuous conditions of the modern world" as the League of Nations Covenant put it, are even more stenuous today than they were in 1919, and are likely to become more so even as still more colonies become independent States in the near future. Few generalizations, however, can be made about the new countries. Not all of them are poor, for example. Kuwait, Israel, Malta, and Iceland at least approach Western European living standards; Nauru, Fiji, Malaysia, Singapore, and the Bahamas are relatively prosperous, more so than many older countries. Many "new" countries are really quite old and only reemerging from a period of colonial domination. India, Lebanon, Syria, Burma, Sri Lanka, Israel, Korea, Vietnam, Kampuchea (Cambodia), Morocco, Ghana, Samoa, Swaziland, Oman, Tonga, and Fiji are not creations of the colonial period, but have histories measured in centuries and even millennia. Others also have deep roots and rich cultures even if they have not existed as independent political entities before. Nevertheless, we must generalize here and confine ourselves to only a few essentials of this adventure in human history, this miraculous birth and nurturing of new political units containing nearly half the world's population.

NATION BUILDING AND STATE BUILDING

A country just emerging from colonial status has a great deal to do. If it has had a reasonable gestation period with gradually increasing degrees of self-government, if its birth is peaceful and orderly, and if it receives adequate postpartum assistance from its former colonial ruler, other countries, and international agencies, then the transition from colony to State is generally smooth and successful. The State-building process, however, can be difficult, even traumatic, under less than ideal conditions. Just the simple mechanics of running a country, especially a poor one, can be most bewildering to people who lack administrative and political experience.

The most extreme case of this kind, of course, was the former Belgian Congo. Other countries have had similar difficulties attending their birth, although not nearly as severe. Indonesia, Algeria, Guinea, Bangladesh, and others have had difficulties building a State where none had existed before. But even more difficult, if less dramatic, is the process of nation building; that is, trying to weld a nation out of disparate tribal and ethnic groups thrown together by colonial rulers who had only their own interests in mind. Some new countries seem to have accomplished this task with considerable success in a relatively short time. Others still appear to be having great difficulties in creating nation-states.

Lebanon, after a long period of peace and prosperity, has dissolved into civil war and possible partition, and there is no assurance that any other new States do not face the same fate. (Not even older countries are immune from communal tensions; Canada faced the very real threat of secession of Quebec; Spain is still fighting Basque separatists; guerrilla warfare continues in Northern Ireland.) Some of the new countries are relatively homogeneous and have had few if any ethnic problems since independence. The rest are still assiduously engaged in nation building.

The nation-building process is even more complex and frequently more painful than State building. It means generating and nurturing nationalism to replace regionalism or tribalism or localism, creating a sense of identification with and loyalty to this new thing called "the State." Often people who had been told all their lives that "the government" was bad and had to be opposed were suddenly being told that "the government" was good and had to be supported. Some new countries have been fortunate in having talented and charismatic leaders to guide them through the transition period and become symbols of the new State around which most factions could rally. When the hero of independence dies or is deposed, however, especially early in the independence period, the effect can be catastrophic unless a new leader emerges very quickly to replace the old one. Lacking wise and trained leadership, with no democratic tradition and little experience in self-government,

Part of the iconography of a new State.   The Black Star Memorial in Accra, built by Ghana's first President, Kwame Nkrumah, to commemorate Ghana's independence. (Harm J. de Blij)

many new countries have soon come under military rule. In this, as in many other aspects of independent statehood, the new countries of Africa and Asia are emulating Latin America. Nowhere, however, have military rulers as a class proven any better than their civilian counterparts.

Fostering nationalism in the new countries involves but goes beyond much of what Jean Gottmann called *iconography*. Depending on local conditions, the following activities have been typical of new governments in recent decades: resurrecting or fabricating a glorious precolonial history; initiating a new educational system in which new national values are stressed; building the army into a highly visible national symbol, often working in roadbuilding, disaster relief, public health, or clearing land for pioneer settlement; instituting a national information service to carry the government's message through every possible medium to the people of the country and abroad; developing national sectoral organizations (peasants, chiefs, teachers, manual workers); and the iconography, or symbolism, of national distinctiveness itself—a flag, an anthem, heroic slogans, a steel mill, a national stadium, a national airline, a national costume, replacement of European place and personal names with local ones, a palace of culture or convention hall in the capital, postage stamps carrying nationalistic messages, and similar devices—all designed to replace anticolonialism as centripetal forces to neutralize or overcome the centrifugal forces long held in check by the colonial rulers.

The very emphasis on fostering national unity, however, often leads to neglect of the rights and interests of minorities, even to their suppression; to the neglect of substance in favor of form or symbol; to investment in extravagant public works, elaborate ceremonies, and conspicuous consumption by the leadership at the expense of essential infrastructure and productive enterprises; even to xenophobia directed against both foreign nationals within the country and other countries themselves. Carried beyond reasonable limits, nationalism can be destructive and retard the development of a new country.

ECONOMIC DEVELOPMENT

In most cases, as we have seen, the colonial countries invested in their colonies whatever they felt was necessary to control and administer the territory and to extract its wealth, whether or not that wealth was shared with the territory. After independence, the old relationships between ruler and ruled may be altered or completely severed or scarcely changed at all, exept symbolically, but almost invariably there is a new emphasis on development of the country for its own benefit. Some countries found themselves at independence far from rich but still with sound economic structures, money in the treasuries, and good credit ratings. Others, while not as well off as the first group at independence, nevertheless had resources that were much in demand in the industrialized countries and could look forward with some confidence to a considerable improvement in their living standards.

At the opposite end of the scale are countries so poverty-stricken that the United Nations has created the category of "least developed among the developing countries." Indeed, some observers feel that some simply can never be self-supporting and will either remain permanent charity cases or go out of business entirely. In between the extremes, the great majority of the new countries are struggling against many handicaps, buffeted by political rivalries and conflicts, caught in situations over which they have little or no control, trying to become a little less poor.

Many of the new countries at independence found themselves with economies geared almost entirely toward the former metropolitan countries. Thus the extractive industries—mines, plantations, sawmills, oil wells—tended to be well developed at the expense of manufacturing and services. Transportation systems were designed to connect the capital with important administrative centers, and seaports with important economic centers. Export crops were emphasized at the expense of subsistence crops and often food had to be imported into countries that could, if permitted, feed themselves. Land tenure was often a serious problem, with the best agricultural land held in plantations and *latifundia*. Soil exhaustion and erosion were severe in some places. Trade agreements had to be negotiated that would help the new countries break out of the "colonial economy" pattern of exporting commodities and importing manufactured goods. Vital imports had to be bought in soft currency markets, on a barter basis, or with borrowed money. Populations were growing rapidly, compounding all their economic, social, and political problems. The catalogue of economic problems is very long, but we are concerned here primarily with what is being done to relieve them.

Nearly every new country has undertaken some degree of national economic planning, even those that have rejected a Marxist approach to economic development, and some have adopted one or another variety of socialism. Some have undertaken land reform programs, attempted to settle nomads, invested in massive infrastructure projects, tried to diversify the economy, begun industrializing (in part for political or nationalistic reasons, even if the industries are uneconomic and agriculture is neglected), and joined with other countries in a variety of ways to achieve together what they could not achieve separately.

If the economic problems of new countries are greater than ever before in history, so is the assistance available to them from outside. In most cases the former metropoles have aid programs for their former colonies, with the French program particularly large. Second, other rich countries have developed aid programs, generally for selected poor countries. Third, there are now a large number of international and regional agencies devoted to assisting developing countries (not just new ones). Finally, developing countries have begun assisting one another, also on both a bilateral and multilateral basis.

Too large a proportion (how much one can only guess) of this aid is wasted on inappropriate projects or equipment or training; too much is squandered on military equipment, supplies, and training; too much is lost through mismanagement and corruption; too much is misapplied because of competition among donors; and even good programs go bad with changes of government or policy or personnel. The "oil crisis" of 1973–1974 and the subsequent quintupling of petroleum prices has dealt a cruel blow to the poor countries of the world, most of whom have suffered far more than the industrialized rich countries. Some have been so badly hurt that the United Nations established a Special Fund to aid those countries "most seriously affected" by this situation.

Another problem that has become prominent in the last few years and is likely to become much more serious is that of debt servicing. A number of poor countries have such heavy burdens of repayment of loans, both principal and interest, that they are actually sending more money to rich countries than they are receiving from them.. Many other economic problems face the new countries, of course, but looming over all of them, both in magnitude and importance, is the incredibly rapid growth of population in most developing countries. In many cases countries that get ample assistance and follow all the rules do indeed increase production, improve their infrastructure, and modernize their economies in general—only to find all of their gains canceled out by an ever-growing population.

But we should not despair. The problems of new poor countries are immense but not overwhelming. They are all surviving somehow. Many are making headway and a few are doing quite well, moving into the middle class, so to speak. Family planning (or population control) has caught on and has in some places been remarkably successful, contributing materially to higher living standards. Development plans are being rethought and reorganized, and showing better results. Aid programs are being better coordinated. Poor countries have at last understood that environmental degradation is a problem for them and not just for the industrialized countries and are beginning to adopt environmental protection measures. Gradually, the developing countries have learned that more is to be gained by cooperation with rich countries than by confrontation with them.

FOREIGN RELATIONS

A new country has not only internal problems to wrestle with, but foreign ones as well. They include establishing formal relations with other countries, joining intergovernmental organizations, reviewing treaties to which it had been made a party, dealing with boundary problems, participating in international conferences, negotiating aid agreements and a multitude of technical matters, defining and articulating positions on international issues, even finding and training personnel to do all these things. These are formidable tasks, to say nothing of the problem of paying for them all. Several small new countries have simply elected not to do all this; they chose instead to keep their foreign relations and economic activities to a minimum. This generally means membership in the United Nations (but little activity therein) and a few other organizations essential to the country, diplomatic relations with the former colonial ruler and perhaps one other great power and one or two neighbors, a few routine treaty relationships, and very little else. A few have no more than nominal diplomatic relations with other countries. Indeed, Western Samoa postponed applying for UN membership for 14 years after independence and Nauru and Tuvalu (formerly Ellice Islands) have not yet applied. Most new

countries, however, are very much involved in world affairs.

The first conference of emerging countries, introduced in Chapter 18, was the Afro-Asian Solidarity Conference held in Bandung, Indonesia, in April 1955. Attending it were the People's Republic of China and 28 independent or soon-to-be-independent developing countries from Gold Coast and Turkey to the Philippines and both Vietnams. It marked the beginning of cooperation among them in economic matters, human rights and self-determination, problems of dependent peoples, and promotion of world peace and cooperation. But cooperation among both new and old poor countries, while much discussed and publicized, was limited primarily to confrontation with the rich countries on decolonization, trade, aid, human rights, and disarmament. Most of the new countries chose to remain "neutralist" or "nonaligned" in the Cold War between the United States and the Soviet Union. (This is the origin of the term "Third World," which has since taken on so many other meanings.) Others, however, regardless of their public statements, were at birth or soon after clearly aligned with one side or the other. "The spirit of Bandung" faded as the world changed rapidly in many ways during the 1960s.

One problem faced by nearly all new States is their boundaries. Even the island States have problems with maritime boundaries, but our concern here is with land boundaries. As we have seen, many of these boundaries were drawn by the colonial rulers with little regard for the needs or interests of the indigenous peoples. Throughout Africa and Asia today there are scores of boundary and territorial disputes, and even Latin America still has its share. The problems are most numerous and acute, however, in Africa. Considering the number and nature of these problems, it is remarkable that there have not been more border wars in Africa. In part, this has been due to the weakness of most of the contestants, to a lingering spirit of pan-Africanism, and to the need for unity in the face of greater perils, both internal and external, all of which have encouraged neighbors to negotiate their differences quietly and settle them by compromise. Many boundary segments have thus been delimited and at least partially demarcated.

Perhaps a more important influence, however, has been the policy adopted by the Organization of African Unity (enshrined in Article III, Paragraph 3 of its 1963 Charter) of "respect for the sovereignty and territorial integrity of each state and for its inalienable right to independent existence." The African States, many founded by revolutionary movements and many with radical domestic and foreign policies, have thus pledged themselves to preserve the *status quo* imposed by the colonial powers. This policy was based on the Roman law principle of *uti possidetis juris*, the same policy adopted by the Spanish American colonies during their wars for independence a century and a half earlier.

Despite the dilemmas posed by several border disputes and attempted secessions, the principle has been upheld: no fragmentation of existing countries; no forcible boundary changes; in a territorial dispute between an African State and a European State, the African is invariably right; all colonial territories must become independent intact. The African States have been unanimous in opposing the "homelands" policy of South Africa in her own territory and in Namibia, and they have refused to recognize Transkei, Bophuthatswana, Venda, and Ciskei as independent. They have also refused to recognize the validity of the 1974 referenda on the Indian Ocean island of Mayotte, in which the mostly Christian inhabitants voted overwhelmingly to remain a French territory, and insist that the island is an integral part of independent and largely Muslim Comoros, with which it was administered by the French until the end of 1975 when the Comoro Islands became independent. Self-determination, it would seem, has its limits in Africa—and elsewhere.

UNITED NATIONS ASSISTANCE

The new States are receiving development assistance of many kinds from a variety of international governmental and nongovernmental agencies. Prominent among them are the Organization for Economic Cooperation and Development (OECD), the European Communities (EC), and the Organization of Petroleum Exporting Countries (OPEC). Most important, however, is the work of the United Nations, its affiliates, and specialized agencies.

Although the General Assembly provides policy guidance and overall supervision of development assistance, most of this particular work comes within the purview of the Economic and Social Council and of the Department of International Economic and Social Affairs of the Secretariat. Apart from lingering problems of apartheid in South Africa and decolonization, the emphasis of the United Nations has, in fact, shifted since the influx of new members in the early 1960s from political and security matters to economic and social matters. This work has become so important, so complex, and handled by so many UN organs that it is currently being reorganized so as to place it within a more rational and efficient structure. Here we can describe only briefly some of the development work of the United Nations.

The Economic and Social Council (ECOSOC) has established five regional economic commissions that have very broad functions within their regions. They are the United Nations Economic Commissions for Europe (ECE), Latin America and the Caribbean (CEPAL), Africa (ECA), and Western Asia (ECWA), and the United Nations Economic and Social Commission for Asia and the Pacific (ESCAP). They do not provide money, but instead concentrate on research, publication, training, advice, coordination, stimulation of economic cooperation and integration within the region, and similar activities. The United Nations Development Programme (UNDP) operates in five areas: surveying and assessing development assets or resources of individual countries, stimulating capital investment, training in a wide range of vocational and professional skills, adapting and utilizing modern technology in development projects, and economic and social planning. Very little of its assistance involves financial aid to countries; most is in the form of experts, fellowships, specialized equipment, and technical services. UNDP also coordinates the work of all UN agencies in the individual developing countries.

More specialized assistance is rendered in their areas of competence by the United Nations Conference on Trade and Development (UNCTAD), the United Nations Industrial Development Organization, United Nations Environment Programme, World Food Council, and International Trade Centre. In addition, nearly all the specialized and affiliated organizations provide assistance. In a few cases, such as the International Labour Organization (ILO) and the Food and Agriculture Organization (FAO), the bulk of their work is devoted to helping developing countries. None of these agencies, except the World Bank Group and the International Monetary Fund (IMF), actually lends or gives money to the poor countries, except small amounts for specific projects from time to time. Another UN agency worthy of mention is UNITAR, the United Nations Institute for Training and Research, a small organ of the General Assembly devoted to enhancing the effectiveness of the organization itself. Its research projects, however, have included some of special interest to developing countries and its training is devoted almost exclusively to the developing countries. The training includes seminars, fellowships, courses, and exercises for diplomats-in-training. This is the kind of quiet, undramatic, and relatively inexpensive work being carried on by UN agencies that is so essential to the new and inexperienced countries of the world.

The International Monetary Fund does provide funds under very strict conditions for the purpose of stabilizing currencies. (The funds must be repaid and are not development funds *per se*; in fact, they are provided to rich countries experiencing temporary foreign exchange problems as well as to poor countries.) The World Bank Group helps developing countries almost exclusively. It consists of three components. The main one is the International Bank for Reconstruction and Development (IBRD or World Bank). It does preinvestment surveys and comprehensive studies of a country's development needs, helps develop overall development plans, and lends money on strict, almost commercial, terms for both specific projects and broader programs. The International Development Association (IDA) lends smaller amounts of money on concessionary terms (long-term, low-interest, long grace period), generally to the poorest countries and for projects least likely to generate actual cash returns. The International Finance Corporation (IFC), smallest of the three, makes loans to private businesses in developing countries that are likely to contribute significantly to their countries' economic development.

To put things in perspective, of the $30.6 billion in development aid to poor countries (new and old) in 1983, only $13.8 billion came from all multilateral institutions, including, but not limited to, those mentioned in this section.* The real value of UN and most other multilateral aid is not its size, which is obviously relatively small, but its high quality and largely nonpolitical nature. Many developing countries, if they have a choice, prefer multilateral aid for these reasons. The UN also utilizes personnel from developing countries who may be more acceptable to the receiving countries than personnel from rich countries. And, as mentioned before, there is grow-

*1983/1984 United nations *Statistical Yearbook*, p. 491.

ing interest in technical cooperation among developing countries, and the United Nations and other multilateral agencies are best equipped to stimulate and assist this encouraging trend.*

NEOCOLONIALISM
AND NEOIMPERIALISM

Few sharp divisions can be drawn between States in the world. There are nearly always some that fall in between two categories or overlap both. The division between rich and poor, as we have seen, is blurred by those not so poor and those that are not so rich. Similarly with imperialism and its victims, with colonialism and decolonization. Some countries are only nominally independent and some former colonies have themselves become imperialists.

The term "neocolonialism" was probably used first by Kwame Nkrumah, President of Ghana, the first black African State to receive independence after being a European colony. When Malaysia became independent five months after Ghana in 1957, Nkrumah ridiculed the move, pointing out that in practice nothing had changed except the flag and some other superficialities. Malaysia was still a British colony because the country had a capitalist economy controlled by the British, British troops were still stationed in Singapore (then a part of Malaysia), and the country was still very much oriented toward Britain. Thus, the arrangement was nothing but a new variety of colonialism. Later on, he and others developed this theme further. Nkrumah later elaborated it in a book whose title emulated Lenin's.† A few excerpts will convey the essentials of the concept.

The essence of neo-colonialism is that the State which is subject to it is, in theory, independent and has all the outward trappings of interna-

tional sovereignty. In reality its economic system and thus its political policy is directed from outside. . . . The result of neo-colonialism is that foreign capital is used for the exploitation rather than for the development of the less developed parts of the world. Investment under neo-colonialism increases rather than decreases the gap between the rich and the poor countries of the world. . . . Neo-colonialism is also the worst form of imperialism. For those who practice it, it means power without responsibility and for those who suffer from it, it means exploitation without redress. . . . Neo-colonialism is based upon the principle of breaking up former large united colonial territories into a number of small nonviable States which are incapable of independent development and must rely upon the former imperial power for defense and even internal security. Their economic and financial systems are linked, as in colonial days, with those of the former colonial ruler. . . . "Aid," therefore, to a neo-colonial State is merely a revolving credit, paid by the neo-colonial master, passing through the neo-colonial State and returning to the neo-colonial master in the form of increased profits. Secondly, it is in the field of "aid" that the rivalry of individual developed States first manifests itself. So long as neocolonialism persists so long will spheres of interest persist, and this makes multilateral aid—which is in fact the only effective form of aid—impossible.*

There is a great deal more to the thesis of neocolonialism, and some of it is exaggerated. Nevertheless, it is undeniably true that many former colonies are still essentially political satellites of the great powers and other rich countries and this factor must be taken into consideration in any analysis of current world relations, including geopolitics and power inventories.

Perhaps the most obvious and important manifestation of neocolonialism is the current pattern of international trade. We discuss this in more detail later, but a brief summary now may be useful. The great bulk of international trade today (both in volume and in value) is carried on among the industrial (rich) countries. They benefit both from buying and selling and

---

*The United Nations sponsored an international conference on this subject in Buenos Aires in 1978.
†K. Nkrumah, *Neo-colonialism; The Last Stage of Imperialism*, London: Thomas Nelson, 1965.

Ibid., Introduction.

from providing shipping, insurance, banking, and other services necessary to the conduct of international trade. Next in importance is trade between the rich countries and the poor countries. This is still primarily a "colonial economy"; that is, the rich countries sell manufactured goods and semifinished goods to the poor countries (mostly ex-colonies) in exchange for commodities (foods, fuels, and raw materials). Since the rich countries control much of the commodity production in the poor countries, they tend to hold down their prices, despite characteristic short-term price fluctuations. And since they also control their own factories, they tend to keep the prices of manufactured goods high. Thus, the poor countries are caught in a squeeze from which it is very difficult to escape. Both UNCTAD and OPEC represent attempts to escape from this squeeze. Regional economic integration and domestic "bootstrap" operations are others. Nationalization of foreign-owned enterprises is another. Whether any of these efforts will be successful in the long run remains to be seen, but clearly they will be much more successful—and perhaps unnecessary—if the rich countries cooperate with the poor ones.

The new countries, however, are not only victims of imperialism, colonialism, and neocolonialism. Some of them have indulged in these practices themselves. Soon after coming to power in Egypt, for example, Colonel Gamal Abdul Nasser began preaching Pan-Arabism, the unity of all the Arabs from the Atlantic to the Persian Gulf. It seems, though, that he envisioned himself at the head of this vast Arab commonwealth, for one by one he tried to gain control over nearly all his neighbors. Before the British left the Anglo-Egyptian Sudan, he worked diligently to get the Sudanese to agree to Egyptian rule. While Libya was still poverty-stricken, he put enormous pressure on old King Idris to join Egypt—until oil was found. Libya suddenly had its own source of revenue, and Egypt was no longer attractive. Nasser also had an ongoing battle with King Hussein of Jordan (to say nothing of his continuing war against Israel), engineered a federation with Syria that only lasted three years (1958–1961), and fought a five-year war in Yemen (1962–1967) with

A case of neoimperialism. President Sukarno of Indonesia tried for years to wrest from the British their territories in Borneo. In August 1963 a United Nations mission visited Jesselton, British North Borneo, to ascertain the feelings of the people about their future. Their feelings are shown quite clearly in this photo. A month later, this territory joined Singapore, Sarawak, and independent Malaya to form the Federation of Malaysia, but Indonesia continued her attacks on the Borneo territories until 1966. British North Borneo is now Sabah and Jesselton is now Kota Kinabalu. (UN)

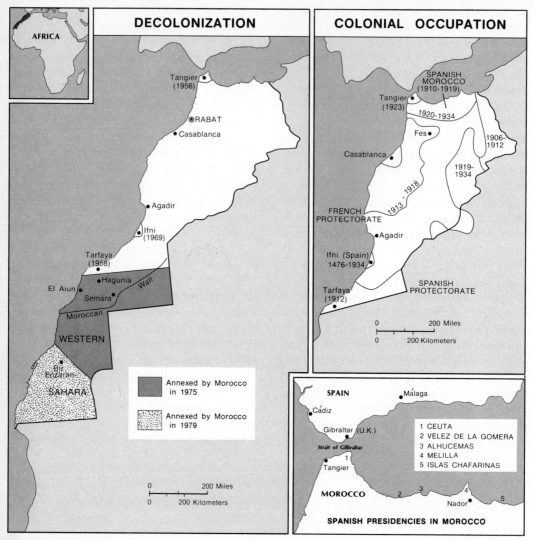

Morocco, from victim of imperialism to independent state to neoimperialist.   This country was gradually conquered after Spain and France agreed to establish protectorates over it and Tangier was made an international city. Decolonization came much more quickly (1956–1958) and Morocco went on to occupy all the former Spanish Sahara. Spain still retains five *plazas de soberania*, or presidios, essentially exclaves of Spain. The three smaller ones are islands. Note the wall the Moroccans have built in the south to keep out the POLISARIO forces.

up to seventy thousand troops and the most modern weapons available to him. The fact that he was not particularly successful does not make Nasser any less imperialistic.

India likewise has been behaving much like an imperialist power from its birth. In 1948, when the Muslim Nizam of Hyderabad acceded to Pakistan, the Indian army simply invaded his large state in southern India and it was absorbed into India. Sim-

ilarly, when the Muslim people of Kashmir clearly wanted to join Pakistan, India attempted to annex it and fought a war with Pakistan over it, occupying the most valuable two thirds of it. France yielded her five colonies in the peninsula to India peacefully between 1952 and 1954, but Portugal stubbornly held on to her four colonies there. India lost patience and in 1961 simply invaded and annexed them. The Indian army went to work again in

1971 when it invaded East Bengal to aid separatists there who were fighting the Pakistani army. The Indians withdrew and Bangladesh emerged from what had been East Pakistan. Even without fighting, India inherited the British role as protector of Sikkim and Bhutan and the dominant influence in Nepal. In 1971, India graciously permitted Bhutan to become nominally independent and even join the United Nations but still keeps her under control in a classic neocolonial relationship. In 1974 she absorbed Sikkim completely as a state.

Indonesia under President Sukarno was also imperialistic, waging a violent propaganda war against the Netherlands for western New Guinea and even dropping paratroops into the territory. He finally won the territory but lost his prolonged "confrontation' with Britain and then Malaysia to secure control of Sarawak, Brunei, and Sabah (formerly British North Borneo). Then, in December 1975, post-Sukarno Indonesia invaded East Timor as the Portuguese withdrew and after heavy fighting against Timorese forces annexed the territory in July 1976.

Morocco and Mauritania ignored the principles both of self-determination and *uti possidetis juris* when they partitioned Western Sahara between them as the Spanish withdrew in 1975. (There is a touch of irony in this partnership since Morocco had a long-standing claim to all of Mauritania as well as the Spanish territory and even delayed Mauritania's admission to the United Nations for a year because of it.) Somalia's irredentist wars against Ethiopia and Kenya, whatever their justification on ethnic grounds, were as imperialistic as Germany's claims to Czechoslovakia's Sudetenland in the 1930s. Vietnam's conquest of Kampuchea in 1978–1979 and Libya's occupation of a huge strip of northern Chad between 1978 and 1987 were only the latest, but almost certainly not the last, cases of what we may call "neoimperialism," a new phase or cycle of imperialism being practiced by States that are not even capitalist or rich. Some theories about imperialism will have to be revised.

The new States of the world are interesting enough individually; collectively they are fascinating. It is instructive also to apply to each of them the theories we have developed through the past century in political geography. No discussion of international relations will be of any value from this point on unless it takes into account the effects of the tripling of the international community in a third of a century and the role of the new States individually and collectively.

REFERENCES FOR PART FIVE

Books and Monographs

A

Abbas, Mikki, *The Sudan Question: The Dispute over the Anglo–Egyptian Condomimium 1884–1951.* New York: Praeger, 1952.

Adams, T.W., and A. Cottrell, *Cyprus Between East and West.* Baltimore: Johns Hopkins Univ. Press, 1968.

Akzin, Benjamin, *New States and International Organization.* Paris: UNESCO, 1955.

al-Baharna, Husain M. *The Arabian Gulf States: Their Legal and Political Status and Their International Problems,* 2nd rev. ed. Beirut: Librarie du Liban, 1975.

Alexander, Yonah and Robert A. Friedlander (eds.), *Self-Determination: National, Regional, and Global Dimensions.* Boulder, CO: Westview, 1980.

Allan, J.A. (ed.), *Libya Since Independence.* New York: St. Martin's Press, 1982.

Allen, Philip M., *Self-Determination in the Western Indian Ocean.* New York: Carnegie Endowment for International Peace, 1966.

Allen, Sir Richard, *Malaysia, Prospect and Retrospect: The Impact and Aftermath of Colonial Rule.* New York: Oxford Univ. Press, 1968.

——, *Imperialism and Nationalism in the Fertile Crescent: Sources and Prospects of the Arab–Israeli Conflict.* New York: Oxford Univ. Press, 1974.

Almond, Gabriel A. and James S. Coleman, *The Politics of the Developing Areas.* Princeton, NJ: Princeton Univ. Press, 1960.

Amin, Abdul Amiz, *British Interests in the Persian Gulf.* Leiden, Neth.: Brill, 1967.

Apter, David E., *The Gold Coast in Transition.* Princeton, NJ: Princeton Univ. Press, 1955.

Aruri, Naseer, *Jordan: A Study in Political Development (1921–1965).* The Hague: Nijhoff, 1972.

Ashford, Douglas E., *Political Change in Morocco.* Princeton, NJ: Princeton Univ. Press, 1961.

Asiwaju, A.I. (ed.), *Partitioned Africans.* New York: St. Martin's Press, 1985.

Augelli, John P., *The Panama Canal Area in Transition.* American Universities Field Staff Reports, North America Series, Part I, 1981, No. 3; Part II, 1981, No. 4.

Austen, Ralph A., *Modern Imperialism: Western Overseas Expansion and Its Aftermath, 1776–1965.* Lexington, MA: Heath, 1969.

Ayearst, Morley, *The British West Indies: The Search for Self-Government.* New York: New York Univ. Press, 1960.

B

Bacon, Elizabeth E., *Central Asians Under Russian Rule.* Ithaca, NY: Cornell Univ. Press, 1966.

Baker Fox, A., *The Power of Small States: Diplomacy of World War II.* Chicago: Chicago Univ. Press, 1967.

Ballard, J.A. (ed.), *Policy-Making in a New State; Papua New Guinea 1972–77.* St. Lucia: Univ. of Queensland Press, 1981.

——, *Uncertain Dimensions: Western Overseas Empires in the Twentieth Century.* Minneapolis: Univ. of Minnesota Press, 1985.

Banks, A.L., *The Development of Tropical and Subtropical Countries with Particular Reference to Africa.* London: Arnold, 1954.

Barber, James, *Rhodesia: The Road to Rebellion.* London: Oxford Univ. Press for the Institute of Race Relations, 1967.

Barone, Charles A., *Marxist Thought on Imperialism: A Critical Survey.* Armonk, NY: Sharpe, 1985.

Beale, Howard K., *Theodore Roosevelt and the Rise of America to World Power.* New York: Collier, 1956.

Beardsley, Charles, *Guam: Past and Present.* Rutland, VT: Tuttle, 1964.

Bender, Gerald J., *Angola Under the Portuguese: The Myth and the Reality.* Berkeley: Univ. of California Press, 1978.

Benedict, Burton (ed.), *Problems of Smaller Territories.* London: Univ. of London, 1967.

Bentwich, Norman, *Israel Resurgent.* New York: Praeger, 1960.

Berard, Victor, *British Imperialism and Commercial Supremacy.* New York: Longmans, 1906.

Betts, R.F., *Assimilation and Association in*

French Colonial Theory (1890–1914). London, 1961.

Bhutto, Zulfikar Ali, *The Myth of Independence*. New York: Oxford Univ. Press, 1969.

Bley, Helmut, *South-West Africa Under German Rule 1894–1914*. Evanston, IL: Northwestern Univ. Press, 1968.

Bloomfield, Richard J. (ed.), *Puerto Rico; The Search for a National Policy*. London: Frances Pinter, 1985.

Boyce, Peter J., *Foreign Affairs for New States*. New York: St. Martin's Press, 1978.

Brausch, George, *Belgian Administration in the Congo*. London: Oxford Univ. Press, 1961.

Brecher, Michael, *The New States of Asia*. New York: Oxford Univ. Press, 1966.

Brett, E.A., *Colonialism and Underdevelopment in East Africa: The Politics of Economic Change 1919–1939*. Portsmouth, NH: Heinemann Educational Books, 1973.

Bretton, H.L., *Power and Stability in Nigeria: The Politics of Decolonisation*. New York: Praeger, 1962.

Brewer, A., *Marxist Theories of Imperialism; A Critical Survey*. London: Routledge & Kegan Paul, 1980.

Brookfield, Harold C., *Colonialism, Development and Independence: The Case of the Melanesian Islands in the South Pacific*. Toronto: Macmillan, 1972.

Buchanan, K., *The Geography of Empire*. Nottingham, Eng.: Spokesman, 1972.

Buchheit, Lee C., *Secession: The Legitimacy of Self-Determination*. New Haven, CT and London: Yale Univ. Press, 1978.

Burrows, Susan, *Self-Determination: An Interdisciplinary Annotated Bibliography*. New York: Garland, 1986.

Burton, Benedict, *Problems of Smaller Territories*, London: Oxford Univ. Press, 1967.

C

Cabral, Amilcar. *Revolution in Guinea*. New York: Monthly Review Press, 1970.

Carrington, C.E., *An Exposition of Empire*. New York: Cambridge Univ. Press, 1947.

——, *Gibraltar*. London: Royal Institute of International Affairs, 1956.

Carroll, Faye, *South West Africa and the United Nations*. Lexington: Kentucky Univ. Press, 1967.

Carter, Gwendolen M., *Independence for Africa*. New York: Praeger, 1960.

Carter, Gwendolen M. and Patrick O'Meara (eds.), *African Independence; The First Twenty-Five Years*. Bloomington: Indiana Univ. Press, 1986.

Chapman, Terry M., *The Decolonisation of Niue*. Wellington, N.Z.: Victoria Univ. Press, 1976.

Chilcote, Ronald H., *Emerging Nationalism in Portuguese African Documents*. Stanford, CA: Hoover Institution Press, 1972.

—— (ed.), *Protest and Resistance in Angola and Brazil: Comparative Studies*. Berkeley: Univ. of California Press, 1972.

Chowdhuri, R.N., *International Mandates and Trusteeship Systems: A Comparative Study*. The Hague: Nijhoff, 1955.

Christopher, A.J., *Colonial Africa*. Totowa, NJ: Barnes & Noble, 1984.

Church, R.J. Harrison, *Modern Colonization*. London: Hutchinson's Univ. Library, 1951.

Clark, G., *The Balance Sheet of Imperialism*. New York: Columbia Univ. Press, 1936.

Clarke, C.G., and A.J. Payne (eds.), *Politics, Security and Development in Small States*. Winchester, MA: Allen & Unwin, 1987.

Clements, Frank, *Rhodesia: A Study of the Deterioration of a White Society*. New York: Praeger, 1969.

Clutterbuck, R., *Guerrillas and Terrorists*. London: Faber & Faber, 1977.

Clyde, P.H., *Japan's Pacific Mandate*. New York: Macmillan, 1935.

Cobban, Alfred, *National Self-Determination*. London: Oxford Univ. Press, 1944.

Cockram, Gail-Maryse, *South West African Mandate*. Cape Town: Juta, 1976.

Cohen, Michael J., *Palestine and the Great Powers, 1945–1948*. Lawrenceville, NJ: Princeton Univ. Press, 1982.

Cohen, Robin (ed.), *African Islands and Enclaves*. Beverly Hills, CA: Sage, 1983.

Coleman, James S., *Nigeria: Background to Nationalism*. Berkeley: Univ. of California Press, 1958.

Coulter, John Wesley, *The Pacific Dependencies of the United States*. New York: Macmillan, 1957.

——, *The Drama of Fiji: A Contemporary History*. Rutland, VT: Tuttle, 1967.

Cromer, Earl, *Ancient and Modern Imperialism*. London: John Murray, 1910.

Crowder, Michael, *Senegal: A Study in French Assimilation Policy*. London: Oxford Univ. Press, 1962.

——, *West Africa Under Colonial Rule*. Evanston, IL: Northwestern Univ. Press, 1968.

Crozier, Brian, *The Morning After; A Study of Independence*. London: Methuen, 1963.

Currie, David P. (ed.), *Federalism in the New Nations*. Chicago: Univ. of Chicago Press, 1964.

D

Darby, Phillip, *Three Faces of Imperialism*. New Haven, CT: Yale Univ. Press, 1987.

Davidson, James Wightman. *Samoa mo Samoa: The Emergence of the Independent State of Western Samoa*. London: Oxford Univ. Press, 1967.

Davis, William Columbus, *The Last Conquistadores; The Spanish Intervention in Peru and Chile, 1863–1866*. Athens: Univ. of Georgia Press, 1950.

Day, John, *International Nationalism; The Extra-territorial Relations of Southern Rhodesian African Nationalists*. London: Routledge & Kegan Paul, 1967.

Delius, Peter, *The Land Belongs to Us: The Pedi Polity, the Boers and the British in the Nineteenth Century Transvaal*. Berkeley: Univ. of California Press, 1984.

De Smith, Stanley A., *Microstates and Micronesia; Problems of America's Pacific Islands and Other Minute Territories*. New York: New York Univ. Press, 1970.

Deutsch, Karl W. *Nationalism and Social Communication*. New York: Wiley, 1953.

Deutsch, Karl W. and William J. Foltz (eds.), *Nation-Building*. New York: Atherton, 1963.

Dietz, James L., *Economic History of Puerto Rico*. Lawrenceville, NJ: Princeton Univ. Press, 1987.

Dore, Isaak I., *The International Mandate System and Namibia*. London: Westview, 1985.

Duffy, James, *Portuguese Africa*. Cambridge: Harvard Univ. Press, 1959.

Dugard, J. (ed.), *The South West Africa/Namibia Dispute: Documents and Scholarly Writings on the Controversy Between South Africa and the United Nations*. Berkeley: Univ. of California Press, 1973.

Duignan, Peter and Lewis Henry Gann (eds.), *Colonialism in Africa 1870–1960*, (Multi-volume.) Cambridge, England: Cambridge Univ. Press, 1975.

Dunlop, Eric W. and Walter Pike, *Australia . . . Colony to Nation*. London: Longmans, 1960.

Dunn, Rose E., *Resistance in the Desert: Moroccan Responses to French Imperialism*. Madison: Univ. of Wisconsin Press, 1977.

E

Edgell, M.C.R. and B.H. Farrell (eds.), *Themes on Pacific Lands*. Western Geographical Series No. 10. Victoria, B.C.: Univ. of Victoria, 1974.

Eisenstadt, S.N., *The Political Systems of Empires*. New York: Free Press, 1963.

El-Ayouty, Yassin, *The United Nations and Decolonization: The Role of Afro–Asia*. The Hague: Nijhoff, 1971.

Emerson, Rupert, *From Empire to Nation*. Boston: Beacon Press, 1962.

——, *Self-Determination Revisited in the Era of Decolonization*. Cambridge: Harvard Univ. Center for International Affairs, 1964.

Emerson, Rupert and others, *America's Pacific Dependencies*. New York: American Institute of Pacific Relations, 1949.

Erlich, Haggai, *Ethiopia and the Challenge of Independence*. London: Frances Pinter, 1986.

Etherington, Norman, *Theories of Imperialism, War, Conquest and Capital*. Totowa, NJ: Rowman & Littlefield, 1984.

F

Falk, Pamela S. (ed.), *The Political Status of Puerto Rico*. Lexington, MA: Lexington Books, 1986.

Fawcett, Charles B., *A Political Geography of the British Empire*. Boston: Ginn, 1933.

——, "Geography and Empire," in Griffith Taylor (ed.), *Geography in the Twentieth Century*, New York: Philosophical Library; London: Methuen; 1951, 418–432.

Fenelon, K.G., *The United Arab Emirates*, 2nd ed. London and New York: Longmans, 1976.

Fetter, Bruce, *Colonial Rule and Regional Imbalance in. Central Africa*. Boulder, CO: Westview, 1983.

Fieldhouse, D.K., *Black Africa 1945–1980*. London: Allen & Unwin, 1986.

Fisk, Ernest Kelvin, *The Political Economy of Independent Fiji*. Canberra: Australian National Univ. Press, 1970.

Fitzsimmons, M.A., *Empire by Treaty; Britain and the Middle East in the Twentieth Century*. Notre Dame, IN: Univ. of Notre Dame Press, 1964.

Forsythe, David P., *United Nations Peace Making; The Conciliation Commission for Palestine*. Baltimore and London: Johns Hopkins Univ. Press, 1972.

Frankel, S. Herbert, *The Concept of Colonization*. Oxford: Oxford Univ. Press, Clarendon, 1949.

Freymond, J., *Small States in International Relations*. Stockholm: Almqvist & Wiksells, 1971.

Furnivall, J.S., *Netherlands India: A Study in Plural Economy*. New York: Macmillan, 1944.

———, *Colonial Policy and Practice: A Comparative Study of Burma and Netherlands India*. New York: New York Univ. Press, 1956.

G

Gale, Roger W., *The Americanization of Micronesia*. Washington: Univ. Press of America, 1979.

Gann, Lewis Henry and Peter Duignan, *White Settlers in Tropical Africa*. Harmondsworth, Middlesex, Eng.: Penguin, 1962.

Garcia Passalacqua, Juan M., *Puerto Rico; Equality and Freedom at Issue*. New York: Praeger, 1984.

Gerteiny, A.G., *Mauritania: A Survey of a New African Nation*. New York: Praeger, 1967.

Geyer, Dietrich, *A History of Southern Rhodesia*. London: Chatto & Windus, 1965.

———, *Russian Imperialism*. New Haven, CT: Yale Univ. Press, 1987.

Geyer, Dietrich and Peter Duignan, *Burden of Empire; An Appraisal of Western Colonialism in Africa South of the Sahara*. New York: Praeger, 1967.

Gibson, Richard, *African Liberation Movements; Contemporary Struggles Against White Minority Rules*. New York: Oxford Univ. Press for the Institute of Race Relations, 1972.

Gifford, Prosser and William Roger Louis (eds.), *France and Britain in Africa: Imperial Rivalry and Colonial Rule*. New Haven, CT: Yale Univ. Press, 1971.

Gill, T.D., *South West Africa and the Sacred Trust 1919–1972*. The Hague: TMC Asser Instituut, 1984.

Goldblatt, I., *The Mandated Territory of South West Africa in Relation to the United Nations*. Cape Town: Struik, 1961.

Goldman, Robert B., *From Independence to Statehood*. New York: St. Martin's Press, 1984.

Good, Robert C., *U.D.I.; The International Politics of the Rhodesian Rebellion*. London: Faber & Faber, 1973.

Grant, G.C., *The African's Predicament in Rhodesia*. Minority Rights Group, Report No. 8, London, 1972.

Green, L.P., and T.J.D. Fair, *Development in Africa: A Study in Regional Analysis with Special Reference to South Africa*. Johannesburg: Witwatersrand Univ. Press, 1962.

Green, Reginald Herbold, *From Sudwestafrika to Namibia; The Political Economy of Transition*. Uppsala, Swed.: Scandinavian Institute of African Studies, 1981.

Grundy, Kenneth W., *Guerilla Struggle in Africa: An Analysis and Preview*. New York: Grossman, 1971.

H

Hammer, Ellen J., *The Struggle for Indochina*. Stanford, CA: Stanford Univ. Press, 1954.

Hancock, W.K., *Wealth of Colonies*. Cambridge: at the University Press, 1950.

Hanna, Willard A., *Sequel to Colonialism; The 1957–1960 Foundations for Malaysia*. New York: American Universities Field Staff, 1965.

Hardy, W.G., *From Sea unto Sea: Canada—1850 to 1910; the Road to Nationhood*. Garden City, NY: Doubleday, 1960.

Harrigan, Norwell, *The Inter-Virgin Islands Conference: A Study of a Microstate International Organization*. Gainesville: Univ. Presses of Florida, 1980.

Harrison, John A., *Japan's Northern Frontier: A Preliminary Study in Colonization and Expansion with Special Reference to Relations of Japan and Russia*. Gainesville: Univ. of Florida Press, 1953.

Hastings, Peter, *New Guinea: Problems and Prospects*. Melbourne: Cheshire for the Australian Institute of International Affairs, 1969.

Hatch, John Charles, *The History of Britain in Africa: From the Fifteenth Century to the Present*. New York: Praeger, 1969.

Hawley, Donald, *The Trucial States*. New York: Humanities Press, 1971.

Healy, D.F., *U.S. Expansionism: The Imperialist Urge in the 1890's*. Madison: Univ. of Wisconsin Press, 1970.

Hearnshaw, F.J.C., *Sea-Power and Empire*. London: Harrap, 1940.

Heine, Carl, *Micronesia at the Crossroads*. Honolulu: Univ. Press of Hawaii, 1974.

Hidayatullah, M., *The South-West Africa Case*. Bombay: Asia Publishing House, 1967.

Hinden, Rita, *The United Nations and the Colonies*. London: UN Association, 1950.

Hoadley, J. Stephen, *The Future of Portuguese Timor; Dilemmas and Opportunities*. Occasional Paper No. 27. Singapore: Institute of Southeast Asian Studies, 1975.

Hobson, J.A., *Imperialism*. London: Allen & Unwin, 1938.

Hodges, Tony, *Western Sahara; the Roots of the Desert War.* Westport, CT: Lawrence Hill, 1983.

Holland, R.F., *European Decolonization 1918–1981.* New York: St. Martin's Press, 1985.

Hope, Kempe Ronald, *Economic Development in the Caribbean.* Westport, CT: Praeger, 1986.

Hoskins, H.L., *British Routes to India.* New York: Longmans, Green, 1928.

———, *European Imperialism in Africa.* New York: Holt, 1930.

Howitz, Irving Louis, *Beyond Empire and Revolution: Militarization and Consolidation in the Third World.* New York: Oxford Univ. Press, 1982.

Hudson, W.J. (ed.), *Australia and the Colonial Question at the United Nations.* Honolulu: Univ. Press of Hawaii, East–West Center Books, 1970.

——— (ed.), *Australia and Papua New Guinea.* Sydney: Sydney Univ. Press, 1971.

———, *Australia's New Guinea Question.* Melbourne: Nelson in association with the Australian Institute of International Affairs, 1975.

Huibregtse, P.K., *Angola, the Real Story,* trans. by N. Buhr. The Hague: Zuid-Hollandsche Uitgeversmaatschappij, 1973.

I

Ingrams, Harold, *Uganda: A Crisis of Nationhood.* London: HMSO, 1960.

Isaacman, Allen and Barbara Isaacman, *Mozambique: From Colonialism to Revolution, 1900–1982.* Boulder, CO: Westview, 1983.

J

James, Harold and Dennis Sheil-Small, *The Undeclared War.* Totowa, NJ: Rowman & Littlefield, n.d.

Jaster, Robert S., *South Africa in Namibia: The Botha Strategy.* Lanham, MD: Univ. Press of America, 1985.

Jennings, Sir Ivor, *Nationalism and Political Development in Ceylon.* New York: Institute of Pacific Relations, 1950.

Johansen, G.K. and H.H. Kraft, *Germany's Colonial Problems.* London: Butterworths, 1937.

John, Sir Rupert, *Pioneers in Nation-Building in a Caribbean Mini-State.* New York: UNITAR, 1979.

Johnson, Harold Scholl, *Self-Determination Within the Community of Nations.* Leyden, Neth.: Sijthoff, 1967.

Johnson, L.W., *Colonial Sunset: Australia and Papua New Guinea 1970–74.* St. Lucia, London, and New York: Univ. of Queensland Press, 1984.

Johnson, Roberta Ann, *Puerto Rico: Commonwealth or Colony?* New York: Praeger, 1980.

Johnston, Sir Harry H., *A History of the Colonization of Africa.* Totowa, NJ: Rowman & Littlefield, 1966.

Jolliffe, Jill, *East Timor: Nationalism and Colonialism.* St. Lucia: Univ. of Queensland Press, 1978.

Jones, F.C., *Japan's New Order in East Asia: Its Rise and Fall, 1937–1945.* New York: Oxford Univ. Press, 1954.

K

Kahler, Miles, *Decolonization in Britain and France.* Princeton, NJ: Princeton Univ. Press, 1984.

Kanduza, Ackson M., *The Political Economy of Underdevelopment in Northern Rhodesia, 1918–1960.* Halifax, N.S.: Dalhousie African Studies Series 5, 1986.

Kann, R.A., *The Hapsburg Empire: A Study in Integration and Disintegration.* New York: Praeger, 1957.

Karnes, Thomas L., *The Failure of Union: Central America, 1824–1960.* Chapel Hill: Univ. of NC Press, 1961.

Kautsky, J.H. (ed.), *Political Change in Underdeveloped Countries: Nationalism and Communism.* New York: Wiley, 1962.

Kawai, T., *The Goal of Japanese Expansion.* Tokyo: Hokuseido Press, 1938.

Kearney, Robert N., *Politics and Modernization in South and Southeast Asia.* New York: Halsted Press, 1975.

Kelly, Ian, *Hong Kong.* Honolulu: University of Hawaii Press, 1987.

Keltie, J.S., *The Partition of Africa.* London, 1895.

Kennedy, Paul M., *The Samoan Tangle.* New York: Harper & Row, 1974.

Khadduri, Majid, *Modern Libya: A Study in Political Development.* Baltimore: Johns Hopkins Univ. Press, 1963.

Khoury, Philip S., *Syria and the French Mandate.* Princeton, NJ: Princeton Univ. Press, 1936.

Kilson, Martin (ed.), *New States in the Modern World.* Cambridge: Harvard Univ. Press, 1975.

King, Frank H.H., *The New Malayan Nation; A Study of Communalism and Nationalism.* New York: Institute of Pacific Relations, 1957.

Kofele-Kale, Ndiva (ed.), *An African Experiment*

in *Nation Building: The Bilingual Cameroon Republic Since Reunification.* Boulder, CO: Westview, 1980.

Kolarz, Walter, *Russia and Her Colonies.* London: George Philip, 1952.

Kostiner, Joseph, *The Struggle for South Yemen.* New York: St. Martin's Press, 1984.

L

Lagerberg, Kees, *West Irian and Jakarta Imperialism.* New York: St. Martin's Press, 1980.

Lee, J.M., *Colonial Development and Good Government; A Study of the Ideas Expressed by the British Official Classes in Planning Decolonization, 1939–1964.* Oxford: Oxford Univ. Press, Clarendon, 1967.

Legge, J.D., *Australian Colonial Policy; A Survey of Native Administration and European Development in Papua.* Sydney: Angus & Robertson, 1956.

Legum, Colin (ed.), *Zambia; Independence and Beyond. The Speeches of Kenneth Kaunda.* London: Nelson, 1966.

Leibowitz, Arnold H., *Colonial Emancipation in the Pacific and the Caribbean: A Legal and Political Analysis.* New York: Praeger, 1976.

LeVine, Victor T., *The Cameroons: From Mandate to Independence.* Berkeley: Univ. of California Press, 1965.

Lewis, Vaughan A. (ed.), *Size, Self-Determination and International Relations; The Caribbean.* Mona, Jamaica: Univ. of the West Indies, Institute of Social and Economic Research, 1976.

Lijphart, Arend, *The Trauma of Decolonization: The Dutch and West New Guinea.* New Haven, CT: Yale Univ. Press, 1966.

Lin, Tzong-biau and others, *Hong Kong: Economic, Social and Political Studies in Development.* Armonk, NY: Sharpe, 1979.

Lofchie, Michael, *Zanzibar: Background to Revolution.* Princeton, NJ: Princeton Univ. Press, 1965.

Loney, Martin, *Rhodesia: White Racism and Imperial Response.* Harmondsworth, Middlesex, Eng.: Penguin, 1975.

Louis, William Roger, *National Security and International Trusteeship in the Pacific.* Annapolis, MD: Naval Institute Press, 1972.

Lucas, C.P., *The Partition and Colonization of Africa.* Oxford, 1922.

Lugard, Lord, *The Dual Mandate in British Tropical Africa.* Edinburgh: Blackwood, 1922.

Lusignan, Guy de, *French-Speaking Africa Since Independence.* New York: Praeger, 1969.

Lutchman, Harold Alexander, *From Colonialism to Cooperative Republic; Aspects of Political Development in Guyana.* Río Piedras: Univ. of Puerto Rico, 1974.

M

Macmillan, W.M., *The Road to Self Rule.* London: Faber & Faber, 1959.

Madariaga, Salvador de, *The Rise of the Spanish American Empire.* New York: Macmillan, 1947.

Mair, Lucy Philip, *Australia in New Guinea,* 2nd ed. Carlton, Victoria: Melbourne Univ. Press, 1970.

Malcolm, George A., *First Malayan Republic: The Story of the Philippines.* Boston: Christopher, 1951.

Mandaza, Ibbo (ed.), *Zimbabwe: The Political Economy of Transition, 1980–1986.* Atlantic Highlands, NJ: Humanities Press International, 1986.

Marcum, John A., *The Angolan Revolution,* 2 vols. Cambridge: MIT Press, 1969.

Mazrui, Ali A., *Nationalism and New States in Africa.* Portsmouth, NH: Heinemann, 1984.

McDonald, A.H. (ed.), *Trusteeship in the Pacific.* Sydney: Angus & Robertson, 1949.

McHenry, Donald F., *Micronesia: Trust Betrayed; Altruism vs. Self-Interest in United States Foreign Policy.* Washington: Carnegie Endowment for International Peace, 1975.

McTague, John J., Jr., *British Policy in Palestine, 1917–1922.* Lanham, MD: Univ. Press of America, 1983.

Meinig, Donald W., "A Macrogeography of Western Imperialism," in F. Gale and G.H. Lawtoki, *Settlement and Encounters.* London: Oxford Univ. Press, 1969.

Meller, Norman, *The Congress of Micronesia; Development of the Legislative Process in the Trust Territory of the Pacific Islands.* Honolulu: University of Hawaii Press, 1969.

Mercer, John, *Spanish Sahara.* London: Allen & Unwin, 1976.

Miller, J.D.B., *The Politics of the Third World.* London: Oxford Univ. Press, 1966.

Miller, Norman N., *Kenya: The Quest for Prosperity.* Boulder, CO: Westview, 1983.

Minkel, Clarence W. and Ralph H. Alderman, *A Bibliography of British Honduras 1900–1970.* East Lansing: Michigan State Univ., Latin American Studies Center, 1970.

Minter, William, *Portuguese Africa and the West.* Harmondsworth, Middlesex, Eng.: Penguin, 1972.

Mitchell, Sir Harold, *Contemporary Politics and Economics in the Caribbean*. Athens: Ohio Univ. Press, 1967.

———, *Caribbean Patterns*, 2nd ed. New York: Wiley, 1972.

———, *Europe in the Caribbean*. New York: Cooper Square Publishers, 1973.

Misra, B.B., *District Administration and Rural Development in India*. Delhi: Oxford Univ. Press, 1983.

Mlambo, E.E.M., *Rhodesia; The Struggle for a Birthright*. London: Hurst, 1972.

Mondlane, Eduardo, *The Struggle for Mozambique 1969*. Harmondsworth, Middlesex, Eng.: Penguin, 1969.

Moon, P.T., *Imperialism and World Politics*. New York, 1933.

Moorson, Richard, *Walvis Bay, Namibia's Port*. London: International Defence and Aid Fund for Southern Africa, 1984.

Morgan, D.J., *The Official History of Colonial Development*. London: Macmillan, 1980.

Morrell, W.P., *Britain in the Pacific Islands*. Oxford: Oxford Univ. Press, Clarendon, 1960.

Mortimer, Edward, *France and the Africans 1944–1960: A Political History*. New York: Walker, 1969.

Mullen, Ken, *Cocos Keeling; The Islands Time Forgot*. Sydney: Angus & Robertson, 1974.

Myers, Ramon H. and Mark R. Peattie, *The Japanese Colonial Empire, 1895–1945*. Princeton, NJ: Princeton Univ. Press, 1987.

N

Nasser, Gamal Abdul, *Egypt's Liberation: The Philosophy of the Revolution*. Washington: Public Affairs Press, 1955.

Neuberger, Benyamin, *National Self-Determination in Post Colonial Africa*. London: Frances Pinter, 1986.

Nevin, David, *The American Touch in Micronesia*. New York: Norton, 1977.

Niebuhr, Reinhold, *The Structure of Nations and Empires*. New York: Scribner's, 1959.

Nkala, Jericho, *The United Nations, International Law, and the Rhodesian Independence Crisis*. New York: Oxford Univ. Press, Clarendon, 1985.

Nogueira, F., *The United Nations and Portugal; A Study of Anti-Colonialism*. London: Sidgwick & Jackson, 1963.

Nufer, Harold F., *Micronesia Under American Rule: An Evaluation of the Strategic Trusteeship (1947–1977)*. Hicksville, NY: Exposition Press, 1978.

Nuseibeh, H.Z., *The Ideas of Arab Nationalism*. Ithaca, NY: Cornell Univ. Press, 1956.

Nyangoni, Christopher and Gideon Nyandoro, *Zimbabwe Independence Movements*. Totowa, NJ: Rowman & Littlefield, 1979.

Nye, Joseph S., Jr., *Pan-Africanism and East African Integration*. Cambridge: Harvard University Press, 1967.

Nyerere, Julius K., *Freedom and Unity; A Selection from Writings and Speeches, 1952–1965*. London: Oxford Univ. Press. 1967.

O

Oduho, J. and William Deng, *The Problem of the Southern Sudan*. London: Oxford Univ. Press, 1963.

Ofuatey-Kodjoe, W., *The Principle of Self-Determination in International Law*. New York: Nellen, 1977.

Osborne, Robin, *Indonesia's Secret War; The Guerilla Struggle in Irian Jaya*. London: Allen & Unwin, 1985.

P

Padmore, George, *Africa: Britain's Third Empire*. London: Dobson, 1949.

———, *Pan-Africanism or Communism? The Coming Struggle for Africa*. New York: Roy, 1956.

Palley, Claire, *The Constitutional History and Law of Southern Rhodesia, 1888–1965, with Special Reference to Imperial Control*. Oxford: Oxford Univ. Press, Clarendon, 1966.

Palmer, Robin, *Land and Racial Domination in Rhodesia*. Berkeley: Univ. of California Press, 1977.

Palmier, Leslie H., *Indonesia and the Dutch*. London: Oxford Univ. Press, 1962.

Peattie, Mark R., *Nan'yō: The Rise and Fall of the Japanese in Micronesia 1885–1945*. Honolulu: Univ. of Hawaii Press, 1987.

Peterson, J.E., *Oman in the Twentieth Century*. Totowa, NJ: Rowman & Littlefield, 1978.

Phelan, John Leddy, *The Hispanization of the Philippines: Spanish Arms and Filipino Responses 1565–1700*. Madison: Univ. of Wisconsin Press, 1967.

Platt, Raye R. and others, *The European Possessions in the Caribbean Area*. New York: American Geographical Society, 1941.

Plischke, Elmer, *Microstates in World Affairs*. Washington: American Enterprise Institute for Public Policy Research, 1977.

Pollock, Norman Charles, *Studies in Emerging Africa*. London: Butterworths, 1971.

Pomerance, Michla, *Self-Determination in Law and Practice: The New Doctrine in the United Nations.* The Hague, Boston, and London: Nijhoff, 1982.

Pratt, Sir John T., *The Expansion of Europe into the Far East.* London: Sylvan, 1947.

Pratt, R. Cranford and D. Anthony Low, *Buganda and British Overrule,* New York: Oxford Univ. Press, 1960.

Prescott, John Robert Victor, "Resources, Policy and Development in West and North Central Africa," in John W. House and Michael G.R. Conzen (eds.) *Northern Geographical Essays.* Newcastle-upon-Tyne, England: Oriel Press, 1966, pp. 292–309.

Price, David Lynn, *The Western Sahara.* Beverly Hills, CA: Sage, 1979.

Pye, Lucian W., *Politics, Personality and Nation Building: Burma's Search for Identity.* New Haven, CT: Yale Univ. Press, 1962.

R

Rao, R.P. *Portuguese Rule in Goa, 1510–1961.* London: Asia Publishing House, 1963.

Rapaport, Jacques and others. *Small States and Territories: Status and Problems.* New York: Arno, 1971.

Reno, Philip, *The Ordeal of British Guiana.* New York: Monthly Review Press, 1964.

Rézette, Robert, *The Western Sahara and the Frontiers of Morocco.* Paris: Nouvelles Editions Latines, 1975.

———, *The Spanish Enclaves in Morocco.* Paris: Nouvelles Editions Latines, 1976.

Rivlin, Benjamin, *The United Nations and the Italian Colonies.* New York: Carnegie Endowment for International Peace, 1950.

Roberts, George O., *The Anguish of Third World Independence: The Sierra Leone Experience.* Lanham, MD: Univ. Press of America, 1982.

Roberts, W. Adolphe. *The French in the West Indies.* New York: Cooper Square Publishers, 1971.

Robinson, A.N.R., *The Mechanics of Independence: Patterns of Political and Economic Transformation in Trinidad and Tobago.* Cambridge: MIT, 1971.

Ronen, Dov, *The Quest for Self-Determination.* New Haven, CT and London: Yale Univ. Press, 1979.

Ross, Angus (ed.), *New Zealand's Record in the Pacific Islands in the Twentieth Century.* Auckland: Longman Paul for the New Zealand Institute of International Affairs, 1969.

Rudin, H.R., *Germans in the Cameroons.* London, 1938.

S

Sadik, Muhammad T., and W.P. Sharely, *Bahrain, Qatar and the United Arab Emirates: Colonial Past, Present Problems and Future Prospects.* Lexington, MA: Lexington Books, 1972.

Sampson, A., *The Arms Bazaar.* London: Hodder & Stoughton, 1977.

Schatzberg, Michael, *The Political Economy of Zimbabwe.* New York: Praeger, 1984.

Schwartz, Frederick A.O., Jr., *Nigeria: The Tribes, the Nation or the Race; the Politics of Independence.* Cambridge: MIT Press, 1965.

Serapiao, Luis B., and Mohamed A. El-Khawas, *Mozambique in the Twentieth Century: From Colonialism to Independence.* Lanham, MD: Univ. Press of America, 1979.

Shinar, Dov, *Palestinian Nation Building and the Role of Communications in the West Bank.* London: Frances Pinter, 1986.

Shorrock, William I., *French Imperialism in the Middle East: The Failure of Policy in Syria and Lebanon.* Madison: Univ. of Wisconsin Press, 1976.

Simmons, Adele Smith, *Modern Maturities: The Politics of Decolonization.* Bloomington: Indiana Univ. Press, 1982.

Smith, T., *The Pattern of Imperialism.* Cambridge: at the University Press, 1981.

Soggott, David, *Namibia; The Violent Heritage,* New York: St. Martin's Press, 1986.

Stahl, Kathleen M., *British and Soviet Colonial Systems.* London, 1951.

Starushenko, G.B., *The Principle of National Self-Determination in Soviet Foreign Policy,* trans. from Russian. Moscow: Foreign Languages Publishing House, 1963.

Steel, Robert Walter, and Charles A. Fisher (eds.), *The Problem of Malayan Unity in its Geographical Setting.* London: George Philip, 1956.

Stillman, Calvin W., *Africa in the Modern World.* Chicago: Chicago Univ. Press, 1955.

Stoecker, Helmuth (ed.), *German Imperialism in Africa.* Atlantic Highlands, NJ: Humanities Press International, 1986.

Strachey, John, *The End of Empire.* London: Gollancz, 1959.

Strange, Ian F., *The Falkland Islands.* Newton Abbot, Devon, Eng.: David & Charles, 1972.

Strausz-Hupe, Robert, and Harry W. Hazard

(eds.), *The Ideas of Colonialism*. New York: Praeger, 1958.

Sureda, A. Rigo, *The Evolution of the Right of Self-Determination; A Study of United Nations Practice*. Leyden, Neth.: Sijthoff, 1973.

SWAPO, Department of Information, *To Be Born a Nation; The Liberation Struggle for Namibia*. Atlantic Highlands, NJ: Humanities Press International, 1981.

Symonds, Richard, *The Making of Pakistan*. London: Faber & Faber, 1950.

T

Tagupa, William, *Politics in French Polynesia, 1945–1975*. Wellington: New Zealand Institute of International Affairs, 1976.

Thomas, Wolfgang H., *Economic Development in Namibia*. New Brunswick, NJ: Transaction Books, 1985.

Thompson, Carol B., *Challenge to Imperialism*. Boulder, CO: Westview, 1986.

Thompson, Virginia and Richard Adloff, *French West Africa*. Stanford, CA: Stanford Univ. Press, 1958.

———, *The Emerging States of French Equatorial Africa*. Stanford, CA: Stanford Univ. Press, 1960.

———, *Djibouti and the Horn of Africa*. Stanford, CA: Stanford Univ. Press, 1968.

———, *The French Pacific Islands: French Polynesia and New Caledonia*. Berkeley: Univ. of California Press, 1971.

———, *The Western Saharans*. Totowa, NJ: Rowman & Littlefield, 1980.

Topping, Donald M., *The Pacific Islands; Micronesia*. American Universities Field Staff Reports, South East Asian Series, Vol. 25, No. 4. 1977.

Torsvik, P. (ed.), *Mobilization, Center-Periphery Structures and Nation-Building*. Oslo: University of Oslo, 1981.

Touval, Saadia, *Somali Nationalism*. Cambridge: Harvard Univ. Press, 1963.

Towsend, M.E., *The Rise and Fall of Germany's Colonial Empire, 1884–1918*. New York: Macmillan, 1930.

Trumbull, Robert, *Paradise in Trust; a Report on Americans in Micronesia, 1946–1958*. New York: William Sloane, 1959.

U

Udechuku, E.C., *Liberation of Dependent Peoples in International Law*, 2nd ed. London: African Publishers Bureau, 1978.

Umozurike, Umozurike Oji, *Self-Determination in International Law*. Hamden, CT: Shoe String Press, Archon Books, 1972.

V

Van Cleve, Ruth G., *The Office of Territorial Affairs*. New York: Praeger, 1974.

van der Kroef, Justus M., *The West New Guinea Dispute*. New York: Institute of Pacific Relations, 1958.

Villard, Henry Serrano, *Libya: The New Arab Kingdom of North Africa*. Ithaca, NY: Cornell Univ. Press, 1956.

Villari, L., *The Expansion of Italy*. London: Faber, 1930.

Vital, D. *The Survival of Small States*. London: Oxford Univ. Press, 1971.

Viviani, Nancy, *Nauru: Phosphate and Political Progress*. Canberra: Australian National Univ. Press, 1970.

W

Wainhouse, David Walter, *Remnants of Empires: The United Nations and the End of Colonialism*. New York: Council of Foreign Relations, Harper & Row, 1964.

Walker, Eric A., *Colonies*. Cambridge: at the University Press, 1944.

Wallerstein, Immanuel, *Africa: The Politics of Independence*. New York: Random House, 1967.

Wambaugh, Sarah, *The Doctrine of Self-Determination*. New York: Oxford Univ. Press, 1919.

———, *A Monograph on Plebiscites*. New York: Oxford Univ. Press, 1920.

———, *Plebiscites Since the World War*. New York: Carnegie Endowment for International Peace, 1936.

Weinberg, A.K., *Manifest Destiny: A Study of Nationalist Expansion in American History*. Chicago: Quadrangle Books, 1963.

Weinstein, Brian, *Gabon: Nation-Building on the Ogooue*. Cambridge: MIT Press, 1966.

Wellington, John H., *South West Africa and Its Human Issues*. Oxford: Oxford Univ. Press, Clarendon, 1967.

Wenner, Manfred W., *Modern Yemen 1918–1966*. Baltimore: Johns Hopkins Univ. Press, 1967.

Wesley-Smith, Peter, *Unequal Treaty, 1898–1997. China, Great Britain and Hong Kong's New Territories*. Hong Kong: Oxford Univ. Press, 1980.

Wheare, Kenneth C., *The Statute of West-*

minster and Dominion Status. Oxford: Oxford Univ. Press, 1953.

White, Osmar Egmont Dorkin, *Parliament of a Thousand Tribes: A Study of New Guinea.* London: Heinemann, 1966.

Wiens, Herold J., *China's March Towards the Tropics.* Hamden, CT: Shoe String Press, 1954.

Wilhelm, Donald, *Emerging Indonesia.* Totowa, NJ: Rowman & Littlefield, 1980.

Williams, William Appleman, *Empire as a Way of Life.* New York: Oxford Univ. Press, 1980.

Wilson, Thomas, *Ulster Under Home Rule: A Study of the Political and Economic Problems of Northern Ireland.* London: Oxford Univ. Press, 1955.

Wriggins, W. Howard, *Ceylon: Dilemmas of a New Nation.* Princeton, NJ: Princeton Univ. Press, 1960.

Wright, Carol, *Mauritius.* Newton Abbot, Devon, Eng.: David & Charles, 1974.

Y

Yansané, Aguibon Y. (ed.), *Decolonization and Dependency: Problems of Development of African Societies.* Westport, CT: Greenwood Press, 1980.

Yeagher, Rodger, *Tanzania: An African Experiment.* Boulder, CO: Westview, 1983.

Young, C., *Politics in the Congo: Decolonization and Independence.* Princeton, NJ: Princeton Univ. Press, 1965.

Young, Kenneth, *Rhodesia and Independence; A Study in British Colonial Policy.* London: Dent, 1969.

Z

Zahlan, Rosemarie Said, *The Origins of the United Arab Emirates,* London: Macmillan, 1978.

———, *The Creation of Qatar.* Totowa, NJ: Rowman & Littlefield; London: Croom Helm; and New York: Barnes & Noble, 1979.

Periodicals

A

Abuetan, Barid, "Eritrea: United Nations Problem and Solution," *Middle Eastern Affairs,* 2, 2 (February 1951), 35–53.

Aitkin, Don and Edward P. Wolfers, "Australian Attitudes Towards the Papua New Guinea Area Since World War II," *Australian Outlook,* 27 (August 1973), 202–214.

Albinski, H.S., "Australia and the Dutch New Guinea Dispute," *International Journal,* 16 (Autumn 1961), 358–382.

Alley, R., "Independence for Fiji. Recent Constitutional and Political Developments," *Australian Outlook,* 24 (August 1970), 178–187.

Andreyanov, Y., "The Oil Monopolies in the Persian Gulf Colonies," *International Affairs* (Moscow): 70 (October 1962), 76–79.

"American Trusteeship in the Pacific Islands," *World Today,* 3 (July 1947), 317–322.

"The Applicability of the Principle of Self-Determination to Unintegrated Territories of the United States: The Cases of Puerto Rico and the Trust Territories of the Pacific Islands; A Panel," *Proceeding, American Society of International Law,* 67 (November 1973), 1–28.

Apter, David E., "The Role of Traditionalism in the Political Modernization of Ghana and Uganda," *World Politics,* 13 (October 1960), 45–68.

Arden-Clarke, Charles, "The Problem of the High Commission Territories," *Optima,* 8, 4 (1958), 163–170.

Armstrong, Arthur John, "The Emergence of the Micronesians into the International Community: A Study of the Creation of a New International Entity," *Brooklyn Journal of International Law,* 5, 2 (1979), 207–261.

———, "The Negotiations for the Future Political Status of Micronesia," *American Journal of International Law,* 74, 3 (July 1980), 689–693.

Armstrong, Arthur John and Howard Loomis Hills, "The Negotiations for the Future Political Status of Micronesia," *American Journal of International Law,* 78, 2 (April 1984), 484–497.

Armstrong, Elizabeth, "The United States and Non-Self-governing Territories," *International Journal,* 3 (Autumn 1948), 327–333.

Azikiwe, Nnamdi, "Essentials for Nigerian Survival," *Foreign Affairs,* 43, 3 (April 1965), 447–461.

B

Bailey, Sydney D., "The Path of Self-Government in Ceylon," *World Affairs*, n.s., 3, 2 (April 1949), 196–206.

Bains, J.S., "Angola, the UN and International Law," *Indian Journal of International Law*, 3 (January 1963), 63–71.

Baker, Randall, "Breath of Change in Spanish Africa," *Africa Quarterly*, 7 (January–March 1968), 343–360.

Baklanov, I., "The United States: Towards Partitioning and Annexing Micronesia," *International Affairs*, 9 (September 1978), 126–129.

Balogh, Thomas, "The Mechanism of Neo-Imperialism: The Economic Impact of Monetary and Commercial Institutions in Africa," *Oxford University Institute of Statistics*, Bulletin 24, 3 (August 1962), 331–346.

Barber, Hollis W., "Decolonization: The Committee of Twenty-four," *World Affairs*, 138, 2 (Fall 1976).

Barber, James P., "Rhodesia; the Constitutional Conflict," *Journal of Modern African Studies*," 4 (December 1966), 457–469.

Barbour, Nevill, "Aden and The Arab South," *World Today* 15, 8 (1959), 302–310.

Barratt, John, "The Internal Agreement in Rhodesia and the Outlook for an International Settlement," *International of Affairs Bulletin*, 2, 1 (1978), 10–22.

Baudet, H., "Netherlands Retreat from Empire," *Internationale Spectator*, 21 (July 8, 1967), 1029–1064.

"The Belize Mediation," *International and Comparative Law Quarterly*, 17 (October 1968), 996–1009.

Bergsman, Peter, "The Marianas, the United States, and the United Nations: The Uncertain Status of the New American Commonwealth," *California Western International Law Journal*, 6 (Spring 1976), 382–411.

Beri, H.M.L., "The Namibia Tangle," *Africa Quarterly*, 18, 2–3 (1979), 22–36.

Best, Alan C.G., "South Africa's Border Industries: The Tswane Example," *Annals, AAG*, 61, 2 (June 1971), 329–343.

———, "Angola: Geographic Background of an Insurgent State," *Focus*, 26, 5 (May–June 1976), 1–8.

Blakeslee, G.H., "The Japanese Monroe Doctrine," *Foreign Affairs*, 11, 4 (1933), 671–681.

Bogan, E.F., "Government of the Trust Territory of the Pacific Islands," *Annals, American Academy of Political and Social Sciences* (January 1950), 164–174.

Bokor-Szego, H., "The Colonial Clause in International Treaties," *Acta Juridica*, 4, 3–4 (1962), 261–294.

———, "The International Legal Content of the Right of Self-Determination as Reflected by the Disintegration of the Colonial System," in *Questions of International Law*. Budapest: International Law Association, 1966, 7–41.

Bose, N.K., "Bengal Partition and After," *Calcutta Geographical Review*, 9, 1–4 (1949), 14–22.

Bourde, A., "Comoro Islands: Problems of a Microcosm," *Journal of Modern African Studies*, 3 (May 1965), 91–102.

Boustead, J.E.H., "Abu Dhabi 1761–1963," *Royal Central Asian Journal*, 50 (July–August 1963), 273–277.

Bowett, D.W., "Self-Determination and Political Rights in the Developing Countries," *Proceedings, American Society of International Law*, 60 (1966), 129–135.

Boyce, Peter J. and Richard A. Herr, "Microstate Diplomacy in the South Pacific," *Australian Outlook*, 28 (April 1974), 24–35.

Braganza, Berta M., "On Neo-Colonialism," *Afro–Asian and World Affairs*, 4 (Summer 1967), 87–111.

Braine, Bernard, "Storm Clouds over the Horn of Africa," *International Affairs*, 34, 4 (1958), 435–443.

Branch, James A., "The Constitution of the Northern Mariana Islands: Does a Different Cultural Setting Justify Different Constitutional Standards?" *Denver Journal of International Law and Policy*, 9, 1 (Winter 1980), 35–67.

Broderick, Margaret, "Associated Statehood—a New Form of Decolonisation," *International and Comparative Law Quarterly*, 17 (April 1968), 368–403.

Broek, Jan O.M., "Diversity and Unity in Southeast Asia," *Geographical Review*, 34, 2 (1944), 175–195.

Brown, N., "Cyprus: A Study in Unresolved Conflict," *World Today*, 23, 9 (1967), 396–405.

Brunhes, Jean and C. Vallaux, "German Colonization in Eastern Europe." *Geographical Review*, 6 (1918), 465–480.

Bryant, Bunyan and others, "Recognition of Guinea (Bissau)," *Harvard International Law Journal*, 15 (Summer 1974), 482–495.

Buchanan, Keith, "Northern Region of Nigeria: The Geographical Background of its Political Duality," *Geographical Review*, 43, 4 (October 1953), 451–473.

——, "Profiles of the Third World," *Pacific Viewpoint*, 5, 21 (1964), 97–126.

Burns, R.D., "Inspection of the Mandates, 1919–1941," *Pacific Historical Review*, 37 (November 1968), 445–462.

Busk, C.W.F., "The North-West Frontier: Pakistan's Inherited Problem," *Geographical Magazine*, 22, 3 (1949), 85–92.

C

Cabral, João, "Portuguese Colonial Policy," *Africa Quarterly*, 5 (October–December 1965), 153–173.

Cakste, Mintauts, "Latvia and the Soviet Union," *Journal of Central European Affairs*, 9, 1 (April 1949), 32–60.

Calvocoressi, Peter, "South-West Africa," *African Affairs*, 65 (July 1966), 223–232.

Carrington, C.E., "Decolonization: The Last Stages," *International Affairs*, (1962), 29–40.

Castles, Alex C., "Law and Politics in the Rhodesian Dispute," *Australian Outlook*, 21 (August 1967), 165–178.

Cefkin, J. Leo, "The Rhodesian Question at the United Nations," *International Organization*, 22, 3 (Summer 1968), 649–669.

Cervin, Vladimir, "Problems in the Integration of the Afghan Nation," *Middle East Journal*, 6, 4 (Autumn 1952), 400–416.

Chang, King-yuh, "The United Nations and Decolonization: the Case of Southern Yemen," *International Organization*, 26, 1 (Winter 1972), 37–61.

Chimango, L.J., "The Relevance of Humanitarian International Law to the Liberation Struggles in Southern Africa—the Case of Moçambique in Retrospect," *Comparative and International Law Journal (Southern Africa)*, 8 (November 1975), 287–317.

Christopher, A.J., "Colonial Land Policy in Natal," *Annals, AAG*, 61, 3 (September 1971), 560–575.

"Citizens of Trust Territory of the Pacific Islands Have Rights Under the Trusteeship Agreement Which Are Judicially Enforceable," *Texas International Law Journal*, 10 (1975), 138–149.

Clark, D.E., "Manifest Destiny and the Pacific." *Pacific Historical Review*, 1 (March 1932), 1–17.

Clark, Roger S., "The 'Decolonization' of East Timor and the United Nations Norms on Self-Determination." *Yale Journal of World Public Order*, 7, 1 (Fall 1980), 2–44.

——, "Self-Determination and Free Association: Should the United Nations Terminate the Pacific Islands Trust?" *Harvard International Law Journal*, 21, 1 (Winter 1980), 1–86.

Clarke, Colin G., "Political Fragmentation in the Caribbean: The Case of Anguilla," *Canadian Geographer*, 15, 1 (1971), 13–29.

Cockram, Ben, "The 'Protectorates'—an International Problem," *Optima*, 13, 4 (December 1963), 176–184.

Cohen, A., "New Africa and the United Nations," *International Affairs*, 36 (October 1960), 476–488.

Coker, Christopher, "Decolonization in the Seventies; Rhodesia and the Dialectic of National Liberation," *Round Table*, No. 274 (April 1979), 122–136.

Coleman, James S., "Problems of Political Integration in Emergent Africa," *Western Political Quarterly*, 8, 1 (March 1955), 44–57.

Collins, B., "'Consultative Democracy' in British Guiana," *Parliamentary Affairs*, 19 (Winter 1965–1966), 103–113.

Collins, B.A.N., "Independence for Guyana," *World Today*, 22 (June 1966), 260–268.

Comyns-Carr, R., "Gibraltar Dispute as Seen from Spain," *World Today*, 21 (September 1965), 364–367.

Connor, Walter, "Self-Determination: The New Phase," *World Politics*, 20 (1967), 30–53.

Coulter, John Wesley, "The United States Trust Territory of the Pacific Islands," *Journal of Geography*, 47 (October 1948), 253–267.

Crawford, John F., "South West Africa: Mandate Termination in Historical Perspective," *Columbia Journal of Transnational Law*, 6 (Spring 1967), 91–137.

Crawford, Robert and Perumala Dayanidhi, "East Timor: A Study in De-Colonization," *India Quarterly*, 33, 4 (October–December 1977), 419–431.

Crawshaw, Nancy, "The Republic of Cyprus, from Zurich Agreement to Independence," *World Today*, 16, 10 (December 1960), 526–540.

Crocombe, R.G., "Development and Regression in New Zealand's Island Territories," *Pacific Viewpoint*, 3, 2 (September 1962), 17–32.

Crowder, Michael, "Two Cameroons or One?" *Geographical Magazine* 32, 6 (November 1959), 303–314.

Crush, Jonathan, "Colonial Coercion and the Swazi Tax Revolt of 1903–1907," *Political Geography Quarterly*, 4, 3 (July 1985), 179–190.

Cumming, Duncan Cameron, "The Disposal of Eritrea," *Middle East Journal,* 7, 1 (Winter 1953), 18–32.

D

Dale, Richard, "The Political Futures of South West Africa and Namibia," *World Affairs,* 134 (Spring 1972), 325–343.

Davidson, Basil, "The Liberation Struggle in Angola and 'Portuguese' Guinea," *Africa Quarterly,* 10 (April–June 1970), 25–31.

Das, Taraknath, "The State of Hyderabad During and After British Rule in India," *American Journal of International Law,* 43, 1 (January 1949), 57–72.

Dean, Vera K., "Geographical Aspects of the Newfoundland Referendum," *Annals, AAG,* 39 (1949), 70.

de Blij, Harm J., "Mozambique: Fragile Independence," *Focus,* 27, 2 (November–December 1976), 9–16.

Decalo, Samuel, "Regionalism, Politics and the Military in Dahomey," *Journal of Developing Areas,* 7 (1973), 449–478.

Dempsey, Guy, "Self-Determination and Security in the Pacific; A Study of the Covenant Between the United States and the Northern Mariana Islands," *New York University Journal of International Law and Politics,* 9 (Fall 1976), 277–302.

Dennett, T., "Japan's 'Monroe Doctrine' Appraised." *Annals, American Academy of Political and Social Science,* 215 (1941), 61–65.

Dillard, Hardy C., "Status of South-West Africa (Namibia)—Separate Opinion," *International Lawyer,* 6 (April 1972), 409–427.

Dingelstedt, V., "Ruling Nations: Considerations on Their Characters," *Scottish Geographical Magazine,* 26 (1911), 291–306.

Dobby, E.H.G., "Malayan Prospect," *Pacific Affairs,* 23, 4 (December 1950), 392–401.

Duffy, James, "Portugal in Africa," *Foreign Affairs,* 39, 3 (April 1961), 481–493.

Dugard, John, "The Revocation of the Mandate for South West Africa," *American Journal of International Law,* 62, 1 (January 1968), 78–97.

———, "Namibia (South West Africa): The Court's Opinion, South Africa's Response, and Prospects for the Future," *Columbia Journal of Transnational Law,* 11 (Winter 1972), 14–49.

Dunlop, John Stewart, "The Influence of David Livingstone on Subsequent Political Developments in Africa," *Scottish Geographical Magazine,* 75, 3 (December 1959), 144–152.

Dymov, G. "Persian Gulf Countries at the Crossroads," *International Affairs,* 3 (March 1973), 53–59.

E

Edwards, K.C., "Luxembourg: How Small Can a Nation Be?" *Northern Geography* (1966), 256–267.

El-Ayouty, Yassin, "The United Nations and Decolonisation, 1960–70," *Journal of Modern African Studies,* 8, 3 (October 1970), 462–468.

El Mallakh, Ragaei, "The Challenge of Affluence: Abu Dhabi," *Middle East Journal,* 24, 2 (Spring 1970), 135–146.

Emerson, Rupert, "American Policy Toward Pacific Dependencies," *Pacific Affairs,* 20 (September 1947), 259–275.

———, "Colonialism, Political Development, and the UN," *International Organization,* 19, 3 (Summer 1965), 484–503.

———, "Self-Determination," *Proceedings, American Society of International Law,* 60 (1966), 135–141.

———, "The United Nations and Colonialism," *International Relations,* 3 (November 1970), 766–781.

F

Fair, T.J.D., "A Regional Approach to Economic Development in Kenya," *South African Geographical Journal,* 45 (1963), 55–77.

Fairfield, Roy P., "Cyprus: Revolution and Resolution," *Middle East Journal,* 13, 3 (Summer 1959), 235–248.

Falk, Richard A., "The South West Africa Cases: An Appraisal," *International Organization,* 21, 1 (Winter 1967), 1–23.

Fawcett, J.E.S., "Gibraltar; the Legal Issues," *International Affairs,* 43 (April 1967), 236–251.

Ferguson, A.J., "Constitutional Development in Portuguese Africa," *Africa Institute Bulletin,* 1 (June 1963), 126–130.

Field, Michael, "Abu Dhabi: Struggle for Stability," *Middle East International,* 6, 43 (September 1971).

Fisher, Charles A., "The Expansion of Japan: A Study in Oriental Geopolitics," *Geographical Journal,* 115 (1950), 1–19, 179–193.

———, "The Changing Political Geography of British Malaysia," *Professional Geographer,* 7, 2 (March 1955), 6–9.

———, "The Geographical Setting of the Proposed Malaysian Federation," *Journal of Tropical Geography,* 17 (1963), 99–115.

———, "The Malaysia Federation, Indonesia and the Philippines: A Study in Political Geography," *Geographical Journal* 129, 3 (September 1963), 311–328.

———, "The Vietnamese Problem in its Geographical Context," *Geographical Journal*, 131, 4 (December 1965), 502–515.

Fitzgerald, C.P., "Chinese Expansion in Central Asia," *Royal Central Asian Journal* (1963), 292–293.

Fletcher-Cooke, John, "Some Reflections on the International Trusteeship System, with Particular Reference to Its Impact on the Governments and Peoples of the Trust Territories," *International Organization*, 13 (Summer 1959), 422–430.

Flugal, R.R., "The Palestine Problem: A Brief Geographical, Historical, and Political Evaluation," *Social Studies*, 48, 2 (1957), 43–51.

Fox, A.B., "The United Nations and Colonial Development," *International Organization*, 4 (May 1950), 199–218.

Franck, Thomas M. "The Stealing of the Sahara," *American Journal of International Law*, 70, 4 (October 1976), 694–721.

Franck, Thomas M. and Paul Hoffman, "The Right of Self-Determination in Very Small Places," *New York University Journal of International Law and Politics*, 8, 3 (Winter 1976), 331–386.

Frazier, E. Franklin, "Urbanization and Its Effects upon the Task of Nation-Building in Africa South of the Sahara," *Journal of Negro Education*, 30, 3 (Summer 1961), 214–222.

Furber, Holden, "The Unification of India, 1947–1951," *Pacific Affairs*, 24, 4 (December 1951), 352–371.

Furnivall, J.S., "Twilight in Burma: Independence and After," *Pacific Affairs*, 22, 2 (June 1949), 155–172.

G

Gale, Roger William, "A New Political Status for Micronesia," *Pacific Affairs*, 51, 3 (Fall 1978), 427–447.

Gann, Lewis Henry, "Portugal, Africa and the Future," *Journal of Modern African Studies*, 13 (March 1975), 1–18.

Gilmore, William C., "Legal Perspectives on Associated Statehood in the Eastern Caribbean," *Virginia Journal of International Law*, 19, 3 (1979), 489–555.

Ginsburg, G., "'Wars of National Liberation' and the Modern Law of Nations; the Soviet Thesis," *Law and Contemporary Problems*, 29 (Autumn 1964), 910–942.

Ginsburg, Norton, "From Colonialism to National Development," *Annals, AAG*, 63, 1 (March 1973), 1–21.

Glaser, Edwin, "The Liquidation of Colonialism and the Progressive Development of International Relations and Legality," *Revue romaine d'études internationales*, No. 12 (1971), 31–60.

Glassner, Martin Ira, "The Foreign Relations of Jamaica and Trinidad and Tobago 1960–1965," *Caribbean Studies*, 10, 3 (October 1979), 116–153.

Good, Kenneth, "Settler Colonialism in Rhodesia," *African Affairs*, 73 (January 1974), 10–36.

Goodman, Elliot R., "The Cry of National Liberation: Recent Soviet Attitudes Toward National Self-Determination," *International Organization*, 14, 1 (Winter 1960), 92–104.

Gopal, Ajit, "The High Commission Territories," *Africa Quarterly*, 6 (July–September 1966), 128–134.

Gorbold, R., "Tanganyika: The Path to Independence," *Geographical Magazine*, 35, 7 (November 1962), 371–384.

Gordon, Edward, "Resolution of the Bahrain Dispute," *American Journal of International Law*, 65, 3 (July 1971), 560–568.

Gott, K.D., "The Australian Trusteeship in New Guinea," *Eastern World* (January–February 1949), 22–23.

Gottmann, Jean, "The Political Partitioning of Our World: An Attempt at Analysis," *World Politics*, 4, 4 (July 1952), 512–519.

Gould, Peter R., "Tanzania 1920–1973: The Spatial Impress of the Modernization Process," *World Politics*, 22, 2 (January 1970), 149–170.

Grahl-Madsen, Atle, "Decolonization; the Modern Version of a Just War," in *German Yearbook of International Law*, 22 (1979), 255–273.

Green, D.M., "America's Strategic Trusteeship Dilemma—Its Humanitarian Obligations," *Texas International Law Journal*, 9 (Summer 1974), 175–204.

Griffin, James, "Papua New Guinea and the British Solomon Islands Protectorate; Fusion or Transfusion?" *Australian Outlook*, 27 (December 1973), 319–328.

Gunthardt, Walter, "Papua–New Guinea en Route to Independence," *Swiss Review of World Affairs*, 23 (June 1973), 17–19.

H

Haas, Anthony, "Independence Movements in the South Pacific," *Pacific Viewpoint*, 11–12 (1970–1971), 97–120.

Haas, Ernst B., "Attempt to Terminate Colonialism; Acceptance of the United Nations Trusteeship System," *International Organization*, 7 (February 1953), 1–21.

Hamdan, G., "The Political Map of the New Africa," *Geographical Review*, 53, 3 (July 1963), 418–439.

Hance, William A., "Economic Potentialities of the Central African Federation," *Political Science Quarterly*, 69, 1 (March 1954), 29–44.

Hargreaves, John, "Pan-Africanism After Rhodesia," *World Today*, 22 (February 1966), 57–63.

Harrison, Tom, "Sarawak in the Whirlpool of Southeast Asia," *Geographical Magazine*, 31, 8 (December 1958), 378–385.

Hasan, Najmul, "Namibia; South-West Africa," *Pakistan Horizon*, 28, 3 (1975), 61–78.

Hastings, P., "New Guinea—East and West," *Australian Outlook*, 14 (August 1960), 147–156.

Heard-Bey, Frauke, "The Gulf States in Transition," *Asian Affairs*, 59 (February 1972), 14–22.

Hellen, J.A., "Independence or Colonial Determinism? The African Case," *International Affairs*, 44, 4 (October 1968), 691–708.

Henderson, W.O., "Economic Aspects of German Imperial Colonization," *Scottish Geographical Magazine*, 54 (1938), 150–161.

Henry, Michael, "The Anglo–Spanish Dispute over Gibraltar," *Contemporary Review*, 212 (April 1968), 169–173.

Herbert, Nicholas, "Federation and the Future of the Gulf," *Middle East*, 8, 6 (1968), 10–14.

Herman, Paula, "British Honduras; a Question of Viability," *World Affairs*, 138 (Summer 1975), 60–68.

Herr, R.A., "A Minor Ornament: The Diplomatic Decisions of Western Samoa at Independence," *Australian Outlook*, 29 (December 1975), 300–314.

Hickey, John, "Keep the Falklands British? The Principle of Self-Determination of Dependent Territories," *Inter-American Economic Affairs*, 31 (Summer 1977), 77–88.

Higgins, Rosalyn, "International Court and South West Africa: The Implications of the Judgment," *International Affairs*, 42, 4 (October 1966), 573–599.

———, "International Law, Rhodesia, and the U.N.," *World Today*, 23 (March 1967), 94–106.

Higgott, Richard, "Structural Dependence and Decolonisation in a West African Landlocked State: Niger," *Review of African Political Economy*, No. 17 (January 1980), 43–58.

"The High Commission Territories; a Remnant of British Africa," *Round Table*, No. 213 (December 1963), 26–40.

Hill, Allan G., "Bahrain Is Independent," *Geographical Magazine*, 44 (September 1972), 846–852.

Hodgkiss, A.G. and Robert W. Steel, "The Changing Face of Africa," *Geography*, 46, 211 (1961), 156–160.

Hofmeyer, Jan H., "Germany's Colonial Claims: A South African View," *Foreign Affairs*, 17, 4 (July 1939), 788–798.

Hopkins, Jack W., "Toward Self-government in the Trust Territory," *Philippine Journal of Public Administration*, 8 (April 1964), 132–135.

Hoselitz, Bert F., "Nationalism, Economic Development, and Democracy," *Annals, American Academy of Political and Social Science*, 30, 5 (1956), 1–11.

Hottinger, Arnold, "Portuguese Africa," *Swiss Review of World Affairs*, 22 (May 1972), 15–23.

Houphouet-Boigny, Felix, "Black Africa and the French Union," *Foreign Affairs*, 35, 4 (July 1957), 593–599.

House, John, "Political Geography of Contemporary Events: Unfinished Business in the South Atlantic," *Political Geography Quarterly*, 2, 3 (July 1983), 233–246.

Howe, Marvine, "The Birth of the Moroccan Nation," *Middle East Journal*, 10, 1 (1956), 1–16.

Huaraka, Tunguru, "Walvis Bay and International Law," *Indian Journal of International Law*, 18, 2 (1978), 160–174.

Huff, David L. and James M. Lutz, "The Contagion of Political Unrest in Independent Black Africa," *Economic Geography*, 50, 4 (October 1974), 352–367.

Hunter, G., "Independence and Development; Some Comparisons Between Tropical Africa and South-East Asia," *Ekistics*, 104 (1964), 3–6.

Hurewitz, J.C., "The UN and Disimperialism in the Middle East," *International Organization*, 19, 3 (Summer 1965), 749–763.

Huxley, Elspeth, "West Africa in Transition,"

Geographical Magazine, 25, 6 (October 1952), 310–320.

I

"Independent Qatar," *Middle East Economic Digest,* 15, 10 (December 1971), 1429–1436.

Iveković, Ivan and Dimitrije Babić, "Changes in Portugal and the Progress of Decolonization," *Socialist Thought and Practice,* 13 (January 1975), 44–73.

J

Jackson, Harold Karan, "Our 'Colonial' Problem in the Pacific," *Foreign Affairs,* 39, 1 (October 1960), 56–66.

Johnson, D.H.N., "Trusteeship: Theory and Practice," in *Yearbook of World Affairs,* London: Stevens, 1951.

Johnson, Richard, "The Future of the Falkland Islands," *World Today,* 33, 6 (June 1977), 223–231.

Jones, N.S. Carey, "The Decolonization of the White Highlands of Kenya," *Geographical Journal,* 131, 2 (June 1965), 186–201.

Judson, W.V., "The Strategy Value of Her West Indian Possessions to the United States," *Annals, AAG,* 19, 2 (1902), 383–391.

K

Kamanu, Onyeonoro S., "Succession and the Right of Self-Determination; an O.A.U. Dilemma," *Journal of Modern African Studies,* 12, 3 (September 1974), 355–376.

Kay, David A., "Politics of Decolonization; the New Nations and the United Nations Political Process," *International Organization,* 21 (Autumn 1967), 786–811.

Kearns, Kevin C., "Prospects of Sovereignty and Economic Viability for British Honduras," *Professional Geographer,* 21, 2 (February 1969), 97–103.

Kerr, J.R., "The Political Future of New Guinea," *Australian Outlook,* 13, 3 (September 1959), 181–192.

Khapoya, Vincent B., "Determinants of African Support for African Liberation Movements; a Comparative Analysis," *Journal of Modern African Studies,* 3 (Winter 1976), 469–489.

Khol, Andreas, "The 'Committee of Twenty-four' and the Implementation of the Declaration on the Granting of Independence to Colonial Countries and Peoples," *Revue des droits de l'homme,* 3, 1 (1970), 21–50.

Kilner, Peter, "Britain, the U.N. and South Arabia," *World Today,* 22 (July 1966), 269–272.

———, "South Arabian Independence," *World Today,* 23 (August 1967), 333–339.

Kimble, George H.T., "The Federation of Rhodesia and Nyasaland," *Focus,* 6, 7 (March 1965).

Knight, David B., "Territory and People or People and Territory?: Thoughts on Post-Colonial Self-Determination," *International Political Science Review,* 6 (1985), 248–272.

Kretzmann, Edwin M.J., "Oil, Water, and Nationalism," *Proceedings, Institute of World Affairs,* 33 (1959), 175–183.

Kroef, Justus M. van der, "Society and Culture in Indonesian Nationalism," *American Journal of Sociology,* 58, 1 (July 1952), 11–24.

———, "Nationalism and Politics in West New Guinea," *Pacific Affairs,* 34 (Spring 1961), 38–53.

———, "The West New Guinea Problem," *World Today,* 17 (November 1961), 489–502.

———, "West New Guinea; the Uncertain Future," *Asian Survey,* 8 (August 1968), 691–707.

Kubiak, T.J., "Rhodesia (Zimbabwe): White Minority Rule in a Black State," *Focus,* 27, 2 (November–December 1976), 1–8.

Kumar, Satish, "The United Nations Committee of 24; Its Origins and Work," *Africa Quarterly,* 5 (October–December 1965), 174–187.

Kundu, Samarendra, "Southern African Liberation and India," *Africa Quarterly,* 18 (July 1978), 20–23.

Kuyper, P.J., "The Limits of Supervision: The Security Council Watchdog Committee on Rhodesian Sanctions," *Netherlands International Law Review,* 25, 2 (1978), 159–194.

L

Lackner, Helen, "Revolution in the Gulf," *Race,* 15 (April 1974), 515–527.

Laing, Edward A., "Independence and Islands: The Decolonization of the British Caribbean," *New York University Journal of International Law and Politics,* 12, 2 (Fall 1979), 281–312.

Laschinger, Michael, "Roads to Independence; the Case of Swaziland," *World Today,* 21 (November 1965), 486–494.

Lateef, Abdul, "Bahrain; Emerging Gulf State," *Pakistan Horizon,* 26, 1 (1973), 10–15.

Latham-Koenig, A.L., "Ruanda-Urundi on the Threshhold of Independence," *World Today,* 18 (July 1962), 288–295.

Lawless, Robert, "The Indonesian Takeover of East Timor," *Asian Survey,* 16, 10 (October 1976), 948–964.

Leibowitz, Arnold H., "The Applicability of Federal Law to Guam," *Virginia Journal of International Law,* 16 (Fall 1975), 21–63.

Leifer, Michael, "Indonesia and the Incorporation of East Timor," *World Today,* 32, 9 (September 1976), 347–354.

Levine, Stephen, "The United States and Micronesia: Towards Political Change," *New Zealand International Review,* 15 (November–December 1976).

Lewis, Gordon K., "British Colonialism in the West Indies: The Political Legacy," *Caribbean Studies,* 7 (April 1967), 3–22.

Lewis, I.M., "Pan-Africanism and Pan-Somalism," *Journal of Modern African Studies,* 1 (1963), 147–162.

Lijphart, Arend, "The Indonesian Image of West Irian," *Asian Survey,* 1, 5 (July 1961), 9–15.

Linebarger, G.C., "The Netherlands–Indonesian Dispute," *World Affairs,* 125 (Spring 1962), 30–35.

Lissitzyn, Oliver J., "International Law and the Advisory Opinion on Namibia," *Columbia Journal of Transnational Law,* 11 (Winter 1972), 50–73.

Louis, William Roger, "The South West African Origins of the 'Sacred Trust,' 1914–1919," *African Affairs,* 66 (January 1967), 20–39.

———, "The United Kingdom and the Beginning of the Mandates System, 1919–1922," *International Organization,* 23, 1 (Winter 1969), 73–96.

Lowenthal, David and Colin G. Clarke, "Island Orphans: Barbuda and the Rest," *Journal of Commonwealth and Comparative Politics,* 18, 3 (December 1980), 293–307.

Luthera, V.P., "Britain and Rhodesia," *Afro–Asian and World Affairs,* 4 (Summer 1967), 112–122.

M

MacDonald, Barrie, "Secession in Defense of Identity: The Making of Tuvalu," *Pacific Viewpoint,* 16 (1975), 26–44.

Machel, Samora, "The People's Republic of Mozambique; the Struggle Continues," *Review of African Political Economy,* 4 (November 1975), 14–25.

MacLaughlin, Jim, "The Political Geography of 'Nation-Building' and Nationalism in Social Sciences: Structural vs. Dialectical Accounts," *Political Geography Quarterly,* 5, 4 (October 1986), 299–329.

Manchanda, S.L., "Aden and South Arabia; Past, Present and Future," *Afro–Asian and World Affairs,* 4 (Spring 1967), 13–26.

Marston, Geoffrey, "Termination of Trusteeship," *International and Comparative Law Quarterly,* 18, 1 (January 1969), 1–40.

Martelli, G., "The Portuguese in Guinea," *World Today,* 21 (August 1965), 345–351.

Martin, L., "The Geography of the Monroe Doctrine and the Limits of the Western Hemisphere," *Geographical Review,* 30 (1940), 525–528.

Mason, Philip, "Partnership in Central Africa," *International Affairs,* 33, 2 (1957), 154–164; 33, 3 (1957), 310–318.

Matthews, Geoffrey and Marc Lee, "The United Nations in Contempt; Denying Gibraltarians Their Rights," *Round Table,* No. 231 (July 1968), 311–315.

"Mauritius, Another Newly Independent Pygmy State," *Swiss Review of World Affairs,* 18 (April 1968), 23–25.

Mazrui, Ali A., "Consent, Colonialism, and Sovereignty," *Political Studies,* 11 (February 1963), 36–55.

———, "Africa's Experience in Nation-Building; Is It Relevant to Papua and New Guinea?" *East Africa Journal,* 7 (November 1970), 15–23.

McAuley, J., "Paradoxes of Development in the South Pacific," *Pacific Affairs,* 27 (June 1954), 138–149.

McDougal, Myres S. and W. Michael Reisman, "Rhodesia and the United Nations; the Lawfulness of International Concern," *American Journal of International Law,* 62, 1 (January 1968), 1–19.

McGee, T.G., "Aspects of the Political Geography of Southeast Asia," *Pacific Viewpoint,* 1, 1 (March 1960), 39–58.

McHenry, D.E., "The United Nations Trusteeship System," *World Affairs Interpreter,* 19 (July 1948), 149–158.

McKitterick, T.E.M., "The End of a Colony— British Guiana 1962," *Political Quarterly,* 33 (January–March 1962), 30–40.

McMaster, David N., "Political Upheavals and Economic Development: Uganda, *Focus,* 26, 5 (May–June 1976), 9–16.

McNickle, R.K., "Pacific Dependencies," *Editorial Research Reports,* (September 9, 1949), 579–595.

Meinig, Donald W., "The American Colonial Era: A Geographic Commentary," *Proceed-*

ings, *Royal Geographical Society of Australia, South Australia Branch*, 59 (1957–1958), 1–22.

Melamid, Alexander, "Political Geography of Trucial 'Oman and Qatar,' " *Geographical Review*, 43, 2 (April 1953), 194–206.

Menon, Lakshmi N., "International Responsibility for Dependent Areas," *India Quarterly*, 11 (July–September 1955), 248–261.

Mercer, John, "The Cycle of Invasion and Unification in the Western Sahara," *African Affairs*, 75 (October 1976), 498–510.

Merrit, Richard L., "Systems and the Disintegration of Empires," *General Systems*, 8 (1963).

Metelski, J.B., "Micronesia and Free Association; Can Federalism Save Them?" *California Western International Law Journal*, 5 (Winter 1974), 162–183.

Metford, J.C.J., "Falklands or Malvinas? The Background to the Dispute," *International Affairs*, 44, 3 (July 1968), 463–481.

Mezerik, A.G. (ed.), "Colonialism and the United Nations," *International Review Service*, 10, 83 (1964), 1–105.

Mihaly, Eugene B., "Tremors in the Western Pacific; Micronesian Freedom and U.S. Security," *Foreign Affairs*, 52, 4 (July 1974), 839–849.

Miller, J.D.B., "Papua New Guinea in World Politics," *Australian Outlook*, 27 (August 1973), 191–201.

Miller, Jake C., "The Virgin Islands and the United States; Definition of a Relationship," *World Affairs*, (Spring 1974), 297–305.

Miller, Joseph C., "The Politics of Decolonization in Portuguese Africa," *African Affairs*, 74 (April 1975), 135–147.

Mink, Patsy T., "Micronesia; Our Bungled Trust," *Texas International Law Forum*, 6 (Winter 1971), 181.

Minter, William, "Joint Ventures in Colonialism; Portugal and Its Allies," *Contemporary Review*, 224 (April 1974), 199–204.

Mittelman, James H., "Collective Decolonisation and the U.N. Committee of 24," *Journal of Modern African Studies*, 14 (March 1976), 41–64.

Monroe, Elizabeth, "Kuwait and Aden: A Contrast in British Policies," *Middle East Journal*, 18, 1 (Winter 1964), 63–74.

Monroe, Malcolm W., "Namibia—the Quest for the Legal Status of a Mandate: An Impossible Dream?" *International Lawyer*, 5 (July 1971), 549–557.

Mori, Roland, "Portugal in Southern Africa," *Swiss Review of World Affairs*, 20 (April 1970), 8–11.

"Moves Toward Autonomy in the Solomon Islands," *Australian Foreign Affairs Record*, 44 (April 1973), 242–250.

Mühlemann, Christoph, "Shaky Independence for Grenada," *Swiss Review of World Affairs*, 23 (March 1974), 22–23.

Murphy, John F., "Whither Now Namibia?" *Cornell International Law Journal*, 6 (Fall 1972), 1–43.

Murphy, Raymond E., "Economic Geography of a Micronesian Atoll," *Annals, AAG*, 40, 1 (March 1950), 58–83.

Mushkat, Marion, "The Process of African Decolonization," *Indian Journal of International Law*, 6 (October 1967), 483–508.

N

"Namibia, South Africa, and the Walvis Bay Dispute," *Yale Law Journal*, 89, 5 (April 1980), 903–922.

Nawaz, M.K., "Colonies, Self-Government, and the United Nations," *Indian Yearbook of International Affairs*, 11 (1961), 3–47.

————, "The Meaning and Range of the Principle of Self-Determination," *Duke Law Journal*, (1965), 82–101.

Neumann, H., "Portugal's Policy in Africa; the Four Years Since the Beginning of the Uprising in Angola," *International Affairs*, 41 (October 1965), 663–675.

Newbury, C.W., "Aspects of French Policy in the Pacific, 1853–1906," *Pacific Historical Review*, 27 (February 1958), 45–56.

Nichertlein, Sue, "The Struggle for East Timor; Prelude to Invasion," *Journal of Contemporary Asia*, 7, 4 (1977), 486–496.

"Niue Approaches Self-Government," *New Zealand Foreign Affairs Review*, 23 (March 1973), 3–11.

Norman, H.E., "The Genyosha: A Study in the Origins of Japanese Imperialism." *Pacific Affairs*, 17, 3 (1944), 261–284.

Nyerere, Julius K., "Rhodesia in the Context of Southern Africa," *Foreign Affairs*, 44, 3 (April 1966), 373–386.

O

Owen, David, "The Importance of Negotiated Settlements in Southern Africa," *International Affairs Bulletin*, 3, 2 (1979), 47–54.

Owen, R.P., "The Rebellion in Dhofar—a

Threat to Western Interests in the Gulf,'' *World Today,* 29, 6 (June 1973), 266–272.

Owiti Ayaga, Odeyo, ''United Nations and African Decolonization; Ineffectiveness of the Security Council,'' *Indian Journal of International Law,* 12 (October 1972), 581–605.

P

Pace, R.P., ''The United Nations and Self-Determination,'' *Tydskrif vir Hedendaagse Romeins–Hollandse Reg,* 30 (1967), 124–136.

Pal Singh, Surendra, ''India and the Liberation of Southern Africa,'' *Africa Quarterly,* 10 (April–June 1970), 4–8.

Papaioannou, Ezekias, ''Neo-Colonialism and Developing Countries,'' *World Marxist Review,* 14 (May 1971), 86–96.

Pélissier, René, ''Spain's Discreet Decolonization,'' *Foreign Affairs,* 43, 3 (April 1965), 519–527.

Pelzer, Karl J., ''Micronesia; a Changing Frontier,'' *World Politics,* 2, 2 (January 1950), 251–266.

Perham, Margery, ''The Sudan Emerges into Nationhood,'' *Foreign Affairs,* 27, 4 (July 1949), 665–677.

Perlman, M.M., ''The Republic of Lebanon,'' *Palestine Affairs,* 2, 11 (November 1947), 109–114.

Perrons, D., ''Ireland and the Break-up of Britain,'' *Antipode,* 11 (1980), 53–65.

Petersen, Glenn, ''Breadfruit or Rice? The Political Economics of a Vote in Micronesia,'' *Science and Society,* 43, 4 (1979–1980), 472–485.

Pick, H., ''Independent Mauritania,'' *World Today,* 17 (April 1961), 149–158.

Pillai, R.V. and Mahendra Kumar, ''The Political and Legal Status of Kuwait,'' *International and Comparative Law Quarterly,* 11, 1 (January 1962), 108–130.

Plishke, Elmer, ''Self-Determination: Reflections on a Legacy,'' *World Affairs,* 140, 1 (Summer 1977), 41–57.

Ponoma, Jean Baptiste, ''The People of Reunion Want Freedom,'' *World Marxist Review,* 21 (July 1978), 80–83.

Potholm, Christian P., ''The Protectorates, the O.A.U. and South Africa,'' *International Journal,* 22, 1 (Winter 1966–67), 68–72.

Poznanski, L., ''Constitutional Development in the British Solomon Islands Protectorate,'' *Parliamentarian,* 54 (April 1973), 93–96.

Premdas, Ralph R., ''Papua New Guinea; Internal Problems of Rapid Political Change,''

Asian Survey, 15, 12 (December 1975), 1054–1076.

————, ''Ethnonationalism, Copper, and Secession in Bougainville,'' *Canadian Review of Studies in Nationalism,* 4 (1977), 247–265.

Prescott, John Robert Victor, ''The Geographical Basis of Kenya's Political Problems,'' *Australian Outlook,* 16 (1962), 270–282.

Purcell, D.C., Jr., ''Economics of Exploitation: The Japanese in the Mariana, Caroline and Marshall Islands, 1915–1940,'' *Journal of Pacific History,* 11, 3–4 (1976), 189–211.

R

Rahmatullah, Khan and Kaur Satpal, ''The Deadlock over South-West Africa,'' *Indian Journal of International Law,* 8 (April 1968), 179–200.

Rama Rao, T.W., ''The Right of Self-Determination: Its Status and Role in International Law,'' in *Internationales Recht und Diplomatie, 1968* (Cologne).: 1969, 19–28.

Ramazani, Rouhollah K., ''The Settlement of the Bahrain Dispute,'' *Indian Journal of International Law,* 12 (1972), 1–14.

Ray, Aswini K., ''De-colonization Through the United Nations; an Analysis of the Soviet Role in the Framing of the Charter,'' *Afro–Asian and World Affairs,* 3, 1 (1966), 22–30.

Richardson, Henry J., III, ''Constitutive Questions in the Negotiations for Namibian Independence,'' *American Journal of International Law,* 78, 1 (January 1984), 76–120.

Riedel, Eibe H., ''Confrontation in Western Sahara in the Light of the Advisory Opinion of the International Court of Justice of 16 October 1975; a Critical Appraisal,'' in *German Yearbook of International Law,* 19 (1976), 405–422.

Riggs, Fred W., ''Wards of the U.N.: Trust and Dependent Areas,'' *Foreign Policy Reports,* 26, 6 (June 1, 1950), 54–63.

Rippy, J. Fred, ''Antecedents of the Roosevelt Corollary of the Monroe Doctrine.'' *Pacific Historical Review,* 9 (1940), 267–280.

Rivlin, Benjamin, ''Unity and Nationalism in Libya,'' *Middle East Journal,* 3, 1 (January 1949), 31–44.

Roberts, Margaret, ''Political Prospects for the Cameroun,'' *World Today,* 16, 7 (July 1960), 305–315.

Robertson, Sir James, ''The Sudan in Transition,'' *African Affairs,* 52, 209 (October 1953), 317–327.

Robinson, G.W.S., "Ceuta and Melilla: Spain's Plazas de Soberanía," *Geography*, 43 (1958), 266–269.

Robinson, K.E., "French West Africa," *African Affairs*, 50, 199 (April 1951), 123–132.

Rothschild, Donald S., "The Politics of African Separatism," *Journal of International Affairs*, 1 (1961), 18–28.

———, "Rhodesian Rebellion and African Response," *Africa Quarterly*, 6 (October–December 1966), 184–196.

Roucek, J.S., "Geopolitics of Portuguese Colonialism," *Contemporary Review*, 205 (June 1964), 284–296.

Rudebeck, Lars, "Political Mobilisation for Development in Guinea–Bissau," *Journal of Modern African Studies*, 10 (May 1972), 1–18.

Rupen, Robert A., "Mongolian Nationalism," *Royal Central Asian Journal*, 45, Pt. 2 (April 1958), 157–178.

S

Sady, E.J., "Department of the Interior and Pacific Island Administration," *Public Administration Review*, 10, 1 (1950), 13–19.

Saul, John S., "Liberation Movements Provide Structure for Emerging States," *International Perspectives*, (November–December 1974), 34–38.

Sayre, Francis B., "American Trusteeship Policy in the Pacific," *Proc., Ac. Pol. Soc.*, 22 (January 1948), 406–416.

———, "Legal Problems Arising from the United Nations Trusteeship System," *American Journal of International Law*, 42 (April 1948), 263–298.

Schaffer, B.B., "The Concept of Preparation: Some Questions About the Transfer of Systems of Government," *World Politics*, 18 (1965), 42–67.

Schaffer, Rosalie P., "The Extension of South African Treaties to the Territories of South West Africa and the Prince Edward Islands," *South Africa Law Journal*, 95 (March 1978), 63–70.

Schurz, Carl, "Manifest Destiny." *Harpers Monthly*, 87 (October 1893), 737–746.

Schweinfurth, Ulrich, "The Future of the Small Islands in the Pacific," *Aussenpolitik*, 28 (1977), 343–356.

"Self-determination and Security in the Pacific: A Study of the Covenant Between the United States and the Northern Mariana Islands," *New York University Journal of International Law and Politics.* 9 (1976), 277–302.

Serra, Jaime, "Portuguese Colonialism Up a Blind Alley," *World Marxist Review*, 15 (May 1972), 78–82.

Shannon, L.W., "Is Level of Development Related to Capacity for Self-Government? An Analysis of the Economic Characteristics of Self-Governing and Non-Self-Governing Areas," *American Journal of Economics and Sociology*, 17 (July 1958), 367–381.

Shaudys, Vincent K., "The External Political Geography of Dutch New Guinea," *Oriental Geography*, 5, 2 (1961), 145–160.

———, "Geographic Consequences of Establishing Sovereign Political Units," *Professional Geographer*, 14, 2 (March 1962), 16–20.

Shaw, Malcom, "The Western Sahara Case," in *British Yearbook of International Law*, 49 (1978), 119–154.

Sheean, Vincent, "The People of Ceylon and Their Politics," *Foreign Affairs*, 28, 10 (October 1949), 68–74.

Shilling, Nancy A., "Problems of Political Development in a Ministate: The French Territory of the Afars and the Issas," *Journal of Developing Areas*, 7 (July 1973), 613–634.

Sibeko, Alexander, "Portugal and Africa; Breaking the Chains," *Af. Communist*, 59 (1974), 29–47.

Sidell, Scott, "The United States and Genocide in East Timor," *Journal of Contemporary Asia*, 11, 1 (1981), 44–61.

Singh, L.P., "The Commonwealth and the United Nations Trusteeship of Non-Self-Governing Peoples," *International Studies*, 5 (January 1964), 296–303.

Skurnik, W.A.E., "France and Fragmentation in West Africa: 1945–1960," *Journal of African History*, 8, 2 (1967), 317–333.

Slonim, S., "The Origins of the South West Africa Dispute; the Versailles Peace Conference and the Creation of the Mandates System," in *Canadian Yearbook of International Law*, 6 (1968), 115–143.

Smith, C. G., "Arab Nationalism: A Study in Political Geography," *Geography*, 43, 4 (November 1958), 229–242.

Smogorzewski, K.M., "The Russification of the Baltic States," *World Affairs*, 4, 4 (October 1950), 468–481.

Solomon, Neal S., "The Guam Constitutional Convention of 1977," *Virginia Journal of International Law*, 19, 4 (Summer 1979), 725–806.

Somerville, J.J.B., "The Central African Federation," *International Affairs*, 39, 3 (July 1963), 386–402.

Spackman, Ann, "Constitutional Development in Trinidad and Tobago," *Social and Economic Studies*, 14 (December 1965), 283–320.

Spengler, J.J., "Economic Development: Political Preconditions and Political Consequences," *Journal of Politics*, 22, 3 (August 1960), 387–416.

Stephen, Michael, "Natural Justice at the United Nations: The Rhodesia Case," *American Journal of International Law*, 67, 3 (July 1973), 479–490.

Stokes, E., "Great Britain and Africa: The Myth of Imperialism," *History Today*, 10 (August 1960), 554–563.

Suret-Canale, J., "The French Community at the Hour of African Independence," *International Affairs*, 7 (January 1961), 32–37; 7 (February 1961), 23–30.

Sutcliffe, Robert B., "Rhodesian Trade Since UDI," *World Today*, 23 (October 1967), 418–422.

Sykes, John, "Colonial Moçambique; Some Unconventional Conclusions," *New Middle East*, No. 31 (April 1971), 29–31.

T

Tagupa, W.E., "Northern Marianas: Secession from Trusteeship and Accession to Commonwealth," *Journal of Pacific History*, 12, 1–2 (1977), 81–85.

Tangri, R.K., "Rise of Nationalism in Colonial Africa: The Case of Colonial Malawi," *Comparative Studies in Society and History*, 10 (January 1968), 142–161.

Tanner, R.E.S., "The Belgian and British Administrations in Ruanda-Urundi and Tanganyika," *Journal of Local Administration Overseas*, (1965), 202–211.

Tata, Robert J., "Poor and Small Too: Caribbean Mini-States," *Focus*, 29, 2 (November–December 1978), 1–11.

Tauber, Laurence, "Legal Pitfalls on the Road to Namibian Independence," *New York University Journal of International Law and Politics*, 12, 2 (Fall 1979), 375–410.

Taylor, Alan R. and Eugene P. Dvorin, "Political Development in British Central Africa (1890–1956)," *Race*, 1, 1 (November 1959), 61–78.

Teague, Michael, "Portugal's Permanence in Africa," *Geographical Magazine*, 28, 9 (November 1955), 326–336.

Tomasek, Robert D., "British Guiana: A Case Study of British Colonial Policy," *Political Science Quarterly*, 74, 3 (September 1959), 393–411.

Tregonning, K.G., "The Partition of Brunei," *Journal of Tropical Geography*, 11 (April 1958), 84–89.

Twitchett, Kenneth J., "Colonialism: An Attempt at Understanding Imperial, Colonial and Neo-Colonial Relationships," *Political Studies*, 13 (October 1965), 300–323.

U

Umozurike, U.O., "International Law and Self-Determination in Namibia," *Journal of Modern African Studies*, 8 (December 1970), 585–603.

———, "International Law and Colonialism in Africa: A Critique," *Zambia Law Journal*, 3/4, 1/2 (1971/1972), 95–124.

"The United Nations, Self-Determination and the Namibia Opinions," *Yale Law Journal*, 82 (January 1973), 533–558.

V

Vance, Robert T., Jr., "Recognition as an Affirmative Step in the Decolonization Process: The Case of Western Sahara," *Yale Journal of World Public Order*, 7, 1 (Fall 1980), 45–87.

van der Veur, Paul W., "The United Nations in West Irian: A Critique," *International Organization*, 18, 1 (Winter 1964), 53–73.

Van Wyk, J.T., "The United Nations, South West Africa and the Law," *Comparative and International Law Journal (Southern Africa)*, 2 (1969), 48–72.

Velazquez, C.M., "Some Legal Aspects of the Colonial Problem in Latin America," *Annals, American Academy of Political and Social Science*, 360 (July 1965), 110–119.

Venkatavaradan, T., "The Question of Southern Rhodesia," *Indian Yearbook of International Affairs*, 13 (1964), 112–150.

Verrier, A., "The Need for Statesmanship in British Guiana," *World Today*, 19 (August 1963), 352–358.

Vevier, Charles, "American Continentalism: An Idea of Expansion 1845–1910," *American Historical Review*, 65, 2 (January 1960), 323–335.

Viviani, Nancy, "Australians and the Timor Issue," *Australian Outlook*, 30 (August 1976), 197–226.

Vladimirov, K., "U.S. Colonial Possessions in the Pacific," *International Affairs*, (January 1962), 71–73.

Von Vorys, Karl, "New Nations: The Problem

of Political Development," *Annals, American Academy of Political and Social Science*, 358 (1965), 1–179.

Voth, A., "Portugal in Africa: Angola and Mozambique," *Contemporary Review*, 212 (March 1968), 141–146.

W

Walker, P.C. Gordon, "The Future of City and Island States," *New Commonwealth*, 29, 8 (April 1955), 369–371.

Watt, D.C., "Britain and the Future of the Persian Gulf States," *World Today*, 20 (November 1964), 488–496.

———, "The Decision to Withdraw from the Gulf," *Political Quarterly*, 39 (July–September, 1968), 310–321.

Weatherbee, Donald E., "Portuguese Timor; an Indonesian Dilemma," *Asian Survey*, 6, 12 (December 1966), 683–695.

Welensky, Roy, "The United Nations and Colonialism in Africa," *Annals, American Academy of Political and Social Science*, 354 (July 1964), 145–152.

Wellings, Paul, "The 'Relative Autonomy' of the Basotho State: Internal and External Determinants of Lesotho's Political Control," *Political Geography Quarterly*, 4, 3 (July 1985), 191–218.

Wellington, John H., "South West Africa: The Facts About the Disputed Territory," *Optima*, 15 (March 1965), 40–54.

Whitaker, Paul M., "The Revolutions of 'Portuguese' Africa," *Journal of Modern African Studies*, 8 (April 1970), 15–35.

Whittlesey, Derwent, "Southern Rhodesia—an African Compage," *Annals, AAG*, 46, 1 (March 1956), 1–97.

Wigny, Pierre, "The Belgian Plan for Democracy in Africa," *Optima*, 8, 1 (March 1958), 23–31.

Wiskemann, Elizabeth, "The "Drang nach Osten" Continues," *Foreign Affairs*, 17, 4 (July 1939), 764–773.

Wolfers, Edward P., "Papua New Guinea on the Verge of Self-Government," *World Today*, 28, 7 (July 1972), 320–326.

Y

Young, Richard, "The State of Syria: Old or New?" *American Journal of International Law*, 56, 2 (April 1962), 482–488.

Z

Zoli, C., "The Organization of Italy's East African Empire," *Foreign Affairs*, 16, 1 (October 1937), 80–90.

Part Six

CONTEMPORARY INTERNATIONAL RELATIONS

Chapter 23

INTERNATIONAL LAW

There are a number of different approaches to international relations. The systems approach is one, in which international society is considered as a system with subsystems and fixed rules binding them together and making them behave in particular ways. One problem with this is that international relations are neither systematic nor predictable. Another view is that international relations are based solely on power: each State acts in accordance with its own perceptions of self-interest, mainly the acquisition and retention of power, and the real decision makers are the most powerful States. But we have seen how difficult it is to define and measure power and how the components of power are continually changing. We prefer to consider international relations as relations among members of a community.

Like most communities, it is organized, but the organization is not systematic. Some members have a large measure of independence and others are still dependent, but all are interdependent. Some members are more powerful than others, however power is defined, but the balance of power is continually shifting and even small and weak States often play important roles in international affairs. The community has grown gradually with no plan and no steady progression of development. The whole is bound together by an intricate network of bilateral and multilateral relationships of all kinds: religious, political, economic, historical, cultural, legal, ideological, and all of these are overlapping and interrelated. In the preceding two Parts, we have discussed several of these relationships under the general headings of geopolitics and of imperialism, colonialism, and decolonization; now we examine more of these relationships that are geographic in nature.

Every society or community must be governed by a set of general principles and specific rules if it is to function at all. The international community is no exception. It is governed by a complex network of principles, treaties, judicial decisions, customs, practices, and writings of experts that are binding on States in their mutual relations. This is what we call international law, and it has been evolving for centuries. While international law is not created by a particular legislature and enforced by an executive with police powers as domestic law is, it is binding nevertheless. And it is usually effective. Despite the scoffing of cynics, the plain fact is that States do recognize their interdependence, do understand the need for world order, and do operate according to international law most of the time. Naturally, as in every society, there are lawbreakers. But in the international community no State can long remain a chronic lawbreaker. Even if formal sanctions are not applied against it, a State that refuses to accept and abide by the rules simply isolates itself from the rest of the community and suffers from the lack of normal intercourse every State needs. Law is the only alternative to anarchy; law is demanded by the community of interests among States.

Having said this, we must recognize that "international law" has been largely created by a relative handful of States, primarily in Western Europe. Although it does contain some components of other traditions, it is basically derived from Roman law, the Anglo-Saxon common law, and Christian theology. When the dismantling of empires began in earnest after World War II, many new States realized that they were expected to accept and adhere to a body of law in whose creation they had played no role. In some cases they had strong indigenous legal traditions that they valued highly and which they wanted to project into the international sphere. In others, Western law represented imperialism (indeed, had often been used to justify and buttress imperialism) and must *ipso facto* be rejected. Despite these feelings, however, on the whole the new States respect, utilize, and help develop international law, while objecting to certain aspects of it that they deem contrary to their interests. These interests are often different from, sometimes contrary to, the community of interests recognized by the older Western States. Conflicts are thus inevitable, but they are being resolved both by traditional methods and through the political process, principally in the United Nations. It is in the United Nations, in fact, where international law is currently being most vigorously developed.

THE UNITED NATIONS AND INTERNATIONAL LAW

Article 13 of the UN Charter charges the General Assembly with responsibility for, among other things, "encouraging the progressive development of international law and its codification." Overall responsibility for legal matters is assigned to the Sixth Committee of the General Assembly, but the actual work is being done chiefly by the International Law Commission (ILC), established in 1947 by the General Assembly. The 21 members of the ILC are distinguished authorities drawn from all major legal traditions who do not represent governments but function in their personal capacities. The new States not only participate actively in the work of the Sixth Committee and the ILC, but look to them to broaden international law so that it becomes truly universal. Among the

---

THE WORLD COURT AND POLITICAL GEOGRAPHY

Among the cases considered by the World Court during its first four decades, the following are of special interest in political geography: the Corfu Channel (**United Kingdom** v. **Albania**, 1948–1949), Norwegian Fisheries (**U.K.** v. **Norway**, 1951), Miniquiers and Ecrehos (**France** v. **U.K.**, 1953), Antarctica (**U.K.** v. **Argentina and Chile**, 1956), Right of Passage over Indian Territory (**Portugal** v. **India**, 1957–1960), Sovereignty over Certain Frontier Land (**Belgium** v. **Netherlands**, 1959), Arbitral Award Made by the King of Spain (**Honduras** v. **Nicaragua**, 1960), Temple of Preah Vihear (**Cambodia** v. **Thailand**, 1962), South-West Africa (**Ethiopia** v. **South Africa; Liberia** v. **South Africa**, 1966), Northern Cameroons (**Cameroon** v. **United Kingdom**, 1963), North Sea Continental Shelf (**Federal Republic of Germany** v. **Denmark; FRG** v. **Netherlands**, 1969), Fisheries Jurisdiction (**U.K.** v. **Iceland; FRG** v. **Iceland**, 1974), Western Sahara, 1975, Aegean Sea Continental Shelf (**Greece** v. **Turkey**, 1978), Continental Shelf (**Libya** v. **Tunisia**, 1982), Gulf of Maine (**United States** & **Canada**, 1984), and Continental Shelf (**Libya** v. **Malta**, 1985). Of these cases, all but the Corfu Channel, Right of Passage, Northern Cameroons, and South-West Africa cases were boundary or territorial disputes, and of these four, all but the Corfu Channel case involved colonies or mandates. Seven cases involved the Law of the Sea. We may expect the Court to be asked to deal even more in the future with cases in these categories.

many topics dealt with in great detail by the ILC in its first four decades, two are of special interest to geographers: international rivers and the Law of the Sea. We discuss international rivers later in this chapter and the Law of the Sea in Chapters Twenty-nine and Thirty.

Another UN organ helping to develop international law is the International Court of Justice, successor to the Permanent Court of International Justice of the League of Nations. The ICJ, or World Court, is composed of 15 judges elected by the General Assembly and the Security Council. Like the members of the ILC, they represent "the main forms of civilization and the principal legal systems of the world." They act in their individual capacities and not as representatives of their governments. The Court sits in The Hague and considers cases brought to it by States either for judgement or for advisory opinions.

Other UN bodies participate, some of them rather marginally, in international lawmaking, including the General Assembly itself and the specialized agencies. More important, however, are the international lawmaking conferences. Most of them have dealt with strictly legal matters, such as diplomatic and consular immunities, State succession, and treaties, but at least four fall largely within the field of political geography. These are the three UN Conferences on the Law of the Sea and the UN Conference on Transit Trade of Land-locked Countries. Unlike conferences of experts gathered to discuss current problems and perhaps pass resolutions, these conferences are designed to produce *conventions* (multilateral treaties) which, if duly ratified, become binding on at least the signatories and part of international law. It is possible that we will see fewer lawmaking conferences in the future, with more of this work being done by the Sixth Committee. But we are unlikely to see any reduction in the overall contribution of the United Nations to the codification and progressive development of international law. On the contrary, as

more States join the international community, as all States become ever more tightly bound together by mutual dependence, and as the need for universal law becomes both more accepted and more urgent, we can expect a great increase in UN activity in this field.

CONFLICTS AND CONFLICT RESOLUTION

There are and probably always will be conflicts and disputes among States. They have many origins and take many shapes. Some are bilateral and some involve groups of States on both sides. Some are ancient, some relatively new. Some are fairly simple, others intricately complex. All, however, can be resolved without resort to war if the parties are willing. Conflict resolution (or, as it is called in international law, *pacific settlement of disputes*) has long been a popular subject for political scientists. Geographers have tended to concentrate on the issues instead of on the process of settlement. If geographers are to understand a conflict and analyze it usefully, however, they must be familiar with the procedures as well as the issues.

By far the most important method of settling a dispute is through *bilateral negotiations*. The parties use diplomatic channels to begin with. Sometimes the normal diplomatic efforts will be supplemented or supplanted by negotiations between cabinet ministers and even heads of government. On occasion imaginative devices will be used: a private citizen acting as go-between, a meeting in a train parked on a bridge straddling an international river, an official going secretly in disguise to talk with a hostile chief of State. Generally, though, the negotiations themselves are quite mundane, even if they are accompanied by flamboyant publicity and ostentatious saber-rattling. Most disputes are settled this way; only if the dispute is particularly intractable or the parties unwilling to talk with one another is a third party called in to help.

There are five standard types of third-

MODES OF REDRESS SHORT OF WAR

There are some less pacific methods of dealing with international grievances or disputes. In international law, these are known as "modes of redress short of war." The mildest response is called **retorsion,** applied when the act or acts complained of are unfriendly but not illegal. Such acts may include discriminatory tariffs, immigration restrictions, port rules, currency control, and the like. Typically, the retort is reciprocity; that is, the complainant responds in kind. A more vigorous response to an alleged unfriendly or illegal act is **retaliation.** A State has a large arsenal of retaliatory weapons available to it, and it may exercise them as it deems appropriate within the limits of its own ability and commitments. It may take such diplomatic action as recalling its ambassador, closing its embassy or breaking diplomatic relations. Economic action might involve raising tariffs or imposing quotas against the goods of the other party or selective or general embargoes or boycotts. And an aggrieved State might take military action; a brief raid into the other party's territory by land, sea, and/or air to destroy selected military targets or take military prisoners, or perhaps capture of one of the other party's vessels at sea. Finally, it may resort to **reprisal,** a form of retaliation far in excess of the acts complained of. Reprisal borders on aggression and sometimes might be considered illegal, though the distinction is unclear. As with the modes of pacific settlement of disputes, the less damaging the mode of redress short of war, the more frequently it is used. Reprisal is a last resort short of war.

party participation: *good offices, conciliation, mediation, arbitration,* and *judicial proceedings.* The disputing parties may choose any one of them, use a combination or variation of them, or progress from one to another. In any case, they determine the third party and establish the ground rules, either in general or in detail.

The simplest form of third-party participation, the one in which the third party is least directly involved in the dispute, is called *good offices.* The third party expedites bilateral negotiations by performing such services for the disputants as providing a neutral site for the negotiations; supplying interpreters, office space, secretarial services, and the like; transmitting messages between the parties; doing basic research and providing factual information to the parties; and even providing entertainment and sightseeing so as to create and maintain a relaxed and friendly atmosphere for the negotiations. This work is done every day around the world, generally quietly with little or no publicity. Many countries, especially Switzerland, provide good offices, but by far the most frequent and useful provider of the service is the United Nations. This is, in fact,

one of its least publicized but most important services to world public order. Even while countries are shooting at each other, they can negotiate quietly at UN headquarters or elsewhere under UN auspices.

A third party can intervene rather modestly in the negotiations by offering *conciliation.* A conciliator will consider the positions of both sides and offer a compromise solution to the problem. He or she does not participate in the negotiations, undertake detailed studies, or pass judgments. The conciliator's function is to facilitate the resolution of a dispute by offering a face-saving solution to the parties. Very close to conciliation, but more formal and active, is *mediation.* A mediator studies the case in more detail, participates actively in the negotiations, and offers a formal proposal for solution of the problem. Usually only the more difficult problems require third-party intervention in the first place and, therefore, the tendency is for the parties that have agreed on such third-party intervention to prefer mediation to conciliation. The number of situations requiring conciliation or mediation is much smaller than those requiring only good offices, but they tend to be more politicized and involve higher stakes for the parties—and sometimes for the world.

If a dispute is more legal than political in nature (and it is frequently difficult to disentangle the two) and if it has been protracted and particularly trying for both parties, they may resort to *arbitration*. This is a more formal, time-consuming, and expensive undertaking and, consequently, is less frequently utilized than any method discussed so far. But many arbitrations over the years have had far-reaching influence not only on the parties, but also on the evolution of international law, even though technically they are not precedents. The parties to the dispute agree in advance whether the results of the arbitration are to be advisory only or actually binding on them. Usually, they agree on binding arbitration. Then they choose an umpire or arbitrator (who may be a sovereign, a distinguished judge, or a tribunal or panel). If there is more than one arbitrator, the parties typically choose one or two each and they choose another to serve as president or chairman.* The parties then submit to the arbitrator(s) a *compromis*, a formal statement of the principles and rules of law that the parties agree are applicable in this particular case. The arbitrator then takes testimony, studies it, and renders a decision. If there is a question of damages or compensation involved, the arbitrator may also issue an award.

Unlike good offices, conciliation, and mediation, arbitration aims at justice, not compromise. It still represents, however, the parties' attempt to arrive at an amicable resolution of a problem under their own rules by arbitrators of their own choice. *Judicial proceedings*, on the other hand, are formal adversary proceedings before a permanent court following established rules. They are typically the last

resort, used after all other methods of pacific settlement of a dispute have been rejected or have failed. The proceedings may take place before a national court, a regional court (such as the Court of Justice of the European Community) or before the International Court of Justice. Naturally, being sovereign States, one or both parties may ignore or reject either arbitral or judicial decisions, but generally they are respected.

INTERNATIONAL RIVERS AND LAKES

Some aspects of international law are of special interest to geographers because of their spatial dimension. We have already discussed a number of them and cover others in subsequent chapters. Here we examine a subject that overlaps many others: international rivers and lakes. Generally, the same principles and rules apply to lakes as to rivers; we mention lakes only when there is a special reason to do so.

International rivers are rivers shared by two or more countries. The international boundary may follow the river or cut across it. If the river serves as a boundary, the actual boundary may be on the left bank (looking downstream), the right bank, or somewhere in between. If it flows across two or more States, there are upstream and downstream riparians that may have different interests in the river. (Sometimes part of a river will serve as a boundary and another part traverse one or more States.) *Internationalized* rivers are those that by treaty or other formal arrangement have been opened to navigation by vessels of States in addition to those of the riparians, even if they lie entirely within the territory of a single country. Among them are the Scheldt, Rhine, Niger, Danube, Congo, Zambezi, Amazon, Plata, and St. Lawrence. The ones most heavily used for international navigation have international commissions to supervise and regulate their use. Navigation has traditionally been the most important use made of interna-

---

*Arbitrators are sometimes chosen from among the judges on call from the Permanent Court of Arbitration, founded at The Hague in 1899. In this case each party selects two from the list, and they select a fifth to be the umpire.

tional rivers, but other uses have become so important in the last half century that the International Law Commission is currently undertaking a thorough study of "the law of the non-navigational uses of international watercourses."

From earliest times, people have used rivers to transport themselves and their goods. Water transport is particularly useful and economical for heavy and bulky goods. Rivers large and small were the world's first highways. With increasing population and industrialization, with cheaper and more efficient land and air transport, the proportion of goods carried on internal waterways (canals as well as rivers) has decreased. The volume of waterborne commerce has nevertheless increased dramatically and is more important than ever.

Over the centuries both natural and political barriers to river navigation have gradually been removed, beginning in Europe with its high density of population, industry, and waterways. The waterways of North America are also intensively utilized and have been reshaped and managed, generally more elaborately and on a larger scale than in Europe. Elsewhere, rivers tend to be used more for local navigation by small vessels, often very small ones, and the engineering works are not as elaborate.

The natural obstacles to international river traffic, however, have been much easier to remove than the artificial ones. States have frequently forbidden nonnationals to navigate on national and international rivers within their territories. Where navigation by foreigners was permitted, it was often subjected to tolls, taxes, and restrictions. As trade grew in importance, the commercial advantages of free navigation gradually assumed greater importance than absolute territorial sovereignty. As early as the eleventh and twelfth centuries, territories in Europe, particularly in Italy, began the internationalization of rivers. By the late eighteenth century, however, tolls and restrictions were still common. The principle of free transit

was reestablished by the armies of revolutionary France, which opened the River Scheldt, closed by the Dutch since 1648, to the traffic of all riparian States. In 1795, Spain and the United States negotiated the Treaty of San Lorenzo el Real that granted Americans the right to navigate the Mississippi and other Spanish rivers to the sea. This gave substance to President Jefferson's claim that "the ocean is free to all men, and their rivers to all their inhabitants." Gradually, the concept of freedom of the seas was applied to rivers, considered by some to be long, narrow arms of the sea.

At the end of the Napoleonic Wars, the Congress of Vienna codified the principles of freedom of navigation on international waterways that had been established in recent years and applied them specifically to the Scheldt, the Meuse, and the Rhine. Through the nineteenth century the principles of free transit on rivers gradually became established and by the end of the century transit duties in Europe had virtually disappeared. In Latin America, the newly independent States opened their territories to the ships and traders of all States. The Congo Act of 1885 opened the Congo River and all its tributaries and the Niger River to the trade of all States. By World War I, most navigable waterways, natural and artificial, in Europe, East Asia, Africa, and the Americas had been internationalized. Today there are few political problems in river navigation; where they do exist, they tend to be part of a much larger pattern of hostility. Some restrictions on foreign navigation on both national and international rivers are still practiced by States such as China and the Soviet Union, even where the rivers have been internationalized by treaty, but these restrictions are often informal and difficult to document. On the whole, however, river navigation is freer now than it has been for centuries.

Nonnavigational uses of rivers are many and varied and becoming more so all the time. The earliest civilizations—Egypt, Mesopotamia, the lower Indus, Southeast

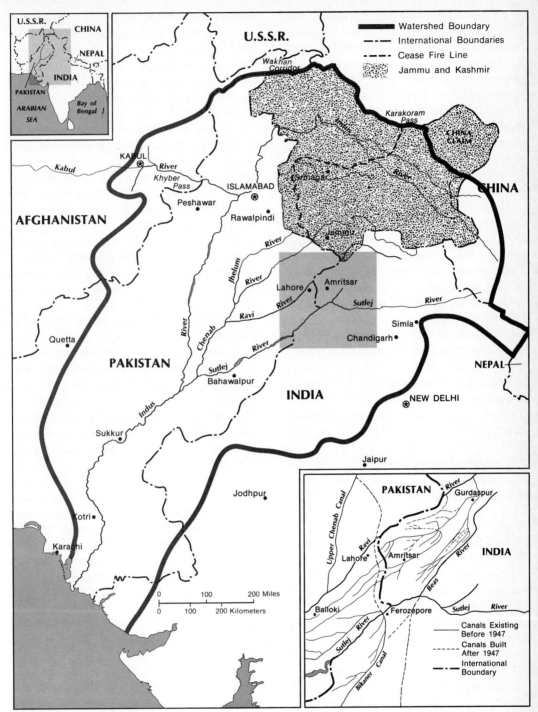

Division of the Indus Rivers. The large map shows the main stream of the Indus River and its principal tributaries. Note the importance of Jammu and Kashmir in this respect as well as its strategic location between China and India and Pakistan. Note also the relationship of the Khyber Pass and Pakistan's northwest frontier province to this area and the location of Pakistan's capital of Islamabad. Amritsar is a center of the Sikh religion and Chandigarh is capital of the Indian states of Punjab and Haryana. The inset map of the irrigation canal system is generalized but shows clearly the effect of the 1947 partition on what had been an integrated system.

Asia, east China—all developed in river valleys using the river water for irrigation. Irrigation has spread around the world from these culture hearths, even to regions that are relatively humid. In some places, such as the southeastern United States, this irrigation merely supplements rainfall and is used largely for luxury crops, but in other areas, such as Egypt, the river water is absolutely essential for the maintenance of even a bare subsistence for the people.

All around the world, even in recent years, there have been disputes over the uses of international rivers. One, which brought two large neighbors to the very brink of war, was over the Indus River. During their rule in India, the British developed a sophisticated multipurpose system for the upper Indus Valley, utilizing the main stream, six major tributaries, and a number of smaller tributaries. When India was partitioned in 1947, the new international boundary cut right through this integrated system, wreaking havoc on irrigation, flood control, hydroelectricity, and other services in the basin. In April 1948, India cut off the flow of water from the eastern tributaries of the Indus into the canals leading into Pakistan. Service was restored a month later, but the dispute over ownership of the Indus waters dragged on until both parties finally agreed in 1960 to a World Bank plan for dividing them between the two countries. Since then, a permanent Indus commission has supervised implementation of the Indus Waters Treaty, but the enormous cost and disruption to India and Pakistan for creating and operating two parallel river-development schemes could have been avoided if the neighbors had formed a joint commission immediately in 1947 to operate and improve the system they inherited from the British.

Similar but less virulent disputes over the waters of the Nile, Colorado, Columbia, Ganges, Euphrates, and other rivers have been settled only after arduous bilateral negotiations and in some cases third-party participation. As the world's population grows larger and richer, it places greater and greater demands on Earth's finite supply of fresh water. Inevitably, as the larger streams are utilized, smaller and smaller streams are likely to be the focus of competing demands for their water.

The Jordan waters dispute between Israel and Jordan, Lebanon and Syria is now more than 40 years old. Both Israel and international experts have repeatedly urged cooperative development of the whole Jordan system for the benefit of all the countries in the basin, but the Arab countries have steadfastly refused to cooperate with Israel for political reasons. As a result Israel and Jordan have developed parallel water-diversion systems for irrigation and other uses.

An even smaller and much more obscure stream, the Río Lauca, was the cause of a major crisis between Chile and Bolivia in the early 1960s. The Lauca rises in Chile high in the Andes and flows for about 225 kilometers into Lake Coipasa in Bolivia. In 1939, Chile announced a plan to divert some of the river's water down into the valley of the Río Azapa to provide irrigation and domestic water for the city of Arica. Bolivia protested then and on a number of occasions thereafter as plans were developed and then executed. In 1962, the diversion actually began. Bolivia broke off diplomatic relations with Chile and requested an urgent special meeting of the Council of the Organization of American States (OAS), charging Chile with "geographic aggression"! Bolivia was dissatisfied with the OAS action and quit the organization for a time. The diversion continues, but the basic issue has not been resolved. Since there is no generally accepted international law on the subject of nonnavigational uses of international rivers (other than a broad principle that the upstream riparian should not use a river in such a way as to harm a downstream riparian), conflicts involving them are bound to continue.

Other nonnavigational uses of rivers include fishing, flood control, hydroelectric

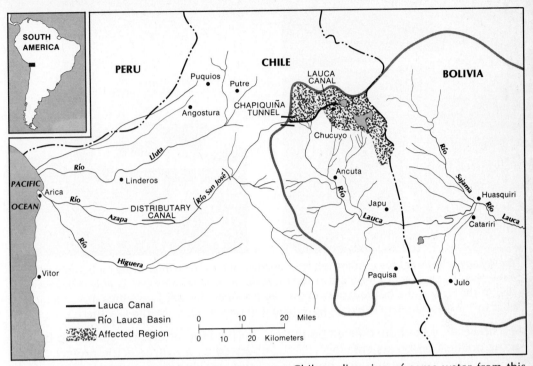

The Río Lauca. Bolivia created a controversy over Chilean diversion of some water from this stream, but soon linked it with her desire for a *salida al mar,* or outlet to the sea. Bolivia never demonstrated any damage done to her territory or people by the diversion.

power generation, recreation, waste disposal, industrial processes, timber floating, domestic consumption, drainage, construction, and aquaculture. Various traditions around the world cover some of these uses, but the traditions often conflict with one another. So do many of the uses of rivers. The problem is serious enough in the United States, where water law in the southwest is very different from water law in the northwest and northeast, but it is far more severe at the international level. In Europe most major problems have been worked out over generations, but in Latin America generations of work by experts have failed to produce a generally accepted code for regulating the use of international river waters. In Africa and Asia there has not even been a serious attempt at such standardization. This is why the International Law Commission has undertaken a comprehensive study with a view to developing a draft convention on the subject, and why its work is so important.

A final point concerns the concept of the river basin, regardless of political boundaries, as an ecological unit that should be developed according to a comprehensive, integrated, multipurpose plan. All the present and potential uses of the main stream, its tributaries and its drainage basin should be accounted for as well as possible so as to minimize conflicts over them. Such conflicts include competing uses, waste of water, soil erosion, water and soil pollution, salinization and waterlogging of overirrigated land, destruction of wildlife habitats, overfishing, unnecessarily destructive floods, and other ills caused by unwise use of this scarce natural resource. This concept is really quite old, but its first modern expression was the Tennessee Valley Authority, still the model for similar schemes the world over. The Papaloapan in Mexico, the Cauca in Colombia, the Damodar in India, the Helmand in Afghanistan, and the Volta in Ghana are just a few examples. Developing *international* river basins in this man-

The Mekong Scheme in Laos.   At the left is a Canadian survey team measuring topographical levels of the Nam Ngum River, a tributary of the Mekong, in 1960. The project is now well advanced. (UN) At the right is the headquarters of the Lao National Mekong Committee in Vientiane, still active in 1979 despite the change of government in Laos in 1975. (Martin Glassner)

ner, however, is far more difficult because of the political problems entailed.

Two ambitious schemes for multipurpose cooperative development of international river basins are those for the Plata and the lower Mekong. The first involves Brazil, Paraguay, Bolivia, Argentina, and Uruguay, which share the basin of the Paraguay–Paraná–Plata river system. In 1941, the five countries met in Montevideo and agreed on a resolution expressing their aspirations for regional development. Nothing more happened until the 1950s when the countries individually and in pairs began serious development work there. Finally, all five countries met in Buenos Aires in 1967 and established an intergovernmental coordinating committee to begin planning a variety of cooperative projects. Since then many bilateral projects have been completed and more are under way, all within the general framework of the Plata Basin program. It has the active support of the OAS and other regional and international agencies, but its course has not been smooth. There have been numerous squabbles among the participants, particularly Brazil, Paraguay, and Argentina, over everything from overall priorities to the location of highway bridges

to the type of electric current to be generated by a hydroelectric project. Nevertheless, it goes on.

Even larger is the project for developing the lower basin of the Mekong, the largest unbridged and undammed river in the world, shared by Thailand, Laos, Vietnam, and Kampuchea. It was initiated by the UN Economic Commission for Asia and the Far East (ECAFE, predecessor of ESCAP) and the Colombo Plan in the mid-1950s. As the years went by and plans developed, more sponsors joined in: 25 countries plus several international agencies and private foundations. Detailed plans and cost estimates were ready by 1960, and field surveys began soon after. Before work had gone very far, the United States got heavily involved in what had been small-scale fighting in South Vietnam. As the scale of the war increased, and it spread through much of Laos and Cambodia (as it was then called), the work slowed down and nearly ceased. The Mekong Committee,* composed of the four riparian States, which does the planning, has not met since 1975. In 1977, Laos, Thailand, and Vietnam agreed

*Officially, the Committee for Coordination of Investigations of the Lower Mekong Basin. Its headquarters is in Bangkok.

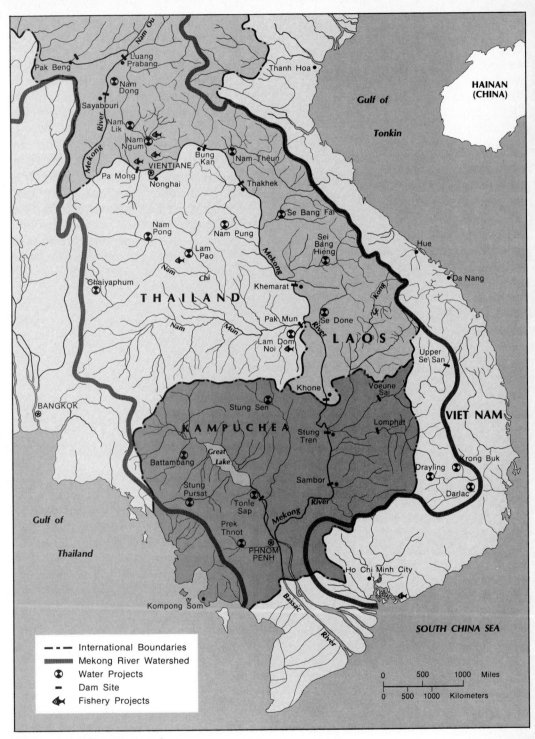

The Lower Mekong Basin Scheme.

to form an interim committee and the pace of the work picked up.

Its work program for 1986 included 93 projects requiring initial funding of approximately US $372.6 million, in addition to the many projects already completed. Development work continues in the fields of hydrology, meteorology, mapping by remote sensing, economic and social studies, social planning and resettlement, environmental studies and planning, irrigation, drainage, flood and salinity intrusion control, hydroelectricity, navigation, agriculture and aquaculture, geological surveys, water-borne diseases, control of erosion and sedimentation, and agro-industrial development. In January 1986, it ran a UNDP Workshop for Nile Basin Countries in Bangkok to illustrate the advantages of the Mekong approach to the Nile riparians. There is still hope that Kampuchea will join in and that full cooperation in the basin, which was so inspiring and productive before the late 1960s, will resume.

Other multinational basin-development organizations include those for the Niger and Senegal rivers and Lake Chad in Africa and, still incipient, the Amazon in South America. We can be reasonably certain that more of these schemes will evolve, for the benefits of such cooperation are so great and so obvious that not even the most ardent nationalists can deny them.

Chapter 24

INTERNATIONAL TRADE

International trade, in a sense, has existed for several thousand years, since the first organized human communities began trading surpluses with other communities. We have both written and archaeological evidence of elaborate trading systems functioning in all of the early culture hearths and of goods being traded among them, often over impressive distances. Geographers, economists, anthropologists, and historians have studied in detail the trading systems of West Africa, South America, the South Pacific, and other regions before their exposure either to Europeans or to modern technology. Modern international trade, however, really began with the Industrial Revolution and the decline of mercantilism.

Mercantilism was the dominant economic theory in Europe during the late Middle Ages and into the eighteenth century. It was based on the notion that wealth consisted only of gold and silver. Naturally, a country could get rich only by accumulating large stores of gold and silver bullion. There were three ways of doing this: first, by stealing it or by conquering countries that had large stores of bullion; second, by finding and exploiting new sources of the precious metals; and third, by exchanging goods and services for gold and silver. The first method was employed from time to time but proved impractical as a general policy; the second led to the great Age of Discovery and the creation of the first large European colonial empires; the third produced stagnation and decay, even for some countries with large empires. Mercantilism was an essentially restrictive economic policy that made a few individuals and countries rich but did nothing to improve the lot of the many millions of ordinary people in the world. Gold and silver produce nothing; only investment produces anything, and it is production of goods and services that improves people's lives, not hoarding of gold and silver. The exchange of surplus production increases the value of the initial investment and multiplies people's real incomes.

The first major attack on mercantilism was launched by the English philosopher David Hume in 1752 when he demonstrated the fallacy of the "favorable balance of trade" concept. His friend Adam Smith followed up in 1776 with an elaborate treatise called *Wealth of Nations* that expounded the concept of international division of labor, or specialization and international trade. This was a direct attack on *autarky* and a plea for free trade, which would bring some benefit to everyone rather than great benefit to a few. Smith was followed by David Ricardo, John Stuart Mill, and other classical English economists who developed further the theory of *comparative advantage*.* This is the theoretical underpinning of free trade. The indisputable advantages of specialization and trade led to the free trade movement of the nineteenth century and the burgeoning world trade that developed from it. Today most people in the world can afford *some* imported product, and *some* of what most people produce is exported. A far cry indeed from the situation three hundred years ago when most people subsisted on what they could produce themselves and only a few rich and powerful people could benefit from international trade.

Not all governments, however, adopted the *laissez faire* policies so important for free trade. Some, in fact, persisted with mercantilist policies for generations. Antonio Salazar, dictator of Portugal and erstwhile economics professor, based his *Estado Novo* on mercantilist principles. It brought stability but not prosperity to Portugal and her empire. Few countries ever adopted a complete *laissez faire* attitude toward international trade; in most, the liberalization of trade was limited by protectionist policies or other considerations. Still, comparative advantage, economies of scale, and differential demand for goods in countries with different cultures were powerful forces generating the expansion of trade. Gradually, the "colonial economy" (the exchange of the manufactured goods of Western Europe for the foods, fuels, and raw materials of Africa, Asia, and the Americas) was surpassed in importance by trade among the industrialized countries.

As the industrializing countries became richer as a result of their control over man-

ufacturing, commodities, and trade, they began demanding and producing ever-more sophisticated and expensive products. They found that the only feasible sources of the goods they wanted were other industrialized countries that, through no coincidence at all, were also the only countries rich enough to buy the new manufactured goods they were producing. The bulk of world trade then began to flow among the rich countries themselves. This pattern was reinforced by the restrictions placed by the colonial States on manufacturing and trade by their colonies, by protective tariffs imposed by some industrializing States, by subsidization of local agriculture, by development of substitutes for some commodities, and by other practices and developments. Today we find this pattern still dominant, even though many poor countries really are "developing" and becoming more important traders in their own right rather than serving simply as appendages of a colonial ruler. Table 24-1 shows this pattern clearly for 1982. Over 63 percent of all goods traded among States and territories originated in the developed market economy (noncommunist) States, with another 9 percent coming from the more developed socialist countries including the Soviet Union. The poorer countries of the world contributed only 26 percent, and this includes enormous quantities of petroleum.

The table shows a number of other interesting things, including who trades with whom. The developed market economy countries, for example, did 68.2 percent of their trading among themselves, and bought only 27 percent of their total imports from the developing countries, including the Asian socialist ones, of which China is by far the largest. Even the poor countries bought almost two thirds of their imported goods from the rich countries, and the major petroleum exporters (in whose cases the terms "poor" and "developing" need new definitions) bought 77.4 percent of their imports of all kinds in the rich industrialized countries—which, of course, are the best cus-

*This theory is best illustrated by the hypothetical example of the best lawyer in town who is also the best typist in town. He and everyone else will benefit most if he specializes in the law and hires a typist. Even though he has an *absolute* advantage over the typist in typing, the typist has a *comparative* advantage in that specialty. So even if a country can produce *everything* more efficiently than *every* other country, it can maximize its assets and grow more rapidly if it specializes in those things in which it has a comparative advantage and imports other things from somewhat less efficient producers.

Table 24–1 Origin and Destination of World Trade (%) 1976 and *1982*

Origin	World	Developed Market Economy Countries	Socialist Countries		Developing Countries	
			Eastern Europe	Asia	Total	Major Petroleum Exporters
World	100.0	100.0	100.0	100.0	100.0	100.0
Developed Market Economy Countries	65.1	68.6	34.7	56.4	67.9	84.0
	63.6	*68.2*	*29.0*	*57.8*	*63.1*	*77.4*
Socialist Countries						
Eastern Europe	8.8	3.7	53.3	32.1	5.2	4.1
	9.1	*4.0*	*59.5*	*21.6*	*4.7*	*4.9*
Asia	0.8	0.4	1.4	—	1.6	0.9
	1.2	*0.8*	*1.5*	—	*2.3*	*0.9*
Developing Countries						
Total	25.3	27.3	10.5	11.5	25.3	11.1
	26.1	*27.0*	*10.0*	*20.1*	*29.8*	*16.7*
Major Petroleum Exporters	14.0	16.0	2.3	0.0	12.8	1.0
	12.2	*13.7*	*1.9*	*0.5*	*12.8*	*2.1*

Source: United Nations Conference on Trade and Development, *Handbook of International Trade and Development Statistics* (Sales No.E/F. 78.11.D.1), pp. 58–59; and E/F. 84.11.D.12, pp.66–67.)

tomers for their principal export. Note also that more than half the total trade of the European countries with centrally planned economies ("socialist" countries) was among themselves; obviously, politics rather than comparative advantage is the determining factor here.

Economics, in fact, is overshadowed by politics in many aspects of international trade. We discuss some of these factors later, but now we will see how the vast mechanism of international trade is regulated.

GATT AND UNCTAD

The system of world trade that had matured in the late nineteenth century was disturbed by World War I, battered by the Great Depression, and destroyed by World War II. All along, however, it had been weakened by autarkic policies adopted by various countries to restrict trade for a number of reasons. There were, of course, surviving mercantilists, but more important were desires to protect infant industries, build up uneconomic industries for strategic reasons, conserve scarce foreign currencies, give

preferences to political allies or former colonies, develop monopolies and cartels to increase profits by manipulating prices, maintain strong merchant fleets, and achieve other limited objectives, all of which brought effectively to an end the limited experiment with free trade promoted by Britain. High protective tariffs, including especially the U.S. Smoot-Hawley Tariff of 1930, were only the most publicized instrument of this economic nationalism that escalated a bank failure in Austria and a stock market crash in the United States into a great worldwide Depression, illustrating the interdependence of much of the world, even then.

Toward the end of World War II, Western businessmen, statesmen, and economists felt the urgent need to rebuild the world economy, including the trading system. In July 1944, they gathered at Bretton Woods, New Hampshire, for this purpose. There they designed the International Bank for Reconstruction and Development (IBRD or World Bank) and the International Monetary Fund (IMF or Fund). As successful as these agencies have been, however, they could not alone promote a return to a system of reasonably free trade and payments in the postwar world. In 1947, two separate but interrelated sets of negotiations began, one to create an International Trade Organization, which would enable private enterprise to play an active role in trade liberalization; the other to develop a tariff agreement among 23 of the most important trading countries. The Havana Charter establishing the ITO was signed in 1948, but only one country (Australia) ever ratified it and it never went into effect. *The General Agreement on Tariffs and Trade (GATT)* had meanwhile been signed in Geneva in 1947; it was ratified and did go into effect. It has been enormously successful in its mission of encouraging and guiding the expansion of world trade.

GATT is essentially two things: a complex network of nearly two hundred interrelated bilateral trade agreements, and a series of rules, all designed to lower tariffs and eliminate nontariff barriers to trade. Its "temporary" secretariat in Geneva has taken on the characteristics of a permanent organization, it has developed loose links with the United Nations, its semiannual meetings serve as forums within which new proposals for trade liberalization can be explored, and it has incorporated many of the provisions of the Havana Charter into its own rules and functions. The original 23 adherents to GATT have been joined by 69 others, and another 30 participate under special arrangements, another sign of its vitality. Even a few "socialist" countries have joined, a sure sign of its indispensable role in facilitating the trade of countries just emerging into the world trading system. The large formal tariff-cutting conferences of GATT, held approximately every five years, have reduced or eliminated tariffs on many thousands of items.* The rules against discriminatory trade practices, dumping (selling export goods below cost), and other nontariff barriers to trade are usually followed—although Japan, the United States, and other countries do vi-

olate them from time to time—and the rules of GATT currently govern some 85 percent of international trade.

During the nineteenth century, a new technique was developed to supplement the traditional principle of reciprocity. This was the "most-favored-nation" clause incorporated into a bilateral trade agreement. This MFN clause provided that each partner would grant to the other whatever tariff and other trade concessions had been granted or would be granted in the future to any third country. This practice helped somewhat in reducing tariffs but did not become generally accepted until GATT made it the foundation of its whole system. The MFN clause and the GATT tariff-cutting rounds have replaced the U.S. Reciprocal Trade Agreements and have been the principal instruments in a general but still quite inadequate worldwide reduction of tariffs. MFN status must be negotiated and is frequently (as in the United States at present) linked to political concessions.

GATT provides for two major exceptions to its rules about tariff reduction and nondiscrimination. It permits developing countries to use protective tariffs to protect infant industries for limited periods, and it permits groups of countries to grant concessions to their members not granted to outsiders, provided the result is an expansion rather than a contraction of trade. Both devices are widely and for the most part successfully used.

There is no doubt that GATT has contributed hugely to the incredible expansion of world trade during its existence. But, as we see in Table 24–1, the benefits of this trade expansion have not been evenly distributed. They have accrued largely to the industrial countries, those that were already rich. The "colonial economy" still hampers to some extent the development of the poor, largely ex-colonial countries of the world. There has long been a general tendency for prices of manufactured goods to rise more rapidly than the prices of commodities in world trade, thus worsening the terms of trade

*The most recent was the Tokyo Round, 1973–1979.

of those countries that are primarily producers of commodities and consumers of manufactured goods; that is, the poor countries, generally former colonies.*

All these trends became evident to the newly emerging countries in the late 1950s and 1960s. They saw that the rich were getting richer, called GATT "a rich man's club," and refused, many of them, to become parties to it. They felt that GATT's mandate was too restricted to help them in development and demanded a new organization that could. As their numbers increased, so did their voting strength in the United Nations. By 1964, there were 75 of them, and they were able to pressure the rich countries (including the USSR) to agree—reluctantly—to the convening of a special UN conference on the subject. We will discuss this conference and its results shortly, but first we should consider some of the major trade issues of the 1980s and 1990s with which the world will have to deal both through GATT and through other mechanisms.

The latest round of trade liberalization negotiations under GATT began with a preliminary meeting of ministers and other officials from more than 70 GATT members in Punta del Este, Uruguay, in September 1986. This "Uruguay Round" is likely to last for several years because it must tackle and, if possible, resolve some exceedingly difficult problems. This is so because of two basic factors: first, the easy things have already been done, and now the residual conflicts must be faced; and second, the world trading pattern has changed considerably since the Tokyo Round and the current negotiators face new and more challenging problems.

The huge growth in international trade is related to four closely interrelated trends, all of which have political causes and/or effects. First, the *mobility of capital* of the current era of history is quite un-

precedented and unexpected even half a century ago. But like trade itself, most of this international investment (some two thirds) has been concentrated in the developed countries. Second, international firms (transnational corporations) have increasingly engaged in *intraindustry trade*; that is, trade in products of the same industrial sector, so that countries have been bound more closely together. Third, there has been a growing *tendency toward oligopoly* in international trade, thus reducing price competition and increasing the importance of technological innovation and other factors in worldwide competition. Fourth, some transnational firms have created a new kind of international division of labor within their own organizations, so that much of the growth of trade is accounted for by *trade among components of the same company*. All four of these trends are linked by the transnational corporation, to which we return later in this chapter.

Other issues with which the Uruguay Round must grapple are:

1. Nearly all of GATT's trade expansion work so far has been in manufactured goods, but the problems of *agricultural production* and distribution (discussed in more detail in Chapter Thirty-five) have become too great to be evaded any longer. Now the negotiators will have to wrestle with the costly, inefficient, and trade-distorting agricultural policies of Western Europe and the United States especially, in particular their more than $50 billion of subsidies to farmers to encourage overproduction and then the dumping of surplus agricultural products on saturated markets, in which those who are really hungry are too poor to buy these surplus foods and fibers.

2. Trade in *services*, such as banking, tourism, insurance, shipping, computer software, communications, and so on, has grown spectacularly, especially among the rich countries,

*Since 1973, of course, petroleum has been a major exception to this general tendency, but few other commodities producers have been able to exercise the same politicoeconomic power as OPEC, the Organization of the Petroleum Exporting Countries.

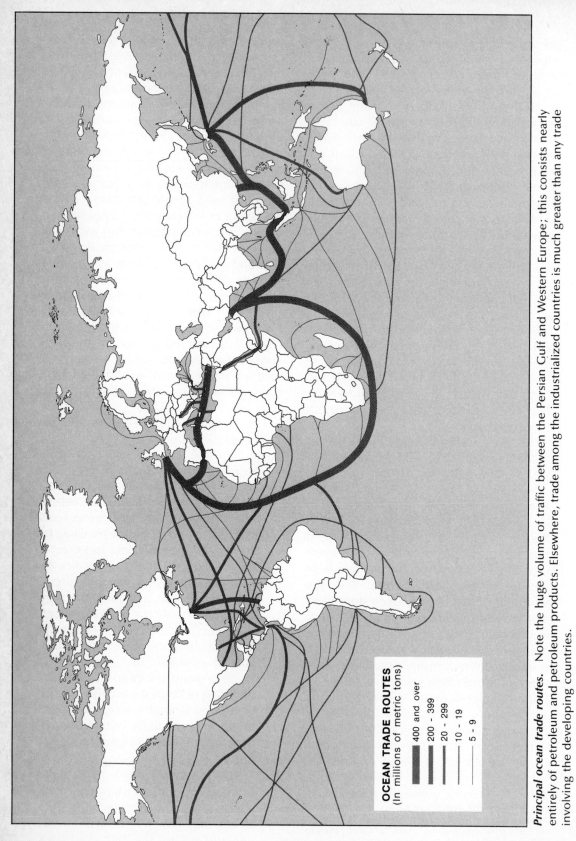

Principal ocean trade routes. Note the huge volume of traffic between the Persian Gulf and Western Europe; this consists nearly entirely of petroleum and petroleum products. Elsewhere, trade among the industrialized countries is much greater than any trade involving the developing countries.

OCEAN TRADE ROUTES
(In millions of metric tons)

— 400 and over
— 200 - 399
— 20 - 299
— 10 - 19
— 5 - 9

even as much manufacturing has passed from them to some newly industrializing countries (such as Brazil, South Korea, Taiwan, Singapore, and Mexico). The rich countries want services brought under the GATT system, but many developing countries do not. In Punta del Este they compromised on a plan to discuss services outside GATT at first and, perhaps in 1991, to bring a report on them into the negotiations on new GATT rules.

3. Largely on the insistence of the United States, the Uruguay Round will include drafting of rules to protect *intellectual property* (e.g., patents, trademarks, copyrights) from piracy and counterfeiting.

4. Again largely because of an American initiative, the negotiators will consider (though not necessarily resolve) distortions of international trade caused by certain *investment policies*, such as local content requirements.

All of these issues are further complicated by such relatively new—and probably temporary but nonetheless vexing—situations as OPEC-dictated crude petroleum prices, far lower than they were in the late 1970s and early 1980s, and the colossal and unprecedented deficit in the U.S. balance of payments, which has led to protectionist pressures not seen in more than half a century. In addition, there are still a wide range of more or less traditional distortions of international trade that we list at the end of this chapter.

But some constructive and hopeful developments are taking place. For example, the powerful entry of China into the world trading system, the growing strength of some economic integration arrangements (discussed in detail in the next chapter), including movements for free trade between Australia and New Zealand and between Canada and the United States; the growing, but still small, trade among developing countries (the so-called "South-South trade"); and efforts by bodies other

than GATT to reduce barriers to trade and to use trade as a stimulus to development. Among these bodies are the OECD (covered in Chapter Twenty-seven) and UNCTAD, to which we now turn.

The United Nations Conference on Trade and Development was held in Geneva from 23 March to 16 June 1964. The rich countries refused to make any significant concessions, but the Conference achieved enough so that it was established in December 1964 as a permanent organ of the General Assembly. Since then UNCTAD has met in New Delhi in 1968, Santiago de Chile in 1972, Nairobi in 1976, Manila in 1979, Belgrade in 1983, and Geneva in 1987. Between sessions of the Conference, its work is carried on by its permanent body, the Trade and Development Board, whose headquarters is also in Geneva. Nearly every country in the world belongs to it (including some that do not belong to the United Nations, such as the Holy See [Vatican City], San Marino, both Koreas, Monaco, Liechtenstein, and Switzerland). The core of its membership, however, is the *Group of 77*, or the developing countries, named for the original 75 plus two later arrivals that were instrumental in convening UNCTAD in 1964. (Today the Group of 77 has approximately 126 members, including Romania and Yugoslavia, and it functions as a caucus or negotiating group within the United Nations.)

UNCTAD's *raison d'être* has been to initiate changes in international relations leading to a new international economic order based on new concepts, among which is the use of trade policy as an instrument of economic development. This is not very different from the policies of the individual European countries that in the past used trade as instruments both of their economic development and of their imperialist policies, but today conditions are very different from those of a century and more ago. Since UNCTAD's mission is so much greater and more complex than GATT's, its secretariat is also much larger and its activities more diverse.

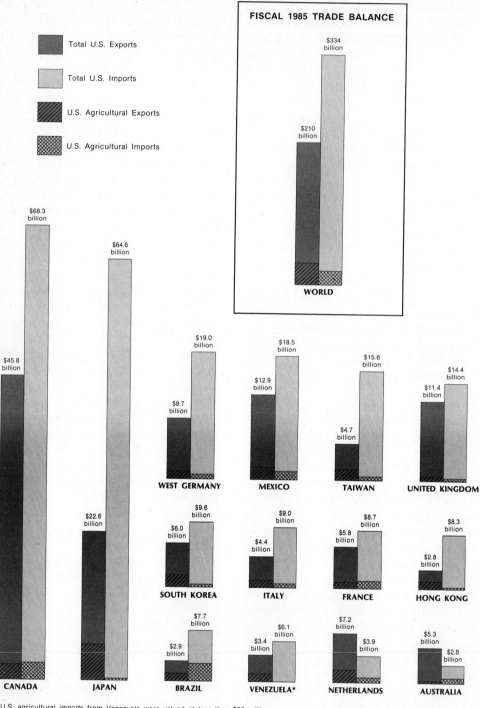

FISCAL 1985 TRADE BALANCE

$334 billion

$210 billion

WORLD

Total U.S. Exports

Total U.S. Imports

U.S. Agricultural Exports

U.S. Agricultural Imports

$68.3 billion

$64.6 billion

$45.8 billion

$22.6 billion

CANADA   **JAPAN**

$19.0 billion

$8.7 billion

WEST GERMANY

$18.5 billion

$12.9 billion

MEXICO

$15.8 billion

$4.7 billion

TAIWAN

$14.4 billion

$11.4 billion

UNITED KINGDOM

$9.6 billion

$6.0 billion

SOUTH KOREA

$9.0 billion

$4.4 billion

ITALY

$8.7 billion

$5.8 billion

FRANCE

$8.3 billion

$2.8 billion

HONG KONG

$7.7 billion

$2.9 billion

BRAZIL

$6.1 billion

$3.4 billion

VENEZUELA*

$7.2 billion

$3.9 billion

NETHERLANDS

$5.3 billion

$2.8 billion

AUSTRALIA

*U.S. agricultural imports from Venezuela were valued at less than $50 million.

Source: Department of Agriculture.
Farmline, September 1986. P. 15.

Aspects of U.S. merchandise trade deficits, 1985.   These graphs indicate clearly both the voracious appetite of the United States for material goods and the broad spread of her trade deficits. Not only is Japan *not* the only country that sells more to the U.S. than it buys from her, but Japan *buys* more from the U.S. than any other country except Canada.

We cannot detail UNCTAD's activities here, but it might be useful simply to summarize some of them in outline form.

I. Restructuring of international trade to reduce dependency of some countries on others.
  A. Increasing the value of exports of the developing countries.
    1. Sponsorship of commodity agreements (between the major producers and major consumers of a commodity to trade fixed quantities of particular types and qualities at fixed prices over a specified number of years).
    2. For manufactures, the Generalized System of Preferences (GSP) under which the developed countries, through bilateral and multilateral agreements, grant certain import preferences to the manufactured and semi-manufactured goods of developing countries;* efforts to control restrictive business practices.
  B. Regulation of trade through principles and policies on trade and related problems of economic development.
    1. The Charter of Economic Rights and Duties of States (adopted by the General Assembly in 1974); work on the Second UN Development Decade (the 1970s) and the Declaration and Programme of Action on the Establishment of a New International Economic Order (adopted by the General Assembly in 1974).
    2. Sectoral regulation, as in the field of shipping, through aiding poor countries to develop merchant marines and helping obtain special freight rates for poor countries.
II. Restructuring of international cooperation to strengthen it and reorient it for development.
  A. Expansion of international relations by promoting trade among States with varying economic and social systems and levels of development.
  B. Strengthening of international relations by devising and implementing policies and methods that are better adapted to the real needs of poor countries.
    1. Both to encourage all forms of technical assistance and to provide technical assistance in certain sectors of trade and development (UNCTAD's only actual operational function).
    2. More appropriate solutions for certain countries, such as promoting special measures in favor of land-locked States, consideration of the special needs of developing island countries, and promotion of special measures to help the least developed countries, of which there are currently 29.*

In its work, UNCTAD cooperates in many ways with other organs of the United Nations and its specialized agencies as well as with other international governmental and nongovernmental organizations. Especially noteworthy here is its cooperation with GATT that has partly, at least, replaced their original rivalry. Not only do they coordinate their work, particularly in multilateral trade relations, but they jointly operate the UNCTAD/GATT International Trade Centre in Geneva. This center provides to developing coun-

*The United States in 1976 became the nineteenth developed country to implement a GSP.

*A "least-developed country" is defined by the United Nations as one with a gross domestic product of under $100 per capita annually, not more than 10 percent of the GDP coming from manufacturing, and a literacy rate not exceeding 20 percent.

CANADA'S FOREIGN TRADE IN 1985

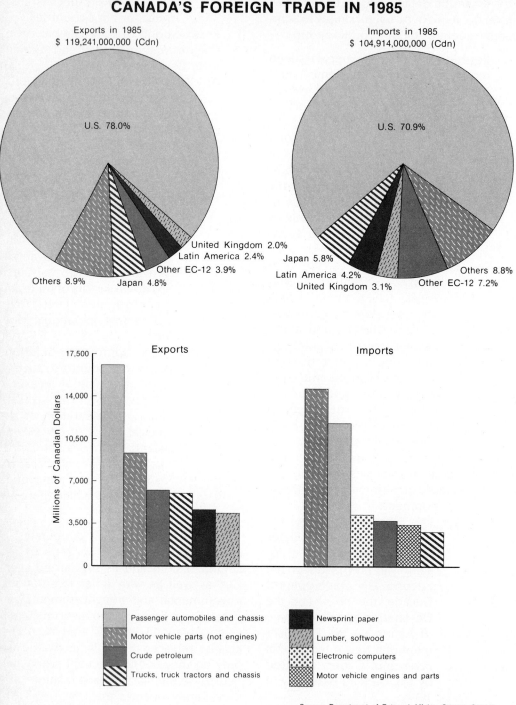

Source: Department of External Affairs, Ottawa, Ontario,
Canada. *Canada Reports*, October 1986 Pp. 2 & 3.

Canada's foreign trade in 1985. Note the overwhelming dependence of Canada on trade with
the United States, and the similarities of the major components of the largest two-way flow of
trade in the world. These factors have encouraged negotiation of a full free trade area between
the two countries.

Palais des Nations, Geneva. This was the headquarters of the League of Nations and currently serves as the European office of the United Nations. UNCTAD headquarters is located in the high-rise annex at the top of the photo. (United Nations)

tries a trade promotion advisory service, a training service for export promotion specialists, and a market development advisory service.

It can not be said that UNCTAD has been a rousing success. The GSP, for example, while having some symbolic value, has not led to the rich countries buying a significantly larger proportion of their imports from poor countries. In fact, as Table 24–1 shows clearly, this flow actually diminished slightly between 1976 and 1982. This is due in part, of course, to the worldwide recession during the period resulting largely from the "oil crises" of 1973–1974 and 1979, but even the very slow recovery in the world since 1982 has not changed the situation significantly. The commodity agreements sponsored or inspired by UNCTAD have fallen on hard times. The International Tin Agreement collapsed in 1985; those for sugar, rubber, and cacao are in trouble; only the one for coffee is reasonably healthy, though under attack by the United States. And the system of negotiating within UNCTAD by groups (The Group of 77; Group B, the developed market economy countries;

Group D, the European "socialist" countries; and China) has become ossified and is no longer generating creative thinking and discussion.

Nevertheless, there is no thought of dissolving UNCTAD and no member has yet seriously considered withdrawing from it. It is still useful as a forum for "North–South" economic negotiations, and it is still performing useful technical services in the areas of shipping, insurance, and aid to land-locked countries in Asia and Africa, among others. Now in its third decade, UNCTAD may well overcome its current problems and change the world to an even greater degree.

ECONOMIC SANCTIONS

In the last chapter, we mentioned economic sanctions as one of the modes of redress short of war available to States with grievances against other States. They deserve more attention, however, and since most economic sanctions involve international trade in some way, this seems like an appropriate place to discuss them.

The most important economic sanctions against an alleged offender are the *embargo* on exports to that country and the *boycott* of imports from the country. Both embargoes and boycotts may be partial or total, may be applied by one or more States or by an intergovernmental organization against one or more States, may or may not involve third States (through secondary and tertiary boycotts), and are enforced and evaded with varying degrees of effectiveness. The same is true of other kinds of sanctions besides embargoes and boycotts. They include such measures as freezing the assets of the object State that are located in other countries, a ban on granting of credit and other financial transactions, prohibiting the planes and ships of the object State to carry goods or passengers of the boycotting State or to call at its seaports and airports, and banning advertising, insurance, patents, licences, technical or managerial assistance, and other activities that could aid the economy of the object of the sanctions.

The United States has imposed embargoes since colonial times, notably in 1807 against Britain, in 1940 against Japan, and since the early 1950s against China, Cuba, and other communist-ruled countries. The League of Nations imposed economic sanctions against Italy, the United Nations against Rhodesia and South Africa, and the Organization of American States and the Organization of African Unity against States in their regions. In fact, economic sanctions seem to have become more common in recent years and are especially favored by the United States and by the smaller States of Africa—though, of course, against very different targets. Surveying the history of such sanctions, one can only reach the conclusion that their results have been mixed.

By far the most comprehensive, sophisticated, polemical, and protracted sanctions in modern times have been those imposed by the Arab countries against Israel. Their object is not simply to redress a grievance or to punish a transgression or even to coerce some desired behavior; their object is to destroy Israel if possible while strengthening themselves. While the secrecy surrounding both the application and evasion of sanctions precludes a definitive evaluation, it is clear that Israel has not been destroyed nor have the Arab countries achieved noteworthy net economic or political gains as a result of the sanctions.

The same is true of the United Nations-sponsored sanctions against Rhodesia. Through the entire period from her unilateral declaration of independence in November 1965 through the Security Council's imposition of sanctions in April 1966 to the final achievement of negotiated, internationally recognized independence under majority rule in April 1980, the sanctions never were implemented effectively. Portugal refused to cooperate until 1975; France and other countries (including the United States) were less than enthusiastic about enforc-

ing the sanctions. Even Britain's famed Beira Patrol, while stopping and searching 52 tankers in the Indian Ocean between 1966 and the day of Mozambique's independence in 1975, never found a single tanker bound for Beira with crude oil destined for Rhodesia. Yet Rhodesia managed to sell her tobacco and chrome, buy petroleum and all manner of other goods, build important domestic industries, maintain disguised diplomatic missions abroad, and otherwise thwart most of the sanctions.

Current proposals to impose stringent sanctions against South Africa are, in light of both the history of sanctions and current realities, unlikely to be adopted and if adopted are unlikely to achieve the desired effects: abolition of apartheid and establishment of majority rule. Selected sanctions imposed in very special circumstances and effectively enforced can influence the behavior of the object State to some degree and can enhance the prestige of the party imposing the sanctions. But there is little evidence that sanctions alone can intimidate any but the smallest and weakest States (against whom sanctions are rarely applied anyway). Nevertheless, sanctions do have sufficient value that they will continue to be invoked for as far into the future as we can see. Most important, by offering an aggrieved country an alternative to war, sanctions may and apparently do prevent war—or at least hasten the conclusion of a war. This alone makes sanctions worthwhile, even if they are futile in other contexts.

TRANSNATIONAL CORPORATIONS*

We have mentioned the role of transnational firms in international trade, but they deserve more attention. They have received considerable attention—mostly

*The term "multinational corporation" has been abandoned by the United Nations in favor of "transnational" for a number of reasons. Outside the United Nations the trend seems also to be in favor of "transnational," though the two terms are sometimes used interchangeably. There is still no precise, generally accepted definition of either term.

unfavorable—from the press and public for their roles in overthrowing and installing governments in Latin America. The United Fruit Company in Central America in the interwar period and the International Telephone and Telegraph Company in Chile in the early 1970s are only the most notorious examples. During the first two decades after World War II, transnational corporations (TNCs), some with a century of experience behind them, proliferated and expanded rapidly. Coinciding with decolonization, this movement conjured up images of a new form of colonialism.

Indeed, the size, skills, and mobility of those new empires were seen by many critics as constituting a threat to the sovereignty of small countries, especially those recently emerging from colonial status without the skill, experience, and power to deal with the corporate giants on anything like equal terms. Many of the larger firms, in fact, are richer, more powerful and even larger by several measures than many of the new States. Demand developed for some devices to control them and to prepare new (or even older) developing countries to stand up to them.

As a result of these new pressures, and of several well-publicized excesses of TNCs, the international community began to take them seriously. Studies were undertaken, principles developed, and codes of conduct proposed. In 1974 ECOSOC, the United Nations Economic and Social Council, created a Commission on Transnational Corporations, which, with its secretariat, the Centre on Transnational Corporations, has been laboring to negotiate a comprehensive code of conduct for TNCs. In 1976 the OECD developed a declaration on international investment and has since been active in the field. In 1977 the International Labor Organization promulgated a declaration of principles on TNCs and social policy. In 1980 UNCTAD adopted a Code of Conduct on Restrictive Business Practices. And the work goes on—and so does the controversy.

There are two basic points of view on

An overseas operation of a transnational corporation. This synthetic detergent plant at Ain-Sebaa, north of Casablanca, Morocco, had 110 employees in 1971, of whom 8 were Europeans and the manager, Jordanian. These proportions vary widely among transnational corporations, but the general trend is for them to employ more and more local personnel, including managerial staff. This plant produced 75 to 80 percent of the detergents used in Morocco, an indication of the kind of influence wielded by foreign-owned corporations in many developing countries. (Martin Glassner)

TNCs, with many variations. One sees them as an instrument of capitalist exploitation, designed to keep the peoples of the poor countries dependent on the rich countries and permanently subjugated to them, while raping their lands of resources and destroying their social fabrics. The other views TNCs as gentle giants, engines of development, bringing the benefits of modern science, technology, finance, and management to poor, benighted souls who otherwise would be at the mercy of harsh geographic conditions, stultifying social patterns, and corrupt and tyrannical governments. The truth, as usual, falls somewhere between the two extremes.

While still formidable, the political power of TNCs is declining and was probably never as great as legend would have it. Under pressures from intergovernmental organizations, national governments, and the general public (aroused more by the Nestlé baby formula scandal than by any government overthrow), TNCs have begun behaving more responsibly. They are also better understood now than they were in the 1960s. We know now, for example, that most transborder capital flows are among countries that are already rich, not from rich to poor countries; that most foreign facilities of TNCs headquartered in rich countries are located in other rich countries; and that, while the United States is the most important investor in other countries, she is also the most important host to foreign investment and locus of foreign branches and subsidiaries of TNCs.

The global economy is so complex now and so tightly interconnected and steadily growing more so, that import and export figures become more difficult to interpret, corporate responsibility becomes more diffuse, heroes and villains more difficult to identify. An American automobile assembled in the United States bearing an American nameplate typically contains components made in two dozen or more countries. American firms operating in Japan not only sell their products to Japanese consumers, but also provide goods and services to Japanese industry and even export to the United States. Despite the strenuous efforts of the U.S. government to destroy the government of Nicaragua, some 40 U.S.-based transnationals were still operating profitably in Nicaragua in late 1986. Japanese firms have established factories south of the U.S. border to take advantage of Mexico's *maquiladora* program and import components from Japan and other countries, assemble them or finish them in Mexico, and ship the finished products to the United States and elsewhere. Examples abound.

Another striking development that has helped to defuse criticism of transnationals is the growth of TNCs based in developing countries, not rich ones. By 1980, there were about one thousand of them operating in 125 countries, and they were proliferating at a rate more than two and a half times that of firms in the OECD countries. Based in such countries as South Korea, Hong Kong, Taiwan, Singapore, Argentina, Brazil, Venezuela, Mexico, and India, they engage in the same

activities as their counterparts from richer countries, and from similar motivations. While still miniscule compared with the OECD transnationals, these "Third World" TNCs already dominate direct foreign investment in Asian developing countries and the textile industry, and in other regions and sectors their influence is growing rapidly. The political implications of this phenomenon, to say nothing of the economic ones, are considerable.

Even the States with centrally planned economies are joining this activity that they once reviled. By 1986, State enterprises of countries belonging to COMECON, the Soviet-led Council for Mutual Economic Assistance, had established over six hundred companies in developing countries and in all but one of the OECD countries. In keeping with their ideological posture, they claim that they are not motivated by a desire for profits and should therefore be exempt from any international regulation. In fact, however, they are profit-oriented and earn large quantities of hard currencies desperately needed at home. Again, the political implications of this relatively new phenomenon may be very great indeed.

The serious, detailed academic study of TNCs has only begun. Like the subject of sanctions, this political/economic/geographic topic requires study and analysis by political geographers. Neither subject has yet been adequately addressed by them, and there is much scope here for energetic, creative work. It can yield very great rewards.

REMAINING DISTORTIONS IN INTERNATIONAL TRADE

While there has been, very broadly speaking, a general movement toward freer trade since World War II, many restrictive practices are still utilized. Some of them have economic motivations, but many are essentially political. Their net effect, whatever their objectives, is to reduce the total volume of trade artificially and to divert it out of its natural channels into artificial ones. The following list of restrictive trade practices is by no means exhaustive. Many others are extremely imaginative, technical, or obscure, but no less harmful.

By far the most common is the *protective tariff*. By discouraging competition (in some cases eliminating it altogether through excessively high tariffs), the government of the importing State also protects the local producer from the normal penalties a free market levies against an inefficient, high-cost, obsolete, careless, or dishonest producer. The consumer then suffers from both high prices and poor quality when denied access to foreign goods in a free competitive system. A protective tariff can be justified, but only for infant industries during the period of their infancy. A case can be made for protection of high-cost, inefficient industries (including primary ones) that bring important social or security benefits to a country if the added cost is considered tolerable by the society. A modest revenue tariff can also be important to a small, poor country for which it may be a principal source of income. But high tariffs imposed to "save local jobs" or "combat low-wage imports" usually result in retaliation. Economists call this a "beggar-thy-neighbor" policy and have demonstrated conclusively that in a trade war resulting from such "protection" everyone is left exposed and no one benefits except the inefficient producers and the intermediaries.

Even if tariffs are kept low—through GATT or other arrangements—there are numerous nontariff barriers to trade and other practices designed to restrict competition to give an unfair advantage to one exporter over another. They include (in no particular order) such things as import and export *quotas*, or quantitative limits on particular items being traded; *preemptive buying*, or buying quantities of something just to keep another party from obtaining it; *licensing*, or requiring importers to obtain (or buy) licences to import military, luxury, or other special goods; *exchange controls*, designed to keep to a minimum the outflow of scarce foreign exchange, especially "hard" currencies; unusual or unnecessarily high *standards* for protection of the importing country's health, safety, and morals; cumbersome import *procedures* and excessive *documentation;* complex and expensive *packaging requirements;* confusing or unreasonable *classification* of imported goods and customs *nomenclature;* extraordinary *export incentives; dumping; taxes* imposed on imports and exports in lieu of tariffs; *State trading*, practiced generally by the socialist countries and to a limited extent by many other countries; *control on exports* of arms, nuclear materials, goods in short supply, and strategic technology and products; *"voluntary" restraints on exports* in fact coerced by importers to reduce competition; and many more.

There are also many illegal, unethical, or just unsavory practices that raise questions about trade statistics and trade agreements. For one thing, there is a great and growing flow from rich countries to poor countries of outmoded, defective, unsafe, and toxic products banned in the home country. This includes pharmaceuticals and agricultural chemicals which can be—and sometimes are—lethal. Then there is a counterflow from poor countries to rich ones of counterfeit, substandard, and even dangerous goods, including imitation name-brand goods of all kinds and defective parts for automobiles and aircraft.

Finally, *smuggling* of all kinds is a major problem around the world. In Chapter Twenty-eight, we discuss in some detail the worldwide clandestine trade in arms and drugs; what we speak of here is what is generally referred to officially as "unrecorded" or "informal" trade. This is used to bypass customs duties or outright bans on imports of certain products. Along borders everywhere in the world, there are exchanges of goods and services by local residents back and forth across these borders. This is generally not a problem; often, in fact, it performs a valuable service where governments are unable to

Smuggler's goods. This jewelry was being smuggled from Hong Kong into Canada concealed in a tight corset called a bodypacker. It was confiscated by alert customs officers in Vancouver. (Canadian Department of National Revenue, Customs and Excise)

provide adequately for their citizens in remote border areas. It also stimulates the local economy, cements ties between relatives and friends and creates international good will. What is of concern, however, is the large-scale transfer of contraband for commercial purposes. This can seriously damage government economic policies—or, if the government is a party to the smuggling, can generate huge profits for corrupt officials. It is estimated, for example, that one quarter to one third of Paraguay's actual foreign trade is unrecorded, consisting of goods smuggled in both directions across the borders of Brazil and Argentina. Regardless of the reasons for smuggling, it does represent a serious distortion of international trade.

GATT, UNCTAD, the UN Commission on International Trade Law, and other agencies have been trying to combat these practices, so far with only limited success. Other practices that affect trade are not necessarily harmful; in fact, if conceived, designed, and implemented wisely, they can actually help to free and expand trade. They can be grouped under the headings of preferences, discussed below, and economic integration, to which we devote the next chapter.

PREFERENTIAL TRADING

Among the economic weapons in the armories of some States is that of giving preferred treatment to particular trading partners. Preferences became important during the nineteenth century but have achieved even more importance in the past half century. Some preference systems have deliberately been *ad hoc,* others have been abandoned as conditions have changed, and still others survive and thrive. Among the minor or obsolete ones of recent times have been the special trading arrangements between the United States and the Philippines, part of the "independence package" that led other ex-colonies to question openly whether the Philippines was really sovereign; the sugar quotas granted by the United States to selected exporting countries and realloted periodically; the various currency blocs that were so prominent in the first decade and a half after World War II (the sterling, ruble, franc, and dollar blocs); and the greatly skewed trading partnerships between South Africa and the landlocked countries of Southern Africa and between India and the Himalayan kingdoms. By far the most important, however, have been the Commonwealth Preference System and COMECON.

The *Commonwealth Preference System* (originally Imperial Preference) was one of the most important features of the Commonwealth from the Imperial Conference in Ottawa in 1932 until the entry of Britain into the European Communities in 1973. Under this system, Britain granted free entry to the goods of Commonwealth countries and imposed tariffs on competing goods of nonmembers, thus abandoning her traditional policy of free trade. On their part, the other Commonwealth countries agreed to assess lower tariffs on many goods they imported from one another than on similar goods imported from nonmembers. There were exceptions to this system, breaches of it, and changes in it over the years, but it survived even the creation of GATT, which made, as we have seen, provision for such preferential arrangements. The dependence of many Commonwealth countries on the protected British market, especially for their agricultural exports, was one of the major factors delaying Britain's entry into the European Economic Community. Not until they had begun developing other markets and Britain was able to negotiate a gradual phasing out of her Commonwealth preference obligations did the merger take place. There is still, however, a strong tendency for Commonwealth countries to trade with one another, and they account for one quarter to one third of total world trade.

Except for China, the countries with centrally planned economies ("socialist" countries) are all members or observers of COMECON, *the Council for Mutual Economic Assistance*. These countries encounter difficulties in trading with countries with market economies because their pricing systems must be arbitrary; their currencies are not convertible into "hard" currencies (such as US dollars, sterling, yen, Swiss francs, or West German marks); and they have both political and economic reasons for turning inward and trading very little with outsiders. Until recently, they had very little to sell to hard currency countries and so could not earn the foreign exchange necessary to buy goods outside their bloc. Therefore, whatever trading they did was largely among themselves by barter. There was also an ideological imperative for this arrangement and a fear (probably legitimate) of being penetrated by capitalism and imperialism if they traded with capitalist countries.

Nevertheless, COMECON never developed a formal, compulsory trading system or any mechanism for enforcing one, and trade among the European members (including the USSR) has dropped steadily from well over 80 percent in the early 1950s to about 66 percent in 1961 to 58 percent in 1973 and just over half in 1984. The comparable figures for the Asian socialist countries (including China) have always been much lower. Intrabloc trade, however, is still greater than the trade among members of most organizations actually attempting economic integration.

Chapter 25

ECONOMIC INTEGRATION

The basic arguments in favor of economic integration are quite simple. Two or more countries can, by combining their economies, reduce their costs of production by producing for a larger market, thus achieving economies of scale; provide a greater variety of goods' and more opportunities for both managers and workers in a larger economy; speak with a louder voice in international economic (and perhaps political) affairs; and achieve a variety of ancillary benefits. These benefits have been realized in varying degrees around the world. Any disadvantages to economic integration seem to be felt primarily by individuals or small sectors of society rather than by the society as a whole. Achieving economic integration, however, whatever its benefits, is exceedingly difficult. There are many obstacles, but the most important is the reluctance of States to surrender any of their sovereignty, any control over economic decision making, any political options. Consequently, economic integration has not made very impressive progress in the past quarter century.

There are four levels, or degrees, of economic integration. A *free trade area* is the simplest. Here, two or more countries agree to eliminate tariffs and other barriers to trade between or among them so that goods flow freely across their mutual boundaries. Each member, however, retains its own tariffs and other trade restrictions in respect of nonmembers (third par-

ties). The free trade area may be limited to certain classes of goods, such as agricultural products, or even to individual products, such as the U.S.–Canada free trade area in automobiles and automobile parts. A *customs union* is a free trade area plus a common external tariff. That is, the members agree not only to trade freely among themselves but also to form one unit for trading with nonmembers. The customs duties received at the ports of entry of all members are commonly pooled and either used for common purposes or apportioned among the members or both.

A *common market* is a customs union plus the free movement within the group of capital and labor as well as goods. This means that an investor may invest capital in any of the member countries without discrimination, a firm may seek loans or investments or workers in any of the member countries, and a worker may seek work and be employed in any of the member countries as long as he or she is a citizen of one of them. Complete economic integration is achieved through an *economic union* in which the members have not only a common market, but also common economic and monetary policies, a common currency, common banking and insurance systems, uniform taxes and corporation laws, and so on. Economic union without political union is probably impossible, except for very small countries, such as Luxembourg or Liechtenstein, since common political deci-

The Liechtenstein-Austria boundary—visual evidence of an economic union. Since Liechtenstein and Switzerland share a single economy, they maintain joint customs posts on the border with Austria, as here in Schaanwald, Liechtenstein. On the other hand, on the Swiss-Liechtenstein border there are no customs or immigration posts at all. (Martin Glassner)

sions must be made before these steps can be taken and this is most difficult with multiple constitutions, parliaments, cabinets, central banks, and pressure groups.

Typically, countries agree to form free trade areas, customs unions, or common markets gradually, in stages, usually according to a fixed schedule. Thus, there are degrees of these forms of integration. A common market may even be working toward an economic union and have economic institutions beyond the free circulation of goods, labor, and capital. Groups may also fail to reach their goals and yet retain their original names. Thus the name of a group may be misleading; it does not necessarily indicate its degree of integration or its success in reaching its goals. Countries that wish to integrate can get considerable help from the UN regional commissions.

UNITED NATIONS
REGIONAL COMMISSIONS

One of the most vital functions of the United Nations is assisting member countries, their dependencies, and even nonmembers with their economic growth and development. Many of the UN organs and specialized agencies are devoted to this effort, as is much of the UN budget. Of all the UN organs working in the economic area, the regional commissions, part of the Economic and Social Council, are among the most important and least known. All of them are concerned to some extent with social as well as economic mat-

ters, and one of them even has "social" in its name. They all have broadly similar organizational systems and activities, but they differ in many ways according to the needs and resources of the regions they serve. All of them, in varying degree, are concerned with economic integration, the one for Latin America being most involved in this activity and the one for Western Asia least of all. This is why they are introduced in this chapter.

The activities of these UN regional commissions are so diverse and so widespread and so intimately integrated into the daily lives of people nearly everywhere in the world that they are at least indirectly associated with nearly every subject covered in this book. Much of the basic theoretical work on economic integration, for example, was done by economists of the commission for Latin America and the Caribbean (still commonly known by its original Spanish initials, CEPAL). The commissions do research and publish important studies, statistics, and analyses of regional and national economic problems; provide guidance and technical assistance in the formulation and execution of development and integration programs; sponsor seminars, workshops, and training programs; sponsor meetings, conferences, and groups of experts to grapple with economic problems; and otherwise facilitate the economic growth and health of their members. Typically, many of these activities are carried out through subsidiary bodies and in subregional offices within the region. The commissions also coop-

UNITED NATIONS REGIONAL COMMISSIONS

UN Economic Commission for Europe (ECE)

Founded: 1947   34 members, including Byelorussia, Ukraine, Albania, Cy-
Headquarters: Geneva   prus, Switzerland, Malta, United Kingdom, and Canada.

UN Economic and Social Commission for Asia and the Pacific (ESCAP)

Founded: 1947   38 Members, including Nauru, Tuvalu, South Korea, France,
Headquarters: Bangkok   Netherlands, USSR, United Kingdom, and United States;
                                     9 associate members, all dependent territories of mem-
                                     bers.

UN Economic Commission for Latin America and the Caribbean (ECLAC; CEPAL)

Founded: 1948   40 members, including Canada, France, Netherlands, Por-
Headquarters: Santiago   tugal, Spain, United Kingdom, United States, and the Ba-
                                     hamas; 4 associate members, all dependent territories of
                                     members.

UN Economic Commission for Africa (ECA)

Founded: 1958   50 members, all independent African States; 1 associate
Headquarters: Addis Ababa   member: Namibia; 3 observers: African National Con-
                                     gress, Pan Africanist Congress, Southwest African People's
                                     Organization.

UN Economic Commission for Western Asia (ECWA)

Founded: 1974   14 original members, all Arab countries of Southwest Asia,
Headquarters: Baghdad   plus Egypt and the Palestine Liberation Organization, both
                                     admitted in 1977.

Africa Hall, Addis Ababa, Ethiopia. This is the headquarters of the United Nations Economic Commission for Africa. Nearby, out of sight at upper left, is the kind of squalid slum that the UN economic commissions are working to eliminate. (United Nations)

erate with other intergovernmental organizations in furthering their work. While some have been more effective than others, all deserve more attention and credit than they have received.

In order that we may understand better the processes, problems, and possibilities of economic integration, we will now examine a few examples from several regions.

EUROPE

The best-known economic group in the world is the EEC, the European Economic Community (popularly but quite inaccurately known as the European Common Market or simply as "the common market," as if there were no others), now only a part of the *European Communities*. The European Communities (informally called the Community and abbreviated EC) is far more than a common market. The 1957 Treaty of Rome that created it clearly envisages and lays the groundwork for *political* as well as economic union of Western Europe, and the members have made very impressive progress in that direction. The institutions of the EC resemble a government and perform many of the functions of a government. Most of its activities are concentrated in its Brussels headquarters—giving Brussels a big advantage in the competition for the honor of being the future "capital of Europe"—but important activities are also located in the other competitors, with the Court of Justice in Luxembourg and the Parliament in Strasbourg. Nearly one hundred countries have accredited diplomatic representatives to the EC. It functions as a unit in international trade negotiations and has signed the United Nations Convention on the Law of the Sea. It formulates common policies on many international issues, though it still has difficulties agreeing on some domestic policies. It is by far the most powerful trading unit in the world and approximately equal to the United States in total economic strength. It has come a long way toward

fulfilling the vision of Jean Monnet, the French "political economist" sometimes called "The Father of Europe."

All this power and unity did not develop overnight, however. The nucleus of the EC is the Economic Union of Belgium and Luxembourg (UEBL), which dates from 1922. In 1944, Belgium, Luxembourg, and the Netherlands signed a Customs Union Treaty in London that became fully effective on 1 January 1948. Then in 1960, the Treaty of Benelux Economic Union came into force. Now *Benelux* functions as a unit within the EC, having its own institutions (including a Court of Justice founded in 1974), and may be moving toward becoming a single State.

After World War II the Marshall Plan provided a catalyst for putting into practice the new ideal of European unity, with Benelux not only as a nucleus, but also as a model. The Organization for European Economic Cooperation (OEEC) was formed in 1948 by 18 European countries to administer Marshall Plan funds in an orderly and rational manner. This gave them the experience and the confidence to attempt economic integration. The first fruit of this new unity drive was the European Coal and Steel Community (ECSC), a product of "the Schuman Plan" (named for French Foreign Minister Robert Schuman), founded in 1951 by France, West Germany, the Netherlands, Italy, Luxembourg, and Belgium. This was a huge pool of these countries' coal and iron ore resources and production and their iron and steel industries, essentially a common market for these products. Its High Authority was located in Luxembourg. The European unity movement gathered strength and in 1957 created both the European Atomic Energy Community (Euratom) and the European Economic Community, with the same membership as the ECSC. The three communities worked successfully side by side. The EEC by 1 July 1968 (18 months ahead of schedule) had eliminated tariffs on intracommunity trade and had established a common external tariff. On 1 July 1967, the three communities merged their institutions and now have one commission, council, parliament, and court, although they still retain their separate identities within the merged Communities. The United Kingdom, Ireland, and Denmark formally joined the Communities on 1 January 1973. The Norwegian government had also wanted to join at that time, but its people rejected membership in a 1972 referendum, largely because they wanted to protect their fishing industry from the competition of other Community fishermen. The Europe of Six became the Europe of Nine. In May 1979, Greece—long an associate member— joined the EC as a full member; Spain and Portugal joined in 1986; Turkey may be next.

If the EC is beginning to seem like a federation, the analogy is not entirely incorrect. But it is a federation-in-process rather than a federation-in-being. A great deal more still must be done, with many more common policies to be hammered out, both internal and external. But look at what has been accomplished! Germany and France, these age-old enemies who fought three wars within one lifetime, are working together toward common goals. The regional policy has contributed greatly toward reducing the economic disparities among the various regions, bringing special aid to the backward areas of southern Italy, western and southwestern France, northern Netherlands, eastern Germany, half of Ireland, and parts of England, Scotland, and Wales. Common policies in many areas have been put into effect, monetary union is very close to realization, and, perhaps most important of all, the EC has avoided the pitfall of becoming an exclusivist, inward-oriented "rich man's club" by signing many special agreements with other countries.

The European Free Trade Association (EFTA) was created through the 1960 Stockholm Convention by seven Western European countries who believed in creation of a single European market but who, for various reasons, were unable to join the EEC. The strongest member was the United Kingdom, who, with her Commonwealth commitments, the use of her currency (the pound sterling) as a worldwide reserve currency, and her reluctance to accept the political provisions of the Treaty of Rome, still needed the advantages of a wider trading group to help her recover the losses sustained by participation in World War II and the loss of most of her empire. The other original members were Austria, Norway, Portugal, Denmark, Sweden, and Switzerland, with Finland as an associate member. EFTA's goals were more modest than those of the EEC and it reached them more rapidly. It achieved free trade in industrial goods by the end of 1966, three years ahead of schedule. It was not designed as a competitor to the EEC, but it stems from the same dreams and aspirations, and from the same experience within the OEEC. From the beginning it aimed at strengthening its members to a point where they could join the EEC individually or collectively.

This policy has been quite effective; Britain, Denmark, and Portugal have joined the EC, and EFTA and the EC have a close treaty relationship.

EFTA's continuing vitality is indicated by its admission of Finland and Iceland as full members, making a current total of six; by its special relationship with Yugoslavia; by its active consultative Committee of Members of Parliament of EFTA members; by the 1984 ministerial meeting between EC and EFTA members; the 1984 meeting of heads of all EFTA governments; and by a continuing dynamic program of stimulating both trade and economic growth of its members.

In the early 1960s, it was common to refer to Europe as being "at sixes and sevens," a reference both to the EEC and EFTA and to the uncertainties then prevailing about the future of Europe in the face of a perceived Soviet threat, the dominance of the United States in the world economy, economic distress in many parts of Europe, declining coal production and increases in energy needs, changes being wrought by decolonization, and political

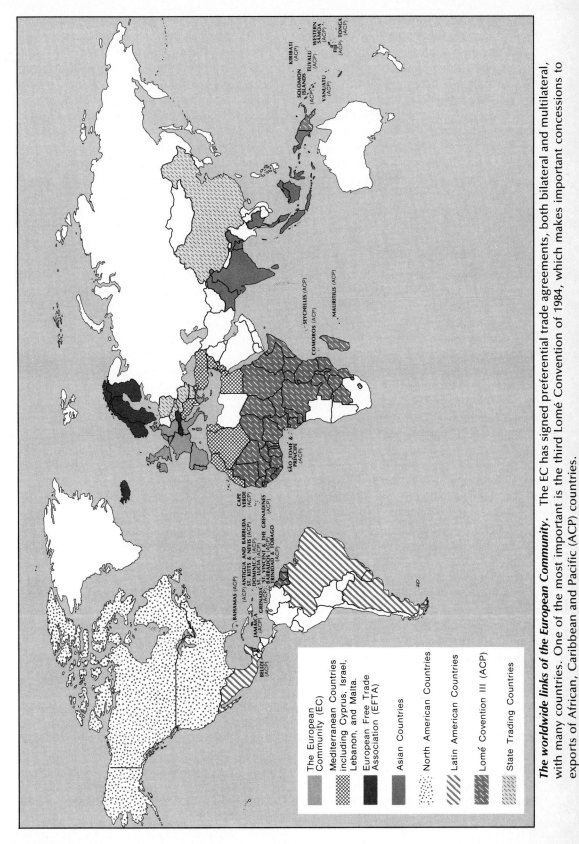

The worldwide links of the European Community. The EC has signed preferential trade agreements, both bilateral and multilateral, with many countries. One of the most important is the third Lomé Convention of 1984, which makes important concessions to exports of African, Caribbean and Pacific (ACP) countries.

The European Community (EC)

Mediterranean Countries including Cyprus, Israel, Lebanon, and Malta.

European Free Trade Association (EFTA)

Asian Countries

North American Countries

Latin American Countries

Lomé Covention III (ACP)

State Trading Countries

KIRIBATI (ACP)

SOLOMON ISLANDS (ACP)

TUVALU (ACP)

VANUATU (ACP)

WESTERN SAMOA (ACP)

FIJI (ACP)

TONGA (ACP)

SEYCHELLES (ACP)

MAURITIUS (ACP)

COMOROS (ACP)

SÃO TOMÉ & PRÍNCIPE (ACP)

CAPE VERDE (ACP)

BAHAMAS (ACP)

ANTIGUA AND BARBUDA (ACP)

ST. KITTS & NEVIS (ACP)

DOMINICA (ACP)

ST. LUCIA (ACP)

ST. VINCENT & THE GRENADINES (ACP)

GRENADA (ACP)

BARBADOS (ACP)

TRINIDAD / TOBAGO (ACP)

JAMAICA (ACP)

BELIZE (ACP)

379

turmoil in more than one country. Since then, however, the EC and EFTA have survived and surmounted crisis after crisis. Today, notwithstanding their weaknesses and problems, these organizations are strong and confident and so obviously beneficial—not only to their members, but also to the world at large—that there is no really strong and organized opposition to either. Unfortunately, the successes of economic integration in Europe have not been replicated elsewhere in the world, despite many attempts to do so.

LATIN AMERICA

Outside Europe, Latin America has made the most progress toward economic integration. Common features in the histories, cultures, and outlooks of most of its countries have led them to many attempts at cooperation and even federation during their century and a half of independence. But isolation, competition, internal politics, and nationalism have conspired to frustrate most of those attempts. Since World War II the emphasis has been on economic development and on the cooperation and integration necessary to achieve it. The UN Economic Commission for Latin America and the Caribbean has been by far the most dynamic of the economic commissions and has been led by some of the world's most talented and articulate economists.

Another regionwide organization devoted, among other things, to promoting economic integration is the *Latin American Economic System* (SELA). One of its immediate forerunners was the Special Committee on Latin American Coordination, established in 1963 by the Organization of American States. In 1974, Mexico and Venezuela promoted a more highly structured economic system. An agreement to create such a system was signed in Panama City in October 1975 by 25 Latin American countries, including Barbados, Jamaica, Guyana, Grenada, and Trinidad and Tobago. Its headquarters is in Caracas. It is responsible for promoting coop-

eration to achieve the integral, self-sustaining, and independent development of the area. The organization aims to create Latin American multinational enterprises; encourage the conversion of raw materials in the region, industrial complementation, intraregional trade, and the export of manufactures; promote the development of transport, communications, and tourism; support the integration processes in the region, coordinate positions and strategies on economic and social matters, among other activities. It is too early to evaluate its achievements, but enthusiasm for it in the region seems to fluctuate according to changing economic and political conditions.

Of the organizations aimed specifically at economic integration and thus bearing directly on international trade, the most ambitious so far has been the *Central American Common Market* with headquarters in Guatemala City. Between 1951 and 1957, the five Central American States of Guatemala, Honduras, El Salvador, Nicaragua, and Costa Rica signed six bilateral free trade agreements among themselves. By stages during the next few years these countries agreed on closer and closer trade relationships until four of them signed the General Treaty on Central American Integration in December 1960, to which Costa Rica acceded in 1962. By 1969, about 95 percent of trade among the members had been liberalized and the common external tariff applied to 97.5 percent of tariff items. They also established 14 autonomous economic cooperation bodies, including the Central American Bank for Economic Integration.

The CACM, however, has not yet adopted a common agricultural policy, its scheme for establishing subregional industries rather than competing industries in each country has not made much progress, and several of its other projects are foundering. The reasons for its failure to make significant progress after an inspiring beginning are many, among them the war of July 1969 between El Salvador and Honduras and the continuing hostility between them. Other reasons include the gravitation of new industries toward the more developed parts of the area, the lack of coordination of development policies, and balance-of-payments problems. All can be attributed to a lack of commitment by the members to subdue their own petty nationalisms and cooperate with one another for their mutual benefit. Perhaps they are simply too poor to bear the financial as well as the political and psychological costs of such integration. European integration, after all, began in earnest only after centuries of rivalry, hostility, and war, and is stimulated and sustained both by real internationalist sentiments and by a good deal of cash, both of which are absent in Central America—and in most of the rest of the world.

The Caribbean Community (CARICOM) is an outgrowth of CARIFTA, the Caribbean Free Trade Association, which itself emerged in 1968 out of the wreckage of the Federation of the West Indies. CARIFTA's goals were quite modest, centering on the expansion of trade among the members under harmonious and equitable terms, and the balanced and progressive economic development of the economies of the area. It had a Council and a Secretariat, located in Georgetown, Guyana. Perhaps because of its modest goals, progress in meeting them was good and its leaders were encouraged to take a further step toward economic integration. The decision to establish a common market and a wider community in the Caribbean was made in 1972 and they came into being with the Treaty of Chaguaramas of 1973.

Currently, CARICOM has 13 members, all of the former members of CARIFTA (which it replaced), and the Caribbean Common Market has 9 members. It has much broader objectives than CARIFTA, both economic and political, and more mechanisms to help the members reach them. They include a common external tariff, harmonization of tax systems, rationalization of agriculture, harmonization of monetary policies, and joint action in relation to industrial development pro-

grams, transport, and tourism. The members also attempt to coordinate their political policies in various international forums and conferences.

The structure of CARICOM is relatively simple. Below the three principal organs are the institutions responsible for various functional areas, such as health, agriculture, and foreign affairs. It has associated institutions, much like the specialized agencies of the United Nations. These include two regional universities, the East Caribbean Common Market, the Regional Shipping Council, and five others. Finally, there are the Community Secretariat (located in Georgetown) and many integration institutions that predate CARICOM. There are two recognized groups of members: the more developed countries (MDCs) are Jamaica, Trinidad and Tobago, Barbados, and Guyana; the rest are the less developed countries (LDCs). Despite all efforts to equalize benefits received from integration by these two groups, it seems that most still gravitate toward the MDCs. In addition, the same centrifugal forces that destroyed the West Indies Federation are still at work and the centripetal forces of shared interests and viewpoints and a common Caribbean identity are still very weak. While there are some hopeful signs that CARICOM will fulfill its promise and unify the Commonwealth Caribbean (and perhaps other Caribbean countries as well), the obstacles to unity are great. As the late Errol Barrow, then Prime Minister of Barbados, observed in 1964, "We live together very well, but we don't like to live together together."

The Latin American Free Trade Association (LAFTA) was the largest of the subregional organizations aimed at economic integration in Latin America, and the least successful. Just as in Central America, the South American countries built up during the early postwar years a network of bilateral agreements that enabled them to expand trade with one another, but they proved inadequate to achieve any lasting benefits and trade among them actually fell appreciably between 1955 and 1961.

Organizations for economic integration in Latin America.

With the aid of CEPAL, Mexico and six South American States formulated and signed in 1960 the Treaty of Montevideo that established LAFTA.

The main objective of the treaty was the establishment of a free trade area among the members by 1973, not a very difficult one to reach, it would seem. Yet it never was reached. The reluctance to cooperate on any but the absolute minimal level is indicated by the relegation of economic integration as a means of accelerating development to a secondary objective with no regulatory provisions. Tariff-cutting negotiations among the members were on an item-by-item basis and painfully slow. Most of the tariff cutting was in manufactured consumer goods and capital goods, products that have never been traded extensively among them. Protection of local industries took precedence over cooperation or the expansion of intraregional trade, partly because of the very real and very considerable differences in the members' levels of economic development. By 1980, intra-LAFTA trade still amounted to only 14 percent of members' total trade. The organization was finally recognized as a failure, and it was allowed to expire in 1980.

LAFTA was immediately succeeded by the *Latin American Integration Association* (ALADI), created by the former members of LAFTA (Mexico and all of the States of South America except Guyana and Suriname) through a new 1980 Treaty of Montevideo, much less ambitious than that of 1960. It lays down no strict rules and no timetables and no restrictions on members' trade arrangements with third countries. It also permits the conclusion of trade agreements with nonmember developing countries; to date ALADI has concluded such agreements with the Central American countries and some Caribbean ones.

It has gotten off to a slow start. Its first five years were dedicated almost completely to the renegotiation of the remnants of LAFTA. This process has so far yielded some 40 bilateral trade agreements, but only a single multilateral one, the Regional Tariff Preference. Even this is not very inspiring since it anticipates only very modest tariff reductions for trade among member countries. Far from an increase in intraregional trade, the period 1982–1986 actually saw a decrease of 40 percent in an already low figure. This was partly due to the regionwide shortage of foreign exchange, and more regional trade talks are scheduled to reverse the trend, but prospects are not bright. As pointed out earlier, Latin America is not Europe and what works in Europe may not necessarily work in Latin America—or in any other part of the world.

AFRICA

Progress in developing trading groups and integrating economies in Africa has been even more difficult and less impressive than in Latin America. All the factors we mentioned earlier in our discussions of imperialism, decolonization, nation building, and the colonial economy have conspired against economic integration in Africa, despite the romantic ideology of Pan-Africanism that was so popular in the late 1950s and early 1960s.

There has been no lack of effort toward integration in Africa, however. For a time, just a decade, in fact, the *East African Community* was considered a model to be followed by groups of countries elsewhere in Africa. Its origin was a customs union between Kenya and Uganda in 1917. After Britain obtained a mandate over Tanganyika, that territory was gradually assimilated into the union, becoming a full member in 1927. Throughout the colonial period Britain encouraged a kind of "functional approach" to economic integration in East Africa, gradually developing a large number of important common services throughout the area. In 1947, the British established in Nairobi the East Africa High Commission (EAHC) to coordinate approaches in the territories toward common problems. It functioned quite well, given the difficulties it faced.

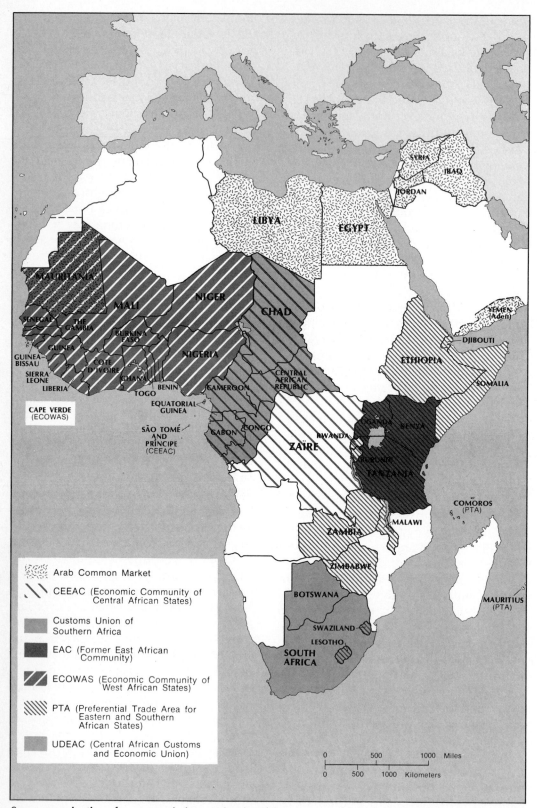

Some organizations for economic integration in Africa.

With the approach of independence and persistence of some problems, however, it was considered wise to replace the EAHC with an East African Common Services Organization (EACSO), which came into existence formally two days after Tanganyika became independent in 1961. EACSO, despite shortcomings and inequities, survived rising African and local nationalisms to provide very good services to the three countries, even after Kenya and Uganda became independent. It was replaced in 1967 by the East African Community, with headquarters in Arusha, Tanzania, and offices of the various common services dispersed from Nairobi around all three countries.

Despite many assets and high hopes, however, the Community went downhill almost from its creation. The many unresolved issues and problems of both the common services and the budding common market simply were not solved. It gradually disintegrated and finally collapsed in 1977.

In Chapter Twenty-seven, we present a chart showing the names and membership of many African organizations with a variety of economic and other objectives, including, in a few cases, economic integration. Of them, perhaps the *Economic Community of West African States* (*ECOWAS*) has the best chance for long-term success. It was founded in 1975 by 15 countries meeting in Lagos, and it began functioning there in 1977, the year of the death of the East African Community. A decade later it was still vigorous, though still very far from achieving true integration. If the provisions of its treaty are faithfully carried out, West Africa will be a far better place in which to live, for it is sweeping and visionary, though tempered by realism.

ARAB STATES, ASIA, AND THE PACIFIC

In the vast area from Morocco across North Africa, through all Asia and across the Pacific, not a single organization has been formed primarily to achieve economic integration or even sharply increased trade among the members. The nearest approach is the so-called *Arab Common Market.** The Agreement on Arab Economic Unity was signed in June 1957 but did not enter into force until June 1964. By May 1976, 12 States had become parties to it, but only 7 of them have since taken even initial steps toward forming a free trade area. These 7 are Libya, Mauritania, South Yemen, Egypt, Iraq, Jordan, and Syria. Despite their abolition of most tariffs and quotas on their mutual trade in 1971 (Jordan excepted because of her special need for protection of industries), many obstacles limit the effectiveness of this move. As a result, bilateral trade agreements have tended to be more flexible and effective than the multilateral tariff reduction system among Arab countries. Some progress has been made among them in mining and livestock development, investment and financial cooperation, but trade among them is still miniscule.

In Asia and the Pacific economic cooperation has made some progress but economic integration has seen none at all. The UN regional commissions may be able to stimulate integration but considering the deep-rooted and bitter hostilities among the countries and lack of any real cooperation among them or even perception of many common interests, it is hard to see how any significant progress can be made in this area in the near future.

The whole field of economic integration among developing countries is a fruitful area for investigation by political geographers. They can explain the links among politics, economics, and the physical environment that tend to inhibit integration and even cooperation and recommend solutions. It would be a most worthwhile contribution.

*The Arab Common Market is supervised by the 13-member (including the PLO) Council of Arab Economic Unity and does not constitute a separate organization. The council has a number of other activities aimed at economic cooperation and development.

Chapter 26

LAND-LOCKED STATES

One of the most perplexing problems in international politicogeographic relations today is access to the sea and its resources for land-locked countries (those with no seacoast at all). States with very short coastlines, such as Zaïre and Jordan and Iraq, often have problems of access, but at least they do have some sovereign territory along the sea that they can utilize constructively. Truly land-locked States have no choice but to transit the territory of another State enroute to and from the sea. If the transit State has adequate transportation and port facilities and if it is friendly and cooperative, this need not be a problem. But frequently, for one reason or another, transit is taxed, impeded, curtailed, or suspended and the interior State is nearly or completely cut off from the world.

If we count Andorra, Liechtenstein, San Marino, and Vatican City as States, then there are at present 30 land-locked States in the world.* Of the 30, nine are in Europe and rather well-to-do by world stan-

dards. Of the remaining 21, which are all poor or "developing" countries, 15 are among the 29 "least developed" countries by UN reckoning: the poorest of the poor. It is no accident that more than half the world's poorest States have no coastline. A location in the interior of a continent, especially one as large as Africa or Asia, isolates a country from the main world flows of goods, people, and ideas. Interior countries lack the "window on the world" that a seaport provides. Once isolation was valued, for it provided some measure of security and protection against a wide range of dangers from hurricanes and tsunamis to pirates and foreign invaders. Now, however, isolation is a disadvantage.

CHARACTERISTICS OF LAND-LOCKED STATES

Those land-locked States closest to the sea are located mostly in Europe, smallest of all the continents and the one with the best circulation system. Only Bolivia, Lesotho, and Malawi on the other continents can be considered reasonably close to the sea. But this advantage is largely canceled out for these States by high transportation costs resulting from difficult terrain or sparsely settled country that neither produces nor consumes very much. Typically, partly because of topographic features, partly because of the poverty of most land-locked *and* transit States, the transport

*Tibet and Sikkim, formerly at least quasi-States, have been absorbed by China and India, respectively. The former Indian State of Jammu and Kashmir has been divided between India and Pakistan and has no international personality. Byelorussia is a member of the United Nations and participates in international conferences concerning land-locked States, but no one outside the Soviet Union pretends that it is sovereign. West Berlin is not a State, but it is surely land-locked and bears all the burdens that interior States do—and many more besides. Since it is governed by very special international agreements, however, much of what is covered in this chapter does not apply to it.

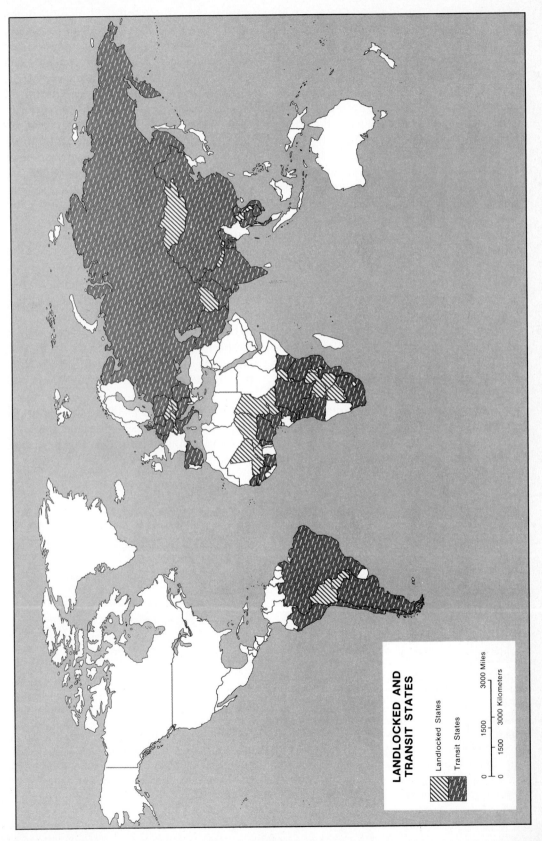

LANDLOCKED AND
TRANSIT STATES

Landlocked States

Transit States

0   1500   3000 Miles

0   1500   3000 Kilometers

systems used by land-locked States are inadequate in facilities, maintenance, and management. Frequently, there are serious imbalances in the direction or seasonal flow of goods as well. Communications, which are essential to keep transit traffic flowing smoothly, often are inadequate or unreliable, making traffic management very difficult. Delays in transit caused by these circumstances not only increase such costs as insurance, storage, interest on loans, penalties, and similar direct charges, but also increase the risk—and the actuality—of loss, theft, damage, and deterioration of goods. Merchants at either end of a trading relationship can lose a great deal of money when delays in transit mean goods arrive too late to be useful or saleable. Construction projects can be delayed and machinery and vehicles immobilized, all for want of materials, equipment, or parts that are lost, damaged, or delayed in transit.

Of all the land-locked States outside Europe, all but two have three or more neighbors with which they must remain on good terms if they are to be reasonably certain of free access to and from the sea. The more coastal neighbors a land-locked

country has, the wider a choice of routes to the sea it should have. In practice, though, few really have more than one or two practicable routes, and only Bolivia and Zambia (and to a lesser extent Afghanistan and Rwanda) can readily choose between ports on two oceans. A few land-locked countries have land-locked neighbors, which might seem to be something of an asset, but the existence of more than one State between the land-locked State and the sea merely complicates the situation.

Even more important is the distance and nature of the terrain between the economic core of a land-locked State and the core areas of its neighbors or between its core and a good seaport. In some cases, notably in South America and West Africa, the core area of a land-locked State is closer to that of another State or to a seaport than to the extremities of the State itself. This is largely due to the European colonial practice of providing transportation and communications between coastal ports and the interior to facilitate the export of cash crops, minerals, ivory, and slaves. It was and still is helpful in terms of access to the sea, but is an overall weakness of

Transit problems of developing land-locked States.   At the left is one of the better portions of the Tribhuvan Rajmarg, the Indian Army-built highway between Kathmandu and the Indian border that is Nepal's principal route to and from the sea. In the "hills" south of Kathmandu, the longest straight stretch of road is less than two kilometers long. In the photo, the two heavily loaded Nepali trucks cannot pass on the road and one of them must pull off the road to let the other go by. On the right is one of the three ferries that carry vehicles across the Mekong River between Thanaleng, Laos (background), near Vientiane, and Nongkhai, Thailand. This is a major bottleneck on the principal route for Lao imports and exports. These tank trucks are the only means for importing petroleum products into Laos. (Martin Glassner)

too many African land-locked States; this weakness is a distinct handicap in bargaining for the right to *use* these facilities.

All the land-locked States may be considered militarily weak, even compared with middle-rank powers, such as Canada or Israel. Few of them are truly modern, united nation-states in complete control of their national territory and guided by the State idea. Fewer still outside of Europe are economically strong. Thus, while some of the land-locked States are useful to the larger coastal States because of their location, minerals, or votes in the United Nations, in terms of the traditional indicators of national power we discussed earlier, they are uniformly weak. Their bargaining power in bilateral negotiations with transit States derives principally from international law and the possibilities of reciprocity and mutual economic benefits. In multilateral negotiations, they could have the advantage of being able to present a solid front but, as we will see, this advantage is more apparent than real. They still must rely on persuasion rather than power.

Among the developing land-locked States, few are able to ship the bulk of their imports and exports over one route by one mode of transport. Usually, there are expensive, time-consuming, and risky breaks-in-bulk and transshipments, often several in one journey. Moreover, with very few exceptions, land-locked States have little or no control over the availability, suitability, or operating efficiency of the transport systems and port facilities outside their borders. Furthermore, even though transit trade can be profitable for the transport systems of transit States, the bargaining power of the land-locked States is considerably reduced in many cases because they compete with their transit States in the production of the same few products, particularly in Africa. As for imports, the profitability of transit trade is often reduced by smuggling. This varies in type, degree, and technique depending on the prevailing economic conditions and

policies in both the land-locked and transit States at a given time, but there is no doubt that smuggling can seriously injure economies that are fragile to begin with.

Few of these problems, however, are peculiar to land-locked countries alone; many are suffered by the interior districts of developing coastal States as well. What compounds these problems for the land-locked States is that between them and the sea is at least one international border, sometimes more. This means that the transit States can, and often do, impose complicated transit and customs formalities, excessive documentation, and even, at times, higher charges for transit traffic than for domestic traffic. Not only do goods in transit, therefore, have to endure cumbersome procedures and high costs at the ports, but also at the border, and this increases the time and expense involved in transit.

These and other problems can be reduced, although never eliminated, if the land-locked State either has several alternate routes to the sea or is on excellent terms with its transit State or States. Happily, as we have seen, one or both these conditions prevails in the case of nearly all land-locked States. There are, however, many examples of transit routes being deliberately blocked for varying periods. Clearly, political as well as economic factors are important in determining the true significance of "land-locked-ness" in individual cases.

There is one more characteristic that distinguishes land-locked States from coastal States: these States are not only dependent on transit States for their access to and from the sea, but also they *know* they are so dependent. There is a distinct psychological factor involved that must always be taken into account when considering land-locked States in terms of history, politics, economics, geography, or international law. The urge to the sea, the feeling of national claustrophobia, has been strongest in the large interior States of Europe and in Bolivia. It was evident in

Ethiopia before it acquired Eritrea and still is to a lesser degree in Afghanistan. It is less conspicuous in Africa, but these countries became land-locked *only* upon the attainment of independence. Whether a feeling of geographic strangulation will develop among the interior States of Africa remains to be seen. Certainly Mali, Uganda, Zambia, Malawi, Lesotho, Zimbabwe and Swaziland have felt the effects of having their routes to and from the sea controlled by hostile powers, and Niger and Chad were innocent victims of the Nigerian civil war.

RELEVANT INTERNATIONAL LAW

From ancient times, land-locked territories have often faced obstructions, restrictions, tolls, or heavy transit fees on goods and persons enroute to or from the sea. There is no "right of innocent passage" on land (or in the air) as there is at sea. Gradually, however, insistence on absolute sovereignty began to give way to a recognition of the advantages of a free flow of trade. By the eleventh and twelfth centuries of the Christian era, territories in Europe, particularly in Italy, were giving treaty rights to land-locked territories and the internationalization of rivers was beginning. Nevertheless, restrictions and heavy tolls were still common by the late eighteenth century.

Gradually, through the nineteenth century the principle of free transit through straits, on rivers and overland became firmly established as recognition grew that such transit was necessary for the development of commerce and industry to the benefit of both land-locked and coastal States. Railroads and other new forms of transportation were instrumental in generating greater competition among industries, necessitating a freer flow of goods, and in reducing the advantages of waterways, hence increasing the bargaining power of the land-locked States. By the end of the century, transit duties in Europe had virtually disappeared, the Latin American States had reversed Spanish trade policies and opened their territories

to unrestricted trade, and most navigable waterways in Europe, Africa, East Asia, and South America had been internationalized. But access to the sea was still not accepted as a specific right of land-locked States.

President Woodrow Wilson of the United States included "a free and secure access to the sea" for Serbia and Poland among his Fourteen Points for a just and permanent peace settlement in Europe after World War I. The League of Nations Covenant for the first time included "freedom of communications and transit" as a worldwide goal and standard. In pursuance of this provision (Article 23 [e]) of the Covenant, a series of conferences during the 1920s produced both multilateral conventions and bilateral treaties aimed at the facilitation of free transit. Of these, the most important were the Barcelona Convention of 1921 and the Convention and Statute on the International Regime of Maritime Ports, signed at Geneva on 9 December 1923.

In the period after World War I a relatively new concept—corridors to the sea—enjoyed a measure of acceptance and a number of States and territories had their boundaries adjusted to allow for very short shorelines on the open sea or a navigable river. Finland, Iraq, Transjordan, Palestine, and Colombia all received such corridors, as had several territories in Africa in the late nineteenth century. But the most famous was the Polish Corridor. Corridors present many problems of self-determination, transit across the corridors themselves, and security, and they no longer seem to be a practical means of assuring access to the sea for a land-locked State. Even the transit conferences of the 1920s did not consider corridors in their deliberations. They did, however, cover in great detail navigable waterways and overland transit.*

*In 1975, Bolivia and Chile briefly negotiated a proposed corridor for Bolivia along the Peruvian border to the sea. The negotiations ended inconclusively in 1979 due to opposition within Bolivia and to the refusal of Peru to assent to the proposed transfer of territory, as required by the 1929 Treaty of Santiago.

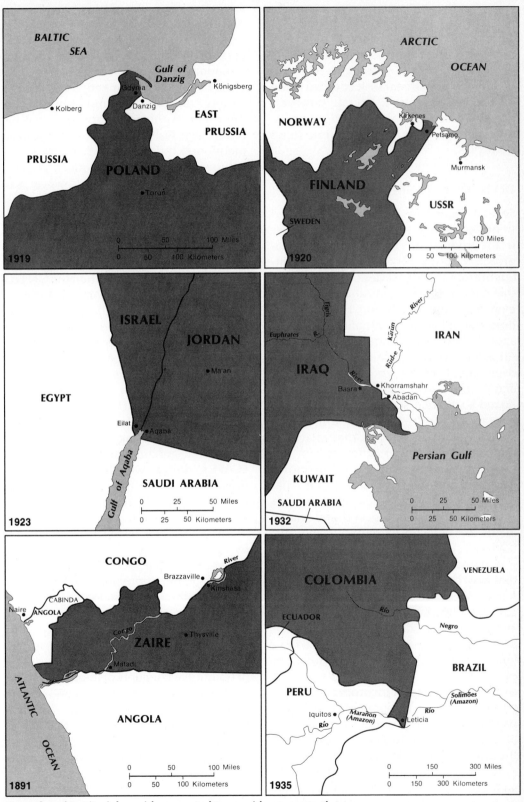

Examples of territorial corridors created to provide access to the sea.

During the interwar period, other important agreements were negotiated concerning railways, power lines, pipelines, ports, and other transport facilities, all in conformity with and tending to reinforce the principles of free transit codified in the Barcelona Convention. Aircraft, however, present special problems, and air transport is a major class of exceptions to these principles.

DEVELOPMENTS SINCE WORLD WAR II

Several important historical events and trends conjoined shortly after World War II to produce an atmosphere and approach toward access to the sea quite different from those just described. For one thing, territorial reorganizations in Europe eliminated the Polish and Finnish corridors and enhanced international cooperation in both water and overland transit. An elaborate system of internationalized rivers and canals, free zones and free ports, and special transit arrangements has produced a situation so adequate that there have been few recent difficulties with transit, even between communist and noncommunist countries (except for the special case of West Berlin). Therefore, Europe is no longer important in discussions of access to the sea, except as an example of how land-locked and transit States can develop harmonious and mutually satisfactory arrangements.

At the same time, the world trading system was being reorganized to meet the needs of reconstruction and economic expansion following the war. Both the GATT and the Havana Charter reaffirmed and in some respects strengthened the Barcelona Convention's provisions for freedom of transit. Article V of GATT and Article 33 of the Havana Charter incorporated transit provisions into an elaborate system for expanding world trade rather than the more limited approaches that had prevailed in the past. Attitudes were changing.

We may identify two basic approaches and one lesser approach to the question of access to the sea, all rooted deeply in international law. The first approach is that of freedom of transit, already discussed in some detail. The second view is that access to the sea derives directly from the "freedom of the seas." These two approaches are closely linked; indeed, some writers consider one simply an aspect of the other. A third concept, one that has never been really important and is scarcely mentioned today, is that sea access constitutes a public law servitude.

The early postwar period, however, saw the rapid evolution of a relatively new concept, embodied in GATT and the Havana Charter, namely that access to the sea is essential for the expansion of international trade and economic development. Indeed, in January 1956, when the UN Economic Commission for Asia and the Far East (ECAFE) considered the problems of its land-locked members (Afghanistan, Laos, and Nepal), its recommendations referred to the "needs" of these States rather than "rights," a term that pervades the classical literature on the subject. Significantly, though, ECAFE did not call for new rules or procedures for the benefit of land-locked States, relying still on existing international law and practice. One result of ECAFE's interest was the beginning of a drive within the United Nations to deal with access to the sea in a comprehensive and definitive manner.

This activity has been stimulated chiefly by the achievement of independence by many former European colonies in Asia and Africa, their entry into the United Nations, and their decision to use the United Nations as a prime vehicle for protecting their new "sovereignty" and supporting their economic development. Most definitely, the mid-1950s saw the fading away of traditional concepts of international law regarding land-locked States—replete with such words as "rights," "obligations," "sovereignty," and "natural law"—and the emergence of newer, more practical arguments for "a free and secure access to the sea."

Early in 1958, the UN Conference on the Law of the Sea assigned the question of

Dakar, capital of Senegal, largest city of francophone Africa and chief transit port for land-locked Mali. (UN)

access to the sea to its Fifth Committee and ultimately adopted the committee's recommendations as Article 3 of the Convention on the High Seas, one of four conventions produced by the conference. Instead of subsiding, however, pressure for a more definitive solution to the question of access to the sea, particularly for the developing land-locked States, increased after 1958.

ECAFE continued to pursue the matter, in Manila in 1963 and Tehran in March 1964. At the latter meeting a resolution was unanimously passed stressing the "critical importance" of access to the sea in economic development and urging that the subject be considered at the forthcoming UN Conference on Trade and Development (UNCTAD). After long and complex discussions of access to the sea, UNCTAD finally adopted eight principles and several recommendations, similar to those adopted in 1958 but with some significant differences. Regional transit agreements were encouraged and exemption from most-favored-nation clauses urged. Other provisions covered special facilities and rights of land-locked States.

THE CONVENTION ON TRANSIT TRADE OF LAND-LOCKED STATES

On the recommendation of UNCTAD, a Committee on Preparation of a Draft Convention Relating to Transit Trade of Land-locked Countries met at UN Headquarters

during October and November 1964 and, as had been the case since 1956, the African and Asian land-locked States led the discussions. Its draft convention was presented the following summer to the UN Conference on Transit Trade of Land-locked Countries. This was the first conference convened under UNCTAD, which had become a permanent agency of the United Nations. The conference was attended by delegates of 58 countries and observers of 10 countries. Their deliberations again covered basic principles and technical details, ancient traditions, and new circumstances. Transit and land-locked States did not form solid blocs, nor did developed and developing States, nor were there any other definable, cohesive blocs. The issues were so complex, in fact, and the participants split so many ways, that the convention produced by the conference was a compromise of compromises. Nevertheless, it is most significant.

For the first time in history, an international lawmaking conference had dealt exclusively with the question of access to the sea, particularly for the developing land-locked States of Africa, Asia, and South America. This certainly gave a measure of "status" to the problems of those States, but did not solve them. The convention did not change the geographical condition of any land-locked country. Although it entered into force in 1967, it still has been ratified or acceded to by very few transit States, and many others refuse to accept it as binding. Despite its importance as a "datum plane," it was only one more step in the continuing drive of the land-locked States to reduce to a minimum the difficulties inherent in their mediterranean ("middle of the land") location.

PROGRESS SINCE 1965

Since 1965, the developing land-locked States have tried to improve their situation in many different ways. We can group these into five major categories.

1. There has been a greater emphasis on *internal development,* with particular stress laid on improved transportation networks and facilities. As improved transport contributes to the general economic development of a country, some handicaps of isolation and internal weakness are reduced, living standards rise (if the improvements are not canceled out by population growth), and the State becomes a more cohesive unit. At the same time its demand for imports increases and it must export more to pay for them so its need for improved transit across coastal States is increased, making it both more valuable as a customer on transit routes and more vulnerable to disruptions in its transit.

2. *Bilateral negotiations* for improved transit remains the most important and most common method of obtaining it. International law provides a framework for such negotiations and sets minimum standards, but does not (at least not yet) guarantee a *right* of free transit. Even if it did, the individual land-locked and transit States would still have to work out detailed arrangements based on local conditions. In these negotiations the coastal State invariably has the advantage. Even when reasonably satisfactory agreements are concluded, they can be breached in times of stress. The Afghan–Pakistan border, to cite an unusual case, was closed in 1950 and 1955 for short periods and again from September 1961 to July 1963. Bilateral arrangements, while necessary, are a slender reed on which to lean.

3. Both transit and land-locked States, as well as the international community in general, have been making concerted efforts in recent years to *improve transport facilities and seaports* for everyone's benefit. Congested seaports are a worldwide problem and improved transport is an essential element in any country's economic development. At the same time some land-locked States are vigorously developing arrangements and facilities for *alternative transit routes* through countries other than the traditional transit State. Here the classic case in modern times is Zambia, which, after Southern Rhodesia declared independence in 1965, began seeking a more secure, though longer and more expensive, route to the sea through Tanzania. Afghanistan has diverted most of her transit trade from Pakistan to the Soviet Union, Rwanda normally uses routes through Uganda and Kenya rather than Zaïre as formerly, and so on. Although time-consuming and expensive, these methods do give the interior State more flexibility, hence more security and greater bargaining power, than reliance on a single transit State or inadequate facilities.

4. As we noted in the last chapter, many postwar efforts at *regional economic integration* have involved land-locked countries. If the degree of integration is great enough, border formalities are greatly reduced or eliminated for goods in transit and the land-locked country may even share in the ownership and control of the port and transport facilities. Uganda had been the best example of this type since she was a full partner in the East African Common Services Organization and the East African Community. But after

the Israeli rescue of hostages from the Entebbe Airport in July 1976, the border between Uganda and Kenya was closed for some time. As relations between the two partners continued to deteriorate and the East African Community disintegrated, so did Uganda's transit arrangements. Economic integration, while highly desirable for many reasons, is still no substitute for political integration as a means of solving the political problem of guaranteed access to and from the sea.

5. The land-locked States have continued their *efforts in the United Nations* and its various organs and specialized agencies both to establish a legal right of transit and to gain special considerations and assistance to ameliorate their special problems. UNCTAD has been especially active in developing "special measures related to the particular needs of the land-locked developing countries." The World Bank, the United Nations Development Programme, ESCAP, CEPAL, the Asian Development Bank, and other agencies have been rendering practical assistance in various ways. In December 1976, the General Assembly established the UN Special Fund for Land-locked Developing Countries. It acted as an agent of the General Assembly in collecting and disbursing voluntary contributions "in order to compensate the land-locked countries for their additional transport costs." Efforts to establish a legal right of transit were concentrated in the Third United Nations Conference on the Law of the Sea (UNCLOS III) and in the Seabed Committee that preceded it.

THE LAND-LOCKED STATES AT UNCLOS III

During the lives of both the Seabed Committee (1967–1973), which acted as a preparatory committee for UNCLOS III, and of the Conference itself (1973–1982), the land-locked States worked diligently to achieve

Land-locked States on stamps. Bolivia issued the top four stamps in 1979 to commemorate the centennial of the loss of her coastal province to Chile at the beginning of the War of the Pacific. Since this loss, she has waged an unremitting struggle to regain some kind of *salida al mar*, or outlet to the sea. The bottom stamp honors the acquisition of coastal Eritrea in 1952 by Ethiopia, giving this landlocked country a sea coast for the first time in modern history. (Martin Glassner)

two objectives: a guaranteed right of free *transit to and from the sea* and the greatest possible *access to the resources of the sea.* United Nations Convention on the Law of the Sea contains a section (Part X) titled "Right of Access of Land-locked States To and From the Sea and Freedom of Transit." The right of access is thus conceded, but not a right of transit, without which access is meaningless. The coastal States are still unwilling to surrender even a small part of their sovereignty to assure free transit to their less fortunate neighbors. To emphasize this point within Part X, one article provides that, "the terms and modalities for exercising freedom of transit shall be agreed between the land-locked States and the transit States concerned through bilateral, subregional or regional arrangements" and that "Transit States, in the exercise of their full sovereignty over their territory, shall have the right to take all measures necessary to ensure that the rights and facilities . . . for land-locked States shall in no way infringe their legitimate interests."

Other articles, among other things, exempt the special rights and facilities of land-locked States from the application of most-favored-nation clauses; prohibit customs duties, taxes, and most charges from being placed on traffic in transit; provide that free zones or other customs facilities may be established at ports for traffic in transit; require that delays or technical difficulties of traffic in transit should be avoided or eliminated; and that "ships flying the flag of land-locked States shall enjoy treatment equal to that accorded to other foreign ships in maritime ports." This last provision, incidentally, is neither facetious nor hypothetical. Switzerland has long had an efficient and profitable merchant fleet operating chiefly out of Genoa; Bolivia, Nepal, Paraguay, Swaziland, Uganda, and other land-locked States also operate or have considered high-seas fleets.

On the whole, considering the gains and losses, this Convention represents a slight net gain for the land-locked States compared with the 1965 Convention. It does not, however, assure the "free, uninterrupted and continuous" movement of traffic in transit referred to in the 1965 Convention.

In December 1970, the General Assembly declared that the seabed and ocean floor beyond the limits of national jurisdiction and its resources are "the common heritage of mankind, irrespective of the geographical location of States, whether land-locked or coastal." Through the debates in UNCLOS III and the successive draft treaty articles considered, this declaration was not challenged. The land-locked States are to share more or less

equitably in both the machinery to be created to manage the international seabed area and in the benefits to be derived from seabed mining. Nevertheless, because of profound legal and economic uncertainties, it seems most unlikely that any developing land-locked State—or any poor country for that matter—will derive any appreciable benefit from seabed mining until well into the twenty-first century. Of much greater practical importance to them is the sharing of benefits of resource exploitation within "the limits of national jurisdiction."

These limits include the continental shelves to the outermost edge of the continental margin, *and* a 200-nautical-mile exclusive economic zone, measured from the coastline. All the resources in these areas, living and nonliving, belong to the coastal States. Thus, within a quarter century, the land-locked States have, in effect, been driven another 188 or 197 nautical miles *back* from the resources of the sea. They have been denied their traditional free access to all the world's known offshore oil and natural gas, all the presently exploitable minerals, most of the readily accessible manganese nodules, and 85 to 95 percent of the world's current fish catch. In UNCLOS III they succeeded only in gaining a token right to a portion of the surplus fish within the exclusive economic zone; that is, the fish that the coastal State cannot or chooses not to harvest up to the allowable catch based on conservation principles. The land-locked States—some 20 percent of the world's independent States and 4 percent of its population— will derive little or no benefit from the coming intensive exploitation of our last great frontier on earth.

Although they failed to gain much from UNCLOS III, the land-locked States will undoubtedly continue their efforts to improve their situation by the methods outlined here: internal development, bilateral negotiations (both for transit and access to the economic zones of coastal States), improved and diversified transit routes and facilities, regional economic integration, and assistance from international agencies. They are also likely to turn more to air transport, joint ventures for marine resource development, and other devices for overleaping the territory that lies between them and the sea and its resources. It is possible also that, as we indicated in our discussion of "strategic location," some of the poorest and most isolated States of today could become the Switzerlands of tomorrow. At this time, however, this prospect seems rather remote.

Chapter 27

INTERGOVERNMENTAL ORGANIZATIONS

We have stressed in this book that today the dominant form of political organization is the State and its ideal aspect the nation-state. We have also discussed other forms of political organization—tribe, city-state, feudal realm. In each case the form did not satisfy completely the needs and aspirations of the participants and they joined together in a larger association, retaining their identities but cooperating and even surrendering some of their sovereignty for the common good. The Iroquois Confederacy, the Hanseatic League, and the Holy Roman Empire of the German Nation are only the most prominent examples of organizations designed to attain goals unattainable by any unit individually. Farther back in history are more examples; the Greek city-states are the best known. None encompassed more than a small portion of the earth's land or population, and none was able to withstand severe pressure for very long. They show us nonetheless that human beings cannot best be served exclusively by parochial political organizations, no matter how grand or powerful. More than a century ago, in 1842, Alfred Tennyson included in his long poem *Locksley Hall* a remarkable prediction of the twentieth century that expresses the aspirations and the dreams of many centuries:

For I dipt into the future, far as human eye could see,
Saw the Vision of the world, and all the wonder that would be;

Saw the heavens fill with commerce, argosies of magic sails,
Pilots of the purple twilight, dropping down with costly bales;

Heard the heavens fill with shouting, and there rain'd a ghastly dew
From the nations' airy navies grappling in the central blue;

Far along the world-wide whisper of the south-wind rushing warm,
With the standards of the peoples plunging thro' the thunderstorm;

Till the war-drum throbb'd no longer, and the battle flags were furl'd
In the Parliament of man, the Federation of the world.

The post-World War II era has been one of unprecedented activity in the field of supranationalism. There is something new in this: in their determination to associate, States are doing what they have not been willing to do on such a scale previously. They are giving up some of their sovereignty in return for security, economic advantage, cultural strength, or whatever benefit they perceive to accrue from their involvement.

THE LEAGUE OF NATIONS

The League of Nations was primarily designed to curb aggressive war and was created to act as an instrument of collective security through joint repudiation of any aggressor. The horrors of World War I contributed to a desire for such an agency, and the United States was among those countries that participated in discussions leading to its establishment. It did not spring up full-blown after the war, however. For generations philosophers, poets, and moralists had been proposing such an organization and during the war even practical politicians and statesmen prepared carefully reasoned proposals for one. During the nineteenth century, a number of peace conferences took place, The Hague Court of Arbitration was founded, and international bureaus, such as the Universal Postal Union and the International Institute of Agriculture, were established to deal with particular work in which international cooperation was essential. During the war the Allies cooperated on a scale hitherto unknown among "sovereign States." They also laid much of the groundwork for the League, so that when the Paris Peace Conference convened in 1919, the League Covenant was drafted in a few days under the leadership of President Wilson.

But the League received its first setback at the very beginning: in the United States, the proposal to join the organization was defeated, mainly through the hostility of Republican Senators to the efforts of President Wilson, who had strongly championed U.S. participation. The members had hoped that the threat of total economic boycott would stifle any would-be aggressor; the absence of the United States from the organization eliminated that possibility and dealt a severe blow to the confidence of the membership. In all, 63 States were members of the League of Nations, though, in fact, the total membership at any single time did not reach 63. But despite these troubles, the League of Nations represented a tremendous leap forward in international relations, for it embodied a principle that until that time had been a matter only for theoretical discussion among political scientists and historians.

The League of Nations was officially terminated in 1946 after the United Nations had begun to function. Its contribution to the latter organization in terms of principles and practices is such that the *United Nations* in many ways is a renewal of the same aspirations possessed by the League.

In evaluating the work of the League of Nations, we can hardly do better than to quote Lord Robert Cecil, one of its founders and staunchest supporters and representative of Britain at the final session of the League Assembly on 9 April 1946, when he said:

It is common nowadays to speak of the failure of the League. Is it true that all our efforts for those twenty years have been thrown away? . . . The work of the League is purely and unmistakably printed on the social, economic and humanitarian life of the world. But above all that, a great advance was made in the international organization of peace. . . . It was not, indeed, a full-fledged federation of the world—far from it—but it was more than the pious aspiration for peace embodied in those partial alliances which had closed many great struggles. . . . We saw a new world centre, imperfect materially, but enshrining great hopes, an Assembly representing some fifty peace-loving nations, a Council, an international civil service, a World Court of International Justice, so often before planned but never created, an International Labour Office to promote better conditions for the workers. And very soon there followed that great apparatus of committees and conferences striving for an improved civilisation, better international co-operation, a larger redress of grievances and the protection of the helpless and oppressed.

Truly this was a splendid programme, the very conception of which was worth all the efforts which it cost . . . but, as we know, it failed in the essential condition of its existence—namely, the preservation of peace—and so, rightly or wrongly, it has been decided to bury it and start afresh. . . .

The League is dead: Long live the United Nations!

THE UNITED NATIONS

The United Nations was founded in 1945 under conditions similar to those attending the birth of the League of Nations. It had the benefit of the League's experience, its successes and failures, and many of its staff and delegates. Its roots are deeper and its foundations firmer than those of its predecessor. Its membership has finally approached universality and no member has yet resigned or been expelled.* Its activities are more varied than those of the League, its structure more complex, its budget and secretariat far larger, its role as a public forum more effectively utilized, and its political and security functions more vigorously carried out. There are many reasons for its relative success and durability, despite many problems and weaknesses, but the most important are undoubtedly the active participation of the United States and the importance of UN multilateral diplomacy and cooperation to the developing countries.

We have already discussed some UN organs and activities—its role in the decolonization process and in assisting the new States; the International Court of Justice and the International Law Commission: GATT and UNCTAD; the regional economic commissions; assistance to landlocked States—and we discuss others later. Many others have a spatial dimension and are thus of interest to geographers. They have varying memberships, fields of activity, and degrees of effectiveness, in all of which political factors are important. Some of these aspects are included in the diagram of the UN system. Others responsible to the Economic and Social Council, not shown here, are the United Nations Fund for Population Activities, Commission on Transnational Corporations, Commission on Human Settlements, and World Tourism Organization.

*President Sukarno of Indonesia, reacting to the UN's failure to support his claims on Malaysia's Borneo territories, informed that body on 20 January 1965 that Indonesia was withdrawing "at this stage and under the present circumstances." After his downfall, the new government on 19 September 1966 announced its decision "to resume full cooperation with the United Nations and . . . participation in its activities." In addition, South Africa has been excluded from participation in the General Assembly since 1974 and from several other UN bodies and specialized agencies.

Activities of the Security Council

The activities of the Security Council of greatest politicogeographical interest are those involving peace-keeping. Since the initiation of the United Nations, the Security Council has dispatched military personnel to perform peace observation or truce supervision functions in Indonesia, northern Greece, Kashmir, around the armistice lines of Israel, Sinai (twice), Lebanon (twice), and Cyprus. Multinational forces under the UN flag fought wars in Korea (1950–1953) and the Congo (1960–1964). In early 1987 nearly twelve thousand military personnel were engaged in UN peacekeeping duties, primarily in Cyprus, the Middle East, and Kashmir. Only 56 UN members have so far participated in these missions, however, and only 19 have joined in two or more major operations. Forces are generally raised on an *ad hoc,* voluntary basis, with States providing troops, supplies, logistical support, and funds as they see fit at the time, although on a few occasions members have been assessed to pay for operations.*

One major activity of the United Nations (as was true of the League of Nations) is to sponsor large international conferences devoted to specific topics. We have already mentioned the lawmaking conference; the other type is a conference of government officials and experts (who are sometimes the same people) who discuss a problem, study reports on it, and ultimately make recommendations to States and to the United Nations on courses of action that should or could be taken to deal with the problem. There have been scores of such conferences and there will be many more. Among those dealing with geographic topics have been those on food, population, the human environ-

*In 1968 Canada and the Nordic countries (Denmark, Norway, Finland, and Sweden) designated units of their armed forces as standby forces to be assigned as needed to UN duty, but so far no other States have followed their lead. It is also noteworthy that by the end of 1986 the five current UN peacekeeping forces had suffered a total of 333 troops and observers killed on active duty.

THE UNITED NATIONS SYSTEM

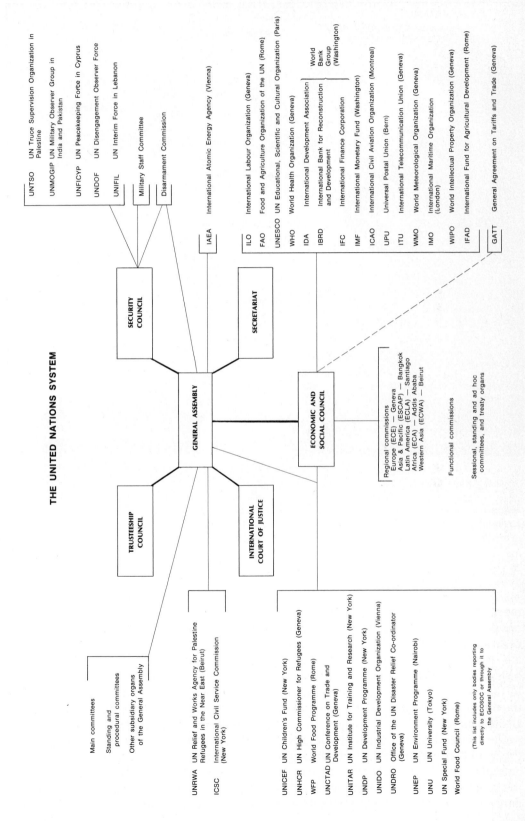

TRUSTEESHIP COUNCIL

GENERAL ASSEMBLY

SECURITY COUNCIL

INTERNATIONAL COURT OF JUSTICE

SECRETARIAT

ECONOMIC AND SOCIAL COUNCIL

Main committees

Standing and procedural committees

Other subsidiary organs of the General Assembly

UNTSO UN Truce Supervision Organization in Palestine

UNMOGIP UN Military Observer Group in India and Pakistan

UNFICYP UN Peacekeeping Force in Cyprus

UNDOF UN Disengagement Observer Force

UNIFIL UN Interim Force in Lebanon

Military Staff Committee

Disarmament Commission

IAEA International Atomic Energy Agency (Vienna)

ILO International Labour Organization (Geneva)

FAO Food and Agriculture Organization of the UN (Rome)

UNESCO UN Educational, Scientific and Cultural Organization (Paris)

WHO World Health Organization (Geneva)

IDA International Development Association

IBRD International Bank for Reconstruction and Development

World Bank Group (Washington)

IFC International Finance Corporation

IMF International Monetary Fund (Washington)

ICAO International Civil Aviation Organization (Montreal)

UPU Universal Postal Union (Bern)

ITU International Telecommunication Union (Geneva)

WMO World Meteorological Organization (Geneva)

IMO International Maritime Organization (London)

WIPO World Intellectual Property Organization (Geneva)

IFAD International Fund for Agricultural Development (Rome)

GATT General Agreement on Tariffs and Trade (Geneva)

UNRWA UN Relief and Works Agency for Palestine Refugees in the Near East (Beirut)

ICSC International Civil Service Commission (New York)

UNICEF UN Children's Fund (New York)

UNHCR UN High Commissioner for Refugees (Geneva)

WFP World Food Programme (Rome)

UNCTAD UN Conference on Trade and Development (Geneva)

UNITAR UN Institute for Training and Research (New York)

UNDP UN Development Programme (New York)

UNIDO UN Industrial Development Organization (Vienna)

UNDRO Office of the UN Disaster Relief Co-ordinator (Geneva)

UNEP UN Environment Programme (Nairobi)

UNU UN University (Tokyo)

UN Special Fund (New York)

World Food Council (Rome)

Regional commissions
Europe (ECE) — Geneva
Asia & Pacific (ESCAP) — Bangkok
Latin America (ECLA) — Santiago
Africa (ECA) — Addis Ababa
Western Asia (ECWA) — Beirut

Functional commissions

Sessional, standing and ad hoc committees, and treaty organs

(This list includes only bodies reporting directly to ECOSOC or through it to the General Assembly)

401

Kashmir, a divided land.   India and Pakistan have fought several bitter wars over this former princely state and its future is still in doubt, although its *de facto* partition along a cease-fire line may be permanent. United Nations peace-keeping forces have been on duty here for more than 20 years. The area north and west of the cease-fire line is occupied by Pakistan; the rest, by far the most valuable portion, by India.

ment, human settlements (Habitat), technical cooperation among developing countries, desertification, water, science and technology for development, energy, outer space, and pollution of the Mediterranean as well as many sponsored by the intergovernmental agencies and other UN affiliates.

The United Nations has neither solved the world's great problems nor brought an end to war, for the UN only represents the collective will of its members. With nationalism still the strongest political force

in the world, it is unreasonable to expect that States would grant the United Nations truly supranational authority. It is very far indeed from being a world government. The significance of the United Nations, in the present context, lies in the fact that it reflects, however imperfectly, the continued desire on the part of States for mutual action, open communications, and a maintenance of international peace and security. States that have faced the censure of the United Nations nevertheless continue to participate and contribute. This

United Nations peacekeeping forces in action. In the top photo, Australian and Uruguayan officers of the UN Military Observer Group in Kashmir confer with Indian and Pakistani officers. They are investigating an incident and interrogating a witness. (UN) The bottom picture shows members of the Fijian contingent of the United Nations Interim Force in Lebanon inspecting a motorbike at a checkpoint south of Tyre. (UN/Inge Lippmann)

surrender of even the smallest measure of sovereignty is a completely new phenomenon on the politicogeographic scene, and a long stride along the road toward "the Parliament of man, the Federation of the world."

THE COMMONWEALTH

We have already mentioned *The Commonwealth* (originally the British Commonwealth of Nations) in connection with decolonization and international trade, but it is such an extraordinary organization that it deserves some additional attention. It is not a federation, not an empire, not an alliance, not a trade group, not, in fact, like any other organization in the world. It is a free association of nearly 50 countries on every continent, as diverse a group of States as can be imagined. All they have in common is their recognition of the British sovereign as Head of the Commonwealth, their former status as British colonies, use of the English language and legal system, and agreement that Commonwealth membership brings them important benefits. These benefits no longer include the Commonwealth preference system, as we have seen, or

NONGOVERNMENTAL ORGANIZATIONS

Over 700 international **non**governmental organizations (NGOs or INGOs) that are in consultative status with the Economic and Social Council play a very important role in the work of the United Nations. They provide a vehicle for citizens, including students, to make their opinions known directly to the UN, to learn about obscure details of UN work in areas of special interest to them, and to help in drafting proposals for UN action. A representative sample of those concerned with geographic matters would include:

International Federation of Agricultural Producers	United Towns Organization
International Planned Parenthood Federation	Inter-American Planning Society
International Union of Local Authorities	International Association for Water Law
Associated Country Women of the World	International Civil Airport Association
Carnegie Endowment for International Peace	International Law Association
International Association of Ports and Harbours	International Road Transport Union
International Catholic Migration Commission	International Union of Railways
International Commission on Irrigation and Drainage	Society for International Development
International Council of Environmental Law	World Population Society
International Union for Conservation of Nature and Natural Resources	Friends of the Earth
American Association for the Advancement of Science	Sierra Club
International Association on Water Pollution Research	International Chamber of Commerce
International Association of Physical Oceanography	International Geographical Union
International Society for Photogrammetry	

free migration into Britain (virtually halted since 1963), or the traditional protection afforded by the Royal Navy. Even the Commonwealth Development and Welfare Fund can no longer be as generous as it once was. Although it is generally considered to have come into existence at the Imperial Conference of 1926, it is actually a product of gradual evolution; it has never been formally organized, has no constitution or charter, and until 1965 never had a headquarters or secretariat. What then is it? How is it organized? What does it do?

There is no longer a hierarchy of membership in the Commonwealth. All members are equal. When a British territory

becomes independent it may apply for full Commonwealth membership and, if it does, it is (almost automatically) admitted as an equal, no matter how small or weak or poor it may be. The members exchange opinions in a friendly, informal, and intimate atmosphere; they do not agree on positions, make collective decisions, or take collective action. Even the Secretariat, located in London, has no career service, is forbidden to perform executive functions, and is formally limited to servicing governmental and other meetings, preparing papers on matters of common interest, giving advice, and assisting in areas of functional cooperation. It has, however, by common consent grown to perform executive functions and generate stronger influence, still informal, within the group. The first secretary-general was from Canada, the second from Guyana.

The principal activity of the Commonwealth is the biennial Heads of Government Meeting, with senior government officials meeting in the alternate years. There are also regular meetings of ministers responsible for education, finance, health, law, food production and rural development, and youth matters. Commonwealth high commissioners (ambassadors) in London form a useful group for special purposes, such as overseeing the sanctions against Rhodesia and supervising the budget of the Secretariat.

There are scores of other significant activities and institutions in agriculture, forestry, commerce, communications, science and technology, and other practical affairs, but it is primarily a framework for consultation, mutual education, and mutual support. Formal activities and organization are relatively new and still not essential to its success. Its durability is astonishing in an era of such rapid and profound change. It has survived depression and world war, the dismantling of the British Empire, squabbles and even wars between members (India–Pakistan, Uganda–Tanzania), Britain's entry into the EC, sharp disagreements about Rhodesia and South Africa, and many other potentially shattering experiences. Through it all, only three countries have withdrawn: Ireland (1949), South Africa (under pressure in 1961), and Pakistan (1972). It is not difficult to imagine the Commonwealth continuing indefinitely, ever-changing, ever-adapting, serving as a halfway house between the loneliness of isolation and the often bewildering complexity of the United Nations. It is truly a singular institution.

THE FRENCH COMMUNITY

The *French Community* has a very different history. It was born with the Fifth Republic, itself largely the product of the se-

ries of disasters that befell France in its efforts to retain a hold over its farflung colonial empire. Prior to the Community, the French Empire consisted of a very complex organizational structure known as the French Union, in which overseas but self-governing territories were represented to some degree in the Paris government. However, so many changes took place in the empire during and after World War II that almost constant revision was necessary. In 1958, with the crisis in Algeria dragging on for several years and "the wind of change" having an impact in other parts of Africa, the Community was erected and French overseas possessions were given a choice: to become independent outside any French Community, to attain independence within the Community framework, or to continue as dependencies as before. When the vote was taken, all African territories, except French Guinea and French Somaliland, chose autonomy and membership in the Community. French Guinea chose complete independence outside the French sphere; French Somaliland chose to remain a French colony.

Under the Fifth Republic, independent States that were members of the Community did not enjoy the representation in the French Parliament possessed by self-governing French overseas territories. In turn, the French president did not have the powers in the member States once enjoyed by the Parliament. Nevertheless, France continued to have considerable influence in her former empire.

But the relationship between France and the countries with which it constituted the Community, as described in the 1958 Constitution, has been replaced by a series of contractual agreements that describe the area of cooperation between each country and France. The institutions of the Community have fallen into disuse, although the Community has never been formally dissolved and it is impossible to determine when the various bodies ceased functioning.

This was no great loss, for the French

Community was never more than a poor imitation of the Commonwealth and was primarily an instrument for retaining French dominance over her former colonies. It had not the long historical evolution, the flexibility of self-government and mutual assistance of the Commonwealth. Perhaps France never really had her heart in it. As a highly centralized State, she probably could not really accommodate herself to the decentralization attendent upon the loss of her empire. It seems unlikely that the Commonwealth will ever be successfully imitated.

ORGANIZATION FOR ECONOMIC COOPERATION AND DEVELOPMENT

In the wake of the devastation of World War II, the United States offered Europe generous economic assistance in rebuilding. Formally known as the European Recovery Program, it was popularly called the Marshall Plan after U.S. Secretary of State George C. Marshall, who proposed it in 1947. The plan was worked out cooperatively with the 18 European countries that accepted the U.S. invitation to join. (All the communist countries rejected the invitation, Poland and Czechoslovakia with regret.) They formed the Organization for European Economic Cooperation (OEEC) to administer the program at their end.*

The OEEC provided Europeans with the first opportunity in history to work cooperatively on a major and continuing peaceful project affecting the vital interests of all of them. It continued functioning well even after the EEC and EFTA divided its members into the Six, the Seven, and the five nonparticipants in either grouping. It laid the groundwork for the UN Economic Commission for Europe, and it carried out its principal mission with efficiency and élan by the end of the 1950s.

It was so successful that everyone concerned felt it would be a pity to scrap it, to break up a winning team. And so in 1961, it was converted into the *Organization for Economic Cooperation and Development (OECD)*.

The fundamental differences between the two organizations are apparent from the change in name. First, it was no longer strictly a European organization; the United States and Canada were charter members; Japan joined in 1964; Finland, which had been an associate, became a full member; and Australia and new Zealand joined as full members.* Second, its new emphases were on development instead of on recovery, including aid of various kinds to the poor countries of the world, and on expansion of world trade. With headquarters in Paris and an elaborate structure, its activities are many and varied. Most are highly technical, but some are of interest to geographers. They include committees on tourism, maritime transport, trade, international investment and transnational enterprises, research, various industries, fisheries, agriculture, and nuclear energy. Perhaps most interesting, and certainly one of the most potent of the OECD organs, is the Development Assistance Committee (DAC), its principal vehicle for channeling, monitoring, and evaluating aid from the OECD to developing countries. The economic forecasts of the OECD may be open to challenge, its technical accomplishments may be uneven, its relationships with the EC, EFTA, and ECE obscure, and the motives and effectiveness of its development assistance questionable, but the OECD remains a powerful and influential force in world affairs.

Unlike GATT, the OECD has no poor members; it is very much a "rich man's club." In its wealth and unity lie its power. In the Conferences on International Economic Cooperation during the mid-1970s, (the so-called "North–South dialogues")

---

*The members were Austria, Belgium, Denmark, France, Germany (West), Greece, Iceland, Ireland, Italy, Luxembourg, Netherlands, Norway, Portugal, Spain, Sweden, Switzerland, Turkey, and the United Kingdom.

*Yugoslavia and the EEC Commission participate in the work of the OECD with a special status.

SOME INTERNATIONAL NONGOVERNMENTAL ORGANIZATIONS

The Inter-Parliamentary Union, founded in 1889 to promote personal contacts among the members of the world's parliaments, currently has 72 member groups and its secretariat is in Geneva. **The International Council of Scientific Unions** was founded in 1931 to coordinate international cooperation in theoretical and applied sciences. The academies or research councils of 66 countries belong as well as 17 international scientific unions. Its headquarters is in Paris. It has committees on Antarctic, oceanic, space, and water research. **The International Air Transport Association (IATA)** has headquarters in Montreal and Geneva. At present over 90 international airlines are full members. IATA was founded in 1948 to promote civil aviation and enable airlines to collaborate on technical and marketing problems. Until 1978 it also set most international airline fares.

---

the OECD countries, largely located in the Northern Hemisphere, were on one side and the Group of 77, primarily found in the Southern Hemisphere, on the other. The conferences were inconclusive. The parties were no more able to reach understandings about such fundamental issues as industrialization of poor countries, trade concessions by the rich to the poor, the flow of aid from the North to the South, the distribution and effective use of all forms of energy, and overall economic relations between North and South than they have been in the United Nations, UNCTAD, GATT, or elsewhere.

OTHER INTERNATIONAL ORGANIZATIONS

Among the numerous other international governmental and nongovernmental organizations, a few that are both geographic and influential should be mentioned briefly. The *Organization of the Petroleum Exporting Countries* is one. It was organized in 1960 by Iran, Iraq, Kuwait, Saudi Arabia, and Venezuela to unify and coordinate members' petroleum policies and to safeguard their interests. The founders have been joined by Algeria, Ecuador, Gabon, Indonesia, Libya, Nigeria, Qatar, and the United Arab Emirates. Its Secretariat is located in Vienna. *OPEC* established a Special Fund in 1976 to provide financial assistance to other developing countries on easy terms as a contribution toward the redistribution of wealth from

the rich countries to the poor, but most aid has gone to Jordan, Syria, and the PLO.

OPEC's quadrupling of oil prices in 1973 and the embargo of 1973–1974 (the "oil weapon") were initially aimed at Israel and her friends; the continuation of OPEC's supply and pricing policies has become part of the drive for a New International Economic Order.*

The Group of 77, basically the developing countries of the world, now numbers approximately 130 members and takes an active role in shaping common positions on many current issues. It is a negotiating group within the United Nations and other global fora, but has no tangible projects or activities. The same is true of the *Nonaligned Movement (NAM),* which grew out of the Bandung Declaration of 1955, was formalized at the Belgrade Conference of 1961, and currently has 101 members, all of them also in the Group of 77. The criteria for membership in both are rather loose; the 77 contains a number of countries that are hardly poor and the NAM contains a number of countries that are clearly aligned with one or another of the superpowers. Nevertheless, both groups are powerful voices speaking for at least the leadership of countries that are small, poor, and weak.

The Organization of the Islamic Con-

*The Organization of Arab Petroleum Exporting Countries (OAPEC) was founded in 1968, currently has 10 members, and its headquarters is in Kuwait. While officially it sponsors a number of joint economic undertakings, its activities are primarily political.

International organizations on stamps. This is a popular theme on stamps, especially the United Nations. These stamps honor the United Nations Development Programme, the International Court of Justice, the World Health Organization, the World Meterological Organization, the International Maritime Organization, the Commonwealth, and the Non-Aligned Movement. The anti-UN stamp at the bottom was issued by Iran in 1982. (Martin Glassner)

ference was organized in 1971 and now has more than 40 members in Africa, Asia, and the Indian Ocean. It has a Secretariat in Jeddah, Saudi Arabia, but no formal structure. Its aims are very broadly to promote Islamic solidarity among members and cooperation in many fields. Despite annual meetings of ministers, however, and numerous resolutions and pronouncements, it has few concrete achievements to its credit.

REGIONAL ORGANIZATIONS

Developing side by side with the international or worldwide organizations just described have been a great many groups of States within particular regions. Regardless of how we classify them functionally, we must recognize that they all perform multiple functions, thus overlapping all categories, and all are fundamentally political, even if their stated purpose is economic or cultural or social. All are groups of political bodies (States) organized as a result of political decisions to achieve political goals.

There are many reasons for the existence of regional organizations. Some observers feel that true world organization is impossible to achieve all at once and that therefore regional organizations can serve as "building blocks" gradually forging greater links among themselves until the ultimate goal is reached. Others feel that regional organization *is* the ultimate goal, that only groups based on a clear community of interests with a common heritage and a common approach to solving problems are necessary or possible. Another view is that regional organizations are a compromise, a "halfway house" between the State and a world government, and are, therefore, to be accepted as experiments or trial runs preparing us for the real thing. Regardless of the theoretical bases for them, however, they have existed for a long time, continue to proliferate, and disappear when no longer needed or desired. In the survey that follows, we include only those regional organizations of some size and durability and those whose functions are broadly geographic. There are hundreds more that are smaller, more technical, or less substantial. There are also scores of organizations of sub-State units: of individual government ministries, of subdivisions of States, of technocrats and nongovernmental professionals and other citizens, all organized on a regional or subregional basis.

ORGANIZATIONS OF GENERAL COMPETENCE

The largest and generally most active of the regional organizations apart from those devoted to economic integration are those commonly called "political' or "political-cultural," but which in reality have economic and social functions as well. Foremost among them in nearly every respect is the *Organization of American States* (OAS), whose headquarters is in Washington.

During the century and a half since most of the Spanish American colonies became independent, a unique network of relationships has developed among them, many of which have traditionally included the United States and many that have recently been joined by the independent countries of the Commonwealth Caribbean. This network is commonly called the Inter-American System. There are skeptics who doubt the efficacy of the system. Many observers recognize the need for changes in it and some question whether the needed changes are possible or even worthwhile. Nevertheless, nowhere else in the world has there ever been such a large and diverse group of independent States that have voluntarily associated themselves in so many ways for so long.

There were a number of unsuccessful attempts to form an organization of the newly independent States of the Western Hemisphere during the nineteenth century, but not until 1890 did the effort bear fruit. At an inter-American meeting in Washington in that year, the Pan American Union was born. During the next half century it developed a number of functional organs, served as a framework for organizing hemispheric defense, aided in the settlement of disputes among its members, and otherwise proved useful to both the United States and the smaller and poorer Latin American States. It was reorganized at Bogotá in 1948 and transformed into the Organization of American States. Its principal functions originally were collective defense and pacific settlement of disputes within the hemisphere. Its principal cornerstones are the Inter-American Treaty of Reciprocal Assistance (the Rio Treaty of 1947), the Charter of the OAS (1948), and the American Treaty on Pacific Settlement (1948). The last of these binds its signatories to utilize all the methods of pacific settlement that we discussed in detail in Chapter Twenty-three. The Rio Treaty organized a military alliance that we discuss briefly later in this chapter. Here we review the structure and work of the OAS stemming from its Charter, as amended by the Protocol of Buenos Aires in 1967.

The OAS today is organized much like a small United Nations. Its supreme organ is the General Assembly, which meets annually to decide general policies and actions of the organization and to coordinate the work of the other organs. The Meeting of Consultation of Ministers of Foreign Affairs is utilized frequently on an *ad hoc* basis "to consider problems of an urgent nature" (OAS Charter) or to serve as an Organ of Consultation in case of a threat to international peace and security (Rio Treaty). Directly responsible to the General Assembly are the Permanent Council, which deals with any matter referred to it by the General Assembly or the Meeting of Consultation and performs certain mandated functions; the Inter-American Economic and Social Council (CIES);* the Inter-American Council for Education, Science and Culture (which is similar to UNESCO); the Inter-American Commission on Human Rights; the Inter-American Juridical Committee; the General Secretariat (formerly the Pan American Union); and the specialized conferences, held as needed to deal with special technical matters or to develop specific aspects of inter-American cooperation. Many conferences have dealt with geographic subjects, including agriculture, ports and harbors, highways, natural resources, and Indian affairs.

The OAS, like the UN, has specialized organizations affiliated with it. Among them are the Inter-American Indian Institute (Mexico City), the Inter-American Institute of Agricultural Sciences (San José), the Pan American Health Organization (Washington; formerly the Pan American Sanitary Bureau and now also functioning as the regional arm of WHO), and the Pan American Institute of Geography and History (Mexico City). The PAIGH has committees on geography, history, geophysics, and cartography. Its geography publications are of high quality and should be more widely known.

Today all the independent States of the Western Hemisphere are members of the OAS, except Canada, which has consistently refused to join, and Guyana and Belize, which are prohibited from membership by the amended Charter until their territorial disputes are resolved. (Cuba was suspended from active participation in 1962 but remains a member.) Its activity in many practical matters, such as development of the Plata Basin and settlement of disputes, has been considerable, but it is far from being central to the everyday concerns of its members. What matters most, however, is probably the perception—irrespective of the reality—of "hemispheric unity," of a "special relationship," among the diverse States and peoples of the hemisphere, a true Inter-American System, epitomized by the OAS. Despite many problems and uncertainties, it is likely to continue indefinitely and be of value to its members and to the world community.

A comparable, but much younger and less developed, organization has been formed in Africa: the *Organization of African Unity* (OAU). Founded in 1963 by 30 States, all but a very few newly independent, it gradually replaced a long list of regional and subregional groups, many of them ephemeral, which had formed during the first decade of decolonization. Its spiritual father was Kwame Nkrumah, President of Ghana and an advocate of Pan-Africanism. Its objectives were furtherance of African unity and international cooperation in Africa; coordination of political, economic, cultural, health, scientific, and defense policies; and the eradication of colonialism in Africa. Its early years were dominated by the anticolonialism struggle, and there is no doubt that it did hasten decolonization.

Another important achievement has been its stabilizing influence in the continent and its assistance in the settlement or containment of many disputes among its members. As we have seen, its insistence on the principle of *uti possidetis juris* has prevented many violent conflicts over the frequently arbitrary boundaries inherited from the colonial powers. Where fighting has broken out between neighbors (Algeria–Morocco, Somalia–Ethiopia, Ghana–Burkina Faso), OAU action has been successful; it has been unsuccessful in cases of civil war (Congo, Nigeria, Sudan) and extracontinental intervention (Shaba, Ogaden, Angola). A third major achievement of the OAU has been to formulate and represent to the world African positions on important world issues.

The headquarters of the OAU is in Addis Ababa, also the headquarters of the UN Economic Commission for Africa, with which it sometimes works very closely. Its principal organ is the Assembly of Heads of State and Government, held annually

*The Alliance for Progress was operated through CIES until the Alliance died in 1974.

and preceded by semiannual ministerial meetings. There are a permanent Secretariat and four specialized commissions responsible to the Assembly directly or indirectly through the Council of Ministers. They deal respectively with mediation, conciliation, and arbitration; defense; economics and transport; and education, science, and culture. Its numerous other organs are responsible to its secretary general; they include functional agencies and a number of regional offices, some of which have rather specialized functions.

The OAU has survived many crises in its nearly three decades of existence, and that is probably its most significant achievement. Beyond survival and the modest successes achieved in decolonization and peacekeeping, the OAU has so far accomplished little. There are a number of reasons for this, of which three are probably most important. First, the diversity of its membership is much greater than that of the OAS, and there is no history of cultural affinity or political cooperation to underpin current efforts. Second, all nonpolitical activities have had to be subordinated to the struggles against colonialism and apartheid. And third, the continent simply lacks the funds, technical skills and experience to deal effectively with the other problems on a continental basis.

The latest—but assuredly not the last—crisis to plague the OAU is the question of the former Spanish Sahara (Western Sahara). After a lengthy and divisive debate, the Sahrawi Arab Democratic Republic was admitted to the OAU in 1982 as the 51st member, even though its existence is largely imaginary. In protest, Morocco withdrew from the OAU in November 1984. Nevertheless, the OAU will very likely limp along, perhaps even gain some strength, and ultimately turn its attention more to economic development, health, science, education, and culture.

A regional group that straddles two continents but consists of a single dominant ethnic group is the *League of Arab States*. Among the 22 current members (including the Palestine Liberation Organization),

some are more Arab than others, some have important non-Muslim minorities, and in some the Arabic language is used regularly only by a small proportion of the population; nevertheless, the members are bound together by perceived common "race," religion, and language. But there are also many things that continue to divide the Arab peoples. Political systems differ greatly, and there is disapproval in certain Arab States of the governments that rule in others. Perhaps the greatest unifying element is the general hostility to the State of Israel and "the forces of Zionism."

During World War I, some Arab leaders hoped for the establishment of a great Arab State, but they saw their hopes destroyed by the many diverse forces that still prevail there: tribalism, poverty, illiteracy, vested interests. With the breakup of the Ottoman Empire and the elimination of Turkish control in the Middle East, Britain and France were put in charge of mandates over various parts of the Arab world. Arab nationalism arose after World War I just as African nationalism did after World War II: the French withdrew from Syria, the British from Iraq. Opposition to a British-sponsored Jewish State grew and Arab unity began to develop.

The Arab League was founded in 1945 by Egypt, Iraq, Syria, Lebanon, Transjordan (now Jordan), Saudi Arabia, and Yemen (North). It has a Council, a number of special committees, and a Permanent Secretariat, all located in Cairo until 1979 when League headquarters was transferred to Tunis as a reaction to the Israel–Egypt peace treaty. Also affiliated more or less loosely with it are the Arab Educational, Cultural and Scientific Organization; the Civil Aviation Council of Arab States; the Arab Centre for Studies of Arid Zones and Dry Lands; the Council of Arab Economic Unity that supervises the Arab Common Market; and perhaps a dozen other groups. Prior to the Yom Kippur War of 1973 (Egypt and Syria against Israel), however, the League did little practical work in any of these fields. There was some mediation of disputes among members and small

Regional organizations on stamps.   Since regional organizations are so important to so many countries, it is not surprising that many of them are honored on stamps. This selection illustrates the Council of Europe, South Pacific Commission, Organization of American States, Arab League, Organization of African Unity, Regional Cooperation for Development (defunct now), Africa Fund, Andean Group, Council of the Entente, North Atlantic Treaty Organization, European Communities, Colombo Plan, European Free Trade Association, Association of Southeast Asian Nations, United Nations Economic Commission for Asia and the Far East, and United Nations Economic Commission for Latin America and the Caribbean. (Martin Glassner)

peacekeeping forces were occasionally employed (such as in Kuwait in 1961), but its energies were primarily devoted to political, economic, and propaganda warfare against Israel.

Recently, however, it has become more constructive, serving as a conduit for petrodollars from its more prosperous members to the poorer ones, sending a peacekeeping force into Lebanon (all but a small portion of it Syrian troops), sponsoring scholarship and training programs for Palestinian Arab refugees, and engaging in some social and cultural activities.

Very different from all of these organizations is the *Council of Europe*. A product of the postwar popular movement for

a united Europe, which spawned many international governmental and nongovernmental organizations in the region, it is primarily a consultative political body. It was founded in 1949 by the Benelux countries, Denmark, France, Ireland, Italy, Norway, Sweden, and the United Kingdom. Since then, 11 other countries have joined, including Malta and Cyprus, making it the largest organization of its kind in Europe. Its organs are a Council of Ministers with powers of decision and of recommendation to governments, and a Parliamentary Assembly, originally composed of members of the parliaments of the member States, but for the first time elected directly by the people of Europe in May 1979. Both are served by a Secretariat in Strasbourg. Its work in the human rights field is outstanding and its European Court of Human Rights can serve as a model for similar courts elsewhere.

Its overriding function is to further European unity; in pursuance of this aim, it has concluded about 80 treaties covering particular aspects of European cooperation. It has no real power or authority, yet it has been influential in many fields and serves to some extent as the "conscience" of Europe. Its value is indicated by the fact that it has not only survived 30 years of rivalries, crises, and changes, but it is still attracting new members: Liechtenstein joined as recently as 1978.

Much smaller but with deeper roots and more day-to-day activity is the *Nordic Council.* The Nordic countries have a long history of cooperation and coordination, although there has not in recent times been any strong move toward economic or political integration of Iceland, Norway, Denmark, Sweden, and Finland. So many joint institutions exist in the region that their exact number is unknown; certainly there are at least 22 permanent committees of national civil servants and 83 other permanent governmental organs besides the council itself. This is probably the most informally integrated region in the world today.

The Nordic Council. The flags of all five Nordic countries fly at the land border posts of the members and at other significant sites. This photograph was taken at the Bella Convention Center in Copenhagen. (Scandinavian National Tourist Boards)

The Nordic Council was established in 1952. It has an elaborate structure consisting of a Council of Ministers, five standing committees, a Secretariat (in Oslo), a Cultural Secretariat (in Copenhagen), and subsidiary organs. The five members' parliaments, secretariats, cabinets, and agencies all participate in its decision making and work. The council is supervised and directed by its Presidium (located in Stockholm). This largely decentralized and intricate network seems to enhance cooperation and coordination in particular fields of interest, including foreign policy, but the various links themselves are not coordinated and the members disagree on many matters. They are still very much individual "sovereign" States.

The *Organization of Central American States* (ODECA) was founded in 1951 as the latest of several attempts to reestablish the unity of Central America that existed in colonial times. Under its new charter of 1965, ODECA's highest authority is the Conference of Foreign Ministers, except when the Heads of Government meet. Under them are executive, legislative, and economic councils and the Central American Court of Justice. Its Secretariat is in San Salvador. ODECA was deeply involved in the dispute between Honduras and El Salvador; it placed a cease-fire supervision unit on the border of the two countries in August 1976 in its first major peacekeeping operation. It remains to be seen whether ODECA will be more successful than its nineteenth-century predecessors in breaking down the parochialism of its members and forging unity among Guatemala, Honduras, Nicaragua, El Salvador, and Costa Rica.

In 1981, six members of the Arab League formed the Cooperation Council for the Arab States of the Gulf, more commonly known as the *Gulf Cooperation Council (GCC).* The six—Bahrain, Kuwait, Oman, Qatar, Saudi Arabia, and the United Arab Emirates—are all oil rich (except for Oman) and ruled by conservative monarchs. While officially dedicated to a broad program of cooperation in trade, industry, agriculture, transport, and a variety of social and cultural affairs, there is little doubt that the GCC's primary interest is in common defense against external threats, most immediately from the fanatically Shiite revolutionary regime of Iran. Many of its activities seem designed to strengthen the partners against a very powerful Iranian challenge, especially should the Iran–Iraq war end somehow in Iran's favor. Nevertheless, there are enough commonalities among the members and enough resources, so that *if* they have a strong enough determination to succeed, the GCC could lay the foundations for the first successful united Arab State in many hundreds of years.

There is no organization in Asia comparable to the OAS, the OAU, or the Arab League. The nearest one to it is the *Association of Southeast Asian Nations* (ASEAN). Founded in 1967 by Indonesia, Malaysia, the Philippines, Thailand, and Singapore, it was joined by newly independent Brunei in 1984. ASEAN replaced the Association of Southeast Asia (Malaysia, the Philippines, and Thailand) and the proposed Maphilindo (Malaysia, Philippines, Indonesia) and has displaced the Asia and Pacific Council (ASPAC), which, with nine members, was the largest Asian political organization. As in Africa, many others have been proposed or formed but few have survived for any length of time. Founded originally as a device for economic cooperation, ASEAN accomplished very little and remained low-key until after the communist takeover of Indochina in 1975. Since then its political, social, and cultural activities have become more important and the organization as a whole has "come alive." It still has no charter, however, and a central Secretariat was finally established in Jakarta only in 1976. It has a number of committees composed of various officials of ASEAN members, most of which have subsidiary bodies.

Its major organ has been its annual conference of foreign ministers and its heads of government did not meet until 1976. Its

committees consider, among other things, matters relating to communications, science and technology, transport, food production, and commerce and industry. There is as yet no dispute-settlement machinery, nor is there, as some members want, a common policy for having the region designated a "zone of peace" or for standing staunchly against communism.

Its chief activities remain economic, but although there have been many agreements in principle and staements of intent, there have been few accomplishments in economic matters. Originally drawn together by common fears and weaknesses, ASEAN has been prevented by mutual distrust and narrow nationalism from achieving its full potential. The distrust seems to be fading after two decades of working together, and ASEAN has recently begun taking positions and actions in regard to such matters as energy and Indochinese refugees (including the "boat people"). It may someday become a vital (and perhaps larger) organization, but as yet it has only incipient, although growing, popular support and few centripetal forces beyond a vague fear of China and Vietnam.

The newest subregional organization of general competence is the *South Asian Association for Regional Cooperation (SAARC)*. It was formally launched in August 1983 on the separate initiatives of Bangladesh and Nepal, and its heads of government first met in Dhaka in December 1985. Its brief history suggests that the leaders of its seven members—Bangladesh, Bhutan, India, Maldives, Nepal, Pakistan, and Sri Lanka—are approaching the novel adventure with varying degrees of optimism and caution. These countries, after all, like those farther east, have no history of regional cooperation, except through the Colombo Plan and ESCAP, both very much larger organizations in which the differences among them can be more readily camouflaged or ignored.

SAARC is organized rather loosely at this early stage, with a Council of Ministers as its highest policy-making body, a Secretariat and nine technical committees responsible for such matters as meteorology, sports, science, transport, agriculture, health, rural development, telecommunications, culture, arts, and postal services. Its headquarters was inaugurated in Kathmandu, Nepal, in January 1987, and a Bangladeshi diplomat took over as Secretary-General for a two-year term. Considering the deep and long-standing frictions among some of the members and the difficulty of some of their bilateral problems, it is remarkable that the organization has come into being at all. It is unusual among such subregional groupings in that it is so completely dominated by one country, India—one, moreover, that has not been particularly enthusiastic about it. The degree of its ultimate success may very well be proportional to the degree of participation and support by India.

In the South Pacific there are three regional organizations: the *South Pacific Commission (SPC), the South Pacific Forum,* and the Forum's subsidiary, the *South Pacific Bureau for Economic Cooperation* (SPEC). Their functions and membership overlap considerably, and in 1987 they were discussing merger. The Commission is the oldest and largest of the three. It originated in 1947 when Australia, France, New Zealand, the Netherlands, the United Kingdom, and the United States signed the Canberra Agreement. The Netherlands withdrew in 1962 after she gave up her half of New Guinea. Since then the organization has evolved dramatically. It now has 27 members of equal status, 22 island States and territories and 5 metropolitan countries. It is now an international technical assistance agency with an advisory and consultative role, providing information, training, assistance, and advice in many social, economic, and cultural fields. It does not, however, have any political functions or coloration, it does not operate large aid programs or common services, and it has only a modest Secretariat staff in Noumea, New Caledonia.

The Commission is guided by the annual South Pacific Conference, its decision making body, in which every member has one vote. In November 1986, it concluded the Convention for the Protection of the Natural Resources and Environment of the South Pacific Region, a potentially important step toward preserving a fragile environment in danger of destruction. All in all, the Commission seems to have profited from the mistakes of the Caribbean Commission (later called the Caribbean Organization and now defunct), and it has not been viewed as an instrument of neocolonialism. It is likely to continue providing valuable services and stimulating a regional consciousness among the peoples of the region.

The Forum is very different. It is the annual meeting of heads of government of the independent and self-governing States of the South Pacific, currently totaling 13, with the Federated States of Micronesia as an observer. It has no formal charter or rules of procedure, and functions by consensus. It provides an opportunity for informal discussions on a wide range of issues and problems. It has tended to emphasize political and security matters since its founding in 1971, including such thorny issues as independence for New Caledonia, French nuclear testing in French Polynesia, and apartheid in South Africa. It has endorsed the establishment of a South Pacific Nuclear-Free Zone and a South Pacific Regional Environment Program, and it deals in various ways with fisheries, petroleum, shipping, telecommunications, and trade problems.

In 1973, it established SPEC, which has headquarters in Suva, Fiji, and has since 1975 served as the Secretariat of the Forum. Its aim is to facilitate cooperation and consultation among the members on trade, economic development, tourism, and related matters. In January 1987, the South Pacific Regional Trade and Economic Cooperation Agreement (SPARTECA) entered into force; under its terms Australia and New Zealand grant many concessions to imports from the other members in or-

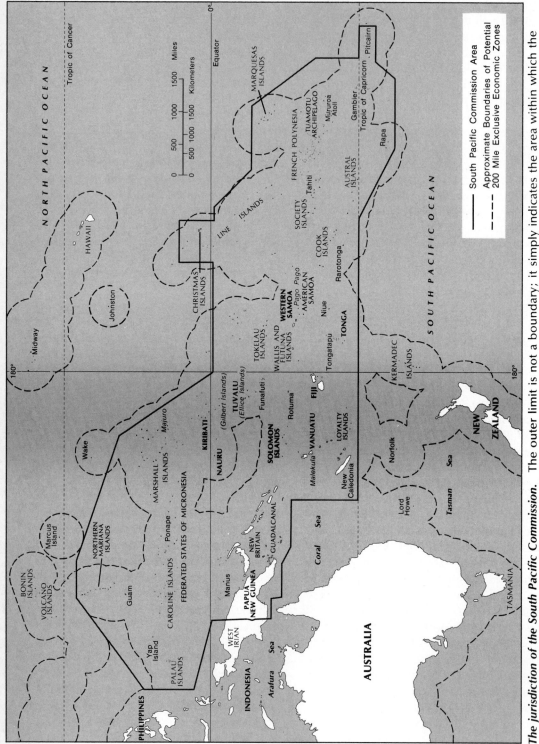

The jurisdiction of the South Pacific Commission. The outer limit is not a boundary; it simply indicates the area within which the islands participate in activities of the commission.

der to help redress a chronic trade imbalance. SPEC is also heavily involved in regional transport services, energy, disaster relief, telecommunications, fisheries, and other matters. It is gradually increasing both the scope and importance of its activities. If it does, indeed, merge with the SPC, the resulting organization could be one of the world's most useful and successful regional intergovernmental organizations.

Africa has not yet developed major organizations of general competence other than the OAU. Instead, there is a large number of subregional organizations, many with overlapping memberships and functions. In time these will probably evolve into fewer and stronger bodies; meanwhile they are difficult to follow. The chart of Selected African Subregional Organizations shows the names and memberships of some of these groups.

ECONOMIC ORGANIZATIONS

We have devoted a whole chapter to organizations designed to foster economic integration in several regions and have discussed several devoted to development of international rivers and lakes, but there are many other organizations with less ambitious economic goals. Among them are the Maghreb Permanent Consultative Committee and a large number of regional and subregional development banks and institutions for financial cooperation. Instead of trying to cover all of them, we will discuss briefly only two of the more prominent ones in an attempt to convey their variety and significance.

The *Council for Mutual Economic Assistance*, generally abbreviated COMECON (or CMEA or CEMA), was founded in 1949 largely as a response to the Marshall Plan and the OEEC. Its headquarters is in Moscow. Original members were the Soviet Union and the "socialist" countries of Eastern Europe, with Mongolia and North Korea as observers. Albania was expelled (de facto) in 1961, Mongolia was admitted in 1962, Yugoslavia became an

associate in 1964, Cuba and Vietnam were admitted in 1972 and 1978.* It was not designed as a device for economic integration; in fact, it had no clear, detailed objective or plan and has gone through a number of stages, each characterized by a different emphasis and organization.

At first Stalin tried to use it for classical colonial purposes: to exploit the resources of each country for the benefit of the Soviet Union, sending them Soviet products in return at very high prices, all on a barter basis. This gave way to an attempt to make each of the Eastern European countries a miniature replica of the USSR, with its own basic and subsidiary industries, agricultural system, and so on. The organization did not really become active until 1956, when it shifted its emphasis to coordination of the individual economic plans of its members. This evolved in the early 1960s into a plan for "the international socialist division of labor," with each country assigned to produce certain specialties for the whole group (East Germany—optical goods and basic chemicals; Hungary—pharmaceuticals and wheat; Poland—coal and steel; Romania—oil-drilling equipment and vegetables). In 1964, Romania rebelled against this plan, declaring that she "did not want to become the vegetable garden of Eastern Europe." This led to other rebellions and a failure of the specialization scheme in particular and multilateral approaches in general. Emphasis shifted to bilateral trading agreements with some efforts at coordination, but with several members undertaking their own internal reforms, this became increasingly difficult.

After the Warsaw Pact invasion of Czechoslovakia in August 1968 and Soviet installation of a more compliant government there, COMECON members reacted more strongly than ever against any supranationalism (except for the Soviet Union), but there was still no agreement on what COMECON should be. As it

---

*Observers in 1987 were Angola, China, Ethiopia, North Korea, and Laos.

SELECTED AFRICAN SUBREGIONAL ORGANIZATIONS

	General Competence						Predominantly Economic										Development-oriented						Other
	African & Mauritian Common Organization*	African & Malagache Union	Mano River Union	Council of the Entente	Union of Central African States	Organization of the Senegal River States	Economic Community of West African States	Preferential Trade Area for Eastern & Southern Africa	Customs Union of West African States	Economic Community of the Great Lakes	Central African Customs & Economic Union	Equatorial Customs Union	Customs Union of Southern Africa	Economic Community of the States of Central Africa	West African Economic Union	Southern African Development Coordination Conference	Organization for the Development of the Senegal River	Authority for Integrated Development of Liptako-Gourma	Organization for the Development of the Gambia River	Niger Basin Authority	Lake Chad Basin Commission	Kagera Basin Organization	CILSS (Club du Sahel)
Angola																X							
Benin	X	X		X			X		X						X					X			
Botswana													X			X							
Burkina Faso	X	X		X			X		X						X			X		X			X
Burundi								X		X				X								X	
Cameroon		X									X			X						X	X		
Cape Verde							X																X
Central African Republic	X	X									X	X		X									
Chad	X	X			X						X	X		X						X	X		X
Comoros								X															
Congo	X	X									X	X		X									
Côte d'Ivoire	X	X		X			X		X						X					X			
Djibouti								X															
Equatorial Guinea											X			X									
Ethiopia								X															
Gabon	X	X									X	X		X									
The Gambia							X												X				
Ghana							X																

418

N.B. The classification of these organizations is somewhat arbitrary; in reality, many of them have multiple functions, many of which are not actually performed. Some of the organizations, in fact, are moribund, whereas there are many smaller organizations that are active. In addition, there are many subregional banks and other financial institutions. Clearly, there is a need for joint action among African States.
*Dissolved in 1985.

bogged down in drift, delay, and debate, members turned more and more to trade with the West. Finally, in the summer of 1971 the Council agreed on a program of "integration," which really means closer cooperation and coordination of economic activities rather than a merger of economies as already described. COMECON still functions primarily as an instrument to encourage cooperation and coordination among its members, but each member is developing its own economy as it chooses and, as we have seen, intragroup trade is falling, though it is still more than half of total trade.

The *Colombo Plan for Cooperative Economic and Social Development in Asia and the Pacific* was established in 1951 as a result of a meeting of Commonwealth foreign ministers held in Colombo in 1950. The original Asian members were British Borneo and Malaya (now joined into Malaysia), Cambodia (now Kampuchea), Ceylon (now Sri Lanka), India, Laos, Pakistan, Singapore, and Vietnam (no longer a member). They have since been joined by 13 others.* There are 6 extraregional members which are the primary donors of aid: United Kingdom, United States, Japan, Canada, Australia, and New Zealand. (Israel was also an active member for a time.) Its purpose is explained in its formal title, but this may be somewhat misleading. The Colombo Plan is not a master plan at all but rather a mechanism for review and coordination of bilateral aid programs and projects worked out between donor and recipient countries. Its highest body is its Consultative Committee, which meets in different countries. The Colombo Plan works closely with ESCAP, with which it has been a partner in the Asian Highway and Lower Mekong Basin scheme, and with other international and regional organizations.

Britain has been the leader and chief donor of funds, but the other donors have also contributed subtantially. A most encouraging feature is that some recipients,

notably India and the Philippines, have become donors as well. Projects include large multipurpose dams and other infrastructure elements, but most aid is in the form of technical assistance and training. Most training takes place outside the region, but local facilities are being developed and increasingly used.

Although the numbers of participants are roughly equal, it would be unreasonable to expect the Colombo Plan to have had the same salutary effect that the OEEC had in Europe because of the very different conditions in the two regions. But the very fact that the food situation has improved somewhat has been a great achievement, and many of the Colombo Plan States have irrigation works, cement factories, and other industrial concerns to show for their participation. Its quiet success is due to its businesslike, nonpolitical functioning, as indicated by the great diversity in its membership and its durability.

MILITARY ALLIANCES

It is very likely that the first inter-State organization in history was a military alliance. Since then there have been hundreds of them, perhaps thousands. Our own observations of contemporary alliances reveal again and again the essential truth of the old adage that a State has no permanent friends or permanent enemies, only permanent interests. But even a cursory glance at the current world situation should make one wonder if States even know what their permanent interests are!

The array of alliances developed for the bipolar world of the 1940s and 1950s is in disarray as new centers of power have emerged. China, Western Europe, Japan, and the Group of 77 have all become powers in their own right because of their economic, demographic, numerical, intellectual, or moral strength, of necessity diluting the dominance of the United States and the Soviet Union and distracting them from their concentration on each other. What has happened, for example, to the ring of alliances the United States built around

*Afghanistan, Bangladesh, Bhutan, Burma, Fiji, Indonesia, Iran, South Korea, Maldives, Nepal, Papua New Guinea, Philippines, and Thailand.

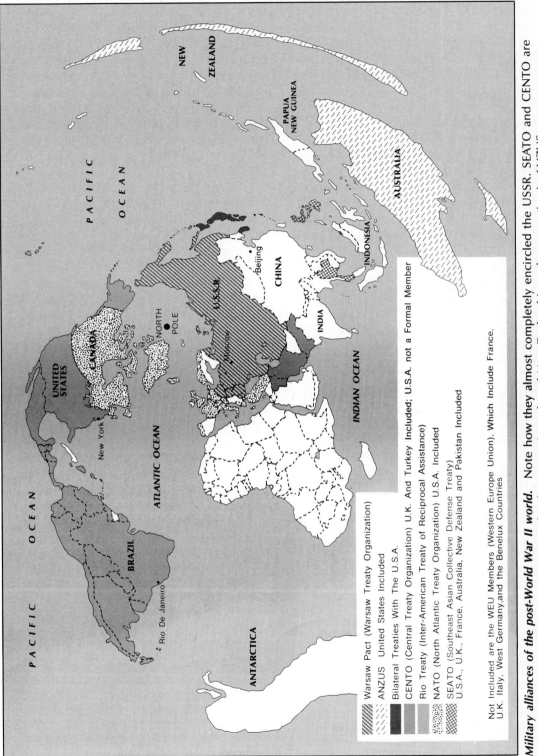

Military alliances of the post-World War II world. Note how they almost completely encircled the USSR. SEATO and CENTO are defunct now, the US alliance with Taiwan has been terminated, and New Zealand is no longer active in ANZUS.

Legend:

Warsaw Pact (Warsaw Treaty Organization)
ANZUS United States Included
Bilateral Treaties With The U.S.A.
CENTO (Central Treaty Organization) U.K. And Turkey Included; U.S.A. not a Formal Member
Rio Treaty (Inter-American Treaty of Reciprocal Assistance)
NATO (North Atlantic Treaty Organization) U.S.A. Included
SEATO (Southeast Asian Collective Defense Treaty)
U.S.A., U.K., France, Australia, New Zealand and Pakistan Included

Not Included are the WEU Members (Western Europe Union), Which Include France,
U.K. Italy, West Germany,and the Benelux Countries

100

421

the "Rimland" of Eurasia to "contain" the USSR and China? Here we discuss only the world's two major military alliances.

The *North Atlantic Treaty Organization* (NATO) was created in 1949 in Washington by the United States, Canada, and 10 Western European countries. It was joined by Greece and Turkey in 1952 and by the Federal Republic of Germany in 1955. Unlike the military alliances created before 1949, it has its own military force composed of national units from all member countries (except Iceland, which has no military forces) under unified command. It regularly carries out joint maneuvers and, notwithstanding deficiencies and problems, is a formidable military power, by far the most powerful in the world. It has withstood three "cod wars" between Britain and Iceland and the far more serious conflict, just barely short of full-scale war, between Greece and Turkey over Cyprus and another over their mutual continental shelf boundary in the Aegean Sea. It has endured communal strife in Belgium and Northern Ireland, repeated government changes in Italy, recession, inflation, communist participation in the governments of several members, and the withdrawal of France in 1966 from its military activities.* Unlike most other military alliances, it has a *raison d'être* stronger than that of simply uniting for collective security. Most of the members have a real community of interest expressed in numerous institutions, traditions, and organizations, some of which we have already discussed. How long will it last? No one can predict, of course, but probably for a long time, although its character may change.†

The *Warsaw Treaty Organization (War-*

saw *Pact)* was created in May 1955 by the Soviet Union and her Eastern European satellites not as a reaction to NATO, as some people believe, but in response to the rearming of West Germany and her admission into NATO, which they perceived as a real potential threat to them. It functions much like NATO and has headquarters in Moscow. Albania, because of her alignment with China after the Sino–Soviet split, was suspended in 1962 and withdrew in 1968. Unlike NATO, the Warsaw Pact has engaged in military action—though not of the kind envisaged in 1955—when token forces from East Germany, Poland, Bulgaria, and Hungary (but not Romania) accompanied the Soviet forces that invaded Czechoslovakia in August 1968 and crushed the new liberal communist government of Alexander Dubcek. Since then the Warsaw Pact has continued many of its activities but at a reduced level and with little apparent enthusiasm. It seems as obsolete as the western military alliances, but will probably survive, even if only as a ghost, as long as NATO does.

This survey of regional organizations suggests a great many topics that deserve serious investigation and analysis by political geographers. International relations is seriously neglected in political geography; it is an exciting area that offers unlimited opportunities for applying every investigative technique and tool at our disposal. We know that political geographers can make valuable contributions to more peaceful and rewarding international relations.

We must avoid undue optimism, however. There is powerful opposition to advancing human organization beyond the State. The official Soviet view, for example, has been expressed this way:

It should . . . be borne in mind that, in the epoch of the peaceful coexistence of states, there continues a competition between different socio-economic systems. This competition precludes any possibility of replacing the sovereignty of states with supranational bodies.*

*This forced NATO to move its headquarters from Paris to Brussels.

†The **Western European Union** (Benelux, France, Britain, Italy, and West Germany) was formed in 1955 as an outgrowth of the 1948 Brussels Treaty of collaboration in economic, social, and cultural matters and for collective self-defense. Its principal purpose is to oversee West Germany's adherence to her undertaking to limit her armaments production. Its social and cultural activities were transferred to the Council of Europe in 1960, and its military functions have been incorporated into those of NATO. It still performs some minor economic and political functions.

*Zhukov, Gennady and Yuri Kolosov, *International Space Law*, New York: Praeger, 1984, p. 187.

Chapter 28

OUTLAWS AND MERCHANTS OF DEATH

In the eight years since the last edition of this book was written, a relatively new element has been added to the array of forces active in international relations discussed in the preceding chapters. Individuals and small groups have been engaging worldwide in a variety of unlawful acts designed to enrich themselves or to achieve some limited political goal but their actions have international ramifications. Piracy, terrorism, drug trafficking, and arms trafficking are not really new. Each has a long and dishonorable history, and there is an abundance of literature on each. None has been of professional interest to geographers, however, certainly not to political geographers, until now. What are new and what now make these activities of interest to political geographers are four factors.

1. The incredible scale of most of these activities and their reach into every sector of society.

2. The worldwide distribution of all of these activities with their locations and transport routes constantly shifting in response to changing circumstances.

3. The significant and growing links among them, forming alliances that increase the destructiveness of each.

4. The very considerable impact they are having on policies and decisions and activities of governments at all levels.

All four of these factors cause distortions in international trade and in tourism and other travel. They divert resources of all kinds, in rich and poor countries alike, from economic and social development to security and law enforcement and, of course, to the illegal activities themselves. They influence government actions, corrupt individuals and erode democratic institutions. For the remainder of this century—and perhaps well beyond—no discussion of international relations can be complete or realistic without the inclusion of outlaws and merchants of death.

PIRACY

Since the late 1970s, there has been a sharp increase of incidents labeled as piracy, including air piracy, by governments, commentators, and news media. Few if any of these incidents are piracy in any legal sense, as we will see, but they do resemble the popular image of piracy in some respects and deserve our attention.

During the period 1980–1984, over four hundred incidents of "piracy" were officially reported, though this figure does not include attacks on refugees from Indochina (the so-called "boat people"), and in any case probably represents only a small portion of such incidents. Many go unreported for a variety of reasons: insurance, fear of retribution, possible exposure of other illegal activity, and so on.

Most acts of "piracy" today are thefts of high-value, low-bulk cargo, such as electronics equipment, from ship cargoes and of cash and valuables from merchant ships, their crews, and passengers. Most take place in port or in territorial waters and are thus indisputably within the jurisdiction of the State in whose waters they occur. Most take place in areas of poverty and high unemployment, where crime is one of the few alternatives to starvation. Much of it is carefully planned and carried out by organized gangs (usually small, but ranging up to 50 or more) operating from shore. Most of the thieves are armed with knives or machetes, rarely guns, and prefer to flee rather than fight if confronted by armed guards. Often the local police cannot or will not deal with the problem, nor can the small, unarmed crews, so some shipowners have taken to hiring private guards from the ports—who often belong to criminal gangs and actually facilitate the thefts.

The character of piracy in individual locations varies in detail from these generalizations. They are most appropriate for West Africa. Nigerian ports were infested with ship thieves until the government began to take decisive action against them in the mid-1980s. They then shifted their activities to other ports farther west. The situation is similar in the Straits of Malacca, where ships must slow down and become easy prey for pirates in canoes and small boats. Since over two hundred ships a day pass through the straits, there are plenty of targets and over 150 incidents were reported from here in 1981–1984 alone. Singapore controls piracy very well, and there are few incidents in her waters, but Indonesia does not and the nearby Riau Archipelago is a major pirate base.

In the Gulf of Siam, the South China Sea, and the Sulu Sea piracy is characterized by great cruelty and repeated violent attacks, especially on helpless refugees in small craft. In many cases Thai fishermen are part-time pirates, but they are also victims of pirates. Off the coast of Brazil, there are occasional reports of piracy, but only in the port of Santos is it more than sporadic and even there it is not a serious problem. Piracy around the Caribbean is

still different. There, gangs involved in narcotics trafficking steal yachts and other small boats to use in their smuggling operations, primarily on the coast of Colombia and among the Leeward and Windward islands.

This brief summary shows a clear pattern: rarely are ships attacked by "pirates" on the high seas outside the jurisdiction of any State. Now that most coastal States have declared exclusive economic zones of up to 200 nautical miles, crimes committed within these zones cannot be considered piracy since the zones have been removed from the classification of high seas. Even aircraft are rarely hijacked over international waters, and thus the term "air piracy" is probably not applicable in most cases. In fact, the term has not yet been used in criminal proceedings. Under the Tokyo Convention of 1963, the State of registry of an aircraft is competent to try such cases, and The Hague Convention of 1970 refers only to "the offense" of unauthorized seizure of an aircraft. The Montreal Convention of 1971 similarly avoids the word "piracy," and there has been no serious attempt since then to define aircraft hijacking legally as piracy.

There is, in fact, no public international law defining piracy, only national laws based on individual policies and situations. The nearest approaches to an international definition were in the 1958 Geneva Convention on the High Seas and in the 1982 United Nations Convention on the Law of the Sea (discussed in more detail in Chapter Thirty), which simply repeats the 1958 formulation verbatim. This definition is exceedingly difficult to interpret and leaves many gaps unfilled; it is therefore almost useless in combating "acts of violence or detention, or any act of depredation, committed for private ends by the crew or passengers of a private ship or aircraft."

Instead of relying on the Law of the Sea to combat piracy, however defined, States, in addition to using their own law enforcement mechanisms, are cooperating internationally. The International Maritime Bureau, sponsored by the International Chamber of Commerce, works with the International Maritime Organization (formerly the Intergovernmental Maritime Consultative Organization) to register and investigate both maritime fraud (a serious and growing problem) and piracy. Various shipowners' associations do similar work. In 1982, the United Nations High Commissioner for Refugees (UNHCR) began an antipiracy campaign, largely by helping Thailand to patrol the Gulf of Siam. In 1983, UNCTAD also began investigating "maritime fraud, including piracy." In 1985, the UNHCR shifted its emphasis to land-based enforcement of criminal laws and the United States established a Regional Anti-Piracy unit within the Refugee Section of her embassy in Bangkok.

All of this antipiracy activity may be having the desired effect, for the rate of pirate attacks dropped in 1986. We shall see whether this is the beginning of a long-term trend or merely a temporary respite. Piracy in either case is likely to continue to interfere with, and increase the cost of, international trade; be used for political purposes at times; and require international cooperation for its suppression. Political geographers could assist in this endeavor with careful studies and analyses of modern piracy.

TERRORISM

Like piracy, terrorism is illegal everywhere but still not adequately defined in international law. Rarely, if ever, is a person charged with "terrorism"; rather, the charge under domestic law is likely to be murder, kidnapping, assault, or some other common crime or crimes. Nevertheless, terrorism is of interest to us because it is politically motivated and is generally carried out across State boundaries. It differs from traditional guerrilla warfare (a topic studied to a very limited degree by a few political geographers) chiefly in that it is very small-scale, sporadic violence committed against civilians, often randomly, in order to intimidate governments and

achieve some political goal, rather than organized attacks by irregular or paramilitary forces against military personnel or facilities in order to achieve a military goal. Terrorists probably contributed to the success of some colonial wars in the period 1945–1980, but nowhere since then have they achieved their own stated objectives.

Terrorists can usefully be divided into two major categories: separatists and ideologists. We have already discussed nationalism and irredentism and later we will discuss religious, linguistic, and ethnic minorities in some detail. Here we only point out that, as we indicated in our discussion of non-State nations in Chapter 11, many aggrieved minority groups around the world have resorted to terrorism in order to call attention to their desire to separate from the State in which they find themselves, or at least to win greater autonomy within the State. The South Moluccans, Croatians, Bretons, Corsicans, Irish nationalists, Kurds, Tamils in Sri Lanka, Philippine Muslims (Moros), Karens and Kachins in Burma, Basques in Spain, Sikhs in Punjab, Puerto Rican nationalists, Palestinians, blacks in southern Sudan, Armenians, numerous groups in Lebanon, and many other minority peoples around the world have spawned terrorists.

The other major category, ideologists, is represented by radical leftists in France, West Germany, Italy, Japan, El Salvador, Guatemala, Peru, the Philippines, and elsewhere, and by radical rightists in El Salvador, Colombia, France, Israel, and other countries. There are also rebels against the national governments of Angola, South Africa, Mozambique, Burma, India, Nicaragua, and other countries, who combine nationalism, political ideology, and probably other motivations in their struggles. Some groups in all categories have been defeated, some are quiescent, some are active now, and some are nascent. Whatever their motivations, size, location, tactics, or other characteristics, they pose many problems not only for govern-

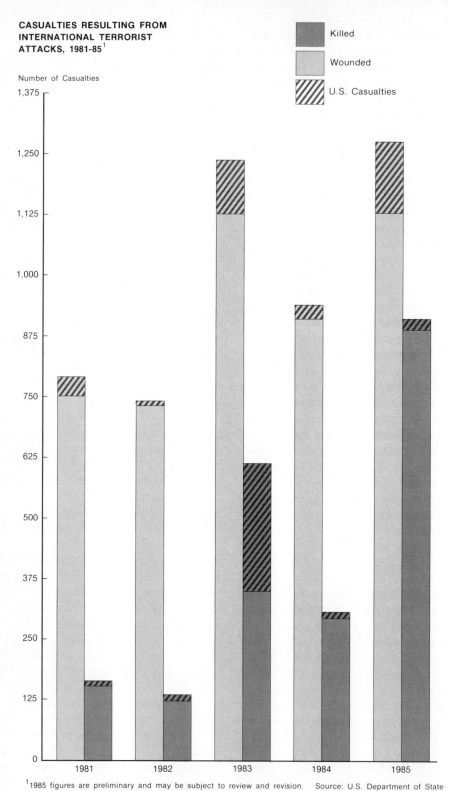

**CASUALTIES RESULTING FROM
INTERNATIONAL TERRORIST
ATTACKS, 1981-85**[1]

Number of Casualties

Killed

Wounded

U.S. Casualties

[1]1985 figures are preliminary and may be subject to review and revision.   Source: U.S. Department of State

Casualties resulting from international terrorist attacks, 1981–85.   Such casualties
are likely to continue to fluctuate from year to year but remain a problem for
some time. International cooperation against terrorism, however, will probably
result in its gradual reduction.

ments, but also for private citizens who become their victims.

While all these terrorists carry out their activities against governments (often indirectly), many of them are sponsored or aided by other governments. This State-sponsored terrorism (euphemistically called "low-intensity conflict") is not a new phenomenon either, but it has become dramatically more significant recently. There is reasonably impressive evidence, for example, that Libya has aided terrorists in Chile, Colombia, El Salvador, Guatemala, New Caledonia, Northern Ireland, Pakistan, the Philippines, Portugal, and Thailand as well as various Palestinian groups. Similarly, Syria, Iran, Cuba, and South Yemen have been accused by the United States of supporting terrorism in various countries around the world, mostly in the Middle East. Government support of terrorism includes such things as training, weapons, cash, passports, transportation, intelligence information, and even diplomatic cover—including use of the diplomatic pouch, diplomatic passports, and protection in embassies. This support complicates an already difficult problem.

Combating terrorists is not easy. Terrorists can attack anyone or anything at any time. But governments cannot protect everyone and everything all the time. Terrorists can also utilize the latest technology, which often enables them to do more damage at less cost with less risk than earlier methods. Modern transportation and communications not only enhance the mobility and security of terrorists, but magnify the impact of their actions through extensive media coverage. State terrorism, that is, terrorism carried out directly by States, is even more difficult to deal with, even if State actions can be legally defined as terrorism and even if the evidence of State terrorism is conclusive, neither of which happens very often. In 1983, North Korean agents in Rangoon killed 16 South Korean officials by setting off a bomb near them. But what kind of sanctions could be carried out against North Korea and by whom? Other countries, including the su-

perpowers, have also been accused of terrorism, but the actions in question are usually excused as self-defense, retaliation, or aid to a friendly government.

There has been some progress in combating terrorism through international cooperation. Some information on terrorists is exchanged through Interpol; there are occasional binational antiterrorist operations; the ICAO conventions (referred to earlier in connection with piracy) have helped reduce aerial hijackings; and the United Nations produced in 1985 the International Convention Against the Taking of Hostages. But mostly it has been a matter of individual States taking such actions individually as they see fit. The most flamboyant of such actions to date was the bombing by the United States of assorted targets in Libya in 1986, but there is no evidence that it had the desired effect. Israel's invasion of Lebanon in 1982 and subsequent expulsion of the Palestine Liberation Organization (PLO) from the country disrupted Palestinian terrorism for a time—but only for a time. It was not until June 1987 that the first high-level antiterrorist meeting of major industrial States (nine of them) took place (in Paris) to discuss collaboration against terrorism. We can be hopeful about the outcome of any collaboration that might result from this meeting—but not optimistic.

NARCOTICS TRAFFIC

Narcotics is big business. It is estimated that in 1985 alone, $50 *billion* worth of drugs, including marijuana, cocaine, heroin, LSD, amphetamines, and other controlled substances, were sold in the United States, the world's biggest market, but certainly not the only one. Worldwide, the figure can only be guessed at, but it is probably triple this amount. Besides this huge amount of money diverted from productive and constructive activities in nearly every inhabited area of the world—apparently including Antarctica!—drug production, processing, and distribution have generated inflation in poor countries, cor-

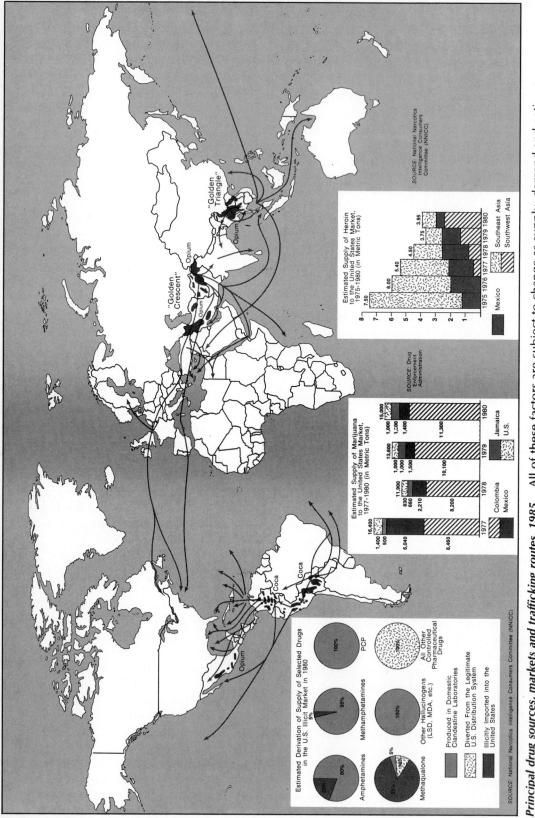

Principal drug sources, markets and trafficking routes, 1985. All of these factors are subject to change as supply, demand and anti-drug activities change. There is no sign yet, however, of a drop in the total demand for or supply of drugs.

rupted governments, ruined development plans, caused innumerable deaths
(not only from the drugs themselves, but
also from bullets and bombs), made a
mockery of many trade figures, and caused
untold suffering and misery.

The principal drug-producing areas of
the world are Mexico, some Caribbean islands, the lower Andean slopes and valleys, the Iran–Afghanistan–Pakistan
"Golden Crescent," and the Burma–Thailand–Laos "Golden Triangle." The chief
consuming areas are the United States and
a number of countries in Western Europe.
While this "North–South" trade does help
ever so slightly to redress the imbalance
of material wealth in the world, its negative effects are far more consequential. The
distinction between producer and consumer, for example, is rapidly disinte-
grating. The United States now produces
a very large proportion of the controlled
drugs it consumes, while everywhere that
there is drug production, processing, or
trafficking, local drug abuse begins. Drugs
are even infiltrating societies once thought
too moral, too tightly controlled, or too
remote to succumb to the lure. Today there
are few places in the world free of drug
abuse.

How can we account for the size and
pervasiveness of the drug business? Fundamentally, of course, it's a product of a
seemingly insatiable demand in societies
that can afford to indulge in a relatively
cheap escape from difficult problems, real
or imagined. We cannot here go into the
sociological causes of drug abuse, but we
can look briefly into the supply side of the
equation.

In some parts of the world, drugs, like
alcoholic beverages, have been used but
seldom seriously abused for centuries;
they are part of the local culture. They
include hallucinogens in northern Mexico
and southwest United States, coca in the
Altiplano and central Andes, hashish in the
Middle East, and opium in Asia from Turkey to China. In addition, very small
amounts of these drugs are used widely
for legitimate medicinal purposes: mor-

phine (derived from opium), marijuana,
and amphetamines, for example. Drug
abuse and drug trafficking became a serious international problem in the nineteenth century. Chinese resistance to European importation of opium into China
led to the Opium War of 1839–1842 between China and Great Britain, and the
first international drug-control treaty, the
International Opium Convention, was
signed in 1912. But only since the 1950s
has narcotraffic assumed overwhelming
proportions.

Coca, marijuana, and opium poppies are
all easy to grow, adaptable to different
physical environments, and immensely
profitable for poor farmers who are ill-pre-
pared to grow legitimate crops profitably.
Coca, for example, the shrub whose leaves
produce cocaine, produces five crops a
year with very little cultivation and generates up to 10 times the profits of other
crops. Peasants in many producing countries, especially in South America, have
resorted to violence to protect their crops
from drug enforcement officers. Drug-
plant eradication and crop-substitution
programs have frequently floundered and
even failed completely because of peasant resistance, little cooperation from
corrupted or intimidated local officials,
environmental concerns, or violent opposition from the traffickers themselves.
There is evidence, however, that these
programs are beginning to be more successful, notably in Turkey, Pakistan, and
Thailand.

But as production is reduced in one
place, it often simply moves to another.
Marijuana production, for example, began in Thailand about 1982, but the government clamped down on it quickly. It
thereupon moved across the Mekong to
Laos, where the government not only protects, but actually helps the growers. Marijuana is now the leading export from Laos
to the United States. In April 1984, the Co-
lombian minister of justice was murdered
by drug chieftains, prompting the government, at last, to mount a reasonably successful campaign against them. As coca

Coca leaves for sale in Bolivia. The peoples of the central Andes have long chewed these leaves to assuage their chronic hunger, thirst, cold, and fatigue and to make their lives endurable in this harsh environment. More recently, however, coca has become a cash crop to satisfy the demand in rich countries for cocaine. (UN)

production began to drop in Colombia, it began in Ecuador. By 1985, less than two years after production began, Ecuador ranked third in the world in coca production.

Similarly with the distribution system. As one route is disrupted by authorities, another is developed. Currently, for example, the traditional route from Pakistan to the United States and Europe through the Middle East is being disrupted by the turmoil in Lebanon, law enforcement efforts in Iran, and perhaps the Gulf War and other factors, so a new route is rapidly developing across Africa from Nairobi or Addis Ababa through West African ports. And as American surveillance and interdiction become more effective in the Caribbean, more drugs are being sent from South America overland and by air through Mexico directly to California.

Slowly, international cooperation is beginning to develop in combating drug production and distribution. The United States has won the reluctant support of some countries, notably Colombia, Bolivia, and Turkey, for bilateral control efforts. Various regional intergovernmental organizations are also beginning to take the problem seriously. In 1986 alone, it was considered by the Council of Arab Ministers of the Interior, ASEAN, the Commission of the EC, the Council of Europe, the OAS, the OECD, the Organization of the Islamic Conference, the SAARC, and others as well as by several regional and interregional conferences convened by the United Nations.

In 1961 and 1971, major UN conventions on drugs were adopted, though they have not yet been effectively implemented by many States. The UN Secretariat has a Division of Narcotic Drugs; there is a UN Fund for Drug Abuse Control and in Vienna the UN International Narcotics Control Board is very active. All three agencies, however, have been adversely affected by the United Nations' financial difficulties.

So far, international efforts to control the diversion of legal drugs into illicit channels have proven quite effective; there is minimal diversion at present. The illegal drugs, however, continue to vex the world. In June 1987, the United Nations held in Vienna an International Conference on Drug Abuse and Illicit Trafficking and its Commission on Narcotic Drugs was preparing a draft convention on the subject.

It is much too soon to see any positive results from these international drug-con-

Crop substitution in the opium-producing area of Thailand. One method of dealing with the drug problem is to assist farmers to substitute other crops for opium poppies, marijuana, and coca. Here the Thai government, with UN support, is delivering seeds by helicopter to Ban Phui, a key village in the roadless hill region of the north. (UNPDAC/Bangkok)

trol efforts. It is likely, however, that before long the international community will come to view narcotraffic as a worldwide scourge, like slavery and piracy used to be, and adopt a vigorous and effective program to suppress it. If not, the world is likely to face a clear and present danger much greater than that of nuclear warfare.

ARMS TRAFFIC

In Chapter Nineteen, we discussed the general problem of the almost unbelievable buildup of armaments in the world, both nuclear and conventional, and the perils of nuclear war. Here we are concerned with the almost equally difficult-to-believe, and perhaps even more frightening, problem of the dispersion of sophisticated conventional arms to virtually every part of the world. Competition among arms producers is so great and control over arms sales so weak that today nearly anyone can get nearly any kind of weapon he or she wants for nearly any purpose at "affordable" prices. Broadly speaking, there are two categories of international arms traffic: government-to-government transfers, including sales, loans, and gifts of weapons, and government approved, brokered and subsidized private sales to other governments; and the private, usually clandestine, generally small-scale, international arms market. We will briefly examine each of them.

According to the *SIPRI Yearbook*, during the period 1981–1985, some $66 *billion* worth of arms were traded more or less openly by governments. Of this total, the United States provided the greatest proportion by far, about 39 percent of total arms exports, followed by the Soviet Union with about 28 percent. Of the total, 64 percent is currently going to developing countries which can ill afford such staggering costs for unproductive imports. Of the arms suppliers to these developing countries, the Soviet Union is first with about 32 percent, followed closely by the United States with about 27 percent. But their combined share of such sales has been declining, from 69 percent in 1978–1982 to 59 percent in 1981–1985, as new suppliers enter the market.

The proliferation of sellers as well as buyers is a major new factor in the arms trade. China now ranks seventh in arms exports to developing countries, followed by Spain, Israel, and Brazil. Developing countries, in fact, are the fastest-growing arms exporters in the world, and most of their customers are other developing countries. It is questionable whether in the long run either buyers or sellers profit from such trade.

The figures given so far are all for "major conventional weapons." To them must be added many, many more billions of dollars worth of smaller personal and crew-served weapons (many extremely deadly), military equipment of all kinds, ammunition and explosives, spare parts, repair and modernization services, and so on. Then there is the export of military personnel. Nepal has traditionally earned significant amounts of hard currency by exporting tough "Gurkha" soldiers to serve in the Indian and British armies, but now Cuba, Pakistan, North Korea, and other developing countries are sending their own regular military personnel abroad to serve as instructors, guards, advisors, and even (as in the case of Cuba) to engage in combat. It is estimated that in 1983, for example, Pakistan had thirty thousand troops serving in 24 countries, mostly in Africa and the Arabian peninsula, twenty thousand of them in Saudi Arabia.

The rate of increase of world military expenditures, especially by developing countries, has slowed since 1982, apparently owing to a number of factors: (1) the drop in oil prices and the worldwide recession of the 1980s have reduced the disposable funds available for arms purchases; (2) many developing countries are already saturated with weapons; and (3) a number of developing countries, as we have seen, have become producers and even exporters of arms. But Iran, Pakistan, India, and a few other countries are still increasing their arms purchases. Iran alone is spending $15 to $20 billion a year on "defense," nearly as much as France or West Germany! Whenever there is discussion of arms control or disarmament, it must be remembered that despite all the rhetoric from the "Third World" or "nonaligned States" or "peace-loving" developing countries about nuclear-free zones or nuclear proliferation, there is little support among them for conventional arms control.

Of great and growing importance is the apparent trend of governments, especially in Western Europe, toward relaxing their own controls over the arms trade and gradually reducing their subsidies and supervision of arms production. Cooperation among Western European governments in arms production and distribution

is growing, but so is competition among private suppliers, to say nothing of the new producers entering the market. There is also a trend now toward the transfer of arms-production technology, for both political and economic reasons, and this accelerates the decentralization of weapons production in the world.

All of these trends reinforce the importance of the private trade in conventional arms. Today, with the proper connections and enough cash, any revolutionary group, any separatist group, any local warlord, any fanatical hate group can buy on the open or black market virtually any conventional weapon short of sophisticated aircraft, naval vessels, armor, and artillery. Weapons are openly advertised in widely circulated magazines, and international arms fairs showcase the latest military equipment available for sale to virtually anyone, with few questions asked. In an already unstable world, such activities can hardly lead to greater stabilization.

What should or can be done about the situation? The decision makers are all working for individual governments, and few of them seem very concerned about it. Perhaps political geographers can increase their awareness of the problem through careful studies and recommendations.

LINKAGES

While each of these types of criminal activity is interesting in itself and in varying degree significant in international relations, what makes them especially susceptible to politicogeographical analysis is the linkages among them. These linkages are becoming so complex and pervasive that, if the trend continues, it may not be long before there is one global network of criminal activity ranging from mugging and petty theft to major revolutionary operations. This is a frightening prospect and for that reason alone likely to be thwarted by governments acting more and more in concert. Meanwhile, however, the criminals are affecting—even upsetting—many of the politicogeographical patterns discussed in this book.

We begin our discussion of linkages with the international traffic in arms, both government-to-government transfers and the commercial market, since all of these criminal activities involve the use of weapons. Generally, modern pirates are lightly armed and do not yet appear to be participating in the arms trade, but terrorists and narcotics traffickers certainly are. Except for small arms and simple explosive devices, most of the weapons they use must be provided directly or indirectly by governments. Some are stolen from military armories and ammunition dumps and some are surplus sold at auction or otherwise openly available. More and more, however, terrorists of various persuasions and purveyors of drugs of various degrees of lethality are deliberately supplied with new and sometimes fairly sophisticated weapons, ammunition, explosives, communications equipment, and other paraphernalia. Particularly in South America and South Asia, there have been numerous all-out battles between narcotics traffickers and politically motivated terrorists, sometimes allied in the same operation, and soldiers and/or police and other law enforcement authorities. In these battles, the bad guys sometimes have more and bigger weapons and the good guys don't always win.

There is even evidence that some governments are engaging in drug traffic themselves. Not just an occasional corrupt official or trafficker infiltrated into the government, but the government as a matter of policy controlling the flow of drugs through its territory and extracting commissions for its services, or the government raising cash by selling drugs, or affording protection to the traffickers in return for cash payments, intelligence about rival gangs, or assistance in other areas. In a number of weak, poor, unstable countries, renegade or mutinous troops have gone into the drug business, taking their weapons with them and sometimes forming alliances with local revo-

lutionaries. Very soon after being expelled from China to Burma in 1949, for example, several divisions of Chinese Nationalist troops fought local rebels, bandits, and opium traffickers until they gradually took over most of the illegal activities there and settled down as warlords in much of northern Burma. Since then similar patterns have been noted in Uganda, Afghanistan, and elsewhere.

Much more important, the Soviet Union and her allies and, as noted earlier, Iran, Syria, and Libya, are known to be supplying material and other assistance to various groups seeking to destabilize unfriendly or neutral governments. The United States also engages in these activities, though perhaps on a smaller scale, and on a still smaller scale, so do, or have, Israel, South Africa, Turkey, Iraq, Taiwan, and perhaps other countries.

The linkages among insurgents, terrorists, drug producers, drug traffickers, and common criminals is well documented. In Burma, Afghanistan, Colombia, Peru, Bolivia, Jamaica, the Philippines, and elsewhere, they exchange drugs for weapons, provide intelligence and protection for one another, help one another "launder" money, and otherwise collaborate, ignoring ideological differences in the joint pursuit of profits. It is doubtful that the average user of cocaine or marijuana or any other drug in the United States or Canada or Western Europe truly understands that a portion of the money he or she pays for a moment's pleasure is used to destroy the lives of other people, not only slowly through the provision of more drugs, but violently through murders of rivals, assassination of government officials, slaughter of uncooperative peasants, and killing of law enforcement officers no older than themselves.

Terrorists have also engaged in actions which could be considered piracy. The Irish Republican Army, the Polisario Front in Western Sahara, the Basque ETA, the Moro Liberation Front and the New People's Army in the Philippines have all attacked ships in port and at sea. In 1985, Arab terrorists seized the Italian cruise ship Achille Lauro and held it for weeks, killing an American passenger and terrorizing the others until they were finally allowed to escape without having achieved any of their objectives, except gaining more publicity. Airplane hijacking has been a popular terrorist action but seems to be declining now because of increased security and diminishing success.

To summarize: International criminal activity, whether labeled as drug running, arms trafficking, terrorism, or piracy, is likely to be closely linked with all of the other categories and aided, actively or passively, by some governments. There are those who try to justify any or all of these activities or excuse them on sociological or ideological grounds, but their impact on civilized society is incalculable and inexcusable. Bringing the question down to an individual level, we can do no better than to paraphrase Golda Meir, late prime minister of Israel, commenting on Arab terrorism: It is possible that somewhere there is a cause worth dying for, but there can surely be no cause worth killing for.

REFERENCES FOR PART SIX

Books and Monographs

A

Abi-Saab, Georges, *The United Nations Operation in the Congo, 1960–1964*. Oxford: Oxford Univ. Press, 1978.

Addona, A.F., *The Organisation of African Unity*. Cleveland: World Publishing, 1969.

Adeniran, Tunde and Yonah Alexander (eds.), *International Violence*. New York: Praeger, 1983.

Agrawala, S.K. and others, *New Horizons of International Law and Developing Countries*. Bombay: N.M. Tripathi, 1983.

Akinsanya, Adeoye A., *The Expropriation of Multinational Property in the Third World*. New York: Praeger, 1980.

———, *Multinationals in a Changing Environment: A Study of Business–Government Relations in The Third World*. New York: Praeger, 1984.

Akzin, Benjamin, *New States and International Organizations*. Paris: UNESCO, 1955.

Alexander, Yonah and others, *Terrorism: Theory and Practice*. Boulder, CO: Westview, 1979.

Alexander, Yonah and Robert A. Friedlander (eds.), *Self-Determination: National, Regional, and Global Dimensions*. Boulder, CO: Westview, 1980.

Alexander, Yonah and Charles K. Ebinger (eds.), *Political Terrorism and Energy: The Threat and Response*. New York: Praeger, 1982.

Alexander, Yonah and Kenneth Myers (eds.), *Terrorism in Europe*. New York: St. Martin's Press, 1982.

Alexander, Yonah and Allan Nanes (eds.), *Legislative Responses to Terrorism*. Dordrecht, Neth.: Nijhoff, 1986.

Alexandrowicz, Charles Henry, *The Law-Making Functions of the Specialised Agencies of the United Nations*. London: Angus & Robertson, 1973.

Allen, Beverly A. and Christian S. Ward (eds.), *The United States, Transnational Business, and the Law*. London, Rome, and New York: Oceana, 1985.

al-Sowayegh, Abdulaziz, *Arab Petro-Politics*. New York: St. Martin's Press, 1984.

Aluko, Olajide and Timothy M. Shaw (eds.), *Southern Africa in the 1980's*. London: Allen & Unwin, 1985.

Alves, Dora, *The ANZUS Partners*. CSIS Significant Issues Series, Vol. 6, No. 8, Lanham, MD: Univ. Press of America, 1984.

Amate, C.O.C., *Inside the OAU; Pan-Africanism in Practice*. New York: St. Martin's Press, 1986.

Anand, R.P., *Asian States and the Development of Universal International Law*. Delhi: Vikas, 1972.

———, *International Law and the Developing Countries*. Dordrecht, Neth.: Kluwer, 1987.

Andemicael, Berhanykun, *Peaceful Settlement Among African States: Roles of the United Nations and the Organization of African Unity*. New York: UNITAR, 1972.

———, *Regionalism and the United Nations*. Dobbs Ferry, NY: Oceana, 1979.

Andic, Fuat M. and others, *A Theory of Economic Integration for Developing Countries, Illustrated by Caribbean Countries*. London: Allen & Unwin, 1971.

Ansair, Javed, *The Political Economy of International Economic Organization*. Boulder, CO: Lynne Rienner, 1986.

Archer, Clive, *International Organizations*. London: Allen & Unwin, 1983.

Arlinghaus, Bruce E., *Arms for Africa: Military Assistance and Foreign Policy in the Developing World*. Lexington, MA; Toronto: Heath, 1983.

Arms Production in the Third World. London: Taylor & Francis, 1985.

Arms Sales: A Useful Foreign Policy Tool? Washington: American Enterprise Institute for Public Policy Research, 1982. Panel moderated by John Charles Daly.

Armstrong, David, *The Rise of the International Organization*. New York: St. Martin's Press, 1983.

ASEAN and Pacific Economic Co-operation. Development Papers No. 2. Bangkok: ESCAP, 1983.

Axline, Andrew W., *Underdevelopment, Dependence and the Politics of Integration: A Caribbean Application*. Ottawa: Univ. of Ottawa, 1977.

———, *Caribbean Integration: The Politics of Regionalism*. London: Frances Pinter; New York: Nichols, 1979.

B

Baehr, Peter R. and others, *United Nations; A Reality and Ideal*. Praeger, 1984.

Bahbah, Bishara and Linda Butler, *Israel and Latin America*. New York: St. Martin's Press, 1986.

Bailey, Martin, *Oilgate; Sanctions Scandal*. London: Hodder & Stoughton, 1979.

Bailey, Sidney D., *The General Assembly of the United Nations*. New York: Praeger, 1964.

Baldwin, Robert E., *Nontariff Distortions of International Trade*. Washington: Brookings Institution, 1970.

Ball, M. Margaret, *The OAS in Transition*. Durham, NC: Duke Univ. Press, 1969.

———, *The Open Commonwealth*. Durham, NC: Duke Univ. Press, 1971.

Barnet, Richard J. and Ronald E. Muller, *Global Reach; The Power of the Multinational Corporations*. New York: Simon & Schuster, 1974.

Becker, Jillian, *The PLO*. New York: St. Martin's Press, 1984.

Behrham, Jack N. and Raymond F. Mikesell, *The Impact of U.S. Foreign Direct Investment on U.S. Export Competitiveness in Third World Markets*. CSIS Significant Issues Series, Vol. 2, No. 1. Lanham, MD: Univ. Press of America, 1980.

Bell, J. Bowyer, *Transnational Terror*. Lanham, MD: Univ. Press of America, 1975.

———, *The Gun in Politics*. New Brunswick, NJ: Transaction Books, 1987.

Bender, Gerald J. (ed.), *International Affairs in Africa*. Newbury Park, CA: Sage, 1987.

Benoit, Emile, *Europe at Sixes and Sevens: The Common Market, the Free Trade Association and the United States*. New York: Columbia Univ. Press, 1962.

Berber, Friedrich J., *Rivers in International Law*. London: Stevens, 1959.

Beres, Louis René, *Terrorism and Global Security: The Nuclear Threat*. Boulder, CO: Westview, 1979.

Bergsten, C. Fred and William R. Cline, *The United States–Japan Economic Problem*. Washington: Institute for International Economics, 1987.

Bergsten, C. Fred and others, *Auction Quotas and US Trade Policy*. Washington: Institute for International Economics, 1987.

Bindoff, S.T., *The Scheldt Question*. London: Allen & Unwin, 1945.

Blackwelder, Brent and Peter Carlson, *Survey of Water Conservation Programs in the Fifty States*. Washington: Island Press, 1984.

Boardman, Robert and others, *Europe, Africa and Lome III*. Dalhousie African Studies Series 3. Lanham, MD: Univ. Press of America, 1985.

Bokor-Szegö, Hanna, *The Role of the United Nations in International Litigation*. Amsterdam, Oxford, and New York: North-Holland, 1978.

Bonn, M.J., *Whither Europe—Union or Partnership?* New York: Philosophical Library, 1952.

Borrmann, Axel and others, *The EC's Generalized System of Preference*. The Hague, Boston, London: Nijhoff, 1981.

Bowett, D.W., *The Law of International Institutions*, 2nd ed. London: Stevens, 1982.

Boxer, Baruch, *Israeli Shipping and Foreign Trade*. Research Paper No. 48, Chicago: Univ. of Chicago, Dept. of Geography, 1957.

Bracewell-Milnes, Barry, *Eastern and Western European Economic Integration*. New York: St. Martin's Press, 1976.

Braillard, Philippe and Mohammad-Reza Djalili, *The Third World and International Relations*. London: Frances Pinter, 1986.

Brewster, H. and C.Y. Thomas, *The Dynamics of West Indian Economic Integration*. Mona, Jamaica: Univ. of the West Indies, Institute of Social and Economic Research, 1967.

Bronckers, M.C.E.J., *Selective Safeguard Measures in International Trade Relations: Issues of Protectionism in GATT*. Deventer, Neth.: Kluwer; The Hague: T.M.C. Asser Instituut, 1985.

Brown, W. Norman, *The United States and India and Pakistan*. Cambridge: Harvard Univ. Press, 1953.

Brumter, Christian, *The North Atlantic Assembly*. Dordrecht, Boston, Lancaster: Nijhoff, 1986.

Brzoska, Michael and Thomas Ohlson, *Arms Transfers to the Third World, 1971–85*. Oxford and New York: Oxford Univ. Press, 1987. A publication of the Stockholm International Peace Research Institute (SIPRI).

Buergenthal, Thomas, *Law-Making in the International Civil Aviation Organisation*. Syracuse, NY: Syracuse Univ. Press, 1969.

Butler, Nick, *The International Grain Trade*. New York: St. Martin's Press, 1986.

Butler, William E. (ed.), *A Source Book of Socialist International Organizations*. Alphen aan den Rijn, Neth.: Sijthoff & Noordhoff, 1978.

C

Cahn, Anne Hessing and others, *Controlling Future Arms Trade.* New York: McGraw-Hill, 1977.

Calderón Cruz, Angel, *Problemas del Caribe Contemporáneo/Contemporary Caribbean Issues.* Río Piedras: Univ. of Puerto Rico, 1979.

Callaghy, Thomas M. (ed.), *South Africa in Southern Africa; The Intensifying Vortex of Violence.* New York: Praeger, 1983.

Camps, Miriam and William Diebold, Jr., *The New Multilateralism: Can the World Trading System Be Saved?* New York: Council on Foreign Relations, 1983.

Cannizo, Cindy (ed.), *The Gun Merchants: Politics and Policies of the Major Arms Suppliers.* Elmsford, NY: Pergamon, 1980.

Carl, Beverly May, *Economic Integration Among Developing Nations: Law and Policy.* Praeger, 1986.

Carlson, Jack and Hugh Graham, *The Economic Importance of Exports to the United States,* CSIS Significant Issues Series, Vol. 2, No. 5. Lanham, MD: Univ. Press of America, 1980.

Carlson, Lucille, *Geography and World Politics.* Englewood Cliffs, NJ: Prentice-Hall, 1958.

Carter, Gwendolen M. and Patrick O'Meara (eds.), *International Politics in Southern Africa.* Bloomington: Indiana Univ. Press, 1982.

Cassen, R.H., *Soviet Interests in the Third World.* Newbury Park, CA: Sage, 1986.

Cervenka, Zdenek (ed.), *Land-locked Countries of Africa.* Uppsala, Swed.: Scandinavian Institute of African Studies, 1973.

Chaliand, Gerard, *Terrorism: From Popular Struggle to Media Spectacle.* Atlantic Highlands, NJ: Humanities Press International, 1987.

Chapman, J.D. (ed.), *The International River Basin.* Vancouver: Univ. of British Columbia, 1963.

Chatterjee, S.K., *Legal Aspects of International Drug Control.* The Hague, Boston, London: Nijhoff, 1981.

Chauhan, Balbir R., *Settlement of International Water Law Disputes in International Drainage Basins.* Berlin: Erich Schmidt Verlag, 1981.

Chen, Samuel Shih-Tsai, *Basic Documents of International Organization,* rev. ed. Dubuque, IA: Kendall/Hunt, 1979.

Chernick, Sydney E., *The Commonwealth Caribbean: The Integration Experience.* Balti-more: Johns Hopkins Univ. Press for the World Bank, 1978.

Chetley, Andrew, *The Politics of Baby Foods: Successful Challenges to an International Marketing Strategy.* London: Frances Pinter, 1986.

Chiang, Pei-heng, *Non-governmental Organizations at the United Nations.* New York: Praeger, 1981.

Clark, Robert P., *The Basque Insurgents: ETA, 1952–1980.* Boulder, CO: Westview, 1984.

Cline, R. and Yonah Alexander, *Terrorism: The Soviet Connection.* London: Taylor & Francis, 1984.

Cline, William R., (ed.), *Trade Policy in the 1980's.* Washington: Institute for International Economics, 1984.

Clout, Hugh, *Regional Variations in the European Community.* New York: Cambridge Univ. Press, 1986.

Cobbe, James H., *Government and Mining Companies in Developing Countries.* Boulder, CO: Westview, 1979.

Coker, Christopher, *NATO, the Warsaw Pact and Africa.* New York: St. Martin's Press, 1986.

Codding, George Arthur, Jr., *The Universal Postal Union.* New York: New York Univ. Press, 1964.

Colby, Charles Carlyle (ed.), *Geographical Aspects of International Relations.* New York: Books for Libraries Press, 1969.

Cordes, Bonnie and others. *Trends in International Terrorism, 1982 and 1983.* Santa Monica, CA: RAND Corp., 1984.

Cottrell, Alvin J. and others, *Arms Transfers and U.S. Foreign and Military Policy.* CSIS Significant Issues Series, Vol. 1, No. 7. Lanham, Md: Univ. Press of America, 1980.

Crassweller, Robert D., *The Caribbean Community: Changing Societies and U.S. Policy.* New York: Praeger, 1972.

Crelinsten, Ronald D. and Denis Szabo, *Hostage-Taking.* Lexington, MA: Heath, 1979.

Cuddington, John T. and others, *Tariffs, Quotas and Trade: The Politics of Protectionism.* San Francisco: ICS Press, 1987.

Curtis, Michael, *Western European Integration.* New York: Harper & Row, 1965.

Curtis, Michael and Susan Aurelia Gitelson (eds.), *Israel in the Third World.* New Brunswick, NJ: Transaction Books, 1977.

D

Dale, William, *The Modern Commonwealth.* London: Butterworths, 1983.

Daoudi, M.S. and M.S. Dajani, *Economic Sanc-*

tions: Ideals and Experience. London, Boston, Melbourne and Henley, Oxfordshire, Eng.: Routledge & Kegan Paul, 1983.

Davies, Rob J., *Trade Sanctions and the Regional Impact in Southern Africa.* London: Africa Bureau 1981.

Day, John C., *Managing the Lower Rio Grande.* Chicago: Univ. of Chicago Press, 1970.

Dean, Vera M., *The Nature of the Non-Western World.* New York: New American Library, Mentor Books, 1957.

Deener, David R., (ed.), *Canada–United States Treaty Relations.* Durham, NC: Duke University Press, 1963.

Degenhardt, Henry W., (compiler), *Treaties and Alliances of the World.* Gale Research Co., 1987.

de Lattre, Anne and Arthur M. Fell, *The Club du Sahel; An Experiment in International Cooperation.* Paris: OECD, 1984.

Dell, Sidney, *Trade Blocs and Common Markets.* New York: Knopf, 1963.

Demas, William Gilbert, *From CARIFTA to Caribbean Community.* Georgetown, Guyana: Commonwealth Caribbean Regional Secretariat, 1972.

———, *Essays on Caribbean Integration and Development.* Jamaica: Univ. of the West Indies, Institute of Social and Economic Research, 1976.

Destler, I.M., *American Trade Politics: System Under Stress.* Washington: Institute for International Economics, 1986.

Deutsch, Karl W. and others, *Political Community and the North Atlantic Area.* Princeton, NJ: Princeton Univ. Press, 1957.

Díaz-Alejandro, Carlos F. and Gerald K. Helleiner, *Handmaiden in Distress: World Trade in the 1980's.* Ottawa: North–South Institute; Washington: Overseas Development Council; London: Overseas Development Institute, 1982.

Diebold, William, *The Schuman Plan: A Study in Economic Cooperation, 1950–1959.* New York: Praeger, 1959.

Donald, Sir Robert, *The Polish Corridor and the Consequences.* London: Butterworths, 1929.

Donaldson, Robert H., *The Soviet Union in the Third World.* Boulder, CO: Westview, 1981.

Dore, Isaak I., *The International Mandate System and Namibia.* Boulder, CO: Westview, 1985.

Dorsey, Gray L., *Beyond the United Nations: Changing Discourse in International Politics*

and Law, Vol. V. Lanham, MD: Univ. Press of America, 1986.

Doxey, Margaret, *The Contemporary World; Maps, Facts and a Guide to International Organizations.* Peterborough, Ontario: Trent Univ., Dept. of Political Studies, 1984.

DuBois, Victor D., *Crisis in OCAM.* American Universities Field Staff Reports, West Africa Series, Vol. 14, No. 2. 1972

Dunning, J.H., *International Production and the Multinational Enterprise.* London: Allen & Unwin, 1981.

Dupree, Louis, *Pakistan 1964–1966, Part IV: The Regional Cooperation for Development.* American Universities Field Staff Reports, South Asia Series, Vol. 10, No. 8. 1966.

Dupuy, René-Jean, *The Future of International Law in a Multicultural World.* The Hague, Boston, London: Nijhoff, 1984.

E

El-Agraa, A.M. (ed.), *The Economics of the European Community,* 2nd ed. New York: St. Martin's Press, 1985.

El-Agraa, A.M. and A.J. Jones, *The Theory of Custom Unions.* New York: St. Martin's Press, 1981.

Elias, T. Olawale, *Africa and the Development of International Law.* Leiden, Neth.: Sijthoff, 1972.

Ellen, Eric F. (ed.), *Violence at Sea; A Review of Terrorism, Acts of War and Piracy, and Countermeasures to Prevent Terrorism.* Paris: ICC Publishing (International Chamber of Commerce), 1986.

Emery, James J., *Technology Trade with The Middle East: Policy Issues.* Boulder, CO: Westview, 1985.

Evans, Ernest, *Calling a Truce to Terror: The American Response to International Terrorism.* Westport, CT: Greenwood Press, 1979.

Evans, John W., *The Kennedy Round in American Trade Policy; The Twilight of the GATT?* Cambridge: Harvard Univ. Press, 1971.

Ezenwe, Uka, *ECOWAS and the Economic Integration of West Africa.* New York: St. Martin's Press, 1983.

F

Fabian, Larry L., *Soldiers Without Enemies: Preparing the United Nations for Peacekeeping.* Washington: Brookings Institution, 1971.

Falk, Richard A. and Saul H. Mendelovitz, *Regional Politics and World Order.* San Francisco: Freeman, 1973.

Farer, Tom J., *The United States and the Inter-*

American System: Are There Functions for the Forms? Washington: American Society of International Law, 1978.

———— (ed.), *The Future of the Inter-American System.* New York, London, Sydney, and Toronto: Praeger, 1979.

Farhi, David, *The Limits of Dissent: Facing the Dilemmas Posed by Terrorism.* Lanham, MD: Univ. Press of America, 1977.

Farley, Philip J. and others, *Arms Across the Sea.* Washington: Brookings Institution, 1978.

Fedorowicz, Jan K., *East–West Trade in the 1980's.* Boulder, CO: Westview, 1987.

Fehrenbach, T.R., *The United Nations in War and Peace.* New York: Random House, 1968.

Feld, Werner J., *Multinational Corporations and U.N. Politics; The Quest for Codes of Conduct.* Elmsford, NY: Pergamon, 1980.

Feld, Werner J. and Robert S. Jordan, with Leon Hurwitz, *International Organizations: A Comparative Approach.* New York: Praeger, 1983.

Fennell, Rosemary, *The Common Agricultural Policy of the European Community.* Totowa, NJ: Rowman & Littlefield, 1980.

Fenwick, Charles G., *The Inter-American Regional System.* New York: McMullen, 1949.

Ferencz, Benjamin B., *Enforcing International Law—A Way to World Peace: A Documentary History and Analysis.* London, Rome, and New York: Oceana, 1983.

Fifer, J. Valerie, *Bolivia: Land, Location, and Politics Since 1825.* Cambridge: At the University Press, 1972.

Fishel, Wesley R., *The End of Extraterritoriality in China.* Berkeley: Univ. of California Press, 1952.

Fisher, Bart S. and Jeff Turner (eds.), *Regulating the Multinational Enterprise: National and International Challenges.* New York: Praeger, 1983.

Fisher, Bart S. and Kathleen M. Harte (eds.), *Barter in the World Economy.* New York: Praeger, 1985.

Florinsky, Michael T., *Integrated Europe?* New York: Macmillan, 1955.

Foltz, William J. and Henry S. Bienen, *Arms and the African.* New Haven, CT: Yale Univ. Press.

Fontaine, Roger W., *The Andean Pact: A Political Analysis.* Washington Papers, Vol. 5, No. 45, Lanham, MD: Univ. Press of America, 1977.

Forman, Brenda, *America's Place in the World Economy.* New York: Harcourt, Brace and World, 1969.

Franck, Thomas M., *East African Unity Through Law.* New Haven, CT: Yale Univ. Press, 1964.

————, *Judging the World Court.* New York: Priority Press, 1986.

Franko, Lawrence G. and Sherry Stephenson, *French Export Behavior in Third World Markets.* CSIS Significant Issues Series, Vol. 2, No. 6. Lanham, MD: Univ. Press of America, 1980.

Freedman, Lawrence and others, *Terrorism and International Order.* London: Routledge & Kegan Paul, 1986.

Freeman, D.B., *International Trade, Migration and Capital Flows.* Research Paper No. 146, Chicago: Univ. of Chicago, Dept. of Geography, 1973.

Freymond, Jean F., *Political Integration in the Commonwealth Caribbean.* Geneva: Institut Universitaire de Hautes Études Internationales, 1980.

Fried, Edward R. and Philip H. Trezise (eds.), *U.S.–Canadian Economic Relations: Next Steps?* Washington: Brookings Institution, 1984.

Friedlander, Robert, *Terrorism: Documents of International and Local Control,* Vols. 1 and 2. Dobbs Ferry, NY: Oceana, 1979.

————, *Terrorism: Documents of International and Local Control,* Vol. 3. London, Rome, and New York: Oceana, 1981.

————, *Terror-Violence: Aspects of Social Control.* London, Rome, and New York: Oceana, 1983.

G

Gal-Or, Naomi, *International Cooperation to Suppress Terrorism.* New York: St. Martin's Press, 1985.

Gandia, Delsie, *The EEC's Generalized Scheme of Preferences and the Yaoundé and other Agreements.* Totowa, NJ: Alanheld, 1981.

Gauhan, Altaf (ed.), *Regional Integration; The Latin American Experience.* London: Third World Foundation for Social and Economic Studies; Boulder, CO: Westview, 1985.

Geiser, Hans J. and others, *Legal Problems of Caribbean Integration; A Study on the Legal Aspects.* Leyden, Neth.: Sijthoff; St. Augustine, Trinidad: Institute of International Relations, 1976.

Gereffi, G., *The Pharmaceutical Industry and Dependency in the Third World.* Princeton, NJ: Princeton Univ. Press, 1983.

Gerson, L.L., *Woodrow Wilson and the Rebirth of Poland.* New Haven, CT: Yale Univ. Press, 1953.

Gervasi, Tom, *Arsenal of Democracy: American Weapons Available for Export*. New York: Grove Press, 1977.

Ginther, K. and W. Benedek (eds.), *New Perspectives and Conceptions of International Law: An Afro–European Dialogue*. Vienna and New York: Springer-Verlag, 1983.

Glassner, Martin Ira, *Access to the Sea for Developing Land-locked States*. The Hague: Nijhoff, 1970.

———, *Bibliography on Land-locked States*, 2nd ed. Dordrecht, Neth.: Nijhoff, 1986.

Glassner, Martin Ira, "CARICOM and the Future of the Caribbean," in Gary S. Elbow, ed. *International Aspects of Development in Latin America: Geographical Perspectives*. Muncie, IN: CLAG Publications, 1977, 111–117.

Glick, Leslie Alan, *Multilateral Trade Negotiations; World Trade After the Tokyo Round*. Totowa, NJ: Rowman & Allanheld, 1984.

Goblet, Y.M., *Political Geography and the World Map*. New York: Praeger, 1955.

Godana, Bonaya Adhi, *Africa's Shared Water Resources: Legal and Institutional Aspects of the Nile, Niger and Senegal River Systems*. London: Frances Pinter; Boulder, CO: Lynne Rienner, 1985.

Golubev, G. and A. Biswas, *Large Scale Water Transfers*. London: Taylor & Francis, 1985.

Goodrich, Leland Matthew, *United Nations*. London, 1960.

———, *The United Nations in a Changing World*. New York: Columbia Univ. Press, 1974.

Gordon, Colin, *The Atlantic Alliance: A Bibliography*. London: Frances Pinter, 1978.

Gray, Peter H., *Free Trade or Protection?* New York: St. Martin's Press, 1985.

The Great Tin Crash; Bolivia and the World Tin Market. London: Latin American Bureau; New York: Monthly Review Press, 1987.

Green, N.A. Maryan, *International Law: Law of Peace*. Plymouth, England: Macdonald & Evans, 1982.

Greenaway, David (ed.), *Current Issues in International Law*. New York: St. Martin's Press, 1986.

Gregg, Robert W. (ed.), *International Organization in the Western Hemisphere*. Syracuse, NY: Syracuse Univ. Press, 1968.

Grewlich, Klaus W., *Transnational Enterprises in a New International System*. Alphen aan den Rijn: Sijthoff & Noordhoff, 1980.

Grieco, Joseph M., *Between Dependency and Autonomy: India's Experience with the International Computer Industry*. Berkeley, Los Angeles, and London: Univ. of California Press, 1984.

Gruhn, Isebill V., *Regionalism Reconsidered: The Economic Commission for Africa*. Boulder, CO: Westview, 1979.

Grunwald, Joseph and others, *Latin American Integration and U.S. Policy*. Washington: Brookings Institution, 1972.

Grunwald, Joseph and Kenneth Flamm, *The Global Factory: Foreign Assembly in International Trade*. Washington: Brookings Institution, 1985.

Grzybowski, Kazimierz, *Soviet Public International Law; Doctrines and Diplomatic Practice*. Leyden, Neth.: Sijthoff, 1970.

H

Haas, Ernst, *The Uniting of Europe*. Stanford, CA: Stanford University Press, 1957.

Hadley, Eleanor M., *Japan's Export Competitiveness in Third World Markets*. CSIS Significant Issues Series, Vol. 3, No. 2. Lanham, MD: Univ. Press of America, 1981.

Haines, C. Grove, *European Integration*. Baltimore: Johns Hopkins Univ. Press, 1957.

Hall, Duncan, *The British Commonwealth of Nations*. London: Methuen, 1920.

Hall, R. Duane, *Overseas Acquisitions and Mergers: Combining for Profits Abroad*. New York, Westport, CT and London: Praeger, 1986.

Hallstein, Walter, *United Europe: Challenge and Opportunity*. Cambridge: Harvard Univ. Press, 1962.

Halperin, Ernst, *Terrorism in Latin America*. Lanham, MD: Univ. Press of America, 1976.

Hamilton, Geoffrey (ed.), *Red Multinationals or Red Herrings? The Activities of Enterprises from Socialist Countries in the West*. London: Frances Pinter, 1986.

Hammond, Paul Y. and others, *The Reluctant Supplier: U.S. Decisionmaking for Arms Sales*. Cambridge, MA: Oelgeschlager, Gunn & Hain, 1983.

Han, Henry H., *Terrorism, Political Violence and World Order*. Lanham, MD: Univ. Press of America, 1984.

Hansen, Roger D., *Central America: Regional Integration and Economic Development*. Washington: National Planning Association, 1967.

Harkavy, Robert E. and Stephanie Newman (eds.), *International Arms Transfers*. New York: Praeger, 1979.

Harrell-Bons, Barbara, *ECOWAS: The Eco-*

nomic Community of West African States. American Universities Field Staff Reports, Africa Series, No. 6, 1979.

Harris, Richard G., Trade, Industrial Policy and International Competition. Toronto: Univ. of Toronto Press, 1987.

Hart, Michael M., Canadian Economic Development and the International Trading System. Toronto: Univ. of Toronto Press, 1987.

Hathaway, Dale E., Agriculture and the GATT: Issues in a New Trade Round. Washington: Institute for International Economics, 1987.

Hawley, James P., Dollars and Borders: U.S. Government Attempts to Restrict Capital Flows, 1960–1980. Armonk, NY: Sharpe, 1986.

Hazari, International Trade. New York: Columbia Univ. Press, New York Univ. Press, 1986.

Henig, Ruth B. (ed.), The League of Nations. New York: Harper & Row, 1973.

Hetzel, Nancy K., Environmental Cooperation Among Industrial Countries: The Role of Regional Organizations. Lanham, MD: Univ. Press of America, 1980.

Higgins, Rosalyn, The Development of International Law Through the Political Organs of the United Nations. London: Oxford Univ. Press, 1963.

———, United Nations Peacekeeping Documents and Commentary, Vol. IV: Europe 1946–1979. New York and Oxford: Oxford Univ. Press, 1981.

Hine, R.C., The Political Economy of European Trade. New York: St. Martin's Press, 1985.

Holzman, Franklyn D., International Trade Under Communism—Politics and Economics. New York: Basic Books, 1976.

Hong, Wontack, Trade, Distortions and Employment Growth in Korea. Honolulu: Univ. of Hawaii Press, 1979.

House, Arthur H., U.N. Civilian Operations in the Congo. Lanham, MD: Univ. Press of America, 1978.

Hoya, Thomas W., East–West Trade: Comecon Law, American–Soviet Trade. London, Rome, and New York: Oceana, 1984.

Hudson, Manley O., The Permanent Court of International Justice. New York: Macmillan, 1934.

Hudson, R. and others, An Atlas of EEC Affairs. London: Methuen, 1984.

Hufbauer, Gary Clyde, Subsidies in International Trade. Washington: Institute for International Economics, 1984.

Hufbauer, Gary Clyde and Jeffrey J. Schott, Economic Sanctions in Support of Foreign

Policy Goals. Washington: Institute for International Economics, 1983.

———, Trading for Growth: The Next Round of Trade Negotiations. Washington: Institute for International Economics, 1985.

———, Economic Sanctions Reconsidered: History and Current Policy. Washington: Institute for International Economics, 1985.

Hufbauer, Gary Clyde, and others, Trade Protection in the United States: 31 Case Studies. Washington: Institute for International Economics, 1986.

Hunter, Shireen, Gulf Cooperation Council; Problems and Prospects. Significant Issues Series Vol. 6, No. 15. Washington: Georgetown Univ., Center for Strategic and International Studies, 1984.

——— (ed.), The PLO After Tripoli. CSIS Significant Issues Series Vol. 6, No. 10. Lanham, MD: Univ. Press of America, 1984.

I

Ince, Basil A. and others (eds.), Issues in Caribbean International Relations. Lanham, MD: Univ. Press of America, 1983.

International Rivers; Some Case Studies. Occasional Paper No. 1. Bloomington: Indiana Univ. Dept. of Geography, 1965.

Isaacs, Arnold H., Dependence Relations Between Botswana, Lesotho, Swaziland and the Republic of South Africa. Leiden, Neth.: African Studies Centre, 1982.

J

Jackson, Marvin, East–South Trade: Economics and Political Economics. Armonk, NY: Sharpe, 1985.

Jackson, Richard L., The Non-aligned, the UN, and the Superpowers. New York: Praeger and Council on Foreign Relations, 1983.

Jacobson, Harold K., Networks of Interdependence: International Organizations and the Global Political System. New York: Knopf, 1979.

James, Preston E., One World Divided: A Geographer Looks at the Modern World. New York and London: Blaisdell, 1964.

Jasteer, Robert, Southern Africa. New York: St. Martin's Press, 1985.

Jenks, Christopher W., The Prospects of International Adjudication. London, 1964.

Johnston, Ronald J., The World Trade System; Some Enquiries into Its Spatial Structure. New York: St. Martin's Press, 1976.

Jones, Stephen B., The Arctic: Problems and Possibilities. New Haven, CT: Yale Institute of International Studies, Memorandum No. 29, 1948.

Jones, Stephen B. and Marion F. Murphy, *Geography and World Affairs*. Chicago, Rand McNally, 1971.

Jorgensen-Dahl, Arnfinn, *Regional Organization and Order in South-East Asia*. New York: St. Martin's Press, 1982.

Jutte, Rudiger and Annemarie Grosse-Jutte (eds.), *The Future of International Organization*. London: Frances Pinter, 1980.

K

Kalijarvi, Thorsten V., *Modern World Politics*. New York: T.Y. Crowell, 1942.

Kaplan, Morton A., *System and Process in International Politics*. New York: Wiley, 1964.

Kapteyn, P.J.G. and others (eds.), *International Organization and Integration: Annotated Basic Documents and Descriptive Directory of International Organizations and Arrangements*. The Hague, Boston, London: Nijhoff, 1982.

Kaufman, Edy and others, *Israeli–Latin American Relations*. New Brunswick, NJ: Transaction Books, 1979.

Kay, David A., *The New Nations in the United Nations, 1960–1967*. New York: Columbia Univ. Press, 1970.

Kennedy, S., *The Pan-Angles: A Consideration of the Federation of the Seven English-Speaking Nations*. New York and London: Longmans, Green, 1914.

Khalidi, Rashid, *Under Siege; PLO Decision-making During the 1982 War*. New York: Columbia Univ. Press, 1985.

Khan, Kabir-ur-Rahman, *The Law and Organisation of International Commodity Agreements*. The Hague, Boston, London: Nijhoff, 1982.

Khan, Khushi M. (ed.), *Multinationals of the South*. New York: St. Martin's Press, 1986; London: Frances Pinter, 1987.

Khong, Cho Oon, *The Politics of Oil in Indonesia: Foreign Company-Host Government Relations*. New York: Cambridge Univ. Press, 1986.

Kidron, Michael, and Ronald Segal, *The New State of the World Atlas*. New York: Simon & Schuster, 1984.

Kihl, Young W., and others, *World Trade Issues; Regime Structure and Policy*. New York: Praeger, 1985.

Kirisci, K., *The PLO and World Politics*. London: Frances Pinter, 1986.

Kitzinger, V.W., *The Politics and Economics of European Integration*. New York: Praeger, 1963.

Klare, Michael T., *Supplying Repression*. New York: Field Foundation, 1977.

Kline, John M., *International Codes and Multinational Business: Setting Guidelines for International Business Operations*. Westport, CT and London: Greenwood Press, Quorum Books, 1985.

Kohona, Palitha Tikiri Bandara, *The Regulation of International Economic Relations Through Law*. Dordrecht, Neth.; Boston, Lancaster, Eng.: Nijhoff, 1985.

Koul, Autar K., *The Legal Framework of UNCTAD in World Trade*. Leyden, Neth.: Sijthoff, 1977.

Krause, Lawrence B., *U.S. Economic Policy Toward the Association of Southeast Asian Nations: Meeting the Japanese Challenge*. Washington: Brookings Institution, 1982.

Krutilla, John V., *The Columbia River Treaty; the Economics of an International River Basin Development*. Baltimore: Johns Hopkins Univ. Press, 1967.

Kumar, Krishna (ed.), *Transnational Enterprises: Their Impact on Third World Societies and Cultures*. Boulder, CO: Westview, 1982.

Kurz, Anat, and Ariel Merari, *Asala*. Boulder, CO: Westview, 1986.

Kuusi, Juha, *The Host State and the Transnational Corporation. An Analysis of Legal Relationships*. Westmead, Eng.: Saxon House, Teakfield, 1979.

L

Labrie, Roger P. and others, *U.S. Arms Sales Policy: Background and Issues*. Lanham, MD: Univ. Press of America, 1982.

Lakos, Amos, *International Terrorism*. Boulder, CO: Westview, 1986.

Lall, Arthur, *The UN and the Middle East Crisis, 1967*. New York and London: Columbia Univ. Press, 1968.

Lammers, J.G., *Pollution of International Watercourses: A Search for Substantive Rules and Principles of Law*. Dordrecht, Neth.; Lancaster Eng.: Nijhoff, 1984.

Langdon, Frank, *The Politics of Canadian–Japanese Economic Relations, 1952–1983*. Vancouver: Univ. of British Columbia Press, 1984.

Langdon, Steven W., *Multinational Corporations in the Political Economy of Canada*. New York: St. Martin's Press, 1981.

Lasok, D. and W.J. Cairns, *The Customs Union—*

A Step Towards West European Integration. Hingham, MA: Kluwer, 1983.

Lauterpacht, Hersh, *Recognition in International Law.* Cambridge, 1947.

———, *The Development of International Law by the International Court.* London: 1958.

Lawrence, Robert Z. and Robert E. Litan, *Saving Free Trade: A Pragmatic Approach.* Washington: Brookings Institution, 1986.

Lawson, Ruth C., *International Regional Organizations: Constitutional Foundations.* New York: Praeger, 1962.

The League of Nations in Retrospect: Proceedings of the Symposium. Berlin and New York: de Gruyter, 1983.

Lee, R. and P. Ogden, *Economy and Society in the EEC: Spatial Perspectives.* Farnborough, Eng.: Saxon House, 1976.

Lefever, Ernest W., *Uncertain Mandate; Politics of the UN Congo Operation.* Baltimore: Johns Hopkins Univ. Press, 1967.

Le Marquand, David G., *International Rivers: The Politics of Cooperation.* Vancouver: Univ. of British Columbia, Westwater Research Center, 1977.

Lerner, Daniel and Morton Gorden, *Euratlantica: Changing Perspectives of the European Elites.* Cambridge: MIT Press, 1969.

Levasseur, Alain A. and Enrique Dahl, *Multinational Corporations; Investments, Technology, Tax, Labor, and Securities: European, North and Latin American Perspectives.* Lanham, MD: Univ. Press of America, 1986.

Leventhal, Paul and Yonah Alexander (eds.), *Preventing Nuclear Terrorism.* Lexington, MA: Lexington Books, 1987.

Levin, Aida Luisa, *The OAS and the UN: Relations in the Peace and Security Field.* New York: UNITAR, 1974.

Lewis, S. and Thomas G. Mathews (eds.), *Caribbean Integration; Papers on Social, Political and Economic Integration.* Río Piedras: Univ. of Puerto Rico, Institute of Caribbean Studies, 1967.

Leyton-Brown, David (ed.), *The Utility of International Economic Sanctions.* New York: St. Martin's Press, 1987.

Lichtheim, George, *The New Europe: Today—and Tomorrow.* New York: Praeger, 1963.

Lillich, Richard B. (ed.), *Transnational Terrorism: Conventions and Commentary; A Compilation of Treaties, Agreements, and Declarations of Special Interest to the United States.* Charlottesville, VA: Michie, 1982.

Lindquist, John and others, *Strategies for River Basin Management.* Boston: Kluwer Academic, 1985.

Lippmann, Walter, *Western Unity and the Common Market.* Boston: Little Brown and Atlantic Monthly Press, 1962.

Lloyd, Peter J., *International Trade Problems of Small Nations.* Durham, NC: Duke Univ. Press, 1968.

Lodge, Janet (ed.), *European Union.* New York: St. Martin's Press, 1985.

Lodge, Juliet, *Terrorism: A Challenge to the State,* New York: St. Martin's Press, 1981.

———, *The European Community and New Zealand.* London: Longwood, 1982.

———, (ed.), *Institutions and Policies of the European Community.* London: Frances Pinter, 1983.

Louscher, David J. and Michael D. Salomone (eds.), *Marketing Security Assistance.* Lexington, MA: Lexington Books, 1987.

Lowenfeld, Andreas F., *Public Controls on International Trade.* New York: Matthew Bender and St. Martin's Press, 1979.

Luard, Evan, *The United Nations.* New York: St. Martin's Press, 1985.

Luciani, Giacomo, *The Oil Companies and the Arab World.* St. Martin's Press, 1984.

M

Maachou, Abdelkader, *OAPEC; Organization of Arab Petroleum Exporting Countries.* New York: St. Martin's Press, 1982.

MacDonald, Robert W., *The League of Arab States.* Princeton, NJ: Princeton Univ. Press, 1965.

Makonnen, Yilma, *International Law and the New States of Africa.* New York: UNIPUB, 1983.

Malmgren, Harald B., *Changing Forms of World Competition and World Trade Rules.* CSIS Significant Issues Series, Vol. 3, No. 3. Lanham, MD: Univ. Press of America, 1981.

Martin, Linda G. (ed.), *The Asean Success Story.* Honolulu: University of Hawaii Press, 1987.

Martz, John D., *Central America: The Crisis and the Challenge.* Chapel Hill: Univ. of North Carolina Press, 1959.

Martz, John D. and Lars Schoultz (eds.), *Latin America, the United States, and the Inter-American System.* Boulder, CO: Westview, 1980.

Mason, Kenneth, *The Geography of Current Affairs.* Oxford: Oxford Univ. Press, Clarendon, 1932.

Mattelart, Armand, *Transnationals and the Third*

World: The Struggle for Culture. South Hadley, MA: Bergin & Garvey, 1983.

Mayne, R., *The Community of Europe.* London: Gollancz, 1962.

Mazrui, Ali A., *Africa's International Relations.* London: Heinemann, 1977.

Mazzeo, Domenico (ed.), *African Regional Organizations.* Cambridge and elsewhere: Cambridge University Press, 1984.

McDonald, Vincent R., *The Caribbean Economies.* New York: MSS Information Corp., 1972.

McGovern, Edmond, *International Trade Regulation: GATT, the United States and the European Community.* Exeter, England: Globefield Press, 1982.

McIntosh, Malcolm, *Arms Across the Pacific.* London: Frances Pinter, 1987.

McIntyre, John R. and Daniel S. Rapp, *The Political Economy of International Technology Transfer.* Westport, CT: Greenwood Press, Quorum, 1986.

McKinlay, R.D. and A. Mughan, *Aid and Arms to the Third World: An Analysis of the Distribution and Impact of US Official Transfers.* London: Frances Pinter, 1984.

McMillan, Carl H., *Multinationals from the Second World.* New York: St. Martin's Press, 1987.

McWhinney, Edward, *United Nations Law Making: Cultural and Ideological Relativism and International Law Making for an Era of Transition.* New York and London: Holmes & Meier; Paris: UNESCO; 1984.

Menon, Rafan, *Soviet Power and the Third World.* New Haven, CT: Yale Univ. Press, 1986.

Merari, Ariel and others, *Inter 85.* Boulder, CO: Westview, 1986.

Merari, Ariel and Shiomi Elad, *The International Dimension of Palestinian Terrorism.* Boulder, CO: Westview, 1987.

Merican, T.S.J., *The Third World and International Law: A Study of the Role of International Law in the Contemporary World,* 2nd ed. London: Sweet & Maxwell, 1981.

Michel, Aloys A., *The Indus Rivers: A Study of the Effects of Integration.* New Haven, CT: Yale Univ. Press, 1967.

Mickolus, Edward F. (compiler), *The Literature of Terrorism: A Selectively Annotated Bibliography.* Westport, CT: Greenwood Press, 1980.

Middleton, Robert, *Negotiating on Non-tariff Distortions of Trade.* New York: St. Martin's Press, 1975.

Mikdashi, Zuhayr, *Transnational Oil: Issues,* *Policies and Perspectives.* London: Frances Pinter, 1986.

Miller, Abraham H., *Terrorism and Hostage Negotiations.* Boulder, CO: Westview, 1980.

Millett, Richard and W. Marvin Will (eds.), *The Restless Caribbean; Changing Patterns of International Relations.* New York: Praeger, 1979.

Mirza, Hafiz, *Multinationals and the Growth of the Singapore Economy.* New York: St. Martin's Press, 1986.

Moskos, C., Jr., *Peace Soldiers: The Sociology of a UN Military Force.* Chicago: Chicago Univ. Press, 1976.

Moss, R., *Urban Guerrillas: The New Face of Political Violence.* London: Temple Smith, 1972.

Msabaha, Ibrahim S.R. and Timothy M. Shaw (eds.), *Confrontation and Liberation in Southern Africa.* Boulder, CO: Westview, 1987.

Mueller, Jerry E., *Restless River; International Law and the Behavior of the Rio Grande.* El Paso: Texas Western Press, 1975.

Multinational Investment in the Economic Development and Integration of Latin America. Washington: Inter-American Development Bank, 1968.

Munger, Edwin S., *Geography of Ocean Outlets for the Belgian Congo.* Elizabethville: Institute of Current World Affairs, 1952.

Murphy, John F., *The United Nations and the Control of International Violence: A Legal and Political Analysis.* Totowa, NJ: Allanheld, Osmun, 1982.

———, *Punishing International Terrorists: The Legal Framework for Policy Initiatives.* Totowa, NJ: Rowman & Littlefield, 1985.

Myers, Kenneth, *Terrorism in Europe.* New York: St. Martin's Press, 1982.

Myrdal, Gunnar, *Rich Lands and Poor: The Road to World Prosperity.* New York: Harper & Row, 1957.

Mytelka, Lynn Krieger, *Regional Development in a Global Economy; The Multinational Corporation, Technology and Andean Integration.* New Haven, CT, and London: Yale Univ. Press, 1979.

N

Naff, Thomas and Ruth C. Matson (eds.), *Water in the Middle East: Conflict or Cooperation?* Boulder, CO, and London: Westview, 1984.

Nakhleh, Emile A., *The Gulf Cooperation Council.* New York: Praeger, 1986.

Nathan, K.S. and M. Pathmanathan (eds.), *Trilateralism in Asia; Problems and Prospects in U.S.–Japan–Asian Relations*. Honolulu: Univ. of Hawaii Press, 1987.

Nelson, Daniel N., *Alliance Behavior in the Warsaw Pact*. Boulder, CO: Westview, 1986.

Neres, P., *French-Speaking West Africa*. London: Oxford Univ. Press, 1962.

Neuberger, Benyamin, *National Self-determination in Post-Colonial Africa*. Boulder, CO: Lynne Rienner, 1986.

Neuman, Stefanie G., *Military Assistance in Recent Wars*. New York: Praeger, 1986.

Neumann, Ambassador Robert G. (ed.), *NATO: The Next Thirty Years*, CSIS Significant Issues Series, Vol. 1, No. 6. Lanham, MD: Univ. Press of America.

Newbigin, Marion I., *Geographical Aspects of the Balkan Problems in Their Relation to the Great European War*. New York: Putnam, 1915.

Nicholas, H.G., *The United Nations as a Political Institution*. London: Oxford Univ. Press, 1968.

Nincic, Miroslav and Peter Wallensteen (eds.), *Dilemmas of Economic Coercion: Sanctions in World Politics*. New York: Praeger, 1983.

Nkonoki, Simon R., *Regional Development Planning of the Kagera River Basin in Eastern Africa Under the Kagera Basin Organisation (KBO); A Case Study of Hydropower Development Planning and Related Environmental Impacts*. Bergen, Norway: DERAP Publications, November 1983.

Nolan, Joanne E., *Military Industry in Taiwan and South Korea*. New York: St. Martin's Press, 1986.

Nove, Alec, *East–West Trade*. Lanham, MD: Univ. Press of America, 1987.

Nsekela, Amon J. (ed.), *Southern Africa: Toward Economic Liberation*. London: Rex Collings, 1981.

Nye, Joseph S., *Pan-Africanism and East African Integration*. Cambridge: Harvard Univ. Press, 1965.

Nystrom, J. Warren and George W. Hoffman, *The Common Market*, 2nd ed. New York: Van Nostrand, 1976.

O

OECD, *The OECD Guidelines for Multinational Enterprises*. Washington, April 1986.

Ogilvie, Alan Grant, *Europe and Its Borderlands*. Edinburgh: Nelson, 1957.

Ojuatey-Kodjoe, W. (ed.), *Pan-Africanism: New Directions in Strategy*. Lanham, MD: Univ. Press of America, 1986.

O'Keefe, D. and H.G. Schermers, *Essays in European Law and Integration*. Hingham, MA: Kluwer, 1982.

Okolo, Julius Emeka and Stephen Wright (eds.), *West Africa*. Boulder, CO: Westview, 1987.

Ontiveros, Suzanne (ed.), *Global Terrorism*. Santa Barbara, CA: ABC–CLIO, 1986.

Osmańczyk, Edmund Jan, *The Encyclopedia of the United Nations and International Agreements*. Philadelphia and London: Taylor & Francis, 1985.

O'Sullivan, Noel K., *Terrorism, Ideology and Revolution*. Boulder, CO: Westview, 1986.

P

Pacini, Deborah and Christine Franquemont (eds.), *Coca and Cocaine; Effects on People and Policy in Latin America*. Cambridge, MA: Cultural Survival, 1986.

Padmore, George, *Pan-Africanism or Communism? The Coming Struggle for Africa*. New York: Roy, 1956.

Palmer, Ransford W., *Problems of Development in Beautiful Countries; Perspectives on the Caribbean*. Lanham, MD: North–South Publishing Co., 1984.

Panikkar, K.M. and others, *Regionalism and Security*. New Delhi: India Council of World Affairs, 1948.

Papadopoulos, Andrestinos N., *Multilateral Diplomacy Within the Commonwealth: A Decade of Expansion*. The Hague, Boston, London: Nijhoff, 1982.

Parker, Geoffrey, *The Logic of Unity: A Geography of the European Community*. London: Longman, 1981.

———, *A Political Geography of Community Europe*. London: Butterworths, 1983.

Payne, Anthony J., *The Politics of the Caribbean Community 1961–79: Regional Integration Among New States*. New York: St. Martin's Press; Manchester, Eng.: Manchester Univ. Press, 1980.

Pearson, Charles S. (ed.), *Multinational Corporations, Environment and the Third World*. Durham, NC: Duke Univ. Press, 1987.

Perusse, Roland I., *A Strategy for Caribbean Economic Integration*. Hato Rey, Puerto Rico: North–South Center Press, 1971.

Pésci, Kálmám, *The Future of Socialist Economic Integration*. Armonk, NY: Sharpe, 1981.

Peterson, Dean F. and A. Berry Crawford (eds.), *Values and Choices in the Development of the Colorado River Basin*. Tucson: Univ. of Arizona Press, 1978.

Pierre, Andrew J. (ed.), *Arms Transfers and American Foreign Policy*. New York: New York Univ. Press, 1979.

————, *The Global Politics of Arms Sales*. Princeton, NJ: Princeton Univ. Press, 1982.

Piper, Don Courtney, *The International Law of the Great Lakes; A Study of Canadian–United States Co-operation*. Durham, NC: Duke Univ. Press, 1967.

Pomfret, Richard, *Mediterranean Policy of the European Community: A Study of Discrimination in Trade*. New York: St. Martin's Press, 1986.

Potholm, Christian P. and Richard A. Fredland (eds.), *Integration and Disintegration in East Africa*. Washington: Univ. Press of America, 1980.

Pounds, Norman J.G. and William N. Parker, *Coal and Steel in Western Europe*. Bloomington: Indiana Univ. Press, 1957.

Poynter, Thomas A., *Multinational Enterprises and Government Intervention*. New York: St. Martin's Press, 1986.

Preeg, Ernest H., *Traders and Diplomats; An Analysis of the Kennedy Round of Negotiations Under the General Agreement on Tariffs and Trade*. Washington: Brookings Institution, 1970.

Preiswerk, Roy (ed.), *Regionalism and the Commonwealth Caribbean*. Trinidad: Institute of International Relations, 1969.

Pryce, Roy, *The Political Future of the European Community*. London: Marshbank & Federal Trust, 1962.

Pryor, Frederic L., *The Communist Foreign Trade System*. Cambridge: MIT Press, 1963.

R

Ra'anan, Uri and others, *Hydra of Carnage; International Linkages of Terrorism*. Lexington, MA: Lexington Books, 1987.

Rajan, M.S., *The Expanding Jurisdiction of the United Nations*. Bombay: N.M. Tripathi; Dobbs Ferry, NY: Oceana, 1982.

Rallo, Joseph C., *The Security of Western Europe*. London: Frances Pinter, 1986.

Raman, K. Venkata (ed.), *Dispute Settlement Through the United Nations*. New York: UNITAR, 1977.

Rees, Judith and Peter Odell, *The International Oil Industry*. New York: St. Martin's Press, 1987.

Regional Industrial Cooperation: Experiences and Perspectives of ASEAN and the Andean Pact. Vienna, Aus.: UNIDO, 1986.

Relations Between the United Nations and Non-UN Regional Intergovernmental Organizations. New York: UNITAR Conference Report No. 3, 1973.

Renninger, John P., *Multinational Co-operation for Development in West Africa*. UNITAR Regional Study No. 5. Elmsford. NY: Pergamon, 1979.

Rey, Louis, *The Challenging and Elusive Arctic Region*. CSIS Significant Issues Series, Vol. 6, No. 5, Lanham, MD: Univ. Press of America, 1984.

Robertson, A.H., *The Relations Between the Council of Europe and the United Nations*. New York: UNITAR Regional Study No. 1, 1972.

Robinson, John, *Multinationals and Political Control*. New York: St. Martin's Press, 1983.

Robock, S.H. and K. Simmonds, *International Business and Multinational Enterprises*. Homewood, IL: Irwin, 1983.

Roosa, Robert V. and others, *East–West Trade at a Crossroads: Economic Relations with the Soviet Union and Eastern Europe*. New York and London: New York Univ. Press, 1982.

Roseneau, James N., *Linkage Politics: Essays on the Convergence of National and International Systems*. New York, 1969.

Rostow, Walt W., *The United States and the Regional Organizations of Asia and the Pacific, 1965–1985*. Austin: Univ. of Texas Press, 1986.

Rotberg, Robert I. and others, *South Africa and Its Neighbors; Regional Security and Self-Interest*. Lexington, MA: Lexington Books, 1985.

Rubin, Seymour J. and Thomas R. Graham, *Environment and Trade: The Relation of International Trade and Environmental Policy*. Totowa, NJ: Allanheld, Osmun; London: Frances Pinter, 1982.

Rubin, Seymour J. and Gary Clyde Hufbauer (eds.), *Emerging Standards of International Trade and Investment: Multinational Codes and Corporate Conduct*. Totowa, NJ: Rowman & Allanheld, 1984.

Rubin, Seymour J. and Thomas R. Graham (eds.), *Managing Trade Relations in the 1980's: Issues Involved in the GATT Ministerial Meeting of 1982*. Totowa, NJ: Rowman & Allanheld, 1984.

Russett, Bruce M., *International Regions and the International System: A Study in Political Ecology*. Chicago: Rand McNally, 1967.

Rustow, Dankwart, *A World of Nations*. Washington: Brookings Institution, 1967.

S

Sachs, M.Y., *The United Nations: A Handbook on the United Nations*. New York: Wiley, 1977.

Saddy, Fehmy (ed.), *Arab–Latin American Relations*. New Brunswick, NJ: Transaction Books, 1983.

Safarian, A.E., *The Performance of Foreign-Owned Firms in Canada*. Washington and Montreal: Canadian–American Committee, 1969.

Safarian, A.E. and Bertin (eds.), *Multinationals, Governments, and International Technology Transfer*. New York: St. Martin's Press, 1987.

Sakensa, K.P., *Cooperation in Development Problems and Prospects for India and Asean*. CA: Sage, 1985.

Saliba, Samir N., *The Jordan River Dispute*. The Hague: Nijhoff, 1968.

Sampson, Anthony, *The Arms Bazaar: From Lebanon to Lockheed*. New York: Viking, 1977.

Sanders, Thomas G., *Andean Economic Integration*. American Universities Field Staff Reports, West Coast South America Series, Vol. 15, No. 2. 1968.

Sandwick, John A., *The Gulf Cooperation Council*. Boulder, CO: Westview, 1987.

Sarkesian, Sam C. and William L. Scully (eds.), *U.S. Policy and Low-Intensity Conflict: Potentials for Military Struggles in the 1980's*. New Brunswick, N.J., and London: Transaction Books, 1981.

Sarna, Aaron J., *Boycott and Blacklist: A History of Arab Economic Warfare Against Israel*. Totowa, NJ: Rowman & Littlefield, 1986.

Saunders, Christopher T. (ed.), *East–West–South*. New York: St. Martin's Press, 1981.

——, *Regional Integration in East and West*. New York: St. Martin's Press, 1983.

Sauvant, Karl P., *The Group of 77; Evolution, Structure, Organization*. London, Rome, and New York: Oceana, 1981.

——, *The Collected Documents of the Group of 77*. Dobbs Ferry, NY: Oceana, 1981.

——, *International Transactions in Services*. Boulder, CO: Westview, 1986.

Savary, Julian, *French Multinationals*. London: Frances Pinter, 1984.

Schaefer, Henry W., *Comecon and the Politics of Integration*. New York: Praeger, 1972.

Seegers, Kathleen Walker, *Alliance for Progress*. New York: Coward-McCann, 1964.

Seers, Dudley and others, *Integration and Unequal Development; the Experience of the EEC*. New York: St. Martin's Press, 1981.

——, *The Second Enlargement of the EEC*. New York: St. Martin's Press, 1982.

Segal, Aaron Lee, *The Politics of Caribbean Economic Integration*. Río Piedras: Univ. of Puerto Rico, Institute of Caribbean Studies, 1968.

Sekiguchi, Sueo, *Japanese Direct Foreign Investment*. Totowa, NJ: Rowman & Littlefield, 1979.

Sesay, Amadu and others, *The OAU After Twenty Years*. Boulder, CO, and London: Westview, 1984.

Seymour, Ian, *OPEC; Instrument of Change*. New York: St. Martin's Press, 1981.

Shaw, Timothy M. and Kenneth A. Heard (eds.), *Cooperation and Conflict in Southern Africa; Papers on a Regional Subsystem*. Washington: Univ. Press of America, 1976.

Shaw, Timothy M. and 'Sola Ojo (eds.), *Africa and the International Political System*. Lanham, MD: Univ. Press of America, 1982.

Shaw, Timothy M. and Yash Tandon (eds.), *Regional Development at the International Level*, Vol. II: *African and Canadian Perspectives*. Lanham, MD: Univ. Press of America, 1985.

Shihata, Ibrahim F.I. and others, *The OPEC Fund for International Development: The Formative Years*. New York: St. Martin's Press, 1983.

Siekmann, Robert C.R., *Basic Documents on United Nations and Related Peace-Keeping Forces*. Dordrecht, Neth.: Nijhoff, 1985.

Sigmund, Paul E., *Multinationals in Latin America: The Politics of Nationalization*. Madison: Univ. of Wisconsin Press, 1980.

Sinjela, A. Mpazi, *Land-locked States and the UNCLOS Regime*. London, Rome and New York: Oceana, 1983.

Smith, David Ingle and Peter Stopp, *The River Basin*. Cambridge: At the University Press, 1979.

Smithers, Sir Peter, *Governmental Control: A Prerequisite for Effective Relations Between the United Nations and Non-United Nations Regional Organizations*. New York: UNITAR Regional Study No. 3, 1973.

Smogorzewski, Casimir, *Poland, Germany and the Corridor*. London: Williams & Norgate, 1930.

Snyder, Jed C., *Defending the Fringe*. Boulder, CO: Westview, 1987.

Sobell, Vladimir, *The Red Market: Industrial Cooperation and Specialisation in Comecon*. Brookfield, VT: Gower, 1984.

Sohn, Louis B. (ed.), *International Organiza-*

tion and Integration; Annotated Basic Documents of International Organizations and Arrangements. Dordrecht, Neth.: Nijhoff, 1986.

Soward, F.H. and A.M. Macarlay, *Canada and the Pan American System*. Toronto: Ryerson, 1948.

Spero, Joan Edelman, *The Politics of the International Economic Relations*. New York: St. Martin's Press, 1985.

Spraos, John, *Inequalising Trade? A Study of North/South Specialisation in the Context of Terms of Trade Concepts*. London and New York: Oxford Univ. Press, 1983.

Sprout, Harold and Margaret Sprout, *Man-Milieu Relationship Hypotheses in the Context of International Politics*. Princeton, NJ: Princeton Univ. Center for International Studies, 1956.

———, *Foundations of International Politics*. Princeton, NJ: Van Nostrand, 1962.

Stairs, Denis and Gilbert R. Winham, *The Politics of Canada's Economic Relationship with the United States*. Toronto: Toronto Univ. Press, 1987.

Stamp, L. Dudley, *The British Commonwealth*. London: Longmans, 1951.

Stern, Robert M. and others, *Perspectives on a U.S.–Canadian Free Trade Agreement*. New York: St. Martin's Press, 1987.

Stoetzer, O.C., *The Organization of American States*. New York: Praeger, 1965.

Stohl, Michael and George A. Lopez (eds.), *The State as Terrorist: The Dynamics of Governmental Violence and Repression*. Westport, CT: Greenwood Press, 1984.

Stokman, F.N., *Roll Calls and Sponsorship: A Methodological Analysis of Third World Group Formation in the UN*. Leiden, Neth.: Sijthoff, 1977.

Stolper, Thomas E., *China, Taiwan, and the Offshore Islands*. Armonk, NY: Sharpe, 1985.

Strack, Harry R., *Sanctions; The Case of Rhodesia*. Syracuse, NY: Syracuse Univ. Press, 1978.

Strom, Gabriele Winai, *Development and Dependence in Lesotho, the Enclave of South Africa*. Uppsala, Swed.: Scandinavian Institute of African Studies, 1978.

Summers, Lionel M., *The International Law of Peace*. Dobbs Ferry, NY: Oceana, 1972.

Suryadinata, Leo, *China and the Asean States: The Ethnic Chinese Dimension*. Athens: Ohio Univ. Press, 1985.

T

Tabory, Mala, *The Multinational Force and Observers in the Sinai*. Boulder, CO, and London: Westview Press, 1986.

Taylor, Michael and Nigel Thrift (eds.), *The Geography of Multinationals*. New York: St. Martin's Press, 1982.

Taylor, Paul (ed.), *International Institutions at Work*. London: Frances Pinter, 1987.

Taylor, Paul and A.J.R. Groom, *International Organisation: A Conceptual Approach*. London: Frances Pinter, 1978.

Taylor, Philip, *Nonstate Actors in International Politics; from Transregional to Substate Organizations*. Boulder, CO, and London: Westview, 1984.

Taylor, T., *Defence, Technology and International Integration*. London: Frances Pinter, 1982.

Teclaff, Ludwik A., *The River Basin in History and Law*. The Hague: Nijhoff, 1967.

———, *Water Law in Historical Perspective*. Buffalo, NY: William S. Hein, 1985.

Ten Years of CARICOM. Washington: Inter-American Development Bank, 1984.

Thakur, Ramesh, *Peacekeeping in Vietnam*. Edmonton, Alta., Can.: Univ. of Alberta Press, 1984.

———, *International Peacekeeping in Lebanon*. Boulder, CO: Westview, 1987.

Thayer, George, *The War Business: The International Trade in Armaments*. New York: Simon & Schuster, 1969.

Thoman, Richard S. and Edgar C. Conkling, *Geography of International Trade*. Englewood Cliffs, NJ: Prentice-Hall, 1967.

Timmermans, C.W.A. and E.L.M. Volker, *Division of Powers Between the European Communities and Their Member States in the Field of External Relations*. Hingham, MA: Kluwer, 1981.

Timmins, David B. and William M. Timmins, *The Inter-National Economic Policy Coordination Instrument: Some Considerations Drawn from the OECD Experience*. Lanham, MD: Univ. Press of America, 1985.

Tondel, L.M., *The Southeast Asia Crisis*. Dobbs Ferry, NY: Oceana, 1966.

Tostensen, Arne, *Dependence and Collective Self-reliance in Southern Africa; the Case of the Southern African Development Coordination Conference (SADCC)*. Uppsala, Swed.: Scandinavian Institute of African Studies, 1982.

Tovias, Alfred, *Tariff Preferences in Mediterranean Diplomacy*. New York: St. Martin's Press, 1978.

Tumlir, Jan, *Protectionism: Trade Policy in Democratic Societies*. Lanham, MD: Univ. Press of America, 1986.

Tung, William L., *International Organization Under the United Nations System*. New York: T.Y. Crowell, 1969.

Tuomi, Helen and Raimo Väyrynen, *Transnational Corporations, Armaments and Development*. New York: St. Martin's Press, 1982.

Turner, Barry, and Gunilla Nordquist, *The Other European Community*. New York: St. Martin's Press, 1982.

Tussie, Diana (ed.), *Latin America in the World Economy*. New York: St. Martin's Press, 1983.

———, *The Less Developed Countries and the World Trading System; A Challenge to the GATT*. London: Frances Pinter, 1986, New York: St. Martin's Press, 1987.

Twitchett, Kenneth J., *Europe and the World; The External Relations of the Common Market*. New York: St. Martin's Press, 1976.

Tyler, William G., *Advanced Developing Countries as Export Competitors in Third World Markets: The Brazilian Experience*. CSIS Significant Issues Series, Vol. 2, No. 8. Lanham, MD: Univ. Press of America, 1980.

U

United Nations, *Development of Water Resources in the Lower Mekong Basin*. Sales No. 57.II.F.8., 1957.

———, *Integrated River Basin Development*. Sales No. E.70.II.A.4., 1970.

United Nations, *Experiences in the Development and Management of International River and Lake Basins*. New York, 1983.

United Nations, *The Blue Helmets; A Review of United Nations Peace-Keeping*. New York, 1985.

United Nations, *The History of UNCTAD 1964–1984*. New York, 1985.

United Nations Economic Commission for Western Asia. *Economic Integration in Western Asia*. London: Frances Pinter, 1985.

Urquhart, Brian, *The United Nations and International Law*. New York: Cambridge Univ. Press, 1986.

Urquidi, Victor L., *Free Trade and Economic Integration in Latin America*. Berkeley: Univ. of California Press, 1964.

V

van Hamel, J.A., *Danzig and the Polish Problem*. New York: Carnegie Endowment for International Peace, 1933.

Venn, Fiona, *Oil Diplomacy in the Twentieth Century*. New York: St. Martin's Press, 1986.

Villar, Roger, *Piracy Today; Robbery and Violence at Sea Since 1980*. London: Conway Maritime Press, 1985.

Vitányi, Béla, *The International Regime of River Navigation*. Alphen aan den Rijn, Neth.: Sijthoff & Noordhoff, 1979.

Völker, E.L.M. (ed.), *Protectionism and the European Community*. Deventer, Neth.: Kluwer, 1983.

W

Wainhouse, David W., *International Peace Observation*. Baltimore and London: Johns Hopkins, Univ. Press, 1969.

Waldmann, Raymond J., *Regulating International Business Through Codes of Conduct*. Washington and London: American Enterprise Institute for Public Policy Research, 1980.

Wallace, Cynthia Day, *Legal Control of the Multinational Enterprise: National Regulatory Techniques and the Prospects for International Control*. The Hague, Boston, London: Nijhoff, 1982.

Wallace, William V., and Roger A. Clarke, *Comecon, Trade and the West*. London: Frances Pinter, 1986.

Wallerstein, Immanuel, *Africa; the Politics of Unity*. New York: Random House, 1967.

Walrond, Grantley W. and Raj Kumar, *Options for Developing Countries in Mining Development*. New York: St. Martin's Press, 1985.

Wardlaw, Grant, *Political Terrorism: Theory, Tactics and Counter-measures*. New York: Cambridge Univ. Press, 1983.

Waterbury, John, *Hydropolitics of the Nile Valley*. Syracuse, NY: Syracuse Univ. Press, 1979.

Weintraub, Sidney, *Free Trade Between Mexico and the United States*. Washington: Brookings Institution, 1984.

Weiss, Leonard, *Trade Liberalization and the National Interest*. CSIS Significant Issues Series, Vol. 2, No. 2. Lanham, MD: Univ. Press of America, 1980.

Weiss, Thomas G., *Multilateral Development Diplomacy in UNCTAD*. New York: St. Martin's Press, 1986.

Wells, Clare, *The UN, UNESCO and the Politics*

of Knowledge. New York: St. Martin's Press, 1987.

Westermeyer, William E. and Kurt M. Shusterich (eds.), *United States Arctic Interests: The 1980's and 1990's.* New York: Springer-Verlag, 1984.

Whalley, John, *Canada and the Multilateral Trading System.* Toronto: Univ. of Toronto Press, 1987.

———, *Canada-United States Free Trade.* Toronto: Univ. of Toronto Press, 1987.

——— and others, *Canadian Trade Policies and the World Economy.* Toronto: Univ. of Toronto Press, 1987.

———, *Canada's Resource Industries and Water Export Policy.* Toronto: Univ. of Toronto Press, 1987.

Wheare, Kenneth C., *The Constitutional Structure of the Commonwealth.* Oxford: Oxford Univ. Press, 1960.

Whitaker, Arthur P., *The Western Hemisphere Idea: Its Rise and Decline.* Ithaca, NY: Cornell Univ. Press, 1954.

Whitaker, P., *Political Theory and East African Problems.* New York and London: Oxford Univ. Press, 1964.

Willemin, Georges and Roger Heacock, *The International Committee of the Red Cross.* Dordrecht, Neth.: Nijhoff, 1984.

Wilson, Robert R., *International Law and Contemporary Commonwealth Issues.* Durham, NC: Duke Univ. Press, 1971.

Winham, Gilbert R., *International Trade and the Tokyo Round Negotiation.* Princeton, NJ: Princeton Univ. Press, 1987.

Wionczek, Miguel S. (ed.), *Economic Cooperation in Latin America, Africa, and Asia.* Cambridge, MA: MIT, 1969.

Wiseman, Henry (ed.), *Peacekeeping: Appraisals & Proposals.* Pergamon, 1983.

Wiskemann, Elizabeth, *Germany's Eastern Neighbours: Problems Relating to the Oder–Neisse Line and the Czech Frontier Regions.* London: Oxford Univ. Press, 1956.

Wonnacott, Paul, *The United States and Canada: The Quest for Free Trade.* Washington: Institute for International Economics, 1987.

Woolcock, Stephen, and others, *Interdependence in the Post-Multilateral Era: Trends in U.S.–European Trade Relations.* Lanham, MD: Univ. Press of America, 1985.

Woronoff, Jon, *Japan's Commercial Empire.* Armonk, NY: Sharpe, 1984.

Wright, Quincy, *The Role of International Law in the Elimination of War.* Manchester, Eng.: Manchester Univ. Press, 1961.

Y

Yannopoulos, George N., *Greece and the EEC.* New York: St. Martin's Press, 1986.

Yeats, Alexander J., *Trade Barriers Facing Developing Countries.* New York: St. Martin's Press, 1979.

Yin, John, *The Soviet Views on the Use of Force in International Law.* Hong Kong: Asian Research Service, n.d.

Yochelson, *Services and the U.S. Trade Policy.* Lanham, MD: Univ. Press of America, 1987.

Yuill, Douglas, and others (eds.), *Regional Policy in the European Community.* New York: St. Martin's Press, 1980.

Z

Zacklin, Ralph and others (eds.), *The Legal Regime of International Rivers and Lakes.* The Hague; Boston, London: Nijhoff, 1981.

Zagaris, Bruce, *Foreign Investment in the United States.* New York: Praeger, 1980.

Zartman, I. William, *Ripe for Resolution: Conflict and Intervention in Africa.* New York: Council on Foreign Relations, 1985.

Zoller, Elisabeth, *Peacetime Unilateral Remedies: An Analysis of Countermeasures.* Dobbs Ferry, NY: Transnational, 1984.

Periodicals

A

Abegunrin, Layi, "Southern African Development Coordination Conference (SADCC): Towards Regional Integration of Southern Africa for Liberation," *A Current Bibliography on African Affairs,* 17, 4 (1985), 363–384.

Aderbidge, A.B. and others, "Symposium on West African Integration," *Nigerian Journal of Economic and Social Studies,* 5, 1 (March 1963), 1–40.

Anglin, Douglas G., "SADCC After Nkomati," *African Affairs,* 84, 335 (April 1985), 163–182.

Armstrong, Hamilton Fish, "The World Is Round," *Foreign Affairs*, 31, 2 (January 1953), 175–199.

Austin, Dennis, "Sanctions and Rhodesia," *World Today*, 22 (March 1966), 106–113.

Austin, Warren R., "The Development of International Law by the United Nations," *Proceedings American Society of International Law*, (1948), 132–141.

Axline, Andrew W., "Integration and Development in the Commonwealth Caribbean; the Politics of Regional Negotiations," *International Organization*, 32, 4 (Autumn 1978), 953–973.

B

Bailey, Sydney D., "Nonmilitary Areas in UN Practice," *American Journal of International Law*, 74, 3 (July 1980), 499–524.

Baral, Lok Raj, "SARC, But No 'Shark': South Asian Regional Cooperation in Perspective," *Pacific Affairs*, 58, 3 (Fall 1985), 411–426.

Barnekov, Christoper C., "Sanctions and the Rhodesian Economy," *Rhodesian Journal of Economics*, 3 (March 1969), 44–75.

Barton, Thomas Frank, "Outlets to the Sea for Land-locked Laos," *Journal of Geography*, 59, 5 (May 1960), 206–220.

Bernstein, R.A. and P.D. Weldon, "A Structural Approach to the Analysis of International Relations," *Journal of Conflict Resolution*, 12, 2 (June 1958), 159–181.

Beukema, Herman, "The Geographical Factor in the Study of International Relations," *Professional Geographer*, 7, (1948), 21–23.

Binaisa, Godfrey L., "Organization of African Unity and Decolonization: Present and Future Trends," *Annals, American Academy of Political and Social Science*, 432 (July 1977), 52–69.

Blackwell, Michael, "Lomé III: The Search for Greater Effectiveness," *Finance and Development*, 22, 3 (September 1985), 31–34.

Boutros-Ghali, Boutros Y., "The Arab League, 1945–1955," *International Conciliation*, No. 498 (May 1954), 387–448.

——, "The Addis Ababa Charter," *International Conciliation*, No. 546 (1964).

Bowen, Robert E., "The Land-locked and Geographically Disadvantaged States and the Law of the Sea," *Political Geography Quarterly*, 5, 1 (January 1986), 63–69.

Bowman, Larry W., "The Subordinate State System of Southern Africa," *International Studies Quarterly*, 12 (September 1968).

Brewster, H., "Caribbean Economic Integration—Problems and Perspectives," *Journal of Common Market Studies*, 9, 4 (June 1971), 282–298.

Briggs, Herbert Whittaker, "Reflections on the Codification of International Law by the International Law Commission and by Other Agencies," *Recueil des Cours de l'Académie de Droit International de La Haye*, 126 (1969), 233–316.

C

Caflisch, Lucius C., "Land-Locked States and Their Access to and from the Sea," *British Yearbook of International Law*, 49 (1978), 71–100.

Cambell, John C., "Diplomacy on the Danube," *Foreign Affairs*, 27, 2 (January 1949), 315–327.

Calvet, A.L., "A Synthesis of Foreign Direct Investment Theories and Theories of the Multinational Firm," *Journal of International Business Studies*, 12 (1981), 43–59.

Caribbean Yearbook of International Relations. Ed. University of the West Indies, Institute of International Relations, Trinidad and Tobago. Alphen aan den Rijn, Netherlands: Sijthoff & Noordhoff.

Catroux, G., "The French Union: Concept, Reality and Prospects," *International Conciliation*, No. 495 (1953), 195–256.

Chhabra, Hari Sharan, "Oil to Rhodesia—a Western Conspiracy," *India Quarterly*, 34 (January–March 1978), 26–38.

Cole, Babalola, "ECOWAS: Problems and Prospects," *A Current Bibliography on African Affairs*. 17, 3 (1985), 267–277.

Coleman, James S., "Problems of Political Integration in Emergent Africa," *Western Political Quarterly*, 8, 1 (1955), 45–57.

Conkling, Edgar C., "Toward an Integrated Approach to the Geography of International Trade," *Professional Geographer*, 33, 1 (February 1981), 16–25.

Coplin, W.D., "International Law and Assumptions About the State System," *World Politics*, 17, 4 (July 1965), 615–634.

Corbridge, Stuart, "Perversity and Ethnoregionalism in Tribal India: The Politics of the Jharkand." *Political Geography Quarterly*, 6, 3 (July 1987), 225–240.

Costanzo, G.A., "The Association of Overseas Countries and Territories with the Common Market," *Civilizations*, 8 (1958), 505–526.

Crary, Douglas D., "Geography and Politics in the Nile Valley," *Middle East Journal*, 3, 3 (July 1949), 260–276.

Cunningham, J.K., "A Politico-Geographical Appreciation of New Zealand Foreign Policy," *New Zealand Geographer*, 14, 2 (October 1958), 147–160.

Cunningham, Susan M., "Multinational Enterprises in Brazil: Locational Patterns and Implications for Regional Development," *Professional Geographer*, 33, 1 (February 1981), 48–62.

D

Dale, Edmund H., "Some Geographical Aspects of African Land-locked States." *Annals, AAG*, 58, 3 (September 1968), 485–505. Follow-up commentaries appear in *Annals, AAG*, 59, 4 (December 1969), 820–822.

Desmond, Annabelle, "The Common Market," *Population Bulletin*, 18, 4 (July 1962), 65–90.

Doherty, Kathryn B., "Jordan Waters Conflict," *International Conciliation*, No. 553 (May 1965).

Doxey, Margaret, *Canada and the Evolution of the Modern Commonwealth. Behind the Headlines*, 40, 2 (1982).

Drysdale, Alasdair, "Political Conflict and Jordanian Access to the Sea," *Geographical Review*, 77, 1 (January 1987), 86–102.

Dubey, M., "International Law Relating to the Transit Trade of Land-locked Countries." *Indian Yearbook of International Affairs*, 14, (1967), 22–44.

E

East, W. Gordon, "The Geography of Land-locked States," *Transactions and Papers, Institute of British Geographers*, No. 28 (1960), 1–22.

Efron, Reuben and Allan S. Nanes, "The Common Market and Euratom Treaties: Supranationality and the Integration of Europe," *International and Comparative Law Quarterly*, 6, 4 (October 1957), 670–684.

Eggleston, Sir F.W., "The Kashmir Dispute and Sir Owen Dixon's Report." *Australian Outlook*, 2 (1951), 3–9.

Eken, Sena, "Breakup of the East African Community," *Finance and Development*, 16, 4 (December 1979), 36–40.

Etzioni, Amitai, "A Paradigm for the Study of Political Unification," *World Politics*, 15, 1 (October 1962), 44–74.

———, "European Unification: A Strategy of Change," *World Politics*, 16, 1 (October 1963), 32–51.

F

Fenwick, Charles G., "The Arms Embargo Against Bolivia and Paraguay," *American Journal of International Law*, 28 (1934), 534–538.

Fisher, Charles A., "South Asia: The Balkans of the Orient?" *Geography*, 47 (1962), 347–367.

———, "The Britain of the East? A Study in the Geography of Imitation," *Modern Asian Studies*, 2 (1968), 343–376.

Fleming, Douglas K., "The Common Market Today," *Journal of Geography*, 66 (1967), 449–453.

Fleure, H.J., "The Geographic Study of Society and World Problems," *Scottish Geographical Magazine*, (1968).

Franck, Thomas M. and others, "The New Poor: Land-locked, Shelf-locked and other Geographically Disadvantaged States." *New York University Journal of International Law and Politics*, 7, 1 (Spring 1974), 33–57.

Freeman, T. Walter and Mary M. Macdonald, "The Arctic Corridor of Finland," *Scottish Geographical Magazine*, 54, 4 (July 1938), 219–230.

Friedman, Wolfgang, "The United Nations and the Development of International Law," *International Journal*, 25 (Spring 1970), 272–286.

G

Galtung, Johan, "On the Effects of International Economic Sanctions; with Examples from the Case of Rhodesia," *World Politics*, 19 (April 1967), 378–416.

Geiser, Hans-Joerg, "The Lomé Convention and Caribbean Integration; a First Assessment," *Caribbean Yearbook of International Relations 1975*, 330–354.

———, "Regional Integration in the Commonwealth Caribbean," *Journal of World Trade Law*, 10, 6 (November–December 1976), 546–565.

Gigax, William R., "The Central American Common Market," *Inter-American Economic Affairs*, 16, 2 (1962), 59–77.

Gilmore, William C., "Legal and Institutional Aspects of the Organisation of Eastern Caribbean States," *Review of International Affairs*, 11, 4 (October 1985), 311–328.

Gladden, E.N., "The East African Common Services Organization," *Parliamentary Affairs*, 16, 4 (Autumn 1963), 428–439.

Glassner, Martin Ira, "The Río Lauca: Dispute Over an International River," *Geographical Review*, 60, 2 (April 1970), 192–207.

———, "International Law Regarding Land-locked States," *Nepal Review,* 11, 8 (June 1970), 388–393.

——, "The Status of Developing Land-locked States Since 1965," *Lawyer of the Americas,* 5, 3 (October 1973), 480–498.

——, "Developing Land-locked States and the Resources of the Seabed," *San Diego Law Review.* 11, 3 (May 1974), 633–655.

——, "Land-locked Nations and Development," *International Development Review,* 19, 2 (September 1977), 19–23.

——, "CARICOM: A Community in Trouble," *Focus,* 29, 2 (November–December 1978), 12–16.

——, "The Land-locked States at the Third United Nations Conference on the Law of the Sea," *Proceedings,* AAG, Committee on Marine Geography, (1978) 119–122.

——, "Transit Rights for Land-locked States and the Special Case of Nepal," *World Affairs,* 140, 4 (Spring 1978), 304–314.

——, "The Transit Problems of Land-locked States; the Cases of Bolivia and Paraguay," *Ocean Yearbook 4.* Chicago: University of Chicago Press, 1983, 366–389.

——, "Land-locked States and the 1982 Law of the Sea Convention," *Marine Policy Reports* (University of Delaware), 9, 1 (September 1986), 8–14.

Gonzalez, Anthony, "Trade Strategies of the Commonwealth Caribbean in a Changing World Economic Order; Special Reference to the Lomé Convention and Caribbean Economic Integration," *Caribbean Yearbook of International Relations 1975,* 355–373.

Gonzalez, Heliodoro, "U.S. Arms Transfer Policy in Latin America: Failure of a Policy," *Inter-American Economic Affairs,* 32, 2 (Fall 1978), 67–89.

——, "Arms-Sales Policy: The Chilean Case," *Inter-American Economic Affairs,* 34, 3 (Winter 1980), 3–24.

Goormaghtigh, John, "European Coal and Steel Community," *International Conciliation,* No. 503 (May 1955), 343–408.

Gottmann, Jean, "Geography and International Relations," *World Politics,* 3, 2 (January 1951), 153–173.

——, "The Political Partitioning of Our World: An Attempt at Analysis," *World Politics,* 4, 4 (July 1952), 512–519.

Govindaraj, V.C., "Land-locked States and Their Right of Access to the Sea." *Indian Journal of International Law,* 14, 2 (April 1974), 190–216.

——, "Land-locked States—Their Right to the Resources of the Sea-bed and the Ocean Floor." *Indian Journal of International Law,* 14, 3–4 (July–December 1974), 409–424.

Grieve, Muriel J., "Economic Sanctions Theory and Practice," *International Relations* 3 (October 1968), 431–443.

Gross, Leo, "The United Nations and the Role of Law," *International Organization,* 19, 3 (Summer 1965), 537–561.

H

Haas, Ernst B., "Regionalism, Functionalism, and Universal International Organization," *World Politics,* 8, 2 (January 1956), 238–263.

Halderman, John W., "Some Legal Aspects of Sanctions in the Rhodesian Case," *International and Comparative Law Quarterly,* 17 (July 1968), 672–705.

Hall, Kenneth O. and Byron W. Blake, "The Emergence of the African, Caribbean, and Pacific Group of States: An Aspect of African and Caribbean International Cooperation," *African Studies Review,* 22, 2 (September 1979), 11–125.

Hardy, O., "South American Alliances: Some Political and Geographical Considerations," *Geographical Review,* 4–5 (1919), 259–265.

Harkema, Roelof C., "The Ports and Access Routes of Land-locked Zambia," *Geografisch Tijdschrift,* 6, 3 (May 1972), 223–231.

Harris, P.B., "Rhodesia; Sanctions, Economics and Politics," *Rhodesian Journal of Economics,* 2 (September 1968), 5–20.

Hartshorne, Richard, "The Polish Corridor," *Journal of Geography,* 36, (1937), 161–176.

——, "The Politico-Geographic Pattern of the World," *Annals, American Academy of Political and Social Sciences,* 218 (1941), 45–57.

Harvey, Heather Joan, "The British Commonwealth: A Pattern of Cooperation," *International Conciliation,* No. 487 (January 1953), 1–48.

Hassan, Tariq, "Third Law of the Sea Conference; Fishing Rights of Land-locked States." *Lawyer of the Americas,* 8, 3 (October 1976), 686–742.

Hasse, Rolf, "Why Economic Sanctions Always Fail—the Case of Rhodesia," *Intereconomics* (Hamburg): 7–8 (July–August 1978), 194–199.

Hawkins, A.M., "The Rhodesian Economy Under Sanctions," *Rhodesian Journal of Economics,* 1 (August 1967), 1–44.

Hill, J.E., Jr., "El Chamizal: A Century-Old Boundary Dispute," *Geographical Review,* 55, 4 (October 1965), 510–522.

————, "El Horcon: A United States–Mexican Boundary Anomaly," *Rocky Mountain Social Science Journal*, 4, 1 (April 1967), 49–61.

Hilling, David, "Routes to the Sea for Landlocked States," *Geographical Magazine*, 44, 4 (January 1972), 252–264.

Hirsch, Abraham M., "Utilization of International Rivers in the Middle East: A Study of Conventional International Law," *American Journal of International Law*, 50, 1 (1956), 81–100.

————, "From the Indus to the Jordan; Characteristics of Middle East International River Disputes," *Political Science Quarterly*, 71, 2 (June 1956), 203–222.

Hoffman, George, "Toward Greater Integration in Europe: Transfer of Electric Power Across International Boundaries," *Journal of Geography*, 55, 4 (April 1956), 165–176.

Huang, Thomas T.F., "Some International and Legal Aspects of the Suez Canal Question," *American Journal of International Law*, 51, 2 (April 1957), 277–307.

Hundley, Norris, Jr., "The Colorado Waters Dispute," *Foreign Affairs*, 42, 3 (April 1964), 495–500.

————, "The Politics of Water and Geography: California and the Mexican-American Treaty of 1944," *Pacific Historical Review*, 36, 2 (1967), 209–226.

Husbands, Jo L., "A World in Arms: Geography of the Weapons Trade," *Focus*, 30, 4 (March–April 1980), 1–16.

I

Inglehart, Ronald, "An End to European Integration?," *American Political Science Review*, 61, 1 (March 1967), 91–105.

International Journal on World Peace. Published quarterly in New York since 1983 by the professors of the World Peace Academy.

Issawi, Charles, "The Bases of Arab Unity," *International Affairs*, 3, 1 (January 1955), 36–47.

J

James, Preston E., "Concepts and World Crises," *Journal of Geography*, 74, 1 (January 1975), 8–15.

Jayaraman, T.K. and Omkar Lal Shrestha, "Some Trade Problems of Landlocked Nepal," *Asian Survey*, 16, 12 (December 1976), 1113–1123.

Johnson, Gilbert R., "United States–Canadian Treaties Affecting Great Lakes Commerce and Navigation," *Inland Seas*, 3, 4 (October 1947), 203–207.

Johnson, R.W., "The Canada–United States Controversy over the Columbia River," *University of Washington Law Review*, 41 (1966), 676–763.

Jones, Stephen B., "Views of the Political World," *Geographical Review*, 45, 3 (July 1955), 309–326.

Journal of European Integration. Saskatoon: University of Saskatchewan, Department of Political Science. Published semiannually since 1977.

K

Kafando, Talata, "The Case of Liptako–Gourma," *Ceres*, 37 (January–February 1974), 45–48.

Kain, Ronald Stuart, "Bolivia's Claustrophobia," *Foreign Affairs*, 16, 4 (July 1938), 704–713.

Kalijarvi, Thorsten V., "Obstacles to European Unification," *Annals, American Academy of Political and Social Science*, (1963), 46–53.

Kalla, Patricia, "The GATT Dispute Settlement Procedure in the 1980's: Where Do We go from Here?" *Dickinson Journal of International Law*, 5, 1 (Fall 1986), 82–101.

Karan, Pradyumma P., "Dividing the Water: A Problem in Political Geography," *Professional Geographer*, 12, 1 (1961), 6–10.

Kirchheimer, Otto, "The Decline of Intra-State Federalism in Western Europe," *World Politics*, 3, 3 (April 1951), 281–298.

Kitzinger, U.W., "Europe: The Six and the Seven," *International Organization*, 14, 1 (Winter 1960), 20–36.

Kliot, Nurit, "Geography of Hostages: The Case of Lebanon," *1983 Yearbook of the Israeli Geographical Association*, (1984).

Kuehnelt-Leddihn, Erik R.V., "The Petsamo Region," *Geographical Review*, 34, 3 (July 1944), 405–417.

Kurey, Gregory S., "GATT and the VRA: Japanese Automobile Imports and Trade Protectionism," *Dickinson Journal of International Law*, 5, 1 (Fall 1986), 51–80.

L

Landau, Georges D., "The Treaty for Amazonian Cooperation: A Bold New Instrument for Development." *Georgia Journal of International and Comparative Law*, 10, 3 (1980), 463–489.

Langer, W.L., "The Struggle for the Nile," *Foreign Affairs*, 14, 2 (1936), 259–273.

Langlands, B.W., "Concepts of the Nile," *Uganda Journal*, 26, 1 (March 1962), 1–22.

Latimer, Paul, "The Law of the Sea: Problems Arising in the Analysis of the Rights of Land-locked States," *International Relations*, 5, 6 (November 1977), 180–197.

Laylin, John G. and Rinaldo L. Bianchi, "The role of Adjudication in International River Disputes," *American Journal of International Law*, 53, 1 (January 1959), 30–49.

"The League of Arab States," *Arab Perspectives*, 1, 1 (April 1980), 31–56.

Lee Yong Leng, "Economic Aspects of Supra-nationalism: The Case of ASEAN," *Political Geography Quarterly*, 2, 1 (January 1983), 21–30.

Leich, Marian Nash, "The Sinai Multinational Force and Observers," *American Journal of International Law*, 76, 1 (January 1982), 181–182.

Lepawsky, Albert, "International Development of River Resources," *International Affairs*, 39, 4 (October 1963), 533–550.

Lerner, Daniel, "Will European Union Bring About Merged National Goals?" *Annals, American Academy of Political and Social Science.* (1963), 34–45.

Lewis, I.M., "Pan-Africanism and Pan-Somal-ism," *Journal of Modern African Studies*, 1 (1963), 147–162.

Lewis, Vaughan A., "Evading Smallness: Regional Integration as an Avenue Towards Viability," *International Social Science Journal*, 30, 1 (1978), 57–72.

Longrigg, Stephen H., "New Groupings Among the Arab States," *International Affairs*, 34, 3 (July 1958), 305–317.

Lowenstein, Karl, "Sovereignty and International Cooperation," *American Journal of International Law*, 48, 2 (April 1954), 222–244.

M

Makil, R., "Transit Rights of Land-locked Countries: An Appraisal of International Conventions." *Journal of World Trade Law*, 4, 1 (January–February 1970), 35–51.

Malan, Theo, "The Preferential Trade Area (PTA) of Southern and Eastern African Countries," *Africa Institute Bulletin*, 13 (1982), 104–106.

Marts, Marion, and W.R.D. Sewell, "Conflict Between Fish and Power Resources in the Pacific Northwest," *Annals, AAG*, 50, 1 (March 1960), 42–50.

McConnell, James E., "Foreign Ownership and Trade of United States High-Technology Manufacturing," *Professional Geographer*, 33, 1 (February 1981), 63–71.

————, "Geography of International Trade," *Progress in Human Geography*, 10, 4 (December 1986), 471–483.

McFadden, Eric J., "The Collapse of Tin: Restructuring a Failed Commodity," *American Journal of International Law*, 80, 4 (October 1986), 811–830.

McNee, Robert B., "Centrifugal–Centripetal Forces in International Petroleum Company Regions," *Annals, AAG*, 51, 1 (March 1961), 124–138.

McRae, D.M. and J.C. Thomas, "The GATT and Multilateral Treaty Making: The Tokyo Round," *American Journal of International Law*, 77, 1 (January 1983), 51–83.

Meinig, Donald W., "Cultural Blocs and Political Blocs: Emergent Patterns in World Affairs," *Western Humanities Quarterly*, 10, 3 (Summer 1956), 203–222.

Meredith, B., "Geo-Politics and the Transnationals," *Contemporary Review*, 227, 1319 (1975), 316–320.

Merritt, Richard L., "Distance and Interaction Among Political Communities," *General Systems*, 9 (1964), 255–263.

Milič, Milenko, "Access of Land-locked States to and from the Sea." *Case Western Reserve Journal of International Law*, 13, 3 (Summer 1981), 501–516.

Mirvahabi, Farin, "The Rights of the Land-locked and Geographically Disadvantaged States in Exploitation of Marine Fisheries," *Netherlands International Law Review*, 26, 2 (1979), 130–162.

Morris, Kenton W., "The St. Lawrence Seaway—Its Development and Economic Significance," *Journal of Geography*, 55, 9 (December 1956), 447–452.

Murphy, John F., "Recent International Legal Developments in Controlling Terrorism," *Chinese Yearbook of International Law and Affairs*, 4 (1984), 97–127.

N

Naghmi, Shafqat Hussain, "Exclusive Economic Zone and the Land-locked States." *Pakistan Horizon*, 33, 1 and 2 (1st and 2nd Quarters 1980), 37–48.

Nijim, Bashir K., "Conflict Potential and International Riparian Dispute: The Case of Jordan," *Iowa Geographer*, 28 (Fall 1971), 8–14.

O

O'Connell, Joseph F., "The Caribbean Community; Economic Integration in the Commonwealth Caribbean," *Journal of Interna-*

tional Law and Economics, 2, 1 (1976), 35–66.

Osala, E.M., "Some Current Aspects of International Commodity Policy," *Journal of Agricultural Economics,* 18 (1967), 27–46.

Ostrander, F. Taylor, "Zambia in the Aftermath of Rhodesian UDI: Logistical and Economic Problems." *African Forum,* 2, 3 (Winter 1967), 50–65.

P

Padelford, Norman J., "Cooperation in the Central American Region: The Organization of Central American States, *International Organization,* 11, 1 (Winter 1957), 41–54.

Padelford, Norman J. and Rupert Emerson, eds., *Africa and International Organization,* Special issue of *International Organization,* 16, 2 (Spring 1962).

Palmer, Norman D. (ed.), "The National Interest—Alone or with Others?" *Annals, American Academy of Political and Social Sciences,* 282 (July 1952), 1–118.

Pathmanathan, M. "The Political Dynamics of ASEAN Cooperation," *Indonesian Quarterly,* 6, 3 (July 1978), 28–41.

Patterson, Ernest Minor (ed.), "Looking Toward One World," *Annals, American Academy of Political and Social Sciences,* 258 (July 1948), 1–123.

——— (ed.), "World Government," *Annals, American Academy of Political and Social Sciences,* 264 (July 1949), 1–123.

——— (ed.), "NATO and World Peace," *Annals, American Academy of Political and Social Sciences,* 288 (July 1953), 1–237.

Payne, Anthony J., "The Rise and Fall of Caribbean Regionalisation," *Journal of Common Market Studies,* 19, 3 (March 1981), 255–280.

Pazzanita, Anthony G., "Legal Aspects of Membership in the Organization of African Unity: The Case of the Western Sahara," *Case Western Reserve Journal of International Law,* 17, 1 (Winter 1985), 123–158.

Perlow, Gary H., "The Multilateral Supervision of International Trade: Has the Textiles Experiment Worked?" *American Journal of International Law,* 75, 1 (January 1981), 93–133.

Petersen, N. and J. Alklit, "Denmark Enters the European Communities," *Scandinavian Political Studies,* 8 (1973), 198–213.

Pobbi-Asamani, Kwadao O., "Self-reliance in West Africa: The Political Economy of the Economic Community of West African States—The First Ten Years, 1975–1985—A

Preview," *A Current Bibliography on African Affairs,* 17, 1 (1984), 5–15.

Polland, V.K., "A.S.A. and A.S.E.A.N. 1961–1967: Southeast Asian Regionalism," *Asian Survey,* 10, 3 (March 1970), 244–255.

Porter, Richard C., "Economic Sanctions: The Theory and the Evidence from Rhodesia," *Journal of Peace Science,* 3, 2 (1978), 93–110.

Pounds, Norman J.G., "A Free and Secure Access to the Sea," *Annals, AAG,* 49, 3 (September 1959), 256–268.

Pye, Lucian W., "The Non-Western Political Process," *Journal of Politics,* 20, 3 (August 1958), 468–486.

R

Ramsaran, Ramesh, "Integration and Underdevelopment in the Commonwealth Caribbean," *Intereconomics* (Hamburg), 7–8 (July–August 1977), 197–200.

———, "CARICOM: The Integration Process in Crisis," *Journal of World Trade Law,* 12, 3 (May–June 1978), 208–217.

Robson, Peter, "The Mano River Union," *Journal of Modern African Studies,* 20, 4 (December 1982), 613–628.

Root, Franklin R., "The European Coal and Steel Community," *Studies in Business and Economics,* University of Maryland, Bureau of Business and Economic Research, 9, 3 (December 1955), 1–19; 10, 1 (June 1956), 1–16.

Rose, William J., "Czechs and Poles as Neighbors," *Journal of Central European Affairs,* 11, 2 (July 1951), 153–171.

Rubin, Alfred P., "Terrorism and Piracy: A Legal View," *Terrorism: An International Journal,* 3, 1–2 (1979) 117–130.

Rubin, Seymour, "Transnational Corporations and International Law: An Uncertain Partnership," *Chinese Yearbook of International Law and Affairs,* 4 (1984), 39–49.

Russett, Bruce M., "Is There a Long-run Trend Toward Concentration in the International System?" *Comparative Political Studies,* 1 (1958), 103–122.

———, "Components of an Operational Theory of International Alliance Formation," *Journal of Conflict Resolution,* 12, 3 (September 1968), 285–301.

S

Sackey, James A., "The Structure and Performance of CARICOM: Lesson for the Development of ECOWAS," *Canadian Journal of African Studies,* 12, 2 (1978), 259–277.

Sarup, Amrit, "Transit Trade of Land-locked

Nepal," *International and Comparative Law Quarterly*, 21, 2 (April 1972), 287–306.

Sautter, Hermann, "LAFTA's Successes and Failures," *Intereconomics* (Hamburg), 5 (1972), 149–152.

Schachter, Oscar, "The Development of International Law Through the Legal Opinions of the United Nations Secretariat," *British Yearbook of International Law*, 25 (1948), 91–132.

Schrodt, P. (ed.), "Arms Transfers: Special Issue," *International Interactions*, 10 (1983), 1–141.

Sewell, W.R.D., "The Columbia River Treaty: Some Lessons and Implications," *Canadian Geographer*, 10, 3 (1966), 145–156.

Shaw, Timothy M., "International Organizations and the Politics of Southern Africa; Towards Regional Integration or Liberation?" *Journal of Modern African Studies*, 3, 1–19.

Shee Poon-Kim, "A Decade of ASEAN, 1967–1977," *Asian Survey*, 17, 8 (August 1977), 753–770.

Simmonds, K.R., "International Economic Organizations in Central America and the Caribbean. Regionalism and Sub-Regionalism in the Integration Process," *International and Comparative Law Quarterly*, 19, 3 (July 1970), 376–397.

Simsarian, J., "The Division of Waters Affecting the United States and Canada," *American Journal of International Law*, (1938), 488–518.

Smith, Harold A., "The Waters of the Jordan: A Problem of International Water Control," *International Affairs*, 25, 4 (October 1949), 415–425.

Sprout, Harold and Margaret Sprout, "Geography and International Politics in an Era of Revolutionary Change," *Journal Conflict Resolution*, 4, 1 (March 1960), 145–161.

Spykman, Nicholas John, "Frontiers, Security and International Organizations," *Geographical Review*, 32, 3 (1942), 436–447.

Steiner, H. Arthur, "Communist China in the World Community," *International Conciliation*, No. 533 (May 1961).

Stephenson, Glenn V., "The Impact of International Sanctions on the Internal Viability of Rhodesia," *Geographical Review*, 65, 3 (July 1975), 377–389.

Sulaiman, M.A., "Free Access: The Problem of Land-locked States and the 1982 United Nations Convention on the Law of the Sea," *South African Yearbook of International Law*, 10 (1984), 144–161.

Sweeney, Jane Chace, "State-Sponsored Terrorism: Libya's Abuse of Diplomatic Privileges and Immunities," *Dickinson Journal of International Law*, 5, 1 (Fall 1986), 133–165.

T

Tabibi, Abdul Hakim, "The Right of Land-locked Countries to Free Access to the Sea." *Österreichische Zeitschrift für Offenttliiches Recht*, 23 (1972), 117–146.

Tarlton, C.D., "The Styles of American International Thought: Mahan, Bryan, and Lippman," *World Politics*, 17, 4 (July 1965), 584–614.

Thee, M., "Third World Armaments: Structure and Dynamics," *Bulletin of Peace Proposals*, 13 (1982), 113–118.

Thompson, Dennis, "The European Economic Community: Internal Developments Since the Breakdown of the British Negotiations," *International and Comparative Law Quarterly*, 13 (July 1964), 830–853.

Travis, Martin B., "The Organization of American States: A Guide to the Future," *Western Political Quarterly*, 10, 2 (1956).

Triska, Jan F. and Howard E. Koch, Jr., "Asian–African Coalition and International Organization: Third Force or Collective Impotence?" *Review of Politics*, 21 (1959), 417–455.

V

Van Cleef, Eugene, "Finland—Bridge to the Atlantic," *Journal of Geography*, 48, 3 (March 1949), 99–105.

Venkatesan, S., "UNCTAD and the Least Developed and Land-locked Countries." *Foreign Trade Review* (New Delhi), 7, 3 (October 1972), 280–289.

Verwey, Wil D., "The International Hostages Convention and National Liberation Movements," *American Journal of International Law*, 75, 1 (January 1981), 69–92.

W

Ward, Michael, Review Essay: "Recognizing a Hegemon," *Political Geography Quarterly*, 4, 4 (October 1985), 343–346.

Wasserman, Ursula, "Economic Sanctions—the Rhodesian Experience," *Journal of World Trade Law*, 9, 5 (September–October 1975), 590–593.

Weisfelder, Richard F., "The Southern African Development Coordination Conference (S.A.D.C.C.)," *South Africa International*, 13, 2 (October 1982), 74–96.

Whitaker, Arthur P., "Development of American Regionalism; the Organization of American States," *International Conciliation*, No. 469 (March 1951), 121–164.

White, Gilbert F., "The Changing Dimensions of the World Community," *Journal of Geography*, 4 (1960), 165–170.

————, "The Mekong River Plan," *Scientific American*, 208, 4 (April 1963), 49–59.

Wigmore, J.H., "A Map of the World's Law," *Geographical Review*, 19 (1929), 114–120.

Williams, Michael and Michael Parsonage, "Britain and Rhodesia: The Economic Background to Sanctions," *World Today*, 29, 9 (September 1973), 379–388.

Wolfe, Alvin W., "The African Mineral Industry: Evolution of a Supranational Level of Integration," *Social Problems*, 11, 2 (Fall 1963), 153–164.

Wright, Esmond, "The 'Greater Syria' Project in Arab Politics," *World Affairs*, 5, 3 (July 1951), 318–329.

Wright, John K., "Geography and the Study of Foreign Affairs," *Foreign Affairs*. 17, 1 (October 1938), 153–163.

Wright, L.A., "A Study of the Conflict Between the Republics of Peru and Ecuador," *Geographical Journal*, 98, 5–6 (November–December 1941), 253–272.

Y

Yansane, Aguibov, "The State of Economic Integration in North West Africa South of the Sahara: The Emergence of the Economic Community of West African States (ECOWAS)," *African Studies Review*, 20, 2 (September 1979), 63–87.

Yearbook of International Organizations. London: Taylor & Francis.

Part Seven

OUR LAST FRONTIERS

Chapter 29

THE TRADITIONAL LAW OF THE SEA

Those who go down to the sea in ships,
Who do business in great waters;
They see the works of the Lord
And His wonders in the deep.
——FROM PSALM 107

So wrote the psalmist more than two thousand years ago when seafaring was already an ancient and respected calling. King Solomon nearly a thousand years earlier maintained a large and profitable merchant navy. The ancient Phoenicians and Polynesians ranged farther over the oceans of the world than we can imagine in what seem to us to be the frailest of sailing craft. The ancient Greeks not only viewed the sea as the binding agent of their fragmented homeland, but concentrated so much on the sea that they became vulnerable to overland attack. The Roman Empire, on the other hand, managed to combine sea power with effective organization on land. The Mediterranean became what the Greeks had never been able to make it, an interior sea, a Roman lake.

The Romans ventured far beyond the Mediterranean, for we have descriptions of Indian Ocean coasts written by Roman sea captains. But in the centuries that followed, the oceans remained an insurmountable barrier to Europe. Chinese vessels reached the eastern coast of Africa while the Dark Ages settled over the European continent. Some may even have reached America. Not until the ascent to power of the Spanish and Portuguese kingdoms on the Iberian Peninsula, almost a dozen centuries after the fall of Rome, did the oceans begin once again to form a link between Europe and other parts of the world. Since then, the link has been permanent, and the oceans have carried European power to all parts of the globe. In search of land and resources, the Spanish and Portuguese vessels returned home with new knowledge of the earth. Soon, there were competitors for the domination of the sea routes to spices, gold, and slaves. The Hollanders successfully challenged the Portuguese; the English challenged the Dutch. Once the oceans had contained Europe—now they were the routes to wealth and power.

The development of overseas "spheres of influence" began almost immediately. So intense was the competition for control in the Americas that Spanish and Portuguese interests were limited by the Treaty of Tordesillas even before the end of the fifteenth century. The treaty, in fact, constitutes a first attempt to define and delimit a geometric boundary. Known as the "Pope's Line," it divided the yet unexplored world into a Spanish and a Portuguese realm, the boundary lying along the meridian 370 leagues west of the Cape Verde Islands.* The Portuguese established themselves along the west coast of Africa and made contact with the Arabs plying the eastern coasts. The Dutch laid claim to major portions of southeast Asia and established settlements and fortifications in a multitude of other places, in-

*Hence Portuguese cultural influence dominates Brazil, while Spanish influences lie to the west. According to the treaty, land to the west of the meridian should go to Spain, land to the east should become Portuguese.

461

cluding Cape Town. Danes, Brandenburgers, English, French, Germans, Italians—all entered the scramble for overseas possessions at one time or another.

While the oceans had thus become avenues of conquest and power, they also became a threat to coastal communities, not only in the colonized world but also in Europe itself. A powerful State that could invade the Americas, Africa, or Asia could also assert its strength in Europe by attacking actual or potential competitors. Hitherto, frontiers had acted as separating agents on land, and it was never difficult to interpret the intentions of any army moving into such frontiers, whether such an army was one of occupation or conquest. But a group of vessels approaching offshore might simply be passing toward some distant land, or might suddenly attack, having approached unmolested to within a few hundred meters of the coast!

Through all these centuries, while some States were stoutly defending the freedom of the seas, others were trying to assert various kinds of jurisdiction or even complete sovereignty over portions of the sea. The Greek city-states, the Roman Republic, Venice, and other Italian city-states were the most active in the Mediterranean in laying the foundations for what was to become the Law of the Sea, whose core was the concept of the territorial sea.

THE TERRITORIAL SEA

In the thirteenth and fourteenth centuries, the Norwegians, Danes, English, and Dutch exercised control over various parts of the North Sea and North Atlantic Ocean. They did so by treaty or ordinance, and usually the areas claimed were not precisely defined. In the fourteenth century, activity in the North Sea greatly increased. The Dutch fishing fleet expanded, and fishing vessels from Flanders, France, and England joined in the search for the best fishing grounds. Sporadic friction occurred, but it was not until the end of the sixteenth century that the question of

countries' rights on neighboring waters developed into a full-scale legal battle.

In the 1590s, the Danes, to whom power had gone from Norway, decided to abandon the "closed sea" practice they had inherited from the Norwegians. Instead, they announced that a belt of water, eight miles in width, lying around their possession of Iceland, constituted Danish territorial waters. Some of northwestern Europe's best fishing grounds lie off the Norwegian, Icelandic, and Faeroe coasts, and soon the Danish crown defined similar belts around all these possessions. It was not long before they increased these widths; in the mid-seventeenth century they claimed as much as twenty-four miles.* The English, meanwhile, had drawn baselines along the British coast from promontory to promontory, cutting off large portions of the coastal waters; some of these baselines were more than 50 miles in length. Furthermore, the English demanded that the Dutch and others fishing near their coasts obtain permits to do so.

Thus began the legal debate that ultimately produced many of the principles to which States still adhere today. The Dutch, who did not have large coastal bays and indentations, argued that closing portions of the sea constituted violation of "customary" international law. Early in the seventeenth century, a Dutch jurist, Hugo Grotius, published a treatise, the significant part of which was called *The Free (Open) Sea*. The British legal expert John Selden soon replied under the title *The Closed Sea*.

As the seventeenth century wore on, the argument grew in intensity. Indeed, it was a major contributor to four naval wars fought between the English and the Hollanders, while the legal debate raged on in the intervening periods. The Dutch objected to restrictions of any kind other than

*Measurements of distances at sea and in the air are customarily expressed in nautical miles, and all references to miles in these two chapters are understood to mean nautical miles. A nautical mile is approximately one minute of arc on the earth's surface, or about 1.15 English or statute miles.

a narrow belt of territorial sea perhaps determined by the cannon-shot principle; the British stood firm on the principle of extensive closed seas. But time was on the side of the Hollanders. Like the Dutch, the English came to depend to an ever greater extent on their growing merchant fleet, and they wanted that fleet to face as few obstacles as possible. Thus by the end of the seventeenth century, the English had begun to realize that their policy of closed seas could not be reconciled with their dependence on free oceanic trade. Thus they yielded to some extent to Dutch desires, and the debate now came to focus on the problem of the manner by which the extent of any belt of territorial sea should be determined.

Again, two schools of thought existed. The question previously had revolved around "free" or "closed" seas; it now concerned the principle of effective occupation as opposed to that of a continuous (if narrow) belt of territorial sea along all coasts. This is the point where Cornelius Van Bynkershoek enters the legal controversy. Van Bynkershoek, the Dutch jurist, was probably the most vocal and energetic legal expert ever to become involved in this question. As a result, many innovations are attributed to him which actually seem to have originated with others. But Van Bynkershoek had the ability to express himself so well and convincingly that his name has become inextricably attached to a number of rules and definitions.

Van Bynkershoek's *De Dominio Maris Dissertatio,* first published in 1702 and revised in 1744, summarized the entire question as it stood at that time, established a terminology that is still in use today, and brought the Dutch standpoint forcefully to the attention of the English. The English favored the concept of a continuous belt of equal width rather than Van Bynkershoek's shore-domination principle, and this time the English idea came to be generally accepted. The Danes, French, and Italians preferred the equal-width zone of territorial water also, and now the problem was one of determining the exact width of such a belt.

The Danes in 1745 announced that they would claim a width of 1 marine league, equal to 4 nautical miles. The Dutch agreed to observe a belt of equal width, but they insisted on using the cannon-shot as their determinant. Abbé Fernando Galiani, an Italian writer, is credited with the proposal that 3 miles be used as the generally acceptable standard width. Galiani apparently based his suggestion on what he viewed as the maximum range of the cannon of that day, but military experts have determined that the range of eighteenth-century cannons probably did not exceed 1 mile. Whatever the origins of the 3-mile limit, after the United States and Britain, around the turn of the nineteenth century, adopted it for their maritime boundaries, it rapidly became the most common claim in many parts of the world.

During much of the nineteenth century and the first half of the twentieth, the situation remained relatively stable by virtue of European colonial control and the application of European claims to overseas possessions. There were disputes over European fishing grounds and several separate treaties were signed among interested States, but until World War II, those States that thrived partly or largely on their overseas trade maintained the 3- or 4-mile limit. In 1951, 80 percent of the merchant-shipping tonnage of the world was registered in countries that subscribed to the 3-mile limit, and another 10 percent to States adhering to the 4-mile limit. Today more and more States are claiming 12 and even more miles as their territorial sea. Bays are being closed off, fishing disputes are growing in number and intensity, and the range of claims and justifications is unprecedented.

OTHER ASPECTS OF THE TRADITIONAL LAW OF THE SEA

Once the international community accepted the concept that every coastal State is entitled to a territorial sea and most

Soviet trawler in the Bering Sea. Vessels of this type, sailing chiefly under Soviet and Japanese flags, are primarily used to process fish caught by smaller vessels carried on board. Their voyages commonly last more than a year and their canned or frozen catches may be sold in ports far from home. (National Marine Fisheries Service)

coastal States had adopted 3 or 4 nautical miles as the breadth of their territorial seas,* the other issues of maritime jurisdiction became either easier to resolve or less important. They included methods of measuring the territorial sea; protection of fishing grounds; jurisdiction over customs, fiscal, immigration, and health matters; neutrality and security jurisdiction outside the territorial sea; and the status of bays and straits. The Law of the Sea, like other branches of international law, continued to evolve at a leisurely pace. All this was changed by World War I.

As we have already seen, World War I destroyed the nineteenth-century world. Out of the wreckage came, among other things, the League of Nations. As part of its effort to develop and codify international law, the League sponsored a number of law-making conferences. One of the most important was the Conference for the Codification of International Law, held at The Hague in 1930. Among the many matters debated there were several issues involving the Law of the Sea. They included: a uniform 3-mile territorial sea; the claims of certain States to broader territorial seas; acceptance of the principle of a contiguous zone within which the coastal State could exercise jurisdiction in

*A major exception was the Ottoman Empire, which claimed a breadth of 6 miles; several other Mediterranean States did likewise.

customs, health, and security matters; proposals for extensive exclusive fishery zones; and a newer "common patrimony" concept. The Conference failed to resolve these matters, but it did perform a useful function by identifying and partially defining them, since they were to grow steadily in importance. The major confrontation at The Hague was between distant-water fishing States, such as Japan and Britain, which wanted a narrow territorial sea and no exclusive fishing zones, and those States that wanted to keep foreign fishing fleets as far from their shores as possible. This signaled a major shift in emphasis in the evolving Law of the Sea from commerce and security to the resources of the sea.

Through the 1930s and throughout World War II the Law of the Sea continued to evolve slowly, despite excessive claims to maritime jurisdiction by a small number of countries, led by the Soviet Union, which in 1927 claimed a 12-mile territorial sea. This deliberate pace was ended abruptly on 28 September 1945. On that day President Harry Truman of the United States issued two proclamations. The first was a moderate one, announcing that the United States would regulate fisheries in those areas of the high seas contiguous to her coasts, but that there would be no restrictions on navigation in these zones. The second proclamation, however, was dramatic. It asserted U.S. jurisdiction over the

Drilling for oil off Australia. Nearly a third of all of the petroleum being produced in the world today comes from wells drilled into the continental shelf and this proportion is likely to increase in the future. Australia has recently become a significant oil producer, mostly from offshore wells. This one is in the Bass Strait off the coast of Victoria State. (UN/Australian Information Service)

resources of the continental shelf and of the subsoil of the shelf. A subsequent policy statement defined the outer limit of U.S. jurisdiction as the 200-meter isobath.* Motivated though it was by a perceived need to tap known and suspected offshore oil deposits so as to reduce dependence on imports for a ravenous oil-based economy and by the need to establish federal control over this resource in the face of claims by several coastal states, it nevertheless set off what we call "The Great Sea Rush" of the twentieth century.

Looking back on it now, the terms of "the Truman Proclamation," as it is now called, seem quite moderate, since they included preservation of all traditional high seas rights in the water above the shelf (superjacent water). It did, however, call the world's attention to the fact that out beyond the coastal zone was something of value besides fish and that there was nothing to prevent a coastal State from simply grabbing it.† If the United States could do it, why couldn't others? They could—and did. Only a month later Mexico also claimed both exclusive fishing and

*An isobath is a line connecting points of equal water depth, an underwater contour line. A depth of 200 meters is roughly 100 fathoms or about 660 feet.

†By this one presidential act, the United States acquired exclusive rights to the vast mineral and living resources of 2.4 million square kilometers of land under the sea (700,000 square miles).

continental shelf rights. Argentina in 1946 claimed not only its extraordinarily broad continental shelf, but also the superjacent waters. In the following year Peru and Chile went even further and claimed *sovereignty* over a belt of sea extending 200 nautical miles from their coasts. They claimed that since they had little or no continental shelf, they were entitled to the rich fisheries of the Humboldt Current as compensation. Since then claims to maritime jurisdiction have multiplied and escalated so rapidly and so far that the Law of the Sea has been transformed.

THE UNITED NATIONS CONFERENCES ON THE LAW OF THE SEA

We may recall that one of the functions of the United Nations is to continue the work of the League in developing and codifying international law. One subject assigned to the International Law Commission was the preparation of draft conventions on several aspects of the Law of the Sea in 1958. Much of the customary Law of the Sea was thus codified into three conventions, on the Territorial Sea and the Contiguous Zone, on the High Seas, and on Fishing and Conservation of the Living Resources of the High Seas; and the Convention on the Continental Shelf codified a new doctrine that had been adopted by so many

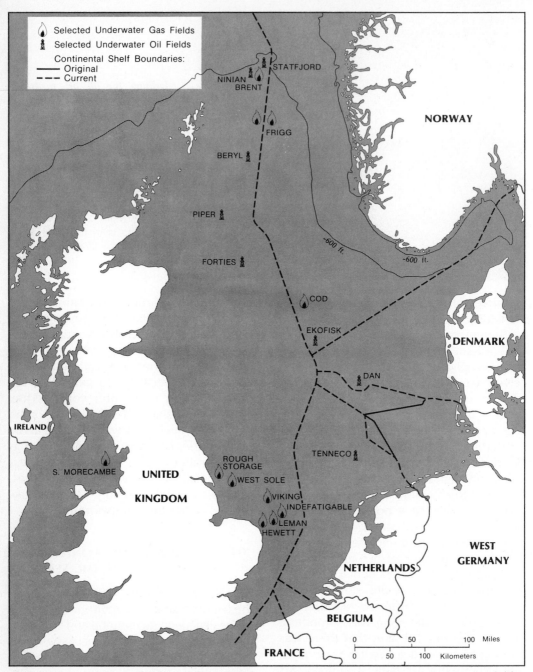

The North Sea continental shelf.   Note the deep trench off the coast of Norway; this was deliberately ignored when the countries surrounding the North Sea decided on median lines as their shelf boundaries. Because of the configuration of her coast, Germany received a very small portion of the shelf and none of the area most promising for hydrocarbon deposits. She then brought the Netherlands and Denmark before the World Court, which ruled that equity as well as the median line should be taken into consideration. The parties then negotiated new boundaries for the German sector. Only selected oil and gas wells are shown; there are many more.

States so rapidly that it has been referred to as an "instant custom."

Why was the 1958 conference so much more successful than The Hague Conference of 1930, even though many of the issues were the same? Much had changed in the interim. The 1930 conference was dominated by lawyers who tended to be doctrinal and legalistic and who felt no sense of urgency. The 1958 conference was attended by 86 States (compared with 47 in 1930), including many new ones in Asia, Africa, and the Middle East, and their delegations included experts in many fields besides law. And the times were different. In 1930, the world was quiet; in 1958, the delegates were faced with the continental shelf doctrine, population pressures, proliferating claims to maritime resources and territory, the relatively greater strength of the United Nations, rapidly advancing technology, increased expectations of the poor countries of benefits from the sea, and, of course, the Cold War between the United States and the Soviet Union. The Law of the Sea was not, has never been, and never can be divorced from the tangle of political, economic, and social developments in the world at large.

Despite its very considerable success, the 1958 conference left a number of matters unsettled, including two of the most controversial ones: the breadth of the territorial sea and the establishment of exclusive fishing zones. In order to resolve these and other issues, the Second United Nations Conference on the Law of the Sea was held in Geneva in the spring of 1960. It was short (six weeks), sharp, and unsuccessful. A U.S. compromise proposal on the territorial sea—a six-mile territorial sea plus a six-mile fishing zone—was defeated by one vote. If only a single country out of the 87 present had voted for it instead of against it, much of the subsequent history of the world would have been very different.

The four 1958 conventions and the 1965 Convention on Transit Trade of Landlocked States went into effect and became the new datum plane for the Law of the

Arvid Pardo. At the signing ceremony of the United Nations Convention on the Law of the Sea, Montego Bay, Jamaica, December 1982, he was an observer but had no official role. (Martin Glassner)

Sea. But the requisite ratifications had scarcely been deposited with the Secretary-General of the United Nations when another climactic event led to its total rethinking and renegotiation. On 17 August 1967, Arvid Pardo, Permanent Representative (ambassador) of Malta to the United Nations, proposed to the Secretary-General that the agenda of the forthcoming twenty-second session of the General Assembly include an item entitled "Declaration and treaty concerning the reservation exclusively for peaceful purposes of the seabed and of the ocean floor, underlying the seas beyond the limits of present national jurisdiction, and the use of their resources in the interests of mankind." In an accompanying memorandum, he suggested that the seabed and ocean floor be declared a "common heritage of mankind." He proposed the creation of an international agency that would exercise complete jurisdiction over this area. Arvid Pardo's proposal set off a chain of events that have transformed the Law of the Sea more in the past two decades than in the thousands of years of its evolution before 1967.

In December 1967, the General Assembly established an *ad hoc* Committee on the Peaceful Uses of the Seabed and Ocean Floor Beyond the Limits of National Jurisdiction, which came to be known as the Seabed Committee. In 1969, the committee was enlarged (ultimately to 91 members) and became essentially a preparatory committee for the Third United Nations Conference on the Law of the Sea (UNCLOS III). The Seabed Committee labored on for four more years. It was unable to produce a draft convention on the Law of the Sea, but it did do a lot of research, its members and the public generally learned a great deal about the sea, and it defined issues and narrowed the range of options for dealing with them. While it was working, interest in the sea was growing both within the United Nations and in the world outside. A revised Law of the Sea was becoming more urgent. Why?

There are a number of reasons. First, the 1958 and 1965 conventions left a number of important matters unsettled, had not been ratified or acceded to by enough States to be firmly established as binding on the entire community of States, and were being outdistanced by events. Second, decolonization had led to the admission to the United Nations of 41 new countries between 1958 and 1967 and the Group of 77 had organized as a pressure group within the UN. They wanted to participate in both the exploitation of new resources and the progressive development of international law. Third, the potential value of manganese nodules, discovered by the British oceanographic research vessel HMS *Challenger* in 1873–1876, was first understood and publicized in the early 1960s.* Fourth, the "population explosion" and the equally spectacular "technology explosion" were putting enormous pressure on finfish, shellfish,

*The nodules are really polymetallic, containing at least 27 elements and at least 14 other constituents, all varying widely in proportions. They are not rocks, but round or potato-shaped lumps of precipitates from the seawater that litter the floors of the continental shelves, the ocean basins, and even some freshwater lakes. The potential value of their manganese, copper, cobalt, and nickel alone is estimated in the trillions of dollars.

marine mammals, and seaweed, with some species being harvested to the brink of extinction.

Fifth, the superpowers were turning the sea into a colossal military playing field. Their nuclear submarines were ranging everywhere and even their surface ships were intruding into waters hitherto free of their rivalries, such as the polar seas and the Indian Ocean, and they were developing techniques for placing on the seafloor, not only sensing devices, but also nuclear and other weapons of mass destruction. Sixth, the ecosystem of the sea was being seriously damaged by modern technology in addition to overfishing, as dramatized by the 1967 *Torrey Canyon* oil-spill disaster, yet there was no international mechanism for adequately handling this problem.

Seventh, the International Geophysical Year (1957–1958), the International Indian Ocean Expedition (1959–1965), and other cooperative ventures in scientific cooperation not only produced a great deal of new information about the sea, but graphically demonstrated how little we know about it even now. They also demonstrated the value of international as well as interdisciplinary cooperation in marine scientific research. Eighth, more countries were extending their jurisdiction farther and farther out to sea, unrestrained by any provisions of international law. Since annexing land of neighbors was no longer feasible, they were turning for expansion to the defenseless and nearly friendless sea. By the end of 1967, about a dozen countries were already claiming some sort of jurisdiction *beyond* 12 nautical miles, and six more joined them by 1970. The trend was clear.

All these factors led the General Assembly to adopt in December 1970 a "Declaration of Principles" that essentially enshrined Arvid Pardo's common heritage proposals. Another resolution authorized UNCLOS III. But already the common heritage concept was being undermined by the very States giving it such vociferous

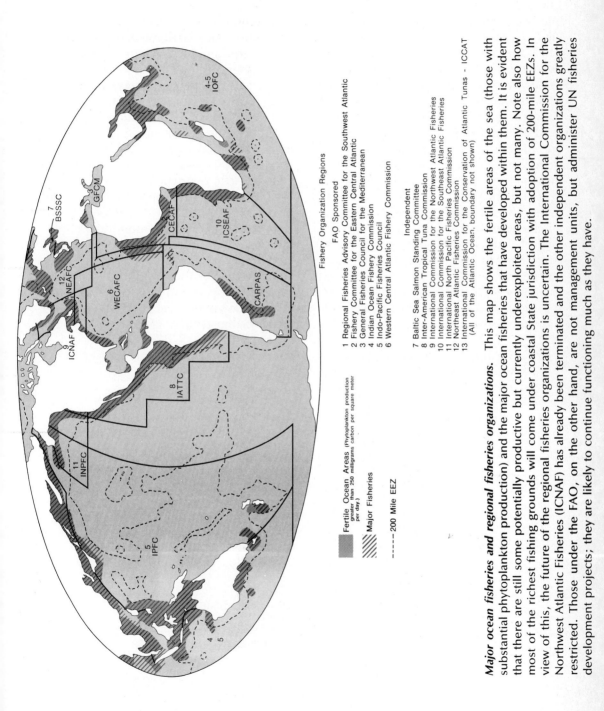

Fishery Organization Regions

FAO Sponsored

1 Regional Fisheries Advisory Committee for the Southwest Atlantic
2 Fishery Committee for the Eastern Central Atlantic
3 General Fisheries Council for the Mediterranean
4 Indian Ocean Fishery Commission
5 Indo-Pacific Fisheries Council
6 Western Central Atlantic Fishery Commission

Independent

7 Baltic Sea Salmon Standing Committee
8 Inter-American Tropical Tuna Commission
9 International Commission for the Northwest Atlantic Fisheries
10 International Commission for the Southeast Atlantic Fisheries
11 International North Pacific Fisheries Commission
12 Northeast Atlantic Fisheries Commission
13 International Commission for the Conservation of Atlantic Tunas - ICCAT
   (All of the Atlantic Ocean; boundary not shown)

Fertile Ocean Areas (Phytoplankton production
greater than 250 milligrams carbon per square meter
per day.)

Major Fisheries

———— 200 Mile EEZ

Major ocean fisheries and regional fisheries organizations. This map shows the fertile areas of the sea (those with substantial phytoplankton production) and the major ocean fisheries that have developed within them. It is evident that there are still some potentially productive but currently underexploited areas, but not many. Note also how most of the richest fishing grounds will come under coastal State jurisdiction with adoption of 200-mile EEZs. In view of this, the future of the regional fisheries organizations is uncertain. The International Commission for the Northwest Atlantic Fisheries (ICNAF) has already been terminated and the other independent organizations greatly restricted. Those under the FAO, on the other hand, are not management units, but administer UN fisheries development projects; they are likely to continue functioning much as they have.

verbal support in the Seabed Committee and the General Assembly. Through 1972 and 1973 a series of regional meetings in Latin America and Africa endorsed the demand of Chile, Ecuador, and Peru for some kind of 200-mile economic zone within which the coastal State would have exclusive rights to *all* resources, living and nonliving. By the time UNCLOS III opened with a short organizational session in New York in December 1973, they had managed to acquire enough support so that one of the major conference issues had already been settled in principle. Nevertheless, the Conference still had a great deal of work to do.

The Third United Nations Conference on the Law of the Sea was the largest, longest, and most complex diplomatic conference in history, and without question one of the most important. We should first say a little about the Conference itself, though a brief review can convey only a glimmer of its magnitude and complexity.

First, the number of States participat-ing—135 to 150 per session—was triple the number at The Hague Conference of 1930 and nearly double the number at Geneva in 1958. Many newcomers were not only inexperienced in creating international law, but had no maritime experts or maritime tradition. Second, the objective of the Conference was to produce a single "generally acceptable" convention covering *all* aspects of the Law of the Sea, old and new; the agenda contained 92 "subjects and issues" that somehow had to be incorporated into a treaty, and their complexity, interrelatedness, and importance were almost overwhelming. Third, the organization of the Conference, its rules of procedure, and the dynamics of conference diplomacy were more than a little unusual. While there was a very general North–South lineup (roughly the rich countries versus the Group of 77 with the socialist States in the middle), chiefly on the question of the seabed, it was not very rigid even there. No single country or group or interest, in fact, dominated the Conference, and traditional alignments of States

Caracas welcomes UNCLOS III. "Welcome to Our Land, People of the Sea" reads this colorful billboard designed for the first substantive session of the Third United Conference on the Law of the Sea. About 5000 people participated in this 10-week session in the magnificent new Parque Central conference facilities, and the Venezuelan government did everything possible to assure its success. (Martin Glassner)

were not often evident. Science and technology, when introduced at all, were used as instruments of national policy.

There were 11 sessions of the Conference, from December 1973 to December 1982. The first substantive session (of 10 weeks), which met in Caracas in the summer of 1974, was the longest. The others were held in New York and Geneva, and a formal signing ceremony was held in Montego Bay, Jamaica. Each session had a different mood and most had different procedures. Nevertheless, the Conference seemed to be dominated by three interlocked themes: a general determination to establish some kind of international regime for the seabed beyond the outermost limits of national jurisdiction; a vigorous nationalism, expressed primarily, but certainly not exclusively, in the drive by coastal States to secure as much of the sea and its resources as they possibly could; and the effort of the Group of 77 to use the emerging new Law of the Sea as an instrument to help them achieve a New International Economic Order.

The Conference participants were organized in a great variety of formal and informal groups, subgroups, working groups, consulting groups, and so on. The most important were the regional groups: the Latin American, Asian, African, and Eastern European Groups, and the Western European and Others Group, which included nearly all the States left over. Other groups overlapped these and one another: the Islamic Group, Arab Group, EC, and NATO, for example, and other overlapping groups were issue-oriented, such as the Archipelagic States, Coastal States Group, Oceanic Group, the Territorialists (those advocating a 200-mile territorial sea), and others.

Most bizarre of all to a geographer was the Group of Land-locked and Other Geographically Disadvantaged States. We have already discussed land-locked States, and both their characteristics and problems seem clear enough. But *what* was (or is) a "geographically disadvantaged State"? No one seems to know. Many definitions have been offered but none has been generally accepted. It seems to mean a country that stands to gain little or nothing from a 200-mile exclusive economic zone, but using all the proposed criteria, that comes to a total of 110 countries! Added to the land-locked States, this potential group would leave only a handful of countries outside. In practice, however, admission to the group was largely based on political criteria, not geographic ones. The Holy See (Vatican City) withdrew from the group in 1976 purportedly because it was becoming

Table 29-1 Group of Land-locked and Geographically Disadvantaged States at the Third United Nations Conference on the Law of the Sea, with Dates of Admission to the Original Group

Afghanistan*	Ethiopia 10 April 1975	Liechtenstein*	Swaziland*
Algeria April 1976	Finland	Luxembourg*	Sweden
Austria*	The Gambia 10 April 1975	Malawi*	Switzerland*
Bahrain	German Democratic	Mali*	Syria June 1977
Belgium	Republic	Mongolia*	Turkey April 1976
Bhutan*	Germany, Federal	Nepal*	Uganda*
Bolivia*	Republic of	Netherlands	United Arab Emirates
Botswana*	Greece April 1976	Niger*	Upper Volta
Bulgaria 7 May 1975	Holy See*†	Paraguay*	(now Burkina Faso)*
Burundi*	Hungary*	Poland	Zäire
Byelorussian SSR*	Iraq	Qatar 10 April 1975	Zambia*
Cameroon June 1977	Jamaica 10 April 1975	Romania April 1979	Zimbabwe*—
Central African	Jordan 10 April 1975	Rwanda*	Observer, then
Republic*	Kuwait	San Marino*	full member,
Chad*	Laos*	Singapore	March 1981
Czechoslovakia*	Lesotho*	Sudan	

*Land-locked
†Withdrew summer 1976

too political. Israel, which qualifies on at least two counts and applied for admission, was rejected for political reasons. Nevertheless, the Land-locked and GDS was at times a potent lobby for their interests and managed to win some useful concessions in the Conference.

The Third United Nations Conference on the Law of the Sea will without question be recorded in the history books of the future as a milestone in political affairs. It produced what is essentially a constitution for the sea, nearly three quarters of the earth's surface. In the next chapter we examine parts of this constitution, those with a spatial dimension and therefore of interest to geographers.

Chapter 30

THE DEVELOPING LAW OF THE SEA

I must go down to the seas again, to the lonely seas and the sky,
And all I ask is a tall ship and a star to steer her by,
And the wheel's kick and the wind's song and the white sail's shaking,
And a gray mist on the sea's face, and a gray dawn breaking.

UNCLOS III

Would that all of us would ask as little from the sea as John Masefield does in his poem "Sea Fever"! Unfortunately, though, our demands are so much greater that we compete with one another, ferociously at times, for the sea's bounty and for strategic control of it. We must therefore have rules to regulate this competition among States and among conflicting uses of the sea. In 1982, UNCLOS III adopted overwhelmingly the United Nations Convention on the Law of the Sea. Only three other States joined the United States in voting against it. On the very first day that it was open for signature (10 December 1982) 119 States and two other entities signed it and 38 more signed within the following two years. By mid-1987, 32 States had ratified it. While it has not yet entered into force, it does represent a consensus on the developing Law of the Sea.

Much of this developing law derives from the traditional concepts and practices discussed in the last chapter. Some of it, in fact, is retained *verbatim* from the 1958 and 1965 conventions. Although in this chapter we tend to stress the innovations, the dramatic depatures from tradition, we must remember that everything being done now in the Law of the Sea is really a continuation of a very long tradition, two thousand years of legal development and six thousand years of seataring.

The Territorial Sea

As we pointed out in the last chapter, the concept of a territorial sea has long been incorporated into international law, but there has never been any agreement on its breadth. Generally speaking, the major maritime and naval powers wanted—and claimed—a narrow territorial sea so as to have as few restrictions as possible on their vessels navigating around the world, while many weaker countries, or those with very small merchant and naval fleets, preferred to have a broad buffer zone between potentially hostile or exploitative foreign vessels and their own coasts. The compromise reached is a uniform 12-mile territorial sea, measured from the applicable baseline. Territorial waters are the sovereign territory of the coastal State just as is the land and, like the land, its sovereignty over the territorial sea extends from the core of the earth to the heavens. There is one major restriction, however, on a coastal State's sovereignty over its territorial waters: it must grant to foreign vessels, including warships, the right of innocent passage. This is also a very old restriction, and an essential one if commerce and navigation are to continue linking together the peoples of the world across the great ocean highways. The definitions of the terms "innocent" and "passage," though, are more elaborate and that of "innocent" much more precise than the

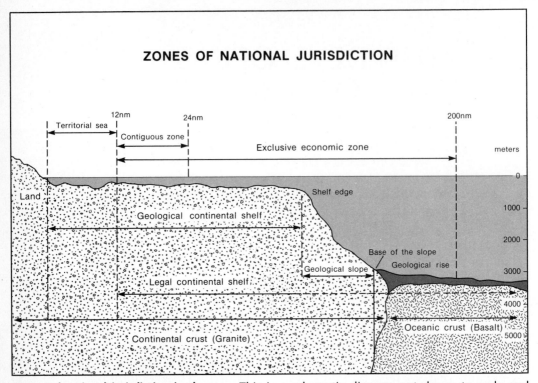

ZONES OF NATIONAL JURISDICTION

Zones of national jurisdiction in the sea. This is a schematic diagram not drawn to scale and designed only to illustrate certain terms and concepts.

traditional one: "Passage is innocent as long as it is not prejudicial to the peace, good order or security of the coastal State." Submarines passing through the territorial sea are still required to navigate on the surface showing their flag. Aircraft still do not have a right of innocent passage.

Many other provisions are included in this section of the text, with several relating to the determination of the baseline from which the breadth of the territorial sea (up to 12 nautical miles) is measured. Normally, this baseline is "the low-water line along the coast as marked on large-scale charts officially recognized by the coastal State." However, "In localities where the coastline is deeply indented and cut into, or if there is a fringe of islands along the coast in its immediate vicinity," straight baselines may be drawn according to rather detailed instructions that take into account bays, estuaries, ports and roadsteads, and other coastal features. Shore-

ward of the baseline, normal or straight, all waters, salt or fresh, are considered *internal waters* (with one exception) in which there is no right of innocent passage. On the whole, these provisions for the territorial sea appear to be a fair and practical compromise between the interests of international navigation and coastal State protection.

A coastal State retains the right to establish a *contiguous zone* on the seaward side of the territorial sea of not more than 24 miles from its baseline. Within this zone "the coastal State may exercise the control necessary to (a) Prevent infringement of its customs, fiscal, immigration or sanitary regulations within its territory or territorial sea; (b) Punish infringement of the above regulations committed within its territory or territorial sea." The contiguous zone is not part of the State's territory, but is instead a portion of the sea within which the State is permitted to exercise limited jurisdiction.

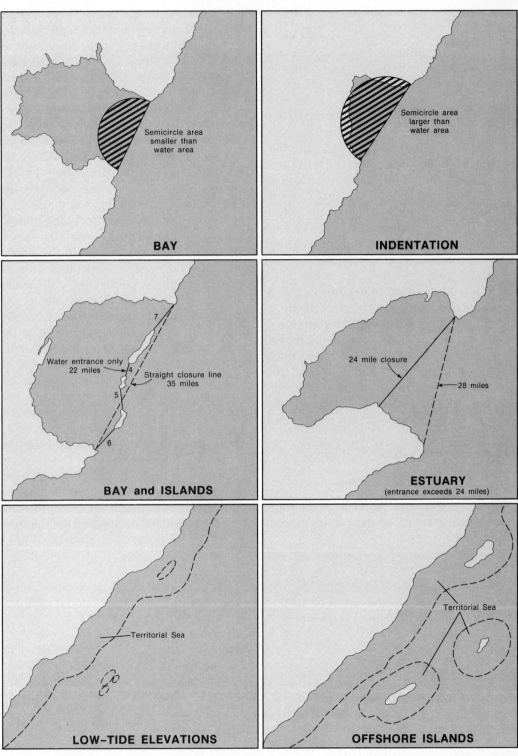

Baselines of the territorial sea. These diagrams show how straight baselines are drawn along irregular coastlines according to the rules spelled out in the 1958 Convention on the Territorial Sea and the Contiguous Zone. These rules have been retained in the United Nations Convention on the Law of the Sea.

Straits Used for International Navigation

One of the major reasons the larger maritime powers opposed expansion of the territorial sea to 12 miles was their concern that more than 100 straits used for international navigation that were between 6 and 24 miles wide would fall under national control and become subject to closure by the straits States. They insisted that high seas rights, especially unrestricted navigation, should be allowed in these straits, while some straits States wanted the rule of innocent passage to prevail in them. The 1958 Territorial Sea Convention has only one sentence on the subject; the UNCLOS has 11 articles spelling out an elaborate new regime called transit passage. It is a real compromise in which the straits States won control over fishing, pollution, traffic movement, and other potential hazards to them, and the maritime States won overflight of aircraft, including military aircraft, and avoided special restrictions on nuclear propulsion, nuclear cargoes, and submarines. The most important of these straits are shown on the front endpaper map.

Archipelagic States

Although discussed at The Hague Conference in 1930, the question of archipelagos was avoided in Geneva in 1958. But Indonesia and the Philippines (and more recently Fiji) lobbied so hard for a new regime for them that UNCLOS III devised one that is really quite new in international law. First, "archipelagic State" is defined as one "constituted wholly by one or more archipelagos and may include other islands," while an archipelago is "a group of islands, including parts of islands, interconnecting waters and other natural features which are so closely interrelated that such islands, waters and other natural features form an intrinsic geographical, economic and political entity, or which historically have been regarded as such." An archipelagic State is entitled to draw straight baselines connecting "the outermost points of the outermost islands and drying reefs of the archipelago provided that within such baselines are included the main islands and an area in which the ratio of the area of the water to the area of land, including atolls, is between one to one and nine to one." Also, generally, "the length of such baselines shall not exceed 100 nautical miles."

These restrictions effectively limit the status of "archipelagic State" to a very few countries (Indonesia, Fiji, the Philippines, the Bahamas, and a few others). The waters enclosed by the baselines (except for internal waters) are designated *archipelagic*

A vital international strait. This waterway connects the Kattegat to the left and the Sound to the right. In the foreground is the Danish port of Helsingor, protected by Kronborg Castle (made famous by Shakespeare as Elsinore in his *Hamlet*). In the background is the larger Swedish city of Helsingborg. (The Danish Tourist Board)

waters over which the State has sovereignty but within which "ships of all States enjoy the right of innocent passage." There are provisions for sea lanes within the archipelagic waters and lists of rights and duties of the State and of ships and aircraft passing through. The territorial sea and contiguous zone are to be measured outward from the baseline. While additional area has been removed from the high seas by these provisions, on the whole they also represent a reasonable compromise.

Islands

We all learned as children that an island is "a piece of land surrounded by water," but such a simple definition could hardly suffice in a world in which over half a million such "pieces of land," no matter how small, might each qualify for its own territorial sea, contiguous zone, continental shelf, and exclusive economic zone. Therefore the definition of "island" in the Convention is more restrictive: "An island is a naturally formed area of land, surrounded by water, which is above water at high tide." Then a new paragraph reads, "Rocks which cannot sustain human habitation or economic life of their own shall have no exclusive economic zone or continental shelf." They may still have territorial seas and contiguous zones, however, and the entire section on islands leaves many questions still unanswered.

Enclosed or Semi-enclosed Seas

One of the more enigmatic parts of the Convention deals with the "enclosed or semi-enclosed sea," defined as "a gulf, basin, or sea surrounded by two or more States and connected to the open seas by a narrow outlet or consisting entirely or primarily of the territorial seas and exclusive economic zones of two or more coastal States." States bordering such bodies of water are exhorted to cooperate in a variety of ways in marine affairs. The definition of the terms is certainly open to question, but a more fundamental question is why the terms are used at all, why any distinction is made between these and any other parts of the sea, why this part is in the Convention at all. It is possible that the reason will become clear in time; it is also possible that the delegates at UN-CLOS III simply turned an academic exercise into a legal issue so as to leave no possible marine feature uncovered by the Convention they were writing.*

The Exclusive Economic Zone

While elements of this concept have roots deep in history, its modern scope is quite new and revolutionary. Each coastal State now is entitled to a zone, including the territorial sea, within which it has "sovereign rights for the purpose of exploring and exploiting, conserving and managing the natural resources, whether living or non-living, of the sea-bed and subsoil and the superjacent waters," as well as jurisdiction over a number of other activities in the zone (EEZ). The breadth of the EEZ is not to exceed 200 nautical miles from the baseline. This part of the Convention has 21 articles, some of them quite long, that describe the numerous detailed rights and few vague responsibilities of the coastal States in their EEZs. They alone have jurisdiction over conservation of fish, marine mammals, and other "living resources" in their zones, including anadromous, catadromous and highly migratory species.† Land-locked States are granted greatly restricted and probably meaning-

---

*The academic exercise referred to is Lewis M. Alexander, "Regionalism and the Law of the Sea: The Case of Semi-enclosed Seas," *Ocean Development and International Law Journal*, 2, 2 (Summer 1974), 151–186.

†Anadromous fish are those, such as salmon, that are born in freshwater streams and migrate out to sea where they spend the greater part of their lives before returning to their birthplaces to spawn. Catadromous fish are those, such as some eels, that are born at sea and migrate into freshwater lakes and streams to live until they return to the sea to spawn. Highly migratory species are those, such as tuna and whales, that range far through the sea, sometimes thousands of kilometers, on a fairly regular schedule.

The United States Exclusive Economic Zone. Most of the EEZs do not yet have boundaries agreed on with the respective neighbors of the United States; those in the Pacific Northwest are still being negotiated with Canada and pose particularly difficult problems.

less rights to certain surplus fish within the EEZ.

The exclusive economic zone has been subtracted from the high seas and the "common heritage of mankind," thus shrinking this historic international area by one third! Within 200 n.m. are all the world's known offshore oil and natural gas deposits, all the presently exploitable minerals, most of the readily accessible manganese nodules, 85 to 95 percent of the world's current fish catch, nearly all marine plants, and nearly all potentially exploitable sources of unconventional en-

ergy (waves, tides, thermal layers). More than half the combined EEZ of the world goes to only 10 countries, of which 6 of the first 7 are already rich (in order of size of maritime territory gained: the United States, Australia, New Zealand, Canada, the USSR, and Japan). All the top 5 gainers of submarine petroleum are also rich (United States, Soviet Union, Britain, Norway, and Australia). Thus, the poor countries, which began the drive for an EEZ and finally won acceptance for it, stand to gain relatively little from it.

Another serious problem with the EEZ

is the very real possibility that it will gradually, on various pretexts, become, in effect, a territorial sea of 200 miles breadth or that it will be extended beyond 200 miles if additional resources are discovered there, or both. It seems highly unlikely at this point that the creation of the EEZ will contribute materially to a reduction of the great and growing gap between rich and poor in the world, or to the wise preservation of the sea as "an indivisible ecological whole," or to the replacement of petty nationalisms with real and substantial international cooperation.

The Continental Shelf

Prior to 1945, the continental shelf was a geological feature known only to geologists and to some fishermen, engineers, and seamen. By 1958, it was not only well known, but so important to so many coastal States that the first United Nations Conference on the Law of the Sea granted the coastal States exclusive rights to the resources of the shelf and its subsoil. The Convention on the Continental Shelf is the shortest of the four Geneva conventions of 1958 and the most tentative. It defines the shelf as "the seabed and subsoil of the submarine areas adjacent to the coast but outside the area of the territorial sea, to a depth of 200 meters or, beyond that limit, to where the depth of the superjacent waters admits of the exploitation of the natural resources of the said area." Both portions of this definition—the obsolete 200-meter isobath and the absurd "exploitability clause"—have been abandoned in the new Convention in favor of a more precise but far more generous (to the coastal States) definition: "the sea-bed and subsoil of the submarine areas that extend beyond its territorial sea throughout the natural prolongation of its land territory to the outer edge of the continental margin, or to a distance of 200 nautical miles." This definition assures that coastal States have exclusive rights to much of what had been considered "the common heritage of mankind."

The remaining nine articles on the continental shelf grant to the coastal States most of the exclusive rights they have obtained in the EEZ, but one article requires coastal States to pay a kind of royalty on nonliving resources produced from their shelves beyond 200 miles, with the money distributed to the poorest countries.

The High Seas

The two thirds of the sea remaining after internal, territorial, and archipelagic waters and the EEZ have been subtracted and assigned to coastal States, continue to be high seas and essentially the property of the international community. Many traditional high seas rights and responsibilities are retained, but the old "freedom of the seas" is dead. This is not necessarily bad, for, in truth, it is obsolete; today it would mean freedom to overfish and freedom to pollute the sea. The new definition of "freedom of the high seas" is both more precise and more narrow than the traditional one expressed in the 1958 High Seas Convention. On the whole, the new provisions preserve the traditional high seas rights that tend to serve community interests, and some have even been enhanced. Although not perfect, from a geographer's standpoint the Convention provisions on the high seas seem satisfactory.

The International Seabed Area

We may recall Arvid Pardo's proposal of 1967 that the "seabed and ocean floor and subsoil thereof beyond the limits of present national jurisdiction" be declared "the common heritage of mankind." It should be used only for peaceful purposes, and its resources exploited for the benefit of the international community, particularly its poorest members. This triggered the massive and thorough review and revision of the Law of the Sea that followed. By far the greatest innovation—and the most complex and controversial issue—is the establishment of a common heritage re-

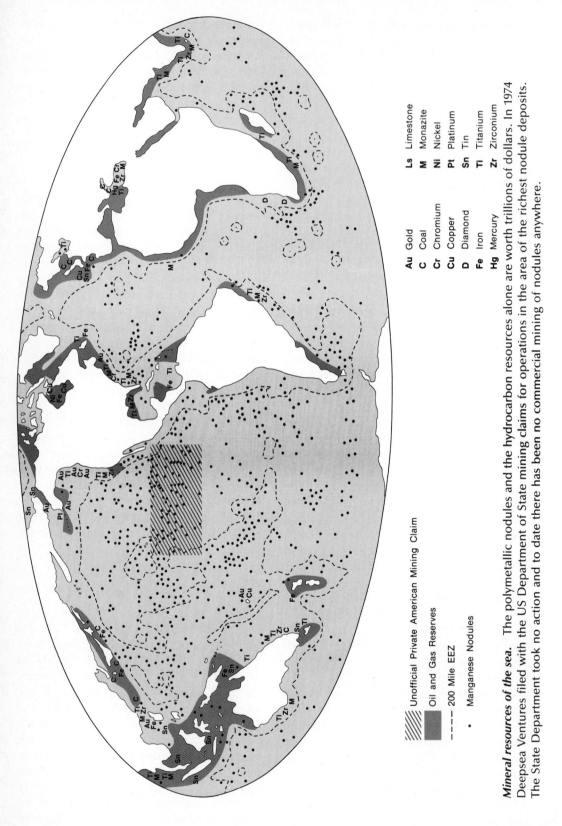

Mineral resources of the sea. The polymetallic nodules and the hydrocarbon resources alone are worth trillions of dollars. In 1974 Deepsea Ventures filed with the US Department of State mining claims for operations in the area of the richest nodule deposits. The State Department took no action and to date there has been no commercial mining of nodules anywhere.

Au Gold   **Ls** Limestone
C Coal   **M** Monazite
Cr Chromium   **Ni** Nickel
Cu Copper   **Pt** Platinum
D Diamond   **Sn** Tin
Fe Iron   **Ti** Titanium
Hg Mercury   **Zr** Zirconium

Unofficial Private American Mining Claim

Oil and Gas Reserves

– – – – 200 Mile EEZ

• Manganese Nodules

Manganese nodules from the sea bed. Nodules are composed of fine-grained oxide material apparently precipitated out of the sea water. They vary widely in their composition as well as in their physical and chemical properties, but all are polymetallic, containing about 26 elements altogether. The most important commercially are manganese, copper, cobalt, and nickel. It is estimated that 22 billion tons of nodules lie on the ocean floor, but the most important concentration apparently is in the northeast Pacific, between Hawaii and Mexico. (UN/S. Lewin)

gime for the International Seabed Area. The main features of this regime are as follows:

1. The Area and its resources cannot be appropriated by any State.

2. The Area is to be used exclusively for peaceful purposes.

3. All rights in the resources of the Area "are vested in mankind as a whole."

4. International cooperation is encouraged in marine scientific research, protection and conservation of the marine environment, and transfer of technology relating to mineral resource exploration and exploitation.

5. Developing countries are assisted to participate effectively in these activities.

6. An International Seabed Authority is created to administer the mineral resources of the Area.

7. The Authority must undertake scientific research in the Area, establish a system of equitable sharing of benefits from the Area, and participate in deep sea mining on its own account

as well as in association with States and private entities.

Much of the Convention is devoted to detailed provisions concerning the structure and operations of the Authority, which do not concern us here. It seems that the poor countries of the world are really going to get very little out of this arrangement until well into the twenty-first century, but at least it should result in the establishment of an international agency that, however unwieldy, will manage the mineral resources of over half the earth's surface for the ultimate benefit of all the peoples of the earth. This might be a giant step toward the replacement of the currently dominant European State system with a form of political organization more suited to our future needs and conditions. Meanwhile, however, it still allows for the expansion of national jurisdiction into the international area and is very far from a common heritage regime that can be effectively implemented.

Preservation of the Marine Environment

Instead of treating ocean space as "an indivisible ecological whole,"* UNCLOS III partitioned it—horizontally, vertically, and functionally—into innumerable international, regional, national, and private jurisdictions with no centralized control over them and no overall plan for the preservation of the integrity of the marine environment. The Conference, partly because of its limited mandate, also dealt not with environmental considerations as a whole, but only with pollution of the sea, and even here only with the 10 to 20 percent of pollution that originates from ships. Although the developing countries have overcome their original unconcern with

*Ocean space includes the surface of the sea, the water column, the seabed and its subsoil. Some would include the airspace over the sea as a part of ocean space as well. The quotation is from Elizabeth Mann Borgese, one of the foremost leaders in the development of the new Law of the Sea and public awareness of it.

A petroleum tanker disaster at sea.   In December 1976, the *Argo Merchant*, a tanker loaded with petroleum, ran aground off the coast of Massachusetts, then broke up and sank. A succession of such disasters spurred UNCLOS III to devise new rules for the preservation of the marine environment, at least insofar as ship-based pollution is concerned. Nationalism and other political considerations, however, have prevented agreement on truly effective measures to protect the "indivisible ecological whole" of ocean space. (NOAA)

environmental matters (and even at times opposition to environmental protection proposals because they feared interference with their economic development in general and industrialization in particular), the Conference was long engaged in the difficult job of allocating responsibility for regulations and enforcement among flag States, port States, and coastal States. The *Torrey Canyon* disaster in 1967 awakened public interest in marine pollution, and the *Amoco Cadiz* disaster in 1978 not far away spurred UNCLOS III into tightening the relevant provisions of the Convention. The result is quite inadequate to assure long-term protection for the marine environment, but it is the best arrangement negotiable at present and does represent real, if modest, progress.

Marine Scientific Research

It is hard for the average person to comprehend how highly politicized marine scientific research has become in much of the world and in UNCLOS III. We tend to accept it unquestioningly as a positive good; it is part of our culture and we have reaped the benefits of it. Yet few of the poor countries of the world (the vast majority of all the countries) have any firsthand experience with marine scientific research or any real understanding of it. They often suspect—and quite understandably, considering their colonial past—that a mysterious ship lurking off their coast doing inexplicable things with mysterious instruments is preparing for either some kind of assault on their "sovereignty and territorial integrity" or on their offshore resources. The struggle in UNCLOS III over a regime for marine scientific research was bewildering. What emerged is a modified consent regime; that is, a State wishing to do any kind of scientific research in any of the areas of coastal State jurisdiction (and remember there are six of them now) must obtain the advance permission of the coastal State, though this consent may be implied if the coastal State does not deny it within a specified time period. There are many other provisions relating to cooperation in such research, sharing of results, and so on. Marine scientific research is being shackled with so many restrictions and cumbersome procedures that in the future it may be less rewarding and far more expensive than it has been until now.

Settlement of Disputes

One of the more hopeful innovations in the UNCLOS is the section on settlement of disputes. It provides for a Law of the Sea Tribunal, *ad hoc* arbitral tribunals, special arbitration procedures, and a conciliation procedure. The mechanism for dispute settlement is probably quite good. It must be, for there are likely to be very many and very serious disputes over marine resources and ocean space as States continue to push their jurisdiction to and beyond the limits contemplated but not specified in the Convention.

OTHER ASPECTS OF THE POLITICAL GEOGRAPHY OF THE SEA

UNCLOS III avoided several aspects of the political geography of the sea and eventually they will have to be addressed through some other agency. First, the status of the Southern Ocean is not defined at all; in fact, there are no limits specified for the international areas—the seabed or the high seas. They are simply whatever is left over after coastal States have taken whatever portions of the sea they wish (since States have considerable freedom in drawing straight baselines). Chile and Argentina have already claimed EEZs for their overlapping claims in Antarctica, but their validity is dubious. We may expect more problems in this region as States begin to exploit its resources more intensively.

The Conference also avoided the whole question of the military uses of the sea. While the phrase "exclusively for peaceful purposes" appears frequently in the Convention, it is nowhere defined. Aside from assuring warships their traditional immun-

ities and permitting submarines to pass submerged through international straits, there is no reference to arms control or disarmament at sea, or the creation of "zones of peace" or "nuclear-free zones" such as that proposed for the Indian Ocean. Although it does not necessarily hinder moves toward disarmament, the Convention does not encourage it either.

Another serious omission, though an understandable one considering the dominance of nationalism in UNCLOS III, is the total absence of any reference to national obligations to preserve the integrity of the marine environment on the continental shelf, in territorial waters, in archipelagic waters, or in the rapidly expanding internal waters, and these are precisely the marine areas most intensively utilized now and will be far into the future. In addition, the failure of a State to meet its obligations relating to the conservation and management of the living resources of the EEZ is excluded from the jurisdiction of the Tribunal, so a State is quite free to ignore even the very limited obligations it undertakes in this area.

Another potential problem might develop as more countries follow the lead of the United States in establishing marine sanctuaries or marine parks. If all vessels are banned from them, except for scientific and tourist vessels, some States might object that freedom of navigation is being unduly restricted. Other jurisdictional and functional problems could arise.

And what about sovereignty over newly formed islands, such as the one that is slowly building up from sediments being dumped into the Bay of Bengal by the Ganges and Brahmaputra rivers? Since the sediments come largely from Nepal and China, perhaps they will have claims to it as valid as those of India and Bangladesh.

No mention is made either of the practice of the superpowers (and others may follow before long) of warning vessels out of enormous tracts of the sea from time to time when they wish to test long-range missiles or other weapons, a blatant violation of the freedom of the high seas. Nor

The final session of UNCLOS III, Montego Bay, Jamaica, 10 December 1982. At the far right, H. E. Mr. Javier Perez de Cuellar, Secretary-General of the United Nations, is addressing the Conference. At the dais are the Conference officials; in the center is H. E. Ambassador Tommy T. B. Koh of Singapore, President of the Conference. In the foreground are some of the delegates. (Martin Glassner)

is there reference to the establishment of Air Defense Identification Zones extending hundreds of kilometers out to sea, which could be construed as a similar violation of the freedom of the skies over the high seas.

The status of ice shelves, such as those adjacent to Antarctica, is not covered either, and the whole question is still unclear.

Even less clear is the legal status of the United Nations Convention on the Law of the Sea. The United States has refused to become a party to it; so have the United Kingdom, West Germany, and a few other important countries. It may not enter into force for many years, if ever. Yet much in it is simply a codification of preexisting customary and treaty law and many other provisions have been so widely accepted that they may be considered customary law even if the Convention never enters into force. All of the seabed provisions, however, are new. A Preparatory Commission is currently thrashing out rules and procedures for the operation of the Authority and the Tribunal pending entry into force of the Convention, but much of this tedious effort may ultimately be wasted.

Nevertheless, the Law of the Sea will continue to evolve, perhaps in ways we cannot yet foresee. Meanwhile, new maritime boundaries and zones are being claimed and/or negotiated around the world, hundreds of them. It is as if the boundary-making process described in Chapter Eight is beginning all over again, but with the added complication of *both* EEZ and shelf boundaries that may not coincide. The United States alone will ultimately have some 30 maritime boundaries with other States and territories, of which only 7 had been negotiated by mid-1987. Of these, only the treaties with Venezuela, New Zealand on behalf of Tokelau, and the Cook Islands are in force, while those with Cuba and Mexico (for 3 boundary segments) await Senate approval. Still to be negotiated are 4 boundary segments with Canada, 3 adjacent to American Samoa, 3 with Kiribati, and 1 each with the British Virgin Islands, Saba, the Dominican Republic, Japan, the Federated States of Micronesia, the Marshall Islands, the Soviet Union, and the Bahamas.

Besides boundaries, many other marine and maritime problems remain to be worked out around the world. And the work must be done well. Political geographers as well as lawyers will have a good deal of work to do in marine affairs for a very long time to come.*

*Besides the professional literature and the UN Office for Ocean Affairs and Law of the Sea, the best source for current information on the developing Law of the Sea is a splendid non-governmental organization:
Council on Ocean Law
1717 Massachusetts Avenue, NW
Suite 302
Washington, DC 20036

Chapter 31

ANTARCTICA

Antarctica is the coldest, windiest, driest, highest, quietest, most remote, and least understood continent on earth. Attached to the continent is a fluctuating apron of shelf and pack ice, a formidable buffer zone making approach to the land far more difficult than to any other continent. Because of its remoteness and inhospitable environment, it was the last continent to be discovered and, except for a few whalers and explorers, it was virtually ignored for more than a century after its discovery. Now its importance in the physical geography of the earth—climate, submarine flows of water through the oceans, biomass—has been recognized, and we are glimpsing its commercial potential.

As with any other region, we have a problem defining the limits of Antarctica. More than with any other continent, the surrounding sea is an integral part of the region and at least a portion of it must be included in our definition. The Antarctic Treaty covers only the region south of latitude 60°S; some observers have suggested latitude 40°S as a boundary. Both of these suggestions have all the advantages and disadvantages of geometric boundaries we have already discussed. The most frequently suggested boundary is the Antarctic Convergence (also called the Polar Frontal Zone), a zone about 50 km (30 miles) wide circling the continent between 55° and 62° South latitude where the cold Antarctic water sinks beneath the warmer waters of the north. Another and perhaps more useful regional delineator is the Subtropical Convergence (also called Subantarctic Convergence), the southernmost reach of the warm, southward-moving currents of the Atlantic, Pacific, and Indian oceans. It lies at approximately 40° South, but displays more latitudinal stability than the Antarctic Convergence. Defining the northern boundary of the Southern Ocean is not just an academic exercise. It will be vitally important in establishing a regime (or regimes) for the management of the Antarctic ecosystem.

Why should we now be concerned with a legal regime for an empty, frozen wilderness so far away from the core area of any country? Because at this stage in their political evolution, human beings simply cannot bear to leave any portion of the earth's surface alone. Every square kilometer must be accounted for—claimed, owned, governed, exploited, and perhaps destroyed. The pristine beauty, the inspiring vastness, the tranquilizing solitude of this magnificent region may not last much longer. Already tourists are intruding into what has been a huge and immensely valuable scientific preserve, and with few legal restrictions on tourism, we can visualize it building to much larger proportions in the near future, bringing to the continent much degradation and little benefit. Even more tempting for commercial exploitation are the mineral

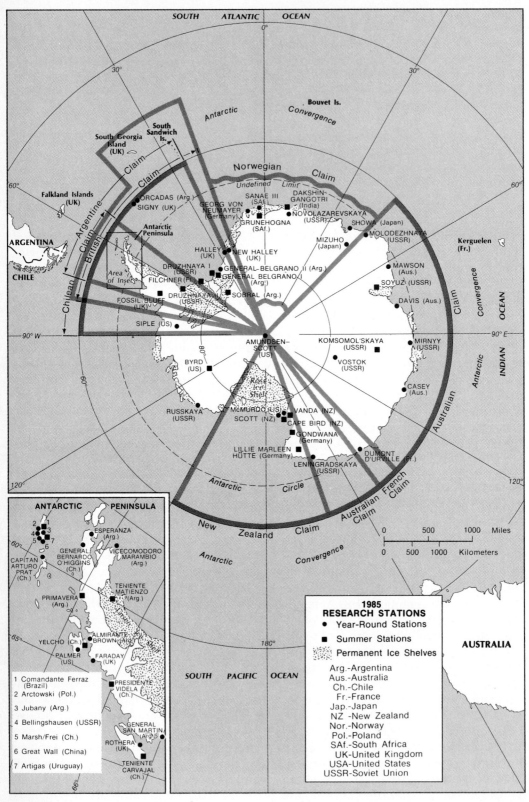

Antarctica

deposits of the continent and the super-abundant marine life of the Southern Ocean.

TERRITORIAL CLAIMS AND GOVERNANCE

Although the continent was postulated by the ancient Greeks and probably seen by Polynesian seafarers centuries ago, specific knowledge of it dates only from the late eighteenth century, when Captain James Cook circumnavigated it (but never saw it) and British and American sealers ventured into Antarctic waters to hunt seals on islands and ice. The continent itself was not spotted until the early nineteenth century when sealers, whalers, and explorers from several countries discovered the Antarctic Peninsula (called Graham Land by the British, the Palmer Peninsula by the Americans, Tierra O'Higgins by the Chileans, and Tierra San Martín by the Argentines). The first formal claim to Antarctic territory was made by Britain in 1908. Since then six other countries have made formal claims: New Zealand, 1923; France, 1924; Australia, 1933; Norway, 1939; Chile, 1940; and Argentina, 1943.

Of all the customary bases for claims to territory under international law, which we discussed in Chapter Seven, the only one with any validity in the Antarctic is discovery and effective occupation, the basis of most existing claims. Since discovery is difficult to prove and occupation difficult to make "effective," other principles have been created and applied here. Best known is the sector principle. This originated in 1907 when a Canadian senator recommended that Canada should declare that she had taken possession of all lands, including islands, lying between her northern coast and the North Pole. Other countries adopted this new concept both in the Arctic (Russia 1916) and the Antarctic (United Kingdom, first in 1917), but it has not been generally accepted as a valid principle of international law. Other theories are the conti-guity doctrine, continuity and hinterland, patrimony and *uti possidetis*, and the region of attraction. All of these have been applied at one time or another, often in combination with the sector principle or discovery and occupation, but none has achieved any significant measure of acceptance by the international community. The matter of sovereignty in Antarctic was so muddled by the 1950s and was becoming so acutely sensitive that there was a need to suspend political competition there.

The opportunity was provided by the International Geophysical Year (IGY), conceived by the International Council of Scientific Unions as a massive, varied, and coordinated study of the earth during a period of great sunspot activity (1957–1958). Neither the Cold War nor competing Antarctic claims prevented collaboration among over 10,000 scientists and technicians from 67 countries working at 2500 IGY stations around the world. Included were 50 stations operated by 12 countries in Antarctica.* They did a great deal of scientific exploration and this international cooperation produced considerable new knowledge about our planet, especially about Antarctica. The international aspect of the operation was epitomized by the exchange of Soviet and American scientists and by the Commonwealth Transantarctic Expedition, which in three months made the first surface crossing of the continent, from Shackleton Base on the Weddell Sea to Scott Base on the Ross Sea via the South Pole. The agreement to allow States working in Antarctica to place scientific stations anywhere on the continent despite prior claims to sovereignty, and the unrestricted exchange of IGY data set precedents for later explorations of the sea and outer space and led directly to the negotiation and success of the Antarctic Treaty.

The Antarctic Treaty was negotiated and signed in Washington in 1959 by the 12

*Argentina, Australia, Belgium, Chile, France, Japan, New Zealand, Norway, South Africa, the United Kingdom, the United States, and the USSR.

States that participated in IGY activities in Antarctica. Since then 25 States have acceded to the Treaty and 8 have been accepted as Consultative Parties (CPs), equal to the original signatories (see box). The treaty went into effect in 1961 and is reviewable after 30 years. It is truly a landmark in international cooperation and in some respects a model for future ventures of this kind in other arenas. It governs the entire continent and has been scrupulously obeyed by all signatories. The spirit of international camaraderie engendered by the IGY has continued to prevail in Antarctica under the treaty. We can hope it will last, but already there are disquieting signs that it may not.

The treaty is short, only 14 articles. To summarize it briefly: Antarctica is to be used for peaceful purposes only; no military activities of any kind are permitted, though military personnel and equipment may be (and are) used for scientific purposes. Freedom of scientific investigation and cooperation shall continue. Scientific program plans, personnel, observations, and results shall be freely exchanged. No prior territorial claim is recognized, disputed, or established and no new claims may be made while the treaty is in force. Nuclear explosions and disposal of radioactive waste are prohibited. All land and ice shelves south of latitude 60°S are covered, but not the high seas of the area. Observers from treaty States have free access to any area and may inspect all stations, installations, and equipment. Treaty States shall meet periodically to exchange information and take measures to further treaty objectives, including the preservation and conservation of "living resources." These consultative meetings shall be open to contracting parties that conduct substantial scientific research in the area. Disputes are to be settled peacefully, ultimately, if necessary, by the International Court of Justice.

Although all the terms of the treaty are being faithfully carried out and we have all benefited from it, some matters the treaty omits are now emerging as problems. For one thing, while the treaty is open for accession by nearly all States, only those doing actual and "substantial" scientific work in the area may participate in the consultative meetings that spell out the rules for operations in the Antarctic. This effectively excludes nearly all the developing countries since they are too poor to bear the heavy costs of Antarctic operations. While the treaty encourages scientific work and bans military activity in the region, it is silent about commercial activities there. As demand grows for Antarctic resources, especially minerals and tourist "destinations," and as technology becomes increasingly able to meet those demands, some regulation (if not outright prohibition) will become essential. The treaty does not come to grips with the question of sovereignty over Antarctic waters: Does ice count as land or water? Are shelf and pack ice to be treated differently? May a State claim territorial waters or other maritime zones around the continent? (Chile and Argentina have already proclaimed 200-mile EEZs adjacent to their overlapping claims in the Antarctic Peninsula.) Even the claims to land territory are only suspended, not resolved.

The sovereignty question may be the most vexing of all. A number of the claimants have organized their claims as colonial territories (Australia, France, UK), complete with their own postage stamps, while others have incorporated their claims into their national territory (Chile, Argentina), complete with administrative structures, and Norway so adamantly rejects the sector theory that she has refused to set northern or southern limits to her claim. Some countries active in Antarctic exploration and research (South Africa, Japan, Belgium, and, most active of all, the United States and the USSR) have made no claims at all and refuse to recognize any existing claims.

It is going to be difficult to get any claimant State to retreat from its position since it has become a matter of prestige for some. New Zealand seems least nationalistic in the matter and may well be

MEMBERS OF THE ANTARCTIC TREATY SYSTEM (1987)

Original Consultative Parties (12)	Date of Ratification or Accession	Non-Consultative (17)	Date of Accession
a. Claimant States (7) (The date in parentheses is the date of the formal claim).		Czechoslovakia	Jun 14 1962
		Denmark	May 20 1965
		Netherlands	Mar 30 1967
		Romania	Sep 15 1971
United Kingdom (1908)	May 31 1960	Bulgaria	Sep 11 1978
Norway (1939)	Aug 24 1960	Papua New Guinea	Mar 16 1981
France (1924)	Sep 16 1960	Peru*	Apr 10 1981
New Zealand (1923)	Nov 1 1960	Spain*	Mar 31 1982
Argentina (1943)	Jun 23 1961	Hungary	Jan 27 1984
Australia (1933)	Jun 23 1961	Sweden*	Apr 24 1984
Chile (1940)	Jun 23 1961	Finland*	May 15 1984
		Cuba	Aug 16 1984
b. Non-claimant States (5)		Rep. of Korea*	Nov 28 1986
		Greece	Jan 8 1987
South Africa	Jun 21 1960	Korea Dem. Rep.	Jan 21 1987
Belgium	Jul 26 1960	Austria	Aug 25 1987
Japan	Aug 4 1960	Ecuador	Sep 15 1987
United States	Aug 18 1960		
Soviet Union	Nov 2 1960		
Later Consultative parties (8) (The date in parentheses is the date of becoming Consultative Party)			
Poland (Jul 29 1977)	Jun 8 1961		
German Dem. Rep. (Oct 5 1987)	Nov 19 1974		
Brazil (Sep 12 1983)	May 16 1975		
Fed. Rep. Germany (Mar 3 1981)	Feb 5 1979		
Uruguay (Oct 7 1985)	Jan 11 1980		
Italy (Oct 5 1987)	Mar 18 1981	*Planning to conduct scientific research in Antarctica and/or seek consultative status in 1988–1989.	
PR China (Oct 7 1985)	Jun 8 1983		
India (Sep 12 1983)	Aug 19 1983		

willing to transfer her claim to an international body, but Argentina and Chile are different. They have gone to very great lengths indeed to secure their claims, erecting various buildings in unused areas, printing maps showing their claims as part of their territory, and more. President Pinochet of Chile spent a week touring the Chilean-claimed sector in January 1977. That topped the Argentine ploy of having a national cabinet meeting on an Antarctic island in August 1973. But then in the summer of 1978, Argentina sent a pregnant woman to their Esperanza Base and she gave birth in January to Emilio Marcos Palma, the first child born in Antarctica. In February, they staged the first wedding in Antarctica (Seargant Carlos Alberto Sugliano and Beatriz Buonamio), also at Esperanza.

Both countries have established permanent civilian colonies there, small groups of dependents of military personnel serving two-year tours of duty (Argen-

Potential Ecuadorian Antarctica claim. While Ecuador has not yet made an Antarctic claim and may not do so, the possibility cannot be ruled out, considering her vigorous pursuit of territorial claims in the Pacific Ocean, the Amazon Basin, and outer space. Such a claim would probably look much like this.

tina on the peninsula in 1977 and Chile on King George Island nearby in 1984). And in October 1984, the Chileans at the latter base (Teniente Marsh) welcomed the first party of tourists to overnight in Antarctica. These are only samples of the intensive activities engaged in by these two countries for a purely nationalistic reason: to strengthen their territorial claims in the event that it becomes possible to reassert them after 1991, despite the fact that legally, such activities, according to Article

IV of the Antarctic Treaty, cannot "constitute a base for asserting, supporting or denying a claim to territorial sovereignty in Antarctica or create any rights of sovereignty in Antarctica."

COMMERCIAL ACTIVITIES

We have already referred several times to the likelihood of commercial, as distinguished from scientific, activity in the Antarctic region. The most important has long been, and likely will continue to be, exploitation of the "living resources" of the Southern Ocean.

The Southern Ocean is home to the world's densest population of marine fauna: 8 species of seal and 12 of whale, some of which have been overharvested almost to extinction; 90 to 100 species of fish, few found elsewhere, but most in small concentrations; large quantities of squid and vast quantities of krill.* Even smaller are the innumerable species of phyto- and zooplankton, produced massively in the mineral-rich Antarctic waters. Of all these animals, the one with the greatest apparent commercial potential is krill, a high-quality food, rich in protein, oil, vitamins, and minerals.

The Soviet Union has been harvesting krill at least since 1971, and Japan followed in the 1972–1973 season. Chile, Taiwan, Korea, Spain, Poland, and East and West Germany are also engaged in the krill industry, but on a smaller scale. Both the Soviets and the Japanese market the krill in various forms, most commonly as a paste used as a food additive and a sandwich spread. Although there are still many problems with the harvesting, processing, and marketing of krill, it would seem that the rapidly growing population of the world, increasing demand for high-quality protein, overfishing of many finfish, exclusion of the major fishing countries from many fisheries by new 200-mile EEZs of coastal States, and other factors will bring more and more krill trawlers to Antarctic waters in the near future. It has been estimated that 50 to 70 million tons of krill could be harvested each year in Antarctic waters without threatening existing stocks. This is about equal to the total catch of all species of fish throughout the world at present.

But this optimistic estimate has already proven false. Not only has the harvest never even approached 2 million tons, but in recent years it has actually been falling. This is undoubtedly due in part to increased feeding by whales saved from slaughter by tightened rules of the International Whaling Commission (IWC), but there is new evidence that there simply is not as much krill in the Southern Ocean as has been thought. Not only fish and marine mammals, but even krill seems to have been overfished.

This problem of overexploitation of the biota of the Southern Ocean led the CPs of the Antarctic Treaty System to convene a conference in Canberra in 1980. There they concluded the Convention on the Conservation of Antarctic Marine Living Resources, which entered into force in 1982. Among other things, it provides that:

1. The convention applies for protection of all living organisms in the sea south of the Antarctic Convergence.

2. Harvesting and associated activities are to be carried out in such a manner as to prevent or minimize long-term risk to the marine ecosystem.

3. A Commission for the Conservation of Antarctic Marine Living Resources (CCAMLR) is established to, among other things, facilitate study of the marine ecosystem, publish relevant data, and formulate conservation measures.

4. Observance of provisions of the convention shall be ensured by an observation and inspection system.

The CCAMLR has headquarters in Hobart, Tasmania, Australia. It holds regular meetings and has begun publishing some useful materials, but to date its actual effectiveness has been limited. The same nationalistic and mercenary impulses that have impeded the IWC and other intergovernmental conservation organizations have been operating here as well. Considering that there have already been violations of the Agreed Measures for the Conservation of Antarctic Fauna and Flora

*Krill is a collective name for many species (11 in Antarctic waters) of large zooplankton—shrimplike crustacea of which *Euphasia superba* is the largest (averaging about 5 cm in length when mature) and most important of the Antarctic species.

(1964), the first major agreement under the ATS, we may well be skeptical about the willingness and/or ability of exploiters of Antarctic resources, whether governments or private parties, to preserve the fragile and valuable ecosystem of the region.

The threat to the Antarctic environment is at least as great from potential exploitation of hydrocarbons and minerals, both on and offshore. Minerals have been found in great variety in Antarctica, but, so far at least, not in concentrations that can be considered commercially exploitable. Fairly sizable deposits of coal and iron ore exist in the Prince Charles Mountains and the Transantarctic Mountains respectively, but they are inaccessible to commercial exploitation. Sand and gravel

are abundant on many raised beaches around the coastline, but they are unlikely to be exported, though they can be used for local construction. More than a dozen other minerals occur on the continent, but because of the enormous costs involved, they are likely to remain unattractive for commercial exploitation for some time to come. The narrow and deep continental shelf may contain commercial quantities of petroleum and natural gas, but the difficulties of extracting and delivering them to market will far exceed those of any other offshore deposits in the world or even those of Alaska's Arctic Slope. Extensive areas of the sea floor around the continent have a covering of scattered manganese nodules and related formations, but they tend to be rather lean in

Antarctica on stamps. The stamps in the top row show the overlapping Chilean and Argentine claims on the continent. The American stamp commemorates the tenth anniversary of the Antarctic Treaty. Those from Australia, Brazil, France, and West Germany symbolize these countries' scientific activities in Antarctica. The Polish stamp illustrates fishing for krill in the Southern Ocean. The four Chilean stamps at the bottom show various Chilean activities and are clearly designed to create sympathy for the Chilean territorial claim. (Martin Glassner)

the most valuable minerals. There is some but probably not much potential for geothermal energy.

In short, it seems that Antarctica, while rich in varieties of minerals, will probably not be an important commercial producer because of the small concentrations and high costs of exploitation. These factors, however, may not necessarily deter some States from permiting, encouraging, or undertaking mining for what they consider to be security or prestige reasons.

Since 1974 the CPs have been considering the question of minerals and began in 1979 to try to develop a minerals regime for the region—in great secrecy. Gradually, however, and only under great pressure from nongovernmental environmental groups around the world and from nonparties to the Antarctic Treaty (especially many of the poorer African and Asian countries), the outlines of successive drafts of a minerals treaty have become known. The same pressures seem to have forced a gradual shift from an emphasis on exploitation of minerals toward somewhat greater regard for environmental considerations, and from exclusive control by ATS members—especially the claimant States and the superpowers—toward greater participation by nonmember States and by nongovernmental organizations. At this point, we cannot yet foresee what kind of minerals regime will ultimately emerge from the negotiations. We can only hope that it preserves the Antarctic ecosystem and tranquility as much as is humanly possible.

A more immediate threat to the Antarctic environment is tourism. At present the continent has an estimated population of about 3000 in summer and 1000 in winter; altogether since the beginning of the IGY in 1957 perhaps 16,000 people have spent some time there. Of them, only a few hundred so far have been tourists, with a few hundred more overflying or sailing near the continent and its islands without actually landing. But already they have disturbed bird nesting areas, left litter behind, and otherwise sullied the environment. More serious, they are getting themselves trapped in the ice and killed in plane crashes. In 1979, an Air New Zealand plane crashed into the continent's highest peak, Mt. Erebus, and all 257 tourists aboard, as well as the crew, were killed. In 1984, on New Year's Eve, 8 wealthy American tourists and their two-man crew were killed when their chartered Chilean plane crashed near Teniente Marsh, to which they had been invited for a party.

Such folly is likely to continue since there is as yet no mechanism for regulating it beyond each country's domestic laws and law enforcement systems. The cost to science of the diversion of scientists, support personnel, and equipment from their vital work, already tightly constrained by time and money, for search-and-rescue operations is very great. Tourists and those in the tourist industry have yet to learn that Antarctica is not Honolulu or Cannes or the Seychelles—an exotic, faraway place that simply *must* be visited. Already 51 American aircraft have crashed in Antarctica while on legitimate scientific missions. Supply ships have been trapped in ice; scientists and others have been stranded for weeks; fires have seriously damaged and even destroyed bases and field camps; and in 1986, the unoccupied Soviet geophysical base Druzhnaya I on the coast of the Weddell Sea vanished—totally, without a trace. Antarctica, while beautiful, is still a dangerous place.

It seems probable that other commercial activities will be attempted in the intermediate future, such as towing icebergs northward to provide pure water for arid areas (already done experimentally), operating scheduled airliner flights between South America and Australia across the continent, providing supplies and services to people working there, and so on. Each of them poses a host of logistical, environmental, and legal problems—and, of course, political ones. Already (in November 1981) New Zealand customs officials in Christchurch have confiscated marijuana and other illegal drugs (in 26 parcels) bound for Americans in Antarctica. This is likely to be—to coin a phrase—only the tip of the iceberg. How can commercial activities be regulated (or prohibited) under the current ATS? There is yet no answer.

ENVIRONMENTAL CONCERNS

We have already mentioned some environmental problems, actual and potential. But there are more. Under the Antarctic Treaty System, 14 Special Protected Areas have been designated in the region to date, and 8 Sites of Special Scientific Interest. But there is no way of legally protecting these areas or of prosecuting violators. The system relies on broad interpretation of the intent of the rules and good-faith compliance, neither of which is likely to survive the growing commercial and strategic interest in the region.

To date the most blatant violation of all the norms, rules, and promises of en-

vironmental protection has been the French airstrip in the Pointe Géologie archipelago near their principal base, Dumont d'Urville, in Terre Adélie. The site is very close to one of the few—and the most accessible—colonies of Emperor penguins, the least common of all species. From an original number of about seven thousand couples before the establishment of the base in the 1950s, the number of couples has declined to an estimated three thousand in 1985. The area is also an important breeding ground for many other species of birds. Yet in January 1983, the French began blasting, bulldozing, and helicoptering to construct an airfield for the use of French Transall C160 aircraft, even though they could cooperate with other countries operating in Antarctica to supply the base without using an airfield at all.

Contrast this with an Australian five-year plan (1985–1990) to clean up all Australian stations. Under this plan over 36 tons of discarded machinery and other de-

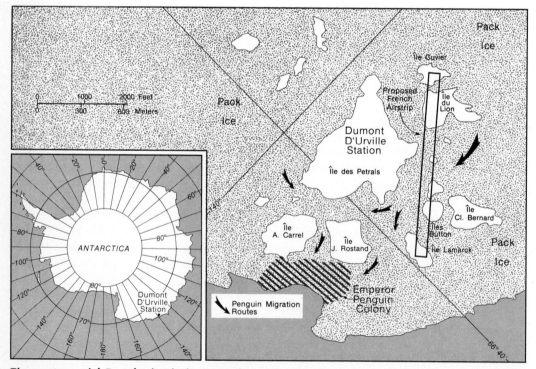

The controversial French airstrip in Antarctica. Like the Argentines, Chileans, and others, the French have not been exactly scrupulous in preserving this fragile and irreplaceable ecosystem.

bris was returned from the Casey Station to Australia in early 1986. Of course, one may wonder why so much junk was allowed to accumulate in the first place.

Despite the best intentions and efforts of most scientists and support personnel, the land and water of Antarctica are threatened. Even the air! Recently scientists discovered over Antarctica a hole in the ozone layer that protects the earth from the ultraviolet rays of the sun. The cause of the hole is not yet certain, but it probably results from the release of manmade chemicals, especially chlorofluorocarbons, into the atmosphere.

If Antarctica is to be spared the less desirable features of modern "civilization," perhaps the best regime for it is some version of a world park. Over the past decade, a number of proposals to turn the entire Antarctic and Southern Ocean, or major portions of the region, into a world park of some kind have been offered by various environmental groups, scholars, and others. Some suggest that it could be operated in some way by the United Nations, others that it could be accommodated within the present Antarctic Treaty System, others that a wholly new organization needs to be developed for it. Although, as we said earlier, the Antarctic Treaty has been scrupulously observed, its environmental provisions are far too weak to be meaningful and the conservation treaties adopted under the ATS are inadequate in a number of respects. Despite the vigorous objections of nearly all governments, some kind of world park may eventually be established in the south polar region.

GEOSTRATEGY

Article I of the Antarctic Treaty states forcefully that "Antarctica shall be used for peaceful purposes only," and apparently it has been. Already, however, there have been disquieting reports of remote sensing devices and satellite tracking facilities functioning within the treaty area. It seems unlikely that nuclear submarines

on patrol or engaged in maneuvers have invariably remained north of 60° South latitude, and the Anglo-Argentine war over the Falkland Islands in 1982 generated other questions about military activities in the treaty area. Yet, generally speaking, Antarctica and the Southern Ocean have remained peaceful, probably because the region is so remote from the world's centers of population and of political, economic, and military activity. Not everyone, however, looks at the region quite so benignly, and many South Americans look at it geopolitically.

In Chapter Eighteen, we pointed out that geopolitics is a lively, respected, and even popular field of study in Latin America, particularly in Brazil and the Southern Cone. It is probably fair to say, in fact, that influential members of the elites, both civilian and military, in this region consider Antarctica and the Southern Ocean to have greater potential geopolitical value than scientific or commercial value. The professional literature in Argentina, Brazil, Chile, and even Peru and Uruguay is laced with discussions of the potential strategic value of the south polar region, especially that portion of it directly south of the South American continent.

The British islands in the South Atlantic—the Falklands and their dependencies—are seen differently from Buenos Aires than from London: not as remote outposts of empire, useful only for weather stations, some fishing, and a bit of science, but as a screen across the South Atlantic protecting (or controlling) the major approaches to Antarctica from the richest, most industrialized, and most powerful countries in the world. They are also seen differently from Santiago, as the Arc of the Southern Antilles, marking a potential eastern limit of a Chilean claim to Pacific Ocean waters. From Brasilia, they are virtually invisible, having no effect whatever on a potential Brazilian claim to a wedge of Antarctic territory based on *defrontação*, the Brazilian version of the sector theory.

The Brazilian geographer Therezinha de

Castro was among the first—in 1958—to urge a Brazilian geopolitical role in the Antarctic. She developed her thesis further in a 1976 book, in which she referred to Antarctica as a future "cornerstone of our destiny . . . a base of warning, interception and departure in . . . the defense of the South Atlantic." Chilean geopoliticians have long argued that Chile should control not only the Straits of Magellan and the Beagle Channel, but also the Drake Passage, which could again become a major world highway if the Panama Canal were interdicted or closed. Argentine geopoliticians argue that Argentine Antarctica, the British-held islands, and all the waters between them ("The Argentine Sea") should be completely incorporated into the country, both to protect shipping lanes and to forestall any Brazilian moves southward.

This is only a fraction of Southern Cone geopolitical thinking about the Antarctic. Much more may be found (in English) in the papers and publications of professor *Jack Child* of the American University in Washington, and in an article by Glassner cited in the references for Part Seven. The important point to be understood, however, is that in the Southern Cone, geopolitics is not just abstract, armchair theorizing, but heavily influences national policy-making. The Argentine invasion of the Falkland Islands and the Argentine–Chilean dispute over three small islands in the Beagle Channel are only two recent examples of this influence. No intelligent discussion of the future of Antarctica can sensibly omit consideration of this viewpoint, regardless of what kind of government each country has at the moment or any other factors. These countries rank the geostrategic value of the Antarctic region much more highly than do either the Europeans or the superpowers. And who knows? They could be right.

MORE VEXING QUESTIONS

Besides the questions of commercial activities, environmental protection, and geostrategic value, others are unanswered by the present legal documents or are raised by political realities. We cannot go into all of them, but two deserve mention at least briefly.

Earlier in this chapter we referred to a number of legal questions regarding sovereignty over Antarctic waters, the status of ice, and so on. These are all just small parts of the Law of the Sea. Antarctica was deliberately excluded, however, from all discussions in UNCLOS III, and the relationship between the Antarctic Treaty and the United Nations Convention on the Law of the Sea is yet to be worked out. Do any of the UNCLOS provisions apply to Antarctica and the Southern Ocean? If so, which ones? To what extent? In a region lacking generally accepted sovereignty, who can accept responsibilities under and derive benefits from the Law of the Sea? The entire matter is incredibly complex and important and will not be settled very soon.

A lesser question but one with many ramifications is the role of South Africa in the Antarctic. South Africa was a signatory of the Antarctic Treaty in 1959 and has been a Consultative Party from the beginning of the Antarctic Treaty System. Yet South Africa has not been active in Antarctic research and has already been excluded from or has withdrawn from many international activities, including a number of specialized agencies of the United Nations. The UN General Assembly in 1985 passed a resolution urging the ATCPs to exclude the country from all of their meetings. That is unlikely to happen without very great pressure being exerted, not only from outside, but from inside the system as well. But if it does happen, or even if the pressure to do so becomes very great, what will be the effects on the system?

THE FUTURE OF ANTARCTICA

In October 1982, the Prime Minister of Malaysia, during the general debate in the UN General Assembly, proposed that the

United Nations focus its attention on Antarctica and convene a meeting which would determine the rights of all countries to uninhabited lands, especially Antarctica. Since then, Antarctica has been discussed during each annual session of the General Assembly, and at its request the Secretariat has produced two important reports on Antarctica. At issue is the demand of a number of developing countries, led by Malaysia, that the common heritage principle be applied to the Antarctic and that in some fashion the developing countries, preferably through the United Nations, be permitted to share in the governance of Antarctica and in its potential "riches." The ATS is viewed as an exclusive club, operating in secrecy for the benefit of a handful of rich (or relatively rich) countries and denying access to a large proportion of the earth's surface to a large proportion of the earth's population. This unfair situation must be rectified in accordance with the overall objectives of a New International Economic Order.

This viewpoint has not won a great many supporters, either in the ATS, in the UN Secretariat, or among the majority of the nonparties to the Antarctic Treaty. But the pressure has continued. As one means of defusing the potentially explosive situation, the U.S. National Science Foundation organized a workshop in Antarctica for a representative group of people. The workshop was held in January 1984 at the Beardmore South Field Camp of the United States at the foot of the Beardmore Glacier in the Transantarctic Mountains. Forty-seven countries were invited and 25 sent representatives, including some from each major grouping of States. In addition, there were two officers of the UN Secretariat and one from the United Nations Environment Programme, representatives of two industries, several university scientists, two reporters, and representatives of seven nongovernmental organizations. The participants read formal papers on various aspects of Antarctica, took field trips (including one to the South Pole), and had many informal discussions.

It is impossible to quantify the results of this workshop, but it is noteworthy that since the beginning of the United Nations debate, and especially since the workshop, more countries have acceded to the Antarctic Treaty, including Papua New Guinea, one of the world's poorest countries, and more have been admitted as CPs, including India and China. In addition, the ATS has opened up to more observers, has shared more information about its activities, and in general has tried to accommodate some of the demands of Malaysia and others without compromising any of the basic principles and practices of the system. It remains to be seen

Beardmore South Field Camp, Antarctica. These are the participants in the January 1984 workshop. (U.S. National Academy of Sciences)

whether this accommodation will be adequate, will have to be extended further, or will fail to prevent broader internationalization of Antarctica under some kind of "common heritage" regime.

The Antarctic Treaty of 1959 represents a very great stride forward in international cooperation. Its provisions for prohibition of military activity, encouragement of science and conservation, banning of nuclear explosions and waste dumping, freezing of territorial claims, periodic consultations, and—perhaps most important—formal inspections to ensure compliance with the treaty, were innova-

tive, far-reaching, and inspiring. We can *hope* that the international community as a whole, not just the parties to the treaty, will resist all attempts at commercial exploitation of the Antarctic region (except for limited and strictly controlled tourism) and preserve it as an international scientific reserve. Only in that way can we bring to life the recognition and the promise of the preamble of the Antarctic Treaty, "that it is in the interest of all mankind that Antarctica shall continue forever to be used exclusively for peaceful purposes and shall not become the scene or object of international discord."*

*Two excellent sources of information about Antarctica are:
International Institute for Environment and Development

3 Endsleigh Street
London WC1H ODD
England

1717 Massachusetts Ave., N.W.
Washington, DC 20036
U.S.A.

c/o CEUR
Corrientes 2835, 7° piso
1193 Buenos Aires
Argentina

Antarctic and Southern Ocean Coalition
(160 NGOs in 30 countries)

P.O. Box 371
Manly, NSW 2095
Australia

1845 Calvert Street, N.W.
Washington, DC 20009
U.S.A.

Chapter 32

OUTER SPACE

The political geography of outer space? Preposterous! How can there be any kind of geography of outer space when geography by definition is the study of the planet Earth and specifically of the inter-relationships between the human inhabitants of the planet and their physical environment? There is no Earth in outer space and no people living there, so geographers can have no professional interest in it. Even if some geographers, such as those specializing in remote sensing, climatology, or transportation, might occasionally glance into outer space in the course of their professional work, why should it be of interest to *political* geographers?

The same arguments were being used as recently as two decades ago with reference to the sea, yet we have seen how the field of marine geography has developed since then—and has even become respectable. While it is certainly premature to attempt to develop a full-blown political geography of outer space, it is not too early to begin laying the foundations, sketching its outlines. And as we do so, we will notice a remarkable resemblance to many aspects of traditional political geography. And why not? Aren't the people who have created political geographical situations on earth and those who have studied them the very people now venturing beyond the atmosphere, even beyond the gravity, of our planet?

Ultimately, new terminology, new theories, new approaches, even new topics will emerge as the new field is developed (perhaps by some of the people reading this book!), but for the present we shall continue with our traditional approach, and merely extend it into a new region.

HOW HIGH IS UP?

We have pointed out that a State's territory extends, according to the traditional international law, from the center of the earth upward "to the heavens." But this raises a serious question that until recently was only a child's riddle: "How high is up?" The answer to the question is that there is no answer, at least, not yet. For more than 60 years, we have had regulations on aerial navigation and the use of airspace, in both domestic and international law, and we discuss some of them in Chapter Thirty-four, but the important point to make here is that none of this large body of air law includes an upper boundary, a limit to a State's jurisdiction and to the applicability of air law. Since the Soviets launched the first artificial satellite into space in 1957, a body of space law has also developed that covers a variety of subjects, but still not defining the lower limit of outer space. This is not simply an academic matter; at some point the vertical limit of national jurisdiction will have to be fixed, just as horizontal limits are, or we most certainly will have boundary disputes over the interface between air and space.

Over the years, discussions within governments, in the professional literature, and especially in the UN General Assembly's Committee on the Peaceful Uses of Outer Space (COPUOS) have produced a great variety of proposals, not very different from those offered in respect of national jurisdiction in the sea and in the Antarctic region. In fact, we will note many analogies to these other frontier regions in our discussion of the political geography of outer space.

One approach to a definition of the air–space boundary has been the functional approach; that is, some people have argued that the limit should be set according to the uses to be made of both air and space. It has even been suggested that there could be several functional limits which could be invoked as appropriate in individual cases. This approach, however, has not gained a great deal of support, and around the world the trend has generally been toward a spatially measured boundary, either limited or unlimited.

Among the many suggested limits, those mentioned most frequently are 100 nautical miles, 80 kilometers, 100 kilometers and 120–150 kilometers. A number of States and scholars have proposed as the boundary the lowest practical level at which space objects could safely remain in orbit. This level is variable and difficult to measure from the earth, but it roughly approximates 100 km. Since among the qualities sought in any boundary are certainty, recognizability, and ease of measurement, the 100 km proposal seems at present to be gaining support. But both States and scholars are still divided on the issue and no one seems in any hurry to settle it.

THE USES OF OUTER SPACE

Undoubtedly, the principal users of space in the first 30 years of the Space Age have been the military establishments of the Soviet Union and the United States, and secondarily, their allies. Space is viewed by them as simply a convenient area in which, and from which, to conduct espionage and communications for military purposes. And, of course, they view it as another potential theater of war. Even now, intercontinental ballistic missiles frequently are launched into space en route to their targets many thousands of kilometers away—so far, only in tests—and both superpowers are trying to develop space-based defense systems against such missiles. The entire subject of the military uses of space, in fact, could very well be placed in Part Four of this book, in our discussion of geopolitics and war. As political geographers and not military geographers, however, we are more concerned with the peaceful uses of space.

Already space has many uses and many more potential ones. All geographers and most geography students are by now familiar, at least to some degree, with remote sensing, the observation and recording by various kinds of instruments mounted on orbiting satellites of various features of the Earth and its atmosphere. Nearly all of us have experienced directly or indirectly radio, television, telephone, or other kinds of messages relayed from one point on the earth's surface (or below it or above it) to another via satellites. Transportation, astronomy, and on-board scientific experiments have all made headlines and in future will make even more. Other uses are still in their earliest stages: space biology and medicine; manufacturing in the weightless vacuum of space of such things as new alloys of metals, pharmaceuticals, glasses, catalysts, and semiconductors; the concentration of solar energy for on-board use and for beaming onward to Earth; mining of the moon, asteroids, and perhaps planets for substances in short supply or even unknown on Earth.

Because of the great commercial as well

as military value of space, over 16,800 objects were launched into space between 1957 and 1987, and most of them are still orbiting the earth.* While all have been launched by national governments or the European Space Agency (ESA), commercial firms are currently preparing to compete with them, and more countries are also developing space-launch capabilities. Even countries, companies, institutions, and individuals that cannot launch objects into space have developed satellites to be launched by others or components of satellites or modules for space shuttles or space stations or experiments to be conducted on board. All of this activity by so many people in so many countries is bound to generate problems and some of them are bound to be political and thus of interest to political geographers.

Unlike activities in Antarctica, activities of any kind in space can be perceived as a potential or actual threat to a State, even one some distance away. Satellites designed and placed into orbit for peaceful scientific or commercial purposes could well be used for military purposes with no one on the ground realizing it. Even perfectly innocent information sent to earth by a satellite can be used to prepare an attack or some other military action. Aside from military potential, remote sensing by or through space vehicles can also give the technologically advanced countries exclusive information about potential economic resources of other countries, particularly developing ones, which could then be used to gain commercial advan-

*Space launches (many of them deploying more than one object), 1957–1982:

USSR	1538	France	10
United States	796	Italy	4
Japan	21	ESA	3
		India	3

Some 6000 orbiting objects are currently being tracked by NORAD (the North American Aerospace Defense Command), of which 22 percent are junk and only 5 percent are satellites working in 1987. Some forty thousand additional bits of orbiting debris at least as big as golf balls have been identified, and there are many times more smaller man-made fragments cluttering near space.

tages for the country possessing the information.

Currently, four major politicogeographical problems connected with space activities are engaging lawyers, diplomats, and scholars.

1. **Direct broadcasting from satellites.** One view of this inchoate technology emphasizes the virtues of a free and unrestricted flow of information; another the necessity for screening and regulation of such broadcasting by governments. Some countries object to commercials coming from satellites into societies free of them. Others are apprehensive about being subjected to pornography, excessive violence, or other controversial material. There is also concern about competing religious or political views being beamed into a society unfamiliar with them or about inappropriate educational material or even about simply being drowned in the materialistic culture of the rich Western countries that alone can afford such expensive space operations.

2. **Space transportation systems.** Transportation has come to be regarded as an activity separate and distinct from scientific and technological missions of space craft. As such, it raises a host of questions of both a national and an international character. Matters of jurisdiction, scheduling, competing uses of space craft and orbits, use of vehicles by nonlaunching States, passage over States during the early stages of launching, and many others are currently receiving the attention of experts. So far the tendency has been to adapt and adjust national legislation to deal with these questions, but very soon it will likely be necessary to develop some new kind of space transportation management and control system.

3. **The environmental impact of space activities.** Human activities have already

harmed space and as they increase in number, variety, scope, and intensity, and damage is likely to increase. The atmosphere of Earth has been polluted by the effluents of high-flying aircraft and rocket boosters; near-earth space is littered with man-made debris. International space law not only provides for compensation for some kinds of environmental harm, but imposes on States an affirmative duty to avoid it. It remains to be seen, however, how effective current rules and practices will be in the future, even the near future.

4. **Use of nuclear power systems in space.** Nuclear energy is currently considered the most efficient power source for certain types of space missions, particularly where the operating time of the mission is longer than a few hundred hours. The potential hazards of a malfunctioning reactor, unplanned reentry into Earth's atmosphere, or intentional release of radioactive substances are all being dealt with in a variety of ways through international cooperation, and to date there is no evidence of damage to Earth or its atmosphere or outer space by any of the hundreds of nuclear power sources that have been used. There is, however, a need to supplement the relevant international space

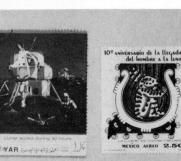

Outer space on stamps. Space activities have been one of the most popular subjects on stamps during the past quarter century. This selection illustrates some of the more geographic aspects of the subject. The two at the top commemorate manned lunar expeditions, the next four, both manned and unmanned space exploration, and the bottom stamp, the Outer Space Telecommunications Congress in Paris. (Martin Glassner)

law on the subject, and that may not be easy.

THE GEOSTATIONARY ORBIT

At present, the most important use of outer space is for communications, and most communications satellites are located in geostationary orbital positions. A satellite that orbits in the same direction that the earth rotates and in a 24-hour period is said to be in a geosynchronous, or synchronous, orbit; that is, it completes a revolution about the planet in exactly the same period as the earth rotates once on its axis. Most such orbits are elliptical or inclined with respect to the equator. If, however, such an orbit is directly over the Equator, it is said to be geostationary. A satellite in a geostationary orbital position is constantly visible from the surface at the same point in the sky, and it can constantly cover a sizable portion of Earth's surface. A satellite can only maintain such a course and speed at an altitude of ±35,787 km.*

The advantages of such a position for observation and communications are obvious. The number of usable positions within the band is constrained by limitations on the use of the radio-frequency spectrum and by the large portions of the band over essentially empty ocean surfaces. Thus, the geostationary orbit is generally recognized as a valuable resource and the most desirable portions of it are already getting crowded. Whether it is a "natural" resource, however, and whether it is truly limited, are highly controversial matters, important to us because of their political implications.

They are likely to become even more important, and more controversial, as very large space stations are introduced into the geostationary orbit to perform functions other than observation and

communications. The gathering and transmission of solar energy, for example, can best be done from such a position. This situation has already produced the first attempt to extend contemporary nationalistic impulses and competition among developed and developing countries on Earth into outer space. The result of this attempt is the Bogotá Declaration.

It is worth quoting at some length this product of the "First Meeting of Equatorial Countries":

The undersigned representatives of States traversed by the Equator* met in Bogota, Republic of Colombia, from November 24 through December 3rd, 1976 with the purpose of studying the geostationary orbit that corresponds to their terrestrial, sea and insular territory and considered as a natural resource. After an exchange of information and having studied in detail the different technical, legal and political aspects implied in the exercise of national sovereignty of States adjacent to said orbit, they have reached the following conclusions. . . .

The Equatorial countries declare that the geostationary synchronous orbit is a physical fact linked to the reality of our planet . . . and that is why it must not be considered part of outer space. Therefore, the segments of the geostationary synchronous orbit are part of the territory over which Equatorial states exercise their national sovereignty. The geostationary orbit is a scarce natural resource . . . therefore, the Equatorial countries meeting in Bogota have decided to proclaim and defend on behalf of their peoples, the existence of their sovereignty over this natural resource. . . .

Fortunately, this absurd attempt at neoimperialism has not attracted much support from nonsigners, and it has been generally ignored by those conducting activities in outer space. It does, however, raise numerous issues that deserve careful attention by political geographers for, while we cannot examine them here, neither the declaration nor the issues it raises will go away.

---

*Actually, because of a number of factors, geostationary satellites move around within a band about 30km deep extending about 150 km north and south of the Equator.

*Brazil, Colombia, Congo, Ecuador, Indonesia, Kenya, Uganda, and Zaïre. Nonsigning equatorial countries are: Gabon, Kiribati, Nauru, Peru, and Somalia.

COOPERATION OR COMPETITION IN SPACE?

Not surprisingly, many of the international rivalries and alignments so prominent on Earth are being projected out into space—but not all. While there is a good deal of competition in space, there is also much cooperation. Essentially, there are currently three groupings of States in respect of space activities similar to the groupings surrounding other issues and activities on land and at sea.

First, the United States and its allies and close associates tend to view space not only as a field for new technological breakthroughs, but for new political and legal breakthroughs as well. They see a large role for private enterprise in space, envision some kind of supranational supervision of space activities, and are willing to share their space technology and the results of their exploitation of space with other countries. Their space pro-

grams, however, tend to be haphazard, ill-coordinated, shortsighted, and competitive in many respects.

The Soviets, on the other hand, have a long-term and well-financed commitment to space activities of all kinds. They are concentrating, for example, on gradually building very large, multipurpose, permanently inhabited space stations rather than the dramatic, one-shot breakthroughs (such as walking on the moon) favored by the United States. They tend, however, to be very conservative in outer space political and legal matters. They do not accept a "common heritage" principle for outer space; they refuse to recognize an outer space law independent of general international law; they do not accept individuals or corporations as subjects of law in space; they oppose an outer space agency with supranational functions; and they oppose any compulsion of States operating in outer space to help those unable to reach it.

INTERGOVERNMENTAL ORGANIZATIONS

International Telecommunication Satellite Organization (INTELSAT)
   106 member countries in 1982; no members from the Soviet bloc
INTERCOSMOS
   10 members of the socialist (Soviet) group
International System and Organization of Space Communications
   (INTER-SPUTNIK) 12 socialist countries
European Space Agency (ESA)
   11 Western European countries, with Austria, Norway, and Canada associated in different ways
International Maritime Satellite Organization (INMARSAT)
   32 members in 1981, from all groupings
Arab Satellite Communication Organization (ARABSAT)
   All 21 States members of the Arab League

NONGOVERNMENTAL ORGANIZATIONS

International Council of Scientific Unions (ICSU)
   18 unions, including the International Geographical Union
Committee on Space Research (COSPAR)
   35 national academies of science from all groupings, and 13 international scientific unions
International Astronautical Federation (IAF)
   60 member societies and educational and research institutions from 36 countries of all groupings

The developing countries are generally wary of being excluded from outer space altogether merely because they lack the scientific, technological, and financial resources to operate there, just as they fear being excluded from the benefits of development of the sea and Antarctica. They view the rich countries—whether socialist or capitalist—as space-resource States; that is, they have the resources to operate in space. They themselves are, then, non-space-resource States. They tend to favor the extension of the common heritage principle to space, restraints on private enterprise in space, supranational instruments for supervision of activities and resolution of disputes in space, some kind of mandatory transfer of space technology to them, and an equitable distribution of the benefits of space activities.

So far, these differing views have not resulted in any serious clashes. The tendency, in fact, has been to compromise, to reconcile differences while dealing with space problems *ad hoc,* as the need becomes apparent. Pragmatism has generally been winning out over ideology (with the notable exception of the Bogotá Declaration), and nearly all countries are cooperating through the United Nations. The United Nations, in fact, has played a leading role in the development of space law and the coordination of space activities almost from the dawn of the Space Age, largely through COPUOS, the Outer Space Division of the Secretariat, and a number of other organs and specialized agencies. In addition, international cooperation in space activities is carried out through an array of organizations (see box).

Besides its day-to-day work in stimulating, coordinating, and utilizing activities in outer space, the United Nations has fostered international cooperation in space through two other mechanisms. First, it has held so far two conferences on the Exploration and Peaceful Uses of Outer Space, the first in 1968, the second in 1982. These conferences studied contemporary and projected space matters and

United Nations leadership in outer space cooperation. This poster calls attention to the second United Nations Conference on the Exploration and Peaceful Uses of Outer Space, held in Vienna in 1982. (UN)

produced lists of conclusions and recommendations directed to member States and to international agencies and bodies. The final reports of these conferences, together with the documentation they produced, serve as important benchmarks of the Space Age.

Second, the United Nations has played a vital role in the development of the international law of outer space. To date, it has sponsored five important treaties.

1. Treaty on Principles Governing the Activities of States in the Exploration and Use of Outer Space, Including the Moon and Other Celestial Bodies (the Principles Treaty, or the Outer Space Treaty), 1967.

2. Agreement on the Rescue of Astronauts, the Return of Astronauts, and the Return of Objects Launched into Outer Space, 1968.

3. Convention on International Liability for Damage Caused by Space Objects, 1972.

4. Convention on Registration of Objects Launched into Outer Space, 1976.

5. Agreement Governing the Activities of States on the Moon and Other Celestial Bodies (the Moon Treaty), 1979.

Of these, the Principles Treaty and the Moon Treaty are of greatest interest to political geographers. The first is generally accepted as being a pioneering venture that entered into force on 10 October 1967, just 10 years and 6 days after the Soviet Union launched the world's first artificial satellite (*Sputnik*) into space and only 6½ years after the first manned spaceship (piloted by the Soviet cosmonaut Yuri Gagarin) was launched. There are objections, however, to many of its provisions (such as those by claimants of segments of the geostationary orbit), and others are imperfectly drafted, subject to widely varying interpretations, or out of date. By 1985, however, it had been ratified or acceded to by 85 States, half the States of the world, and it is still recognized as a pillar of space law.

The Moon Treaty, on the other hand, is much more adventurous and may be ahead of its time. It entered into force on 11 July 1984. By 1985, only 11 States had signed it and only 5 had ratified it. Most of the controversy surrounding it centers on paragraph 1 of Article II: "The moon and its natural resources are the common heritage of mankind. . . ." This extension of the common heritage principle from the seabed to outer space, like its extension to Antarctica, is still vigorously opposed by both the socialist and capitalist States and may not be generally accepted for many years. Nevertheless, both treaties are worthy of careful study by anyone interested in any aspect of outer space, especially its political geography.

BEYOND EARTH ORBITS

So far our discussion has concentrated on near-space activities, chiefly those involving orbiting satellites, shuttles, and other objects. But a number of instrumented and a few manned vehicles have already been sent beyond all normal orbital paths around our planet, to the moon and to and beyond other planets of our solar system. These have produced nothing of politicogeographical interest beyond the refusal of the United States, in accordance with the UN Principles Treaty, to claim the moon or any portion of it. But we can easily envision a complex politicogeographical picture emerging in both the near and distant reaches of outer space during coming decades.

There may well be, for example, numerous occupied orbiting space stations, colonies of earthlings living permanently far from their home planet, organized transportation and communications among these space stations and colonies, and trade among them, to say nothing of colonies on the moon, on asteroids, and on other planets and their moons. These will inevitably generate questions of nationality and nationalism and sovereignty, of ownership and use of resources, of the distribution of costs and benefits, of social stratification and cultural differences, of law and loyalties and rivalries and politics, of frontiers and boundaries and power, and perhaps colonial empires and wars of independence.

Political geographers in the future will have plenty to study and analyze and explain and try to improve beyond Earth's atmosphere and beyond Earth orbits. Meanwhile, there is a great deal to learn about what is being done and what is being planned in space so that we can apply our approach and our techniques to their analysis.*

---

*Two good sources of information, besides the United Nations and individual governments, are:

Institute and Centre of Air and Space Law	International Institute of Space Law
McGill University	Leestraat 43
Montreal, P.Q.	Baarn
Canada	The Netherlands

Although the age-old problem of nationalism leading to territorial claims and boundary disputes is being carried into outer space, we may still be able to contain it. In keeping with Article 3 of the 1967 Principles Treaty, it is still possible that exploration, use, and scientific investigation of outer space will be carried out "in accordance with international law, including the Charter of the United Nations, in the interest of maintaining international peace and security and promoting international cooperation and understanding."

REFERENCES FOR PART SEVEN

Books and Monographs

A

Ademuni-Odeke, *Protectionism and the Future of International Shipping.* Dordrecht, Neth.; Boston, Lancaster: Nijhoff, 1984.

Akaha, Tsuneo, *Japan in Global Ocean Politics.* Honolulu: Univ. of Hawaii Press, 1985.

Alexander, Lewis M., *Offshore Geography of Northwestern Europe; The Political and Economic Problems of Delimitation and Control.* Chicago: Rand McNally, 1963; London: John Murray, 1966.

———, *Marine Regionalism in the Southeast Asian Seas.* Honolulu: East–West Center, 1982.

Alexander, Lewis M. and Lynne Carter Hanson (eds.), *Antarctic Politics and Marine Resources: Critical Choices for the 1980's.* Kingston: University of Rhode Island, Center for Ocean Management Studies, 1984.

Alexandersson, Gunnar, *The Baltic Straits.* The Hague, Boston, London: Nijhoff, 1982.

Amacher, Ryan C. and Richard J. Sweeney, *The Law of the Sea: U.S. Interests and Alternatives.* Washington: American Enterprise Institute for Public Policy Research, 1976.

American Society of International Law, *Policy Issues in Ocean Law.* St. Paul: West, 1975.

Amin, Sayed Hassan, *Marine Pollution in International and Middle Eastern Law.* Glasgow: Royston, 1986.

Anand, Ram Prakash, *Legal Regime of the Seabed and the Developing Countries.* Leyden, Neth.: Sijthoff, 1976.

——— (ed.), *Law of the Sea: Caracas and Beyond. Developments in International Law.* The Hague, Boston and London: Nijhoff, 1980.

———, *Origin and Development of the Law of the Sea: History of International Law Revisited.* The Hague, Boston and London: Nijhoff, 1983.

Anderson, Lee G. (ed.), *Economic Impacts of Extended Fisheries Jurisdiction.* Ann Arbor, MI: Ann Arbor Science, 1977.

Andrassy, Juraj, *International Law and the Resources of the Sea.* New York: Columbia Univ. Press, 1970.

Anninos, P.C.L., *The Continental Shelf and Public International Law.* The Hague: De Swart, 1953.

Aprieto, Virginia L., *Fishery Management and Extended Maritime Jurisdiction: The Philippine Tuna Fishery Situation.* Honolulu: East–West Center, 1981.

Archer, Clive and David Scrivener (eds.), *Northern Waters; Security and Resource Issues.* Totowa, NJ: Barnes & Noble, 1986.

Armstrong, Terence and others, *The Circumpolar North.* London: Methuen, 1978.

Auburn, F.M., *The Ross Dependency.* The Hague: Nijhoff, 1972.

———, *Antarctic Law and Politics.* Bloomington: Indiana Univ. Press, 1982.

B

Barkenbus, Jack N., *Deep Seabed Resources, Politics and Technology.* New York: Free Press, 1979.

Barston, R.P. and Patricia Birnie, *The Maritime Dimension.* Winchester, MA: Allen & Unwin, 1980.

Bartell, Joyce J. (ed.), *The Yankee Mariner and Sea Power: America's Challenge of Ocean Space.* Los Angeles: Univ. of California Press, 1982.

Baxter, Richard R., *The Law of International Waterways.* Cambridge: Harvard Univ. Press, 1964.

Beck, Peter J., *The International Politics of Antarctica.* New York: St. Martin's Press, 1987.

Bell, Frederick W., *Food from the Sea: The Economics and Politics of Ocean Fisheries.* Boulder, CO: Westview, 1978.

Benko, Marietta and others, *Space Law in the United Nations.* Dordrecht, Neth., Boston and London: Nijhoff, 1985.

Birnie, Patricia (ed.), *International Regulation of Whaling: From Conservation of Whaling to Conservation of Whales and Regulation of Whale-Watching,* 2 Vols. London, Rome, and New York: Oceana, 1985.

Blake, Gerald (ed.), *Maritime Boundaries and Ocean Resources.* Beckenham, Eng.: Croom Helm, 1987.

Bloomfield, Lincoln M., *Egypt, Israel and the Gulf of Akaba.* Toronto: Carswell, 1957.

———, *Outer Space: Prospects for Man and Society*. Englewood Cliffs, NJ: Prentice-Hall, 1962.

Bonner, W.N. and R.I. Lewis Smith, *Conservation Areas in the Antarctic*. Cambridge, England: Scientific Committee on Antarctic Research, 1985.

Borgese, Elisabeth Mann, *The Future of the Oceans: A Report to the Club of Rome*. Montreal: Harvest House, 1986.

Bouchez, Leo J., *The Regime of Bays in International Law*. London: Sijthoff, 1964.

Bowett, Derek W., *The Legal Regime of Islands in International Law*. Dobbs Ferry, NY: Oceana, 1978; Alphen aan den Rijn, Neth.: Sijthoff & Noordhoff, 1979.

Brooks, Douglas L., *America Looks to the Sea; Ocean Use and the National Interest*. Boston and Woods Hole, MA: Jones & Bartlett, 1984.

Brown, E.D., *Sea-bed Energy and Mineral Resources and the Law of the Sea,* 3 Vols. London: Graham & Trotman, 1986.

Burke, William T. and others, *National and International Law Enforcement in the Ocean*. Seattle: Univ. of Washington Press, 1975.

Bush, W.M., *Basic Documents on Antarctica*. London, Rome, and New York: Oceana, 1982.

Butler, William E., *International Straits of the World. Northeast Arctic Passage*. Alphen aan den Rijn, Neth.: Sijthoff & Noordhof, 1978.

———, *The Law of the Sea and International Shipping; Anglo–Soviet Post-UNCLOS Perspectives*. New York: Oceana, 1985.

Buzan, Barry, *Seabed Politics*. New York: Praeger, 1976.

C

Carlisle, Rodney P., *Sovereignty for Sale; The Origins and Evolution of the Panamanian and Liberian Flags of Convenience*. Annapolis, MD: Naval Institute Press, 1981.

Center for Ocean Management Studies, *Comparative Marine Policy*. New York: Praeger, 1980.

Červenka, Zdenek, *Land-locked Countries of Africa*. Uppsala, Swed.: Scandinavian Institute of African Studies, 1973.

Chang, Pao-min, *The Sino–Vietnamese Territorial Dispute*. New York: Praeger, 1986.

Charney, Jonathan I. (ed.), *The New Regionalism and the Use of Common Spaces; Issues in Marine Pollution and the Exploitation of Antarctica*. Totowa, NJ: Allanheld, Osmun, 1982.

Chayes, Abram and Paul Laskin, *Direct Broadcasting from Satellites: Policies and Problems*. Washington: American Society of International Law, 1975.

Chipman, Ralph (editor for UN), *The World in Space; a Survey of Space Activities and Issues*. Englewood Cliffs, NJ: Prentice-Hall, 1982.

Christie, E.W. Hunter, *The Antarctic Problem: An Historical and Political Study*. London: Allen & Unwin, 1951.

Christol, Carl Q., *The Modern International Law of Outer Space*. New York and elsewhere: Pergamon, 1982.

Churchill, Robin R. and Alan V. Lowe, *The Law of the Sea*. Manchester, Eng.: Manchester Univ. Press, 1983, 1985.

Colombos, C. John, *The International Law of the Sea*, 5th rev. ed. London: Longmans, 1962.

Crutchfield, James A. and Giulio Pontecorvo, *The Pacific Salmon Fisheries: A Study of Irrational Conservation*. Baltimore: Johns Hopkins Univ. Press, 1969.

Cuyvers, Luc, *Ocean Uses and Their Regulation*. New York: Wiley, 1984.

———, *The Strait of Dover*. Dordrecht, Neth., Boston, Lancaster, Eng.: Nijhoff, 1986.

D

Dahmani, M., *The Fisheries Regime of the Exclusive Economic Zone*. Dordrecht, Neth., Boston, Lancaster, Eng.: Nijhoff, 1987.

Dalhousie Ocean Studies Programme. Halifax, N.S.: Dalhousie Univ. Publishes studies on various aspects of the Law of the Sea.

DeCesari, P. and others (eds.), *Index of Multilateral Treaties on the Law of the Sea*. Milan: Dott Giuffré Editore, 1985.

Deudney, Daniel, *Space: The High Frontier in Perspective*. Worldwatch Paper 50. Washington: August 1982.

Diederiks-Verschoor, I.H., *An Introduction to Air Law*. Deventer, Neth.: Kluwer, 1985.

Dubner, Barry Hart, *The Law of Territorial Waters of Mid-Ocean Archipelagos and Archipelagic States*. The Hague: Nijhoff, 1976.

Durante, Francesco and Walter Rodino, *Western Europe and the Development of the Law of the Sea*. Dobbs Ferry, NY: Oceana, 1979.

Durch, William J. (ed.), *National Interests and the Military Use of Space*. Cambridge, MA: Ballinger, 1984.

Dyer, Ira and Chryssostomos Chryssastomidis (eds.), *Arctic Technology and Policy*. Washington and elsewhere: Hemisphere Publishing, 1984.

E

Eckert, Ross D., *The Enclosure of Ocean Resources: Economics and the Law of the Sea.* Stanford, CA: Hoover Institution Press, 1979.

Edeson, William R. and Jean-François Pulvenis, *The Legal Regime of Fisheries in the Caribbean Region.* Berlin and elsewhere: Springer-Verlag, 1983.

El-Hakim, Ali A., *The Middle Eastern States and the Law of the Sea.* Syracuse, NY: Syracuse Univ. Press, 1979.

English, T. Saunders (ed.), *Ocean Resources and Public Policy.* Seattle and London: Univ. of Washington Press, 1973.

Extavour, Winston Conrad, *The Exclusive Economic Zone; a Study of the Evolution and Progressive Development of the International Law of the Sea.* Geneva: Institut Universitaire de Hautes Études Internationales, 1979.

F

Fawcett, J.E.S., *Outer Space: New Challenges to Law and Policy.* Oxford: Oxford Univ. Press, Clarendon; New York: Oxford Univ. Press, 1984.

Finch, Edward Ridley, Jr. and Amanda Lee Moore, *Astrobusiness: A Guide to the Commerce and Law of Outer Space.* New York: Praeger, 1985.

Fincham, Charles and William van Rensburg, *Bread upon the Waters; the Developing Law of the Sea.* Jerusalem: Turtledove, 1980.

Francioni, F. and T. Scovazzi (eds.), *International Law for Antarctica.* Milan: Dott Giuffré Editore, 1987.

Friedman, Wolfgang, *The Future of the Oceans.* New York: Braziller, 1971.

Friedrich, Christoph, *Germany's Antarctic Claim: Secret Nazi Polar Expeditions.* Toronto: Samisdat, 1978.

Fulton, T.W., *The Sovereignty of the Sea.* Edinburgh: Blackwood, 1911.

G

García Amador, F.V., *The Exploitation and Conservation of the Resources of the Sea.* Leyden: Sijthoff, 1959.

Gold, Edgar, *Maritime Transport: The Evolution of International Marine Policy and Shipping Law.* Lexington, MA, and Toronto: Heath, 1981.

Goldman, Nathan C., *Space Commerce; Free Enterprise on the High Frontier.* Cambridge, MA: Ballinger, 1985.

Goodwin, H.L., *Space: Frontier Unlimited.* Princeton, NJ: Van Nostrand, 1962.

Górbiel, Andrzej, *Legal Definition of Outer Space.* Lódź, Pol.: Uniwersytet Lózki, 1980.

――――, *International Organizations and Outer Space Activities.* Lódź, Pol.: Redake a Naczelna, Uniwersytet Lódski, 1984.

――――, *Outer Space in International Law.* Lódź: Uniwersytet Lódzki, 1984.

Gorove, Stephen (ed.), *Space Shuttle and the Law.* University: Univ. of Mississippi Law Center, 1980.

――――(ed.), *United States Space Law: National and International Regulation.* London, Rome and New York: Oceana, 1982.

Gravesand, J.W.E. Storm Van's and A. van der Veen Vonk (eds.), *Air Worthy: Liber Amicorum Honoring Professor Dr. I.H.Ph. Diedriks-Verschoor.* Deventer, Neth.: Kluwer, 1985.

Greenfield, Jeannette, *China and the Law of the Sea, Air, and Environment.* Alphen aan den Rijn, Neth.: Sijthoff & Noordhoff, 1979.

Gullion, Edmund A. (ed.), *Uses of the Seas.* Englewood Cliffs, NJ: Prentice-Hall, 1968.

H

Hakapää, Kari, *Marine Pollution in International Law. Material Obligations and Jurisdiction, with Special Reference to the Third United Nations Conference on the Law of the Sea.* Helsinki: Suomalainen Tiedeakatemia, 1981.

Hargreaves, Reginald, *The Narrow Seas.* London: Sidgwick & Jackson, 1959.

Hargrove, Eugene C., *Beyond Spaceship Earth; Environmental Ethics and the Solar System.* San Francisco: Sierra Club, 1987.

Hargrove, John Lawrence (ed.), *Who Protects the Ocean? Environment and the Development of the Law of the Sea.* St. Paul, MN: West, 1975.

Harris, Stuart (ed.), *Australia's Antarctic Policy Options.* CRES Monograph 11. Canberra: Australian National Univ., 1984.

Hauser, Wolfgang, *The Legal Regime for Deep Seabed Mining Under the Law of the Sea.* Deventer, Neth.: Kluwer; Frankfort on the Main, W. Ger.: Alfred Metzner Verlag, 1983.

Henderson, Hamish McN. and others, *Oil and Gas Law—the North Sea Exploitation.* Dobbs Ferry, NY: Oceana, 1979.

Herr, Richard A., *Issues in Australia's Marine and Antarctic Policies.* Hobart: Univ. of Tasmania, Dept. of Political Science, 1982.

Hollick, Ann L., *U.S. Foreign Policy and the Law of the Sea.* Princeton, NJ: Princeton Univ. Press, 1981.

Hollick, Ann L. and Robert E. Osgood, *New Era*

of Ocean Politics. Baltimore: Johns Hopkins Univ. Press, 1974.

Hood, Valerie and others, *A Global Satellite Observation System for Earth Resources: Problems and Prospects.* Washington: American Society of International Law, 1977.

Howard, Harry N., *Turkey, the Straits and U.S. Policy.* Baltimore and London: Johns Hopkins Univ. Press, 1974.

Hudson, Heather, *New Dimensions in Satellite Communications; Challenges for North and South.* Dedham, MA: Artech House, 1985.

I

Inamura, Kobo, *Satellite Broadcasting: Its Prehistory, Present and Future as a New Technology.* Bangkok: Dear Book Co., 1981.

J

Jados, Stanley S., *Consulate of the Sea and Related Documents.* University: Univ. of Alabama Press, 1975.

Jansky, Ronald and others, *Communication Satellites in the Geostationary Orbit.* Dedham, MA: Artech, 1983.

Jasani, Bhupendra (ed.), *Space Weapons—The Arms Control Dilemma.* London and Philadelphia: Taylor & Francis, 1984.

Jasentuliyana, Nandasiri (ed.), *Maintaining Outer Space for Peaceful Uses.* Tokyo: UN Univ., 1984.

Jayaraman, K., *Legal Regime of Islands.* New Delhi: Marwah, 1982.

Jessup, Philip C., *The Law of Territorial Waters and Maritime Jurisdiction.* New York: Jennings, 1927.

Jessup, Philip C. and Howard J. Taubenfeld, *Control for Outer Space and the Antarctic Analogy.* New York: Columbia Univ. Press, 1959.

Jhabvala, Farrokh (ed.), *Maritime Issues in the Caribbean.* Miami: Univ. Presses of Florida, 1983.

Johnson, Barbara and Mark W. Zacher (eds.), *Canadian Foreign Policy and the Law of the Sea.* Vancouver: Univ. of British Columbia Press, 1977.

Johnston, Douglas M., *The International Law of Fisheries,* New Haven, CT: Yale Univ. Press, 1965.

——— (ed.), *The Environmental Law of the Sea.* Gland, Switz.: International Union for the Conservation of Nature and Natural Resources, 1981.

———, *Environment Management in the South China Sea: Legal and Institutional Developments.* Honolulu: East–West Center, 1982.

———, *Canada and the New International Law of the Sea.* Toronto: Univ. of Toronto Press, 1987.

Jónsson, Hannes, *Friends in Conflict: The Anglo–Icelandic Cod Wars and the Law of the Sea.* London: Hurst; Hamden, CT: Shoe String Press, Archon Books, 1982.

K

Karas, Thomas, *The New High Ground; Strategies and Weapons of Space Age Wars.* New York: Simon & Schuster, 1983.

Katz, James Everett, *People in Space.* New Brunswick, NJ: Transaction Books, 1985.

Kent, George, *The Politics of Pacific Island Fisheries.* Boulder, CO: Westview, 1980.

Knight, H. Gary (ed.), *The Future of International Fisheries Management.* St. Paul, MN: West, 1975.

Koh, Kheng-Lian, *Straits in International Navigation: Contemporary Issues.* London, Rome and New York: Oceana, 1982.

Kronmiller, Theodore G., *The Lawfulness of Deep Seabed Mining.* London, Rome, and New York: Oceana, 1980.

L

Lachs, Manfred, *The Law of Outer Space; an Experience in Contemporary Law-making.* Leiden, Neth.: Sijthoff, 1972.

Lapidoth, Ruth, *Freedom of Navigation with Special Reference to International Waterways in the Middle East.* Jerusalem Papers on Peace Problems No. 13–14. Jerusalem: Hebrew University, 1975.

Lapidoth-Eschelbacher, Ruth, *The Red Sea and the Gulf of Aden.* The Hague, Boston and London: Nijhoff, 1982.

Laursen, Finn (ed.), *Toward a New International Marine Order.* The Hague, Boston and London: Nijhoff, 1982.

———, *Superpower at Sea; U.S. Ocean Policy.* New York: Prager, 1983.

The Law of the Sea Institute at the University of Hawaii publishes its annual conference proceedings, occasional papers, and workshop reports that are most valuable.

Lay, S. Houston and Howard J. Taubenfeld, *The Law Relating to Activities of Man in Space.* Chicago and London: Univ. of Chicago Press, 1970.

Lee Yong Leng, *Southeast Asia and the Law of the Sea.* rev. ed. Singapore: Singapore Univ. Press, 1980.

Leonard, L. Larry, *International Regulation of Fisheries.* Washington: Carnegie Endowment for International Peace, 1944.

Li, Kuo Lee, *World Wide Space Law Bibliography*. Montreal: McGill Univ., 1978.

Leifer, Michael, *International Straits of the World. Malacca, Singapore and Indonesia*. Alphen aan den Rijn, Neth.: Sijthoff & Noordhoff, 1978.

Lien, Jon, and Robert Graham (eds.), *Marine Parks and Conservation; Challenge and Promise*. Toronto: National and Provincial Parks Association of Canada, 1985.

Logan, Rod M., *Canada, the United States, and the Third Law of the Sea Conference*. Montreal and Washington: Canadian–American Committee, 1974.

Logue, John J. (ed.), *The Fate of the Oceans*. Villanova, PA: Villanova Univ. Press, 1972.

Lovering, J.F. and J.R.V. Prescott, *Last of Lands—Antarctica*. Melbourne: Melbourne Univ. Press, 1979.

Luard, Evan, *The Control of the Sea-bed; Who Owns the Resources of the Oceans?* rev. ed. London: Heinemann, 1977.

Lumb, Richard Darrell, *The Law of the Sea and Australian Off-shore Areas*, 2nd ed. St. Lucia: Univ. of Queensland Press, 1978.

M

Ma, Ying-jeou, *Legal Problems of Seabed Boundary Delimitation in the East China Sea*. Occasional Papers/Reprints Series in Contemporary Asian Studies No. 3. Baltimore: Univ. of Maryland School of Law, 1984.

MacDonald, Charles G., *Iran, Saudi Arabia, and the Law of the Sea: Political Interaction and Legal Developments in the Persian Gulf*. Westport, CT: Greenwood Press, 1980.

Mangone, Gerard J., *Marine Policy for America*. Lexington, MA: Heath, 1977.

Mankabady, Samir, *The International Maritime Organization*. London & Sidney: Croom Helm, 1984.

Mason, C.M. (ed.), *The Effective Management of Resources; the International Politics of the North Sea*. London: Frances Pinter; New York: Nichols, 1979.

Masterson, W.E., *Jurisdiction in Marginal Seas*. New York: Macmillan, 1929.

Matte, Nicholas Mateesco, *Space Policy and Programmes Today and Tomorrow*. Montreal: McGill Univ., Institute and Centre of Air and Space Law, 1980.

———, *Aerospace Law: Telecommunications Satellites*. Toronto: Butterworths, 1982.

McCune, Shannon, *Islands in Conflict in East Asian Waters*. Hong Kong: Asian Research Service, 1984.

McDorman, Ted L. and others, *Maritime Boundary Delimitation: An Annotated Bibliography*. Lexington, MA, and Toronto: Heath, 1983.

McDougal, Myres S. and William T. Burke, *The Public Order of the Oceans; a Contemporary International Law of the Sea*. Dordrecht, Neth: Nijhoff, 1987.

McFee, W., *The Law of the Sea*. Philadelphia: Lippincott, 1950.

Mericq, Luis H., *Antarctica: Chile's Claim*. Washington: National Defense Univ. Press, 1986.

M'Gonigle, R. Michael and Mark W. Zacher, *Pollution, Politics, and International Law; Tankers at Sea*. Berkeley, Los Angeles, and London: Univ. of California Press, 1979.

Miles, Edward, *International Administration of Space Exploration and Exploitation*. Denver: Univ. of Denver Monograph Series in World Affairs, Vol. 8, No. 4 (1971).

Mitchell, Barbara and R. Sandbrook, *The Management of the Southern Ocean*. London: International Institute for Environment and Development, 1980.

Morgan, R., *World Sea Fisheries*. London: Methuen, 1956.

Morris, Michael A., *International Politics and the Sea; the Case of Brazil*. Boulder, CO: Westview, 1979.

Mouton, M.W., *The Continental Shelf*. The Hague: Nijhoff, 1952.

Myhre, Jeffrey D., *The Antarctic Treaty System; Politics, Law and Diplomacy*. Boulder, CO, and London: Westview, 1986.

N

Nordquist, Myron S. and others (eds./compilers), *New Directions in the Law of the Sea*. London, Rome, and New York: Oceana, 1980.

O

Oceans Policy Study Series. Charlottesville: Univ. of Virginia, Center for Oceans Law and Policy. London, Rome, and New York: Oceana, 1985. Monographs published irregularly.

O'Connell, D.P., *The International Law of the Sea*, 2 Vols. New York and Oxford: Oxford Univ. Press, 1984.

Oda, Shigeru, *The International Law of the Ocean Development; Basic Documents*. Leiden, Neth.: Sijthoff, 1977.

————, *International Law of Resources of the Sea*. Alphen aan den Rijn Neth.: Sijthoff & Noordhoff, 1979.

Oerding, J.B., *Frozen Friction Point: A Geopolitical Analysis of Sovereignty in the Antarctic Peninsula*. Gainesville: University of Florida Press, 1977.

Ogley, Roderick, *Internationalizing the Seabed*. Brookfield, VT.: Gower, 1984.

Okidi, C. Odidi, *Regional Control of Ocean Pollution*. Alphen aan den Rijn, Neth.: Sijthoff & Noordhoff, 1978.

Orrego Vicuña, Francisco (ed.), *Antarctic Resources Policy; Scientific, Legal and Political Issues*. Cambridge, Eng.: Cambridge University Press, 1983.

————, *The Exclusive Economic Zone: A Latin American Perspective*. Boulder, CO, Westview, 1984.

Oxman, Bernard H., *From Cooperation to Conflict: The Soviet Union and the United States at the Third U.N. Conference on Law of the Sea*. Seattle: Univ. of Washington, Washington Sea Grant Project., 1985.

P

Padelford, Norman J., *Public Policy for the Seas*. Cambridge; MIT Press, 1970.

Padelford, Norman J. and Jerry E. Cook, *New Dimensions of U.S. Marine Policy*. Cambridge: MIT Sea Grant Project Office, 1971.

Papadakis, Nikos, *The International Legal Regime of Artificial Islands*. Leyden, Neth.: Sijthoff, 1977.

Papadakis, Nikos and Martin Ira Glassner (eds.), *The International Law of the Sea and Marine Affairs: A Bibliography*. The Hague: Nijhoff, 1984.

Park, Choon-ho, *East Asia and the Law of the Sea*. Seoul: Seoul National Univ. Press, 1983; Honolulu: Univ. of Hawaii Press, 1984.

Pharand, Donat, in association with Leonard H. Legault, *The Northwest Passage: Arctic Straits*. Dordrecht, Neth.; Boston; and Lancaster, Eng.: Nijhoff, 1984.

————, *The Waters of the Canadian Arctic Archipelago in International Law*. New York: Cambridge Univ. Press, 1987.

Pinochet de la Barra, Oscar, *Chilean Sovereignty in Antarctica*. Santiago de Chile: Editorial del Pacífico, 1954.

Piradov, A.S. (ed.), *International Space Law*. Moscow: Progress Publishers, 1976.

Platzöder, Renate (compiler/ed.), *Third United Nations Conference on the Law of the Sea: Documents*. Dobbs Ferry, NY: Oceana, 1982 et. seq.

Polar Research Board, National Research Council, *Antarctic Treaty System: An Assessment*. Washington: National Academy Press, 1986. Proceedings of a Workshop held at Beardmore South Field Camp, Antarctica, January 7–13, 1985.

Polomka, Peter, *Ocean Politics in Southeast Asia*. Singapore: Institute of Southeast Asian Studies, 1978.

Pontecorvo, Giulio (ed.), *Fisheries Conflicts in the North Atlantic: Problems of Management and Jurisdiction*. Cambridge, MA: Ballinger, 1974.

————, *The New Order of the Oceans; the Advent of a Managed Environment*. New York: Columbia Univ. Press, 1986.

Post, Alexandra Merle, *Deepsea Mining and the Law of the Sea*. The Hague, Neth.; Boston and London: Nijhoff, 1983.

Prescott, John Robert Victor, *The Political Geography of the Oceans*. New York: Halsted Press, 1975.

————, *Maritime Jurisdiction in Southeast Asia: A Commentary and Map*. Honolulu: East–West Center, 1981.

————, *Australia's Maritime Boundaries*. Canberra: Australian National Univ., Dept. of International Relations, 1985.

————, *The Maritime Political Boundaries of the World*. London and New York: Methuen, 1985.

Q

Quartermain, L.B., *New Zealand and the Antarctic*. Wellington: New Zealand Government Printers, 1971.

Queeney, Kathryn M., *Direct Broadcast Satellites and the United Nations*, Alphen aan den Rijn, Neth.: Sijthoff & Noordhoff, 1978.

Quigg, Philip W., *A Pole Apart; the Emerging Issue of Antarctica*. New York and elsewhere: McGraw-Hill, 1983.

————, *Antarctica: The Continuing Experiment*. Foreign Policy Association Headline Series No. 273. New York: Foreign Policy Association, 1985.

R

Ramazani, Rouhollah K., *The Persian Gulf and the Strait of Hormuz*. Alphen aan den Rijn, Neth.: Sijthoff & Noordhof, 1979.

Rao, P. Chandrasekhara, *The New Law of Maritime Zones*. New Delhi: Milind, 1983.

Ray, John B. (ed.), *The Oceans and Man*. Dubuque, IA: Kendall/Hunt, 1975.

Reijnen, Gijsbertha C.M., *Utilization of Outer Space and International Law*. Amsterdam and New York: Elsevier Scientific Publishing Co., 1981.

Rembe, Nasila S., *Africa and the International Law of the Sea: A Study of the Contribution of the African States to the Third United Nations Conference on the Law of the Sea*. Alphen aan den Rijn: Sijthoff & Noordhoff, 1980.

Rodgers, Patricia Elaine Joan, *Midocean Archipelagos and International Law*. New York: Vantage Press, 1981.

Ross, William M., *Oil Pollution as an International Problem*. Western Geographical Series Vol. 6. Victoria, B.C.: Univ. of Victoria, 1973.

Rozakis, Christos L. and Constantine A. Stephanou (eds.), *The New Law of the Sea: Selected and Edited Papers of the Athens Colloquium on the Law of the Sea, September 1982*. Amsterdam, New York and Oxford: North-Holland, 1983.

Rozakis, Christos L. and Petros N. Stagos, *The Turkish Straits*. Dordrecht, Neth.: Nijhoff, 1987.

S

Samuels, Marwyn S., *Contest for the South China Sea*. New York and London: Methuen, 1982.

Sanger, Clyde, *Ordering the Oceans; the Making of the Law of the Sea*. Toronto: Univ. of Toronto Press, 1987.

Schmidhauser, John R. and George O. Totten, *The Whaling Issue in U.S.–Japan Relations*. Boulder, CO: Westview, 1978.

Sebek, Victor, *Eastern European States and the Development of the Law of the Sea*. Dobbs Ferry, NY: Oceana, 1979.

Shapley, Deborah, *The Seventh Continent: Antarctica in a Resource Age*. Washington: Resources for the Future, 1986.

Shusterich, Kurt Michael, *Resource Management and the Oceans; The Political Economy of Deep Seabed Mining*. Boulder, CO: Westview, 1982.

Singh, Nagendra, *Maritime Flag and International Law*. Leyden, Neth.: Sijthoff, 1978.

Smith, Delbert D., *Space Stations. International Law and Policy*. Boulder, CO: Westview, 1979.

Smith, H.A., *The Law and Custom of the Sea*. London: Stevens, 1948.

Smith, Robert W., *Exclusive Economic Zone Claims*. Dordrecht, Neth.: Nijhoff, 1986.

Stares, Paul B., *The Militarism of Space: US Policy, 1945–1984*. Ithaca, NY: Cornell Univ. Press, 1985.

———, *Space and National Security*. Washington: Brookings Institution, 1987.

Steinhart, Carol and John Steinhart, *Blowout: A Case Study of the Santa Barbara Oil Spill*. North Scituate, MA: Duxbury, 1972.

Stone, Jeffrey C. (ed.), *Africa and the Sea*. Aberdeen: Univ. African Studies Group, 1985.

Sugden, David, *Arctic and Antarctic; A Modern Geographical Synthesis*. Totowa, NJ: Barnes & Noble, 1982.

Symmons, Clive Ralph, *The Maritime Zones of Islands in International Law*. The Hague, Boston, and London: Nijhoff, 1979.

Székely, Alberto, *Latin America and the Development of the Law of the Sea*. Dobbs Ferry, NY: Oceana, 1976.

T

Tangsubkul, Phiphat, *ASEAN and the Law of the Sea*. Singapore: Institute of Southeast Asian Studies, 1982.

———, *The Southeast Asian Archipelagic States: Concept, Evolution and Current Practice*. Honolulu: East–West Center, 1984.

Theutenberg, Bo Johnson, *The Evolution of the Law of the Sea; a Study of Resources and Strategy with Special Regard to the Polar Areas*. Dublin: Tycooly, 1984.

Truver, Scott C., *The Strait of Gibraltar and the Mediterranean*. Alphen aan den Rijn, Neth.: Sijthoff & Noordhoff, 1980.

U

Underdal, Arild, *The Politics of International Fisheries Management: The Case of the Northeast Atlantic*. Oslo, Bergen, and Tromso: Univ. of Oslo Press, 1980.

V

van Bogaert, E.R.C., *Aspects of Space Law*. Hingham, MA: Kluwer, 1987.

VanderZwaag, David L., *The Fish Feud: The U.S. and Canadian Boundary Dispute*. Lexington, MA, and Toronto: Heath, 1983.

Vertzberger, Yaacov Y.I., *Coastal States, Regional Powers, Superpowers and the Malacca–Singapore Straits*. Berkeley: Univ. of California, Institute of East Asian Studies, 1984.

Veur, Paul W., *Search for New Guinea's Boundaries: From Torres Strait to the Pacific*. Canberra: Australian National Univ. Press, 1966.

W

Wadegaonkar, Damodar, *The Orbit of Space Law*. London: Stevens; Bombay: N.M. Tripathi; 1984.

Wall, Patrick (ed.), *The Southern Ocean and the Security of the Free World*. London: Stacey International, 1977.

Wenk, Edward, Jr., *The Politics of the Ocean*. Seattle: Univ. of Washington Press, 1972.

Westermeyer, William E., *The Politics of Mineral Resources Development in Antarctica*. Boulder, CO: Westview, 1984.

Wolfrum, Rüdiyer, *Antarctic Challenge*. Berlin: Duncker & Humblot, 1984.

Y

Young, Oran, *Resource Management at the International Level: The Case of the North Pacific*. London: Frances Pinter, 1977.

Z

Zhukov, Gennady and Yuri Kolosov, *International Space Law*. New York: Praeger, 1984.

Zumberge, James H., *Possible Environmental Effects of Mineral Exploration and Exploitation*. Cambridge, England; Scientific Committee on Antarctic Research, 1979.

Periodicals

A

Abrahamsson, Bernhard J., "The Law of the Sea Convention and Shipping," *Political Geography Quarterly*, 5, 1 (January 1986), 13–17.

Air and Outer Space Law, Thesaurus Acroasium (Thessaloniki), 10, (1981).

Alexander, Frank C., Jr., "A Recommended Approach to the Antarctic Resource Problem," *University of Miami Law Review*, 33 (December 1978), 371–423.

Alexander, Lewis M., "The Expanding Territorial Sea," *Professional Geographer*, 11, 4 (July 1959), 6–8.

———, "Offshore Claims and Fisheries in Northwest Europe," *Yearbook of World Affairs*, 14 (1960), 236–260.

———, "Geography and the Law of the Sea," *Annals, AAG*, 58, 1 (March 1968), 177–197.

———, "The Delimitation of Maritime Boundaries," *Political Geography Quarterly*, 5, 1 (January 1986), 19–24.

Allen, Edward W., "Territorial Waters and Extraterritorial Rights," *American Journal of International Law*, 47, 3 (1953), 478–480.

Almond, Harry H., Jr., "Demilitarization and Arms Control: Antarctica," *Case Western Reserve Journal of International Law*, 17, 2 (Spring 1985), 229–285.

Andrews, John, "Antarctic Geopolitics," *Australian Outlook*, 11, 3 (1951), 3–9.

Annals of Air and Space Law. Montreal: McGill Univ., Institute and Centre of Air and Space Law, published irregularly.

Antarctic. Christchurch. New Zealand Antarctic Society. Published quarterly since 1956.

Antarctic Bibliography. U.S. Superintendent of Documents, Washington. Published irregularly since 1965 (Library of Congress).

Antarctic Journal of the United States. U.S. Superintendent of Documents, Washington. Published quarterly since 1966.

"Antarctic Riches—for whom?" *Mazingira*, (August 1977), 71–77.

Archdale, H.E., "Territorial Waters: What Are They?" *Australian Outlook*, 10, 1 (1956), 42–45.

B

Barrie, G.N., "The Antarctic Treaty: Example of Law and Its Sociological Infrastructure," *Comparative and International Law Journal of Southern Africa*, 8 (July 1975), 212–224.

Baty, Thomas, "The Three Mile Limit," *American Journal of International Law*, 22, 3 (July 1928), 503–538.

Bernhardt, J. Peter A., "Sovereignty in Antarctica," *California Western International Law Journal*, 5 (Spring 1975), 297–349.

Birch, W.F., "Antarctica: Sovereignty and Stewardship," *New Zealand Foreign Affairs Review*, 29 (1979), 35–39.

Blake, Gerald, "Flash-point Through Which

Middle East Oil Must Pass," *Geographical Magazine,* 53, 1 (October 1980), 50–52.

———, "Coveted Waterway of Bab el Mandeb," *Geographical Magazine,* 54, 4 (January 1981), 233–235.

Bloomfield, Lincoln P., "The Arctic: Last Unmanaged Frontier," *Foreign Affairs,* 60, 1 (Fall 1981), 87–105.

Blum, Yehuda Z., "The Gulf of Sidra Incident," *American Journal of International Law,* 80, 3 (July 1986), 668–678.

Boczek, Boleslaw A., "Global and Regional Approaches to the Protection and Preservation of the Marine Environment," *Case Western Reserve Journal of International Law,* 16 (1984), 39.

———, "The Soviet Union and the Antarctic Regime," *American Journal of International Law,* 78, 4 (October 1984), 834–858.

Boggs, S. Whittemore, "Delimitation of the Territorial Sea," *American Journal of International Law,* 24 (1930), 541.

———, "Problems of Water Boundary Definition: Median Lines and International Boundaries Through Territorial Waters," *Geographical Review,* 27, 3 (July 1937), 445–456.

———, "Delimitation of Seward Areas Under National Jurisdiction," *American Journal of International Law,* 45, 21 (April 1951), 240–266.

———, "National Claims in Adjacent Seas," *Geographical Review,* 41, 21 (April 1951), 185–209.

Borgese, Elisabeth Mann, "The Law of the Sea." *Scientific American,* (March 1983), 42–49.

Bowett, Derek W., "The Second U.N. Conference on the Law of the Sea," *International and Comparative Law Quarterly,* 9 (1960), 415–435.

Boyle, Alan E., "Marine Pollution Under the Law of the Sea Convention," *American Journal of International Law,* 79, 2 (April 1985), 347–372.

Brittin, Burdick H., "International Law Aspects of the Acquisition of the Continental Shelf by the United States," *Proceedings, United States Naval Institute,* 74, 550 (December 1948), 1541–1543.

Broder, Sherry and Jon Van Dyke, "Ocean Boundaries in the South Pacific," *University of Hawaii Law Review,* 4, 1 (1982), 1–59.

Burmester, H., "The Torres Strait Treaty: Ocean Boundary Delimitation by Agreement," *American Journal of International Law,* 76, 2 (April 1982), 321–349.

Burton, Steven J., "New Stresses on the Antarctic Treaty: Toward International Legal Institutions Governing Antarctic Resources," *Virginia Law Review,* 65 (April 1979), 421–512.

Butler, Shirley Oakes, "Owning Antarctica: Cooperation and Jurisdiction at the South Pole," *Journal of International Affairs,* 31 (Spring–Summer 1977), 35–51.

Buzan, Barry, "The Neglected Ocean," *Proceedings, United States Naval Institute,* 84, 11 (November 1958), 54–61.

C

Cagle, Malcom W., "The Gulf of Aqaba—Trigger for Conflict," *Proceedings, United States Naval Institute,* 85, 1 (1959), 75–81.

Cerone, Rudy J., "Survival of the Antarctic: Economic Self-Interest v. Enlightened International Cooperation," *Boston College International and Comparative Law Journal,* 2 (1978), 115–129.

Chao, K.T., "East China Sea: Boundary Problems Relating to the Tiao-yu-ta'i Islands," *Chinese Yearbook of International Law and Affairs,* 2 (1982), 45–97.

Charney, Jonathan I., "Ocean Boundaries between Nations: A Theory for Progress," *American Journal of International Law,* 78, 3 (July 1984), 582–606.

Child, Jack, "Antarctica and Argentine Geopolitical Thinking," *CLAG Yearbook 1986.* Baton Rouge: Louisiana State University, Department of Geography and Anthropology, 12–16.

———, "Geopolitical Thinking in Latin America," *Latin American Research Review,* 14, 2 (1979), 89–111.

Chiu, Hungdah, "Some Problems Concerning the Delimitation of the Maritime Boundary Between the Republic of China and the Philippines," *Chinese Yearbook of International Law and Affairs,* 3 (1983), 1–21.

———, "Political Geography in the Western Pacific After the Adoption of the 1982 United Nations Conference on the Law of the Sea," *Political Geography Quarterly,* 5, 1 (January 1986), 25–32.

Christol, Carl Q., "The Moon Treaty Enters into Force," *American Journal of International Law,* 79, 1 (January 1985), 163–168.

Clingan, Thomas A., Jr., "US–Mexican Maritime Relations in the Aftermath of UNCLOS III," *Political Geography Quarterly,* 5, 1 (January 1986), 57–62.

Cohen, Saul B., "The Oblique Plane Air Boundary," *Professional Geographer*, 10, 6 (1958), 11–15.

Colborn, Paul A., "National Jurisdiction over Resources of the Continental Shelf," *Bulletin, Pan American Union*, 82, 1 (January 1948), 38–40.

Coll, Alberto R., "Functionalism and the Balance of Interests in the Law of the Sea," *American Journal of International Law*, 79, 4 (October 1985), 891–911.

Colombos, C. John, "Territorial Waters," *Transactions, Grotius Society*, 9 (1923), 89.

Colson, David, "The Antarctic Treaty System: The Mineral Issue," *Law and Policy in International Business*, 12 (1981), 841–902.

Couper, A.D., "The Marine Boundaries of the United Kingdom and the Law of the Sea," *Geographical Journal*, 151, 2 (July 1985), 228–236.

D

Dean, Arthur H., "The Geneva Conference on the Law of the Sea: What Was Accomplished," *American Journal of International Law*, 52, 4 (October 1958), 607–628.

———, "Freedom of the Seas." *Foreign Affairs*, 37, 1 (October 1958), 83–94.

———, "The Second Geneva Conference on the Law of the Sea: The Fight for Freedom of the Seas," *American Journal of International Law*, 54, 4 (October 1960), 751–790.

De Saussure, Hamilton, "The Impact of Manned Space Stations on the Law of Outer Space," *San Diego Law Review*, 21, 5 (September–October 1984), 985–1014.

De Vorsey, Louis and Megan C. De Vorsey, "The World Court Decision in the Canada–United States Gulf of Maine Seaward Boundary Dispute: A Perspective from Historical Geography," *Case Western Reserve Journal of International Law*, 18, 3 (1986), 415–442.

E

Edmonds, David C., "The 200-Miles Fishing Rights Controversy: Ecology or High Tariffs?" *Inter-American Economic Affairs*, 26, 4 (Spring 1973), 3–18.

El Baradei, Mohamed, "The Egyptian–Israeli Peace Treaty and Access to the Gulf of Aqaba: A New Legal Regime," *American Journal of International Law*, 76, 3 (July 1982), 532–554.

Evensen, Jens, "The Anglo–Norwegian Fisheries Case and Its Legal Consequences," *American Journal of International Law*, 46, 4 (October 1952), 609–630.

F

Feldman, Mark B., "The Tunisia–Libya Continental Shelf Case: Geographic Justice or Judicial Compromise?" *American Journal of International Law*, 77, 2 (April 1983), 219–238.

Feldman, Mark B. and David Colson, "The Maritime Boundaries of the United States," *American Journal of International Law*, 75, 4 (October 1981), 729–763.

Fiske, Clarence, "Territorial Claims in the Antarctic," *Proceedings, U.S. Naval Institute*, 85, 1 (1959), 82–91.

G

Glassner, Martin Ira, "The Status of Developing Land-locked States Since 1965," *Lawyer of the Americas*, 5, 3 (October 1973), 480–498.

———, "Israel's Maritime Boundaries," *Ocean Development and International Law*, 1, 4 (Winter 1974), 303–313.

———, "The Illusory Treasure of Davy Jones' Locker," *San Diego Law Review*, 13, 3 (March 1976), 533–551.

———, "The View from the Near North: South Americans View Antarctica and the Southern Ocean Geopolitically," *Political Geography Quarterly*, 4, 4 (October 1985), 329–342.

———, "Regionalism and the New Law of the Sea," *Prinosi Za Poredbeno Proučavanje Prava i Medunarodno Pravo* (Zagreb), 18, 21 (1985), 409–428.

———, "The New Political Geography of the Sea," *Political Geography Quarterly*, 5, 1 (January 1986), 6–8.

Gorman, Stephen M., "The High Stakes of Geopolitics in Tierra del Fuego," *Parameters*, 8, 2 (June 1978), 45–56.

Greig, D.W., "Territorial Sovereignty and the Status of Antarctica," *Australian Outlook*, 32 (August 1978), 117–129.

Groom, Sidney M., "Our Newest Frontier: The Outer Continental Shelf," *Our Public Lands*, 5, 1 (January 1955), 12ff.

Gross, Leo, "The Geneva Conference on the Law of the Sea and the Right of Innocent Passage Through the Gulf of Aqaba," *American Journal of International Law*, 53, 3 (1959), 564–594.

Grzybowski, Kazimir A., "The Soviet Doctrine of Mare Clausum and Policies in Black and Baltic Seas," *Journal of Central European Affairs*, 14, 4 (January 1955), 339–353.

Guill, James H., "The Regimen of the Seas," *Proceedings, U.S. Naval Institute*, 83, 12 (December 1957), 1308–1319.

Gushue, Raymond, "The Territorial Waters of Newfoundland," *Canadian Journal of Economic and Political Sciences*, 15 (1949), 344–352.

H

Hage, Robert E., *The Third United Nations Conference on the Law of the Sea: A Canadian Retrospective, Behind the Headlines*, 40, 5 (1983).

Hailbronner, Kay, "Freedom of the Air and the Convention on the Law of the Sea," *American Journal of International Law*, 77, 3 (July 1983), 490–520.

Hambro, Edvard, "Some Notes on the Future of the Antarctic Treaty Collaboration," *American Journal of International Law*, 68, 2 (April 1974), 217–226.

Hanessian, John, "Antarctica: Current National Interests and Legal Realities," *Proceedings, American Society of International Law*, 52 (1958), 145–164.

Harry, Ralph L., "The Antarctic Regime and the Law of the Sea Convention: An Australian View," *Virginia Journal of International Law*, 21, 4 (Summer 1981), 727–744.

Hayton, Robert D., "The 'American' Antarctic," *American Journal of International Law*, 50, 3 (July 1956), 583–610.

———, "Polar Problems and International Law," *American Journal of International Law*, 52, 4 (1958), 746–765.

———, "The Antarctic Settlement of 1959," *American Journal of International Law*, 54, 2 (April 1960), 349–371.

Henrikson, Alan K., "Space Politics in Historical and Futuristic Perspective," *Fletcher Forum*, 5, 1 (Winter 1981), 106–114.

Hodgson, Robert D. and Robert W. Smith, "Boundary Issues Created by Extended National Marine Jurisdiction," *Geographical Review*, 69, 4 (October 1979), 423–433.

Honnold, E., "Thaw in International Law? Rights in Antarctica Under the Law of Common Spaces," *Yale Law Journal*, 87 (1978), 804–859.

I

International Organizations and The Law of the Sea: Documentary Yearbook 1985. Edited by the Netherlands Institute for the Law of the Sea. Dordrecht, Neth: Nijhoff, 1987.

J

Jacobson, Jon L., "International Fisheries Law in the Year 2010," *Louisiana Law Review*, 45 (1985), 1161–1198.

Jamali, Mohammed Fadhel, "Supranational Resources: Sea and Space as Areas for International Conflict and Cooperation," *International Journal of World Peace*, 3, 3 (July 1986), 47–64.

Jessup, Philip C. and Howard J. Taubenfeld, "Outer Space, Antarctica, and the United Nations," *International Organization*, 13, 3 (Summer 1959), 363–379.

Joesten, Joachim, "The Second U.N. Conference on the Law of the Sea," *World Today*, 16, 6 (June 1960), 249–257.

Joyner, Christopher C., "Anglo–Argentine Rivalry After the Falklands/Malvinas War: Law, Geopolitics and the Antarctic Connection," *Lawyer of the Americas*, 15, 3 (Winter 1984), 467–502.

———, "The Southern Ocean and Marine Pollution: Problems and Prospects," *Case Western Reserve Journal of International Law*, 17, 2 (Spring 1985), 165–194.

——— (ed.), *Polar Politics in the 1980's*. Special Issue of *International Studies Notes*, 11, 3 (Spring 1985), six articles.

K

Kane, Francis X., "Space Age Geopolitics," *Orbis*, 14 (Winter 1971), 911–933.

Katchen, Martin H., "The Spratly Islands and the Law of the Sea: 'Dangerous Ground' for Asian Peace," *Asian Survey*, 17, 12 (December 1977), 1167–1181.

Kawakami, Kenzo, "The Continental Shelf and Its Geographically Controversial Points," *Geography Journal* (Tokyo), 63, 2 (1954), 53–59.

Kemp, Geoffrey, "Geopolitics, Remote Frontiers and Outer Space," *Fletcher Forum*, 5, 1 (Winter 1981), 115–119.

Kent, George, "Regional Approaches to Meeting National Marine Interests," *Contemporary Southeast Asia*, 5, 1 (June 1983), 80–94.

Kent, H.S.K., "The Historical Origins of the Three Mile Limit," *American Journal of International Law*, 48, 4 (1954), 537–553.

Khlestov, O. and Golitsyn, Vladimir, "The Antarctic: Arena of Peaceful Cooperation," *International Affairs* (Moscow) (August 1978), 61–65.

Kopal, Vladimir, "The Question of Defining Outer Space," *Journal of Space Law*, 8, 2 (Fall 1980), 154–173.

Kumar, C.K., "International Waterways: Strategic International Straits," *India Quarterly*, 14, 1 (January–March 1958), 87–94.

Kunz, Josef L., "Continental Shelf and International Law: Confusion and Abuse," *American Journal of International Law*, 50, 4 (October 1956), 828–853.

L

Lamontagne, Michele R., "United States Commercial Space Policy: Impact on International and Domestic Law," *Syracuse Journal of International Law and Commerce*, 13, 1 (Fall 1986), 129–154.

Lapidoth, Ruth, "The Strait of Tiran, the Gulf of Aqaba, and the 1979 Treaty of Peace Between Egypt and Israel," *American Journal of International Law*, 77, 1 (January 1983), 84–108.

Lee Yong Leng, "Malacca Strait, Kra Canal, and International Navigation." *Pacific Viewpoint*, 19, 1 (1978), 65–74.

Legault, L.H. and Blair Hankey, "From Sea to Seabed: The Single Maritime Boundary in the Gulf of Maine Case," *American Journal of International Law*, 79, 4 (October 1985), 961–991.

Lillie, H.R., "Antarctica in World Affairs," *Canadian Geographical Journal*, 36 (1948), 282–.

Lundquist, Thomas R., "The Iceberg Cometh?: International Law Relating to Antarctic Iceberg Exploitation," *Natural Resources Journal*, 17 (January 1977), 1–41.

M

Ma, Ying-Jeou, "The East Asian Seabed Controversy Revisited: Relevance (or Irrelevance) of the Tiao-yu-t'ai (Senkaku) Islands Territorial Dispute," *Chinese Yearbook of International Law and Affairs*, 2 (1982), 1–44.

Marcoux, J. Michael, "Natural Resource Jurisdiction on the Antarctic Continental Margin," *Virginia Journal of International Law*, 11 (1971), 374–405.

Marine Policy. London: Butterworths. Published quarterly since 1976.

Maritime Policy and Management. London: Taylor & Francis, published quarterly since 1973.

Marvyama, Magoroh, "Social and Political Interactions Among Extraterrestrial Human Communities: Contrasting Models," *Technological Forecasting and Social Change*, 9, 4 (1976), 349–360.

Melamid, Alexander, "The Political Geography of the Gulf of Aqaba," *Annals, AAG*, 47, 3 (September 1957), 231–240.

———, "Legal Status of the Gulf of Aqaba," *American Journal of International Law*, 53, 2 (April 1959), 412–413.

———, "The Division of Narrow Seas," *Political Geography Quarterly*, 5, 1 (January 1986), 39–42.

Mitchell, Barbara and Lee Kimball, "Conflict over the Cold Continent," *Foreign Policy*, No. 35 (Summer 1979), 124–141.

Molde, Jörgen, "The Status of Ice in International Law," *Nordisk Tidsskrift for International Ret*, 51, 3–4 (1982), 164–178.

Moneta, Carlos J. "Antarctica, Latin America and the International System in the 1980's," *Journal of Interamerican Studies and World Affairs*, 23 (February 1981), 29–68.

Moore, John Norton, "The Regime of Straits and the Third United Nations Conference on the Law of the Sea," *American Journal of International Law*, 74, 1 (January 1980), 77–121.

N

Nanda, Ved P., "The Exclusive Economic Zone," *Political Geography Quarterly*, 5, 1 (January 1986), 9–11.

Nash, Marian L., "U.S. Maritime Boundaries with Mexico, Cuba, and Venezuela," *American Journal of International Law*, 75, 1 (January 1981), 161–162.

O

Ocean Development and International Law. New York: Taylor & Francis. Published quarterly since 1973.

Ocean Management. Amsterdam: Elsevier Scientific Publishing Co., Published quarterly since 1974.

Ocean Yearbook. Chicago: University of Chicago Press; International Ocean Institute of Malta.

Oda, Shigeru, "Fisheries Under the United Nations Convention on the Law of the Sea," *American Journal of International Law*, 77, 4 (October 1983), 739–755.

Oxman, Bernard H., "The Antarctic Regime: An Introduction," *University of Miami Law Review*, 33 (December 1978), 285–297.

———, "The Third United Nations Conference on the Law of the Sea: The Eighth Session (1979)," *American Journal of International Law*, 74, 1 (January 1980), 1–47.

———, "The Third United Nations Conference on the Law of the Sea: The Ninth Session (1980)," *American Journal of International Law*, 75, 2 (April 1981), 211–256.

———, "The Third United Nations Conference on the Law of the Sea: The Tenth Session (1981)," *American Journal of International Law*, 76, 1 (January 1982), 1–23.

P

Panel discussion on the new political geography of the sea, *Political Geography Quarterly*, 5, 1 (January 1986), 33–38.

Payne, Richard J., "Southern Africa and the Law of the Sea: Economic and Political Implications," *Journal of Southern African Affairs*, 4, 2 (April 1979), 175–186.

Pardo, Arvid, "Who Will Control the Seabed?" *Foreign Affairs*, 47, 1 (October 1968), 123–137.

———, "Sovereignty Under the Sea: The Threat of National Occupation," *Round Table*, No. 232 (1968), 341–355.

Pardo, Arvid and Richard Young, "The Legal Regime of the Deep Sea Floor," *American Journal of International Law*, 62, 3 (July 1968), 641–653.

Pearcy, G. Etzel, "Geographical Aspects of the Law of the Sea," *Annals, AAG*, 49, 1 (March 1959), 1–23.

———, "Hawaii's Territorial Sea," *Professional Geographer*, 11, 6 (November 1959), 2–6.

———, "The Continental Shelf: Physical Versus Legal Definition," *Canadian Geographer*, 5, 3 (1961), 26–29.

Peterson, M.J., "Antarctica: The Last Great Land Rush on Earth," *International Organization*, 34, 3 (Summer 1980), 377–403.

———, "Antarctic Politics Today," *Current World Leaders*, 28, 7 (1985).

Pinto, M.C.W., "The International Community and Antarctica," *University of Miami Law Review*, 33 (December 1978), 475–487.

Polar Times. Rego Park, NY: American Polar Society. Published semiannually since 1935.

Pounds, Norman J.G., "The Political Geography of the Straits of Gibraltar," *Journal of Geography*, 51, 4 (April 1952), 165–170.

Powers, R.D. and Leonard R. Hardy, "How Wide the Territorial Sea," *Proceedings, U.S. Naval Institute*, 87, 2 (1961), 68–73.

Prescott, John Robert Victor, "Australia's Maritime Claims and the Great Barrier Reef," *Australian Geographical Studies*, 19, 1 (April 1981), 99–106.

Puri, Rama, "Exclusive Political Zone: A New Dimension in the Law of the Sea," *Indian Political Science Review*, 14, 1 (January 1980), 39–54.

R

Reeves, J.S., "The Codification of the Law of the Territorial Waters," *American Journal of International Law*, 24 (1930), 486.

Reisman, W. Michael, "The Regime of Straits and National Security: An Appraisal of International Lawmaking," *American Journal of International Law*, 74, 1 (January 1980), 48–76.

Rhee, Sang-Myon, "Equitable Solutions to the Maritime Boundary Dispute Between the United States and Canada in the Gulf of Maine," *American Journal of International Law*, 75, 3 (July 1981), 590–628.

———, "Sea Boundary Delimitations Between States Before World War II," *American Journal of International Law*, 76, 3 (July 1982), 555–588.

Rose, Julia, "Antarctic Condominium: Building a New Legal Order for Commercial Interests," *Marine Technology Society Journal*, 10, 1 (January 1976), 19–27.

Rothblatt, Martin A., "Satellite Communication and Spectrum Allocation," *American Journal of International Law*, 76, 1 (January 1982), 56–77.

Roucek, Joseph S., "The Geopolitics of Antarctica and the Falkland Islands," *World Affairs Interpreter*, 22 (April 1951), 44–56.

S

San Diego Law Review. Annual issue on the Law of the Sea.

Sea Changes; the Developing Régime of the Sea in Law and State Practice. Cape Town: University of Cape Town, Institute of Marine Law. Published semiannually.

Sharma, Surya P., "International Law of the Outer Space: A Policy-Oriented Study," *Indian Journal of International Law*, 17 (April–June 1977), 185–202.

Smith, Brian, "Innocent Passage as a Rule of Decision: Navigation v. Environmental Protection." *Columbia Journal of Transnational Law*, 21, 1 (1982), 49–102.

Smith, Hance O., "The Role of the Sea in the Political Geography of Scotland." *Scottish Geographical Magazine*, 100, 3 (December 1984), 138–150.

Smith, Robert W., "Trends in National Maritime Claims," *Professional Geographer*, 32, 2 (May 1980), 216–223.

Somogyi, Joseph de, "The Question of the Turkish Straits," *Journal of Central European Affairs*, 11, 3 (1951), 279–290.

Sorensen, Max, "Law of the Sea," *International Conciliation*, No. 520 (1958), 195–256.

Space Policy, London: Butterworths. An international quarterly journal published since 1985.

Sullivan, Walter, "Antarctica in a Two-Power

World," *Foreign Affairs,* 36, 1 (October 1957), 154–166.

Svarlien, O., "The Legal Status of the Arctic," *Proceedings, American Society of International Law,* (1958), 136–144.

Symmons, Clive R., "Maritime Boundary Disputes in the Irish Sea and Northeast Atlantic Ocean," *Marine Policy Reports* (University of Delaware), 9, 1 (September 1986), 1–7.

Symonides, Janusz, "Geographically Disadvantaged States and the New Law of the Sea," *Polish Yearbook of International Law,* 8 (1976), 55–73.

Symposium: "The Role of Private Enterprise in Outer Space: International Legal Implications," *Houston Journal of International Law,* 2, 1 (August 1979), eleven articles.

Symposium: "The International Legal Regime for Antarctica," *Cornell International Law Journal,* 19, 2 (Summer 1986), eight articles.

T

Taijudo, Kanae, "Japan and the Problems of Sovereignty over the Polar Regions," *Japanese Annual of International Law,* 3 (1959), 12–17.

Tangsubkul, Phiphat and Frances Lai Fung-Wai, "The New Law of the Sea and Development in Southeast Asia," *Asian Survey,* 23, 7 (July 1983), 858–878.

Taubenfeld, Howard J., "A Treaty for Antarctica," *International Conciliation,* No. 531 (January 1961).

Toma, Peter A., "The Soviet Attitude Toward the Acquisition of Territorial Sovereignty in Antarctica," *American Journal of International Law,* 50 (July 1956), 611–626.

Treves, Tullio, "Military Installations, Structures, and Devices on the Seabed," *American Journal of International Law,* 74, 4 (October 1980), 808–857.

Triggs, Gillian, "The Antarctic Treaty Regime: A Workable Compromise or a 'Purgatory of Ambiguity'?" *Case Western Reserve Journal of International Law,* 17, 2 (Spring 1985), 195–228.

V

Valencia, Mark J., "The South China Sea: Constraints to Marine Regionalism." *Indonesian Quarterly,* 8, 2 (April 1980), 16–38.

Van Dyke, Jon and Susan Heftel, "Tuna Management in the Pacific: An Analysis of the South Pacific Forum Fisheries Agency," *University of Hawaii Law Review,* 3, 1 (1981), 1–65.

W

Wang, Erik B., *Canada–United States Fisheries and Maritime Boundary Negotiations: Diplomacy in Deep Waters. Behind the Headlines* (Toronto), 38, 6 (1981); 39, 1 (1981).

Wasserman, Ursula, "The Antarctic Treaty and Natural Resources," *Journal of World Trade Law,* (March–April 1978), 174–179.

Williams, M.H., "Russia and the Turkish Straits," *Proceedings, U.S. Naval Institute,* 78, 5 (1952), 479–485.

Wilson, Gregory P., "Antarctica, the Southern Ocean and the Law of the Sea," *JAG Journal,* 30 (Summer 1978), 47–83.

Z

Zumberge, James H., "Mineral Resources and Geopolitics in Antarctica," *American Scientist,* 67 (January 1979), 68–77.

Part Eight

THE POLITICAL GEOGRAPHY OF EVERYDAY LIFE

Chapter 33

THE POLITICS OF RELIGION, LANGUAGE, AND ETHNIC DIVERSITY

In our discussion of the term "nation," we observed that it properly refers to a group with a common culture, of which two major components are religion and language. Much has and will continue to be written about these two cultural elements, but we are concerned here with their political aspects and relationships and in particular those that are linked with geography. One problem we face is trying to separate the two. There is certainly no necessary correlation or linkage between religion and language, yet many nations (or peoples or ethnic groups) have both distinctive religions *and* languages. Thus an ethnic minority in a country may demand cultural autonomy that would include both or may simply be distinguishable from other groups because of both. The Québecois, for example, are distinguished from their fellow Canadians not only by being predominantly French-speaking, but also by being overwhelmingly Roman Catholic in an otherwise largely Protestant country, although language was the basis of the separation movement. Thus, in this chapter, as we refer to an ethnic group (or people or nation), we may be referring to linguistic or religious factors or both.*

*The term "ethnic" is often used loosely to include inherited physical features that some people imply in the term "race." While physical characteristics may be important in the identification of some groups, including self-identification, there is no necessary connection between "race" and either language or religion. We therefore ignore "racial" factors here, even though they are sometimes linked with cultural factors in people's minds.

RELIGION

No longer, as in centuries past, is religion a dominant force in the determination of State boundaries or even in the creation of States. In the twentieth century, the partitions of Ireland, Syria, India, and Palestine were all designed to create separate political units for Catholics and Protestants, Christians and Muslims, Muslims and Hindus, and Jews and Arabs, respectively. In each case, however, while the political map of the world was altered and the course of history changed accordingly, minority groups were included in the new units and some of them have continued to pose problems of integration into the national system. These partitions, furthermore, have led to the creation of only four new States out of more than a hundred born in this century.

No longer are new religions sweeping across the face of the earth, gaining adherents by the millions, often under threat of the sword. The only religions gaining significant numbers of new adherents today are Islam, primarily in the northern parts of East and West Africa, and Christianity, in much of Africa. There is also a steady growth of a number of Protestant denominations in nominally Roman Catholic Latin America. Generally speaking, however, we are in a period of relative religious stability, in which the major changes are taking place *within* religions.

Although the religious wars that characterized Europe for centuries are fortunately behind us and we are living in an era of religious toleration, there has been no shortage of local religious wars and domestic conflicts in which religion has played a significant role. In Northern Ireland, for example, the conflict is not clearly between Catholics and Protestants; that is, it is not a war over religion but rather of different positions and orientations of the various communities within the province. Many members of the large Catholic minority want the province to be reunited with the dominantly Catholic Republic of Ireland, while most of the Protestants want to remain united with Great Britain. Catholics over the centuries have been discriminated against in employment and even some Protestants feel discriminated against by English control of the Ulster economy. But while the religious lines in Ulster are not rigidly drawn and economic and ideological factors are important components of the conflict, it is the religious component that attracts attention and tends to polarize people not only in Ulster but also in Britain, Ireland, and elsewhere.

The Nigerian civil war, which lasted for 30 months in the late 1960s, was particularly bloody. Again, the war had many causes, but because the Ibo nation, largely Roman Catholic, felt that they were not being, and could not be, treated fairly in largely Muslim and animist Nigeria, they broke away to form the State of Biafra. The Ibos lost the war and Biafra was obliterated, but it seems that they are being treated more fairly now in reorganized Nigeria.

A similar uprising, but one more drawn out and less well organized, is the continuing one of the Moro National Liberation Front in Mindanao and the Sulu Archipelago against the government of the Philippines. This group represents many of the Muslims who form a majority in 5 of the 13 provinces of these islands, and they have been fighting for autonomy in a chiefly Christian country.

Almost lost in the avalanche of news and opinion about the war in Vietnam and neighboring countries from the early 1960s

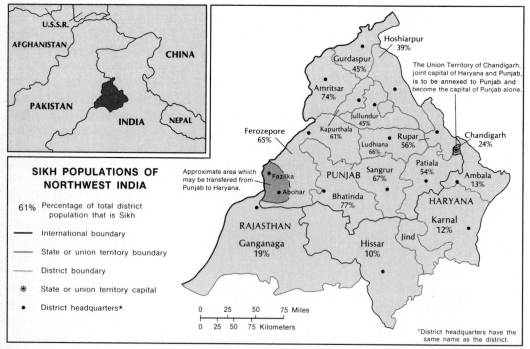

SIKH POPULATIONS OF NORTHWEST INDIA

61% Percentage of total district population that is Sikh

──── International boundary

──── State or union territory boundary

┈┈┈ District boundary

⊛ State or union territory capital

• District headquarters*

The Union Territory of Chandigarh, joint capital of Haryana and Punjab, is to be annexed to Punjab and become the capital of Punjab alone.

Approximate area which may be transfered from Punjab to Haryana.

0   25   50   75 Miles
0   25   50   75 Kilometers

*District headquarters have the same name as the district.

Proposed settlement of Sikh separatists claims in Punjab. In July 1985, Prime Minister Rajiv Gandhi and Sikh leader Sant Harchand Singh Longowal agreed on territorial transfers to settle Sikh claims for a state of their own. However, Longowal was assassinated, opposition to the agreement is strong among both Hindu and Sikh extremists and implementation of the agreement has been postponed indefinitely.

to 1975 was the fact that two indigenous religious groups, the Hoa Hao and the Cao Dai, controlled several important provinces. They figured importantly in the domestic political maneuvering, and they maintained their own armies that at various times fought the Japanese, French, Saigon government, and Viet Cong. While armed and militant religious groups of this type are no longer very common in the world, their quarrels can and sometimes do cause great hardship and suffering. The clearest recent example is Lebanon—once a peaceful, prosperous, cosmopolitan, and reasonably democratic country, but now devastated by years of civil war, fragmented into bitterly hostile factions, and occupied by foreign troops.

Although its roots go back to ancient Phoenicia, Lebanon under the Ottoman Empire was simply Mount Lebanon in Syria. When the French gained control of Syria after World War I they added to Mount Lebanon the coastal strip and the Bekaa Valley and on 1 September 1920 proclaimed there the State of Greater Lebanon—still under French mandate. The people of Mount Lebanon were predominantly Christian (mostly Maronites) and pro-French; the other areas, mainly populated by Muslims, were included within Lebanon to enhance the new country's political and economic position, and the total population had only a slight Christian majority. Tensions between Christians and Muslims grew until 1943 when they formulated the "National Pact." This provided for proportional representation by religious community in the Chamber of Deputies, as shown in Table 33–1. It was also agreed to continue the tradition of a Maronite president, Sunni prime minister, and Shia speaker of the Chamber of Deputies. Informally, cabinet posts, army and civil service jobs, and other positions were allocated according to religious affiliation,

Table 33-1 Lebanon: Estimated Proportions of Religious Groups in Total Population and Allocation of Seats in Chamber of Deputies

Religious Groups	Percentage (1970)*	Deputies (1970)	Estimated percentage of population (1980)
Christians (54%)			Christians (40%)
Catholics	38%	38%	
Maronites	75%	30	23
Greek Catholics	20	6	5
Armenian Catholics	2.0	1	
Roman Catholics	1.5	1	
Syriacs	1.0	0	
Chaldeans	0.5	0	
Greek Orthodox	11	11	7
Armenian Orthodox	4	4	
Protestants	1	1	All other Christians 5
Muslims (39%)			Muslims (53%)
Sunni Muslims	20	20	26
Shia Muslims	19	19	27
Others (7%)			Others (7%)
Druzes	6.0	6	7
Jews	0.5	0	
'Alawites	0.5	0	
	100%	99 deputies	

*Embedded within these figures are perhaps 1 percent of the total population who are Ismaili Muslims, Syrian Orthodox, Bahai, and other groups too small to be represented as such in the Chamber of Deputies. Druzes and 'Alawites are sometimes considered as Muslims whether or not they are aligned politically with Muslims. There has been no census in Lebanon since 1932, and all population figures, including those of the 1932 census, are very likely wrong, perhaps by a considerable margin. Nevertheless, the political system is based on this enumeration.

but not necessarily proportionately. This precarious arrangement was upset in 1958 by a number of internal and external factors, and it completely disintegrated after 1975, when civil war broke out among the private militias maintained by many religious and political groups.

The war continues since none of the fundamental problems has been solved. Underlying the many causes of the war—social, political, personal, and economic—is a basic fact of demography: the Christians are no longer a majority but refuse to surrender their positions of power based on their former majority status. Until the Lebanese people of all factions are able to arrive at a new *modus vivendi*, the country is likely to remain a tinderbox that could set off a big war involving not only Lebanese, Palestinians, Syrians, Israelis, and UN troops, but perhaps others from outside the immediate region as well.

In the United States, religious groups are not armed and militant, and they do not oppose one another politically, but there is no doubt that religion influences politics in many areas of the country. The Mormon influence in Utah and neighboring states, the Catholic majorities or near-majorities in southern New England and the Middle Atlantic states, the "Bible Belt" of fundamentalist Protestants in the South, the concentration of Jews in the major cities, especially in the Northeast and Midwest—these are just a few religious factors to consider when trying to explain or predict electoral behavior in the United States. Religious doctrines, while no longer as important as they once were in determining candidates for public office and the outcome of elections, are still influential in referenda and in legislative action on certain issues.

Religious minorities often play roles, make contributions, and exercise influence in societies far out of proportion to their numbers. This includes both indigenous and immigrant minorities. Among them are Jews, Parsees, Catholic Ibos, Mormons, Quakers, Armenians, Sikhs, and others. While they tend to focus on the economic sphere and are sometimes excluded from politics, these groups are always political factors, especially in areas where they are concentrated. In some countries, especially in Europe, South Asia, and Israel, there are political parties formed by, and representing, particular religious groups. The Christian Democratic parties of Europe and Latin America are generally associated, however loosely, with the Catholic Church, although the influence of the Vatican on them is often exaggerated by their opponents. The National Religious [Jewish] Party in Israel has been a coalition partner in every government since the founding of the State in 1948, and there are other Jewish and Muslim parties as well. The Komeito in Japan and the Jan Sangh in India are similarly parties representing the orthodox wings of the dominant religions, Buddhism and Hinduism, respectively.

Although religious groups are active in politics around the world, nowhere today is there a true theocracy. The last one was eliminated when the Chinese occupied Tibet in the 1950s, drove the Dalai Lama (the spiritual and temporal ruler) into exile, and destroyed the Lamaist Buddhist monasteries that were an integral part of the governing system. The nearest thing to a theocracy today is probably Saudi Arabia, in which the puritanical Wahhabi sect is still influential and the government takes seriously its role as protector of the holiest places of Islam, but it is still a secular State. Similarly, Libya, Iran, and Pakistan—other countries in which Islamic law has been incorporated into the civil and criminal codes—continue to be governed by lay persons and remain secular States.* Other countries have established churches that often receive some degree of financial and other support from the government, but the churches do not control the governments and generally religious freedom is granted to other denominations. In Japan the official religion until the end of World

---

*Vatican City may be considered a theocracy since the Pope is both spiritual and temporal ruler and all officials are clergymen, but this is a very special case. Iran is now arguably a theocracy, though too little is known of her inner workings to be certain.

Religious minorities far from their places of origin.   Top: These Hutterite women hoeing a field in Alberta are among the some 14,000 members of the Hutterian Brethren living in about 115 communal farm settlements in the Canadian prairie provinces, the Dakotas, Montana, and Washington. This Mennonite sect originated in Austria in 1528, fled to Germany after persecution in Switzerland, then to Bohemia, to Hungary, and eventually to Russia in the eighteenth century. About 400 members of the sect came from Russia to South Dakota in 1874–1877 to begin the North American settlements. (National Film Board of Canada) Bottom: Rihaniya is a Circassian village in Upper Galilee, Israel, one of two in the country. These people are Sunni Muslims from the Caucasus who fled Russian persecution in the nineteenth century to settle in Turkish Palestine. They have nearly abandoned their own language in favor of Arabic, but retain their religion and pastoral farming economy. (Martin Glassner) Both of these minority peoples live in peace within the dominant cultures of their refuges, but others around the world are not so fortunate.

War II was Shinto, a nationalistic, militaristic variety of Buddhism, but it was disestablished under the American occupation after the war and the emperor renounced the doctrine of his divine descent.

What pattern emerges from all these observations about the politics of religion to-day? Generally, it is a picture of a world in which religion no longer dominates the lives of people as it once did. Despite survivals of past orthodoxies and occasional revivals of religious feeling and activity (such as in Islam today), we are living in a secularizing world in which religion remains important to many individuals and is locally important politically, but is no longer a force that shapes history. Religious wars and massacres still break out from time to time in some countries, but seldom do these spill over into other countries and not for a long time have they generated massive crusades. Nationalism, it would seem, is a more powerful influence on most people in the world than religion, though it will not necessarily remain so.

LANGUAGE

Throughout this book we have repeatedly referred to the strength of nationalism in the world today. One of the most important components of nationalism has long been language. The European nation-state evolved out of a desire of people to be ruled by people who spoke their language; that is, who were of the same nation. Other cultural components, of course, go into the blend of symbols and feelings and attitudes that bind together a nation, but historically, few even approach language as the most important single element.

Ethnic minorities within many countries cling to their native languages in the face of cultural imperialism of the majority or even the obsolescence of the language in the face of technological change. It is a badge of identification, of belonging to a distinctive group with a proud tradition of its own. Cornish has died out, but Gaelic (both Scottish and Irish varieties), Welsh, and Breton of the old Celtic languages are preserved still and used by ardent nationalists as touchstones of their movements for greater autonomy or even independence. Of all the ancient languages, however, only Hebrew has been revived in modern times, updated, and firmly established as the national language of a modern State, in everyday use by the great majority of the population of Israel.

One reason for the survival of Hebrew through nearly two thousand years of the dispersion of its users throughout the world was its retention as a liturgical language. Other liturgical languages, such as Latin, have also served as unifying elements among peoples who share a religion but little else. No other, however, has retained the almost mystical loyalty of a people for so long as Hebrew among the Jewish people. It never really died out as an everyday language in its homeland either, as clusters of Jews have always lived there, using the ancient language regularly. When they were joined by East European Zionists beginning in the nineteenth century, the newcomers adopted the traditional tongue. By 1948, when the State of Israel was born, the debate over a national lan-

Survival of an ancient language. Around the world many linguistic minorities are trying to preserve their languages in the face of enormous pressures to abandon them in favor of the dominant languages of the countries. Use of the Manx language, for example, has declined drastically in the past half century but survives as a badge of distinctiveness within the United Kingdom. This bilingual sign is an example. (British Tourist Authority)

guage among the six hundred thousand Jews of Palestine had been settled: Hebrew had won out over Yiddish, English, and other candidates. It was a symbol of the rebirth of the ancient Jewish State, a matter of enormous political significance.

The multitude of languages in the world and the tendency of all languages to change through time have created a need for *linguae francae*, languages used in trade and general communication among peoples who speak other languages of their own. Some of these evolved naturally and have been accepted by the people as, at least in part, their own. Examples are Swahili in East Africa, Hausa in West Africa, Pidgin in the South Pacific, and Tupí-Guaraní in Brazil and nearby areas. Others, however, have been introduced by conquerors and either adopted willingly by natives wishing to advance under the new rulers or imposed by the rulers as instruments of cultural imperialism. Quechua, Arabic, French, Latin, Spanish, and English are only a few of the languages spread far and wide in this manner. Since World War II English has replaced French as the nearest approximation of a worldwide *lingua franca*, not because of any inherent virtues it might have, but because the United States emerged from the war as the world's most powerful country, with her troops, business people, teachers, scholars, scientists, and tourists penetrating the farthest reaches of the globe—and often unwilling to learn the local language. American influence, added to British influence within the Commonwealth and beyond, has made English the world's most important language—for the present. Already Spanish has replaced French as the most popular foreign language for Americans to learn and the Spanish-speaking population of the United States is growing rapidly. Some observers estimate that early in the twenty-first century the United States will be a bilingual country.

A *bilingual* country is one in which two languages are dominant rather than one. Examples are Belgium (French and Flemish), Canada (English and French), Sri Lanka (Sinhalese and Tamil), and Paraguay (Spanish and Guaraní). In nearly all bilingual countries other languages are also spoken, but generally by small numbers of people. In a great many countries, the overwhelming majority in fact, many languages are spoken by the native peoples. In most of these *multilingual* countries, one or two languages have achieved dominance, either because they are spoken by the largest groups of the population or because they are the languages of the ruling elites. Sometimes these dominant languages obscure the basic multilingual character of the country. Mexico, Yugoslavia, Nigeria, South Africa, India, China, and the Soviet Union are only a few of the more prominent multilingual States.

In some bilingual and multilingual States, attempts to introduce (or impose) an official "national" language or to suppress a local language have led to vigorous pro-

Area claimed by Tamil separatists in Sri Lanka. Whatever the final settlement of the Tamil–Sinhalese conflict may be, it is unlikely to include actual partition of the country or even Tamil control over this much territory. The major port/naval base of Trincomalee is particularly sensitive.

tests, bloody riots, and even full-scale rebellion or attempted secession, as in India, Sri Lanka, Belgium, Spain, and South
Africa. In most of them, however, a *modus
vivendi* has been reached among the various language groups and the selection of
a national language or languages has been
accomplished peacefully.

In some countries, such as Norway and
Britain, social groups are identified by the
dialect of the national language they speak.
In others, regional dialects label people
and sometimes restrict their participation

in political life. In still other societies secret or ritual languages confer power or
prestige on the few initiates. In others,
mastery of a classical or literary language
or dialect is essential for advancement in
society. Nearly everywhere in the world
language has political as well as cultural
significance.

If language is such an important component of nationalism and if nationalism
is as powerful a force in the world today
as we have indicated earlier, why do the
political boundaries of the world not con-

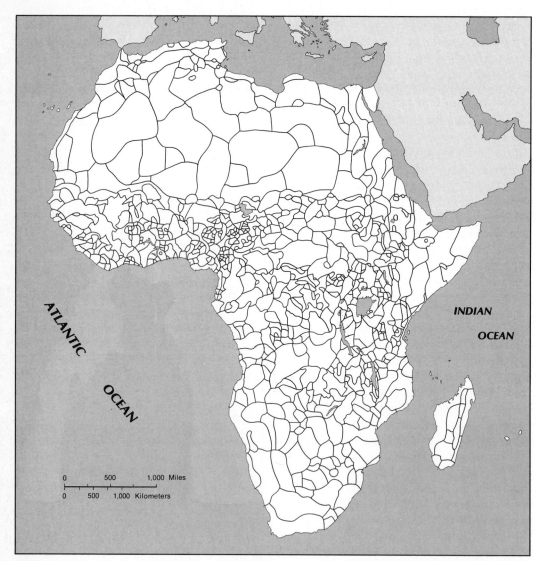

Ethnic groups of Africa. The almost unbelievable cultural fragmentation of the continent contributes greatly to its political instability. It is nearly impossible to draw political boundaries along
strictly ethnic lines that would make sense in other respects.

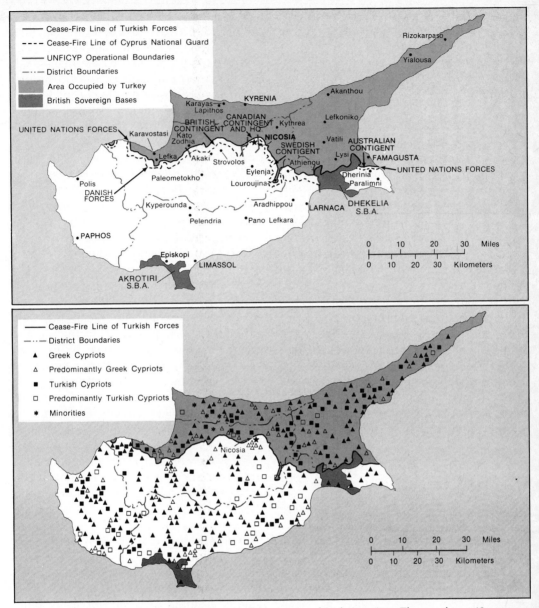

Cyprus. The top map shows the present politicogeographical situation. The northern 40 percent of the island State is occupied by Turkish troops since their invasion in 1974 and the United Kingdom still retains two sovereign base areas in accordance with the 1959 agreements on independence for Cyprus. The rest of the island is governed by Greek Cypriots. UN peacekeeping forces have been separating contending Greek and Turkish Cypriot communities pending a political settlement to the communal dispute. The bottom map is a simplified depiction of the distribution of the Cypriot population in 1960 by ethnic group. It is evident that no partition plan based on ethnic criteria as of 1960 was feasible; since 1974, however, there has been an *ad hoc* exchange of populations, and now the area occupied by Turkey is predominantly Turkish Cypriot. All the large cities had mixed but predominantly Greek Cypriot populations. The "minorities" are primarily Armenians and Maronites. There are a number of Greek Cypriot enclaves within the British Sovereign Base Areas.

form to linguistic boundaries? Part of the explanation lies in the mobility of people. Throughout history peoples have migrated, often over considerable distances, carrying their cultural baggage with them. Although they may have returned to their homelands or amalgamated with the local people, their languages—or traces of them—remain to distinguish regions that otherwise may not differ greatly. Language then becomes only one element, and sometimes a subordinate element, in the culture. Even culture as a whole may not determine boundaries. As we have seen, military, economic, dynastic, physiographic, and other factors are often determinative in boundary making, and language may be ignored. In twentieth century Asia and Africa, anticolonialism has often been the strongest component of nationalism, a rallying cry that unites most people in a colony, regardless of the language they speak. And, as we have also seen, most colonial boundaries have remained intact after independence as a matter of policy, even though they frequently cut across linguistic lines.

ETHNIC MINORITIES AND POLYCULTURAL STATES

There are few true nation-states in the world today. Most countries are home for two, three, or numerous cultural or ethnic groups. It is not just the largest countries, such as the Soviet Union or China, that are polycultural or plural societies, but even many of the smallest, such as Mauritius, Fiji, Comoros, and Trinidad and Tobago. In 1978, Freedom House in New York published a list of 91 minority ethnic groups that allegedly were being denied self-determination in 46 countries on every continent. The list is certainly open to question, yet it does indicate that cultural pluralism is not only worldwide but also a prominent element in political unrest nearly everywhere.

Nigeria and Lebanon are not the only countries in which a census is more than just a useful source of routine statistical information but a sensitive political issue as well. In 1976, for example, the Slovenes of Carinthia in Austria, supported by Yugoslavia and opposed by German-speaking Austrian nationalists, vigorously protested a proposed ethnic census in their province. And Saudi Arabia has yet to conduct a formal census and publish the figures, at least in part because it would inevitably show that the native population is quite small and the number of foreigners, mostly Palestinian and other Arabs, large.

But cultural pluralism need not lead to political instability. In few countries, in fact, are the ethnic groups as bitterly hostile as in Cyprus or Lebanon or Sudan. In most, ethnic differences have been at least temporarily sublimated or suppressed by nationalist or repressive governments. Where the harmony is only superficial and ethnic animosities simmer close to the surface, we may witness in the near future more outbreaks of communal fighting based on racial, religious, linguistic, tribal, or other cultural rivalries. But if such outbreaks can be postponed long enough for a true sense of national identity to develop, for people to be identified and judged as individuals and as members of a larger society instead of as members of a particular ethnic group, we can find a gradual reduction in intercommunal tensions and thus in civil and international strife.

Perhaps the world's best current example of a plural society that has welded itself into a nation-state is Switzerland. With four major languages and two major religions, it could be a land of strife, but is not. There are a number of reasons for this. First, there is no correlation between language, religion, place of residence, and socioeconomic status. German- or French-speaking Catholics, for example, live in both cities and countryside and are found in all socioeconomic levels. Thus, people identify with one another in a number of ways depending on the situation, but always as Swiss, clearly distinct from those across the borders who might share their language or religion. Second, the peoples of Switzerland came together (usually vol-

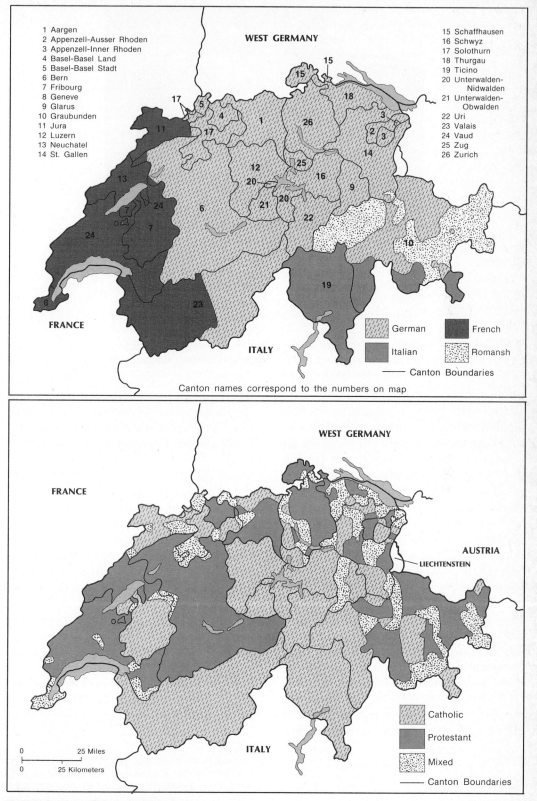

1 Aargen
2 Appenzell-Ausser Rhoden
3 Appenzell-Inner Rhoden
4 Basel-Basel Land
5 Basel-Basel Stadt
6 Bern
7 Fribourg
8 Geneve
9 Glarus
10 Graubunden
11 Jura
12 Luzern
13 Neuchatel
14 St. Gallen

15 Schaffhausen
16 Schwyz
17 Solothurn
18 Thurgau
19 Ticino
20 Unterwalden-Nidwalden
21 Unterwalden-Obwalden
22 Uri
23 Valais
24 Vaud
25 Zug
26 Zurich

WEST GERMANY

FRANCE

ITALY

German
Italian
French
Romansh
Canton Boundaries

Canton names correspond to the numbers on map

WEST GERMANY

FRANCE

AUSTRIA
LIECHTENSTEIN

ITALY

0   25 Miles
0   25 Kilometers

Catholic
Protestant
Mixed
Canton Boundaries

Switzerland—religions and languages.

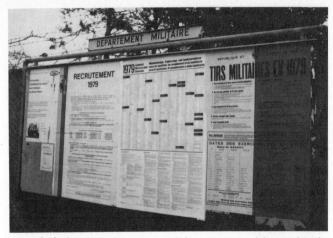

A symbol of Swiss unity. One of the centripetal forces binding together the people of Switzerland is compulsory military service for males beginning with basic training at age 20, followed by service in the active reserves until age 50 (55 for officers). This billboard in Geneva carries announcements of the reserve units being called up for training and other duties for specified periods, as well as other notices concerning military service. The Swiss are rightfully proud of their "armed neutrality," regardless of their language or religion. (Martin Glassner)

untarily) over a long period of time, from 1291 to 1815, partly to protect themselves against more powerful countries surrounding them and partly to share the advantages of a central location in Europe. In order to preserve these advantages, to avoid taking sides in outside conflicts that might prove internally divisive, Switzerland has since 1815 maintained a stance of permanent, armed neutrality. Third, the country is not just a federation, but a confederation, in which the cantons retain a large measure of authority, both by exercising primary responsibility in certain areas and by executing many federal laws. Since religion and language often cut across cantonal lines, many cantons have to reach compromises in most political matters. Finally, Switzerland, more than any other country in the world, practices direct democracy, involving frequent use of the initiative and referendum (even permitting referenda on certain types of treaties), so the likelihood of any ethnic group imposing anything on any other is very small.*

We are not implying here that Switzerland is a model to be emulated by other plural societies. Switzerland is the product of unique circumstances and of seven hundred years of trial and error. Nevertheless, she is a living example of a country whose people live in harmony despite differences of language and religion. These differences can be tolerated in order to attain mutually beneficial objectives, not only here but elsewhere as well.

Considering that the politics of religion and language touch the lives of most people in the world at some time or another, it is remarkable how little serious work had been done in this area by political geographers until recently. We could use many more studies, especially for areas outside Europe.

*Although the Swiss are generally quite conservative, their system does have some flexibility. In 1975, the French-speaking population of predominantly German Berne canton won the right to secede from Berne and in 1978 the new canton of Jura was admitted into the confederation.

Chapter 34

THE POLITICS OF TRANSPORTATION AND COMMUNICATIONS

In our discussion of the State as a political entity, we pointed out that one of its characteristics is a circulation system, a system of transportation and communications that permits a flow of goods, people, and ideas within the State and—equally essential today—between the State and other parts of the world. While such a system has, of course, its social, economic, and technological aspects, even its very existence may be political in nature. The great highway systems of the Incas and the Romans were designed and maintained primarily to facilitate administration of their vast empires, while the rulers of early nineteenth century Japan and Paraguay (among many others) deliberately kept their internal circulation systems primitive and their countries isolated from the rest of the world for similar reasons. A decision to build a road from one place to another or to develop a national merchant fleet or restrict the ownership of shortwave radios is frequently a political decision. Circulation, therefore, is of great interest to the political geographer.

TRANSPORTATION

Simple paths or tracks served as the principal transportation arteries for the great majority of the world's people until quite recently. They still lace together farms, homes, and settlements in much of the world. Gradually, however, they are being replaced by streets, roads, and highways.

The result is increased mobility and expanded horizons for individuals and greater economic, cultural, and political unity for societies. Even warfare has changed as new vehicles traveling over relatively flexible road systems have given armies greater mobility and adaptability than railroads did earlier.

Roads

Nearly everywhere road building and maintenance are the responsibility of government, generally with various levels of government responsible for specific types of roads. Every phase of the process, from the decision to build a road or a network of roads to the allocation of funds for repairing and maintaining roads, has a political component. New superhighways built in urban areas of the United States, for example, are generally routed through slums or working-class areas, disrupting the communities and displacing many people rather than through the much less densely populated wealthy suburbs, largely because of the distribution of political power. In hundreds of towns in southeast United States, streets in sections inhabited by blacks or what used to be called "poor white trash" are still unpaved and essentially unserviced. The Interstate and Defense Highway System, begun in the 1950s and still uncompleted, is, like the *autobahnen* of Hitler's Germany, designed for tanks, heavy artillery, and other military traffic, though of course civilians benefit from it also.

Roads through mountain passes have special significance around the world. Sometimes they open up new areas for settlement or connect important population centers or provide invasion routes. Khyber Pass, Brenner Pass, Cumberland Gap, Mitla Pass, Dzungarian Gates, Cilician Gates, South Pass, the names ring through history, recalling epic struggles between people and nature and among peoples. A highway connecting China and Pakistan through some of the most rugged and forbidding terrain in the world, opened in 1971 after six years of construction, closely follows an ancient caravan route, but its strategic significance is quite modern.

Roads of continental dimensions have also had enormous political significance since the old Silk Road connected the Chinese and Roman empires. Today the Asian Highway reaches from Istanbul to Singapore and Ho Chi Minh City. Even longer is the Pan-American Highway, linking Fairbanks, Alaska with Puerto Montt, Chile, and Buenos Aires, Argentina.* Another international highway is the Carretera Marginal Bolivariana de la Selva (commonly known as "El Marginal") running along the eastern foothills of the Andes through Colombia, Peru, and Bolivia. Designed to link the rivers and roads running east and west through the Andes and to stimulate settlement of the region, its construction slowed considerably after the overthrow by the army of its principal promoter, President Fernando Belaunde Terry of Peru. There is no comparable highway in Africa, reflecting the political fragmentation of the continent as well as physio-

*Latin Americans insist on calling it the Inter-American Highway, believing that "Pan-American" smacks too much of Yankee imperialism. When a bridge replaced a ferry across the Panama Canal in 1962, linking major segments of the highway, Latin Americans stridently protested naming it the Thatcher Ferry Bridge. Reluctantly, the United States agreed to call it the Bridge of the Americas. And the Darien Gap in the highway has still not been closed, at least in part because of cool relations between Panama and Colombia.

graphic, demographic, and economic factors. Some are being planned and constructed, however, coordinated by the staff of the UN Transportation and Communications Decade for Africa.

Railroads

Possibly even more dramatic and long-lasting has been the political significance of railroads, even though they are barely a century and a half old. Railroads are far more efficient (thus cheaper) than roads for carrying low-value bulk products long distances. They are also very efficient for carrying large numbers of people short distances through densely settled country. Consequently, railroads have been instrumental in opening up vast new areas of the world for agriculture, mining, and forestry as well as in stimulating and servicing the urban sprawl so typical of most industrial societies. The political significance of both of these movements is clear.

Most of the world's railroads are government owned and nearly all privately owned railroads are government regulated. Even if privately built, railroads are subject to political as well as economic influences. The land-grant program of the U.S. government to spur railroad construction in the West in the mid-nineteenth century is only the most obvious of these influences. Towns all through the country were born, thrived, shriveled, or died according to whether the railroad came through them or passed them by, and many sprang up along the railroad as the track was laid. Even in Russia, the growth of Tomsk was stunted when the Trans-Siberian Railroad was built far to the south.

In the Soviet Union even today, railroads are given priority by the government over roads, and truck haulage is still rudimentary. The Trans-Siberian has been double-tracked, rerouted north of, instead of through, Manchuria, and improved in other ways. Perhaps the most important and difficult railroad construction in recent times was the 4300-kilo-

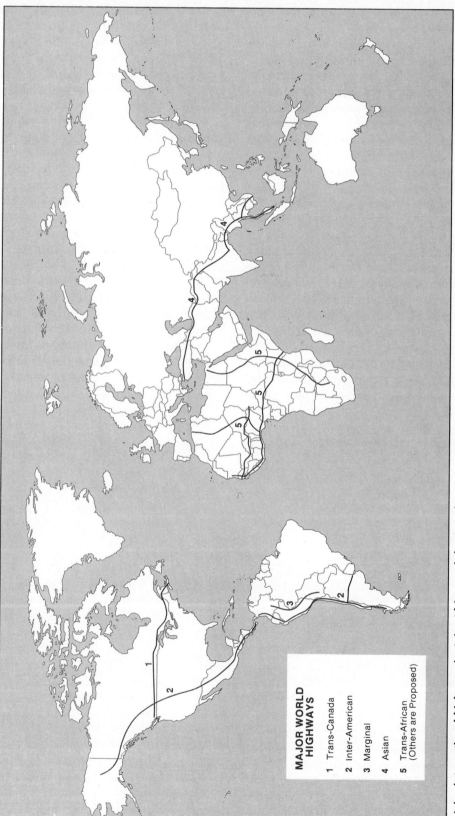

MAJOR WORLD
HIGHWAYS

1 Trans-Canada
2 Inter-American
3 Marginal
4 Asian
5 Trans-African
  (Others are Proposed)

Major international highways in Asia, Africa, and the Americas.

meter Baikal–Amur Mainline Railroad (BAM), paralleling the Trans-Siberian in Yakutia but much more comfortably north of the Chinese border. It was laid in some of the coldest, bleakest, remotest, and most rugged areas on earth at the rate of some 300 miles per year, for reasons partly economic and partly strategic. Another railroad laid largely for strategic reasons was the Tanzam Railroad, built by China to link the rail nets of Zambia and Tanzania after Rhodesia declared her independence in 1965. Zambia was placed in the politically uncomfortable position of continuing to use the Rhodesian Railways for her vital copper exports or of boycotting the Rhodesian Railways and suffering the inevitable economic consequences. The decision was to sever links with southern Africa and establish new ones with East Africa by means of a railroad and pipeline to Dar es Salaam.

The Tanzam Railroad, while new, is not a new idea. It was included by Cecil Rhodes in the nineteenth century as part of his projected Cape-to-Cairo Railway, a dream of British empire builders through much of the scramble for Africa. The dream was never realized, however, as Germany managed to hold on to German East Africa (Tanganyika), thus blocking the most feasible routes. After Britain obtained a mandate over Tanganyika, she welded together an efficient and practical rail system with more than 3000 miles of track in her territories of Tanganyika, Uganda, and Kenya. This system was operated by all three territories through the East African Railways and Harbours. It continued functioning after independence as a service of the East African Community. But the Community collapsed in late 1976 and EAR&H broke up in February 1977. Then each of the former partners had to reorganize its share of the rails and equipment and try to operate its own national rail system.

Another uncompleted transcontinental railroad project has also been the victim of politics. A railroad linking Arica, Chile with Santos, Brazil was proposed in 1928,

The Tan-zam Railway terminal in Dar es Salaam. The railway and its facilities are operated by TAZARA—the Tanzania-Zambia Railway Authority. Its fortunes wax and wane depending on the political situation in southern Africa. (Martin Glassner)

and over the years various segments of it have been put into service. There is still a gap, however, in the mountainous area between Cochabamba and Santa Cruz, Bolivia. Bolivia has accorded this stretch a low priority and construction proceeds very slowly. Part of Bolivia's reluctance to complete the line may be attributed to Brazil's interest in its completion; Bolivia, with considerable historical justification, is suspicious of Brazil's motives.

The role of railways as political instruments cannot be overstressed. Canada came into existence in 1867 only because the British acceded to the demand of the Maritimes and built the Intercolonial Railway linking St. John, New Brunswick, with Montreal. The line was uneconomic but its purpose was political and it was successful. Later, British Columbia demanded to be linked with Canada by rail before she would join the confederation, and did so in 1871 after the Canadian Pacific Railway was built. Similar examples of the unifying effect of railways can be found on every continent. So can examples of railways as instruments of colonial policy, and of political jealousies and suspicions inhibiting the development of railroads or standardization of gauges and equipment, and of political factors in setting freight rates, and of railroads being built primarily for military use, and of political battles between rail and road for the carriage of urban and commuting passengers. Indeed, it would be difficult to exaggerate the political factors in rail transport.

Waterways

Inland waterways similarly have been subject to political influence and have had their effects on politics. Rivers have been great natural highways for millennia and have linked peoples more often than they have divided them. Still, as we have seen, international rivers have frequently been foci of contention among not only riparian States, but among other users as well. More than one country has felt it necessary to use force to open up a river for international navigation. And unlike highways and railways, rivers have other uses besides transportation. In fact, the nonnavigational uses of rivers are growing in importance and the various uses are frequently in conflict with one another. As mentioned earlier, the International Law Commission is currently working on this problem.

Artificial waterways give rise to similar problems. There are numerous tales that can be told of canals proposed and constructed; of rivers straightened, deepened, and canalized, all embroiled in political controversy both domestic and international. The United States abounds in them: the Houston Ship Channel, the Cross-Florida Barge Canal, and the proposal to open a route to Oklahoma City for oceangoing ships, to mention only a few recent examples. But the all-time classic story has to be that of the St. Lawrence Seaway, opened in 1959 after half a century of complex and bitter battles.

There were rivalries between Montreal and New York for seaborne trade, between Canada and the United States, between railroad and shipping interests, between hydropower and navigation needs, between advocates of state and federal action, between Canadian prime ministers and provincial premiers, and on and on. In the end, the United States concurred in its construction only after Canada threatened to build a seaway entirely on its own territory and charge American ships very high tolls to use it! Two considerations made the seaway seem desirable to the U.S. government: the discovery of iron ore on the Ungava Peninsula near the Quebec–Labrador boundary that could replace the depleted iron ranges of Minnesota in feeding midwestern steel mills, and the belief that the seaway would make a material contribution to the defense of the continent. Now it is profitable and almost too successful, cited as a model of practicality, of wise multipurpose devel-

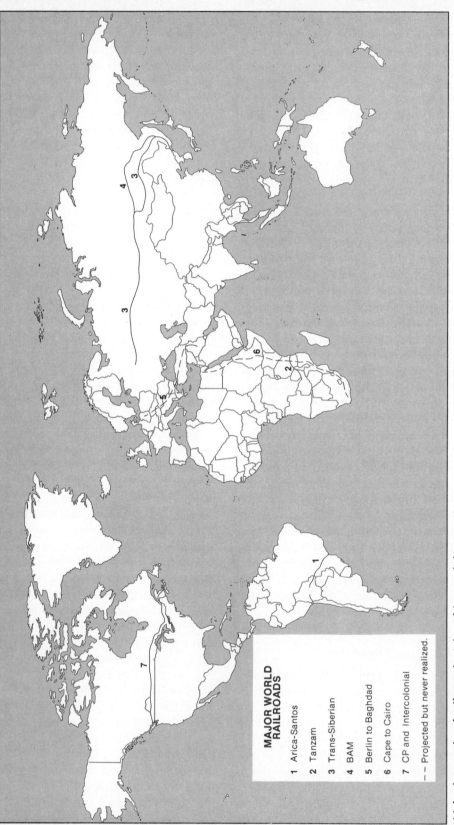

MAJOR WORLD RAILROADS

1 Arica–Santos
2 Tanzam
3 Trans-Siberian
4 BAM
5 Berlin to Baghdad
6 Cape to Cairo
7 CP and Intercolonial
-- Projected but never realized.

Major international railways in Asia, Africa, and the Americas.

opment of an important resource, and of international cooperation!

Pipelines

Pipelines in their modern form are relatively new, though their ancestors are the numerous aqueducts of antiquity. Now every continent is laced with pipelines carrying all manner of petroleum and petroleum products, natural gas, water, wastes, and even coal slurry. Some pipelines have been of heroic proportions, both in length and in difficulties overcome. The Trans-Arabia Pipeline (Tapline) from the Persian Gulf oilfields of Saudi Arabia to the Mediterranean in Lebanon; the Friendship Pipeline carrying petroleum products from the Volga–Urals fields of the USSR to Poland, Czechoslovakia, and East Germany; and the Alaska Pipeline from Prudhoe Bay on the Arctic Ocean to Valdez on the Pacific are only a few better-known ones. Each has its obvious strategic advantages, but each is uniquely vulnerable to attack by saboteurs or aircraft.

The construction and routing of pipelines are no less political than those of any other mode of transport. Witness the Sumed Pipeline built by Egypt in the late 1970s from Suez to the Mediterranean in direct competition with the lines built by Israel from Eilat to the Mediterranean a decade earlier. And the drawn-out debate over the best route for a gas pipeline from Prudhoe Bay to Calgary and the midwestern United States, any of which would require settlement of Canadian native land claims.

The Soviet Union's 1981–1985 plan included a provision for six new pipelines to carry natural gas from Western Siberia to Western Europe. They sought to purchase some Western equipment and technology for one of them. The United States strenuously opposed these sales on political and security grounds and instituted economic sanctions against all potential suppliers. This policy was unsuccessful. The equipment was supplied, the pipeline opened ahead of schedule in late 1983,

and Western Civilization has somehow managed to survive! But we may expect more such rows over pipelines in the future. Now undersea pipelines are becoming common, generally linking oilfields or storage tanks with offshore tanker terminals, but destined to become longer and probably more politically—as well as environmentally—controversial.

Bridges and Tunnels

The past two decades have seen the construction of major bridges and tunnels around the world, many of them crossing strategic waterways and many having major geostrategic as well as economic importance. To name only a few: the bridge across the Bosporus at Istanbul to link Europe with Asia, the bridge–tunnel connecting Honshu and Kyushu in Japan, and the bridge from Puerto Stroessner, Paraguay, across the Río Paraguay to Brazil. Others are planned or proposed: Britain and France have agreed (again) to build a rail tunnel under the English Channel between them; the United Nations is studying construction of a "Europe–Africa permanent link through the Strait of Gibraltar," probably to be a bridge–tunnel combination; the Gulf Cooperation Council is considering an eventual bridge–tunnel across the Persian Gulf, and Italy has been probing a possible bridge across the Strait of Messina between Sicily and the mainland. Eventually, probably, some version of all these links will be built, lacing together even more tightly islands and continents and the peoples who inhabit them. They should help, among other things, to break down the insularity of many countries and peoples and thus reinforce even more the world's interdependence.

Maritime Transport

We have already discussed many features of maritime transport in our chapters on geopolitics, international trade, and the sea. More needs to be said, however, as we are becoming more, not less, dependent on seaborne trade with the passage of time. The increase in number and size of vessels sailing the world's sea lanes has necessitated more stringent rules of the road, tighter controls on ship design and construction, establishment of one-way lanes and traffic control systems in many straits, and other reductions in the traditional freedom of the seas. All this is quite apart from the extension of internal waters and the territorial seas farther out to sea and the creation of archipelagic waters.

The major maritime powers made it quite clear from the outset of the Third UN Conference on the Law of the Sea (UNCLOS III) that they would tolerate no unjustified interference with freedom of navigation. As a result of this position and the realization by most other States that they too benefit from a free flow of international commerce, agreement was reached early in the conference on regimes of transit through straits, archipelagos, territorial waters, and exclusive economic zones. Other agreements on restrictive measures necessary to prevent collisions at sea, marine pollution, and other hazards have been adopted not only in UNCLOS III, but also in IMO, the International Maritime Organization. International cooperation is essential and generally maintained in international shipping, but this does not mean that there are or will be no problems.

Changing technology of maritime transport, changes in the types of goods carried, entry of new countries into the shipping industry, development of new seaports, new sources of fuels and raw materials, and differential rates of economic growth are some factors that are leading to changes in shipping routes and hence the political significance of certain waterways and nearby coasts and ports. Despite the reopening of the Suez Canal, for example, traffic around the Cape of Good Hope has not diminished as much as had been expected and is likely to increase again despite the hazards presented by formidable storms and currents. Unless, of course, South Africa be-

comes less inviting to vessels as a result of political changes.

Some waterways are governed by special regimes. These include the Panama, Suez, and Kiel canals, and the Turkish Straits (the Dardanelles and the Bosporus). In all of them international trade is protected, but under different conditions, and the rules are sometimes interpreted by the State controlling the waterway to suit itself. The 1888 Convention of Constantinople, for example, states explicitly that the Suez Canal is to be free and open to the ships of all countries without distinction both in peace and war. Yet Egypt prohibited passage of Israeli ships and of any ships bound to or from Israel through the canal for a quarter of a century, until late in the 1970s. And under the Montreux Convention of 1936 that governs the Turkish Straits, aircraft carriers are forbidden passage in either direction, yet in July 1976 Turkey permitted the *Kiev*, a new Soviet aircraft carrier, to pass from the Black Sea into the Mediterranean, accepting at face value the Soviet claim that the ship was really a cruiser!

Before leaving the subject of maritime transport, it would be well to reiterate something of the maritime problems of land-locked countries. Lacking seaports, those marvelous windows on the world, these countries are handicapped initially by being isolated, to a greater or lesser extent, from the swirls and eddies of human intercourse along the coasts and across the seas. They are further handicapped in participating in the shipping industry by their lack of home ports and of opportunities to train seamen and masters. Switzerland is unique among landlocked States in having successfully operated a merchant fleet for some time now, using Basel on the Rhine River and the Italian port of Genoa as its home ports. Bolivia again has a merchant fleet, composed of one ship serving Europe and South America. And the Royal Nepal Shipping Corporation operated one chartered ship on one voyage from Calcutta to Bremen in 1972. Land-locked States, espe-

The Panama Canal: critical international waterway. In addition to straits, there are several interoceanic canals that are considered vital for international maritime transport and considered "choke points" by naval strategists. The Panama Canal is no longer as important as it once was because of the increasing size and speed of vessels and the development of a rail "land bridge" across the United States, but the Canal is still vital for the international trade of many countries. (UN)

cially the poorer ones, are even more at the mercy of shipping companies than coastal States without fleets of their own. This is one more reason that they need special help.

Air Transport

Air transport is very likely the most politicized of all modes of transport for a number of reasons. First, air space, unlike ocean space, is not free for anyone's use. It belongs to the State under it and no commercial or military aircraft has a right of innocent passage. States have absolute jurisdiction over use of their air space and ground facilities. Second, air travel is relatively new, developing only after the modern State system was well established and flowering during the period of decolonization. As a result, it has developed not through custom and practice over a long period, but very quickly and by the agreement of a relative handful of States. Third, the technology of aviation is now so costly and so attractive that States are willing to go into debt in order to acquire and maintain in air fleet both for legitimate

security reasons and purely for prestige. Indeed, a national airline, like a seat in the United Nations or a steel mill, is today part of the iconography of new States.

As a result, such matters in international air transport as routes flown, cities served, frequency of flights and types of aircraft used are thrashed out by goverment representatives primarily on the basis of reciprocity and spelled out carefully in bilateral agreements. There is no most-favored-nation clause relating to air transport and few blanket agreements covering multiple countries and arrangements. Even the friendliest of States, such as the United States and the United Kingdom, may engage in arduous and protracted negotiations, perhaps offering to swap an extra flight a week into Hong Kong for an American carrier for the right of a British carrier to fly from London to Denver nonstop. Package deals are sometimes worked out that contain provisions unrelated to air transport. This helps to account for the seemingly irrational practices so common in the international air transport industry.

Since overflight rights must be specifically granted by States, the lack of such

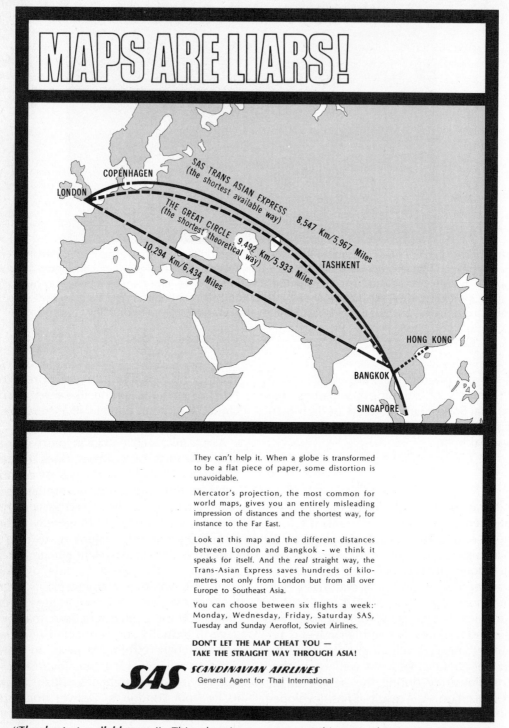

"The shortest available way." This advertisement appeared in several European periodicals in late 1971 and early 1972. It shows not only how most maps distort distances, shapes, and directions, so the charted route is seldom what it appears to be, but how it is nearly impossible to fly a true great circle route for very long because of both technical and political considerations. (Used by permission of SAS)

rights may force planes to travel extraordinarily long distances at great cost to avoid overflying particular countries. Only recently have the Soviet Union and China begun allowing overflights and even landing by selected noncommunist aircraft; most other countries still forbid overflights by aircraft of countries deemed unfriendly. The map of world airline routes shows large areas with a very low density of international flights. On the other hand, Aeroflot, the Soviet State airline, operates extensive international service nearly everywhere, except Australia and parts of South America. This service is maintained

chiefly to obtain convertible currencies and to carry the Soviet presence as far as feasible. The actual operations of Aeroflot abroad, such as marketing, public relations, and flight service, are allegedly marked by inefficiency, unethical sales practices, and a general lack of professionalism. This is probably an extension of the Soviet domestic system that stresses political and strategic goals rather than commercial operation, and its network cannot be justified by the amount of traffic carried. It is, nevertheless, generally accepted by many Western governments as being similar to other airlines, but they

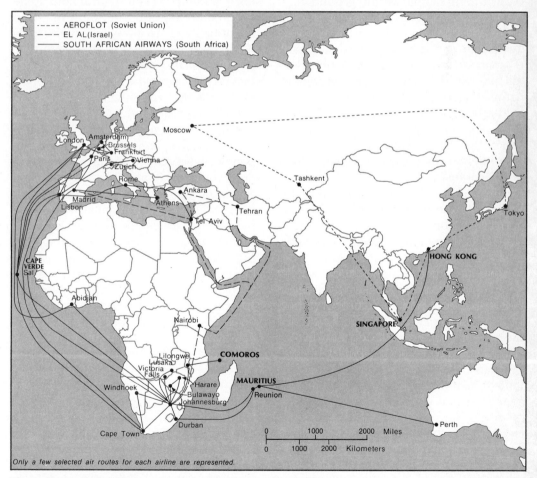

The politics of air transport.   This map shows selected routes of three major international air carriers. Note how El Al and SAA must fly costly circuitous routes in order to avoid over-flying hostile territory. The Aeroflot route was proposed during the 1960s, when Sino-Soviet tension was at its peak. (Manila was considered a possible substitute for Hong Kong.) It never actually went into effect. Today Aeroflot has routes directly across China between Moscow and Peking and Pyongyang, North Korea. El Al no longer serves Tehran.

have not insisted on the customary reciprocity—probably for political reasons.

Other airline routes and service patterns are distorted by such factors as some countries' desire to serve their colonies and former colonies, even if there is little local demand for air transport; a desire to preempt carriers of other countries; and hostility toward certain countries that are denied overflight for their carriers. Many countries prohibit flights by foreign aircraft over particular areas or even permit overflights only in certain designated corridors. The reasons are not limited to the safety of air navigation or the possibility of disturbing people on the ground.

Another vital factor in the airline industry is the predominance of government ownership and subsidy. There are many reasons for this, such as the desire to attract tourist trade and foreign investment, national security, need to conserve foreign exchange, need to serve small or remote places that generate little revenue, and so on. While these reasons individually might be defensible, they contribute to the confusion.

Fares and standards of service were until recently controlled for its members by the International Air Transport Association (IATA), a nongovermental industry association whose members include many government-owned airlines and whose rulings require government approval. In order to restrict competition among international carriers, IATA even regulated the charges for in-flight films and the types of meals that could be served as part of the plane fare, down to the size and content of sandwiches! IATA's stranglehold over fares was broken, in part, by nonmembers, such as Ecuatoriana, Loftleidir, and Aeroflot; the increasingly popular lowfare charter flights; and such innovations as Laker's very low fare–no frills scheduled transatlantic flights and other airlines' lowfare-standby–no-reservation systems.

Airline deregulation, begun in the United States in the late 1970s, in Canada in 1986, and later in Europe, is complicating, not simplifying, this picture, at least at the be-

ginning. We shall see whether it results in affordable, safe and efficient air travel for the average person.

COMMUNICATIONS

Like it or not, perhaps even without realizing it, everyone in the world is enmeshed in an intricate web of messages emanating from innumerable sources and carried by many means from the silent gesture to the most sophisticated devices of modern technology. We are all inevitably, unavoidably influenced by what we see, hear, and feel. Sometimes the influence is benign, even helpful, and sometimes quite the opposite, but even if not quantifiable, it is always significant. We have seen examples of this influence throughout this book but more needs to be said, not about the messages but about the media that carry them.

Taken together, telecommunications—telephone, telegraph, radio, and television—have had an incalculable impact on people everywhere. Modern transportation would be impossible without them. So would democratic elections, efficient government and destructive wars. Radio has been a particularly effective means of entertaining, informing, educating, and persuading even illiterate people in all but the remotest of locations, for by now the loudspeaker and the transistor radio are ubiquitous. In most of the world telecommunications are wholly or partially government owned and everywhere they are at least regulated by government. Some governments are quite responsible in exercising their power in the public interest, whereas others are much less so.

Even more than transportation, telecommunications have an international character and impact that is unmistakable. Radio and television signals leap international boundaries without permission or interference, and jamming them is very costly and only partially effective. Radio Moscow, the Voice of America, and the British Broadcasting Corporation broadcast by shortwave around the world, supplemented by such largely propaganda stations as Radio Free Europe and Radio Liberty. Radio Beijing and the Voice of the Arabs (from Cairo) are two other powerful shortwave stations beaming heavy doses of propaganda far afield. Even the United Nations Radio Service in 1978 began producing 15-minute programs in four languages denouncing *apartheid* and supporting self-determination for the native African peoples, programs broadcast through the facilities of member States.

But Radio RSA, the South African government's external shortwave service, responds in kind with 208 hours a week of broadcasts in English, French, German, Portugese, Spanish, Dutch, Afrikaans, Swahili, Chichewa, Lozi, and Tsonga plus another 20 hours a day in six languages designed exclusively for listeners in Africa.

News and entertainment programs can be influential without any deliberate attempt to persuade. American music, fashions, humor, and attitudes infiltrate even the most closed societies on earth, inadvertently influencing attitudes of others toward them. The process works in reverse also: Radio broadcasts from London during the Battle of Britain in World War II and on-the-scene television coverage of the Vietnam War demonstrably influenced

International transportation and communications on stamps. Although most countries routinely pay tribute to elements of their circulation systems on stamps, far fewer recognize *international* circulation. These stamps illustrate the importance of the St. Lawrence Seaway, the Bridge of the Americas over the Panama Canal, the International Telecommunications Union, World Communications Year, International Civil Aviation Organization, East African Airways (now defunct), ASEAN Submarine Cable Network, Asia-Pacific Broadcasting Network, Radio Moscow, Universal Postal Union, Asian-Oceanic Postal Union, the TANZAM Railway, and the Nairobi-Addis Ababa Highway. (Martin Glassner)

American attitudes toward those conflicts. Even political enemies listen to one another's broadcasts and watch one another's telecasts: East and West Germans, Jordanians and Israelis enjoy one another's television programs designed primarily for domestic consumption.

New technology is constantly intensifying the telecommunications mesh. Undersea cables that carry hundreds of messages simultaneously, teletype and telex, messages carried by laser beam, and programs of all kinds relayed by orbiting satellites and even broadcast directly from satellites—all these and others are not futuristic fantasies but are here now, in current use. They raise a host of political, legal, and ethical questions we can only mention here.

"Pirate" or unauthorized broadcasting beamed at a country from ships in international waters, a practice begun in the 1950s in the North Sea but later brought under international control, made the world debate proposals to regulate the content as well as the frequencies of radio and television broadcasts that cross State boundaries. Now the debate focuses on communications satellites. The United Nations has been working since 1977 on "an international convention on principles governing the use by States of artificial earth satellites for direct television broadcasting." At issue, as noted in Chapter Thirty-two, are not only blatant propaganda messages, but even programs tending to condone or encourage racial bigotry, religious programs received in countries where atheism is the official religion, even commercials received in countries where noncommercial broadcasting is the norm. These are not easy problems to solve, and they are unlikely to be solved soon.

The print media—newspapers, magazines, books, pamphlets and broadsides, trade and professional journals—are also widespread and influential, in some countries even more so than the broadcast media. They also are subject to government ownership or at least control. And to an even greater extent than broadcast media, they serve as mouthpieces of special interest groups of all kinds: labor unions, political parties, churches, ethnic groups, manufacturers, farmers, veterans, retirees, and so on. Even in illiterate societies, the written word may be a powerful force, if only because it may be endowed with mystical qualities, and those who can read are automatically in advantageous positions.

Government censorship, both before and after publication, is widespread in the world; a free press is quite rare, in fact. Even where the press is not overtly owned, controlled, or censored by government, it may be cowed into docility by harassment, threats, revokable licences, control of paper supplies, and still more subtle devices.

Why do governments go to such lengths to control the printed word? Partly because, sometimes for the best of motives, they wish to channel people's thinking along particular lines, to guide the political and social development of their countries as well as economic development, and partly because such control has been repeatedly demonstrated to be both feasible and effective. Not completely effective, of course—witness the *samizdat*, or mimeographed newsletters clandestinely produced and circulated among the intelligentsia in the USSR, and the forbidden books smuggled into South Africa. Nevertheless, it does work and people's thinking is influenced by it.

Many agencies, public and private, are working around the world to encourage and maintain greater freedom in the print media. The Inter-American Press Association, for example, annually rates the countries of the Western Hemisphere according to their degrees of press freedom. The Pan-American Institute of Geography and History (PAIGH) and the United Nations Educational, Scientific and Cultural Organization (UNESCO) are sponsoring long-range international programs for writing "objective" histories of parts of the world. And the American Civil Liberties

Union stoutly defends the First Amendment to the U.S. Constitution guaranteeing freedom of the press, religion, and assembly. But these and other agencies face an uphill struggle.

CIRCULATION

In 1977, the General Assembly of the United Nations endorsed the recommendation made earlier by the Economic Commission for Africa and proclaimed 1978–1988 as the United Nations Decade on Transport and Communications in Africa. It authorized an extensive program designed to develop these facilities in Africa and to mobilize the technical and financial resources required for that purpose. This is fitting, for Africa is the continent with, overall, the poorest circulation system. In part, this is a result of the practices of the colonial powers, who developed multiple and unconnected circulation systems designed to tap the human and natural resources of the areas under their control, to facilitate public administration, and to maintain contact with the home countries. When the colonies became independent, they found these facilities useful, of course, but quite inadequate for the development of modern States and modern economies. They were even less suited for the development of regional and subregional cooperation and normal intercourse. Even now, there is still no continuous overland route connecting the countries of West Africa from Senegal to Cameroon. Between Sierra Leone and Somalia, one still flies via Rome or Harare. And until recently, in order to place a telephone call between Accra in the former British Gold Coast (now Ghana) and Lomé in formerly French Togo, only 125 km apart, it was necessary to route it through Paris and London!

Circulation, in fact, is part of the infrastructure absolutely essential to any economic development program and is being strongly emphasized in Latin America and Asia as well as Africa. But as circulation improves in these regions and worldwide, time and distance will shrink still more rapidly every aspect of our lives, not least the political aspects. No place in the world can any longer remain safe or isolated. Even now there is no place on earth where one can go to "get away from it all," for intercontinental ballistic missiles can reach any place from any other place, and the entire Earth is constantly being surveyed and recorded by remote sensing devices mounted in orbiting artificial satellites.

Who can predict the long-term implications of a world so seemingly shrunken by technology that almost literally everyone will be everyone else's neighbor? Can our present multitude of "sovereign" States survive? Or our multitude of languages? Or belief systems? Or political systems? Will the human species become homogenized culturally and intellectually as well as physically? If so, what are the implications of these changes? If not, what will prevent them? And what will be the implications of a failure of the species to adjust its thinking to its technology? Would it be better to adjust our technology to our cultural and ethical values? We cannot answer these questions, only ask them. But we can say with absolute assurance that inevitably we are all being drawn closer together, we are becoming steadily more interdependent, and we must begin thinking of ourselves as members of one species and citizens of one world.

Chapter 35

THE POLITICS OF POPULATION, MIGRATION, AND FOOD

Ever since Thomas Robert Malthus published his celebrated essay on population in 1798, population has been of greater interest, if not concern, to people in many walks of life around the world. Population geography has been a recognized specialization within our discipline for some time, and the newer field of demography is even more specialized. Population is important in political geography also, and political geographers must be concerned with matters in addition to population density and distribution. Thus, there are demographic questions to consider, such as literacy and education, health and age structure, skills and abilities. All have something to do with the people's mobility, awareness, and political consciousness.

POPULATION

Population Quality

The problem of illiteracy has been a major one facing the emergent ex-colonial States, for it is impossible to transfer allegiances from tribe and region to nation and State without the tools of understanding. Thus, many of these emergent States are spending much of their revenue on education. Unfortunately, the consequences are not always salutary; increased education in the village often causes an emigration from rural areas to the urban centers, where thousands of unemployed must be ac-

commodated, for jobs are few. These unemployed form a major political threat, and several countries have taken steps to limit migration to the cities.

Of equal or even greater importance is the health condition of the population, and this question is related directly to that of food supply. Comparisons between Western European States and some Asian countries in terms of life expectancy, daily calories of food available, dietary balance, and incidence of disease quickly indicate why available energies differ in quantity and are expended in different directions. A population that is largely engaged in day-to-day survival under a subsistence form of agriculture cannot be expected to make any great contribution to the State, economically or otherwise. The struggle for food is a major factor in political geography, affecting as it does the internal condition of the State as well as its international relationships.

When we establish a nation-state "model," we argue that the nation-state must possess a people of sufficient size and quality and consider themselves to be a nation. Our references to quality include considerations of health and well-being, education and skills. But other qualities have much to do with the forging (or failing) of a nation. We in North America, and especially in the United States, only need to look around ourselves to see the detrimental impact of racial and related inequities on this nation. Millions of people have come from Eastern and Western Europe to become U.S. citizens in the real sense of the word, emotionally as well as legally. But the people who came from other continents, especially Africa, have been assimilated far less completely and effectively. This is the sort of situation that fosters what might be called "retribalization," and there are examples in world history to emphasize the dangers to any State inherent in such a situation.

One manner in which the population of any State can be quickly characterized is by means of the age/sex pyramid. The figures indicate the number of persons within various age groups for both sexes living in the country at a given time so that the ratios can be recognized at a glance. Although these pyramids do not reveal anything regarding the spatial distribution of the people involved, they do indicate much concerning longevity, changing male/female ratios (which can be severely disturbed by war or migration), and perhaps the effects of epidemics. They do not provide much help in detail, but they do form

one of those tools used by the political geographer who wishes to gain an insight into the problems of any State as a whole.

Population Growth

World population, which presently totals well over 5 billion, is undergoing a rapid expansion in which several distinct stages can be recognized. Various States are at different stages in this process, known as the *demographic transition*, which are closely related to the introduction of modern amenities in such fields as health and education, and are reflected by the State's degree of urbanization, industrial development, literacy rates, and so forth.

The demographic transition functions essentially in this manner: during the first stage (in which most of the world found itself prior to the revolutions of the late 1700s) both birth rates and death rates are high, resulting in a fluctuating population that increases, but slowly. In time of famine and plague, severe setbacks occur, and then the population gains for a period of

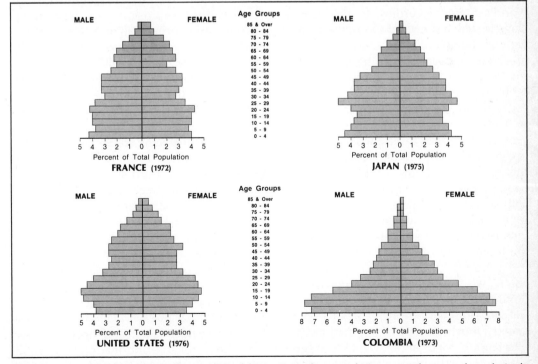

Age-Sex pyramids. Note the effects of World War II on the population growth rates, the relatively large number of older people in France and the "population explosion" in Columbia.

time. During the second stage, as various aspects of development take effect, death rates are reduced. Sanitary standards are raised, diseases and epidemics are fought successfully, food production and distribution improve. The result is an ever-increasing excess of births over deaths (the *demographic gap*) and an "exploding" population. This is largely true because while health conditions get better, socioeconomic conditions remain virtually unchanged. It is still desirable to have a large family to help with the farm work and to contribute income. In many colonial areas, the Western world brought better hygienic conditions, but did not stimulate social change. The third stage, as exemplified by most of Western Europe and North America, is one of renewed stability, and "leveling off" of the rate of increase. As people become educated and urbanization and industrialization take effect, there are social and financial restraints to having a large family. Thus both birth and death rates are very low and population growth is again very slow.

Although it is difficult to classify the world's States according to the growth stage in which they find themselves—and various regions of States may actually be at different stages in the transition—the matter is, nevertheless, a crucial one in political geography. It is probably true that every group of people who became united in a State—either in recent centuries in Western Europe or thousands of years ago in the Middle East—underwent changes in this respect. Probably the most dramatic changes ever to occur took place as a result of the technological revolutions of Europe, but it is worth remembering that rapid technological progress also took place in the Indian States of the Americas and the indigenous States of Africa. Unfortunately, we have no censuses to prove the point, but those early heartlands of political development probably also experienced a mushrooming of population. The Inca Empire, for example, was able to send thousands of residents from the central parts of the State to newly incorporated

outlying districts, with the aim of pacifying the local peoples and introducing modern agricultural techniques. If the conditions of the first stage of the demographic transition had prevailed in that region at that time, such emigration could not have been sustained.

When today a State experiences the impact of modern change, of "development," are the changes that occur similar to those recorded in Europe after 1700? Can the lessons of geography and politics be applied in the presently emerging world? The answer may well be no. Those States that are today still in the first stage of the demographic transition (some African and Asian States) are in many ways unlike preindustrial Europe. Population densities are higher today than they were three centuries ago, so that these emergent States have a different point of departure. Europe's mortality rates declined slowly as the changes caused by the Industrial Revolution took effect; the Western contribution to its colonial realms brought far more sudden declines in death rates. Thus, the rates of population increase (the "explosion") in developing, urbanizing, industrializing Western Europe in its second stage were probably less than they are and will be in many developing States now experiencing these conditions. If these arguments are valid (that conditions during Western Europe's first and second stages both differ importantly from the same stages currently prevailing elsewhere), then there is reason to believe that the third stage in, say, India may not be identical to that of France or Great Britain. In other words, the thesis that the ultimate fate of the population growth pattern is stability, based on the European experience, may well be invalid.

Demographic data are vitally important to political geography, but they also present some of the greatest problems. In many parts of the world, few or no censuses have been taken, and those that have may be unreliable and inadequate. In order to evaluate such factors as those just discussed in any consideration of a State, such as Pakistan or Nigeria, detailed information regarding wage levels, schooling, terms of residence in towns and villages, and many other data are necessary. More often than not, these data are not available, and interpretations and conclusions become subjective and doubtful. Clearly, if errors of great magnitude can prevail in the relatively simple matter of counting heads, the reliability of any other data concerning, for example, wages, taxes, religious preferences, and languages spoken, must be open to question.

POPULATION POLICIES

Several States have population policies that may reveal much concerning their internal demographic conditions. Those States that have expansive population policies desire an increase in the growth rate, and reward large families by tax reduction and even elimination, birth premiums, loans and allowances, and even public praise. The Soviet Union, which desires growth in view of the vastness of its territory and the need for eastward spread of effective occupance, is an example. Bulletin boards in Soviet cities often display photographs of local women who have given birth to their tenth child, with a commending caption and an announcement of an award. The German Democratic Republic (East Germany) has long been hampered by a labor shortage and has adopted a number of policies designed both to bring more women into the labor force and to encourage them to bear more children. Many excellent and cheap nursery school/day-care centers; generous maternity leaves at full pay (a year for the second and subsequent children); and a system of baby bounties, including $500 bonuses for each child, monthly allowances, and loans of up to $2500 that do not have to be paid back if enough children are produced—these are the major features of these efforts. Canada, Israel, and France are other countries that have been pursuing "pronatalist" policies.

Restricting Growth

While some countries pursue expansive population policies, encouraging births and immigration, others pursue *restrictive* policies, encouraging limitation of births and even in a few cases, such as that of the exceedingly densely populated Netherlands, encouraging emigration. Unfortunately, there is not yet sufficient correspondence between the States with restrictive policies and those that need them. A cursory glance at current population figures and trends is sufficient to reveal that population growth is most dramatic—and most alarming—in the poorer countries of the world and the poorer segments of the populations of the richer countries. That is, the population is growing most rapidly in those areas that can least afford to support more people. It is as true now as it ever was that, in the words of the old song, "the rich get richer and the poor have babies."

During the period 1970–1976, according to the 1977 UN *Statistical Yearbook*, the population of the world grew at an average annual rate of 1.9 percent. As usual, the average masks the true range of figures that spread from a low of 0.4 percent in Northern Europe to a high of 3.3 percent in Central America. A growth rate of 1.9 percent is serious enough, but a rate of 3.3 percent is staggering, and even this average hides local growth rates of as high as 3.9 percent per year. And when we consider that some States and parts of others are still in the first stage of the demographic transition—that is, where the death rates are still very high (on the order of 3.0 percent)—we face the frightening prospect of enormous populations when the death rate drops to levels approaching those of the industrialized countries (on the order of 0.9 percent).*

A most unfortunate coincidence in this period of history occurs between skin color and wealth. With the major exception of

*As a point of reference, we might keep in mind that a population growth rate of 3.0 percent per year will result in a nineteenfold increase in a century.

Emotional resistance to family planning. A slogan painted on the wall around a derelict house in Kingston, Jamaica reads, "Birth control; Plan to kill blacks." The picture was taken in 1974. Thirteen years earlier, the author saw a slogan nearby that read, "Birth control is a plot to kill Negroes." The fear persists despite semantic changes, but in Jamaica, as in most developing countries, family planning is gradually being accepted by large segments of the population as both necessary and desirable. (Martin Glassner)

Japan and a few minor exceptions, the societies with the highest per capita incomes and the lowest population growth rates are predominantly light-skinned, whereas the rapidly growing poor populations are predominantly dark-skinned. This has led many in the latter to reject the urgings of the former that they take urgent and effective measures to limit births, viewing this as demographic warfare, even genocide. There is a tendency to view population growth as an asset in the struggle to redress the balance between rich and poor and to overcome the disabilities deriving from colonialism.

Fortunately, this attitude has begun to change. Perhaps it was the example of Japan that stimulated the change when that country demonstrated incontrovertibly the spectacular gains to be achieved from vigorous action to limit population growth. Perhaps it was the sudden realization in such pronatalist countries as Brazil and China that the growth of population was not only gobbling up the economic gains achieved through great effort and sacrifice, but it was even threatening to reduce an already low level of living. Perhaps it was the erosion around the world of religion's hold on the minds and habits of people. Very likely it was a combination of all these factors and others that resulted in a slowing of the world's population

growth rate in 1977 for the first time in memory. Yet even here, the slight drop was due largely to the reversal of China's population policy, and in 1978 China experienced setbacks in her new campaign to reduce the birth rate.

Still, it is politically as well as sociologically significant that most populations that have recently experienced sharp and perhaps permanent drops in their birth rates are both poor and dark-skinned: Mauritius, Singapore, Hong Kong, Taiwan, Barbados, Trinidad and Tobago, Puerto Rico are outstanding in this respect, along with Chile, Argentina, Costa Rica, and Uruguay, all of which are less poor and less dark. Other countries that have made notable progress in this area are Sri Lanka, South Korea, and Cuba. All this is encouraging, but it is sobering to observe that these are all small countries whose combined population is little more than 100 million. We cannot be sanguine about the population problem until we can see comparable—and sustained—progress in such countries as China, India, Indonesia, Pakistan, Brazil, Mexico, Nigeria, and Egypt.

Nor are population problems confined to the poor countries. Even rich and middle-class countries contribute to the excessive demands we are placing on our abundant but finite store of "natural re-

Family planning campaign in a developing country. This billboard outside Port-au-Prince, Haiti is typical of government efforts throughout Africa, Asia, and Latin America to encourage couples to have no more than two children. These campaigns have had varying degrees of success. (Martin Glassner)

sources." After all, poor people, for all their numbers, do not consume very much, while the United States, with some 5 percent of the world's population, consumes about 50 percent of the world's goods and services. This situation raises moral questions as well as political ones. It cannot be sustained forever, and possibly not for very much longer.

International Migration

A few countries are stimulating population growth by encouraging immigration. Today such countries as Brazil, Canada, Israel, and Australia are actually subsidizing immigrants who are expected to become permanent residents. Of course, such immigration is highly selective. No country except Israel (and certainly not the United States) as a matter of policy is willing any longer to welcome "your tired, your poor, your huddled masses yearning to breathe free, the wretched refuse of your teeming shore," as Emma Lazarus so poignantly expressed it in her sonnet inscribed on the base of the Statue of Liberty. The immigrants most desired today are those who are educated, healthy, skilled, financially independent, and easily assimilable; that is, precisely the ones most needed in their own countries.

Most countries, as we have seen, have heterogeneous populations and typically the various ethnic groups within the population display differential rates of growth. Not infrequently, governing elites will attempt to improve their demographic position by forbidding intermarriage among groups, by discrimination of various kinds against the less favored groups, ethnically selective immigration policies, and even, in extreme but by no means rare cases, genocide. The concerns of the rulers about being outnumbered by rival ethnic groups are based on solid demographic, cultural, and historical facts. Flemings have multiplied faster than Walloons in Belgium, for example, and now outnumber them; Muslims are no longer a minority in Christian-dominated Lebanon; and the Slavic rulers of the Soviet Union are perplexed by the rapid growth of the Muslim population of Soviet Central Asia.

Both the Rhodesian and South African governments, representing very small minorities of their populations, attempted to strengthen their positions by encouraging immigration of white Europeans. In both cases the effort failed, in part because of potential immigrants' reluctance to enter a potentially dangerous situation. The South African government virtually abandoned its policy after it discovered that the overwhelming majority of new immigrants were joining the English-speaking rather than the Afrikaans-speaking sector of the population and held distinctly more liberal views than those of the governing elite.

A policy adopted by some countries to ease what appear to be temporary or spot labor shortages is that of the importation of temporary or specialized workers. This situation has been most prominent in the past few decades in Western Europe and Southern Africa, where many millions of people are involved. The recent economic recession in Europe, however, has prompted a recession of the tide of migration, back to Turkey, Greece, Yugoslavia, Italy, Spain, Portugal, back in many cases to joblessness and despair. Smaller movements of this type in this century include those of Basque sheepherders and Mexican farm laborers into the United States, Jamaican laborers to Panama to build the canal, Pacific islanders to Australia to harvest sugar cane, and Egyptian teachers to the Persian Gulf States. This type of movement was much more common in the nineteenth century when, for example, Italian "golondrinas (swallows)" migrated seasonally between Argentina and Italy to work on farms during the Southern and Northern Hemisphere growing seasons.

Selected immigrant groups are also invited to settle in sparsely populated areas of a country, to help expand the ecumene and contribute to the economy. Such movements, generally very small, are typified by Okinawans and Japanese settling in the eastern lowlands of Bolivia and Mennonites in the Gran Chaco of Paraguay and the Mexican state of Chihuahua.

In the past 30 years or so, while international migration for farm labor and agricultural colonization has remained stable or even declined, migration across international boundaries for temporary work in factories, mines and other industrial settings, and in other nonagricultural occupations has vastly increased. And it has become more politicized. We have already mentioned the huge temporary worker programs in Europe and Southern Africa, but they deserve more attention.

In the prosperous countries of postwar Northwest Europe, labor shortages developed quickly following reconstruction. Individually, countries began developing programs for bringing unemployed or underemployed workers from the poorer areas of Southern Europe, North Africa, and Turkey to as far north as Sweden. Some of the workers became modern *golondrinas,* migrating northward for short periods of work and returning home seasonally or when the crisis in the factory had passed or when they were displaced by machines or when they had earned and saved enough to meet their immediate needs—and doing this repeatedly. Others, however, particularly Arabs in France and Turks in Germany, settled down for longer periods, perhaps permanently, sending for their families or starting new families in their new homelands. These are the people who have been hurt most by the economic hard times of the 1980s, and these are the ones who have had the greatest cultural, as well as economic, impact on the landscape. Mosques have been built in Germany, whole neighborhoods in major cities now house Muslim workers and their families, and some social frictions have developed. In France, Switzerland, and other countries, antimigrant demagogues with unconcealed racist attitudes have been urging expulsion of the swarthy foreigners and these proposals have become major political issues. One wonders what will happen in these countries when their economies again expand rapidly and again there is a need for swarthy labor.

In southern Africa the situation is similar but different in important ways. Here the Republic of South Africa is the giant magnet attracting desperate workers from all of the poverty-stricken countries of the region. In mid-1985 there were nearly four hundred thousand of them officially registered, with possibly four times that number in the country. Unlike Europe, South Africa at least tries very hard to control this migration strictly. Migrants are forbidden to bring their families, must live in barracks provided for them, work at specific jobs for specified periods, have no hope of settling permanently or of entering the mainstream of society or even of getting any education or advanced training. Little or nothing is done to reduce tribal, ethnic, or national frictions among the migrants. The money they send home is not just beneficial, it is essential to their families and their countries' economies as well. The threat of expulsion of these foreign workers in the event of imposition of sanctions is a powerful weapon in the hands of the government. The economy of Lesotho would probably collapse, those of Mozambique and Malawi would be crippled, and Botswana, Swaziland, and Zimbabwe would be badly hurt. Of course, South Africa itself would inevitably suffer from the sudden departure of most of the people who do the dirty, hard, and dangerous work, but the cost-benefit ratio of such expulsions is almost impossible to calculate.

Currently, there are other significant international labor migration flows. Among them are those into Argentina from all of her neighbors, into the United States from Mexico and the Caribbean, and into Senegal from Mali and Mauritania. Just as the rise in oil prices in the 1970s triggered a recession in Europe that generated an outflow of temporary workers, so the collapse of oil prices in the 1980s triggered recessions in many of the oil-exporting countries that forced migrants back from Venezuela to Colombia, from Nigeria to a number of countries to the north and west, and from Saudi Arabia and the Gulf States to many Arab countries, Korea, the Philippines, and South Asia. When oil prices rise again, will the flow again be reversed? Is this the best way to provide needed workers? Is economic efficiency worth the political and social costs?

Not all migrants from poor countries to rich ones, however, are uneducated, unskilled workers. The "brain drain" has been a major concern of many developing countries since the late 1940s, and it is constantly growing. Britain, France, Germany, Canada, Australia, and especially the United States are attracting—and welcoming—educated and trained people from developing countries who are seeking higher salaries, better working conditions, better research facilities, more opportunities for advancement, and perhaps more political freedom. Even the character of labor migration from Puerto Rico and Mexico to the United States has changed as more managers, skilled technicians, and professionals migrate northward and blend readily into the general population. Recent immigrants from many Asian, Latin American, and even African countries now occupy important positions in North American and European laboratories, hospitals, computer centers, universities, and corporations.

This form of "foreign aid" to the United States and other rich countries from some of the poorest countries in the world is rarely discussed in the public media or the professional geographic literature. It has been of concern, however, to the United Nations, which has created an Inter-Agency Group on Reverse Transfer of Technology to consider its nature, scale, and effects and to recommend ways of dealing with it. We would do well to do likewise.

Refugees

Not all migrants who cross international boundaries do so voluntarily. The history of the world from biblical times on is repeatedly punctuated by horrific tales of mass expulsions of peoples, of people forced to flee from famine and pestilence and natural disasters. Whole populations have been uprooted by war, revolution, and boundary changes, fleeing or being driven to foreign and sometimes hostile lands. Never before in history, however, have we experienced such a lengthy period of massive and sustained refugee flows as we have in the twentieth century. To recall but a few of them: the partition of India; the breakup of the Ottoman Empire; the creation of Israel and Bangladesh; the civil wars in Sudan and Nigeria; the independence of Algeria; the westward migration of Poland after World War II; Stalin's eastward evacuation of ethnic minorities; the outpouring of political refugees from Hungary in 1956, Cuba in the 1960s, and Indochina in the 1970s; Paraguayans settling in Argentina, Ugandans in Kenya, Jews from Arab countries in Israel; the millions of people forced back and forth across Europe by the Nazis and the dazed survivors who were called "displaced persons." The list can go on and on, and is likely to extend far into the future.

Not only is the magnitude of refugee flows far greater than ever before in history, but they are no longer simply incidental to the climactic political events or natural disasters that generate them. Ref-

The flow of refugees never ceases. This man was one of more than 190,000 people who fled from Hungary to Austria and Yugoslavia during the winter of 1956–1957 after the Soviet Union had crushed the new liberal Communist regime established by Hungarians. He was brought to the United States, along with many of his compatriots, under a special program for political refugees, but he eventually settled in Canada. (UN)

ugees have always evoked some measure of sympathy and support, and, more recently, organized humanitarian aid, but what is needed now is detailed study of the geographic, political, economic and social causes of refugee movements, their nature and characteristics, and their effects on the sending and receiving countries and on the global community as well. Here we can only give a brief introduction to the subject.

The first problem we face is a definition of the term "refugee." According to the 1951 UN Convention Relating to the Status of Refugees and its 1967 Protocol, a refugee is

any person who, owing to a well-founded fear of being persecuted for reasons of race, religion, nationality, membership of a particular social group or political opinion, is outside the country of his nationality and is unable or, owing to such fear, is unwilling to avail himself of the protection of that country, or who, not having a nationality and being outside the country of his former habitual residence, is unable or, owing to such fear, is unwilling to return to it.

This definition, with some variations, has been adopted by a number of countries, including the United States, but it is far from satisfactory. How does one, for example, define *and* prove "well-founded fear" or "persecuted"? Is it really necessary for every single member of a group to prove that he or she is in fact a "refugee"? Much more important, however, this definition does not include people escaping war and civil strife or those who do not cross international boundaries. The Organization of African Unity adopted in 1969 a Convention Governing the Specific Aspects of Refugee Problems in Africa that includes the first category but not the second; even in Africa, for the most part, a person does not become a refugee until he or she crosses a border. There are many other definitional problems, but the important point is that *all* such definitions are inherently political.

Even if we confine our consideration to

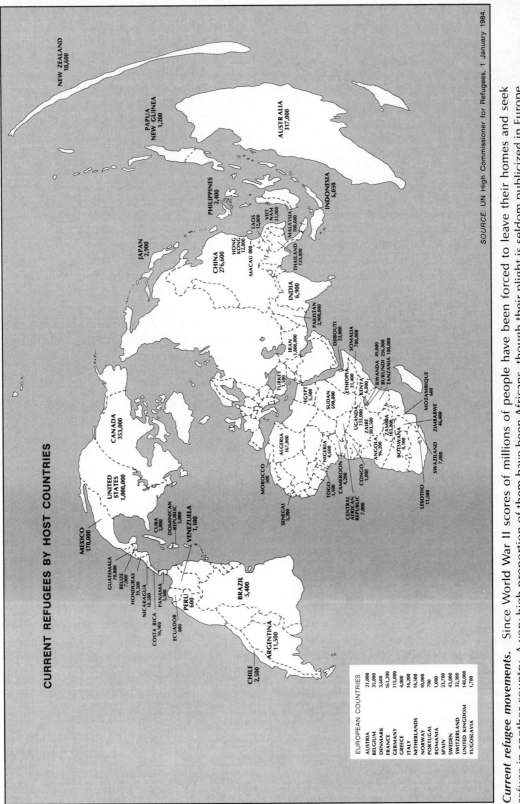

CURRENT REFUGEES BY HOST COUNTRIES

NEW ZEALAND
10,600

PAPUA
NEW GUINEA
1,200

AUSTRALIA
317,000

INDONESIA
6,050

PHILIPPINES
2,400

VIET
NAM
21,000

LAOS
12,000

MALAYSIA
100,000

THAILAND
133,000

HONG
KONG
12,800

MACAU 800

JAPAN
2,900

CHINA
276,600

INDIA
6,900

PAKISTAN
2,900,000

IRAN
1,800,000

DJIBOUTI
23,000

SOMALIA
700,000

TURKEY
1,500

EGYPT
5,500

SUDAN
690,000

ETHIOPIA
31,400

KENYA
6,800

RWANDA 49,000

BURUNDI 256,000

TANZANIA 180,000

MOZAMBIQUE
600

UGANDA
133,000

ZAIRE
303,500

ZAMBIA
103,000

ANGOLA
96,200

BOTSWANA
4,100

ZIMBABWE
46,400

SWAZILAND
7,000

LESOTHO
11,500

ALGERIA
167,000

MOROCCO
500

NIGERIA
4,600

TOGO
1,500

CAMEROON
4,250

CENTRAL
AFRICAN
REPUBLIC
7,000

CONGO
1,000

SENEGAL
5,200

CANADA
353,000

UNITED
STATES
1,000,000

MEXICO
170,000

GUATEMALA
70,000

BELIZE
7,000

HONDURAS
19,500

NICARAGUA
18,500

PANAMA
1,500

COSTA RICA
16,900

CUBA
2,000

DOMINICAN
REPUBLIC
5,000

VENEZUELA
1,100

ECUADOR
800

PERU
600

BRAZIL
5,400

ARGENTINA
11,500

CHILE
2,500

EUROPEAN COUNTRIES	
AUSTRIA	21,000
BELGIUM	35,000
DENMARK	3,600
FRANCE	161,200
GERMANY	115,000
GREECE	4,000
ITALY	14,200
NETHERLANDS	14,500
NORWAY	10,000
PORTUGAL	700
ROMANIA	1,000
SPAIN	23,700
SWEDEN	43,000
SWITZERLAND	32,300
UNITED KINGDOM	140,000
YUGOSLAVIA	1,700

SOURCE: UN High Commissioner for Refugees, 1 January 1984.

Current refugee movements. Since World War II scores of millions of people have been forced to leave their homes and seek refuge in another country. A very high proportion of them have been Africans, though their plight is seldom publicized in Europe or North America. The map shows only refugees so classified and registered by the UNHCR. Several million more are being assisted by UNRWA and national programs.

genuine political refugees, ignoring those who flee economic deprivation and natural calamities, we face other problems. A refugee loses this status, for example, when once given citizenship in another country or when he or she returns to the home country. More important is the fundamental UN principle of *non-refoulement*, the right of a refugee not to be forcibly repatriated to the country of origin. States of asylum are often all-too anxious, often for domestic or external political reasons, to get rid of refugees as rapidly as possible and ignore this right. Then there are the questions arising when persons flee *anticipated* persecution or violence without having actually suffered it, when refugees are interdicted en route to a country of refuge and turned away, or when refugees are interned for political reasons or to discourage further influxes of refugees.

Probably the first major refugee flows in modern times resulted from the expulsion in the late fifteenth and early sixteenth centuries of the Jews and Muslims who had been living in Spain for centuries. This was followed by the expulsion of Protestants from what is now Belgium, of Scots and Irish and heretics from the British Isles, of Huguenots from France, and many more. More recently, wars of independence inspired by the nation-state ideal have generated numerous outflows of people, before, during, and after the wars. After the success of the American War of Independence, for example, thousands of British Loyalists fled to Canada, the Bahamas, the West Indies, and elsewhere to continue living under the British flag. In our century, similar and more massive flows followed the creation of independent Indonesia, Algeria, Kenya, Vietnam, Mozambique, and Israel among others. While decolonization has largely been completed, its legacy lingers on, and parts of Africa and Asia are still reeling from the aftershocks.

Whether in Central America, the Horn of Africa, Pakistan's Northwest Frontier, or anywhere else, the problem of the dis-

position of refugees is difficult, largely because of political considerations. Basically, refugees can be removed from that status by any of three means: voluntary repatriation to their countries of origin, integration into the countries of refuge, or resettlement in third countries. All of these depend on political decisions by governments with many domestic and external considerations influencing them. Meanwhile, the refugees need help.

The two intergovernmental organizations with primary responsibility for refugees are the Office of the United Nations High Commissioner for Refugees (UNHCR) and the Intergovernmental Committee for Migration (ICM), both with headquarters in Geneva. Their mandates are different and complementary, and they generally work as a team. The UNHCR is chiefly responsible for protecting the legal rights of refugees and providing them with physical safety and security, assistance with documentation and supervision, and coordination of emergency assistance, including education and training programs, but it is not itself an operational agency. Instead, it relies on other agencies in planning and implementing programs. These include many of the organs and specialized agencies of the United Nations, the OAU, and the OAS as well as numerous private voluntary organizations, of which the most important are the International Committee of the Red Cross and Red Crescent, and the League of Red Cross Societies.

The ICM implements programs of voluntary repatriation and third-country resettlement. Its mandate is broader than that of the UNHCR in that it may assist persons not classified as refugees under UNHCR definitions. Thus, it aids asylum seekers, economic migrants, labor migration, family reunion, return of individual self-exiles, and transfer of highly qualified personnel among its Latin American member countries. In 1974, it also launched a "return of talent" program to help reverse the brain drain.

International agencies, however, can-

These Kampuchean refugees were learning trades and skills at the Galang Processing Centre in Indonesia in 1984 and earning some money while learning. (UNHCR/R. Burrows)

not solve refugee problems; only governments can. They can do this in two ways, first, by helping refugees *in situ* and by terminating their refugee status by the three methods described above. It is noteworthy that while the United States is the world's largest helper of refugees in absolute terms, both in terms of money expended on refugee aid and of refugees accepted as immigrants, its contributions as a proportion of GNP and as a proportion of either total immigration or of total population, respectively, are miniscule. This is illustrated in Table 35.1 and Table 35.2.

Second, and far more important than

refugee relief, is elimination of the causes of refugee flows in the first place. This is far more difficult and costly in political terms than sending food and medicine to refugees or even inviting them in as permanent residents. Not only are governments failing to consider potential refugee problems when they undertake military action, for example, whether within or across their boundaries, but they often deliberately uproot innocent civilians and evacuate them or simply drive them out of their homes. Even domestic security laws, conscription, persecution of minorities, and suppression of rebellions cause peaceful people to become refugees, and these conditions can only be rectified by governments. To date there is no evidence of a firm commitment by States to prevent refugee problems. As we said earlier, it looks as if we will be experiencing them far into the future.

Table 35.1 Refugees Resettled and Persons Granted Asylum in the United States

Origin	1983	1984
Africa	2,490	3,396
(Ethiopia)	(2,274)	(3,037)
East Asia	39,853	53,148
(Vietnam)	(21,465)	(25,132)
(Kampuchea)	(13,163)	(20,551)
East Europe	13,605	12,391
Latin America	1,496	1,602
Near East and South Asia	12,296	13,611
(Iran)	(7,922)	(10,470)
Other	63	6
TOTAL	69,803	84,154

Table 35.2 Refugees Resettled and Persons Granted Asylum in West European Countries and Canada

	1983	1984
Western Europe	36,187	37,531
Canada	13,642	15,322
TOTAL	49,829	52,853

Source: World Refugee Report. September 1985. Department of State, Bureau for Refugee Programs.

Since many of the world's refugee problems, especially at the moment in Africa, are related to food shortages, we now turn our attention to the difficulty of feeding a hungry world.

THE PROBLEM OF FOOD

There are many people around the world who argue, sometimes very cogently and forcefully, that there is *no* population problem. Many deny that there is a food problem. They base their arguments on interpretations of available statistics (which frequently are incomplete or unreliable) if they are inclined toward quantification, or on simple but firm faith, whether the faith be in God, Marxism, Fate, or Science. Since it is most unwise to rely on either statistics or faith alone, we must try to understand at least the political aspects of population and food, using a combination of approaches. This kind of analysis leads us to believe that there *is* a world population problem and that it is both a cause and a result of a world food problem; in fact, both are really aspects of a single problem. But let us not be misled; the problem is *not* one of too many people or too little food in the world. It is far more complex than that, and here we can only point out some of its more important political aspects.

The United Nations recognized the linkage between the two when it convened in 1974 both the World Population Conference in Bucharest and the World Food Conference in Rome. In both these conferences political factors were prominent and subsequent UN attempts to follow up these conferences with actions have been limited or even frustrated by politics. We can perhaps understand the political aspects of population just described, but the politics of food are less well known and perhaps more difficult to comprehend.

World Food Supply

The 1974 World Food Conference took place in an atmosphere of grave concern.

Only months before, the OPEC oil embargo and quadrupling of crude petroleum prices had generated a near-disaster in agriculture as tractors ran out of gas, the prices of petroleum-based fertilizers and pesticides increased alarmingly, and many other products made from petroleum for use in agriculture became either too costly for farmers to buy or unavailable altogether. Combined with the lingering Sahel drought and adverse weather conditions elsewhere, the embargo and price rise led to localized famines and threats of much greater ones. The specter of Thomas Malthus was ominously present at the conference.

By 1976, however, the world food situation had improved, responding to the drawing down of food stocks for emergency aid, better weather, the UN's program of assistance to the countries most seriously affected by the oil price rise, and other factors. Malthus was forgotten by many. Governments, international agencies, academics, and even Lester R. Brown, one of the most perceptive, prolific, and dynamic observers of world ecological problems and currently President of Worldwatch Institute, were optimistic about our chances of increasing agricultural production greatly enough and rapidly enough to keep up with, or even get ahead of, population growth. But in 1978, the World Food Council reported that "beyond the forseeable recovery of food production in 1976 and 1977, little real progress has taken place toward achieving the [World Food] conference objectives." Then in 1979, Worldwatch Institute reported that worldwide per capita production of fish, wool, beef, oil, mutton, wood, and cereals "has leveled off or begun to decline" since 1960, and Lester Brown, who wrote the UN-financed report, argued for concerted efforts to curb population growth.

Optimism and pessimism about food production have alternated and intermingled just like good harvests and bad harvests, like favorable and unfavorable weather. The reasons are not hard to find. First, we really know little about such things as food production totals, human nutritional needs, food consumption, food preferences and taboos, and many other factors essential in any analysis or projection. Second, even what we do know is interpreted in widely different ways according to the background, attitudes, and motives of the interpreters. Third, projections of food sufficiency or scarcity are based on a variety of projections, for example, of population growth, technological development, and climatic change—all of which are decidedly imprecise. Finally, few analyses or projections give sufficient weight to political factors. Before reviewing some of these political factors it would be well to make a few very brief and very general observations about the present world food situation.

1. Through four-fifths of the twentieth century, total world food production has kept ahead of population increase—but only just.

2. This pattern has left a great many people in the world chronically hungry and many more (about 650 million) suffer chronically from malnutrition.

3. The carrying capacity of the planet is still very much greater than current production; that is, food production can still be increased substantially.

4. Experience has demonstrated that increasing food production substantially and permanently is a very costly and long-range process.

5. Even when a remarkable breakthrough is made, such as the development of high-yielding varieties of wheat in Mexico and of rice in the Philippines that introduced the "Green Revolution" in 1965, its impact is unlikely to be felt very widely, very soon, or for very long.*

*In his acceptance speech on receiving the Nobel Peace Prize in 1970 for his work in initiating the Green Revolution, Norman Borlaug, an American agronomist, emphasized that all he and his colleagues had done was "to buy a little time" and urged population control as the only real solution to the problem.

6. The most dramatic increases in food production have been in the industrialized countries, not in the largely agricultural developing countries; during the 1970s per capita food production actually *declined* in 69 (more than half) of the developing countries.

7. The proportion of foods exported by developed countries to other developed countries has been steadily increasing, so that now it accounts for more than half of all world food trade; poor countries are getting less and less of the food surpluses of the rich countries.

8. Extraordinary quantities—perhaps one quarter to one third in many poorer countries—of all food produced never gets to the people for whom it is intended because of spillage, spoilage, fouling and consumption by insects and rodents, fire and flood damage, and other factors.

9. Despite UN urging and efforts, there is still no worldwide program for the production, storage, and distribution of food, not even a world food reserve.

10. Food aid has never been adequate in either quantity or type and the situation is worsening. Food aid, moreover, is all too frequently used as a tool or a weapon for political purposes.

11. There are grounds for hope that we can solve the food problem. Among them are:

(a) The seriousness of the situation is at last beginning to be understood.
(b) Some developing countries have dramatically increased their food output recently.
(c) There is now an array of experienced agencies providing technical and financial assistance to agriculture.
(d) More and better fertilizers have become available recently.
(e) For the first time in history serious attention is being given to improving productivity of subsistence crops in the humid tropics.

(f) There have recently been great advances in the production of chemical and biological controls for some major diseases and insect pests.

(g) Recent improvements in transport and communications can be applied to the improvement of food production and distribution.

(h) There is increasing awareness of the need for integrated rural development rather than simply increases of food production.

12. The hope is yet to be realized and will not be realized until there is a firm and permanent *commitment* on the part of governments everywhere to reorder their priorities and cooperate to assure everyone on earth an adequate diet.

Food Policies

The last point is perhaps the most important, for it means far more than simply tinkering with gadgets and chemicals and budgets. It means profound changes in the political concepts and practices that may inhibit better food production and distribution, even more than a shortage of water, soil erosion, biological limits to the yield of cows and soybeans, or any of the other very real ecological constraints.

It means, for example, genuine land reform as part of a genuine program of agrarian reform. The deficiencies of the *latifundia* and *minifundia* systems that still characterize much of the agricultural world are well known. But land ownership in many countries is the basis of both economic and political power and those who possess such power do not give it up easily. We discuss land reform and agrarian reform in more detail in the next chapter; here we need only point out that they not only increase food production, but also stem the flow of people to the cities, expand the market for locally manufactured goods, and improve chances of developing a democratic society.

It means shifting much land, labor, and capital from the production on plantations of industrial crops and luxury foods destined for sale in the rich countries to the production of staple foods for the local people. But the plantation system is a vital and persistent element of colonialism that has brought wealth and power not only to the owners (usually foreign), but also to the local elites with whom they are allied.

It means shifting to less complex and more appropriate technology in areas unsuited for giant machines and potent chemicals. But this means less profit and power for the (usually foreign) manufacturers and (often local) distributors and service personnel.

It means reducing the number of middlemen and the amount of packaging and processing, all of which add substantially to the cost of food to the consumer without necessarily helping the farmer. But this again would weaken the positions of those who profit from high-priced food.

It means changing the systems of agricultural subsidies that prevail in the industrialized countries to new ones that would encourage more production rather than simply maintain high incomes for farmers. But this would shift political power and disturb entrenched bureaucracies.

It means developing a rational system for utilizing living marine resources so as to protect the ecosystem of the sea while still maximizing the production of high-quality protein for human consumption instead of for animal feed. But we have already seen that this is unlikely to happen very soon, if ever, as nationalism and exploitation continue to dominate our relations with the sea.

It means shifting financial, material, and human resources away from the production of luxury manufactures into research and development of food products, especially high protein and tropical foods. But are the American people, for example, willing to give up their electric carving knives and their motor boats and their hair dryers?

It means organizing food aid so that the right kinds of foods go to the right people in the right quantities in emergencies, with more emphasis on incentives for local food production where appropriate. But which countries are willing to cease using food aid programs as means to dump surplus agricultural products and obtain political rewards in return?

It means shifting the emphasis in the developing countries from rapid industrialization to improving food production. But how many corrupt and authoritarian leaders are willing to give up either the modern status symbols that swell their egos or the deals with (usually foreign) manufacturers that swell their bank accounts?

Perhaps the most essential and potentially effective method of increasing the food supply and of increasing the overall quality of life for all human beings would be to rechannel money, natural resources, time, and talent currently expended on military equipment and weapons and invest it in *homo sapiens*. We referred to this in Chapter Nineteen and take it up again in Chapter Thirty-Seven, but it bears repeating here.

Finally, somehow, governments must be educated or compelled to treat their own people decently during times of stress brought on by food shortages. The behavior of the Ethiopian government during the drought and famine that afflicted the Sahel and most of eastern and southern Africa during the 1980s was less than admirable and must not be repeated anywhere, ever!

One might be tempted to look for models for expanded food production and more equitable food distribution to the "socialist" countries, those with centrally planned economies. But the first of them—the Soviet Union—has still, 70 years after the revolution, not been able to achieve what we are suggesting. Marxism simply does not seem to work in agriculture, whatever virtues it might have elsewhere. Those socialist countries sometimes cited as models because they have been able to feed their people adequately, such as China, Cuba, and Hungary, are precisely the ones doing the greatest violence to Marxist principles. Cuba continues to export tobacco, sugar, and citrus while importing nearly half of its rice, its basic staple food. China has since 1949 emphasized food production and a tight organization for allocating the scarce resources necessary for high agricultural productivity. And in Hungary nearly all the agricultural land is privately owned and there is nearly a free market economy in agricultural goods. Finally, the Warsaw Pact countries alone accounted in 1983 for 25 percent of all military expenditures and 24 percent of all arms transfers, mostly to developing countries.*

This is not to suggest that socialism has been a total failure in agriculture. Certainly, the socialist countries can, on the basis of their own experience, contribute some useful ideas and techniques to the pool that will also draw from experts of all kinds in the industrialized market economy countries, from leaders in the developing countries, and even, perhaps especially, the dirt farmers themselves, including the peasant farmers of the poorest countries. Nothing less than a rational and sustained cooperative effort can bring population and food once more into balance. Despite the very real obstacles and handicaps, it can be done.

SIPRI Yearbook 1986, pp. 324 and 341.

Chapter 36

THE POLITICS OF ECOLOGY, ENERGY, AND LAND USE

There is little need to elaborate on the quite well-known fact that our physical environment around the world is being rapidly and perhaps irreparably damaged by intense and growing depredations of people. This degradation of the environment is the product of four factors we have already discussed: rapid population growth, increasing wealth leading to increased per capita consumption, ever more sophisticated technology, and accelerating urbanization. These trends are all interrelated and in varying degree interdependent. They are all, furthermore, influenced and shaped in some degree by politics—by positions, policies, and decisions based on political considerations.

Concern with the environment and its relation to politics is not new. We have discussed, for example, concepts of environmental determinism whose origins go back at least to Hippocrates and Aristotle. Ibn Khaldun, as we pointed out in Chapter One, theorized about the effect of the environment on the political units of his time and the life cycle of the State. Montesquieu, Ratzel, Kjellén, and Huntington are only a few others who have followed in this tradition. Toynbee, Wittfogel, and the Sprouts have more recently developed less deterministic concepts of "the impress of the environment on politics," as pointed out by Kasperson and Minghi.* Study of "the impress of politics

upon the environment" does not reach back nearly so far, but Ratzel and Huntington both recognized that it was important and half a century ago Whittlesey examined it in some detail. Kasperson and Minghi suggest that a study of this kind of relationship might be organized into four major components: "political goals, agents of impress, process and effects."* They also consider a third type of environment–politics relationship, "the public management of the environment." Here they review some of the work on this subject by Barrows, Colby, White, Burton, and themselves, all during the twentieth century. They organize their discussion under three main headings—environmental policy and planning, resource allocations, and spatial linkages and area repercussion. These are most useful concepts and can help us understand the three types of relationships as we examine three aspects of the environment–politics linkage.

ECOLOGY†

Before the mid-1960s, "ecology" was a word used commonly by geographers, biologists, and other scientists but seldom heard or seen by the general public. Since

*Ibid., p. 429.
†We prefer the term "ecology" to "environment" because ecology is the study of the mutual relations between organisms and their environment. When it becomes a public issue, however, or when linked with human activity, environment serves just as well, and we therefore use the terms more or less interchangeably.

*Roger E. Kasperson and Julian V. Minghi, eds., *The Structure of Political Geography*, Chicago: Aldine, 1969, pp. 423–435.

then it has been adopted by a new generation concerned about rapid worldwide destruction of our planet's environment. They have broadened and fortified the conservation movement and impressed the public with both the urgency and the practicality of protecting our physical environment from further destruction and actually reversing the trend and restoring the environment if possible to its original state. The euphoria and élan of the early days have largely disappeared, however, and the ecology movement has settled down into a persistent, dogged and frequently successful battle against entrenched interests, rigid thinking, and obsolete laws. Heartening also has been the spread of the new ecology movement from the United States to other countries.

In only one generation Americans have lived through an economy of scarcity during the Great Depression, an economy of abundance in the post-World War II period, and an economy of waste at present. Through it all they have never lost faith in the eternal bounty of nature and the virtue of exploiting it enthusiastically. Now we must shift gears and reverse all this, but every stage, every step involves a political struggle. In only the past quarter-century we have experienced titanic battles in the Congress and the press over the Glen Canyon Dam (Arizona), the Dickey–Lincoln project (Maine), the Cross-Florida Barge Canal, the Central Arizona Project, the Alaska Pipeline, and dozens of other proposals, large and small, for modifying our physical environment in order allegedly to improve our economic and perhaps social environment. We have come to expect government at all levels (but particularly the federal government) to subsidize projects of this nature, but we are still unwilling to accept their regulation and control by government. Worse, we have yet to develop a national consensus on environmental matters that can be expressed in a national plan or at least guidelines.

The Environmental Policy Act of 1970 was a landmark, a giant step in the right direction. It spelled out goals and policies to guide all federal actions that would have an impact on the quality of our environment. It made a concern for environmental amenities and values a part of the mandate of every federal agency. It established the Council on Environmental Quality to identify the policy issues and alternatives for environmental administration. Finally, it required an annual report on the quality of the environment. This act was followed by other, more specialized legislation, such as the Clean Water Act and the Clean Air Act, all the products of diligent and persistent efforts of citizens individually and collectively working with—and on—legislators.

These have helped rectify two of the three basic problems that had prevented the federal government from playing an effective role in long-term environmental planning. When long-range planning has been undertaken, it has generally been intended to deal with problems posed by projected trends rather than to achieve desirable goals, and public policies too often have been defined and carried out in fragmented, narrow programs by mission-oriented agencies. Now there is somewhat more order in both setting and reaching goals, but there is still considerable scope for improvement, both in the planning and in the execution of plans. Both are made more difficult than necessary, however, by the failure so far to deal with a third basic problem: the fact that public administration in general is geared to annual appropriations that tend to favor short-term considerations.

It would be difficult indeed to find someone who would speak forcefully in favor of deliberately destroying our environment. Yet many argue in favor of postponing decisions on ecological problems or insist that they are not really so serious or that other problems should be attended to first or that we really cannot afford to protect wildlife or clean up streams or restore land eroded by careless farming practices or reduce noise pollution. In fact, all these arguments have some merit and it is unwise to ignore them. Since we cannot do everything first or well or at all, we must make choices. These choices are, in part, moral—but they are largely political. Decision making in these cases is always difficult.

The types of decisions to be made fall into three categories.

1. *Priorities of resource allocation.* There is so much to be done. As abundant as our resources are, they are limited and must be distributed among the various tasks to be accomplished. Money, energy, talent, and time must be allocated on some basis other than simply greasing the squeaky wheels, yet we have not yet devised such a system.

2. *Distribution of costs.* We all recognize by now that nothing is free; everything costs, even clean water and air. But who is to pay for achieving and maintaining a livable environment? Ultimately, of course, we all pay for everything, but the real questions are whether we pay now or later; through higher taxes or higher prices; in cash or in kind or simply by foregoing luxuries and reducing our consumption of material things; and, of course, who is to bear what proportion of the costs?

3. *Distribution of benefits.* Should the benefits of an improved environment be distributed evenly throughout the entire country, through all socioeconomic levels and among all ethnic groups? That would seem to be very democratic but we may question whether it is practical or desirable. Should not special attention be given to the physically and mentally handicapped, to deprived minorities, to low-paid workers, and the unemployed? Should the wealthy get subsidized marinas and rural people beautiful parks?

Clearly there are no easy answers to any of these questions. Two contemporary but long-standing problems may serve to illustrate these points.

Strip Mining

There are at present over three thousand surface mines in the United States, spread widely across the midsection of the country from the Appalachians to the Rockies, most of them coal mines and most of them relatively small. Until adoption of the federal Strip Mining Control Act of 1977, regulation of strip mining was largely left to the states. Long after the severe ecological damage of strip mining had been amply documented, its control was spotty and inadequate. Kentucky, the country's leading producer of strip-mined coal, provides a good example.*

Kentucky's efforts to control strip mining have followed a most peculiar course. The period 1947–1967 was marked by a continual increase in the strength and scope of the regulatory effort which culminated in a tremendous burst of initiative in the years 1965 and 1966. In that short span of time, the control program was vastly expanded and improved. Stringent and detailed administrative regulations were promulgated. A new strip mine control law, considered to be the toughest in the nation, was passed. Aggressive enforcement of both the regulations and the statute was undertaken. The available geological data for those years indicate that the improved control program actually achieved a palpable reduction in strip mine-related ecological damage.

The strip mine control program reached a peak of effectiveness in 1967. Then, just as the rest of the country was commencing a period of unprecedented concern for environmental quality, it entered a period of decline.

In this case apparently neither pressure on the Kentucky political system by outside forces nor scandal or other crisis mobilizing public opinion was significant in achieving stricter controls. Rather, "the crucial political actor in the initiation of the program, as well as in its passage by the legislature and its successful implementation, was the Governor, Edward T. Breathitt." But his achievement did not

*The material on Kentucky is based on Marc Karnis Landy, *The Politics of Environmental Reform*, Washington: Resources for the Future, 1976. All quotations are from Chapter 1, pp. 1–16.

survive his term of office: "The deterioration [in strip mine control] was the result of the loss of political autonomy which resulted from electoral defeat suffered by the administration faction in 1967 and its subsequent failure to return to power in the gubernatorial election in 1971."

Because of the tenuous and uncertain nature of state control of strip mining, the Congress was finally persuaded, after decades of effort by ecologists and others, to pass federal legislation. It is too early to evaluate the effectiveness of the 1977 Strip Mining Control Act, but already there are reports of abuse and evasion of its provisions, notably illegal strip mining on federal land. But rigorous enforcement of legislation requires not only money but also sufficient evidence to permit successful prosecution when necessary, and this is often difficult to obtain in regions where strip mining plays such an important role in the lives of people. The problems are likely to increase, not decrease, if demand for coal rises to compensate for shortfalls in petroleum supplies or problems with nuclear energy.

Water Projects

One of the traditions of American politics since the founding of the republic has been the regular congressional appropriations to support what were once termed "internal improvements," more recently referred to as "rivers and harbors projects" and informally known as "pork-barrel" legislation. In April 1977, President Carter, less than three months in office, decided to break with tradition and slashed more than $7 billion worth of water projects from the budget. Despite the ample justification for most of these cuts on grounds of both economy and ecology, congressmen and senators from nearly all sections of the country raised a storm of protest. Their pet projects, those designed to benefit their constituents and win votes, were threatened, and they fought to protect them. In the end, there was a compromise and the president signed

a bill that still provided over $10 billion for public works, including some big and expensive projects he had opposed. The president and the public have learned how hard it is to overcome local demands for federally financed "improvements."

Ecology in Other Countries

Concern with environmental matters has been manifested mostly among the educated and well-to-do. Poor people have other and higher priorities. The same is true of States. Generally, as we have pointed out in our discussion of marine pollution, the poorer countries of the world have only recently begun to realize that their environments are also in danger and that ecological damage may well cost more than industrial development will earn. All over the world, major development projects are being reexamined in light of new understanding of ecological principles. The Jonglei Canal project in Sudan and the Trans-Amazonian Highway in Brazil are examples. Even Egypt's pride, the Aswan High dam, completed only in 1970, has already caused so much ecological damage in the Nile Valley and the eastern Mediterranean that there is talk of dismantling it. But the destruction goes on: the large wild animals of East and Central Africa are rapidly being exterminated, the forests of Southeast Asia are disappearing at a frightening rate, overirrigation continues to destroy cropland through salinization and waterlogging in North Africa and Southwest Asia, and soil erosion continues almost unchecked in the highlands of Latin America. It will be difficult to overcome the suspicion born of colonialism and foreign exploitation and convince the people in these regions that conservation is in their interest.

Even the Soviet Union in recent years has begun to take action on environmental matters, turning slightly from its customary emphasis on heavy industry and massive development projects. In early 1979, the main governmental agency on the environment was upgraded to the status of a State committee and intensified a six-year-old program of cooperation with the United States in 11 major environmental areas, including air and water pollution and earthquake prediction. They have issued a *Red Data Book* listing endangered species of plants and animals, are reducing chemical pollution, moving factories out of the central areas of major cities, removing sulphur emissions of hydroelectric plants, and in general demonstrating a serious commitment to ecology. While we have no detailed information, we might suppose that the political infighting that led to this commitment was at least as brisk as in the United States and is probably still going on.

ENERGY

The "oil crisis" of 1973–1974 suddenly and painfully brought to the attention of the American public what scholars, government officials, ecologists, and others had been saying for many years: that the supply of hydrocarbon fuels, although very large and still unknown, is limited, whereas demand generally rises at an accelerated rate; that the United States and other countries have become too dependent on such fuels; and that petroleum can be used as a political weapon. The spate of books and articles on the politics of oil that had appeared during the 1950s and 1960s was suddenly engulfed by a flood of new analyses and exhortations. The result has been a new line of thinking not only in the industrialized countries, but around the world, and new plans of action to forestall another such "crisis."

For the first time, Americans (and others) began thinking seriously about the total energy picture instead of isolated portions of it, about the folly of building a society based on the expectation of unlimited supplies of cheap energy, about the ease with which mighty States can be held hostage by a few countries poor in technology but rich in fuel. We are still in the midst of this rethinking and it may be a long time before a national consensus emerges on these issues. Meanwhile, we can observe some relationships between politics and energy.

In the United States, more than 90 percent of the natural gas and nearly 75 percent of the crude oil produced domestically come from only five states—Texas, Louisiana, Oklahoma, New Mexico, and Kansas. Despite the undoubted wealth and vaunted political power of the "oil lobby," the fact remains that Americans still enjoy huge amounts of relatively cheap energy. It is clear, however, that this situation cannot last long; it is much too fragile. The fall of the Shah of Iran early in 1979 initiated events still not clearly understood that led to severe shortages of gasoline in much of the United States reminiscent of 1973–1974. Such situations may well recur more frequently and more seriously until the country adopts and maintains a policy based on conservation of energy. We need an appropriate mix of fossil fuels, nuclear power, and such "unconventional" sources of energy as the sun, wind, tides, geothermal steam, alcohol, and thermal layers in the sea, and we must have a rational means of paying for energy.

It is probably both unwise and impractical to move toward autarky in energy, as it is in most other commodities. Despite the massive effort made by the federal government in "Project Independence" immediately after the "oil crisis," the United States is now importing a higher proportion of its crude oil and petroleum products than ever before. Oil shale and tar sands are still not viable sources of the fuel, exploration for oil off the coasts of New England and New Jersey has been disappointing, oil from Alaska's Arctic Slope has not had quite the impact that had been anticipated, and research in controlled thermonuclear energy has so far been unrewarding. Real energy security may well lie in international cooperation rather than competition. But other coun-

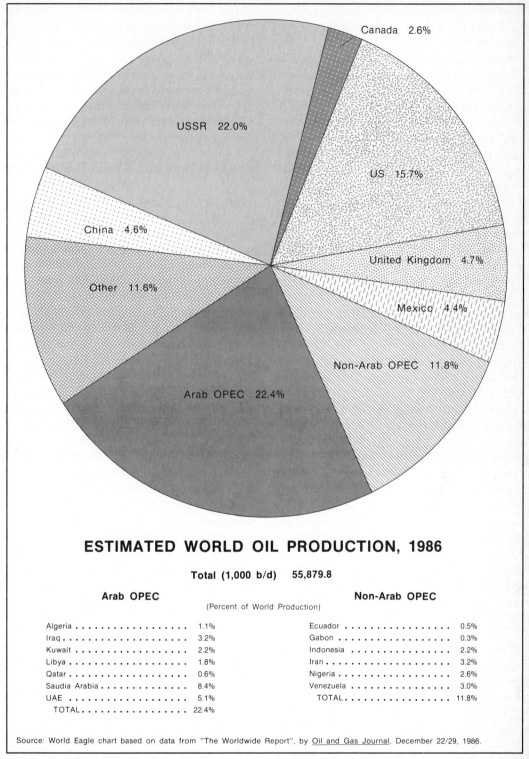

ESTIMATED WORLD OIL PRODUCTION, 1986

Total (1,000 b/d)   55,879.8

Arab OPEC		Non-Arab OPEC	
(Percent of World Production)			
Algeria	1.1%	Ecuador	0.5%
Iraq	3.2%	Gabon	0.3%
Kuwait	2.2%	Indonesia	2.2%
Libya	1.8%	Iran	3.2%
Qatar	0.6%	Nigeria	2.6%
Saudia Arabia	8.4%	Venezuela	3.0%
UAE	5.1%	TOTAL	11.8%
TOTAL	22.4%		

Source: World Eagle chart based on data from "The Worldwide Report", by Oil and Gas Journal, December 22/29, 1986.

Estimated world oil production, 1986. Note that the Soviet Union and the United States are still the world's largest producers. Production is currently being restrained because of relatively low demand, but this situation could easily change quite quickly.

tries' energy pictures are not the same as ours and both their goals and their policies will differ accordingly.

Western Europe, where the Industrial Revolution began, using first water power and then coal, is today much more dependent on external sources for its energy than the United States. Petroleum and natural gas under and around the North Sea have relieved this situation somewhat, and still newer reserves in the form of tar sands in France and heavy crude under the Adriatic Sea may in the future be important, but the basic energy dependence of Western Europe remains unchanged. Consequently, nuclear energy is proportionately more important there than anywhere else in the world, nearly all the hydroelectric potential has been developed, and the world's only large-scale tidal power project functions in the estuary of the Rance River in France. Nevertheless, Western Europe is still vulnerable to oil blackmail and the OPEC oil embargo of 1973–1974 did influence the policies of most States in the region (with the notable exception of the Netherlands) toward the Arab–Israel dispute, though not necessarily to the extent hoped for by the Arabs.

Japan is even more vulnerable, having a much smaller domestic energy stock, no friendly neighbors on whom to call for assistance (nothing like the electricity intertie system binding most of the continental Western European countries, for example) and located at the end of a very long oil supply line vulnerable itself to interdiction at a number of strategic points. Japan is thus, despite her close links with the United States, forced to pay some deference to the political demands of her principal suppliers of petroleum while trying to develop new sources of supply outside the Middle East.

The developing countries present a more complex picture. Some are exporters of petroleum and members of OPEC, and therefore do not have to worry about sources of energy. But as they invest their oil revenues in industrialization and modernization in general, they become more

dependent on imports of all manner of consumer and capital goods, technology, and skills obtainable only in the developed countries. At the same time, their domestic energy needs grow, making a smaller proportion of production available for export (barring a sharp increase in production that is unlikely because of the desire to conserve finite reserves). The result is likely to be decreased bargaining power and a restoration of something approaching a community of interests.

The non-oil exporting developing countries still present a varied picture. Some—the least developed countries—consume so little inanimate energy that for the present, at least, local supplies of wood, cow dung, agricultural wastes, coal, peat, and other traditional organic fuels are adequate to meet present needs. But these materials could also be used for fertilizers, chemicals, and industrial raw materials if other fuels were reliably available at reasonable prices. Other countries are well along the road to industrialization and, thus, more energy-dependent than the poorest countries. Others have attained middle-class status, largely on the basis of the export of agricultural and nonfuel mineral commodities. Both of these groups of countries have been hard hit by increasing oil prices but can get along because they have more financial resources of their own, better credit ratings, and more bargaining power than the poorest countries. Those most seriously affected are generally those whose infant industries, budding transport systems, and nascent urbanization are heavily dependent on foreign energy sources but whose economies cannot yet bear the new costs. These are the countries receiving the most aid from the United Nations, the OECD, and even some of the OPEC countries.*

* Consumer countries have formed organizations to counterbalance OPEC. The International Energy Agency, composed of 20 major consumers, mostly OECD countries, and the Latin American Energy Organization are trying to work out programs to conserve petroleum, use petroleum more efficiently, and develop alternate energy sources. Their prospects of success are still uncertain.

A factor that must be considered, though it cannot yet be evaluated, is the entry of new suppliers of petroleum and natural gas into the market. Mexico is already self-sufficient and has become an important exporter. Recent reports indicate she may have one of the world's largest oil reserves. Brazil, Vietnam, Egypt, and other developing countries may become important producers in the future, with inevitable though unpredictable political consequences.

We have emphasized petroleum here because worldwide it is, and is likely to remain for a long time, the world's most important fuel and the most important fuel in international trade. Coal and natural gas seem unlikely to increase substantially in international trade, although of course locally they may well become more important. Nuclear energy, despite the reevaluation undertaken around the world following the nuclear power plant accident at Chernobyl in the Soviet Ukraine in April 1986, will very likely play a greater role in the world's energy picture than it does now—though probably less of a role than had been predicted two decades ago. One reason for the caution, of course, is concern over the safety of nuclear energy; another is the enormous increase in capital costs of nuclear facilities. But caution is also warranted by the uncertainties of nuclear power. Neither commercial-grade uranium nor advanced nuclear technology is as abundant or widespread as fossil fuels and the technology necessary to utilize them. The possibilities of nuclear blackmail are therefore at least as great for the suppliers of either the fuel or the technology.

Here is still another example of the intricate interdependence of all the States and peoples of the world and of the need for worldwide cooperation rather than competition.

LAND USE

Increasing population is putting severe pressure on another finite resource: land.

Land ownership and land use have been matters for concern and study in many parts of the world for a long time but have never been more important than right now. At all levels, from the individual house lot to whole countries and regions, they present problems so difficult that at times one despairs of trying to solve them. They can be solved, probably, but not very soon and probably not very satisfactorily.

Zoning

One of the most interesting examples of conflicts over land use, and one attracting the attention of some geographers, is that of local zoning. Systematic land zoning within urban areas of the United States dates only from 1916, and even today many urbanizing areas have only rudimentary zoning systems or none at all. In some areas where urban sprawl has become a serious problem, the old zoning regulations have proven inadequate to cope with new situations. Suburbanization has tended to rob central cities of their centrality, thus undercutting the fundamental premise on which zoning was originally based. Now suburbs are themselves becoming central cities, surrounded not by other suburbs but by other central cities. The long-term implications of the suburbanization of the hinterland and the urbanization of suburbia are incalculable, though we have already referred to some of them in our chapters on civil divisions and special purpose districts. Here we can only introduce some specific problems relating to zoning in urban and suburban areas.

The concept of zoning is based on the notion that society in general and most individuals benefit from the separation of various types of land use, generally divided into industrial, business, and residential, and subdivided further according to local perceptions. Those who have lived in countries where zoning is unknown or rare, however, and have been able to live in quiet, charming neighborhoods within easy walking distance of employment,

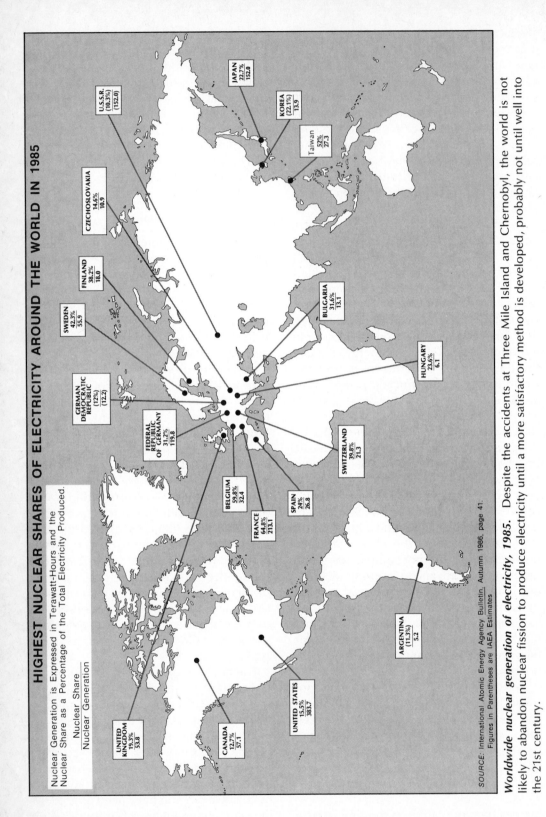

HIGHEST NUCLEAR SHARES OF ELECTRICITY AROUND THE WORLD IN 1985

Nuclear Generation is Expressed in Terawatt-Hours and the
Nuclear Share as a Percentage of the Total Electricity Produced.

Nuclear Share
Nuclear Generation

JAPAN
22.7%
152.0

KOREA
(22.1%)
13.9

Taiwan
52%
27.3

U.S.S.R.
(10.3%)
(152.0)

CZECHOSLOVAKIA
14.6%
10.9

FINLAND
38.2%
18.0

BULGARIA
31.6%
13.1

SWEDEN
42.3%
55.9

HUNGARY
23.6%
6.1

GERMAN
DEMOCRATIC
REPUBLIC
(12%)
(12.2)

FEDERAL
REPUBLIC
OF GERMANY
31.2%
119.8

SWITZERLAND
39.8%
21.3

BELGIUM
59.8%
32.4

FRANCE
64.8%
213.1

SPAIN
24%
26.8

ARGENTINA
(11.3%)
5.2

UNITED
KINGDOM
19.3%
53.8

CANADA
12.7%
57.1

UNITED STATES
15.5%
383.7

*SOURCE: International Atomic Energy Agency Bulletin, Autumn 1986, page 41.
Figures in Parentheses are IAEA Estimates*

Worldwide nuclear generation of electricity, 1985.   Despite the accidents at Three Mile Island and Chernobyl, the world is not likely to abandon nuclear fission to produce electricity until a more satisfactory method is developed, probably not until well into the 21st century.

shopping, and services, all of which are sprinkled throughout the city rather than confined to particular zones, may question the validity of this basic concept.

More serious from a political standpoint are the social effects of zoning. One of the popular ideas of the 1920s, when zoning spread rapidly across the country, was that cities should be surrounded by "garden cities," spacious, quiet, neighborhoods, green with vegetation surrounding single-family homes on large lots facing broad and sometimes deliberately curved streets. This "ideal" suburban community has had distinct advantages for those who live in them and has provided some needed relief from the "asphalt jungle" of the city. But it has created some problems as well. For one thing, as whites became more affluent and moved into these suburbs, they tended to be replaced by blacks, Chicanos, Puerto Ricans, and other minorities, leading to patterns of black cities surrounded by white suburbs. Zoning that includes minimum lot sizes of 1 or 2 acres, prohibits multifamily dwellings, severely restricts the location of businesses providing goods and services, and provides other benefits for the well-to-do family with more automobiles than children has had the effect of keeping the minorities bottled up in the cities with little opportunity to follow the whites out to the suburbs even when they acquire the inclination and means to do so. Environmental zoning can become, by design or chance, exclusionary zoning.

Another effect of the insistence on low-density suburbs has been the acceleration of urban sprawl, with huge areas of often productive farmland given over to subdivisions. The ecological effects of urban sprawl are at least as serious as the social and political ones. Most states by now have adopted some laws protecting open spaces around cities from the depredations of the "developers," but such laws drive up the price of developable land, deprive municipalities of additional revenue-producing land, intensify traffic congestion in the urban areas, and have

other undesirable side effects. This is not to say that it is wrong to insist on green belts around cities, only that better planning is necessary to permit them to perform effectively the tasks for which they are designed.

Another type of zoning that is beginning to be taken more seriously is hazard zoning; that is, prohibition or regulation of land use in areas subject to frequent natural hazards. So far this has been used mainly to exclude residences, businesses, and most industries from floodplains (clearly a public good), but there are still many problems with such zoning even where it is in force. We still have, moreover, virtually no zoning for areas subjected to other natural hazards. An example of the need for such zoning is provided by the Los Angeles area. It has become fashionable since World War II to build homes on the slopes of the canyons in the Santa Monica Mountains and the Hollywood Hills in the northern part of the city and in other nearby areas. Nearly every summer and fall the drought dries up the grass and brush cover that is then set afire by natural or human action. Powerful winds spread the fires through the canyons endangering and even destroying houses. Then, with the vegetation cover removed, the slopes generate massive floods and mudslides during the winter and spring rains, endangering and even destroying still more houses. The hazards are well known, yet the taxpayers are annually called upon to provide emergency services for people who insist on living with them. Similar examples abound in various physical environments around the country.

A still larger problem is one that has been scarcely, if ever, mentioned publicly. It can be bluntly summarized in a single question: How long will it be before we begin to restrict the occupation of desert areas that require enormous expenditures of public funds for water, roads, power, and other facilities to make them habitable, to say nothing of the effects large-scale human habitation and public works have on fragile desert ecosystems? Zoning of this type on a national scale might seem unthinkable now, but in fact there is ample precedent for it.

Land Reform

Contrary to the situation in the United States, it is private, not public, land ownership that is controversial in many other countries. During the eighteenth and nineteenth centuries and well into the twentieth the dominant form of land tenure throughout Latin America, North Africa, Southern and Eastern Europe, and most of Asia was the *latifundio,* the large family-owned estate known by a variety of local names, of which the most common is *hacienda.* Under the *latifundia* system,

a large proportion of the land is owned by a very small percentage of the people. Before the 1789 revolution in France, for example, 40 percent of the land was owned by only 3 percent of the people. In Bolivia 10 percent of the farmers owned nearly 95 percent of the land in 1950. In industrial, urbanized societies, these figures might not be alarming, but in a traditional society in which land plays such a great role in the lives of people, they are serious indeed.

Land reform has been an essential prelude to, or component of, every important economic and social revolution in recent history. Land reform in Japan before industrialization began generated a doubling of agricultural production between 1870 and 1914. Similar increases both in

Contrasting patterns of land use.   While the boundaries visible on these two photographs are not political boundaries they nevertheless tell us much about the politicohistorical development of the regions involved. Left: In the central parts of the United States, the Township and Range System has not only produced the unmistakable, regular pattern of fields and roads, but has to a considerable extent also controlled the county subdivision of the region, which shows similar straight-line characteristics. Right: In Western Ireland, on the other hand, the pattern is less regular—it shows subsequent rather than antecedent characteristics (see Chapter 8). Political and cultural factors everywhere are reflected not only in field patterns but also in the uses to which the land is put and in attitudes toward land in general. (Harm J. de Blij)

agricultural production and farmers' incomes have followed (after an interval of disorganization and uncertainty) the revolutions of 1910 in Mexico and 1952 in Bolivia. Land reform has been fundamental in the programs of the new communist governments and of many democratic governments as well. In fact, it has been amply demonstrated that *latifundismo* retards economic and social progress, while more equitable distribution of land encourages it.

Maldistribution of land also inhibits political democracy. Quite commonly, the large landowners are aligned with the local and national military and religious authorities to form a triumvirate that controls the State. Each reinforces and protects the other two (see Chapter Twelve). This is the major reason why meaningful land reform is so difficult to initiate and sustain without a violent or at least radical revolution. Where it has been attempted peacefully, as in Venezuela, the process does not seem to be so effective. In Venezuela, from the initiation of serious land reform in 1960 to 1973, 75 percent of the land redistributed came from the public domain; only abut 12 percent of the country's private estate land had been affected. But land reform began in Venezuela when economic development, fueled by oil money, had already begun and only a third of the labor force was in agriculture. Expropriated lands were paid for generously, again out of oil revenues, and after a brief early period of expropriation of private land, the reform evolved into what is essentially a colonization program. These three factors together permitted an increase in agricultural production as part of an overall economic development under a stable democratic system, but these three factors are seldom present together. There are, furthermore, still grave imbalances and inequities in the country's socioeconomic picture, and there is no assurance that the present economic development and political stability will continue for very much longer without real land reform.

Land reform is not enough, however. Agricultural production cannot be increased and social inequities redressed without comprehensive programs of *agrarian reform* and *rural development*. These include such things as access to, and better utilization of, land, water, forests, and other natural resources; the development of a rural infrastructure, including electric power, farm-to-market roads, storage and transport facilities, irrigation and drainage projects, pure water, schools and medical facilities; provision of agricultural inputs (such as seeds, fertilizers, pesticides, and machinery), crop insurance, generous credit, and agricultural education and extension training; development of nonagricultural activities in rural areas; and greater participation of rural people, especially women, in rural development. All this costs money, takes time, and requires a reordering of political priorities. And it inevitably has profound and often unpredictable political consequences. Yet in the long run, it is not only desirable but essential to have such agrarian reform and rural development if even the roughly 1 billion rural people in the world today living below an absolute poverty line of $200 per capita income per year are to experience any betterment of their condition (to say nothing of the hundreds of millions who will be joining them in the next generation).

INTERNATIONAL ENVIRONMENTAL PROBLEMS

The most important lesson to be learned from our experience with ecology, energy, and land use has yet to be learned by mankind as a whole: that the planet Earth and its atmosphere constitutes one unified ecosystem and that damage to any part of it inevitably results in damage to the whole. Everything is connected to everything else. Statesmen rarely give much thought to the long-range environmental consequences of their decisions and actions, and politicians do so even more rarely. Even though, as we pointed out earlier, an environmental consciousness has begun to spread around the world, it has not spread far enough or fast enough. Pollution, for example, knows no boundaries, yet there is still no adequate mechanism for preventing, containing, or reversing pollution at the international level.

The nearest thing to a global environmental agency is the United Nations Environment Programme (UNEP), with headquarters in Nairobi. Its Regional Seas Programme has generated and coordinated environmental protection and cleanup activities in designated portions of the global sea. Its Mediterranean Action Programme has had considerable success in cleaning up the Mediterranean Sea, one of the most polluted large bodies of water in the world. Progress has also been made in the Baltic, where the problem is at least as serious. Other international and regional organizations have expressed interest in environmental matters, but seldom do they rank very high on their priority lists. Even the World Bank has come under heavy fire for ignoring ecological considerations when funding huge infrastructure projects in developing countries, and it is beginning to reorder its priorities as a result.

Meanwhile, agricultural pesticides applied in the rich countries of the Northern Hemisphere wash down rivers into the global sea and eventually end up in the bodies of penguins in Antarctica. Industrialized countries in Western Europe are already sending ten to twenty thousand shipments of hazardous wastes every year to Eastern Europe for disposal and are seeking disposal sites in developing countries. Acid precipitation caused by the mixing of atmospheric water with airborne industrial and automotive pollutants is devastating lakes and forests in North America and Northern and Central Europe and is threatening farms and monuments in India, Mexico, Indonesia, Zambia, Brazil and other developing countries. The United States and Canada have cooperated in restoring Lake Erie to reasonably good health after it very nearly "died," but numerous transboundary environmental problems still haunt the two countries.

And Western Europe, rich, sophisticated, and proud, was caught completely off guard in November 1986 when a series of accidental spills of toxic chemicals in Switzerland flowed down the Rhine River, killing hundreds of thousands of fish, contaminating drinking water for millions of people, and otherwise damaging the environment of four countries. This calamity called to mind the words of the English Romantic poet Samuel Taylor Coleridge:

The river Rhine, it is well known,
Doth wash your city of Cologne;
But tell me, nymphs, what power divine
Shall henceforce wash the river Rhine?

What power indeed? Perhaps some global plan of action—backed by a worldwide consensus, determination, sustained effort, and plenty of cash—will emerge as a result of the 1987 report of the World Commission on Environment and Development. This commission, composed of 23 people from all parts of the world and headed by Gro Harlem Brundtland, Prime Minister of Norway, was created by the United Nations but was independent of it. It labored for three years before issuing its massive report on an enormous range of contemporary world problems. Its recommendations, if followed urgently and faithfully, would certainly result in a better world for all of us. But this is unlikely to happen because the political will to do so is still weak or absent altogether. Even given this political will, the problems would still be immensely difficult to solve, as we shall see.

CONFLICTS IN ECOLOGY, ENERGY, AND LAND USE

In this chapter, we have pointed out a number of problems relating to environmental politics from the urban neighborhood to the global one. Each is difficult enough in itself, but their interrelatedness compounds the complexity. Here we suggest a few interrelationships only to illustrate the point.

1. In the rush to "energy independence," we are considering a return to coal as a primary fuel because it is domestically abundant, but we forget the reasons that we converted from coal to oil and gas in the first place: coal is bulky and dirty and, except for expensive anthracite, not very efficient. Underground coal mining, moreover, is hazardous and unhealthful while surface mining is ecologically ruinous.

2. Expansion of the land under cultivation to increase food production means in most of the world using marginal or submarginal land, which would require a great deal of energy and can have most unfortunate ecological consequences.

3. Dispersal of industry to improve the urban environment and provide rural employment can simply accelerate urban sprawl, increase transport costs and energy consumption, and transfer the less desirable features of industry to rural areas, making these areas less attractive than they are now.

4. An increase in energy consumption means additional pipelines through fragile environments, oil spills, and blowouts; unsightly and sometimes dangerous electricity transmission lines; more rail, road and water transport requiring more energy and raw materials to produce and operate; uncertainties about nuclear energy; and changes in the composition of the earth's atmosphere.

5. Increased raising of livestock for food means increased competition with wildlife for forage and destruction of their habitats.

6. Landfills in urban areas or even for offshore facilities, such as airports and oil terminals, can have adverse effects on the ecology of the coastal zone.

7. Economic development, considered a desirable political goal, inevitably requires intensified land use, ever-increasing energy supplies, and ecological damage.

8. Orderly economic growth today requires broad policies and plans for land use, energy, and ecology that can only be developed and administered by government at the expense of some limitations on free enterprise, private ownership of land, and individual behavior.

9. A growing population would require economic growth merely to maintain the present unsatisfactory levels of consumption of goods and services for the great majority of the people of the world, so that cutbacks in production can only make a bad situation worse.

10. Political democracy can best be achieved and maintained in a society that is reasonably prosperous and in which wealth is reasonably equitably distributed, but these conditions can only be developed over a period of time at some environmental cost.

In these 10 points alone, without even considering other summary points that could be made or our more elaborate discussion that led up to them, we can see all the components suggested by Kasperson and Minghi: "political goals, agents of impress, processes and effects." But using these components in linear fashion to analyze environment–politics relationships could quickly lead to analytical dead ends. They, themselves, are interrelated in complex ways. The agents of impress, for example, may select the political goals they wish to reach but find that they have a very limited range of processes from which to choose, while the effects of their choices are felt by people who played no role in the selection of goals, agents, or processes and may object to all of them.

Truly the political geography of everyday life is as wondrous as the political geography of international affairs of which it is so intimately a part.

REFERENCES FOR PART EIGHT

Books and Monographs

A

Adamson, David, *The Kurdish War*. London: Allen & Unwin, 1964.

Agrarian Reform in Latin America: An Annotated Bibliography. Madison: Univ. of Wisconsin Press, 1974.

Alexiev, Alexander R. and S. Enders Wimbush (eds.), *Ethnic Minorities in the Red Army*. Boulder, CO: Westview, 1988.

Alston, P. and K. Tomaševski (eds.), *The Right to Food*. The Hague: Nijhoff, 1984.

Amy, Douglas J., *The Politics of Environmental Mediation*. New York: Columbia Univ. Press, 1987.

Andersen, Walter K. and Shridhar D. Damie, *The Brotherhood in Saffron*. Boulder, CO: Westview, 1987.

Anderson, Irvine H., *Aramco, the United States and Saudi Arabia*. Princeton, NJ: Princeton Univ. Press, 1987.

Anderson, Terry L. (ed.), *Water Rights: Scarce Resource Allocation, Bureaucracy, and the Environment*. Washington: Island Press, 1983.

Arfa, Hassan, *The Kurds: An Historical and Political Study*. London: Oxford Univ. Press, 1966.

Arnold, Guy, *Strategic Highways of Africa*. New York: St. Martin's Press, 1977.

Ashworth, G. (ed.), *World Minorities*, 2 Vols. London: Quartermaine House, 1977, 1978.

Auer, Peter (ed.), *Energy and the Developing Nations*. New York: Pergamon, 1981.

B

Bahbah, Bishara and Linda Butler, *Israel and Latin America*. New York: St. Martin's Press, 1986.

Balaam, David N. and Michael J. Carey (eds.), *Food Politics: The Regional Conflict*. Totowa, NJ: Allanheld, Osmun, 1981.

Banuaziz, Ali and Myron Weiner (eds.), *The State, Religion and Ethnic Politics*. Syracuse, NY: Syracuse Univ. Press, 1986.

Baumbach, Richard O., Jr. and William E. Borah, *The Second Battle of New Orleans: A History of the Vieux Carré Riverfront–Expressway Controversy*. University: Univ. of Alabama Press, 1981.

Beer, William R. and James E. Jacob (eds.), *Language Policy and National Unity*. Totowa, NJ: Rowman & Allanheld, 1985.

Benard, Cheryl and Zalmay Khalilzad, *The Government of God*. New York: Columbia Univ. Press, 1984.

Benningsen, Alexandre and Marie Brosup, *The Islamic Threat to the Soviet State*. New York: St. Martin's Press, 1983.

Berry, Brian J.L. and L.P. Silverman, *Population Redistribution and Public Policy*. Washington: National Academy of Sciences, 1980.

Besmeres, John, *Socialist Population Politics: The Political Implications of Demographic Trends in the USSR and Eastern Europe*. Armonk, NY: Sharpe, 1980.

Best, Alan C.G., *The Swaziland Railway: A Study in Politico-Economic Geography*. East Lansing: Michigan State Univ. Press, 1966.

Bhatt, S., *Aviation, Environment and World Order*. Atlantic Highlands, NJ: Humanities Press, 1980.

Blackburn, John O., *The Renewable Energy Alternative*. Durham, NC: Duke Univ. Press, 1987.

Blinn, Keith W. and others, *International Petroleum Exploration and Exploitation Agreements: Legal, Economic and Policy Aspects*. London: Euromoney Publications, 1986.

Boal, Frederick W. and J.N.H. Douglas (eds.), *Integration and Division: Geographical Perspectives on the Northern Ireland Problem*. London: Academic Press, 1982.

Bohi, Douglas R. and William B. Quandt, *Energy Security in the 1980's; Economic and Political Perspectives*. Washington: Brookings Institution, 1984.

Böhning, W.R., *Studies in International Labor Migration*. New York: St. Martin's Press, 1985.

Bothe, Michael (ed.), *Trends in Environmental Policy and Law/Tendances Actuelles de la Politique et du Droit de l'Environnement*. Gland, Switz.: International Union for the Conservation of Nature and Natural Resources, 1980.

Bowler, Ian R., *Government and Agriculture*. New York: Longman, 1979.

Brass, Paul (ed.), *Ethnic Groups and the State*. London: Croom Helm, 1984.

Brower, David J. and Daniel S. Carol, *Managing Landuse Conflicts*. Durham, NC: Duke Univ. Press, 1987.

Brown, Lester R. and others, *State of the World*, New York: Norton, published annually.

Brown, Peter G. and Henry Shue, *The Border That Joins: Mexican Migrants and U.S. Responsibility*. Totowa, NJ: Rowman & Littlefield, 1983.

Browne, William P. and Don F. Hadwiger, *World Food Policies: Toward Agricultural Interdependence*. London: Frances Pinter, 1986.

Bruchis, Michael, *One Step Backward, Two Steps Forward*. New York: Columbia Univ. Press, 1982.

Burhenne, W.E., *International Environmental Law; Multilateral Treaties*, 3 Vols. Berlin: Erich Schmidt Verlag, 1974–1975.

Burton, Dudley J., *The Governance of Energy*. New York: Praeger, 1980.

C

Cairns, A. and C. Williams, *The Politics of Gender, Ethnicity, and Language in Canada*. Toronto: Univ. of Toronto Press, 1987.

Caldwell, Lynton K., *International Environmental Policy; Emergence and Dimensions*. Durham, NC: Duke Univ. Press, 1984.

Carlozzi, Carl A. and Alice A. Carlozzi, *Conservation and Caribbean Regional Progress*. St. Thomas, U.S.V.I.: Antioch Press, 1968.

Carroll, John E., *Environmental Diplomacy: an Examination and a Prospective of Canadian–U.S. Transboundary Environmental Relations*. Ann Arbor: Univ. of Michigan Press, 1983.

Cathie, John, *The Political Economy of Food Aid*. New York: St. Martin's Press, 1982.

Chadwick, H.M., *The Nationalities of Europe*. Cambridge, England: Cambridge Univ. Press, 1945.

Chami, Joseph G., *Days of Tragedy: Lebanon 1975–76*. New Brunswick, NJ: Transaction Books, 1980.

———, *Days of Wrath: Lebanon 1977–1982*, Vol. II. New Brunswick, NJ: Transaction Books, 1983.

Chaney, Rick, *Regional Emigration and Remittances in Developing Countries: The Portuguese Experience*. New York; Westport, CT; and London: Praeger, 1986.

Chazan, Naomi and Timothy M. Shaw, *Coping with Africa's Food Crisis*. London: Frances Pinter, 1987.

Chiswick, Barry R., *The Gateway: U.S. Immigration Issues and Policies*. Washington and London: American Enterprise Institute for Public Policy Research, 1982.

Chouri, Nazli with Vincent Ferraro, *International Politics of Energy Interdependence*. Lexington, MA: Lexington Books, 1976.

CIMADE, INODEP, MINK, *Africa's Refugee Crisis: What's to Be Done?* Humanities Press International, 1986.

Cioffi-Revilla, Claudio and others (eds.), *Communications and Interaction in Global Politics*. Newbury Park, CA: Sage, 1987.

Clark, Gordon L., *Interregional Migration, National Policy and Social Justice*. Totowa, NJ: Rowman & Allanheld, 1983.

Clarke, Colin and others (eds.), *Geography and Ethnic Pluralism*. London and Winchester, MA: Allen & Unwin, 1984.

Clarke, John I. and Leszek A. Kosinski, *Redistribution of Population in Africa*. Portsmouth, NH: Heinemann, 1982.

Clay, Jason W. and Bonnie K. Holcomb, *Politics and the Ethiopian Famine 1984–1985*. New Brunswick, NJ: Transaction Books, 1987.

Cobban, Helena, *The Making of Modern Lebanon*. Boulder, CO: Westview, 1985.

Codding, George A., Jr. and Anthony M. Rutkowski, *The International Telecommunications Union in a Changing World*. Dedham, MA: Artech House, 1982.

Cohn, Theodore H., *Canadian Food Aid: Domestic and Foreign Policy Implications*. Denver, CO: Univ. of Denver, 1979.

Coleman, William, *The Independence Movement in Quebec 1945–1980*. Toronto: Univ. of Toronto Press, 1984.

Colombo, Furio, *God in America*. New York: Columbia Univ. Press, 1984.

Conant, Melvin A., *The Geopolitics of Energy; A Mediterranean Case Study*. American Universities Field Staff Reports, North America Series Vol. 5, No. 4. 1977.

———, and Fern Racine Gold, *The Geopolitics of Energy*. Boulder, CO: Westview, 1978.

Croll, Elisabeth and others (eds.), *China's One-Child Family Policy*. New York: St. Martin's Press, 1985.

Currey, Bruce and Graeme Hugo (eds), *Famine as a Geographical Phenomenon*. Reidel, 1984.

D

Dando, William A., *The Geography of Famine*. New York: Wiley, 1980.

Das Gupta, Jyotirindra, *Language Conflict and National Development; Group Politics and*

National Language Policy in India. Berkeley: Univ. of California Press, 1970.

Dean, Robert W., *Nationalism and Political Change in Eastern Europe: The Slovak Question and the Czechoslovak Reform Movement.* Denver, CO: Univ. of Denver, 1973.

de Castro, Josué, *The Geography of Hunger.* Boston, MA: Little, Brown, 1952.

Deese, David A. and Joseph S. Nye (eds.), *Energy and Security.* Cambridge, MA: Ballinger, 1981.

de Janvry, Alain, *The Agrarian Question and Reformism in Latin America.* Baltimore and London: Johns Hopkins Univ. Press, 1981.

Dekmejian, *Islam in Revolution,* Syracuse, NY: Syracuse Univ. Press, 1985.

Dellenbrant, Jan Åke, *The Soviet Regional Dilemma: Planning, People and Natural Resources.* Armonk, NY: Sharpe, 1986.

Dempsey, Paul Stephen, *Law and Foreign Policy in International Aviation.* Ardsley-on-Hudson, NY: Transnational, 1987.

Denktash, Rauf R., *The Cyprus Triangle.* London: Allen & Unwin, 1982.

de Vos, George and L. Romanucci-Ross (eds.), *Ethnic Identity; Cultural Continuities and Change.* Palo Alto, CA: Mayfield, 1975.

Dew, Edward, *The Difficult Flowering of Surinam; Ethnicity and Politics in a Plural Society.* The Hague: Nijhoff, 1978.

Dofny, Jacques and A. Akinowo, *National and Ethnic Movements.* Palo Alto, CA: Sage, 1980.

Dominian, Leon, *The Frontiers of Language and Nationality in Europe.* New York: Holt, 1917.

Dorner, Peter (ed.), *Land Reform in Latin America; Issues and Cases.* Madison, WI: *Land Economics* for the Univ. of Wisconsin, Land Tenure Center, 1971.

Dowall, David E., *The Suburban Squeeze: Land Conversion and Regulation in the San Francisco Bay Area.* Berkeley: Univ. of California Press, 1984.

Dowty, Alan, *Closed Borders.* New Haven, CT: Yale Univ. Press, 1987.

Drakakis-Smith, D. and S. Wyn Williams (eds.), *Internal Colonialism: Essays Around a Theme.* Edinburgh: Univ. of Edinburgh and the Institute of British Geographers, 1983.

Drury, Bruce and others (eds.), *Agrarian Reform in Reverse.* Boulder, CO: Westview, 1987.

Dupuy, René-Jean (ed.), *The Future of the International Law of the Environment.* Dordrecht, Neth.; Boston, London: Nijhoff, 1985.

E

Eastman, Clyde, *Immigration Reform and New Mexico Agriculture.* El Paso: Univ. of Texas, 1984.

Ebinger, Charles K., *International Politics of Nuclear Energy.* Washington Papers, Vol. 6, No. 57. Lanham, MD: Univ. Press of America, 1978.

————, *The Critical Link: Energy and National Security in the 1980's.* Cambridge, MA: Ballinger, 1982.

Ehrlich, Thomas, *Cyprus 1958–1967.* New York and London: Oxford Univ. Press, 1967.

Elegant, Robert S., *The Dragon's Seed: Peking and Overseas Chinese.* New York: St. Martin's Press, 1959.

Elmallakh, Dorothea H., *The Slovak Autonomy Movement, 1935–1939.* New York: Columbia Univ. Press, 1979.

Ender, Richard L. and John Choon Kim (eds.), *Energy Resources Development: Politics and Policies.* Westport, CT: Greenwood Press, Quorum Books, 1987.

Englebert, Ernest A. and Ann F. Scheuring (eds.), *Competition for California Water.* Berkeley: Univ. of California Press, 1982.

Esman, Milton J. (ed.), *Ethnic Conflict in the Western World.* Ithaca, NY: Cornell Univ. Press, 1977.

Esposito, *Islam and Politics.* Syracuse, NY: Syracuse Univ. Press, 1984.

F

Faaland, Just, *Population and the World Economy in the 21st Century.* New York: St. Martin's Press, 1982.

Falk, Pamela S., *Petroleum and Mexico's Future.* Boulder, CO: Westview, 1986.

Fawcett, James, *The International Protection of Minorities.* Report No. 41. London: Minority Rights Group, 1979.

Ferris, Elizabeth (ed.), *Refugees and World Politics.* New York: Praeger, 1985.

Fischer, Eric, *Minorities and Minority Problems.* New York: Vantage Press, n.d.

Fishman, Joshua A., *The Rise and Fall of the Ethnic Revival.* Berlin: de Gruyter, 1985.

Flinterman, Cees and others, *Transboundary Air Pollution.* Dordrecht, Neth.: Kluwer, 1986.

Folk-Williams, John A., *Water in the West; What Indian Water Means to the West,* Vol. I. Washington: Island Press, 1982.

Foner, Nancy, *New Immigrants in New York.* New York: Columbia Univ. Press, 1987.

Foon, Chew Sock, *Ethnicity and Nationality in Singapore.* Athens: Ohio Univ. Press, 1987.

Francisse, Anne E., *The Problems of Minorities in the Nation-Building Process: The Kurds, the Copts, the Berbers*. New York: Vantage Press, 1971.

Frankel, Edith Rogovin, *The Soviet Germans*. New York: St. Martin's Press, 1986.

G

Gardiner, C. Harvey, *Pawns in a Triangle of Hate: The Peruvian Japanese and the United States*. Seattle and London: Univ. of Washington Press, 1981.

Gasteyger, Curt (ed.), *The Future for European Energy Security*. London: Frances Pinter, 1986.

Geisler, Charles C. and Frank J. Popper (eds.), *Land Reform, American Style*. Totowa, NJ: Rowman & Allanheld, 1984.

Gellner, Ernest, *Nations and Nationalism*. Ithaca, NY: Cornell Univ. Press, 1983.

Ghai, Dharam and Lawrence D. Smith, *Agricultural Prices, Policy and Equity in Sub-Saharan Africa*. London: Frances Pinter, 1987.

Ghose, Ajit Kumar, *Agrarian Reform in Contemporary Developing Countries*. New York: St. Martin's Press, 1983.

Gilmour, David, *Lebanon*. New York: St. Martin's Press, 1983.

Gittinger, J. Price and others (eds.), *Food Policy*. Baltimore: Johns Hopkins Univ. Press, 1987.

Glantz, Michael H., *Drought and Hunger in Africa*. New York: Cambridge Univ. Press, 1987.

Glaser, William A., *The Brain Drain; Emigration and Return*. New York: Pergamon, 1978.

Glasstone, Samuel and Walter H. Jordan, *Nuclear Power and Its Environmental Effects*. American Nuclear Society, 1980.

Glazer, Nathan (ed.), *Clamor at the Gates: The New American Immigration*. San Francisco: ICS Press, 1985.

Goldman, Robert B. and A. Jeyaratnam Wilson (eds.), *From Independence to Statehood: Managing Ethnic Conflict in Five African and Asian States*. London: Frances Pinter, 1984.

Golubev, G. and A. Biswas, *Large Scale Water Transfers; Emerging Environmental and Social Issues*. London: Taylor & Francis, 1985.

Goodwin-Gill, Guy S., *The Refugee in International Law*. Oxford: Oxford Univ. Press, Clarendon, 1983.

Gormley, William T., Jr., *The Politics of Public Utility Regulation*. Pittsburgh: Univ. of Pittsburgh Press, 1983.

Gorove, Stephen, *Law and Politics of the Danube*. The Hague: Nijhoff, 1964.

Guha, Amalendu and others, *Immigrant Women and Children in Industrial Europe*. Hong Kong: Asian Research Service, n.d.

———, *LDC Immigrants in Nordic Countries*, Hong Kong: Asian Research Service, n.d.

H

Hall, Raymond L. (ed.), *Ethnic Autonomy: Comparative Dynamics of the Americas, Europe and the Developing World*. New York: Pergamon, 1979.

Hall, Richard and Hugh Peyman, *The Great Uhuru Railway: China's Showpiece in Africa*. London: Gollancz, 1976.

Hammer, Tomas, *European Immigration Policy*. Cambridge Univ. Press, 1985.

Hanna, Willard A., *The Kra Isthmus Canal*. American Universities Field Staff Reports, Southeast Asia Series, Vol. 15, No. 12, 1967.

Hansen, Art and Della E. McMillan (eds.), *Food in Sub-Saharan Africa*. London: Frances Pinter, 1986.

Harrison, Selig S., *In Afghanistan's Shadow: Baluch Nationalism and Soviet Temptations*. New York: Carnegie Endowment for International Peace, 1981.

Hauser, Philip M. (ed.), *Population and World Politics*. Glencoe, IL: Free Press, 1958.

Hauser, Philip M. and others, *Urbanization and Migration in ASEAN Development*. Honolulu: Univ. of Hawaii Press, 1985.

Haywood, Keith, *International Collaboration in Civil Aerospace*. London: Frances Pinter, 1986.

Hechter, Michael, *Internal Colonialism: The Celtic Fringe in British National Development, 1536–1966*. London: Routledge & Kegan Paul, 1975.

Heisler, *Transnational Migration: From Foreign Workers to Settlers*. Sage, n.d.

Helmreich, Jonathan E., *Gathering Rare Ones*. Princeton, NJ: Princeton Univ. Press, 1986.

Hill, Brian E., *The Common Agricultural Policy; Past, Present and Future*. New York: Methuen, 1984.

Hodges, Tony and Richard Lawless, *War and Refugees: The West Sahara Conflict*. London: Frances Pinter, 1987.

Hoffman, Thomas and Brian Johnson, *The World Energy Triangle*, Cambridge, MA: Ballinger, 1981.

Hofstetter, Richard R., *U.S. Immigration Policy*. Durham, NC: Duke Univ. Press, 1984.

Hossain, Kamal and Subrata Roy Chowdhury, *Permanent Sovereignty over Natural Resources in International Law*. London: Frances Pinter, 1984.

Hoyt, Edwin C., *National Policy and International Law; Case Studies from American Canal Policy*. Denver: Univ. of Denver Monograph Series in World Affairs (1966).

Husbands, C., *Racial Exclusionism and the City; the Urban Support of the National Front*. London: Allen & Unwin, 1983.

I

ICIHI, *Refugees: The Dynamics of Displacement*. Humanities Press International, 1986.

Illyes, Elemer, *National Minorities in Romania*. New York: Columbia Univ. Press, 1982.

Irland, Lloyd, *Wilderness, Economics and Policy*. Lexington, MA: Lexington Books, 1979.

Ismael, Tareq Y. and Jacqueline S. Ismael, eds., *Government and Politics in Islam*. New York: St. Martin's Press, 1985.

J

Jancar, Barbara, *Environmental Management in the Soviet Union and Yugoslavia*. Durham, NC: Duke Univ. Press, 1987.

Janowsky, Oscar I., *Nationalities and National Minorities*. New York: Macmillan, 1945.

Jay, Martin, *Permanent Exiles; Essays on the Intellectual Migration from Germany to America*. New York: Columbia Univ. Press, n.d.

Jentleson, Bruce W., *Pipeline Politics: The Complex Political Economy of East–West Energy Trade*. Ithaca, NY: Cornell Univ. Press, 1986.

Jones, Richard C. (ed.), *Patterns of Undocumented Migrations: Mexico and the United States*. Totowa, NJ: Rowman & Allanheld, 1984.

Jönsson, Christer, *International Aviation and the Politics of Regime Change*. New York: St. Martin's Press, 1987.

Jordan, Wayne R. (ed.), *Water and Water Policy in World Food Supplies*. College Station: Texas A&M Univ. Press, 1987.

K

Kalt, Joseph P. and Frank C. Schuller (eds.), *Drawing the Line on Natural Gas Regulation*. Westport, CT: Greenwood Press, Quorum Books, 1987.

Karn, Anil, *Politics of Population Influx and Labour Problems*. Delhi: UDH, 1986.

Karim, M.B. (compiler) *The Green Revolution: An International Bibliography*. Westport, CT: Greenwood Press, 1986.

Karklins, Rasma, *Ethnic Relations in the USSR*. London: Allen & Unwin, 1985.

Kapar, *Sikh Separatism*. London: Allen & Unwin, 1985.

Katō, Ichirō and others (eds.), *Environmental Law and Policy in the Pacific Basin Area*. Tokyo: Univ. of Tokyo Press, 1981.

Kaufman, Robert R., *The Politics of Land Reform in Chile 1950–1970*. Cambridge: Harvard Univ. Press, 1972.

Keeler, Theodore E., *Railroads, Freight and Public Policy*. Washington: Brookings Institution, 1983.

Kelly, William J. and others, *Energy Research and Development in the USSR*. Durham, NC: Duke Univ. Press, 1986.

Kent, George, *The Political Economy of Hunger; the Silent Holocaust*. New York: Praeger, 1984.

Khalaf, Samir, *Lebanon's Predicament*. New York: Columbia Univ. Press, 1987.

Kinnane, D., *The Kurds and Kurdistan*. London: Oxford Univ. Press, 1964.

Kirk, John and others, *Studies in Linguistic Geography*. London: Longwood, 1985.

Kliot, Nurit and Stanley Waterman (eds.), *Pluralism and Political Geography; People, Territory and State*. Beckenham, England: Croom Helm, 1983.

Kofele-Kale, Ndiva, *Tribesmen and Patriots: Political Culture in a Poly-ethnic African State*. Lanham, MD: Univ. Press of America, 1981.

Krannich, Ronald L. and Caryl Rae Krannich, *The Politics of Family Planning Policy: A Case of Successful Implementation*. Monograph Series No. 19. Lanham, MD: Univ. Press of America, 1983.

Kuhn, Raymond, *The Politics of Broadcasting*. New York: St. Martin's Press, 1985.

L

Ladas, S.P., *The Exchange of Minorities; Bulgaria, Greece and Turkey*. New York, 1932.

LaFeber, Walter, *The Panama Canal; the Crisis in Historical Perspective*. New York: Oxford Univ. Press, 1978.

Lamm, Vanda, *The Utilization of Nuclear Energy and International Law*. Budapest: Akadémiai Kiadó, 1984.

Laponce, J.A., *Languages and Their Territories*. Toronto: Univ. of Toronto Press, 1987.

Lawrence, Peter, *World Recession and the Food Crisis in Africa*. Boulder, CO: Westview, 1987.

Lee, Changsoo and George de Vos, *Koreans in Japan*. Berkeley: Univ. of California Press, 1982.

Leiss, William (ed.), *Ecology Versus Politics in Canada*. Toronto: Univ. of Toronto Press, 1979.

Leive, David M., *The Future of the International*

Telecommunication Union. Washington: American Society of International Law, 1972.

Lemon, Anthony and Norman Pollock, *Studies in Overseas Settlement and Population*. London: Longman, 1980.

Leo, Christopher, *Land and Class in Kenya*. Toronto: Univ. of Toronto Press, 1984.

Leonard, H. Jeffrey, *Are Environmental Regulations Driving U.S. Industry Overseas?* Washington: Island Press, 1984.

Levine, Barry B. (ed.), *The Carribean Exodus*. New York: Praeger, 1986.

Lewis, I.M., *A Pastoral Democracy: A Study of Pastoralism and Politics Among the Northern Somali of the Horn of Africa*. London: Oxford Univ. Press, 1961.

Liebman, Charles S. and Eliezar Don-Yehiya, *Religion and Politics in Israel*. Bloomington: Indiana Univ. Press, 1984.

Lowe, P. and J. Goyder, *Environmental Groups in Politics*. London: Allen & Unwin, 1983.

Luebke, Frederick C. (ed.), *Ethnicity on the Great Plains*. Lincoln: Univ. of Nebraska Press, 1980.

Luter, James P. and Ann Bowman (eds.), *The Politics of Hazardous Waste Management*. Durham, NC: Duke Univ. Press, 1984.

Lyster, Simon, *International Wildlife Law*. Cambridge: Grotius, 1985.

M

Macartney, C.R., *National States and National Minorities*. London: Oxford Univ. Press, 1934.

Madgwick, Peter and Richard Rose (eds.), *The Territorial Dimension in the United Kingdom*. London: Macmillan, 1982.

Mance, Sir Osborne, *International River and Canal Transport*. London: Oxford Univ. Press, 1945.

Manners, Ian R., *North Sea Oil and Environmental Planning: The United Kingdom Experience*. Austin: Univ. of Texas Press, 1982.

Manogran, Chelvadurai, *Ethnic Conflict and Reconciliation in Sri Lanka*. Honolulu: University of Hawaii Press, 1987.

Marrus, Michael R., *The Unwanted: European Refugees in the Twentieth Century*. New York and Oxford: Oxford Univ. Press, 1985.

Marshall, Dawn I., *The Haitian Problem: Illegal Migration to the Bahamas*. Kingston, Jamaica: Institute of Social and Economic Research, Univ. of the West Indies, 1979.

Martovych, O.R., *National Problems in the USSR*. Edinburg: Scottish League for European Union, 1953.

Martz, John D., *Regime, Politics, and Petroleum*. New Brunswick, NJ: Transaction Books, 1987.

McBride, George McCutchen, *The Land Systems of Mexico*. New York: Octagon, 1971.

McIntosh, C. Alison, *Population Policy in Western Europe: Responses to Low Fertility in France, Sweden, and West Germany*. Armonk, NY: Sharpe, 1983.

McNeill, William H., *Polyethnicity and National Unity in World History*. Toronto: Univ. of Toronto Press, 1985.

McPhail, Thomas L., *Electronic Colonialism*. Newbury Park, CA: Sage, 1987.

Melander, Göran and Peter Nobel, *International Legal Instruments on Refugees in America*. Uppsala, Swed.: Scandanavian Institute of African Studies, 1979.

Mercer, Lloyd J., *Railroads and Land Grant Policy: A Study in Government Intervention*. New York: Academic Press, 1982.

Michaelson, Karen L., *And the Poor Get Children: Radical Perspectives on Population Dynamics*. New York and London: Monthly Review Press, 1981.

Millward, Hugh, *Regional Patterns of Ethnicity in Nova Scotia: A Geographical Study*. Halifax, NS: St. Mary's Univ., International Education Centre, 1981. (Ethnic Heritage Series, Vol. 6.)

Minault, Gail, *The Khilafat Movement: Religious Symbolism and Political Mobilization in India*. New York: Columbia Univ. Press, 1982.

Misiunas, Romuald J. and Rein Taagepera, *The Baltic States: Years of Dependence, 1940–1980*. Berkeley: Univ. of California Press, 1983.

Missakian, J.A., *Searchlight on the Armenian Question (1878–1950)*. Boston: Hairenik, 1950.

Moriyama, Alan T., *Imingaisha*. Honolulu: Univ. of Hawaii Press, 1985.

Morris, Milton D., *Curbing Illegal Immigration*. Washington: Brookings Institution, 1982.

Morris-Jones, W.H. (ed.), *Collected Papers on the Politics of Separation*. London: Univ. of London, Institute of Commonwealth Studies, 1976.

Mosely, George (ed.), *The Party and the National Question in China*. Cambridge, MA: MIT Press, 1966.

Murdock, George P., *Ethnographic Atlas*. Pittsburgh: Univ. of Pittsburgh Press, 1967.

Murphy, Brian, *International Politics of New Information Technology*. New York: St. Martin's Press, 1986.

Murphy, Francis C., *Regulating Flood-Plain Development*. Chicago: Univ. of Chicago Press, 1958.

Mutukwa, Kasuka Simwinji, *Politics of the Tanzania–Zambia Railway Project. Washington, DC: Univ. Press of America, 1979.

N

Nalven, Joseph, Is There a Need for a Guest-Worker Program? El Paso: Univ. of Texas Press.

Nanda, Ved P. (ed.), World Climate Change; the Role of International Law and Institutions. Boulder, CO: Westview, 1983.

Nejatigil, Zaim M., The Turkish Republic of Northern Cyprus in Perspective. Nicosia, Cyprus; 1985.

Nivola, Pietro S., The Politics of Energy Conservation. Washington: Brookings Institution, 1986.

Norton, Robert, Race and Politics in Fiji. New York: St. Martin's Press, 1978.

O

O'Ballance, Edgar, The Kurdish Revolt 1961–1970. London: Faber & Faber, 1973.

Oberai, A.S., ed., State Policies and Internal Migration. New York: St. Martin's Press, 1983.

Ogden, P.E., Migration and Geographical Change. New York: Cambridge Univ. Press, 1984.

Olcott, Martha B. and Lubomyr Hajda (eds.), The Soviet Multinational State. Armonk, NY: Sharpe, 1986.

Organski, Katherine and A.F.K. Organski, Population and World Power. New York: Knopf, 1961.

Orleans, Leo A., Chinese Approaches to Family Planning. Armonk, NY: Sharpe, 1979.

Özgür, Özdemir A., Apartheid: The United Nations and Peaceful Change in South Africa. Dobbs Ferry, NY: Transnational, 1982.

P

Paarlberg, Robert L., Food Trade and Foreign Politics. Ithaca, NY: Cornell Univ. Press, 1985.

Palmer, Robin, Land and Racial Domination in Rhodesia. Berkeley: Univ. of California Press, 1977.

Pastor, Robert A., Migration and Development in the Caribbean. Boulder, CO: Westview, 1985.

Patrick, Richard A., Political Geography and the Cyprus Conflict 1963–1971. Waterloo, Ont., Canada: Univ. of Waterloo, Dept. of Geography, 1976.

Peach, V. and others, Ethnic Segregation in Cities. London: Croom Helm, 1981.

Pearson, Raymond, National Movements in Eastern Europe, 1848–1945. New York: St. Martin's Press, 1983.

Peterson, Dean F. and A. Berry Crawford (eds.), Values and Choices in the Development of the Colorado River Basin. Tucson: Univ. of Arizona Press, 1978.

Petulla, Joseph M., American Environmentalism; Values, Tactics, Priorities. College Station: Texas A&M Univ. Press, 1986.

Pilat, J.F., Ecological Politics; the Rise of the Green Movement. Lanham, MD: Univ. Press of America, 1980.

Pirages, Dennis, The New Context for International Relations: Global Ecopolitics. North Scituate, MA: Duxbury, 1978.

Popper, Frank J., The Politics of Land-Use Reform. Madison: Univ. of Wisconsin Press, 1981.

Prabhakara, N.R., Internal Migration and Population Redistribution in India—Some Reflections. New Delhi: Concept, 1986.

Prindle, David F., Petroleum Politics and the Texas Railroad Commission. Austin: Texas Univ. Press, 1981.

Pringle, D.G., One Island, Two Nations? A Political Geographical Analysis of the National Conflict in Ireland. Letchworth, Hertfordshire, Eng.: Wiley, 1985.

Pryor, Robin J., Migration and Development in South–East Asia: A Demographic Perspective. New York: Oxford Univ. Press, 1979.

R

Ramazani, Rouhollah, Revolutionary Iran. Baltimore: Johns Hopkins Univ. Press, 1986.

Ramberg, Bennet, Nuclear Power Plants as Weapons for the Enemy; an Unrecognized Military Peril. Berkeley, Los Angeles and London: Univ. of California Press, 1985.

Rao, M.S.A., Studies in Migration: Internal and International Migration in India. Delhi: Manohar, 1986.

Raphael, Ray, Tree Talk: The People and Politics of Timber. Washington: Island Press, 1981.

Redclift, M., Development and the Environmental Crisis: Red or Green Alternatives? London and New York: Methuen, 1984.

Rockett, Rocky, Ethnic Nationalities in The Soviet Union. New York: Praeger, 1981.

Rogge, John R., Too Many Too Long; Sudan's Twenty Year Refugee Dilemma. Totowa, NJ: Rowman & Allanheld, 1985.

Rorlich, Azade-Ayse, The Volga Tatars; a Profile in National Resistance. Stanford, CA: Hoover Institution Press, 1986.

Rose, Richard, Understanding the United Kingdom: The Territorial Dimension in Government. London: Longman, 1982.

Rosenbaum, Walter, *Environmental Politics and Policy*. Washington: Congressional Quarterly, CQ Press, 1984.

Rosenfeld, Stanley B., *The Regulation of International Commercial Aviation: The International Regulatory Structure*. Dobbs Ferry, NY: Oceana, 1984.

Rothchild, Donald and Victor A. Olorunsola (eds.), *State Versus Ethnic Claims; African Policy Dilemmas*. Boulder, CO: Westview, 1983.

Rowles, James P., *Law and Agrarian Reform in Costa Rica*. Boulder, CO: Westview, 1985.

Roys, R.L., *The Political Geography of the Yucatan Maya*. Washington: Carnegie Institution of Washington, 1957.

Rüster, Bernd and B. Simma (compilers), *International Protection of the Environment: Treaties and Related Documents*. London, Rome, and New York: Oceana, 1975.

Rywkin, Michael, *Moscow's Muslim Challenge: Soviet Central Asia*. Armonk, NY: Sharpe, 1982.

S

Said, Abdul A. and Luis R. Simmons (eds.), *Ethnicity in an International Context: The Politics of Disassociation*. New Brunswick, NJ: Transaction Books, 1976.

Salter, M.J., *Studies in the Immigration of the Highly Skilled*. Canberra: Australian National Univ. Press, 1978.

Sandbach, Francis, *Environment, Ideology, and Policy*. Totowa, NJ: Rowman & Littlefield, 1980.

Saunders, John, *Population Growth in Latin America and U.S. National Security*. London: Allen & Unwin, 1986.

Scheffer, Walter F. (ed.), *Energy Impacts on Public Policy and Administration*. Norman: Univ. of Oklahoma Press, n.d.

Schiller, Herbert I., *Communications and Cultural Domination*. Armonk, NY: Sharpe, 1976.

Schloss, Aran, *The Politics of Development: Transportation Policy in Nepal*. Monograph Series No. 22. Lanham, MD: Univ. Press of America, 1983.

Schneider, Jan, *World Public Order of the Environment: Towards an International Ecological Law and Organization*. Toronto, Buffalo, and London: Univ. of Toronto Press, 1979.

Schneider, William, *Food, Foreign Policy, and Raw Materials Cartels*. New York: Crane, Russak, 1976.

Sealy, K.R., *The Geography of Air Transport*. London: Hutchinson's University Library, 1957.

Shaw, R. Paul, *Land Tenure and Rural Exodus in Chile, Colombia, Costa Rica and Peru*. Gainesville, Fla.: Univ. Presses of Florida, 1976.

Sheffer, Gabriel, *Modern Diasporas in International Politics*. New York: St. Martin's Press, 1986.

Shiels, Frederick L. (ed.), *Ethnic Separatism and World Politics*. Lanham, MD: Univ. Press of America, 1984.

Short, K.R., *Western Broadcasting Over the Iron Curtain*. New York: St. Martin's Press, 1986.

Silvers, Arthur and others, *Rural Development and Urban Bound Migration in Mexico*. Washington: Johns Hopkins Univ. Press, 1980.

Simmonds, G.W. (ed.), *Nationalism in the USSR and Eastern Europe in the Era of Brezhnev and Kosygin*. Detroit, MI: Univ. of Detroit Press, 1977.

Simon, Rita J. and others, *International Migration; the Female Experience*. New York: Barnes & Noble, 1985.

Smil, Vaclav, *The Bad Earth: Environmental Degradation in China*. Armonk, NY: Sharpe, 1984.

Smith, A.D., *The Ethnic Revival in the Modern World*. Cambridge: At the University Press, 1981.

Smith, Anthony, *Geopolitics of Information: How Western Culture Dominates the World*. Oxford, England: Oxford Univ. Press, 1981.

Smith, David M. (ed.), *Living Under Apartheid: Aspects of Urbanization and Social Change in South Africa*. London: Allen & Unwin, 1982.

Smith, T. Lynn (ed.), *Agrarian Reform in Latin America*. New York: Knopf, 1965.

Soley, Lawrence C., and John C. Nichols, *Clandestine Radio Broadcasting*. New York: Praeger, 1986.

Springer, Allen L., *The International Law of Pollution: Protecting the Global Environment in a World of Sovereign States*. Westport, CT, and London: Greenwood Press, Quorum Books, 1983.

Sprout, Harold H. and Margaret Sprout, *The Ecological Perspective on Human Affairs*. Princeton, NJ: Princeton Univ. Press, 1965.

Stack, John F., Jr., *The Primordial Challenge: Ethnicity in the Contemporary World*. Westport, CT: Greenwood Press, 1986.

Stamp, L. Dudley, *Land for Tomorrow: The*

Underdeveloped World. New York: American Geographical Society, 1952.

Starkie, D.N.M., *The Motorway Age: Road and Traffic Politics in Britain 1896–1970.* Oxford: Pergamon, 1982.

Steenland, Kyle, *Agrarian Reform Under Allende; Peasant Revolt in the South.* Albuquerque: Univ. of New Mexico Press, 1977.

Stevens, Paul, *International Gas.* New York: St. Martin's Press, 1986.

Stevenson, Garth, *The Politics of Canada's Airlines.* Toronto: Univ. of Toronto Press, 1987.

Stinner, William R., *Return Migration and Remembrances; Developing a Caribbean Perspective.* R11E5 Occasional Papers No. 3. Washington: Smithsonian Institution, 1982.

Strand, Paul J. and Woodrow Jones, Jr., *Indochinese Refugees in America.* Durham, NC: Duke Univ. Press, 1985.

Suksamran, Somboon, *Political Buddhism in Southeast Asia.* New York: St. Martin's Press, 1977.

Super and Wright, *Food, Politics and Society in Latin America.* Lincoln: Univ. of Nebraska Press, 1985.

T

Tanzer, Michael, *The Political Economy of International Oil and the Underdeveloped Countries.* Boston: Beacon Press, 1969.

Tapper, Richard, *The Conflict of Tribe and State in Iran and Afghanistan.* New York: St. Martin's Press, 1983.

Teclaff, Ludwik A., *Water Law in Historical Perspective.* Buffalo, NY: William S. Hein, 1985.

Terrie, Philip G., *Forever Wild: Environmental Aesthetics and the Adirondacks Forest Preserve.* Philadelphia: Temple Univ. Press, 1985.

Thiesenhusen, William C., *Latin American Agriculture: Structure and Reform.* London: Allen & Unwin, 1987.

Thompson, Dennis, *Ethnicity, Politics and Development.* London: Frances Pinter, 1986.

Tickell, Crispin, *Climatic Change and World Affairs.* Lanham, MD: Univ. Press of America, 1986.

Tornaritis, Criton G., *The Evolution of the Law Relating to Land In Cyprus.* Nicosia, Cyprus, 1982.

Treverton, Gregory, *Energy and Security.* Totowa, NJ: Rowman & Littlefield, 1981.

Tyson, James L., *U.S. International Broadcasting and National Security.* New York: Ramapo Press, 1983.

U

United Nations, *Air Pollution Across Boundaries.* New York, 1985.

Utton, Albert E. (ed.), *Pollution and International Boundaries; United States–Mexican Environmental Problems.* Albuquerque: Univ. of New Mexico Press, 1973.

V

Van Lier, Irene H., *Acid Rain and International Law.* Toronto; Alphen aan den Rijn, Neth.: Sijthoff & Noordhoff, 1981.

Van Zandt, J. Parker, *The Geography of World Air Transport.* Washington: Brookings Institution, 1944.

Verney, Douglas V., *Three Civilizations, Two Cultures, Canada's Political Traditions.* Durham, NC: Duke Univ. Press, 1986.

Vietor, Richard H.K., *Environmental Politics and the Coal Coalition.* College Station: Texas A&M Univ. Press, 1986.

Vig, Norman and Michael Kraft, *Environmental Policy in the 1980's: Reagan's New Agenda.* Washington: Congressional Quarterly, CQ Press, 1984.

Vocke, Harald, *The Lebanese War.* New York: St. Martin's Press, 1978.

Volkan, Vamik D., *Cyprus—War and Adaptation; a Psychoanalytic History of Two Ethnic Groups in Conflict.* Charlottesville: Univ. Press of Virginia, 1979.

W

Warnock, John W., *The Political Economy of Hunger.* London: Routledge & Kegan Paul, 1987.

Wassenbergh, H.A. and H.P. van Fenema (eds.), *International Air Transport in the Eighties.* Deventer, Neth.: Kluwer, 1981.

Watson, J. Wreford and Timothy O'Riordan, *The American Environment: Perceptions and Policies.* New York: Wiley, 1976.

Weaver, R. Kent, *The Politics of Industrial Change; Railway Policy in North America.* Washington: Brookings Institution, 1985.

Wetstone, Gregory S. and others, *Acid Rain in Europe and North America: National Response to an International Problem.* Washington: Environmental Law Institute, 1983.

Wheeler, G., *Racial Problems in Soviet Muslim Asia.* London: Oxford Univ. Press, 1962.

Williams, Colin H., *National Separatism.* Cardiff: Univ. of Wales Press, 1982.

Wilcher, Marshall E., *Environmental Cooperation in the North Atlantic Area.* Lanham, MD: Univ. Press of America, 1980.

Wimbush, Enders, *Soviet Nationalities in Strategic Perspective*. New York: St. Martin's Press, 1985.

Wionczek Guzman, Miguel and others (eds.), *Energy Policy in Mexico: Problems and Prospects for the Future*. Boulder, CO: Westview, 1985.

Wirsing, Robert G. (ed.), *Protection from Ethnic Minorities*. New York: Pergamon, 1981.

Woolcock, Helen R., *Rights of Passage; Emigration to Queensland in the Nineteenth Century*. London: Tavistock, 1986.

Worldwatch Papers. Washington: Worldwatch Institute. Monographs, topics include environment, energy, food, population, national security, health, migration, and related topics.

X

Xydis, S.G., *Cyprus: Conflicts and Conciliation 1954–1958*. Columbus: Ohio State Univ. Press, 1967.

Y

Yager, Joseph A., *International Cooperation in Nuclear Energy*. Washington: Brookings Institution, 1981.

Yang, Martin M.C., *Socio-Economic Results of Land Reform in Taiwan*. Honolulu: University of Hawaii Press, 1970.

Young, Oran R., *Natural Resources and the State*. Berkeley: Univ. of California Press, 1981.

Z

Zachariah, K.C. and Julien Condé, *Migration in West Africa; Demographic Aspects*. New York: Oxford Univ. Press, 1981.

Zamosc, Leon, *The Agrarian Question and the Peasant Movement in Colombia*. New York: Cambridge Univ. Press, 1986.

Zandt, J.P. van, *The Geography of World Air Transport*. Washington: Brookings Institution, 1944.

Zaslavsky, Victor and Robert J. Brym, *Soviet–Jewish Emigration and Soviet Nationality Policy*. New York: St. Martin's Press, 1983.

Ziegler, *Environmental Policy in the USSR*. Amherst: Univ. of Massachusetts Press; London: Frances Pinter, 1987.

Zucker, Norman L. and Naomi Flink Zucker, *The Guarded Gate; the Reality of American Refugee Policy*. New York: Harcourt Brace Jovanovich, 1987.

Periodicals

A

Agnew, John A., "Political Regionalism and Scottish Nationalism in Gaelic Scotland," *Canadian Review of Studies in Nationalism*, 8 (1981), 115–129.

———, "Place and Political Behaviour: The Geography of Scottish Nationalism," *Political Geography Quarterly*, 3, 3 (July 1984), 191–206.

Alexander, A., "Scottish Nationalism: Agenda Building, Electoral Process, and Political Culture," *Canadian Review of Studies in Nationalism*, 7 (1980), 372–385.

Anderson, James, "Regions and Religions in Ireland: A Short Critique of the 'Two Nations' Theory," *Antipode*, 11 (1980), 44–53.

Aristova, T., "Kurds of Turkmen SSR," *Central Asian Review*, 18, 4 (1965), 302–309.

B

Baker, Oliver E., "The Population Prospect in Relation to the World's Agricultural Resources," *Journal of Geography*, 46, 6 (September 1947), 203–220.

Barker, Mary L., "National Parks, Conservation, and Agrarian Reform in Peru," *Geographical Review*, 70, 1 (January 1980), 1–18.

Beckingham, C.F., "The Cypriot Turks," *Royal Central Asian Journal* (1956), 126–130.

Bedir-Kahn, Kamuran Ali, "The Kurdish Problem," *Royal Central Asian Journal*, 136 (July–October 1949), 237–248.

Bilder, Richard, "Controlling Great Lakes Pollution: A Study in United States–Canadian Environmental Cooperation," *Michigan Law Review*, 70 (January 1972), 473–556.

Blowers, Andrew, "The Triumph of Material Interests—Geography, Pollution and the Environment," *Political Geography Quarterly*, 3, 1 (January 1984), 49–68.

Boal, Frederick W., "Two Nations in Ireland," *Antipode*, 11 (1980), 38–44.

Bole, Janice J., "Feast or Famine: Do Ethiopians Have a Choice?" *Dickinson Journal of International Law*, 5, 1 (Fall 1986), 103–131.

Brookfield, Harold C., "Some Geographic Implications of the Apartheid and Partnership Policies of Southern Africa," *Transactions,*

Institute of British Geographers, 23 (1957), 225–247.

Brown, Lawrence A. and Victoria A. Lawson, "Migration in Third World Settings, Uneven Development, and Conventional Modeling: A Case Study of Costa Rica," *Annals, AAG,* 75, 1 (March 1985), 29–47.

C

Caroe, Olaf, "The Geography and Ethnics of India's Northern Frontiers," *Geographical Journal,* 126, 3 (September 1960), 298–309.

Cerwin, Vladimir, "Problem in the Integration of the Afghan Nation," *Middle East Journal,* 6, 4 (1952), 400–416.

Chamie, Joseph, "The Lebanese Civil War: An Investigation into the Causes," *World Affairs,* 139, 3 (Winter 1976), 171–188.

Chaudhuri, Manoranjan, "Geopolitical Aspects of World Population," *Calcutta Geographical Review,* 11, 3–4 (1949), 48–54.

Cohen, Ronald, "Ethnicity: Problems and Focus in Anthropology," *Annual Review of Anthropology,* 7 (1978), 379–403.

Condominas, G., "Aspects of a Minority Problem in Indochina," *Pacific Affairs,* 24, 1 (1951), 77–82.

Connor, Walker, "The Politics of Ethnonationalism," *Journal of International Affairs,* 27 (1973), 1–21.

Constantinou, Stavros T. and Nicholas D. Dimantides, "Modeling International Migration: Determinants of Emigration from Greece to the United States, 1820–1980," *Annals, AAG,* 75, 3 (September 1985), 352–369.

D

Desbarats, Jacqueline, "Thai Migration to Los Angeles," *Geographical Review,* 69, 3 (July 1979), 302–318.

———, "Indochinese Resettlement in the United States," *Annals, AAG,* 75, 4 (December 1985), 522–538.

Douglas, Neville, Review Essay: "Historical Myth and Materialist Reality: Nationalist and Structuralist Perspectives on the Conflict in Ireland," *Political Geography Quarterly,* 6, 1 (January 1987), 89–101.

E

Edmonds, G.J., "The Place of the Kurds in the Middle Eastern Scene," *Royal Central Asian Journal,* 45 (April 1958), 141–153.

———, "The Kurds and the Revolution in Iraq," *Middle East Journal,* 13, 1 (Winter 1959), 1–10.

———, "The Kurdish War in Iraq: A Plan for Peace," *Royal Central Asian Journal,* 54 (1967), 10–22.

———, "The Kurdish National Struggle in Iraq," *Asian Affairs,* 58, 2 (June 1971), 147–158.

———, "Kurdish Nationalism," *Journal of Contemporary History,* 6, 1 (1971), 87–107.

Ethnic and Racial Studies. London: Routledge & Kegan Paul. Published quarterly since 1977.

F

Falah, G., "The Development of the 'Planned Bedouin Settlement' in Israel, 1964–1982," *Geoforum,* 14, 311–323.

Fenwick, C.G., "Freedom of Communication Across National Boundaries," *American Journal of International Law,* (1950), 107.

Food Policy. London: Butterworths. Published quarterly since 1975.

From Foreign Workers to Settlers? Transnational Migration and the Emergence of New Minorities. Special issue of the *Annals of the American Academy of Political and Social Science,* 485, (May 1986).

G

Garber, Larry, and Courtney M. O'Connor, "The 1984 UN Sub-Commission on Prevention of Discrimination and Protection of Minorities," *American Journal of International Law,* 79, 1 (January 1985), 168–180.

Global Environmental Problems: A Legal Perspective. Annual Symposium, Syracuse Journal of International Law and Commerce.

The Global Political Economy of Food. Special Issue of *International Organization,* 32, 3 (Summer 1978).

"Great Britain and Iraq: 1914–1958; the Kurds," *Round Table,* No. 49 (June 1959), 277–279.

Griffiths, Ieuan L., "The Tazama Oil Pipeline," *Geography,* 54, 2 (April 1969), 214–217.

H

Hantke, Jeanette, "The 1982 Session of the UN Sub-Commission on Prevention of Discrimination and Protection of Minorities," *American Journal of International Law,* 77, 3 (July 1983), 651–662.

Harbeson, Robert W., "Transportation: Achilles Heel of National Security," *Political Science Quarterly,* 74, 1 (1959), 1–20.

Hauser, Philip M., "Demographic Dimensions of World Politics," *Science,* 131 (1960), 1641–1647.

Hoyle, Brian Stewart, "African Politics and Port Expansion at Dar es Salaam." *Geographical Review,* 68, 1 (January 1978), 31–50.

Huang, Thomas T.F., "Some International and Legal Aspects of the Suez Canal Question," *American Journal of International Law*, 51, 2 (1957), 277–307.

J

Jones, Huw R., "Modern Emigration from Scotland to Canada," *Scottish Geographical Magazine*, 95, 1 (April 1979), 4–12.

Jones, Richard C., "Undocumented Migration from Mexico: Some Geographical Questions," *Annals, AAG*, 72, 1 (March 1982), 77–87.

K

Katzenellenbogen, Simon E., "Zambia and Rhodesia: Prisoners of the Past; a Note on the History of Railway Politics in Central Africa," *African Affairs*, 73, 290 (January 1974), 63–66.

Kedourie, Elie, "Continuity and Change in Modern Iraqi History," *Asian Affairs*, 62 (June 1975), 140–146.

Kliot, Nurit, "Lebanon—a Geography of Hostages," *Political Geography Quarterly*, 5, 3 (July 1986), 199–220.

Kotis, Linda L. (ed.), "Feast or Famine: Issues, Problems, and Procedures Relating to Massive Relief Efforts with a Focus on the African Crisis," *Vanderbilt Journal of Transnational Law*, 19, 2 (Spring 1986), 33–464.

Krishan, Radha, "Geopolitics of Petroleum," *Indian Geographer*, 6, 1 (1961), 90–110.

"The Kurds of Transcaucasia and Persia," *Central Asian Review*, 7, 2 (1959), 163–201.

Kurganov, I., "The Problem of Nationality in Soviet Russia," *Russian Review*, 10, 4 (1951), 253–267.

L

Lambert, R.D., "Factors in Bengali Regionalism in Pakistan," *Far Eastern Survey*, 28, 4 (1959), 49–58.

Laponce, J.A., "The French Language in Canada: Tensions Between Geography and Politics," *Political Geography Quarterly*, 3, 2 (April 1984), 91–104.

———, "More About Languages and Their Territories: A Reply to Pattanayak and Bayer," *Political Geography Quarterly*, 6, 3 (July 1987), 265–267.

Lee, Luke T., "The UN Group of Governmental Experts on International Co-operation to Avert New Flows of Refugees," *American Journal of International Law*, 78, 2 (April 1984), 480–484.

Lee Yong Leng, "Language and National Cohe-sion in Southeast Asia," *Contemporary Southeast Asia*, 2, 3 (December 1980), 226–240.

Loya, A., "Radio Propaganda of the United Arab Republic: An Analysis," *Middle East Journal* (1962), 98–110.

M

MacLaughlin, James G. and John A. Agnew, "Hegemony and the Regional Question: The Political Geography of Regional Industrial Policy in Northern Ireland, 1945–1972," *Annals, AAG*, 76, 2 (June 1986), 247–261.

Magraw, Daniel Barston, "Transboundary Harm: The International Law Commission's Study of 'International Liability,' " *American Journal of International Law*, 80, 2 (April 1986), 305–331.

Marinelli, O., "The Regions of Mixed Populations in Northern Italy," *Geographical Review*, 7 (1919), 129–148.

Mayer, Harold M., "Politics and Land Use: The Indiana Shoreline of Lake Michigan," *Annals, AAG*, 54, 4 (December 1964), 508–523.

Michel, Aloys S., "The Canalization of the Moselle and West European Integration," *Geographical Review*, 52, 4 (October 1962), 475–491.

Mitchell, James K., "Social Violence in Northern Ireland," *Geographical Review*, 69, 2 (April 1979), 179–201.

Moseley, G., "China's Fresh Approach to the National Minority Question," *China Quarterly*, 24 (1965), 15–27.

Mountjoy, A.B., "The Suez Canal, 1935–1955," *Geography*, 52 (1957), 186–190.

N

Newman, David, "The Development of the Yishuv Kehillati: Political Process and Settlement Form," *Tijdschrift voor Economische en Sociale Geografie*, 75 (1984).

———, "Ideological and Political Influences on Israeli Rurban Colonization: The West Bank and Galilee Mountains," *Canadian Geographer*, 28, 2 (Summer 1984).

O

O'Loughlin, John, "Distribution and Migration of Foreigners in German Cities," *Geographical Review*, 70, 3 (July 1980), 253–275.

Osman, Abdillahi Said Osman, "International Cooperation to Avert the Flows of Refugees in Africa," *Nordisk Tidsskrift for International Ret*, 51, 3–4 (1982), 179–188.

Owens, W.H., "International Roads for West

Africa," *New Commonwealth*, 40, 6 (June 1962), 363–366.

P

Pallis, A.A., "Racial Migrations in the Balkans 1912–1924," *Geographical Journal*, 66, 4 (October 1925), 315–331.

Pattanayak, D.P. and J.M. Bayer, "Laponce's 'The French Language in Canada: Tensions Between Geography and Politics—A Rejoinder." *Political Geography Quarterly*, 6, 3 (July 1987), 261–263.

Philip, George, "Mexican Oil and Gas: The Politics of a New Resource," *International Affairs*, 56, 3 (Summer 1980), 474–481.

Pirie, Gordon H., "The Decivilizing Rails: Railways and Underdevelopment in Southern Africa," *Tijdschrift voor Economische en Sociale Geografie*, 73, 4 (1982), 221–228.

———, "Race Zoning in South Africa: Board, Court, Parliament, Public," *Political Geography Quarterly*, 3, 3 (July 1984), 207–221.

Pringle, D.C., "The Northern Ireland Conflict: A Framework for Discussion," *Antipode*, 11 (1980), 28–38.

R

Rosencranz, Armin, "The ECE Convention of 1979 on Long-Range Transboundary Air Pollution," *American Journal of International Law*, 75, 4 (October 1981), 975–982.

Ryan, Claude, "The French Canadian Dilemma," *Foreign Affairs*, 43, 3 (April 1965), 462–474.

S

Segal, Aaron, "The Politics of Land in East Africa." *Africa Report*, 12, 4 (1967), 46–50.

Segal, B., "Geopolitics of Broadcasting," *Washington Quarterly*, 6, 2, 140–148.

Shrestha, Nanda R., "The Political Economy of Underdevelopment and External Migration in Nepal," *Political Geography Quarterly*, 4, 4 (October 1985), 289–306.

Smith, Anthony D., "Ethnic Identity and World Order," *Millennium: Journal of International Studies*, 12 (1982), 149–161.

Smith, C.G., "The Great Ditch Across the Suez Isthmus," *Geographical Magazine*, 41 (1969), 259–270.

Smogorzewski, K.M., "The Russification of the Baltic States," *World Affairs*, 4, 4 (1950), 468–481.

Soffer, Arnon, "Geographical Aspects of Changes Within the Arab Community of Northern Israel," *Middle Eastern Studies*, 19 (1983), 213–243.

Stephenson, Glenn V., "Pakistan: Discontiguity and the Majority Problem," *Geographical Review*, 58, 2 (April 1968), 195–213.

Sweet, J.V., "The Problem of Nationalities in Soviet Asia," *Ukranian Quarterly*, 10 (1954), 229–235.

T

Thomas, Trevor M., "World Energy Resources: Survey and Review," *Geographical Review*, 63, 2 (April 1973), 246–258.

V

Van Dyke, Vernon, "Self-determination and Minority Rights," *International Studies Quarterly*, 13 (September 1969), 223–253.

Velikonja, Joseph, "Postwar Population Movements in Europe," *Annals, AAG*, 48, 4 (December 1958), 458–472.

Voskuil, Walter H., "Coal and Political Power in Europe," *Economic Geography*, 18, 3 (July 1942), 247–258.

W

Wallerstein, Immanuel, "Ethnicity and National Integration in West Africa," *Cahiers d'études africaines*, 1 (1960), 129–139.

Wanklyn, Harriet G., "The Middle People: Resettlement in Czechoslovakia," *Geographical Review*, 38, 1 (January 1948), 28–42.

Watts, Michael J., "Politics, the State and Agrarian Development: A Comparative Study of Nigeria and the Ivory Coast," *Political Geography Quarterly*, 5, 2 (April 1986), 103–125.

Wenner, Lettie M., "Arab–Kurdish Rivalries in Iraq," *Middle East Journal*, 17, 1–2 (Winter–Spring 1963), 68–82.

Westermann, W.L., "Kurdish Independence and Russian Expansion," *Foreign Affairs*, 24 (July 1946), 675–686.

Wilkinson, H.R., "Yugoslav Kosmet: The Evolution of a Frontier Province and Its Landscape," *Transactions and Papers, Institute of British Geographers*, 21 (1955), 171–193.

Williams, Colin H., "Ethnic Separatism in Western Europe, *Tijdschrift voor Economische en Sociale Geografie*, 71 (1980), 142–158.

———, "Ideology and the Interpretation of Minority Cultures," *Political Geography Quarterly*, 3, 2 (April 1984), 105–125.

Wise, M.J., "The Impact of a Channel Tunnel on the Planning of South Eastern England," *Geographical Journal*, 131, 2 (June 1965), 167–185.

Part Nine

LOOKING AHEAD

Chapter 37

POLITICAL GEOGRAPHY FOR THE FUTURE

Throughout this book, as we have explored most of the highways and some of the byways of political geography, we have had glimpses of topics new and undeveloped, of ideas that need investigation, of problems that political geographers have little studied but which could be moved toward resolution by their contributions. We have called attention to some of these opportunities as we have gone along. Now it seems appropriate to peer into the future and suggest some other subjects that are virtually virgin territory for political geographers. Some have already been mentioned but require additional discussion; others we have saved until now. All are offered here to suggest how much there is left to do and how political geographers can help to do it. These suggestions are by no means exhaustive.

THE UNITED STATES

Stanley Brunn, in his provocative book *Geography and Politics in America*, suggests six major developments "as potential characteristics of the political geography in the United States in the year 2000." They are (1) the erosion of boundaries, (2) the emergence of city-states, (3) the rise of new political cultures, (4) the reorientation of voting patterns, (5) centralized government planning, and (6) the politicalization of the environment. He then suggests the consolidation of our first or-

der civil divisions into 16 states, one of which would include Puerto Rico and the Virgin Islands.* He discusses each development in turn, and each is worth more space than we can devote to it here. We can, however, add some suggestions to his list.

(1) Considering the continuing heavy immigration of Spanish-speaking people into the United States mainland and their high rate of natural population increase, it is likely that by the year 2000 they will be playing a much larger role than now in our society, economy, and politics. (2) The gradual changeover from reliance on fossil fuels to other sources of energy will very likely reduce the political power of legislators from the oil, coal, and gas-producing regions. (3) We may well see by the end of the century a strong and perhaps successful movement to transfer the capital of the country from Washington to a more central location, not for ease of access, but for social, political, and psychological reasons. (4) If world disarmament, even incomplete disarmament, becomes a reality and not just a dream, it will have profound effects on the distribution of political power in the country since military bases and production tend to be regionalized.

The United States alone offers countless topics for theses and dissertations in po-

*S.D. Brunn, *Geography and Politics in America*, New York, Harper & Row, 1974, p. 411.

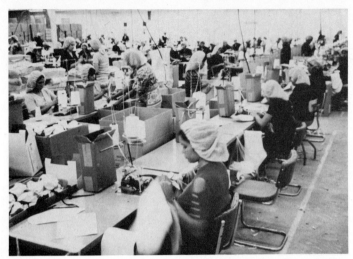

Surgical mask sewing operation in Ciudad Juarez, Mexico. Of the
hundreds of free zones around the world, the zone along the
U.S. border in Mexico is among the most unusual. In 1965 the
Border Industrial Program was inaugurated to promote border
industrialization, stimulate Mexican supporting industries and
reduce unemployment along the border. In 1986, there were
806 textile and assembly plants called *maquiladoras* within the
20 km (12.2 mi) deep zone. They are mostly owned by U.S. and
Japanese firms, employing more than 236,000 people, mostly
women. There were also 59 *maquiladoras* employing nearly 14,000
people located in other free zones in the interior of Mexico. (El
Paso Industrial Development Corporation)

litical geography, but the rest of the world
offers even more.

VERY SMALL PLACES

We have devoted two chapters to discussions of minor civil divisions, municipalities, and special purpose districts. Elsewhere throughout the book we have
referred to the territories of street gangs,
to exclaves of States, neutral zones, international straits, capital cities, and other
very small places, most of which have considerable politicogeographical importance. But other very small places in the
world deserve analysis by political geographers. We offer a few suggestions here.

In our discussion of decolonization we
raised the question of independence for
very small fragments of colonial territory,
and in our discussion of the size of States
we pointed out that there is no real correlation between size and success. Small
can be not only beautiful but practical as

United States foreign trade zones. These are called free trade zones or free ports in other countries.
They are secured areas legally outside the customs territory of the country. Their purpose is to
attract and promote international trade and they are normally located in or near customs ports
of entry. In the United States they are operated as public utilities by states, political subdivisions,
or qualified corporations under Customs supervision. Foreign and domestic merchandise may
be moved into the zones for storage, exhibition, assembly, disassembly, packing, repacking,
sorting, grading and cleaning, and may be manufactured or otherwise processed without payment
of customs duties unless and until the goods enter the customs territory for domestic consumption. Such zones are very popular around the world and are still proliferating; they are making
a great contribution to the increase and free flow of trade.

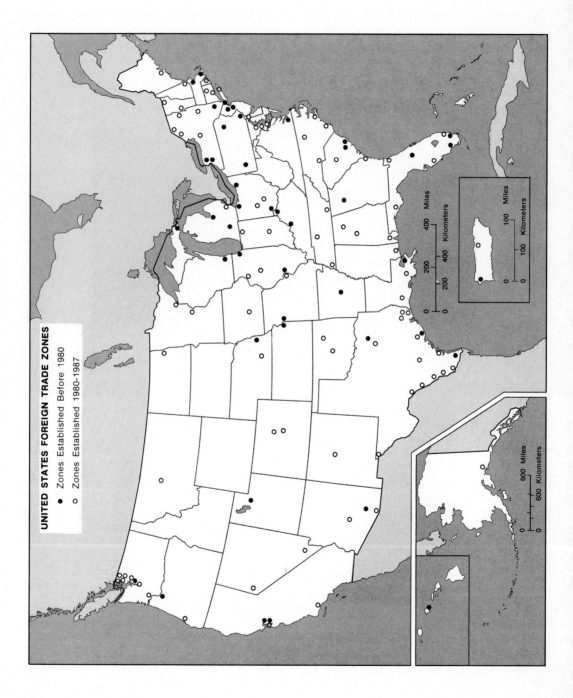

UNITED STATES FOREIGN TRADE ZONES

● Zones Established Before 1980

○ Zones Established 1980-1987

well. But is there no lower limit to the size of States? Must every remaining scrap of colonial empire become a separate State? And what about tiny States that are already independent? If they cannot bear the costs of independence, must they forever be maintained as international charity cases? Is there some realistic and acceptable al-

ternative to statehood for the people who live in very small places?

A widespread and growing practice in the world is the establishment of economic enclaves within countries. These include foreign-owned mines and plantations and their "company towns" as well as free ports and free trade zones—all of

Life in centrally planned economies. Top: The central office of a Soviet collective farm (kolkhoz) near Kharkov. Soviet efforts to increase the annual output of agricultural products have focused on the collective, where various inducements (including slogans, exhortations, public praise, and prizes) are employed to ensure the maximum effort of every worker. (Harm J. de Blij) Bottom: Similar techniques are used in industry. This sign in Wroclaw, Poland boasts of a steady increase in tractor production. Wroclaw is in the territory taken from Germany by Poland after World War II and was called Breslau. (Renée Glassner)

which have been mentioned before in other contexts. Free zones have become especially important. At the turn of the century there were only 11 around the world; in 1985 there were over 400 to be found on every continent except Antarctica. Of particular importance in political geography are export-processing zones and transit zones for land-locked countries. In terms of numbers, distribution, and impact on the host countries, they may well be extremely important.

STATES WITH CENTRALLY PLANNED ECONOMIES

At present there are 16 countries in Europe, Asia, and Latin America frequently referred to as "communist" but more accurately described as having centrally planned economies. Perhaps another score, mostly in Africa, practice some variety of socialism. They have scarcely been considered by political geographers except in terms of boundaries, ethnic minorities, military alliances, and other traditional studies. But political geographers can help determine whether there is, in fact, a "second world" by classifying all these countries in a rational way. They can compare international relations among all or any of these groups with relations among countries with capitalist or mixed systems. They can analyze the role of ideology in decision making in such matters as ecology, economic development, foreign trade, and population. They can compare the centrifugal and centripetal forces operating within them with those operating in other States. Studies of this nature will surely help us to understand them better.

NONGOVERNMENTAL ORGANIZATIONS

In our chapter on international organizations, we called attention to the role played by international nongovernmental organizations (INGOs) in many matters of concern to geographers, especially political geographers. In individual countries, particularly in North America, Western Europe, Israel, Australia, and New Zealand, nongovernmental organizations (NGOs) are enormously important in coordination of activities; transmitting information to their members and the general public; setting professional, technical, and ethical standards; providing development aid to poor countries and to poor sectors of rich countries; influencing policies of governments; and forecasting economic, political, social, demographic, and ecological trends.

Here is an important subject only skimmed by political scientists (principally in studies of "pressure groups") and virtually ignored by geographers, even political geographers interested in decision-making and politicogeographical behavior. Much can be revealed about the importance of NGOs and INGOs in matters of concern to political geographers by studies of, say, the International Planned Parenthood Federation or the Society for International Development or the Sierra Club or the roles of various unofficial groups in particular events and processes. Cooperation of political geographers with political scientists, sociologists, and psychologists could be most rewarding in understanding the interactions of these groups and the attitudes and behavior of their members and nonmembers. Individuals, acting through such groups, may be more influential than we realize.

PUBLIC POLICY

There is currently a great deal of interest among political geographers in decision making at all levels of government with regard to what has come to be called "welfare geography." This is sometimes defined as the study of who gets what, where, and why. This includes the study of the allocation of funds for various purposes by the central government to the various civil divisions; the distribution of military bases and defense contracts, the service areas of various government offices, the diffusion of innovation by governments at various levels, and similar matters. Not yet studied adequately by political geographers are such things as agricultural subsidies of all kinds, the governmental role in the distribution of health care facilities, and the allocation of public resources by cities of various sizes and types. Careful studies of these and related topics can help us understand how our tax money is spent and how such decisions affect not only the landscape, but the kind of society we will have in coming generations.

INDIVIDUAL COUNTRIES AND REGIONS

A staple of political geography during the first half of this century was the study of the political geography of an individual country or of a region or subregion. Indeed, most political geography textbooks and anthologies were organized regionally rather than systematically, as this one is. While the world is changing too rapidly to make regional political geography practical for texts now, there is a real need to return to the production of articles and monographs that will help us understand contemporary events. Would not a serious political geography of Israel be useful? Or of Nicaragua or the Arctic or the Middle East or Africa? In their quest for ever-new behavioral, statistical, or theo-

retical subjects, many contemporary political geographers seem to be forgetting their roots. If they are, indeed, geographers—not social historians, political scientists, economists, or sociologists—then they must be concerned with *places*, where things are happening. Those concerned with political units smaller than a State would perform a useful service if they would produce political geographies of Punjab, of the Midlands, of the South African "homelands," of New York City, and of some of the component units of the United Arab Emirates. Naturally, to be scientific and practical, they would have to go beyond description and use the best tools of modern political geography for analysis. This would help us all to understand better the stories behind the headlines.

INDIGENOUS PEOPLES

In Chapter Fifteen, we discussed the Indian reservation as a kind of special purpose district; in Chapter Thirty-Three we discussed ethnic minorities within States; and scattered through the book are other references to particular problems of aboriginal peoples. But political geographers have simply not done careful studies of the number, distribution, and characteristics of indigenous or aboriginal peoples and their relations with the dom-

Bedouin in Israel. Like nomads everywhere, often under government pressure or even coercion, Israel's Bedouin are settling down. Whether still nomadic, as are these people, settled, or in a transitional stage, Israel's Bedouin have played an important role in the country's political and security affairs since before the founding of the State in 1948. Nomads in other countries have also been politically significant and warrant study by political geographers. (Israeli Consulate General, New York)

Inuit in Canada. The Inuit (Eskimos) have been severely impacted by the imposition of alien cultures on their societies, and they are taking political action to protect their legal rights and way of life. The first Inuit Circumpolar Conference of representatives from Alaska, Canada, and Greenland was held in Barrow, Alaska in June 1977. The Conference is now affiliated with the United Nations. In September 1985 representatives of Scandinavian Sami (Lapps) joined the Inuit in Montreal for an Arctic Policy Conference to discuss broad issues of indigenous peoples in the Arctic. (Canadian Consulate General, New York)

inant ethnic group, their roles in the political and economic life of the States which have grown up around them, and their relations with their fellows across national boundaries.

In order to understand what's going on now in Central America, we should know more about the Miskito and other Indians who live in the Nicaraguan lowlands. Japan is generally considered to have a "homogeneous" society, yet this view completely ignores the Ainu. The treatment of the Papuans of Irian Jaya by the Indonesian authorities should be better understood. What is the role played in Egypt by the Copts, descendants of the original Egyptians who were overwhelmed by the Arab invaders? How do the Canadian and Australian aboriginal reserves differ from the American ones? We may be certain that by the end of this century we will be hearing much more about indigenous peoples around the world through our media of mass communication. Political geographers should provide a deeper understanding of them than these media can.

A NEW INTERNATIONAL ECONOMIC ORDER

Just as decolonization was one of the momentous processes in international affairs during the quarter-century following World War II, the construction of a New International Economic Order is likely to dominate the quarter-century following that. The attainment of political independence is being followed by a struggle in many ways more difficult, the one for economic independence. In both cases, concurrently with independence comes interdependence. The hope is, however, that the interdependence will be based on equality, not on dominance/dependence.

We have already discussed some individual elements in this movement for a New International Economic Order: the role of the United Nations in economic and social development, the OPEC oil embargo and price hikes, the drive to reform international trade, the activities of regional organizations in development, the shifting control of marine resources, and

the importance of transnational corporations. All of these, and more, are leading
to radical transformations in the world's
economic system that must inevitably have
profound consequences for every man,
woman, and child on earth. And if the
world economic system is to change rad-
ically over a period of time, then the political system must inevitably change also.

The most important component of these
transformations would probably be *closing the gap between rich and poor countries,* not by making the rich poorer but
by making the poor less poor. Although
the gap is still growing, not shrinking, we
have witnessed in recent years the emergence of an international middle class, primarily the newly industrializing countries
(NICs) of Asia and Latin America. Their
increasing prosperity is encouraging—but
precarious. Certainly the greatest development effort must be made by the poor
countries themselves, but they need help.
So far, they have gotten relatively little.
Some figures, compiled from several official sources, might be useful:

1. Average ODA (official development
    assistance) from OECD countries in
    1985 equaled 0.49 percent of com-
    bined GNP.

ODA (1985)	Percentage of GNP
United States	0.24%
Denmark	0.80
Sweden	0.86
Netherlands	0.91
Norway	1.03

2. In 1984, the OPEC countries donated
    an average of 1.16 percent of GNP to
    poorer countries.

3. The United States ranked seventeenth
    among OECD donors of aid as a per-
    centage of GNP in 1984.

4. In terms of constant (1950) dollars, U.S.
    "foreign aid" in 1983 was about the
    same as it was during the period 1955–
    1970.

5. In 1984, U.S. "foreign aid" amounted
    to:
    $4,885,400,000 for economic assist-
    ance
    4,692,000,000 for budgeted military
    assistance
    9,348,000,000 for all military assist-
    ance

6. Of the total of some $14 billion in 1984
    aid, $5 billion went to just two coun-
    tries, Israel and Egypt.

7. In the 1987 United Nations Develop-
    ment Programme pledging confer-
    ence, the United States pledged to
    UNDP $0.50 per capita; Norway
    pledged $13.50 per capita.

8. Consistently, 70 to 75 percent of U.S.
    "foreign aid" is spent *in the United
    States.*

These figures should raise some interesting questions about priorities. Here are
some more:*

1. During the period 1960–1985, world
    military expenditures per capita rose
    about 125 percent; world per capita
    GNP rose only about 52 percent.

2. World military expenditures at pres-
    ent are about $1.7 million *a minute.*

3. 1986 estimated stockpile of nuclear
    weapons worldwide is 60,000 units.

4. In 142 countries:
    The total of military personnel was
    26,461,000
    The total number of physicians was
    4,567,400
    The total number of teachers was
    33,520,000

Note that these figures tell us nothing
about quality. Today's nuclear weapons,
for example, are vastly more powerful than
these of 1945; soldiers today are generally
far better trained than teachers. Nevertheless, the figures alone raise more interesting and vital questions, questions
that political geographers should address.

*Ruth Leger Sivard, *World Military and Social Ex-
penditures,* 1986.

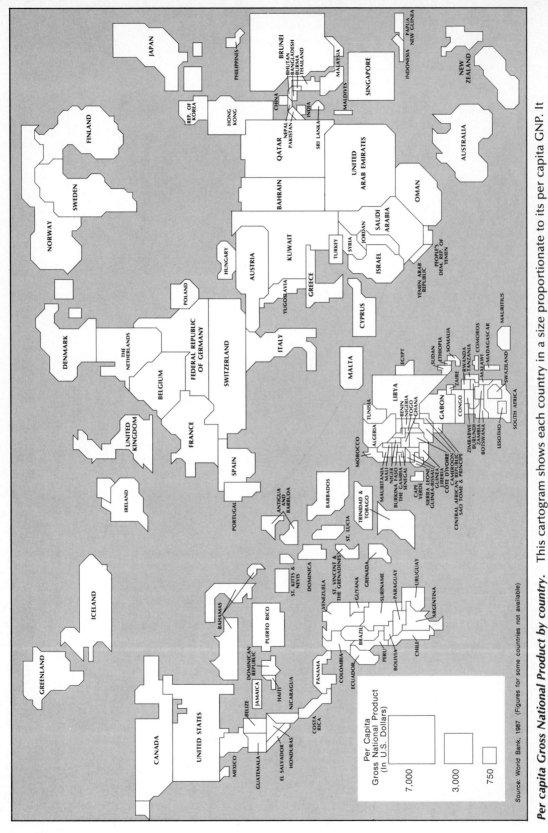

Per capita Gross National Product by country. This cartogram shows each country in a size proportionate to its per capita GNP. It is a very popular measure of prosperity or of economic development, but when used this way it has serious defects. The most important shortcoming is its failure to indicate the distribution of the product within the society. One of the goals of a New International Economic Order is a more equitable distribution of wealth in the world.

Map labels (clockwise and within map):

JAPAN, PHILIPPINES, BRUNEI, BHUTAN, BANGLADESH, BURMA, THAILAND, MALAYSIA, SINGAPORE, INDONESIA, PAPUA NEW GUINEA, NEW ZEALAND, REP. OF KOREA, HONG KONG, CHINA, NEPAL, PAKISTAN, INDIA, SRI LANKA, MALDIVES, AUSTRALIA, FINLAND, QATAR, UNITED ARAB EMIRATES, SWEDEN, NORWAY, BAHRAIN, OMAN, SAUDI ARABIA, KUWAIT, JORDAN, ISRAEL, SYRIA, TURKEY, PEOPLE'S DEM. REP. OF YEMEN, YEMEN ARAB REPUBLIC, HUNGARY, AUSTRIA, GREECE, YUGOSLAVIA, CYPRUS, POLAND, DENMARK, THE NETHERLANDS, FEDERAL REPUBLIC OF GERMANY, SWITZERLAND, ITALY, MALTA, BELGIUM, EGYPT, SUDAN, ETHIOPIA, SOMALIA, RWANDA, TANZANIA, COMOROS, MALAWI, MADAGASCAR, MAURITIUS, FRANCE, LIBYA, ZAIRE, CONGO, GABON, ZIMBABWE, BURUNDI, ZAMBIA, BOTSWANA, LESOTHO, SWAZILAND, SOUTH AFRICA, SPAIN, PORTUGAL, TUNISIA, ALGERIA, MOROCCO, BENIN, NIGERIA, TOGO, GHANA, CAMEROON, CENTRAL AFRICAN REPUBLIC, SAO TOMÉ & PRINCIPE, MAURITANIA, MALI, NIGER, BURKINA FASO, THE GAMBIA, SENEGAL, CAPE VERDE, SIERRA LEONE, GUINEA-BISSAU, GUINEA, LIBERIA, CÔTE D'IVOIRE, IRELAND, UNITED KINGDOM, ICELAND, GREENLAND, ANTIGUA AND BARBUDA, BARBADOS, ST. LUCIA, TRINIDAD & TOBAGO, ST. KITTS & NEVIS, DOMINICA, ST. VINCENT & THE GRENADINES, VENEZUELA, GUYANA, SURINAME, GRENADA, PARAGUAY, URUGUAY, ARGENTINA, BAHAMAS, PUERTO RICO, DOMINICAN REPUBLIC, HAITI, JAMAICA, BELIZE, COLOMBIA, ECUADOR, PERU, BOLIVIA, BRAZIL, CHILE, PANAMA, COSTA RICA, NICARAGUA, HONDURAS, EL SALVADOR, GUATEMALA, MEXICO, CANADA, UNITED STATES

Per Capita
Gross National Product
(In U.S. Dollars)

7,000
3,000
750

Source: World Bank, 1987 (Figures for some countries not available)

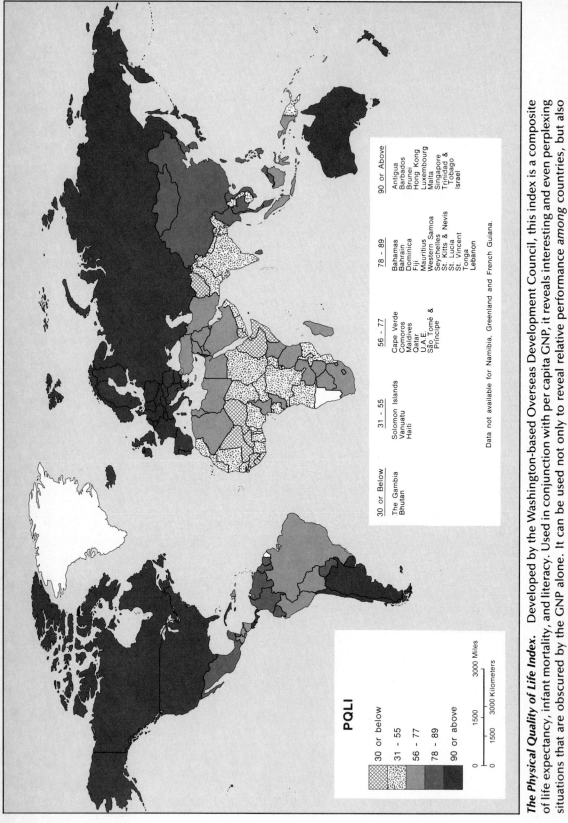

PQLI

	30 or below
	31 – 55
	56 – 77
	78 – 89
	90 or above

0   1500   3000 Miles

0   1500   3000 Kilometers

30 or Below	31 – 55	56 – 77	78 – 89	90 or Above
The Gambia	Solomon Islands	Cape Verde	Bahamas	Antigua
Bhutan	Vanuatu	Comoros	Bahrain	Barbados
	Haiti	Maldives	Dominica	Brunei
		Qatar	Fiji	Hong Kong
		U.A.E.	Mauritius	Luxembourg
		São Tomé &	Western Samoa	Malta
		Príncipe	Seychelles	Singapore
			St. Kitts & Nevis	Trinidad &
			St. Lucia	Tobago
			St. Vincent	Israel
			Tonga	
			Lebanon	

Data not available for Namibia, Greenland and French Guiana.

The Physical Quality of Life Index. Developed by the Washington-based Overseas Development Council, this index is a composite of life expectancy, infant mortality, and literacy. Used in conjunction with per capita GNP, it reveals interesting and even perplexing situations that are obscured by the GNP alone. It can be used not only to reveal relative performance *among* countries, but also to examine various sectors *within* countries.

606

THE ROLE OF POLITICAL GEOGRAPHERS IN THE FUTURE

In this book, we have suggested many topics that need study by political geographers. The clear implication has been that if political geographers investigate these and similar topics, analyze them using all available tools and methodologies and devise new ones where necessary, and make their conclusions known through theses, dissertations, lectures, and publications, the studies would be not only interesting, but useful to those exposed to them.

Political geographers should and will continue in their primary roles as objective observers, students, and critics, remaining apart from events and contributing their insights to the world at large. They have proved their value in such roles in the past and will continue to do so. We suggest, however, that political geographers can also make valuable contributions to society by entering the fields of political combat, testing their theories in the crucible of hard experience, offering practical solutions to real and immediate problems.

This suggestion for the practical application of the concepts of political geography, even ignoring geopolitics, is neither new nor unique. Several distinguished political geographers have served as The Geographer in the U.S. Department of State, including S. Whittemore Boggs, G. Etzel Pearcy, Robert Hodgson, Lewis Alexander, and the incumbent, George Demko. All have been advisors to the government as well as collectors, organizers, and analysts of politicogeographical information. Other political geographers have served the government in various capacities. Isaiah Bowman, one of our most outstanding political geographers, was an advisor to the American Delegation to Negotiate Peace at the Paris Peace Conference in 1919 and later became an advisor and confidant of President Franklin D. Roosevelt. More recently, one political geographer was deeply involved in the dispute between Maine and New Hampshire over their maritime boundary; another was very active in redrawing congressional and other political district boundaries in the State of Washington; another was advisor to Nepal in her negotiations with India for a new transit treaty.

Political geographers can serve as advisors, technicians, and expert witnesses for governments, international agencies, transnational corporations, courts, even political parties. They can advise on drawing boundaries of electoral districts and planning regions, on locations of seats of government, on mapping of political phenomena, on regulations for international rivers, and myriad other matters within their purview.

In applying their approaches, knowledge, techniques, and insights, however, political geographers, like other scholars, must avoid being too closely identified with particular policies, thus losing their status as experts and becoming simply partisans. They must also recognize that their contributions are only being added to the contributions of other experts in various fields, including the art of politics, and are not likely to be decisive. Finally, they must not get so involved in applications that they neglect study and research; that is, they can be scholar–practitioners but not solely practitioners. Within these limits, political geographers can make real contributions to the formulation and execution of policy at every level—local, national, and international. It is an opportunity and a challenge they should not ignore.

REFERENCES FOR PART NINE

Books and Monographs

A

Adeeji, Adebayo, and others, *Economic Crisis in Africa: African Perspectives on Development Problems*. Boulder, CO: Lynne Rienner, 1985.

Agarwala, P.N., *The New International Economic Order: An Overview*. New York: Pergamon, 1983.

Agbor-Tabi, Peter, *U.S. Bilateral Assistance in Africa: The Case of Cameroon*. Lanham, MD: Univ. Press of America, 1984.

Akinrinade, Clusola and J. Kurt Barling, *Paths to African Development*. London: Frances Pinter, 1987.

Apter, David E., *Rethinking Development*. Newbury Park, CA: Sage, 1987.

Atkinson, Michael M. and Marsha A. Chandler (eds.), *The Politics of Canadian Public Policy*. Toronto: Univ. of Toronto Press, 1983.

B

Baidoria, Pedro L., *Brief Review of the Legal System of Free Zones in Latin America*. Washington: Pan-American Union, General Secretariat of the Organization of American States. Document 321, 1968.

Beard, C.A. and George Radin, *The Balkan Pivot: Jugoslavia*. New York: Macmillan, 1929.

Bedjaoui, Mohammed, *Towards a New International Economic Order*. New York: Holmes & Meier, 1979.

Bennett, Robert, *Geography of Public Finance; Welfare Under Fiscal Federalism and Local Government Finance*. New York: Methuen, 1980.

Berberoglu, Berch, *The Internationalization of Capital*. New York: Praeger, 1987.

Betts, Raymond R., *Central and South East Europe, 1945–1948*. London: Royal Institute of International Affairs, 1950.

Black, Jan Knippers, *The Dominican Republic; Politics and Development in an Unsovereign State*. London: Allen & Unwin, 1986.

Blacksell, Mark, *Post-War Europe; a Political Geography*. Boulder, CO: Westview, 1977.

Boateng, E.A., *A Political Geography of Africa*. Cambridge: at the University Press, 1978.

Brandt, Willy, *North–South: A Programme for Survival*. Cambridge: MIT Press, 1980.

Brown, Lawrence D. and others, *The Changing Politics of Federal Grants*. Washington: Brookings Institution, 1984.

Browning, Clyde E., *The Geography of Federal Outlays*. Studies in Geography. Chapel Hill: University of North Carolina, Dept. of Geography, 1973.

Bulajić, Milan, *Principles of International Development Law*. Dordrecht, Neth.; Boston; London: Nijhoff; Belgrade: Exportpress; 1986.

C

Campbell, Robert D., *Pakistan: Emerging Democracy*. Princeton, NJ: Van Nostrand, 1963.

Catsambas, Thanos, *Regional Impacts of Federal Fiscal Policy,* Lexington, MA: D.C. Heath, 1978.

Caviedes, César, *The Politics of Chile: A Socio-Geographical Assessment*. Boulder, CO: Westview, 1979.

Chisholm, Michael, *Modern World Development*. Totowa, NJ: Rowman & Littlefield, 1982.

Clark, M.A.V. and Eric G. Moore (eds.), *Residential Mobility and Public Policy*. Beverly Hills, CA: Sage, 1980.

Cox, Kevin R., *Conflict, Power and Politics in the City: A Geographic View*. New York: McGraw-Hill, 1973.

Cox, Kevin R. and Ronald J. Johnston (eds.), *Conflict Politics and the Urban Scene*. London: Longman, 1982.

D

Dale, Richard, *The Regulation of International Banking*. Cambridge, Eng.: Woodhead-Faulkner; Englewood Cliffs, NJ: Prentice-Hall; 1986.

Das, Parimal Kumar, *The Troubled Region*. Newbury Park, CA: Sage, 1987.

Dear, Michael J. and S.M. Taylor, *Not on Our Street: Community Attitudes to Mental Health Care*. London: Pion, 1982.

Dolman, Antony J., *Resources, Regimes, World Order*. New York: Pergamon, 1981.

Dopfer, Kurt, *The New Political Economy of Development*. New York: St. Martin's Press, 1980.

Drysdale, Alasdair and Gerald H. Blake, *The Middle East and North Africa*. New York: Oxford Univ. Press, 1985.

F

Faaland, Just and J.R. Parkinson, *The Political Economy of Development*. London: Frances Pinter, 1986.

Feinberg, Richard E. and Valeriana Kallab (eds.), *Adjustment Crisis in the Third World*. New Brunswick, NJ: Transaction Books, 1984.

Fields, Rona M., *Northern Ireland*. New Brunswick, NJ: Transaction Books, 1981.

Fitzgerald, W., *The New Europe: An Introduction to Its Political Geography*. New York: Harper & Brothers, 1946.

Flowerdew, Robin (ed.), *Institutions and Geographical Patterns*. New York: St. Martin's Press, 1982.

Forbes, D.K., *The Geography of Underdevelopment*. London: Croom Helm; Baltimore: Johns Hopkins Univ. Press; 1984.

Fordwor, Kwame Donkoh, *The African Development Bank: Problems of International Cooperation*. New York: Pergamon, 1981.

Fossett, James W., *Federal Aid to Big Cities: The Politics of Dependence*. Washington: Brookings Institution, 1983.

Fothergill, Stephen and Graham Gudgin, *Unequal Growth, Urban and Regional Employment Change in the UK*. London: Heinemann Educational Books, 1982.

Fraser-Tyler, Sir W. Kerr, *Afghanistan; a Study of Political Development in Central Asia*. London: Oxford Univ. Press, 1950.

Friters, Gerard M., *Outer Mongolia and Its International Position*. Baltimore: Johns Hopkins Univ. Press, 1949.

G

Gathorne-Hardy, G.M. and others, *The Scandinavian States and Finland; a Political and Economic Survey*. London: Royal Institute of International Affairs, 1951.

Ghurye, G.S., *The Scheduled Tribes of India*. New Brunswick, NJ: Transaction Books, 1980.

Gibbons, A., *The New Map of South America*. New York and London: Century, 1928.

Giddens, A., *The Constitution of Society: Outline of the Theory of Structuration*. Cambridge: Policy Press, 1984.

Goldenstein, L. and M. Seabra, "International Division of Labour and New Regionalization," in *Brazil: Spatial Organization*. Rio de Janeiro: National Commission of Brazil, IGU, 13–64.

Gopalakrishnan, R., *The Geography and Politics of Afghanistan*. New Delhi: Concept, 1986.

Green, L.P. and T.J.D. Fair, *Development in Africa; a Study in Regional Analysis with Special Reference to Southern Africa*. Johannesburg: Witwatersrand Univ. Press, 1962.

Guess, George, *The Politics of United States Foreign Aid*. New York: St. Martin's Press, 1987.

Gumbert, Marc, *Neither Justice Nor Reason; a Legal and Anthropological Analysis of Aboriginal Land Rights*. St. Lucia, London, New York: Univ. of Queensland Press, 1985.

H

Hammer, Richard M. and others (eds.), *North–South Dialogue: A New International Economic Order*. Thessaloniki, Greece: Institute of Public International Law and International Relations of Thessaloniki, 1982.

Hammond, Thomas Taylor, *Yugoslavia—Between East and West*. New York: Foreign Policy Association, 1954.

Hanna, A.J., *The Story of the Rhodesias and Nyasaland*. London: Faber & Faber, 1960.

Hansen, Roger D. (ed.), *The "Global Negotiation" and Beyond; Toward North–South Accommodation in the 1980's*. Austin: Univ. of Texas Press, 1981.

Harloe, Michael (ed.), *New Perspectives in Urban Change and Conflict*. London: Heinemann Educational Books, 1981.

Harris, Nigel, *Of Bread and Guns: The World Economy in Crisis*. Harmondsworth, Middlesex, Eng.: Penguin, 1983.

Harrison, Lawrence E., *Underdevelopment Is a State of Mind: The Latin American Case*. Lanham, MD: Univ. Press of America, 1985.

Hart, Jeffrey A., *The New International Economic Order*. New York: St. Martin's Press, 1983.

Heilleiner, Gerald K., *International Economic Disorder*. Toronto: University of Toronto Press, 1983.

Henig, Jeffrey R., *Public Policy & Federalism*. New York: St. Martin's Press, 1985.

Hollist, W. Ladd and F. LaMond Tullis, *An International Political Economy*. Boulder, CO: Westview, 1985.

Hossain, Kamal and Subrata Roy Chowdhury (eds.), *Permanent Sovereignty over Natural Resources in International Law: Principle and Practice*. New York: St. Martin's Press, 1984.

House, John W., *Frontier on the Rio Grande: A Political Geography of Development and*

Social Deprivation. New York: Oxford Univ. Press, 1982.

Hughes, Steven W. and Kenneth J. Mijeski, *Politics and Public Policy in Latin America*. Boulder, CO: Westview, 1984.

J

Jackson, W.A. Douglas, *The Russo–Chinese Borderlands*. Princeton, NJ: Van Nostrand, 1962.

Johnston, Ronald J., *The Geography of Federal Spending in the United States of America*. Letchworth, Hertfordshire, Eng.: Wiley, 1980.

———. *Residential Segregation, the State and Constitutional Conflict in American Urban Areas*. New York: Academic Press, 1984.

K

Kakwani, Nanak, *Analyzing Redistribution Policies*. New York: Cambridge University Press, 1986.

Karan, Pradyumna P. and William M. Jenkins, *The Himalayan Kingdoms: Bhutan, Sikkim, and Nepal*. Princeton, N.J.: Van Nostrand, 1963.

Kasperson, Roger E., *The Dodecanese: Diversity and Unity in Island Politics*. Chicago: University of Chicago Press, 1966.

Katz, James Everett (ed.), *The Implications of Third World Military Industrialization*. Lexington, Mass.: Lexington Books, 1986.

Kingsbury, Patricia and Robert C., *Afghanistan and the Himalayan States*. Garden City, N.Y.: Doubleday, 1960.

Kirby, Andrew and Stephen Pinch (eds.), *Public Provision and Urban Politics; Papers from the IBG Annual Conference, January 1982*. Geographical Paper No. 80. Reading, Berkshire, Eng.: Univ. of Reading, 1982.

———, *The Politics of Location; An Introduction*. London: Methuen, 1982.

Kirby, Andrew and others, *Public Service Provision and Urban Development*. New York: St. Martin's Press, 1984.

Kohn, Hans, *The Future of Austria*. New York: Foreign Policy Association, 1955.

L

Lattimore, Owen, *Nationalism and Revolution in Mongolia*. New York: Oxford Univ. Press, 1955.

Laufer, Leopold, *Israel and the Developing Countries: New Approaches to Cooperation*. New York: Twentieth Century Fund, 1967.

Lee Yong Leng, *Southeast Asia: Essays in Political Geography*. Singapore: Singapore Univ. Press, 1982.

Lenczowski, George, *Russia and the West in Iran, 1918–1948*. Ithaca, NY: Cornell Univ. Press, 1949.

Little Bear, Leroy and others (eds.), *Pathways to Self-Determination*. Toronto: Univ. of Toronto Press, 1984.

Looney, Robert E., *The Political Economy of Latin American Defense Expenditures; Case Studies of Venezuela and Argentina*. Lexington, MA: Lexington Books, 1986.

Lozoya, Jorge and others, *Alternative Views of the New International Economic Order*. Elmsford, NY: Pergamon, 1979.

Luard, Evan, *The Management of the World Economy*. New York: St. Martin's Press, 1983.

M

Martz, John D., *Central America: The Crisis and the Challenge*. Chapel Hill: Univ. of North Carolina Press, 1959.

Massey, D. and R. Meegan (eds.), *Politics and Method: Contrasting Studies in Industrial Geography*. London: Methuen, 1985.

Mazour, Anatole Grigorevich, *Finland Between East and West*. Princeton, NJ: Van Nostrand, 1956.

McCloud, Donald G., *System and Process in Southeast Asia*. Boulder, CO: Westview, 1986.

Meagher, Robert F., *An International Redistribution of Wealth and Power: A Study of the Charter of Economic Rights and Duties of States*. Elmsford, NY: Pergamon, 1979.

Mollenkopf, John H., *The Contested City*. Princeton, NJ: Princeton Univ. Press, 1983.

Morss, Elliot R. and Victoria A. Morss, *The Future of Western Development Assistance*. Boulder, CO: Westview, 1986.

Muñoz, Heraldo (ed.), *From Dependency to Development*. Boulder, CO: Westview, 1981.

Murphy, T.P., *Science, Geopolitics and Federal Spending*. Lexington, MA: Lexington Books, 1971.

N

Nawaz, Tawfique (compiler), *The New International Economic Order: A Bibliography*. London: Frances Pinter, 1980.

Nissman, David B., *The Soviet Union and Iranian Azerbaijan*. Boulder, CO: Westview, 1987.

Nordenskjöld, O., *The Geography of the Polar Regions*. New York: American Geographical Society, 1928.

O

Odle, Maurice A., *Multinational Banks and Underdevelopment*. Elmsford, NY: Pergamon, 1981.

Ogilvie, Alan Grant, *Europe and Its Borderlands*. Edinburgh: Nelson, 1957.

Olson, Robert K., *US Foreign Policy and the New International Economic Order*. London: Frances Pinter, 1981.

Onwuka, Ralph I. and Amadu Sesay, *The Future of Regionalism in Africa*. New York: St. Martin's Press, 1985.

Onwuka, Ralph I. and Olajide Aluko, *The Future of Africa and the New International Economic Order*. New York: St. Martin's Press, 1986.

Osborne, Harold, *Bolivia, a Land Divided*. London and New York: Royal Institute of International Affairs, 1954.

Osei-Kwame, Peter, *A New Conceptual Model for Study of Political Integration in Africa*. Lanham, MD: Univ. Press of America, 1980.

P

Paddison, Ronan, *The Political Geography of Power*. New York: St. Martin's Press, 1983.

Papp, Daniel, *Soviet Perceptions of the Developing World in the 1980s; The Ideological Basis*. Lexington, Mass.: Lexington Books, 1985.

Peet, Richard, (ed.), *International Capitalism and Industrial Restructuring*. Winchester, MA.: Allen & Unwin, 1987.

Pinch, Steven, *Cities and Services: The Geography of Collective Consumption*. Boston: Routledge and Kegan Paul, 1985.

Pounds, Norman J.G., *A Historical and Political Geography of Europe*. London: Harrap, 1949.

———, *Europe and the Mediterranean*. New York: McGraw-Hill, 1953.

———, *Europe and the Soviet Union, 2nd ed.* New York: McGraw-Hill, 1966.

Prescott, John Robert Victor, *The Reciprocal Relations between Geography and National Policy*. Melbourne: Australian and New Zealand Association for the Advancement of Science, 1967.

———, *The Geography of State Policies*. Chicago: Aldine, 1969.

R

Reitsma, Henrik-Jan A. and others, *The Third World in Perspective*. Totowa, NJ: Rowan & Littlefield, 1985.

Rolef, S.H., *The Political Geography of Palestine*. New York: Middle East Review Special Studies No. 3. (1983).

Ross-Barnett, J. and John Mercer, *Urban Political Analysis and New Directions in Political Geography*. Cedar Rapids: Univ. of Iowa, 1973.

Rostow, Walt Whitman, *Rich Countries and Poor Countries*. Boulder, CO: Westview, 1987.

Roucek, Joseph S., *Balkan Politics: International Relations in No Man's Land*. Stanford, CA: Stanford Univ. Press, 1949.

S

Sakti, Mukherjee, *Outstanding Issues in International Economic Law*. Calcutta: World Press, 1985.

Sanford, Jonathan E., *U.S. Foreign Policy and Multilateral Development Banks*. Boulder, CO: Westview, 1982.

Schmiegelow, Michée (ed.), *Japan's Response to Crisis and Change in the World Economy*. Armonk, NY: Sharpe, 1986.

Scott, Allan and Michael Storper (eds.), *Production, Work, Territory*. Winchester, MA: Allen & Unwin, 1986.

Scott, John, *Africa, World's Last Frontier*. New York: Foreign Policy Association, 1959.

Seegers, Kathleen Walker, *Alliance for Progress*. New York: Coward-McCann, 1964.

Seligson, Mitchell A. (ed.), *The Gap Between Rich and Poor: Contending Perspectives on the Political Economy of Development*. Boulder, CO: Westview, 1984.

Shabad, Theodore, *China's Changing Map; a Political and Economic Geography of the Chinese People's Republic*. New York: Praeger, 1956.

Short, John R., *The Urban Arena: Capital, State and Community in Contemporary Britain*. London: Macmillan, 1984.

Siegfried, André, *Canada: An International Power*. London: J. Cape, 1949.

Simmie, James, *Power, Property and Corporatism: The Political Sociology of Planning*. London: Macmillan, 1981.

Simon, Reeva S., *Iraq Between the Two World Wars*. New York: Columbia Univ. Press, 1986.

Skousgaard, C.J., "Party Influence on Local Spending in Denmark," in K. Newton (ed.), *Urban Political Economy*. London: Frances Pinter, 1976, 48–62.

Smith, David M., *Where the Grass is Greener; Living in an Unequal World*. Baltimore: Johns Hopkins Univ. Press, 1979.

Stanislawski, Dan, *The Individuality of Portugal: a Study in Historical–Political Geography*. Austin: Univ. of Texas Press, 1959.

Staris, Dennis and Gilbert R. Winham, *Canada and the International Political/Economic Environment*. Toronto: Univ. of Toronto Press, 1987.

Steinberg, Ira S., *The New Lost Generation*. New York: St. Martin's Press, 1982.

Stern, H. Peter, *The Struggle for Poland*. Washington: Public Affairs Press, 1953.

Sukhwal, Bheru L., *India; a Political Geography*. New Delhi: Allied Publishers, 1971.

Sweet-Escott, Brickham, *Greece; a Political and Economic Survey, 1939–1953*. London and New York: Royal Institute of International Affairs, 1954.

T

Tayyeb, Ali, *Pakistan: A Political Geography*. London: Oxford Univ. Press, 1966.

Thoman, Richard S., *Free Ports and Foreign Trade Zones*. Cambridge, MD: Cornell Maritime Press, 1956.

Tinbergen, Jan (ed.), *RIO: Reshaping the International Order*. New York: Dutton, 1976.

Tsoukalis, Loukas, *The Political Economy of International Money*. Newbury Park, CA: Sage, 1985.

V

VanDijk, P. and others, *Supervisory Mechanisms in International Economic Organisations*. Deventer, Neth.: Kluwer; The Hague: T.M.C. Asser Instituut; 1984.

Vizulis, I. Joseph, *Nations Under Duress: The Baltic States*. Port Washington, NY: Associated Faculty Press, 1985.

W

Wallerstein, Immanuel, *Modern World System*, Vol I, *Capitalist Agriculture and the Origin of the European World Economy in the Six-* *teenth Century*. New York: Academic Press, 1974.

——, *Modern World System*, Vol II, *Mercantilism and the Consolidation of the European World Economy 1600–1750*. New York: Academic Press, 1980.

——, *Historical Capitalism*. London: Verso, 1983.

——, *The Politics of the World Economy*. London: Cambridge Univ. Press, 1984.

Ward, Peter M., *Welfare Politics in Mexico*. London: Allen & Unwin, 1986.

Watkins, Michael (ed.), *Dene Nation: The Colony Within*. Toronto: Univ. of Toronto Press, 1977.

Weaver, Sally M., *Making Canadian Indian Policy*. Toronto: Univ. of Toronto Press, 1980.

Weinstein, Warren (ed.), *Chinese and Soviet Aid to Africa*. New York: Praeger, 1975.

Whalley, John, *Domestic Policies and the International Economic Environment*. Toronto: Univ. of Toronto Press, 1987.

White, John, *The Politics of Foreign Aid*. New York: St. Martin's Press, 1974.

Whitelegg, J., *Inequalities in Health Care: Problems of Access and Provision*. Straw Barnes, 1982.

Wilson, A. Jeyaratnam and Dennis Dalton (eds.), *The States of South Asia*. Honolulu: University of Hawaii Press, 1983.

Wilson, Thomas (ed.), *Ulster Under Home Rule: A Study of the Political and Economic Problems of Northern Ireland*. London: Oxford Univ. Press, 1955.

Woronoff, Jon, *Asia's "Miracle" Economies*. Armonk, NY: Sharpe, 1986.

Y

Yoffie, David B., *Power and Protectionism; Strategies of the Newly Industrializing Countries*. New York: Columbia Univ. Press, 1983.

Periodicals

A

Ackerman, Edward A., "Public Policy Issues for the Political Geographer," *Annals, AAG*, 52, 3 (September 1962), 292–298.

Alfredsson, Gudmundur, "International Law, International Organizations and Indigenous Peoples," *Journal of International Affairs*, 36, 1 (1982), 113–125.

Archer, J. Clark, "The Geography of Federal Fiscal Policies in the United States of America: An Exploration," *Government and Policy*, 1 (1983), 377–400.

Azikiwe, N., "Essentials for Nigerian Survival," *Foreign Affairs*, 443, 3 (April 1965), 447–461.

B

Badcock, Blair, "Removing the Spatial Bias from State Housing Provision in Australian Cities," *Political Geography Quarterly*, 1, 2 (April 1982), 137–157.

Baldoria, Pedro L., "Political Geography of the Philippines," *Philippine Geographical Journal*, 1, 1 (January 1953), 15–23.

Barbour, Nevill, "Aden and the Arab South," *World Today*, 15, 8 (August 1959), 302–310.

Barone, C.A., "Dependency, Marxist Theory and Salvaging the Idea of Capitalism in South Korea," *Review of Radical Political Economics*, 15 (1983), 43–70.

Barrett, R.E. and M.K. Whyte, "Dependency Theory and Taiwan," *American Journal of Sociology*, 87 (1983), 1064–1089.

Barsh, Russel Lawrence, "Indigenous Peoples: An Emerging Object of International Law," *American Journal of International Law*, 80, 2 (April 1986), 369–385.

Bassett, Keith and Anthony Hoare, "Bristol and the Saga of Royal Portbury: A Case Study in Local Politics and Municipal Enterprise," *Political Geography Quarterly*, 3, 3 (July 1984), 223–250.

Bennett, Sari and Carville Earle, "Socialism in America: A Geographical Interpretation of Its Failure," *Political Geographical Quarterly*, 2, 1 (January 1983), 31–55.

Borchert, John R., "Geography and State–Local Public Policy," *Annals, AAG*, 75, 1 (March 1985), 1–4.

Bowman, Isaiah, "The Geographical Situation of the United States in Relation to World Politics," *Geographical Journal*, 112, 4–6 (1948), 129–145.

Buchanan, K., "Northern Region of Nigeria: The Geographical Background of its Political Duality," *Geographical Review*, 43, 4 (October 1953), 451–473.

Bunge, William, "The Cave of Coulibistrie," *Political Geography Quarterly*, 2, 1 (January 1983), 57–70.

Busk, C.W.F., "The North-West Frontier; Pakistan's Inherited Problem," *Geographical Magazine*, 22, 3 (July 1949), 89–92.

C

Caroe, Sir Olaf, "The India-Tibet-China Triangle," *Asian Review*, 56, 205 (January 1960), 3–13.

Chakrabongse, Chula, Prince of Thailand, "The Political and Economic Background in Thailand," *Journal of the Royal Central Asian Society*, 42, 2 (April 1955), 116–127.

Chase-Dunn, Christopher, "Interstate System and Capitalist World-Economy: One Logic or Two," *International Studies Quarterly*, 25, 1 (March 1981), 19–42.

Chase-Dunn, Christopher and Joan Sokolovsky, "Interstate Systems, World-Empires and the Capitalist World–Economy: A Response to Thompson," *International Studies Quarterly*, 27, 3 (September 1983), 357–367.

Church, R.J. Harrison, "The Islamic Republic of Mauritania," *Focus*, 12, 3 (November 1961).

Clark, Gordon L., "Urban Impact Analysis: A New Tool for Monitoring the Geographical Effects of Federal Policies," *Professional Geographer*, 32, 1 (February 1980), 82–85.

Clarke, John I., "Economic and Political Changes in the Sahara," *Geography*, 46 (April 1961), 102–119.

Corbridge, Stuart, "Political Geography of Contemporary Events V: Crisis, What Crisis?" *Political Geography Quarterly*, 3, 4 (October 1984), 331–345.

Cox, B.A. and C.M. Rogerson, "The Corporate Power Elite in South Africa: Interlocking Directorships Among Large Enterprises," *Political Geography Quarterly*, 4, 3 (July 1985), 219–234.

Cox, Kevin R., "Residential Mobility, Neighborhood Activism and Neighborhood Problems," *Political Geography Quarterly*, 2, 2 (April 1983), 99–117.

Cunningham, J.K., "Maori–Pakeha Conflict, 1858–1885: A Background to Political Geography," *New Zealand Geographer*, 56, 1 (1956), 12–31.

D

Demas, William G., "The Caribbean and the New International Economic Order," *Journal of Interamerican Studies*, 20, 3 (August 1978), 229–263.

Dening, B.H., "Greater Syria; a Study in Political Geography," *Geography*, 35 (June 1950), 110–123.

Deshpande, C.D., "India Recognized; a Geographical Evaluation," *Indian Geography*, 2, 1 (August 1957), 164–169.

Duncan, S.S. and M. Goodwin, "The Local State: Functionalism, Autonomy and Class Relations in Cockburn and Saunders," *Political Geography Quarterly*, 1, 1 (January 1982), 77–96.

Dunleavy, Patrick, Comments: "Class, Consumption and Radical Explanations in Urban Politics: A Rejoinder to Hooper, *Political Geography Quarterly*, 1, 2 (April 1982), 187–192.

Dutta, R., "Uganda: A Geopolitical Study." *Modern Review*, 111, 1 (January 1962), 17–28.

E

Environment and Planning C: Government and Policy. London: Pion. Quarterly journal published since 1982.

F

Fall, Bernard B., "The Laos Tangle," *International Journal*, 16, 2 (1961), 138–157.

Fawcett, Charles Bungay, "Marginal and Interior Lands of the Old World," *Geography*, 32, 1 (1947), 1–12.

Filipp, Karlheinz, "Political Geography Around the World III: Facing the Political Map of Germany," *Political Geography Quarterly*, 3, 3 (July 1984), 251–258.

Fincher, Ruth, "The State Apparatus and the Commodification of Quebec's Housing Cooperatives," *Political Geography Quarterly*, 3, 2 (April 1984), 127–143.

Fisher, Charles A., "The Malaysian Federation, Indonesia and the Philippines: A Study in Political Geography," *Geographical Journal*, 129, 3 (September 1963), 211–328.

Floyd, Barry N., "Pre-European Political Patterns in Sub-Sahara Africa," *Bulletin, Ghana Geographical Association*, 8, 2 (July 1963), 3–11.

Fryer, Donald W., "Economic Aspects of Indonesian Disunity," *Pacific Affairs*, 30, 3 (September 1957), 195–208.

G

Gilliland, H.B., "An Approach to the Problem of the Government of Nomadic Peoples," *South African Geographical Journal*, 29, 2 (1947), 43–58.

Ginsburg, Norton S., "China's Changing Political Geography," *Geographical Review*, 42, 1 (January 1952), 102–117.

Gonen, A. and S. Hasson, "The Use of Housing as a Spatio-Political Measure: The Israeli Case," *Geoforum*, 14 (1983), 103–109.

Gradus, Yehuda and Shaul Krakover, "The Effect of Government Policy on the Spatial Structure of Manufacturing in Israel," *Journal of Developing Areas*, 11 (1977), 393–409.

Grundy, Kenneth W., "African Explanations of Underdevelopment; the Theoretical Basis for Political Action," *Review of Politics*, 28 (January 1966), 62–75.

H

Hall, Peter, "The New Political Geography: Seven Years On," *Political Geography Quarterly*, 1, 1 (January 1982), 65–76.

Haupert, John S., "The Impact of Geographic Location upon Sweden as a Baltic Power," *Journal of Geography*, 58 (1959), 1–14.

Helin, Ronald A., "United the Wings of Pakistan: A Matter of Circulation," *Professional Geographer*, 20 (1968), 251–256.

Hoare, Anthony G., "Dividing the Pork Barrel: Britain's Enterprise Zone Experience," *Political Geography Quarterly*, 4, 1 (January 1985), 29–46.

Hoffman, George W., "South Tyrol: Borderland Rights and World Politics," *Journal of Central European Affairs*, 7 (1947), 258–308.

———, "The Netherlands Demands on Germany: A Post-War Problem in Political Geography," *Annals, AAG*, 42, 2 (June 1952), 129–152.

———, "The Political Geography of a Neutral Austria," *Geographical Studies*, 3, 1 (1956), 12–32.

———, "East Europe: A Study in Political Geography," *Texas Quarterly*, 2, 3 (1959), 57–88.

Hoggart, Keith, "Political Parties and Local Authority Capital Investments in English Cities," *Political Geography Quarterly*, 3, 1 (January 1984), 5–32.

———, "Geography, Political Control and Local Government Policy Outputs," *Progress in Human Geography*, 10, 1 (March 1986), 1–23.

J

Janda, Kenneth and Robin Gillies, "How Well Does 'Region' Explain Political Party Characteristics?" *Political Geography Quarterly*, 2, 3 (July 1983), 179–203.

Johnson, James H., "The Political Distinctiveness of Northern Ireland," *Geographical Review*, 52, 1 (January 1962), 78–91.

Johnston, Ronald J., "Congressional Committees and Inter-State Distribution of Military Spending," *Geoforum*, 10 (1979), 151–162.

———, "Congressional Committees and Department Spending: The Political Influence on the Geography of Federal Expenditure in the United States," *Transactions, Institute of British Geographers*, n.s., 4 (1979), 373.

Jones, Bryan D., Review Essay: "Government and Business: The Automobile Industry and the Public Sector in Michigan," *Political Geography Quarterly*, 5, 4 (October 1986), 369–384.

Jones, Emrys, "Problems of Partition and Segregation in Northern Ireland," *Journal of Conflict Resolution*, (1960), 96–105.

K

Kasperson, Roger E., "Toward a Geography of Urban Politics: Chicago, A Case Study," *Economic Geography*, 41, 2 (April 1965), 95–107.

———, "Ward Systems and Urban Politics," *Southeastern Geographer*, 9, 2 (November 1969), 17–25.

King, L.J. and G.L. Clark, "Government Policy and Regional Development," *Progress in Human Geography*, 2 (1978), 1–16.

Kirby, Andrew, Review Essay: "A Tribute to John House," *Political Geography Quarterly*, 3, 4 (October 1984), 347–348.

Krueger, Ralph, "Changes in the Political Geography of New Brunswick," *Canadian Geographer*, 19 (1975), 121–134.

Kuehnelt-Leddihn, Erik R.V., "The Petsamo Region," *Geographical Review*, 34, 3 (1944), 405–417.

L

Lattimore, Owen, "Origins of the Great Wall of China," *Geographical Review*, 27, 4 (October 1937), 529–549.

———, "The Geographical Factor in Mongol History," *Geographical Journal*, 91, 1 (January 1938), 1–20.

Lattimore, Owen, "The New Political Geography of Inner Asia," *Geographical Journal*, 119, 1 (March 1953), 17–32.

Levi, Werner, "Bhutan and Sikkim: Two Buffer States," *World Today*, 15, 12 (December 1959), 492–500.

Lutz, James M., "The Spatial and Temporal Diffusion of Selected Licensing Laws in the United States," *Political Geography Quarterly*, 5, 2 (April 1986), 141–159.

M

Mair, Andrew, "The Homeless and the Post-Industrial City," *Political Geography Quarterly*, 5, 4 (October 1986), 351–368.

Malecki, Edward J., "Government-Funded R&D: Some Regional Implications," *Professional Geographer*, 33, 1 (February 1981), 72–82.

"The Manaus Free Zone and Western Amazonia," *Bank of London and South America Review*, (March 1969), 146–150.

Mansergh, Nicholas, "Ireland: The Republic Outside the Commonwealth," *International Affairs*, 28, 3 (July 1952), 277–291.

McCarthy, J.J. and M. Swilling, "South Africa's Emerging Politics of Bus Transportation," *Political Geography Quarterly*, 4, 3 (July 1985), 235–249.

McColl, Robert W., "Political Geography of Revolution: China, Vietnam and Thailand," *Journal of Conflict Resolution*, 21 (1967), 153–167.

McGee, T.G., "Aspects of the Political Geography of Southeast Asia," *Pacific Viewpoint*, 1, 1 (1960), 39–58.

McIntyre, Alistair, "Caribbean: Free Ports Among the Islands," *Ceres*, 1, 6 (November–December 1968), 38–41.

McLafferty, Sara, "Urban Structure and Geographical Access to Public Services," *Annals, AAG*, 72, 3 (September 1982), 347–354.

———, "Constraints on Distributional Equity in the Location of Public Services," *Political Geography Quarterly*, 3, 1 (January 1984), 33–47.

McQuillan, D. Aidan, "Creation of Indian Reserves on the Canadian Prairies 1870–1885," *Geographical Review*, 70, 4 (October 1980), 379–396.

Mead, W.R., "Finnish Karelia: An International Borderland," *Geographical Journal*, 118, 1 (March 1952), 40–57.

Melamid, Alexander, "Political Geography of Trucial Oman and Qatar," *Geographical Review*, 43, 2 (April 1953), 194–206.

———, "Political Geography of Economic Underdevelopment," *Geographical Review*, 56, 2 (April 1966), 293–294.

Mohan, John, "State Policies and the Development of the Hospital Service of North-east England, 1948–1982," *Political Geography Quarterly*, 3, 4 (October 1984), 275–295.

Morisett, Jean, "The Aboriginal Nation: The Northern Challenge and the Construction of Canadian Unity," *Canadian Review of Studies in Nationalism*, 7 (1980), 237–249.

N

Nicholson, Norman L., "Some Aspects of the Political Geography of the District of Keewatin," *Canadian Geographer*, 3 (1953), 73–83.

P

Park, Richard L., "East Bengal: Pakistan's Troubled Province," *Far Eastern Survey*, 23, 5 (May 1954), 70–74.

Peet, Richard (ed.), *Restructuring in the Age of Global Capital*. Special Issue of *Economic Geography*, 59, 2 (April 1983). Six articles.

———, "The New International Division of Labor and Debt Crisis in the Third World," *Professional Geographer*, 39, 2 (May 1987), 172–178.

Perry, Peter, "Towards a Political Geography of Corruption," *New Zealand Geographer* (April 1985), 2–7.

Planhol, Xavier de, "Geography, Politics and Nomadism in Anatolia," *International Social Science Journal*, 11, 4 (1959), 525–531.

"Political Geography—Research Agendas for the Nineteen Eighties," *Political Geography Quarterly*, 1, 1 (January 1982), 1–17.

Pollock, N.C., "The Political Geography of Nigeria," *Bulletin of Oxford*, (1968), 195–199.

Proudfoot, Malcolm J., "Chicago's Fragmented Political Structure," *Geographical Review*, 47, 1 (January 1957), 106–117.

Q

Quereshi, Khalida, "The United Arab Emirates," *Pakistan Horizon*, 26, 4 (1973), 3–27.

R

Randall, Richard R. and H.R. Wilkinson, "Political Geography of the Kagenfurt Basin," *Geographical Review*, 47, 3 (July 1957), 406–419.

Reitsma, Hendrik-Jan A., "Malawi's Problem of Allegiance," *Tijdschrift voor Economische en Sociale Geografie*, 65, 6 (1974), 421–429.

———, "Development Geography, Dependency Relations, and the Capitalist Scapegoat," *Professional Geographer*, 34, 2 (May 1982), 125–130.

———, "Geography and Dependency: A Rejoinder," *Professional Geographer*, 34, 3 (August 1982), 337–342.

"Research Agendas for the Nineteen Eighties: Comments, Additions, and Critiques," *Political Geography Quarterly*, 1, 2 (April 1982), 167–180.

Reynolds, David R. and Fred M. Shelley, "Procedural Justice and Local Democracy," *Political Geography Quarterly*, 4, 4 (October 1985), 267–288.

Robinson, K.W., "The Political Influence in Australian Geography," *Pacific Viewpoint*, 3, 2 (July 1962), 21–24.

Romer, E., "Poland: The Land and the State," *Geographical Review*, 4 (1917), 6–25.

Rowley, Gwyn, "Political Geography of Contemporary Events IV: Local Government or Central Government Agency?—the British Case," *Political Geography Quarterly*, 3, 3 (July 1984), 265–268.

S

Sadler, David, "Works Closure at British Steel and the Nature of the State," *Political Geography Quarterly*, 3, 4 (October 1984), 297–311.

Salisbury, Howard G., "The State Within a State: Some Comparisons Between the Urban Ghetto and the Insurgent State," *Professional Geographer*, 23, 2 (April 1971), 105–112.

Saunders, P., Comments: "Urban Politics," *Political Geography Quarterly*, 1, 2 (April 1982), 181–186.

Savage, Robert L., "Diffusion Research Traditions and the Spread of Policy Innovations in a Federal System," *Publius; The Journal of Federalism*, 15, 4 (Fall 1985), 1–27.

Sen, D.K., "China, Tibet and India," *India Quarterly*, 7, 2 (April–June 1951), 112–132.

Seth, Vijay K., "State and Spatial Aspects of Industrialization in Post-Independence India," *Political Geography Quarterly*, 5, 4 (October 1986), 331–350.

Shaudys, Vincent K., "The External Aspects of the Political Geography of Five Diminutive European States," *Journal of Geography*, 61, 1 (January 1962), 20–31.

Shreevastava, M.P., "Political Problems of India: A Functional Approach," *Indian Geographical Journal*, 37, 2 (April–June 1962), 35–44.

Smith, Graham E., "Political Geography and the Theoretical Study of the East European Nation," *Indian Journal of Political Geography*, 40 (1979), 59–83.

Smith, Joseph B., "The Koreans and Their Living Space: An Attempted Analysis in Terms of Political Geography," *Korean Review*, 2, 1 (September 1949), 45–52.

Smith, Neil, "Theories of Underdevelopment: A Response to Reitsma," *Professional Geographer*, 34, 3 (April 1982), 332–337.

Solly, Marion B., "Yugoslavia; a Case Study in Political Geography," in *New Zealand Geographical Society*. Proceedings of the First Geographical Conference, 1955, 61–66.

Sommers, Lawrence M. and Ole Gade, "The Spatial Impact of Government Decisions on Postwar Economic Change in North Norway," *Annals, AAG*, 61, 3 (September 1971), 522–536.

T

Taylor, Peter J., Review Essay: "Chaotic Conceptions, Antinomies, Dilemmas and Dialectics: Who's Afraid of the Capitalist World-Economy," *Political Geography Quarterly*, 3, 4 (October 1984), 87–93.

Thrall, Grant Ian, "Three Pure Planning Scenarios and the Consumption Theory of Land Rent," *Political Geography Quarterly*, 2, 3 (July 1983), 219–231.

V

Valauskas, Charles C., "China Special Economic Zones in Perspective: A Contextual Discussion with Emphasis on the Shekou Industrial Zone," *Hastings International and Comparative Law Review*, 9, 2 (Winter 1986), 149–234.

Vance, James E., "Areal Political Structure and Its Influence on Urban Patterns," *Yearbook of the Association of Pacific Coast Geographers*, 22 (1960), 40–49.

Van Der Kroef, Justus M., "Disunited Indonesia," *Far Eastern Survey*, 27 (1958), 49–63,73–80.

————, "Indonesia: Sources of Disunity," *Orbis*, 2, 4 (1959), 478–491.

Van Valkenburg, Samuel, "Current Political Geographical Problems of Jordania," *Professional Geographer*, 5, 6 (November 1953), 18.

Varma, S.N., "National Unity and Political Stability in Nigeria," *International Studies*, 4, 3 (January 1963), 265–280.

W

Wallerstein, Immanuel, "Dependence in an Interdependent World: The Limited Possibilities of Transformation Within the Capitalist World Economy," *African Studies Review*, 17, 1 (April 1972).

White, Gilbert, "Geographers in a Perilously Changing World," *Annals, AAG*, 75, 1 (March 1985), 10–16.

Whitelegg, John, "Transport Policy, Fiscal Discrimination and the Role of the State," *Political Geography Quarterly*, 3, 4 (October 1984), 313–329.

Whittlesey, Derwent, "Lands Athwart the Nile," *World Politics*, 5, 2 (January 1953), 214–241.

Wilbanks, Thomas J., "Geography and Public Policy at the National Scale," *Annals, AAG*, 75, 1 (March 1985), 4–10.

Wilber, Donald N., "Afghanistan, Independent and Encircled," *Foreign Affairs*, 31, 3 (April 1953), 486–494.

Wilkinson, J.C., "The Oman Question: The Background to the Political Geography of South-East Arabia," *Geographical Journal*, 131, 3 (September 1971), 361–371.

Wirt, F., "The Political Sociology of American Suburbia: a Reinterpretation," *American Sociological Review*, 25 (1960), 514–526.

Wulff, H.E., "Laos," *Australian Geographer*, 7, 4 (February 1959), 141–148.

Z

Zaidi, I.H., "Toward a Measure of the Functional Effectiveness of a State: The Case of West Pakistan," *Annals, AAG*, 56, 1 (March 1966), 52–67.

Name Index

The index pages are designed to supplement the Table of Contents and List of Maps and do not repeat the information contained therein. Boldface page numbers indicate illustrations other than maps.

Alexander, Lewis, 3, 607
Alland, A., 14
Ardrey, Robert, 10
Aristotle, 4

Bertelsen, Judy, 119
Boggs, S. Whittemore, 607
Borgese, Elizabeth Mann, 481n
Borlaug, Norman, 565
Bowman, Isaiah, **5**, 607
Brawer, Moshe, 7
Brown, Lester R., 564
Brunhes, Jean, 61
Brunn, Stanley, 173n, 597
Bunge, William, 243
Burghardt, Andrew, 94

Carneiro, Robert L., 46n
Child, Jack, 242, 496
Christaller, Walter, 9
Cohen, Saul B., 3, 64–65, 235, 283
Cornish, Vaughan, 103–104

Darwin, Charles, 60–61
Davis, William Morris, 61n
de Castro, Therezinha, 495–496
de la Blache, Vidal, 61
Demko, George, 607
de Seversky, Alexander P., 231–233
Deutsch, Karl, 62–63

Frankel, S. H., 283

Galiani, Abbé Fernando, 463
Goode, J. Paul, 23
Gottman, Jean, 10
Gradus, Yehuda, 7
Grotius, Hugo, 462

Hall, Edward T., 9
Harrison, Richard Edes, 23
Hartshorne, Richard, 3, 5, **62**
Haushofer, Karl, 229–230
Herz, John H., 63–64
Hodgson, Robert, 607
Hooson, David J. M., 233
Hume, David, 357

Ibn Khaldun, Abd-al Rahman, 4

Jackson, W. A. Douglas, 3
James, Preston, 61
Jefferson, Mark, 102
Johnston, Ronald J., 200–201
Jones, Stephen B., 5, 63

Kasperson, Roger, 3, 17, 568
Kelly, Philip, 242
Key, V. O., Jr., 200
Kjellén, Rudolf, 224–225
Kliot, Nurit, 7

Lin Piao, 236

McColl, Robert, 119, 121
Mackinder, Halford J., **226**–229
Mahan, Alfred Thayer, 225–**226**
Malmberg, Torsten, 14
Maull, Otto, 231
Meinig, Donald W., 233
Minghi, Julian, 3, 568
Modelski, G., 257
Montesquieu, Charles Louis de Secondat, 4

Nkrumah, Kwame, 238, 316, 410

O'Loughlin, John, 256

Pardo, Arvid, **467**
Pearcy, G. Etzel, 173n, 607
Petty, William, 4
Philbrick, Allan K., 9
Pittman, Howard, 242
Pounds, Norman J. G., 3, 94

Ratzel, Friedrich, **5**, 61, 94, 223–224
Ritter, Carl, 5
Robinson, G. W. S., 77
Robinson, K. W., 112
Rosenthal, L. D., 3, 64–65

Sack, Robert, 14
Selden, John, 462
Semple, Ellen Churchill, 61, 224
Siegfried, André, 200
Smith, Adam, 357
Soffer, Arnon, 7
Soja, Edward, 14
Sommer, Robert, 9
Spate, O. H. K., 102, 104
Spykman, Nicholas John, 229
Strabo, 4

Taylor, Griffith, 61
Taylor, Peter, 200–201
Truman, Harry S., 464–465

Van Bynkershoek, Cornelius, 463
van der Wusten, Herman, 256
Van Valkenburg, Samuel, 61n

Wallerstein, Immanuel, 257
Waterman, Stanley, 7
Weigert, Hans, 3
Whittlesey, Derwent, 5, 61
Wolpert, Julian, 61

Subject Index

The index pages are designed to supplement the Table of Contents and List of Maps and do not repeat the information contained therein. Boldface page numbers indicate illustrations other than maps.

Accretion and avulsion, 67
Africa, 51–53, 383–385
Age-sex pyramids, **553**
Air transport, 545–548
Andorra, 118
Angkor, **51**
Antarctica, 119
Antarctic Treaty, 487–490
Antipode, 7
Arab Common Market, 385
Arab League, *see* League of Arab States
Archipelagic States, 476–477
Argentina, **92**–93, 487, 488, 489–490, **492**, 559
Arms traffic, 431–434
Association of American Geographers, 3, 7
Association of Southeast Asian Nations (ASEAN), 246, **412**, 414, 430, **549**
Australia, 18, 47, 241, 253, 294, **465**, 494–495
Aztecs, 53–54

Bab el Mandeb, Strait of, **233**
Bandung Conference (1955), 238, 313
Bantustans, in South Africa, **125**, 314
Basques, 41
Beagle Channel, **92**
Beira Patrol, 368
Belgium, 115, 288–289
Belgrade, 104
Belmopan, 102
Berlin, West, 76–77, **86**, 386n
Biafra, **125**, 168, 526
Brasilia, **101**
Buffer States and zones, 236–238
Bushmen, 46–47

Campione, Italy, **76**
Canada, 186n, **366**, 400n, 410, 487, 525, **529**, 541, 548, 563, **603**
Caribbean, Commonwealth, 241
Caribbean Basin, 251–253
Caribbean Community (CARICOM), 381
Central American Common Market (CACM), 380–381
Chauvinism, 41
Chile, **92**–93, 488, 489–490, **492**, 493
China, 50–51, 161, 238, 297
Colombo Plan, **412**, 420
Comintern, 247n, 286
Common markets, 374, 377–380, 380–381, 385
Commonwealth, The, 372–373, 403–404, **408**
Comoros, 314
Condominium, territorial, 121
Conflict resolution, 347–349, 409, 483, 488
Contiguous Zone, 474
Continental shelf, 464–467, 479
Cooperation Council for the Arab States of the Gulf, *see* Gulf Cooperation Council
Council of Europe, **412**–413, 430
Council for Mutual Economic Assistance (COMECON), 370, 373, 417–420
Cuban Missile Crisis (1962), 19, **20**
Customs unions, 374, 377, 383
Cyprus, 42, 82, **92**, 115–116, 117

Dakar, **393**
Danzig, **92**, 124
Darwinism, Social, 60–61, 223
Delaware River Basin Commission, 176

Demographic transition, 553–554
Dispute settlement, *see* Conflict resolution
"Domino theory", 239

East African Community (EAC), 383–385, 540, **549**
Economic Community of West African States, 385
Economic unions, 374, **375**, 377
Ecumene, 70, 99
Egypt, **48**, 49–50, 317–318, 604
Enclosed and semi-enclosed seas, 477
Environment, marine, 481–**482**, 483
Environmental determinism, 60–61
European Communities (EC), 105, 314, 377–378, **412**, 430
European Free Trade Association (EFTA), 378–380, **412**
Exclaves and enclaves, 76–77
Exclusive economic zone (EEZ), 477–479

Falkland (Malvinas) Islands, 90–**91**, **92**, 495
Family planning, **556**, 557
Fertile Crescent, 49
Feudal Europe, 57–58
Fisheries, **464**, 465, 477
France:
  Antarctica, **492**, 494
  civil divisions, 160–161
  colonial policies, 289–290, 297
  French Union/French Community, 405–406
  nuclear weapons testing, 245–246
  unitary State, 107
Free cities, 124–125

Free trade areas, 374, 378–380, 381–383
Free zones and free ports, 601
Functional approach in political geography, 62

Gaza Strip, 118
General Agreement of Tariffs and Trade (GATT), 359–363, 392
Geographically disadvantaged States, 471–472
Geostationary orbit, 503
Germany, 293, **492**
Gerrymandering, 204–205
Greece, ancient, 56
Group of 77, 363, 407, 471
Gulf Cooperation Council, 413–414, 543
Gulf Federation, 122

Helgoland, **70**
High seas, 393, 424, 479
Hottentots, 47

Iconography, 10, 20, **310**, 311
Incas, 54–56
India, 166–168, 318–319
Indian Ocean, 254–256, 468
Indian reservations, U.S., 194–199
Indonesia, 42, **317**, 319, 400n
Indus River dispute, 352
Institute of British Geographers, 7
Insurgent States, 119–121
Intergovernmental Committee for Migration, 562
International Air Transport Association (IATA), 407, 548
International Energy Agency, 574n
International Geographical Union, 7, 404, 504
International Geophysical Year (1957–1958), 468, 487–488
International nongovernmental organizations, **407, 504**
International Seabed Area, 479, 481
Iran–Iraq war, 251
Iroquois Confederacy, 170–171
Irredentism, 40, 42, **43, 44,** 45
Islands, 477, 483
Isle of Man, **125,** 159, **530**
Israel:
  Arab boycott, 368
  Bedouin, **602**
  Hebrew language, 530–531
  Law of the Sea, 472
  occupied areas, 118
  Palestine mandate, 301–302

political geography in, 7
religious minorities, **529**
U.S. aid, 604
Italy, 114–115, 293, 304

Japan, 124, 293–294, 528–529, 574, 578
Jordan River, 352

Knights of Malta, 118n, **125**
Kurds, 41

Land-locked States, 366, 467, 471–472, 477–478, 601
Land reform, 566, 578–579
Laos, **388**
latifundia, 312, 578
Latin America:
  ancient civilizations, 52–56
  Antarctica, 495–496
  Cuban Missile Crisis, 19
  economic integration, 380–383
  geopolitics in, 241–242
  Inter-American Highway, 538n
  Latin American Economic System (SELA), 380
  Latin American Energy Organization, 574n
  Latin American Free Trade Association/Integration Association, 381–383
  nuclear-free zone, 245–246
  socio-political structure, **128–129**
  transnational corporations in, 369
Lauca, Rio, 352
League of Arab States (Arab League), 411–412
League of Nations:
  Codification of International Law, 464
  economic sanctions, 368
  general, 398–399
  mandate system, 298–300
  transit conventions, 390
Lebanon, 527–528
London, England, 101

Macau, **289**
Macedonia, 41, 111
Maghreb, 112, 417
Mandates, *see* League of Nations
Maps:
  distorted, 21–23
  mental, 17–20
  propaganda, 24–26
Maquiladora program, in Mexico, 370, **598**
Maritime transport, 543–545
Mayas, 53–54

Mekong Scheme, **355**–356
Mercantilism, 58, 357–358
Mesopotamia, 49
Microstates and ministates, 71
Middle East, 249–251
Minerals, strategic, 247–249
Mombasa, 9–10
Morocco, 85n, **86,** 91, 297, **369**
Multinational corporation, *see* Transnational corporations

Namibia, South-West Africa, 123–124, 249, 285, 302–304, **305**
Narcotics traffic, 427–**431,** 432–434
Nationalism, 13, 40–41, 58–59, 300
Nauru, **302**
Neoimperialism and neocolonialism, 316–319
Nepal, 87, 161–163, 246, **388,** 432, 544
Netherlands, 109, 293
Neutral zones, 121–122
New International Economic Order, 247, 363, 471, 497, 603–606
New Zealand, 294
Nigeria, 168, 526, 559
Nodules, polymetallic (manganese), 468, **481**
Non-Aligned Movement, 238, 313, 407, **408**
Nongovernmental organizations (NGOs), 404, 484n, 498n, 504, 601
Non-state nation, 119–121
Nordic Council, **413**
North Atlantic Treaty Organization (NATO), **412,** 422
Norway, 163, 488
Nuclear-free zones, 245–246, 415

Organization of African Unity (OAU), 91, 314, 410–411, **412,** 560
Organization of American States (OAS), 355, 409–410, **412,** 430
Organization of Arab Petroleum Exporting Countries (OAPEC), 407
Organization of Central American States (ODECA), 413
Organization for Economic Cooperation and Development (OECD), 314, 369, 406–407, 430, 604

Organization for European
  Economic Cooperation
  (OEEC), 406
Organization of the Islamic
  Conference, 407–408, 430
Organization of the Petroleum
  Exporting Countries (OPEC),
  133, 247, 314, 317, 407–411,
  573–574, 604

Pacific Basin/Pacific Rim, 253–254
Pacific Islands, Trust Territory of
  the, 124, 301
Palestine Liberation Organization
  (PLO), 427
Panama Canal, 67–69, **545, 549**
Pan American Union, 409, 410
Pipelines, 543
Piracy, 423–425, 432–434
Plata, Rió de la, 355
Point Roberts, Washington, **76**
Poland, **600**
Political Geography Quarterly, 7
Port Authority of New York and
  New Jersey, 175–176
Portugal, 285, 287–288, 358
Power factors, **131**
Prague, 104
Primate city, 102
Proxemics, 9
Public lands, U.S., 191–194
Pushtunistan, 45

Railroads, 538–541
Refugees, 425, 559–**563**
Research, marine scientific, 482–
  483
Rhodesia, 368
Rivers, international, 349–356
Roads, 537–538
Roman empire, 56–57

St. Lawrence Seaway, 541–542,
  549
Sanction, economic, 367–368,
  543, 559
School districts, U.S., 188–191
Servitudes, international, 69
Singapore, **226**
Smuggling, 371–**372**
Social Darwinism, *see*
  Darwinism, Social
Somalia, 42–45
South Africa, Republic of, 76,
  103, 118, 123–124, 291, 368,
  400, 496, 549, 557–559
South Asian Association for
  Regional Cooperation
  (SAARC), 414–415, 430
Southern Ocean, 483, 485, 491–
  492

South Pacific Commission (SPC),
  412, 415–417
South Pacific Forum, 415–417
South Pacific Regional Trade and
  Economic Agreement
  (SPARTECA), 415–417
Spain, 91, 115, 293
Straits used for international
  navigation, **476,** 544
Strategic minerals, 247–249
Suez Canal, 544

Tangier, 124
Tanzam Railroad, **540, 549**
Tennessee Valley Authority (TVA),
  176–178
Territorial sea, 462–464, 473–474
Terrorism, 425–427, 432–434
Texas, 171n
Thalweg, 84
Theocracies, 528
Third World, 313
Tokyo, 102–103
Tordesillas, Treaty of, 284, 461
Transnational corporations, 361,
  368–370
Trieste, 124
Trust Territory of the Pacific
  Islands, *see* Pacific Islands,
  Trust Territory of the

Uganda, 300
Unified field theory, 63
Union of Soviet Socialist
  Republics:
  Aeroflot, 547–548
  Antarctica, 487, 488, 491, 493
  arms traffic, 431
  civil divisions, 164–166
  ecology, 571–572
  federalism, 116
  gas pipelines, 543
  kolkhoz, **600**
  outer space, **502,** 504, 506
  Pacific Ocean, 253
  Radio Moscow, 548, **549**
  railroads, 538–539
  rings of defense, 237–238
United Kingdom, 114, 159–160,
  290–291, 297, 526
United Nations:
  Antarctica, 496–497
  assistance to new States, 314–
    316
  brain drain, 559
  Decade on Transport and
    Communication in Africa,
    551
  decolonization, 301–308
  Europe–Africa permanent link,
    543

general, 309–403, **408**
good offices, 348
Indian Ocean, 256
International Court of Justice
  (ICJ, World Court), 346–347,
  408
international law, 346–347
International Law Commission,
  346–347
International Monetary Fund,
  315
land-locked States, 395, 467,
  471–472
membership, 72, 308, 313
most seriously affected
  countries, 312
narcotics, 430–**431**
outer space, 500, **505**–506
peacekeeping, 400, **403**
piracy, 424–425
protocol, 12n
refugees, 560–563
regional commissions, 375–
  377, 392, 393, **412**
sanctions, economic, 368
satellite broadcasting, 550
Seabed Committee, 467–468
terrorism, 427
transnational corporations, 369
United Nations Conference on
  the Law of the Sea:
  I (1958), 392–393, 465–467
  II (1960), 467
  III (1973–1982), 395–396,
    468–**484,** 544
United Nations Conference on
  Trade and Development
  (UNCTAD), 363–**367,** 369,
  393–394
United Nations Development
  Programme (UNDP), 315,
  408, 604
United Nations Environment
  Programme (UNEP), 580
United Nations Institute for
  Training and Research
  (UNITAR), 315
United Nations Radio Service,
  549
United Nations Temporary
  Executive Authority (UNTEA,
  New Guinea), 124, 308
University for Peace, 258
World Bank Group, 315, 352,
  580
World Food Conference, 564
zones of peace, 246
United States of America:
  Antarctica, 487, 488, **492**
  arms traffic, 431
  colonial policies, 291–293

United States of America
    (*Continued*)
  ecology, 568–571
  energy, 572–574
  foreign aid, 604
  highways, 537–538
  immigration, 559–560
  Law of the Sea, 484
  outer space, **500,** 504
  religion, 528
  sanctions, economic, 368
  trade deficits, **364**
  zoning of land use, 575–578
uti possidetis juris, 91, 314, 487

Vatican City, 118, 471, 528n
Venezuela, 65, **92**

Warsaw Treaty Organization
    (Warsaw Pact), 422
Waterways, 541–543
Western European Union,
    422n
Western Sahara, 85n, 118–**119,**
    319, 411
West Indies, Federation of the,
    104–105, **113**

World Commission on
    Environment and
    Development, 580

Zambia, 239–241
Zanzibar, 45, 116
Zimbabwe, ancient, **53**
Zionism, 13
Zones of national jurisdiction,
    sea, 473–479
Zones of peace, 246
Zoning of land use, 575–578

MAJOR WORLD POLITICAL UNITS

- Location of Capital

Countries Numbered on Map

1	BELGIUM
2	THE NETHERLANDS
3	LUXEMBOURG
4	FEDERAL REPUBLIC OF GERMANY
5	GERMAN DEMOCRATIC REPUBLIC
6	SWITZERLAND
7	DENMARK
8	AUSTRIA
9	YUGOSLAVIA
10	ALBANIA
11	BULGARIA
12	ROMANIA
13	HUNGARY
14	CZECHOSLOVAKIA
15	SENEGAL
16	THE GAMBIA
17	GUINEA-BISSAU
18	COTE D'IVOIRE
19	BURKINA FASO
20	GHANA
21	BENIN
22	TOGO
23	CENTRAL AFRICAN REPUBLIC
24	UGANDA
25	RWANDA
26	BURUNDI
27	MALAWI
28	ZIMBABWE
29	DJIBOUTI
30	BAHRAIN
31	QATAR
32	UNITED ARAB EMIRATES
33	KUWAIT
34	JORDAN
35	BANGLADESH
36	SINGAPORE